Ch-

I [illegible] you
for [illegible] one
the [illegible].

During that time, you have shown how very capable you are. I know you will be successful in all that you do. You were my student. You are now my friend. In the future I look forward to also calling you my peer.

All the best in your graduate career,

Erl C. Wasi[illegible]

# INORGANIC ELECTRONIC STRUCTURE AND SPECTROSCOPY

## Volume II: Applications and Case Studies


*Edited by*

**Edward I. Solomon**

*Monroe E. Spaght Professor of Chemistry*
*Stanford University*
*Stanford, California*

**A.B.P. Lever**

*Distinguished Professor of Chemistry*
*Faculty of Pure and Applied Science*
*York University*
*Toronto, Ontario, Canada*



WILEY-INTERSCIENCE

A JOHN WILEY & SONS, INC., PUBLICATION

Published by John Wiley & Sons, Inc., Hoboken, New Jersey.
Published simultaneously in Canada.



For general information on our other products and services or for technical support, please contact our Customer Care Department within the United States at (800) 762-2974, outside the United States at (317) 572-3993 or fax (317) 572-4002.

Wiley also publishes its books in a variety of electronic formats. Some content that appears in print may not be available in electronic format. For information about Wiley products, visit our web site at www.wiley.com.

***Library of Congress Cataloging-in-Publication Data is available.***

ISBN-13 978-0-471-97114-6
ISBN-10 0-471-97114-6

Printed in the United States of America.

10 9 8 7 6 5 4 3 2 1

# CONTENTS

# PREFACE

Spectroscopy has had a huge impact in Inorganic Chemistry and indeed most researchers in the field make significant use of a combination of spectroscopic methods and bonding concepts to understand molecular properties on a fundamental level. The sophistication of the instrumentation available, our understanding of the spectroscopic parameters obtained in terms of electronic structure, and our insight into the contribution of electronic structure to physical properties and chemical reactivity have all made great advances in recent years. Unfortunately, most of the key texts in inorganic spectroscopy and electronic structure that were published before 1970 are out of print and often no longer available. The purpose of these volumes is to provide an up to date text covering the key spectroscopic and electronic structure methods and to develop their application and define their impact on a number of important topics in Inorganic Chemistry. Each chapter is written by an internationally recognized leader in that field, with the level of presentation assuming an advanced undergraduate background with a basic understanding of quantum chemistry (at the level of Pauling and Wilson) and group theory (at the level of Cotton). Each chapter begins with the basics, gives examples and evolves into timely topics in the field.

Volume I presents the different spectroscopic methods and levels of electronic structure calculations for transition metal systems. We start with a summary presentation of the key concepts in ligand field theory and its applications to the physical properties of metal complexes. This is a topic where many of the original texts are "out of print". This chapter serves as a foundation for many of the spectroscopic methods which follow. Ground state methods: electron paramagnetic resonance and Mössbauer; excited state methods: polarized absorption, resonance Raman, luminescence, laser spectroscopy; high energy methods: photoelectron and X-ray absorption spectroscopy. This volume concludes with chapters on the modern "computational approaches" such as "ab initio" and "density functional theory" utilized in the calculation of the electronic structure of transition metal complexes.

Volume II contains two parts. The first presents key topics in Inorganic Chemistry where electronic structure and spectroscopy have had a huge impact: bioinorganic chemistry, electron transfer, mixed valence compounds, electrochemistry and photochemistry and photophysics.

The final section of these volumes presents case studies of major classes of inorganic compounds: metal-metal bonds, metal carbonyls and metallocenes, metal nitrosyls, heme sites, spin crossover compounds and magnetic materials.

This presentation of the methods, applications and case studies in inorganic electronic structure and spectroscopy should provide modern researchers with a strong foundation in physical-inorganic chemistry. We thank all the contributors for their excellent and timely contributions to the field.

EDWARD I. SOLOMON
*Monroe E. Spaght Professor of Chemistry*
*Stanford University*
*Stanford, CA*

A.B.P. (BARRY) LEVER
*Distinguished Research Professor of Chemistry*
*York University*
*Toronto, Ontario, Canada*

# CONTRIBUTORS, VOLUME II

P. DAY, The Royal Institution, 21 Albemarle Street, London W1X 4BS, UK E-mail: pday@ri.ac.uk

E.S. DODSWORTH, Department of Chemistry, York University, 4700 Keele St., Toronto, Ontario M3J 1P3, Canada E-mail: dod@yorku.ca

J.F. ENDICOTT, Department of Chemistry, Wayne State University, Detroit, Michigan 48202 E-mail: jfe@fourcroy.chem.wayne.edu

J.H. ENEMARK, Department of Chemistry, University of Arizona, Tucson, AZ 85721-0041 E-mail: jenemark@u.arizona.edu

H.B. GRAY, Division of Chemistry and Chemical Engineering, California Institute of Technology, Pasadena, CA 91125 E-mail: hgcm@cco.caltech.edu

N.E. GRUHN, Department of Chemistry, University of Arizona, P.O. Box 210041, Tucson, AZ 85721

P. GÜTLICH, Fachbereich Chemie, Institut für Anorganische Chemie, Johannes-Gutenberg-Universität Mainz, Staudinger Weg 9, D-55099 Mainz, Germany E-mail: p.guetlich@uni-mainz.de

M.A. HANSON, Department of Chemistry, Stanford University, S.G. Mudd Building, Stanford, California 94305-5080 E-mail: melissa@chem.stanford.edu

A. HAUSER, Département de Chimie Physique, Université de Gèneve, 30, Quai Ernest-Ansermet, CH-1211 Genève 4, Switzerland

M.D. HOPKINS, Department of Chemistry, University of Pittsburgh, Pittsburgh, PA 15260

A.B.P. LEVER, Department of Chemistry, York University, 4700 Keele St., Toronto, Ontario M3J 1P3, Canada E-mail: blever@yorku.ca

D.L. LICHTENBERGER, Department of Chemistry, University of Arizona, P.O. Box 210041, Tucson, AZ 85721-0041 E-mail: dlichten@arizona.edu

G.H. LOEW, Molecular Research Institute, 845 Page Mill Rd., Palo Alto, CA 94304 E-mail: loew@montara.molres.org

V.M. Miskowski, Department of Chemistry, University of Pittsburgh, Pittsburgh, PA 15260

D.E. Richardson, Department of Chemistry, University of Florida, Gainesville, FL 32611-7200 E-mail: der@chem.ufl.edu

P.N. Schatz, Department of Chemistry, University of Virginia, Charlottesville, VA 22901 E-mail: pns@virginia.edu

E.I. Solomon, Department of Chemistry, Stanford University, S.G. Mudd Building, Stanford, CA 94305-5080 E-mail: solomon@chem.stanford.edu

H. Spiering, Fachbereich Chemie, Institut für Anorganische Chemie, Johannes-Gutenberg-Universitat Mainz, Staudinger Weg 9, 6500 Mainz, Germany

B.L. Westcott, Department of Chemistry, Central Connecticut State University, New Britain, CT 06050 E-mail: westcottb@ccsu.edu

J.R. Winkler, Division of Chemistry and Chemical Engineering, California Institute of Technology, Pasadena, CA 91125

# CONTENTS, VOLUME I

# CONTRIBUTORS, VOLUME I

G.M. BANCROFT, Department of Chemistry, University of Western Ontario, London, Ontario M6A 5B7, Canada E-mail: scigmb@uwoadmin.uwo.ca

A. BENCINI, Dipartimento di Chimica, Università Degli Studi di Firenze, Via Marigliano 75/77, 50144 Firenze, Italy E-mail: sandro@chim1.unifi.it; bencini@dada.it

T.C. BRUNOLD, Department of Chemistry, Stanford University, Stanford, CA 94305 E-mail: tbrunold@chem.stanford.edu

D.A. CASE, Department of Molecular Biology, MB1, The Scripps Research Institute, 10666 N. Torrey Pines Rd., La Jolla, CA 92037 E-mail: case@scripps.edu

R. CZERNUSZEWICZ, Department of Chemistry, University of Houston, Houston, TX 77204 E-mail: roman@uh.edu

J. ENSLING, Fachbereich Chemie, Institut für Anorganische Chemie, Johannes-Gutenberg-Universität Mainz, Staudinger Weg 9, 6500 Mainz, Germany E-mail: ensling@iacgu7.chemie.uni-mainz.de

D. GATTESCHI, Dipartimento di Chimica, Università Degli Studi di Firenze, Via Maragliano 75/77, 50144 Firenze, Italy E-mail: gattesch@chim1.unifi.it

H.U. GÜDEL, Institut für Anorganische und Physikalische Chemie, Universitat Bern, Freiestrasse 3, CH-3000 Bern 9, Switzerland E-mail: Guedel@iac.unibe.ch

P. GÜTLICH, Fachbereich Chemie, Institut für Anorganische Chemie, Johannes-Gutenberg-Universität Mainz, Staudinger Weg 9, 6500 Mainz, Germany E-mail: p.guetlich@uni-mainz.de

B. HEDMAN, Stanford Synchrotron Radiation Laboratory, Stanford University, Stanford, CA 94305-5080 E-mail: hedman@ssrl.slac.stanford.edu

M.A. HITCHMAN, Department of Chemistry, The University of Tasmania, Box 252C, G.P.O. Hobart, Tasmania 7001, Australia E-mail: Michael.Hitchman@utas.edu.au

K.O. HODGSON, Department of Chemistry, Stanford University, Stanford, CA 94305-5080 E-mail: hodgson@ssrl.slac.stanford.edu

Y.F. HU, Department of Chemistry, University of Western Ontario, London, Ontario M6A 5B7, Canada E-mail: yhu@julian.uwo.ca

E. KRAUSZ, Research School of Chemistry, The Australian National University, G.P.O. Box 4, Canberra A.C.T. 2601, Australia

A.B.P. LEVER, Department of Chemistry, York University, 4700 Keele St., Toronto, Ontario M3J 1P3, Canada E-mail: blever@yorku.ca

J. LI, Department of Molecular Biology, MB1, The Scripps Research Institute, 10666 N. Torrey Pines Road, La Jolla, CA 92037

C.H. MARTIN, Department of Chemistry, University of Florida, Gainesville, FL 32611

L. NOODLEMAN, Department of Molecular Biology, MB1, The Scripps Research Institute, 10666 N. Torrey Pines Road, La Jolla, CA 92037
E-mail: lou@scripps.edu

H. RIESEN, Research School of Chemistry, The Australian National University, G.P.O. Box 4, Canberra, A.C.T. 2601, Australia

M.J. RILEY, Department of Chemistry, University of Queensland, St. Lucia, Queensland 4072, Australia E-mail: riley@chemistry.uq.edu.au

E.I. SOLOMON, Department of Chemistry, Stanford University, S.G. Mudd Building, Stanford, CA 94305 E-mail: solomon@chem.stanford.edu

T.G. SPIRO, Department of Chemistry, Princeton University, Princeton, NJ 08544
E-mail: spiro@chemvax.princeton.edu

M.C. ZERNER, Department of Chemistry, University of Florida, Gainesville, FL 32611 E-mail: zerner@qtp.ufl.edu

H.H. ZHANG, Department of Chemistry, Stanford University, Stanford, CA 94305-5080

# 1 Bioinorganic Spectroscopy

EDWARD I. SOLOMON AND MELISSA A. HANSON
Department of Chemistry
Stanford University
Stanford, CA 94305-5080, USA
E-mail: solomon@chem.stanford.edu
melissa@chem.stanford.edu



*Inorganic Electronic Structure and Spectroscopy, Volume II: Applications and Case Studies.*
Edited by E. I. Solomon and A. B. P. Lever.
ISBN 0-471-32682-8. 

## 1 INTRODUCTION

Researchers from many areas of Science including Inorganic Chemists, Biochemists, Biophysicists, Molecular Biologists, and Spectroscopists have been drawn to the field of Bioinorganic Chemistry. While they have diverse backgrounds and somewhat different focuses they share the common interest of understanding how the properties of a metalloprotein active site relate to function, and many of the tools utilized to achieve this key goal involve different types of spectroscopy. The spectroscopic methods employed in Bioinorganic Chemistry cover over ten orders of magnitude in photon energy. As can be seen from Table 1, different energy regions provide different complementary insights into the properties of a transition metal active site. By choosing the right combination of methods coupled with ligand field and molecular orbital theories, one can achieve for any system the goals of 1) defining active site geometric and electronic structure and their contributions to catalysis, and 2) elucidating the reaction mechanism on a molecular level. This, however, sometimes involves developing new spectroscopic methods and is often complicated by the fact that many classes of active sites in Biology exhibit unique spectral features when compared to small molecule complexes of the same metal ion. These unique spectral features reflect novel geometric and electronic structures (resulting from the protein ligand, see Reference 1) that can activate the metal ion for catalysis.

Novel geometric and electronic structures in metallobiomolecules result from either unusual ligation or the interaction between metal ions in binuclear and polynuclear sites. Certain ligand–metal bonds can be extremely covalent (e.g., thiolate, phenolate, oxo, sulfido, peroxo, and porphyrin) which will produce low energy intense charge transfer transitions in the absorption spectrum and greatly affect the ground state spin Hamiltonian parameters (*vide infra*). Bridging ligation can allow electron

**Table 1 Spectroscopic Methods**

| Method | Parameters | Information Content |
|---|---|---|
| *Ground State* | | |
| Magnetism | molar magnetic susceptibility ($\chi_m$) | spin of ground state |
| | effective magnetic moment ($\mu = 2.83\sqrt{\chi T}$) | |
| | axial ($D$) and rhombic ($E$) zero field splitting (ZFS) | low symmetry splitting of $d$ orbitals and their relative covalency |
| | exchange coupling constant ($J$) | covalent interaction of bridging ligand with two or more metal ions |
| Mössbauer | isomer shift ($\delta$) | |
| | metal quadrupole splitting ($E_Q$) | oxidation and spin state of $^{57}Fe$ |
| | metal hyperfine ($A^M$) | orbital angular momentum and metal covalency of ground state |
| | electron Zeeman effect ($g_i$, $i = x, y, z$) | |
| | ZFS | see above |
| Electron Paramagnetic Resonance (EPR) | $g_i$, $A_i^M$, $D$, $E$ | see above |
| | ligand superhyperfine coupling ($A^{L_i}$) | covalency of L–M bond |
| Electron Nuclear Double Resonance (ENDOR) | $A^{L_i}$ | see above |
| | nuclear Zeeman effect ($g_N \beta_N H$) | identification of ligand |
| | ligand quadrupole coupling ($P$) | electric field gradient at ligand nuclei |
| Electron Spin Echo Envelope Modulation (ESEEM) | same as ENDOR but for smaller $A^{L_i}$ | |
| *Ligand Field Excited States* | | |
| Electronic Absorption | energies | LF splitting of $d$ orbitals = geometry |
| | intensities ($f$ = oscillator strength) | distortion of site from centrosymmetric |
| | bandshapes | excited state distortions—probes change in bonding on excitation |

**Table 1 (*Continued*)**

| Method | Parameters | Information Content |
|---|---|---|
| Polarized Single Crystal Absorption | parallel ($\pi$), perpendicular ($\sigma$), and axial ($\alpha$) polarized spectra | selection rules for transitions allows assignments |
| Circular Dichroism (CD) | energies/bandshapes | see above |
| | intensities ($R$ = rotational strength) | resolution of bands, magnetic dipole character of transition |
| Magnetic Circular Dichroism (MCD) | energies/bandshapes | see above |
| | intensities (A, B, C terms) | probes nature of excited states, defines metal character in a transition |
| | variable temperature variable field (VTVH) MCD | probes ground state parameters as described above |
| *Charge Transfer Excited States* | | |
| ABS/CD/MCD | see above | sensitive probe of L–M bond—quantitative ligand donor interactions |
| Resonance Raman (rR) | depolarization ratio ($\rho$) | assign vibrations |
| | enhancement profiles (A and B terms) | as with bandshapes gives excited state distortion |
| *Core Excited States* | | |
| Metal K Edge X-ray Absorption Spectrosopy (XAS) | energy | effective nuclear charge on metal ion |
| | intensity | quadrupole and $4p$ mixing due to distortion of site from centrosymmetric |
| L-Edge XAS | energy dependence of intensity | multiplet splitting gives oxidation and spin state |
| | intensity | covalency of metal ion |
| Ligand K-Edge | energy | effective charge on ligand and metal |
| | intensity | covalency of L–M bond |

delocalization between mixed valence metal centers and strong magnetic coupling between homovalent metal ions.

Copper and iron sites are present in a majority of the redox active metalloproteins. These can be classified by their spectral features which, as emphasized above, reflect specific structural features. This characterization is given in Table 2. Relative to normal (or Type II) copper sites which have a mononuclear copper center coordinated to innocent ligands in a tetragonal geometry, blue copper (or Type I) sites have thiolate–copper bonds in a distorted tetrahedral geometry (Figure 1A)[2], coupled binuclear (or Type III) centers have two copper(II)s coupled by a bridging ligand (Figure 1B)[3], and $Cu_A$ centers also have two coppers but with bridging thiolate ligation and in a mixed oxidation state, formally [Cu(II), Cu(I)], with the extra electron strongly delocalized between the two metal centers.[4] It should be noted that for copper proteins, the unique spectral features are present in the oxidized site [Cu(II) $S = 1/2$], while the reduced Cu(I) site has a closed shell $d^{10}$ electron configuration and thus must be studied by high-energy spectroscopy as described in Section 6.

The non-heme iron enzymes in Table 2 often cycle between the high-spin Fe(II) $S = 2$ and high-spin Fe(III) $S = 5/2$ redox states in catalysis. The mononuclear iron enzymes that exhibit unique spectral features include a) the intradiol dioxygenases, which have a phenolate–Fe(III) bond (Figure 2A)[5], b) the heme enzymes, which have a highly delocalized porphyrin ligand (Figure 2B), and c) rubredoxin which has four thiolate–Fe bonds in a $D_{2d}$ distorted $T_d$ structure (Figure 2C). In addition, there are a number of non-heme iron enzymes which have either oxygen bridged binuclear ferric centers in their oxidized state and binuclear ferrous sites bridged by hydroxide and/or carboxylate ligands in their reduced state (Figure 2D)[6], or sulfide bridged two, three, or four iron clusters with terminal thiolate ligands (Figure 2E). The latter sites often have mixed oxidation states which are localized in the two iron-sulfur clusters and delocalized in the three and four iron-sulfur clusters.[7]

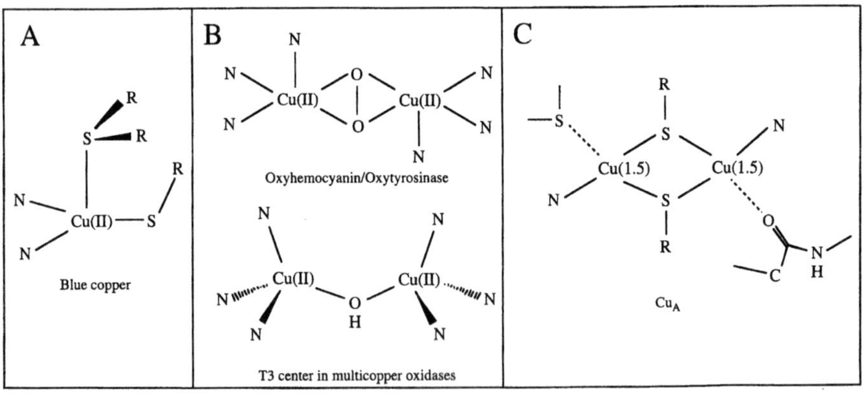


**Figure 1.** Unique sites in copper enzymes. A) Blue copper; B) oxyhemocyanin and oxytyrosinase and the Type III center in the multicopper oxidases; C) the $Cu_A$ center in cytochrome *c* oxidase.

**Table 2 Classification of Copper and Iron Active Sites in Biology**

| *TYPE OF COPPER SITE* | *FUNCTION* | *PROTEIN* |
|---|---|---|
| 1. normal copper (Type II) | oxidase | galactose oxidase |
| 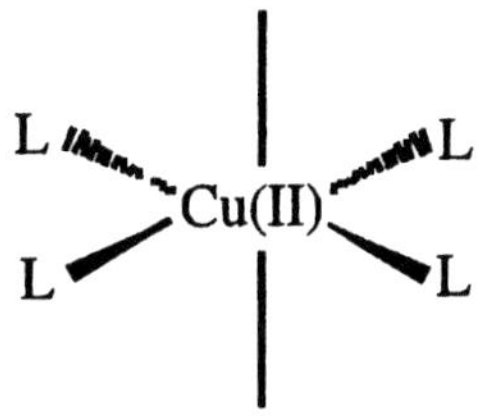 | | amine oxidase |
| | reductase | nitrite reductase |
| | superoxide dismutase | copper/zinc superoxide dismutase |
| | monooxygenase | dopamine $\beta$-hydroxylase |
| | | peptidylglycine $\alpha$-hydroxylating monooxygenase |
| 2. blue copper (Type I) | electron transfer | plastocyanin |
| | | azurin |
| | | stellacyanin |
| 3. coupled binuclear and trinuclear copper (Type III) | dioxygen binding | hemocyanin |
| | monooxygenase | tyrosinase |
| | oxidase | laccase |
| | | ceruloplasmin |
| | | ascorbate oxidase |
| 4. $Cu_A$ | electron transfer | cytochrome *c* oxidase |
| *TYPE OF IRON SITE* | *FUNCTION* | *PROTEIN* |
| 1. mononuclear non-heme | superoxide dismutase | Fe superoxide dismutase |
| | extradiol dioxygenase | catechol 2,3-dioxygenase |
| | hydroperoxidase | lipoxygenase |
| | hydroxylase | phthalate dioxygenase |
| | | $\omega$-hydroxylase |
| | | phenylalanine hydroxylase |
| | | clavaminate synthase |
| | oxidase | clavaminate synthase |
| | | isopenicillin N synthase |
| | | bleomycin |
| 2. phenolate–Fe(III) | intradiol dioxygenase | protocatechuate 3,4-dioxygenase |
| 3. heme | electron transfer | cytochromes |
| | dioxygen binding | hemoglobin |
| | | myoglobin |
| | monooxygenase | cytochrome P450 |
| | dioxygenase | *l*-tryptophan 2,3-dioxygenase |
| | | indoleamine 2,3-dioxygenase |
| | peroxidase | horseradish peroxidase |
| | catalase | bovine liver catalase |
| | oxidase | cytochrome *c* oxidase |
| 4. thiolate–Fe | electron transfer | rubredoxin |
| 5. coupled binuclear non-heme | dioxygen binding | hemerythrin |
| | hydroxylase | methane monooxygenase |
| | oxidase | ribonucleotide diphosphate reductase |
| | desaturase | acyl carrier protein $\Delta^9$ desaturase |
| | hydrolase | purple acid phosphatase |
| 6. iron-sulfur clusters | | |
| $Fe_2S_2(RS)_2$ | electron transfer | plant ferredoxin |
| $Fe_3S_4(RS)_3$ | | *Desulfovibrio gigas* ferredoxin II |
| $Fe_4S_4(RS)_4$ | | bacterial ferredoxin |
| | | high-potential iron protein (HiPIP) |

In this review, we will go over many of the spectroscopic methods listed in Table 1, in order of increasing energy, focusing on these key classes of copper and iron enzymes. The concepts developed here directly extend to many other metalloprotein systems (e.g., Mo(V)=O and low spin Fe(III) $S = 1/2$, Mn(II) $S = 5/2$, Mn(III) $S = 2$, etc.). For each method we will define the spectroscopic parameters and give their information content from ligand field theory, focusing on complexes with normal spectral features. We then explore the origin of the unique spectral features of the centers in Figures 1 and 2 in terms of their novel electronic structures. The relation of electronic and geometric structure to reactivity has been developed in Reference 1.

## 2 GROUND STATE SPECTROSCOPIC METHODS: Cu(II) $S = 1/2$

### 2.1 Normal Spectral Features (Type II Copper)

#### 2.1.1 *Electron Paramagnetic Resonance (EPR)*

2.1.1.1 *$g$ and $A^M$ Values* When a complex with an $S = 1/2$ ground state is placed in a magnetic field ($H$) the $M_S = \pm 1/2$ two-fold degeneracy of the spin (deg. $= 2S + 1$) is split by an amount $g\beta H$ where $\beta$ is the Bohr magneton ($4.6686 \times 10^{-5}$ cm$^{-1}$/gauss) which relates the electron angular momentum to its magnetic moment [$\mu = -\beta(L+2S)$, where $L$ is the orbital and $S$ is the spin angular momentum] and the $g$ value is the fundamental property of interest.

In the standard EPR experiment (see Chapter 2, Volume I, this series) the sample is in an oscillating field of microwaves which have a fixed frequency (X-band: $h\nu \sim$ 9 GHz $= 0.30$ cm$^{-1}$), and a perpendicular magnetic field is varied. As shown in Figure 3, the splitting between the $M_S = +1/2$ and $-1/2$ sublevels of the ground state increases linearly with magnetic field until the resonance condition, $h\nu = g\beta H$, is satisfied and microwave absorption occurs. This leads to a peak in the microwave absorption spectrum at the specific magnetic field which defines the $g$ value ($g = h\nu/\beta H = 0.71448\times$ microwave frequency in MHz/magnetic field in gauss) (Figure 3A). The $g$ factors are often anisotropic which means that they have different values when the complex is oriented in different directions relative to the magnetic field, and consequently the EPR experiment shows absorption of microwaves at different magnetic field values. For a tetragonal complex having axial symmetry (i.e., $z \neq x = y$) as shown in Figure 3, the $g$ value defining the splitting of the $M_S \pm 1/2$ sublevels with the magnetic field along the unique $z$ axis is $g_\parallel$ (Figure 3A), while that for the molecule oriented with the magnetic field in the $x, y$ plane is $g_\perp$ ($= g_x = g_y$) (Figure 3B).

Normally EPR studies are performed on frozen solutions which contain all orientations of the complex relative to the magnetic field and thus microwave absorption will occur over all fields between that associated with $g_\parallel$ and the field corresponding to $g_\perp$. (The $g$ value for a specific orientation $\theta$ of the molecular $z$ axis relative to the magnetic field is given by the directional cosine formula, $g^2 = g_\parallel^2 \cos^2\theta + g_\perp^2 \sin^2\theta$.) This absorption intensity must then be weighted by the relative number of molecules having a specific orientation; this increases rapidly toward the $g_\perp$ direction as shown

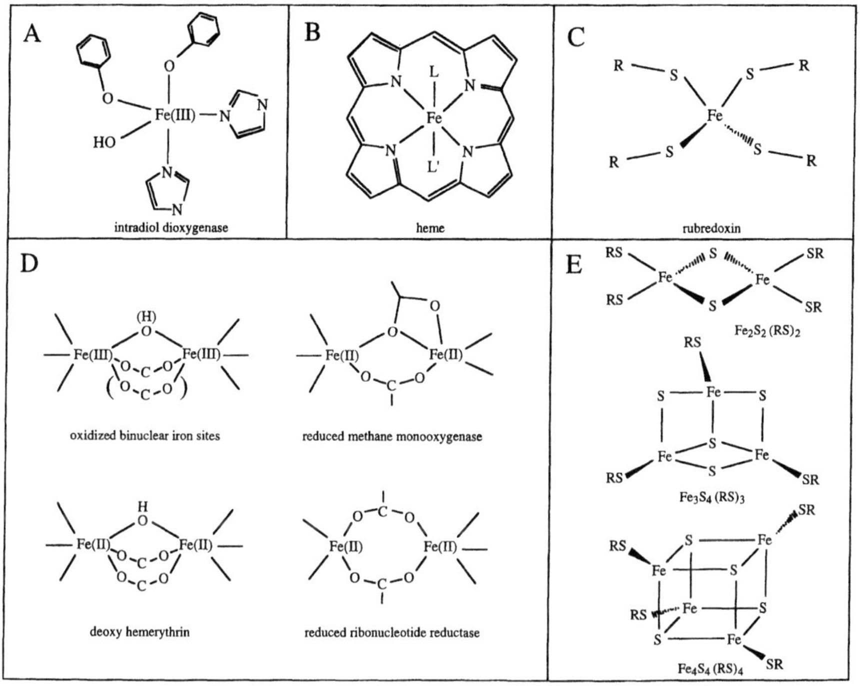


**Figure 2.** Unique sites in iron enzymes. A) Phenolate–Fe(III) ligation in intradiol dioxygenases; B) porphyrin ligation in heme enzymes; C) thiolate–Fe bonds in rubredoxin; D) oxo, hydroxo, and carboxylate bridging ligands in binuclear non-heme iron enzymes; E) sulfido bridges and terminal thiolate ligands in clusters of two, three, or four irons.

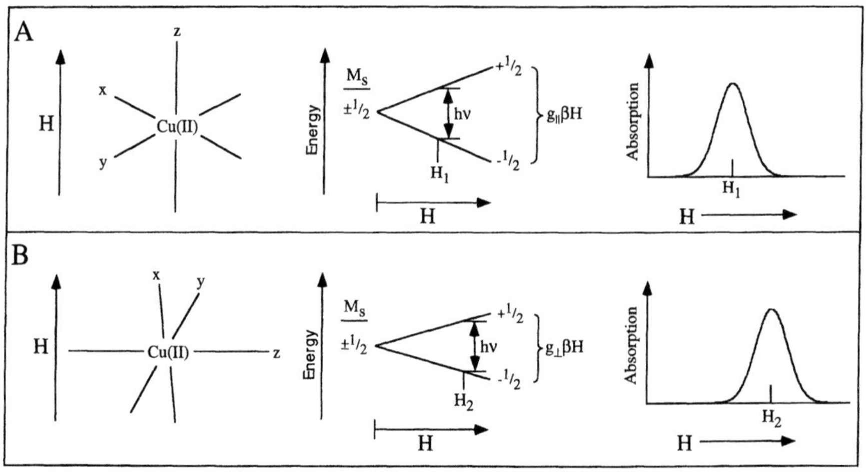


**Figure 3.** EPR experiment for a tetragonal Cu(II) complex A) with $H \parallel z$ and B) with $H \perp z$. Left: orientation of complex; center: associated energy splitting in a magnetic field; right: EPR absorption spectrum.

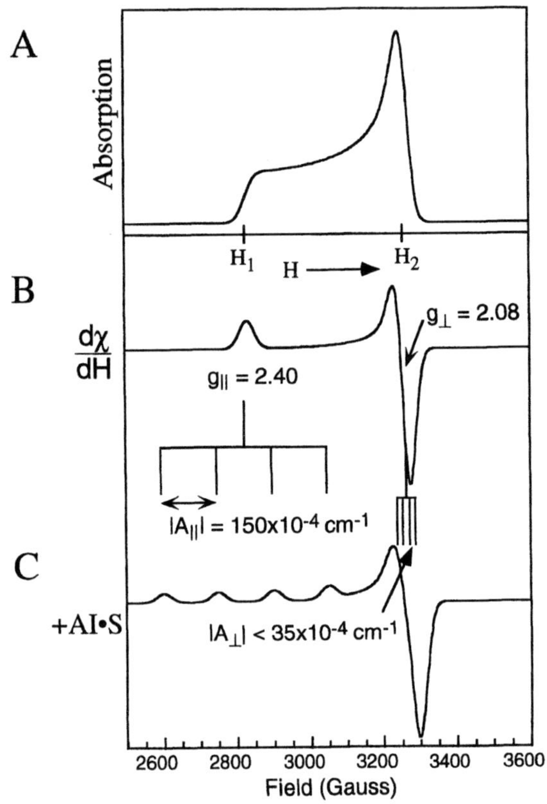


**Figure 4.** Simulated EPR spectrum of a normal copper complex, tetragonal $Cu(H_2O)_6^{2+}$ (at X-band, $\nu = 9.50$ GHz). A) EPR absorption; B) first derivative spectrum without and C) with copper hyperfine splitting.

in Figure 4A. In the EPR experiment the sensitivity is enhanced by taking the first derivative of the microwave absorption spectrum. As shown in Figure 4B this results in a peak at a magnetic field corresponding to the $g_{\|}$ value and an intense derivative shaped feature with its crossover point at the $g_{\perp}$ magnetic field value.

Copper has a nuclear spin ($I = 3/2$) which can couple to the electron spin and produce a hyperfine splitting ($= A\hat{S} \bullet \hat{I}$) of the EPR spectrum into $2I + 1 = 4$ peaks separated by the metal hyperfine coupling constant $A^M$ [where $A$ (cm$^{-1}$) $=$ $A$ (gauss) $\times g \times 4.6686 \times 10^{-5}$ cm$^{-1}$/gauss]. The metal hyperfine splitting of the EPR signal occurs for all orientations of the magnetic field relative to the molecular axes, and from Figure 4C it is seen that the hyperfine coupling constant can also be anisotropic with $A_{\|}$ $(= A_Z) > A_{\perp}$ $(= A_X = A_Y)$ for tetragonal copper(II) complexes.

Many metal complexes have lower than axial symmetry (i.e., $x \neq y$, called a rhombic distortion) which can result in $g_x \neq g_y$ and $A_x \neq A_y$ (and, in principal, these tensors will no longer be collinear, i.e., $A_z^M$ not aligned with $g_Z$). As shown

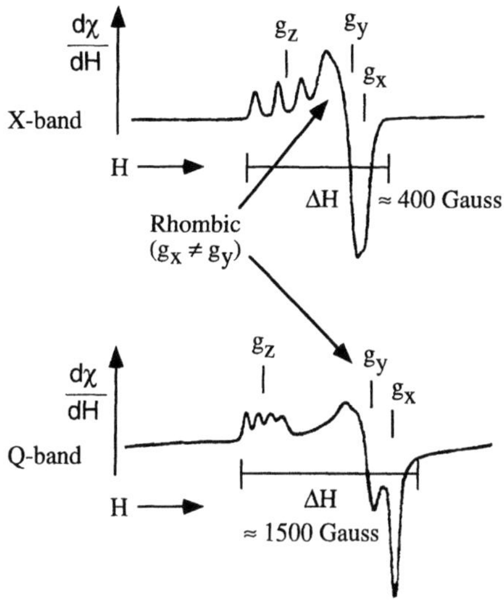


**Figure 5.** EPR spectra of half-met hemocyanin–$NO_2^-$, with rhombic symmetry. A) X-band spectrum ($h\nu = 9$ GHz) which shows broadening in the $g_\perp$ region; B) Q-band spectrum ($h\nu = 35$ GHz) resolves the three different $g$ values. Adapted from Reference 8.

in Figure 5A, this leads to a broadening in the $g_\perp$ region and $\Delta g_\perp = g_x - g_y$ is generally not resolvable at X-band microwave energy unless the rhombic splitting is large. However, if a higher microwave energy is used (usually Q-band, ~35 GHz = 1.17 $cm^{-1}$), the resonance condition for each orientation of the complex will be satisfied over a much larger spread in magnetic fields and the features associated with the different $g$ values can be directly observed (Figure 5B). Note that the hyperfine splittings are independent of magnetic field and thus do not change value between X-band and Q-band but will often broaden and become unresolvable at Q-band due to g strain (*vide infra*).

The EPR spectra in Figure 5 are generally described by a phenomenological spin Hamiltonian of the form given in Equation 1 for an $S = 1/2$ metal ion with a non-zero nuclear spin. The experimental parameters $g_i$ and $A_i$ are obtained directly from the spectrum as described above.

$$\mathcal{H} = \beta \vec{H} \bullet (g_z S_z + g_x S_x + g_y S_y) + A_z^M I_z S_z + A_x^M I_x S_x + A_y^M I_y S_y \qquad (1)$$

2.1.1.2 *Ligand Field Theory* The general expression for the $g$ value of a free metal ion is given in Equation 2 which shows that it depends on $S$, $L$, and $J$, the spin, orbital, and total ($J = |L + S|, |L + S - 1| \ldots |L - S|$) angular momenta of the ground state, respectively.

$$g = 1 + \frac{[J(J+1) - L(L+1) + S(S+1)]}{2J(J+1)} \qquad (2)$$

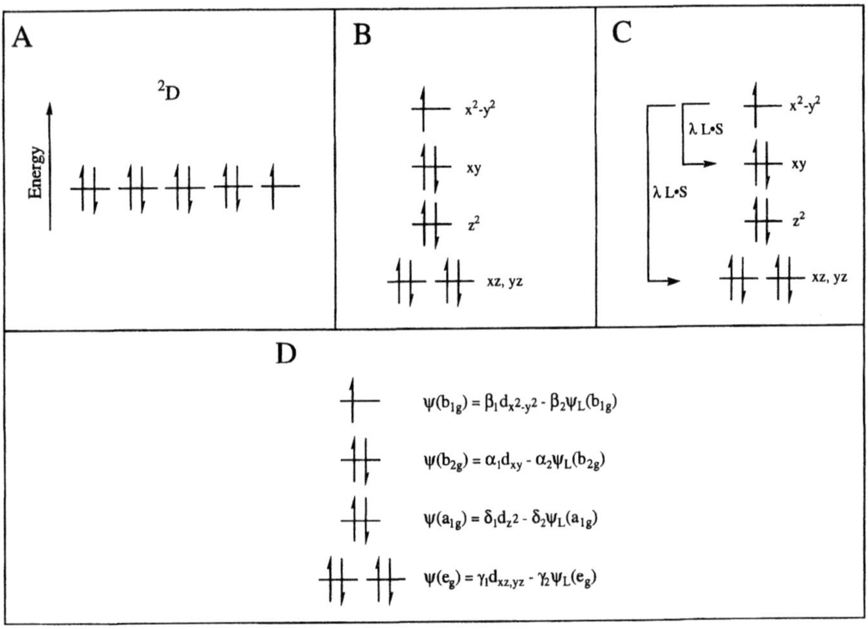


**Figure 6.** Effect of a ligand field on the *d* orbitals of Cu(II) ($d^9$ configuration). A) Five degenerate *d* orbitals in the free ion; B) ligand field splitting of the *d* orbitals in a tetragonal Cu(II) complex, C) spin-orbit coupling allows excited *d* levels to mix into the $d_{x^2-y^2}$ ground state; D) modified ligand field wavefunctions which include covalent interactions with the ligands.

The orbital angular momentum relates to orbital degeneracy and for the free metal ion the five *d* orbitals are degenerate. For Cu(II) ($d^9$ configuration) this produces a $^2D_{5/2}$ ground state with $S = 1/2$, $L = 2$, and $J = 5/2$ giving a $g = 1.200$ (Figure 6A). The ligand field splitting of the *d* orbitals in a tetragonal Cu(II) complex is given in Figure 6B which shows that the ground state now has an unpaired electron in a $d_{x^2-y^2}$ orbital. Because this is non-degenerate, there should be no orbital angular momentum, and the $g$ value (Equation 2, with $L = 0$, $J = S$) should be 2 (to be precise, 2.0023).

As shown in Figure 6C, ligand field theory allows the excited *d* states to mix into the $d_{x^2-y^2}$ ground state through spin-orbit coupling ($\lambda\hat{L} \bullet \hat{S}$ where $\lambda[\mathrm{Cu(II)}] = -830\ \mathrm{cm}^{-1}$). (Note that here the ligand field excited states are designated by the associated hole in the *d* orbital.) This mixes some orbital angular momentum into the ground state and leads to the expression for the $g_i$ values given in Equation 3, where the last term is the orbital contribution which allows the $g$ values to deviate from 2.0023.[9]

$$g_i = 2 - 2\lambda \sum_{n \neq 0} \frac{\langle \Psi_0 | L_i | \Psi_n \rangle \langle \Psi_n | L_i | \Psi_0 \rangle}{E_n - E_0} \tag{3}$$

**Table 3 Effect of orbital angular momentum operators on the real *d* orbitals***

| | | |
|---|---|---|
| $\hat{L}_x d_{xz} = -id_{xy}$ | $\hat{L}_y d_{xz} = id_{x^2-y^2} - i\sqrt{3}d_{z^2}$ | $\hat{L}_z d_{xz} = id_{yz}$ |
| $\hat{L}_x d_{yz} = i\sqrt{3}d_{z^2} + id_{x^2-y^2}$ | $\hat{L}_y d_{yz} = id_{xy}$ | $\hat{L}_z d_{yz} = -id_{xz}$ |
| $\hat{L}_x d_{xy} = id_{xz}$ | $\hat{L}_y d_{xy} = -id_{yz}$ | $\hat{L}_z d_{xy} = -2id_{x^2-y^2}$ |
| $\hat{L}_x d_{x^2-y^2} = -id_{yz}$ | $\hat{L}_y d_{x^2-y^2} = -id_{xz}$ | $\hat{L}_z d_{x^2-y^2} = 2id_{xy}$ |
| $\hat{L}_x d_{z^2} = -i\sqrt{3}d_{yz}$ | $\hat{L}_y d_{z^2} = i\sqrt{3}d_{xz}$ | $\hat{L}_z d_{z^2} = 0$ |

*Reference 9.

$\Psi_0$ is a ground state, $\Psi_n$ are the ligand field excited states, $E_n - E_0$ is the energy of ligand field excited state $n$ above the ground state, and $L_i$ is the $i = x, y$, or $z$ component of the orbital angular momentum operator, the matrix element of which is evaluated between the ground state and the excited state. The effect of the $L_i$ on the real $d$ orbitals is tabulated in Table 3. For tetragonal Cu(II) with a $d_{x^2-y^2}$ ground state and ligand field excited state energies on the order of 12,000 $cm^{-1}$ (see Section 4.1.1.2), Equation 3 gives $g_\parallel (= g_z) = 2.55$, $g_\perp (= g_x = g_y) = 2.14$. Note that the $g$ values are now anisotropic because different orientations of the complex in the magnetic field project out specific components of the orbital angular momentum operator $L_i$ $(i = x, y, z)$ in Equation 3, and from Table 3 these couple the $d_{x^2-y^2}$ ground state to different excited states. These calculated $g$ values are considerably larger than those obtained experimentally (e.g., $CuCl_4^{2-}$: $g_\parallel = 2.221, g_\perp = 2.040$).[10] Ligand field theory treats the $d$ orbitals as having pure metal character, however, covalent interactions with the ligands modify the ligand field wavefunctions as shown in Figure 6D, where $\Psi_L(\Gamma_i)$ indicates a ligand molecular orbital with $\Gamma_i$ symmetry. This ligand character reduces the orbital angular momentum content of these wavefunctions resulting in a reduction of the $g$ values. The last term in Equation 3 should therefore have a "Stevens orbital reduction factor" $k_i^2 < 1$ $(i = x, y, z)$ included to account for the effect of ligand covalency.[11] Modified expressions for the $g$ values which include the ligand and metal coefficients of all the wavefunctions in Figure 6D are given in Reference 12.

There are three contributions to the metal hyperfine coupling constant, $A_i^M$. These are given by Equation 4 for an unpaired electron in the $d_{x^2-y^2}$ ground state of a tetragonal Cu(II) complex.[13] (Note that although the sign of the hyperfine coupling constant can be positive or negative, only its magnitude is measured in the EPR spectrum in Figure 4C).

$$\begin{aligned} A_\parallel &= P_d\left(-\kappa - \tfrac{4}{7} + \left[(g_\parallel - 2.0) + \tfrac{3}{7}(g_\perp - 2.0)\right]\right) \\ A_\perp &= P_d\left[-\kappa + \tfrac{2}{7} + \tfrac{11}{14}(g_\perp - 2.0)\right] \end{aligned} \tag{4}$$

The first term in Equation 4 describes the Fermi contact interaction, $-P_d\kappa$, which is isotropic [the same for all orientations of the complex relative to the magnetic field, i.e., $A_\parallel(\text{Fermi}) = A_\perp(\text{Fermi})$]. This involves the interaction of the unpaired electron spin density at the nucleus with the nuclear spin. An electron in a $d$ orbital has no

**Table 4 Individual Contributions to Metal Hyperfine Coupling**

| | Calculated[14] ($\times 10^{-4}$cm$^{-1}$) | | | | Experimental[10] ($D_{4h}$ $CuCl_4^{2-}$) |
|---|---|---|---|---|---|
| | $A^F$ | $A^{\text{Spin Dip}}$ | $A^{\text{Ind. Dip}}$ | Total | |
| $A_\parallel$ | −172 | −229 | +95 | −306 | −164 |
| $A_\perp$ | −172 | −114 | +13 | −45 | −35 |

probability at the metal nucleus; however, it can interact with core electrons in filled $1s$, $2s$, and $3s$ orbitals and spin polarize these through exchange coupling to produce a net negative spin density at the metal nucleus. A reasonable estimate of this term is given in Table 4 (first column).[14] The second term is the spin dipolar interaction which is anisotropic [$A_\parallel$(spin dipolar) $= -(4/7)P_d$, $A_\perp$(spin dipolar) $= +(2/7)P_d$ where $P_d$[Cu(II)] $= 400 \times 10^{-4}$ cm$^{-1}$]. This interaction involves the electron spin vector, averaged over the metal $d_{x^2-y^2}$ orbital, dipolar coupling with the nuclear spin on the metal. The third term is also orientation dependent [$A_\parallel$(orbital dipolar) $=$ $P_d(g_\parallel - 2.0) + (3/7)(g_\perp - 2.0)$, $A_\perp$(orbital dipolar) $= P_d(11/14)(g_\perp - 2.0)$] and results from the dipolar interaction of the orbital angular momentum spin-orbit mixed into the $d_{x^2-y^2}$ ground state with the nuclear spin and is thus related to the deviations of the g values from 2.0023.

As shown in Table 4, reasonable estimates of all three terms give calculated $A^M$ values which are considerably larger in magnitude than those observed experimentally (square planar $CuCl_4^{2-}$ is given as an example).[10] Covalent delocalization of the electron spin onto the ligands reduces its interaction with the nuclear spin on the metal ion. This covalency has already been included into the orbital dipolar term through the use of experimental $g$ values which are reduced from those of pure $d$ orbitals by covalency (*vide supra*). However, both the Fermi contact and spin dipolar terms must be reduced by $\beta_1^2 < 1.0$ (see $d_{x^2-y^2}$ derived molecular orbital in Figure 6D) to account for the covalent delocalization of the electron spin away from the metal center. For $CuCl_4^{2-}$, $\beta_1^2$ is 0.61 (*vide infra*).[15]

2.1.1.3 *Superhyperfine Interactions* The ligands ($L_i$) can also have nuclear spins $I^{L_i}$ which couple to the electron spin to produce an additional, smaller "superhyperfine" splitting of individual hyperfine lines, each into $2\ I^{L_i} + 1$ components spaced by $A^{L_i}$, the superhyperfine coupling constant of ligand $L_i$ (Figure 7A). This can also be anisotropic ($A_z^{L_i} \neq A_x^{L_i}, A_y^{L_i}$), in this case depending on the orientation of the magnetic field relative to the specific $L_i$–$M$ bond direction (i.e., the $L$–$M$ bond defines the $z_i$ axis of $A^{L_i}$). Note that the ligand superhyperfine coupling constants are measured in the $g$ tensor coordinate system (Figure 7A). Thus, from Figure 7B, the $A_\parallel^{L_i}$ measured in the $g_\parallel$ region of the EPR spectrum of a tetragonal Cu(II) complex corresponds to the $A_{x_i}^{L_i}$ or $A_{y_i}^{L_i}$ component of the ligand superhyperfine tensor; $A_{z_i}^{L_i}$ contributes to the $g_\perp$ region of the EPR spectrum.

From ligand field theory, there are two contributions to the superhyperfine coupling of a specific ligand nuclear spin to the electron spin. The first, which does not require covalency of the $L_i$–$M$ bond, is the dipolar coupling of the electron spin on

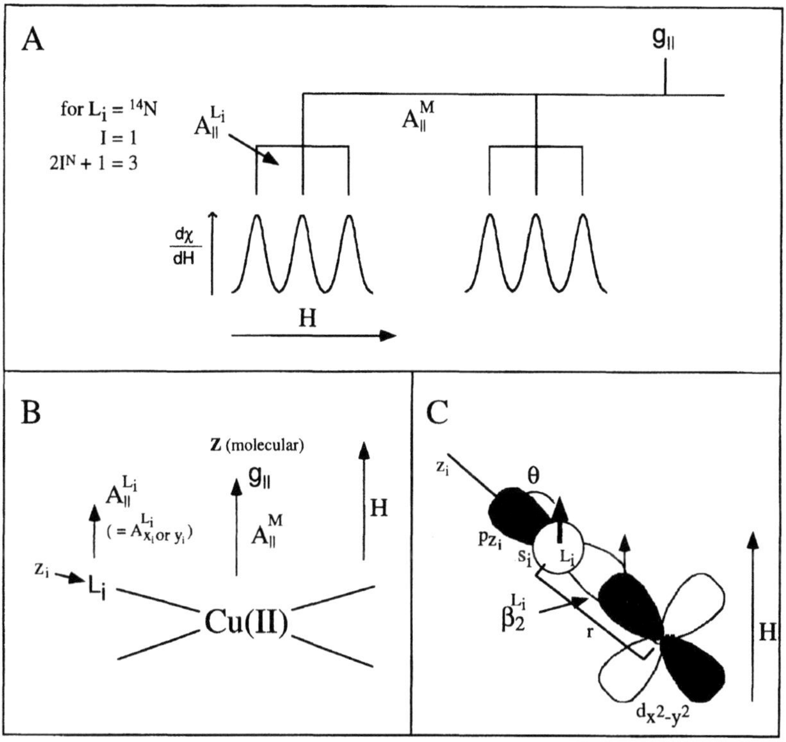


**Figure 7.** Superhyperfine coupling. A) Ligand superhyperfine splitting of metal parallel hyperfine lines; B) ligand $A_{\parallel}$ value in the $g$-tensor coordinate system corresponds to the $A_{x_i}$ or $A_{y_i}$ component in the ligand$_i$–metal bond coordinate system; C) electron spin on the Cu ion is delocalized onto the ligand $i$, allowing it to interact directly with the ligand nuclear spin.

the metal with the nuclear spin on the ligand separated by the $L$–$M$ bond length, $r$. This is given by Equation 5 where $g_N$ is the nuclear $g$ value for ligand $L_i$ given in Table 5, $\beta_N$ is the nuclear magneton [$= (1/1{,}836) \times \beta$] and $\theta$ is the angle of the $L_i$–$M$ bond relative to the magnetic field direction.

$$A^{L_i}(\text{dipolar}) = g\beta g_N \beta_N \left( \frac{3\cos^2\theta - 1}{r^3} \right) \tag{5}$$

This contribution is anisotropic (i.e., dependent on the orientation of the magnetic field relative to the $L_i$–$M$ bond) and, in the absence of covalency, its experimental value can be used to accurately estimate the $L$–$M$ distance.

If significant covalency is present, this will dominate the superhyperfine coupling. As shown in Figure 7C, delocalization of the electron spin onto ligand $i$ by an amount $(\beta_2^{L_i})^2$ allows it to directly interact with the nuclear spin on that ligand. Generally, the ligand orbital involved in bonding to the metal ion is a hybrid having $(1 - n^2)$ ligand $p_z$ and $n^2$ ligand $s$ character. An unpaired electron in an $s$ orbital on the ligand will

**Table 5 Magnetic Properties and Hyperfine Coupling Constants[16] for Biologically Important Ligand Nuclei**

| Ligand | $I$ | Natural Abundance (%) | $g_N$ | $A^{L_i}_{(iso)}$ (gauss) | $A^{L_i}_{\parallel(aniso)}$ (gauss) |
|---|---|---|---|---|---|
| $^{1}H$ | 1/2 | 99.9844 | 5.58536 | 508 | — |
| $^{2}H$ | 1 | 0.0156 | 0.857386 | 78 | — |
| $^{13}C$ | 1/2 | 1.108 | 1.40440 | 1119 | 63.8 |
| $^{14}N$ | 1 | 99.635 | 0.40358 | 557 | 33.5 |
| $^{15}N$ | 1/2 | 0.365 | −0.56608 | −781 | −46.6 |
| $^{17}O$ | 5/2 | 0.037 | −0.75720 | −1659 | −102 |
| $^{33}S$ | 3/2 | 0.74 | 0.42849 | 975 | 56.6 |
| $^{35}Cl$ | 3/2 | 75.4 | 0.54727 | 1672 | 102 |
| $^{37}Cl$ | 3/2 | 24.6 | 0.45553 | 1391 | 84.8 |

have electron spin density at the $L_i$ nucleus and produce a Fermi contact contribution to the superhyperfine coupling constant. This is given in the fifth column of Table 5 [$A^{L_i}$(iso)] for a number of important ligand nuclei in Bioinorganic Chemistry. An unpaired electron in a $p_z$ orbital on the ligand ($z$ here is usually the $L_i$–$M$ bond axis) will only have a spin dipolar interaction with the $L_i$ nuclear spin averaged over the electron's position in the $p_z$ orbital. This will be anisotropic with $A^{L_i}_{\perp}(\text{aniso}) = -1/2A^{L_i}_{\parallel}(\text{aniso})$. Values for $A^{L_i}_{\parallel}$(aniso) for an unpaired electron in the $p_z$ orbital of specific ligands are given in the last column of Table 5. The above considerations lead to Equation 6, which relates the values of the superhyperfine coupling observed experimentally to both the covalency of the $L_i$–$M$ bond, $\beta_2^{L_i}$, and the hybridization of the $L_i$ valence orbital, $n$.

$$\begin{aligned} A^{L_i}_{z}(\text{covalency}) &= \left(\beta_2^{L_i}\right)^2\left[n^2 A^{L_i}_{\text{iso}} + \left(1-n^2\right)A^{L_i}_{\text{aniso}_\parallel}\right] \\ A^{L_i}_{x,y}(\text{covalency}) &= \left(\beta_2^{L_i}\right)^2\left[n^2 A^{L_i}_{\text{iso}} - \tfrac{1}{2}\left(1-n^2\right)A^{L_i}_{\text{aniso}_\parallel}\right] \end{aligned} \tag{6}$$

Although superhyperfine coupling is a very powerful probe of covalency, it is often difficult to obtain experimentally because hyperfine lines are inhomogeneously broadened and splittings of less than a few gauss [$A(\text{cm}^{-1}) = A(\text{gauss}) \times g \times 4.6686 \times 10^{-5}$, therefore for $g = 2.2$, 1 gauss $\cong 1 \times 10^{-4}$ cm$^{-1}$] will generally be unresolvable in a standard [i.e., X-band, continuous wave (cw), single resonance] EPR experiment.

**2.1.2** ***Electron Nuclear Double Resonance (ENDOR)*** ENDOR spectroscopy combines the EPR experiment with an NMR experiment which probes the nuclear spin splittings within an electron spin sublevel and thus provides a high resolution method of obtaining nuclear hyperfine couplings to an electron spin. At low temperature and high microwave power, an EPR transition at a fixed magnetic field corresponding to a specific $g$ value of the site will saturate (i.e., lose signal intensity through population equalization of the spin sublevels involved in the transition due

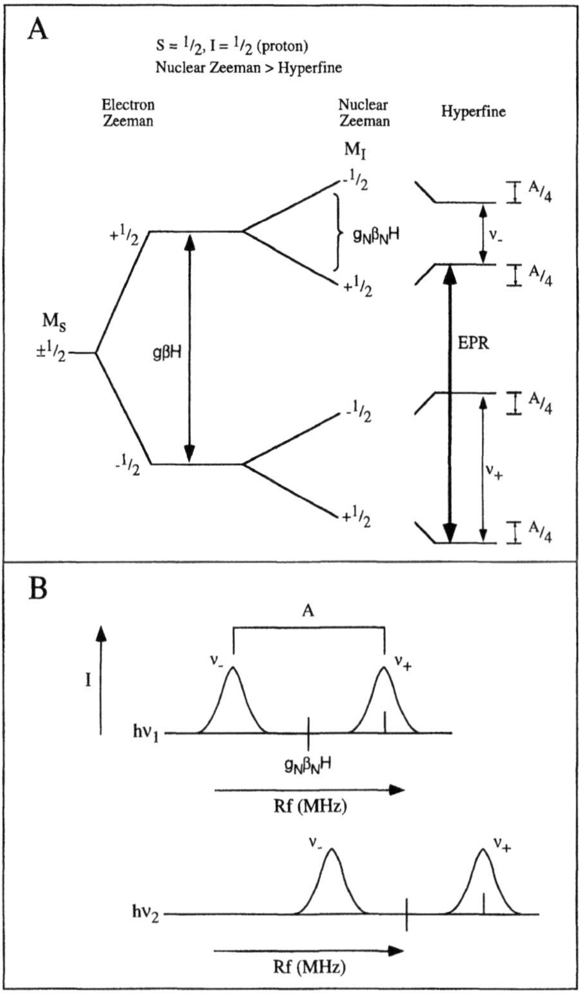


**Figure 8.** Proton ENDOR transitions. A) Energy level diagram depicting splittings for the nuclear Zeeman effect larger than the hyperfine coupling ($\Delta M_S = 1$, $\Delta M_I = 0$ EPR transitions indicated by dark arrow, $\Delta M_S = 0$, $\Delta M_I = 1$ allowed ENDOR transition indicated as $\nu_+$ and $\nu_-$); B) ENDOR spectrum shows two peaks split by the hyperfine coupling constant, A, and centered around the proton Larmor frequency. Increasing the microwave energy to $h\nu_2$ increases the field and shifts of the proton nuclear Zeeman splitting.

to slow spin relaxation). Under these conditions, the system is scanned in energy (MHz) with a radio-frequency (rf) source until nuclear transitions come in resonance (i.e., a double resonance condition is satisfied). Saturation is then reduced due to the presence of additional relaxation pathways, and microwave absorption is observed. Thus in ENDOR spectroscopy the EPR intensity is used to sensitively detect low energy transitions between nuclear spin sublevels.[17]

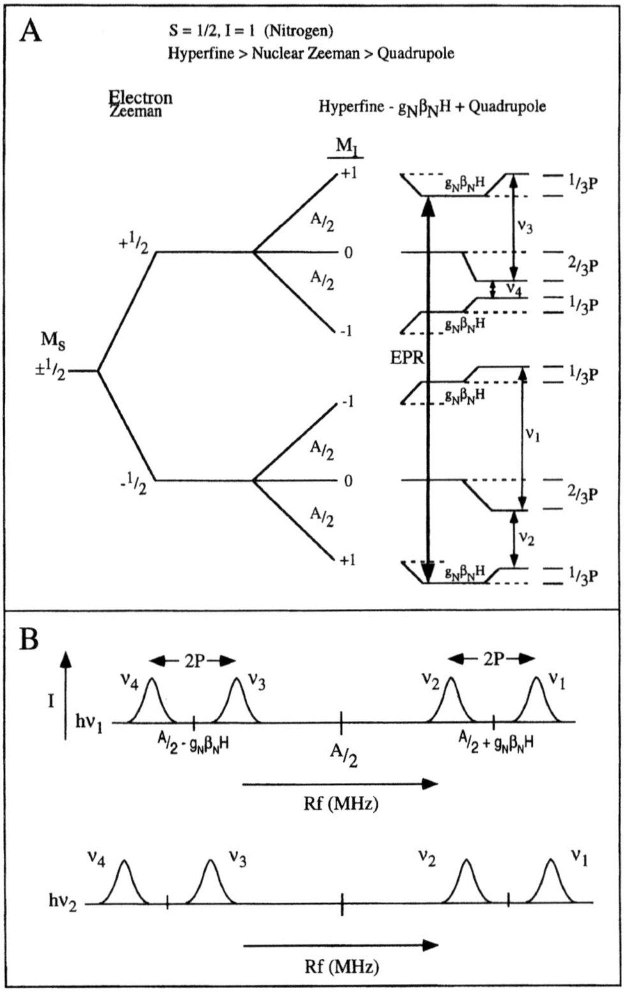


**Figure 9.** ENDOR transitions for most other nuclei using nitrogen $I = 1$ as an example. A) Energy level diagram depicting splittings for the nuclear Zeeman effect smaller than the hyperfine coupling (here A and P are assumed to be positive); (Allowed EPR transition is given by dark arrow, allowed ENDOR transitions indicated as $\nu_1$ to $\nu_4$.) B) ENDOR spectrum shows two sets of two peaks split by twice the nuclear Zeeman energy and centered at one half the hyperfine coupling constant, A. Increasing microwave energy ($h\nu_2 > h\nu_1$) increases the splitting of the two sets of peaks about $A/2$.

The Hamiltonian required to analyze the observed ENDOR frequencies is given in Equation 7 (note that ENDOR spectra are given in MHz where $1 \times 10^{-4}$ cm$^{-1}$ = 3 MHz).

$$\mathcal{H} = g\beta H \bullet S + AS \bullet I - g_N \beta_N H \bullet I + P\left[I_z^2 - \tfrac{1}{3}I(I+1)\right] \qquad (7)$$

The first term describes the electron Zeeman effect and is large relative to the other terms, and the second term describes the hyperfine coupling interaction. These terms have been discussed above. The third term describes a nuclear Zeeman effect which is dependent on the nuclear $g_N$ values given in Table 5, and the last term describes the quadrupole interaction. A nucleus with $I \geq 1$ will have an electric quadrupole moment which interacts with the electric field gradient in the vicinity of the nucleus (i.e., the local chemical environment when the symmetry is less than $O_h$ or $T_d$). Two limits of Equation 7 need to be considered. For protons ($I = 1/2$) the nuclear Zeeman effect is large relative to the hyperfine coupling due to its large $g_N$ value (Table 5). As shown in Figure 8, protons exhibit two peaks in the ENDOR spectrum centered around their nuclear Zeeman energy splitting ($= g_N \beta_N H$: for $g \sim 2$ at X-band, $H \sim 3200$ gauss and this proton Larmor frequency is $\nu_H = 14$ MHz) and split by the hyperfine coupling constant. (Note that in the ENDOR experiment transitions are observed between $M_I$ components of a given $M_S$ sublevel with the selection rule $\Delta M_I = \pm 1$). Alternatively, for most other nuclei, the hyperfine coupling is larger than the nuclear Zeeman term and Figure 9 is appropriate. For a nitrogen ligand ($I = 1$), two quadrupole split peaks are centered at half the hyperfine coupling constant and split by twice the nuclear Zeeman interaction (for $g \sim 2$ at X-band, $H \sim 3200$ gauss and $\nu_N = 1$ MHz).

In order to distinguish between possible assignments of peaks it is often advantageous to perform the ENDOR experiment at a second microwave frequency.[17b] This changes the magnetic field required to satisfy the electron resonance condition for a given $g$ value and thus shifts the nuclear Zeeman resonance frequency as this is dependent on the applied magnetic field. Alternatively, the hyperfine coupling is not magnetic field dependent and does not change with microwave frequency (see bottom of Figures 8B and 9B). Further, the anisotropy in the hyperfine coupling constant can be measured by obtaining ENDOR spectra at the different magnetic fields associated with the different electron g values of the metal site. Finally, it should be noted that ENDOR intensities generally decrease toward the low frequency side of the spectrum. To study the low energy spectral region (associated with small hyperfine interactions) with high sensitivity, one utilizes pulsed EPR methods as described below.

**2.1.3** ***Electron Spin Echo Envelope Modulation (ESEEM)*** Pulsed EPR methods provide a very sensitive mechanism of detecting nuclear spins that are weakly interacting with a paramagnetic metal center (i.e., very low superhyperfine coupling constants).[18] As shown in Figure 10(A and B), an intense microwave pulse is applied to a compound in an external magnetic field ($H$ along the $z$ axis) with the pulse perpendicular to the external field (along $x$ of the rotating frame) and of $\pi/2$ [in units of $1/\nu$(Hz)] duration. The spins centered around the resonance field are rotated 90° into the $y$ axis. The short pulse will excite spin packets with a range of frequencies $\omega_i$ which can be greater ($\omega_+$) or less ($\omega_-$) than the central resonance frequency ($\omega_0$). These start to dephase after the pulse (the Free Induction Decay, FID, Figure 10B3). After a time $\tau$ a second intense perpendicular (along $x$) $\pi$ microwave pulse is applied which inverts the spins (Figure 9B4) such that their time behavior is re-

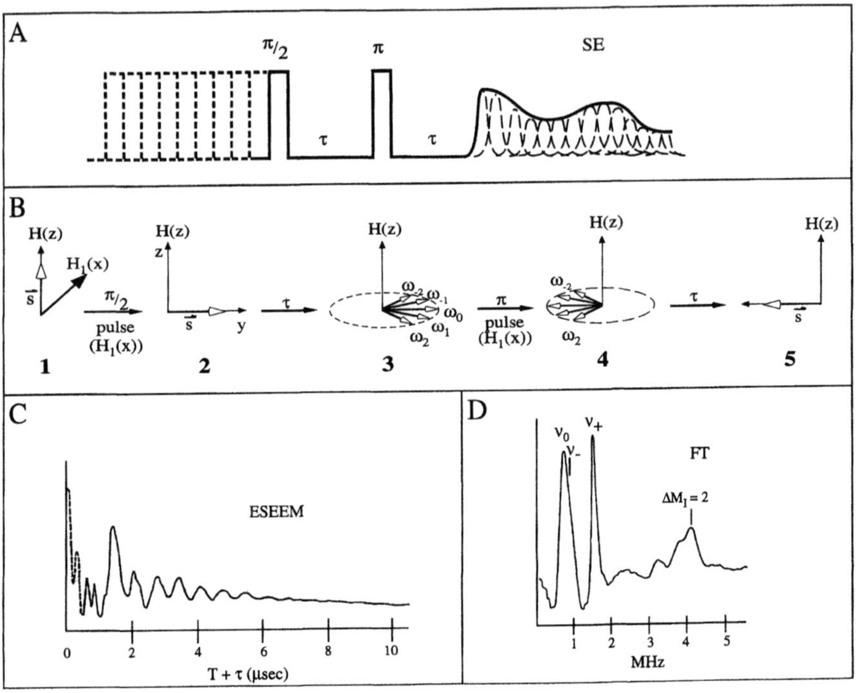


**Figure 10.** The ESEEM experiment. A) Applied microwave pulses generate an envelope of spin echoes; B) pulse sequence and effect on electron spin ($\vec{s}$) packet centered at $\omega_0$; C) decay of the spin echo is modulated by interference patterns due to nearby nuclear spins. ESEEM spectrum of stellacyanin obtained with the three pulse method (see text); D) Fourier transform of time-dependent spectrum of stellacyanin (Figure 10C). Adapted from Reference 18a–c.

versed and they now refocus into the central frequency. Another time period $\tau$ after the second pulse these are back in alignment producing a spin echo (Figure 10B5). The amplitude of the electron spin echo is then detected as a function of the time $\tau$ between pulses (Figure 10A).

The amplitude of the echo decays rapidly with time due to spin relaxation. However, as shown on the right in Figure 10A and in Figure 10C, this decay is modulated by constructive and destructive interferences associated with nearby nuclear spins interacting with the electron spin.[18] This interference occurs because the microwave pulses cover an energy range extending over the $M_I$ components of a given spin sublevel (Figures 8 and 9). At low hyperfine couplings these wavefunctions become somewhat mixed, such that formally forbidden $\Delta M_S = \pm 1$, $\Delta M_I = \pm 1$ microwave transitions become weakly allowed, and these can interfere with the allowed $\Delta M_S = \pm 1$, $\Delta M_I = 0$ transitions. The oscillation frequencies of Figure 10C are dependent on the energy spacings between the different $M_I$ sublevels in Figures 8 and 9. The Fourier transform of this time dependence produces an energy spectrum (Figure 10D) which corresponds to the ENDOR splittings given by Equation 7 ob-

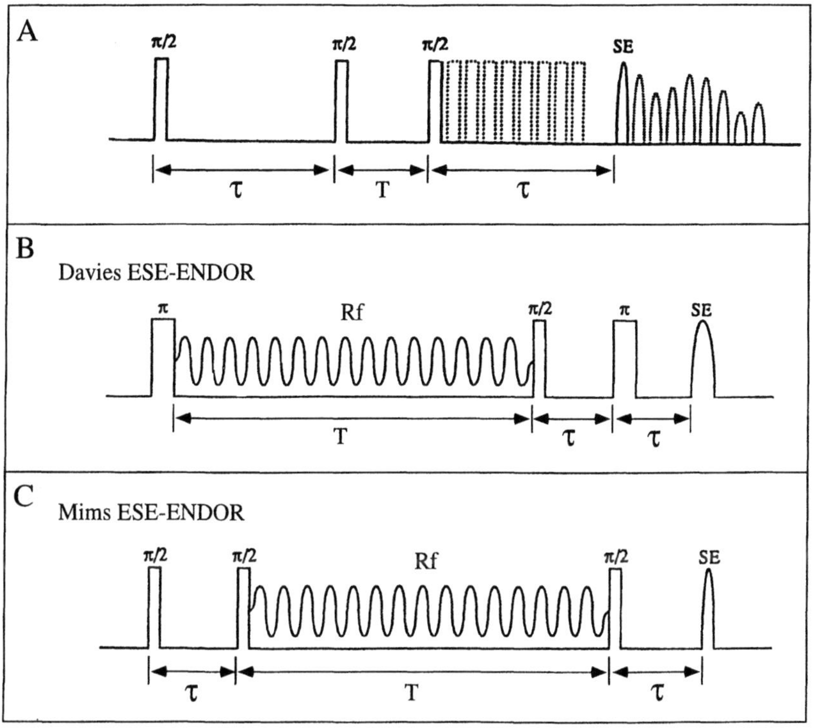


**Figure 11.** Alternative ESE pulse sequences. A) The three pulse stimulated echo experiment. The time $\tau$ between the first and second pulses is maintained while the time T (between the second and third) is incremented; B) Davies ESE-ENDOR sequence; C) Mims ESE-ENDOR sequence. Taken from Reference 18d.

tained with high sensitivity in the weak coupling, $\leq 5$ MHz, region. In practice the two pulse echo decays too rapidly for an accurate Fourier transform and additionally has complications due to nuclear energy sums and differences which also contribute to the envelope modulation in Figure 10C. These complications are not present in three pulse stimulated echo experiments (see Figure 11A), and consequently Fourier transform spectra are generally obtained by the three pulse method.[18]

The ESEEM spectrum shown in Figure 10D is that of the non-coordinated nitrogen of a histidine ligand coupled to the electron spin on a Cu(II) center.[18c] (The coordinated histidine–nitrogen is too strongly coupled to contribute to the ESEEM spectrum but is directly observed in ENDOR, see Section 2.1.2.) The appropriate energy diagram for N ($I = 1$) coupling to an electron spin ($S = 1/2$) is given in Figure 9A. Note that in one $M_S$ manifold the Zeeman splitting opposes that associated with the hyperfine interaction, while in the other $M_S$ manifold they add. It is found that the small hyperfine coupling of the non-coordinated N of histidine to Cu(II) is very close in magnitude to the nuclear Zeeman splitting of N at X-band. Thus in one $M_S$ manifold these cancel, and all that is left is the nuclear quadrupole coupling, which is the quantity observed in the Fourier transformed ESEEM spectrum in Figure 10D (the intense peaks at $< 2$ MHz). The nuclear transitions in the other $M_S$ manifold in Figure 9A have large splittings due to the addition of the hyperfine and nuclear Zeeman terms. These splittings are anisotropic and generally lead to broad contributions to the ESEEM spectrum (region above 3 MHz in Figure 10D) except for the $\Delta M_I = 2$ transition which is generally sharp. As has been previously indicated, it is important to perform ESEEM experiments at different microwave frequencies and hence different magnetic fields so that the nuclear Zeeman splitting can be adjusted to cancel the hyperfine interaction and allow the quadrupole splittings to be observed for other nuclei and ligands.[18b]

The quadrupole term in Equation 7 derives from the electric quadrupole moment of the nucleus interacting with the non-spherical electron charge distribution in the vicinity of that atom [i.e., $P = 3e^2Qq/(4I(2I - 1))$ where $Q$ is the nuclear quadrupole moment, $q$ the electric field gradient, and $e^2Qq$ the quadrupole coupling constant]. This is quite sensitive to the chemical environment of the nucleus, allowing definitive ligand assignments. Equation 7 and Figure 9A assume that the atom with $I \geq 1$ has axial symmetry, in which case one would expect to observe two overlapping $\Delta M_I = \pm 1$ transitions (one peak) at a value of $P$ MHz. However, the ESEEM spectrum in Figure 10D shows two intense peaks at 0.7 and 1.4 MHz. The N nucleus being observed has lower than axial symmetry, and this introduces a new term in the spin Hamiltonian (Equation 7), $(P/3)\ [-(\eta/2)(I_+^2 + I_-^2)]$, where $0 \leq \eta \leq 1$ is the asymmetry parameter which quantitates the rhombic symmetry $[\eta = (q_{xx} - q_{yy})/q_{zz}]$. The appropriate energy diagram for the quadrupole splitting of an $I = 1$ nucleus in rhombic symmetry is shown on the right in Figure 12. There

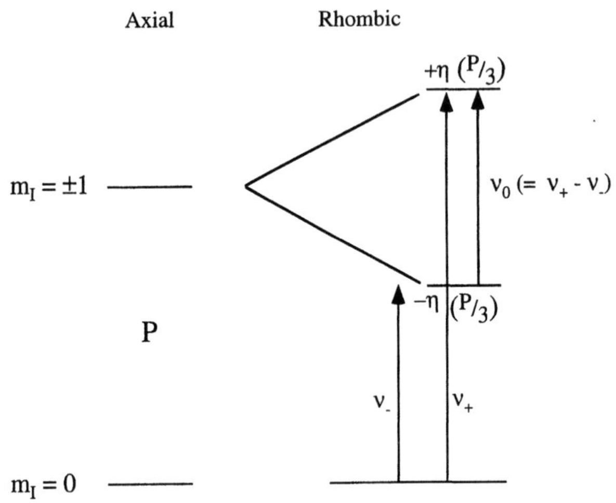


**Figure 12.** Energy level diagram depicting the quadrupole splitting of an $I = 1$ nucleus resulting from A) axial and B) rhombic symmetry, showing the two $\Delta M_I = \pm 1$ allowed transitions, $\nu_-$ and $\nu_+$, and the formally $\Delta M_I = 2$ transition, $\nu_0$, which becomes observable in rhombic symmetry.

are now three allowed $\Delta M_I = \pm 1$ transitions observed in the ESEEM spectrum, $\nu_-$ and $\nu_0$, which are close to coincident at ~0.7 MHz, and $\nu_+$ at ~1.4 MHz. These observed transitions can then be used to obtain the quadrupole coupling constant and asymmetry parameter, $e^2Qq = (2/3)(\nu_+ + \nu_-)$, $\eta = 3\nu_0/(\nu_+ + \nu_-)$. The values obtained from Figure 10D, $e^2Qq = 1.49$ MHz and $\eta = 0.94$, are very similar to those measured by nuclear quadrupole resonance (NQR) spectroscopy for the protonated nitrogen of free imidazole[19] that led to the above assignment.

Finally, it should be noted that electron spin echoes are now being used in combination with scanned rf sources to sensitively detect ENDOR signals (ESE ENDOR) without passage or other effects that can complicate cw ENDOR. The two different pulse sequences employed (Mims and Davies) are given in Figures 11B and 11C.[18d]

## 2.2 Unique Spectral Features

**2.2.1** ***Blue (Type I) Copper Proteins*** Since the early characterization of copper proteins based on their EPR spectra, the blue copper proteins have been recognized as being unique due to their very small parallel hyperfine splitting (Figure 13, plastocyanin: $63 \times 10^{-4}$ cm$^{-1}$; $CuCl_4^{2-}$: $164 \times 10^{-4}$ cm$^{-1}$).[20] As shown in Figure 1A, the blue copper sites are mononuclear with an approximately tetrahedral geometry and a

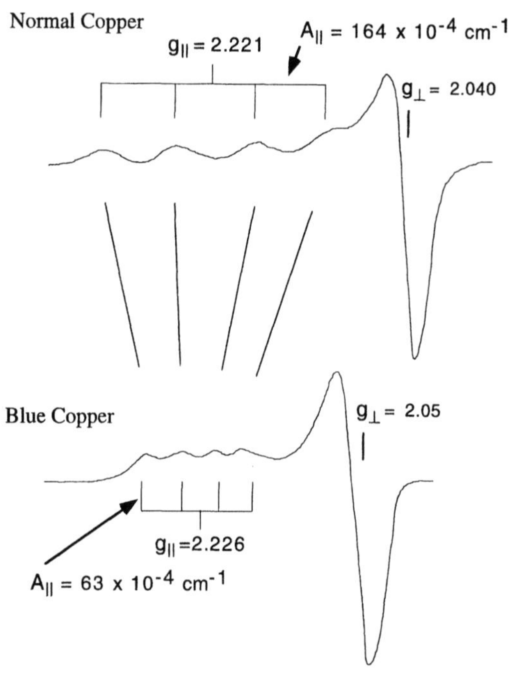


**Figure 13.** EPR spectra of A) normal copper, $CuCl_4^{2-}$ and B) the blue (Type I) copper center in plastocyanin which shows the characteristic extremely small parallel hyperfine splitting. Taken from Reference 2.

thiolate–copper bond.[21] Initially the small hyperfine splitting was ascribed to the distorted tetrahedral structure of the center.[22] In low symmetry the Cu 4$p$ orbitals can mix with the $d_{x^2-y^2}$ orbital, and as shown in Figure 14A, the dipole interaction of an electron spin in a Cu 4$p_z$ orbital with the nuclear spin on the copper would oppose that of an electron spin vector in a $d_{x^2-y^2}$ orbital. If a significant amount of Cu 4$p_z$ (>12%) mixes into the $d_{x^2-y^2}$ orbital in the blue copper site, this would reduce the $A_{\|}$ value to that observed experimentally. However, X-ray absorption spectroscopy (XAS) studies at the copper K-edge (see Section 6) have shown that it is the Cu 4$p_x$, $p_y$ orbitals that mix into the $d_{x^2-y^2}$ ground state of the blue copper site.[23] As shown in Figure 14B, the spin dipolar interaction of an electron averaged over the $p_x$, $p_y$ orbitals with the nuclear spin of the copper would complement that of an electron in the $d_{x^2-y^2}$ orbital and *increase*, not decrease, the hyperfine coupling.

The Cu K-edge XAS studies eliminated the Cu 4$p_z$ mixing model for the small $A_{\|}$ value of blue copper centers and shifted attention to the possibility of a highly covalent active site. This was addressed through quantitative analysis of the EPR $g$ values using self-consistent field X-alpha scattered wave (SCF–X$\alpha$–SW) density functional calculations.[14] The results are given in Figure 15 which show that the blue copper site is significantly more covalent than $CuCl_4^{2-}$ (42% vs. 61% Cu $d_{x^2-y^2}$ character) and that the covalency of the blue copper center is anisotropic involving the thiolate–Cu(II) bond. These results have been experimentally confirmed through copper L-edge[24] and sulfur K-edge XAS studies[23] (see Section 6). The wavefunctions in Figure 15 give the contributions to the metal hyperfine interaction (Equation 4) listed in Table 6. Note that the values for $CuCl_4^{2-}$ given in Table 4 are modified in Table 6 due to covalency. From Table 6, both the Fermi contact and spin dipolar terms are reduced in magnitude in plastocyanin relative to $CuCl_4^{2-}$. This is due to the increased delocalization of the electron spin onto the thiolate ligand (Figure 15)

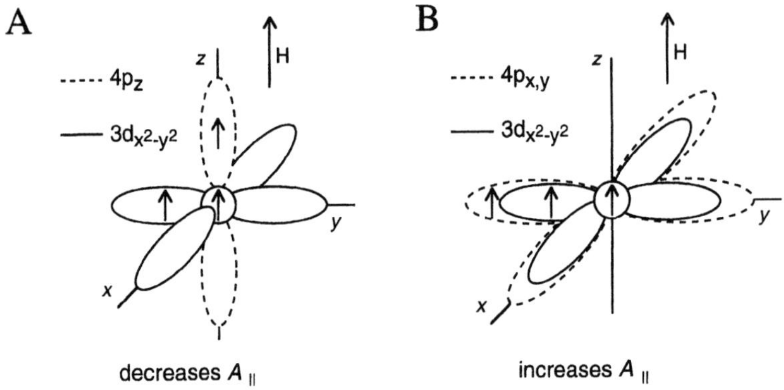


**Figure 14.** Effects of Cu 4$p$ mixing with Cu $d_{x^2-y^2}$ orbital on hyperfine coupling. A) $A_{\|}$ is reduced by Cu 4$p_z$ mixing with the $d_{x^2-y^2}$ orbital due to opposing electron spin dipole interactions with the nuclear spin on Cu; B) $A_{\|}$ is increased by the spin dipolar interaction between the nuclear spin of the copper and an electron spin averaged over the Cu $p_x$ and $p_y$ orbitals mixed with the $d_{x^2-y^2}$ orbital. Taken from Reference 2.

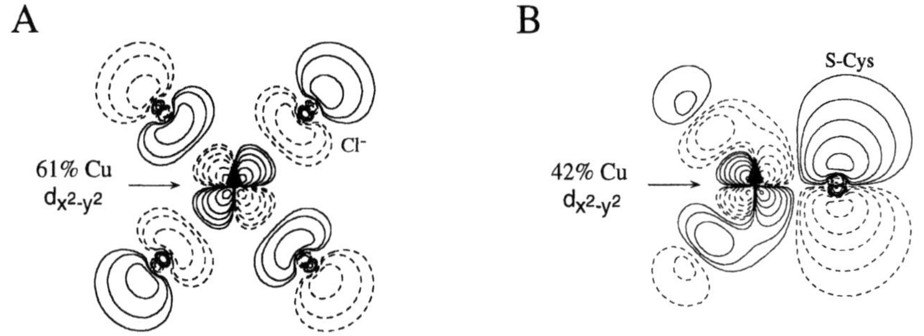


**Figure 15.** Contour plots of ground state wavefunctions generated from SCF–Xα–SW calculations of A) $CuCl_4^{2-}$ and B) plastocyanin, the latter showing significantly more covalency. Adapted from Reference 2.

**Table 6 Contributions[14] to Metal Hyperfine Coupling for Blue Copper as Compared to $D_{4h}$ $CuCl_4^{2-}$***

| | | Calculated ($\times 10^{-4}$ cm$^{-1}$) | | | | Experimental |
|---|---|---|---|---|---|---|
| | | $A^F$ | $A^{\text{Spin Dip}}$ | $A^{\text{Ind. Dip}}$ | Total | |
| $D_{4h}$ $CuCl_4^{2-}$ | $A_\parallel$ | −123 | −155 | +114 | −164 | \|164\| |
| | $A_\perp$ | −123 | +78 | +17 | −28 | \|35\| |
| Plastocyanin | $A_z$ | −79 | −90 | +106 | −65 | \|63\| |
| | $A_{x,y}$ | −79 | +45 | +18 | −15 | \|<17\| |

*Modified from Table 4 to include covalency.

which reduces the electron spin's interaction with the nuclear spin on the copper. The high covalency of the thiolate–Cu(II) bond strongly couples this redox center into the protein which plays a key role in the rapid long range electron transfer function of this site in Biology (see Reference 25).

**2.2.2** ***Coupled Binuclear (Type III) Copper Proteins*** These proteins have two $S = 1/2$ Cu(II)s separated by a distance of ~3.5 Å, which exhibit no EPR signal.[3] The lack of EPR signal has been attributed[26] to a strong antiferromagnetic coupling between the two copper ions through a bridging ligand, peroxide for hemocyanin[27] and tyrosinase and hydroxide for the type III center in the multicopper oxidases[28] (Figure 1B).

2.2.2.1 *Antiferromagnetism* If two Cu(II)s, each with $S = 1/2$, are separated by a distance of ~3.5 Å, their spins will spin–spin dipolar couple producing a broad EPR signal whose shape can be analyzed to extract the metal–metal distance (Figure 16A, see Reference 29 for details). If these coppers are brought closer together ($\lesssim$2.5 Å) their $d_{x^2-y^2}$ orbitals will directly overlap, producing a bonding interaction between the metal ions leading to spin pairing and an $S_{\text{total}} = 0$ ground state (Figure 16B) and

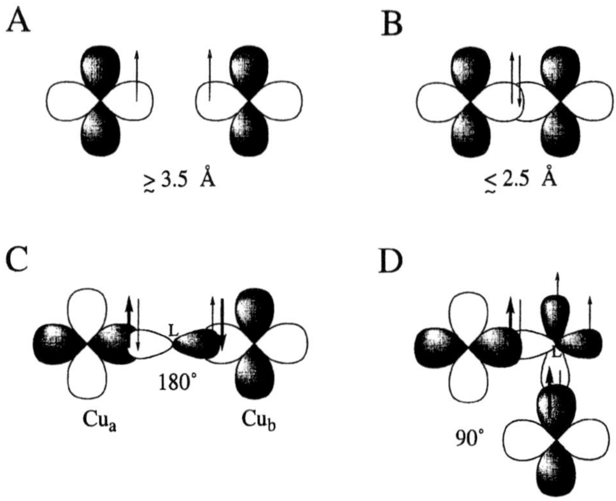


**Figure 16.** Electron interactions in binuclear cupric complexes. A) Spin–spin dipolar coupling occurs between two Cu(II) ions separated by >3.5 Å; B) spin pairing due to the direct overlap of the copper $d_{x^2-y^2}$ orbitals at small Cu–Cu distances; C) spin pairing is a result of the bonding interaction between two half-occupied $d_{x^2-y^2}$ orbitals and the occupied $p_z$ orbital of a bridging ligand in the case of a 180° M–L–M angle. The result is strong antiferromagnetic coupling of the two Cu spins with $S_{\text{total}} = 0$; D) in the case of a M–L–M angle of 90°, spin pairing occurs between each Cu spin and an electron in each of two orthogonal ligand p orbitals. The net effect is weaker ferromagnetic alignment of the two Cu spins, giving $S_{\text{total}} = 1$.

no EPR signal. Such a spin paired ground state is also accessible at distances $\gtrsim$2.5 Å if there is a bridging ligand. As shown in Figure 16C the bonding interaction of the $d_{x^2-y^2}$ orbital on Cu$_a$ with the filled ligand $p$ orbital causes a pairing of the spin on this metal with one of the electron pair on the bridge. The second electron in this pair is polarized toward the opposite side of the $p$ orbital and available for overlap and spin pairing with the electron in the $d_{x^2-y^2}$ orbital on Cu$_b$. The net effect is that the electron spin on Cu$_a$ is antiparallel to that on Cu$_b$ producing an $S_{\text{total}} = 0$ ground state that is EPR inactive. This bonding interaction with the electron pair on the bridge provides a *superexchange* mechanism for strong antiferromagnetic coupling at large metal–metal distances. Note that Figure 16C considers a 180° M–L–M angle and thus the involvement of one ligand $p$ orbital electron pair. At 90° (Figure 16D), each Cu spin pairs with one electron of a pair in each of two orthogonal degenerate $p$ orbitals on the ligand. This case favors the parallel alignment of the polarized spins in each of the two $p$ orbitals on the ligand due to electron exchange and thus a net parallel alignment, or ferromagnetic coupling of the spins on the two Cu ions to produce a spin triplet, $S_{\text{total}} = 1$ ground state. This angle dependence of the exchange coupling (180°: strong antiferromagnetism with an $S = 0$ ground state; 90°: weak ferromagnetism with an $S = 1$ ground state) is generally observed experimentally and given by the Goodenough–Konamori rules.[30]

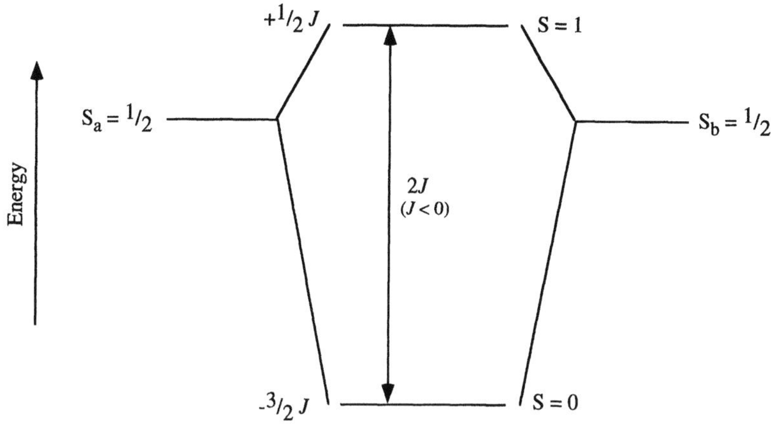


**Figure 17.** Antiferromagnetic coupling of two $S = 1/2$ spins resulting in an $S_{total} = 0$ ground state and an $S_{total} = 1$ excited state split by $2J$ (shown for negative $J$ values). Ferromagnetic coupling of two spins would result in the $S_{total} = 1$ state lowest in energy (for positive $J$ values).

The interaction of two spins on adjacent metal centers is generally described phenomenologically by the Heisenberg, Dirac, Van Vleck (HDVV) spin Hamiltonian given in Equation 8. This Hamiltonian is sometimes written without the minus sign and sometimes without the factor of 2; the form given here is that of Van Vleck in Reference 31, where $J$ is the exchange parameter which describes the magnitude of the coupling of the spins.

$$\mathcal{H} = -2J\hat{S}_a \bullet \hat{S}_b \tag{8}$$

This term couples two spins to produce $S_{total} = S_a + S_b, S_a + S_b - 1, \ldots, |S_a - S_b|$ states of the dimer which are split in energy by an amount given by Equation 9.

$$E_{S_{total}} = -J\left[S_{total}(S_{total} + 1) - S_a(S_a + 1) - S_b(S_b + 1)\right] \tag{9}$$

[Note for Equation 9, $\hat{S}^2_{total} = (\hat{S}_a + \hat{S}_b)^2 = \hat{S}^2_a + \hat{S}^2_b + 2\hat{S}_a\hat{S}_b$, therefore $2\hat{S}_a \bullet \hat{S}_b = \hat{S}^2_{total} - \hat{S}^2_a - \hat{S}^2_b$, and $S^2_i = S_i(S_i + 1)$.] For two $S = 1/2$ ions this produces $S_{total} = 1$ and 0 states split by $2J$ (Figure 17) where $J < 0$ gives the $S_{total} = 0$ state lowest (antiferromagnetism) and $J > 0$ corresponds to $S_{total} = 1$ as the ground state (ferromagnetism). The antiferromagnetic situation is most generally observed and is appropriate for the coupled binuclear copper proteins.

Insight into the origin of this antiferromagnetic coupling can be obtained from molecular orbital theory. As shown in Figure 18A, $d_{x^2-y^2}$ orbitals on each copper (each with one unpaired electron) may be combined to form + and – combinations for the binuclear site [$\phi_1 = (1/\sqrt{2})(d_{x^2-y^2_a} + d_{x^2-y^2_b})$, $\phi_2 = (1/\sqrt{2})(d_{x^2-y^2_a} - d_{x^2-y^2_b})$]. From this figure it can be seen that a $p_z$ orbital on a bridging ligand can

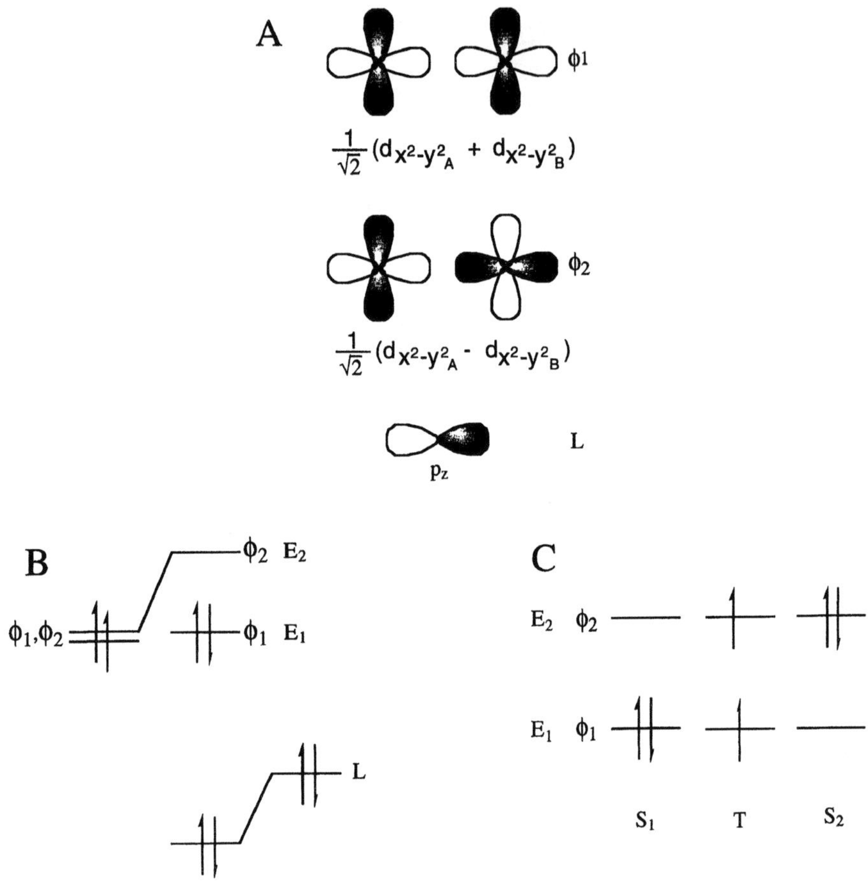


**Figure 18.** Molecular orbital origin of antiferromagnetic coupling. A) Singly occupied $d_{x^2-y^2}$ orbitals on each Cu(II) form symmetric and antisymmetric combinations in a dimer. The $p_z$ orbital of the bridging atom has a bonding interaction only with $\phi_2$, due to non-zero overlap; B) this results in an energy splitting of $\phi_1$ and $\phi_2$; C) singlet (S) and triplet (T) states formed from the two valence electrons on $\phi_1$ and $\phi_2$. $S_1$ and $S_2$ are the doubly occupied singlet states and $T$ is the $(\phi_1)^1(\phi_2)^2$ triplet. The $(\phi_1)^1(\phi_2)^1$ singlet does not affect the analysis due to symmetry restrictions and is not shown.

participate in a bonding interaction only with $\phi_2$ (i.e., non-zero overlap), and this interaction will therefore split the energies of $\phi_2$ and $\phi_1$ (Figure 18B). Inclusion of the two valence electrons (one from each copper) into this molecular orbital diagram leads to the $S_1$ and $T$ states in Figure 18C, where it is qualitatively observed that when $\phi_1$ and $\phi_2$ are close in energy the triplet should be the ground state due to electron repulsion. Increasing the $\phi_2/\phi_1$ energy splitting due to bonding differences with the bridging ligand (i.e., superexchange) will overcome the electron–electron repulsion and lead to spin pairing and the antiferromagnetic ($S_{total} = 0$) ground state.

Quantitatively, the correct singlet ground state wavefunction $\Psi(S)$ has almost equal mixtures of $S_1$ and $S_2$ (Figure 18C right, due to configurational interaction (CI), $\Psi(S) = (1/\sqrt{2})[\Psi(S_1) + \Psi(S_2)]$, while the third singlet $S_3$ derived from the $(\phi_1)^1(\phi_2)^1$ configuration is of the wrong symmetry for CI) and the energy of this singlet, $E_S$, relative to the triplet $(-2J)$ is given by Equation 10.[32]

$$E_T - E_S = -2K_{ab} + \frac{(E_2 - E_1)^2}{J_{aa} - J_{ab}} \tag{10}$$

$K_{ab}$ is the ferromagnetic contribution to the exchange and is expected to be small ($<50\ cm^{-1}$), while the second term usually dominates. $J_{aa}$ and $J_{ab}$ are the one and two site Coulomb repulsion integrals, respectively, with their energy difference estimated to be in the range of 5 to 10 eV for transition metal complexes, and $E_2 - E_1$ is the energy difference of the $\phi_2$ and $\phi_1$ molecular orbitals shown in Figure 18. Thus the larger the energy difference of $\phi_2$ and $\phi_1$, the greater the antiferromagnetic coupling, and it is the differences in the bonding interaction with the bridging ligand (i.e., superexchange) that causes this energy splitting. With more than one bridging ligand, the bonding interactions with the two metal ions can increase this splitting, leading to stronger antiferromagnetism, or decrease the $\phi_1/\phi_2$ splitting, thus reducing the antiferromagnetic coupling[33] if the valence orbitals on the different bridging ligands interact differently with $\phi_1$ and $\phi_2$. The magnitude of the singlet/triplet splitting in Figure 17 ($2J$) is most generally determined through magnetic susceptibility experiments.

2.2.2.2 *Magnetic Susceptibility* ($\chi$) When a sample is placed in a magnetic field $H$, the magnetic field inside the sample is different from the applied field due to the magnetization of the sample, $M$, which is associated with its magnetic moment. This quantity can be measured at the high sensitivity required for metalloprotein systems using either a Faraday balance or, preferably, a superconducting quantum interference device (SQUID) susceptometer (see Reference 34 for details on this instrument). The magnetic moment of a mole of sample is proportional to the applied magnetic field; the proportionality constant is the susceptibility of the sample to magnetic polarization, $\chi$ ($cm^3$/mol in cgs emu units) (Equation 11).

$$M = \chi \bullet H \tag{11}$$

In magnetochemistry, we are interested in relating the molar susceptibility to the magnetic moments of the individual molecules, $\mu$, which in turn relate to the energies of their ground state sublevels in a magnetic field.[35] Following Van Vleck[31], the energies of the individual sublevels $E_n$ of a molecule are expanded in a power series in the magnetic field (Equation 12). The magnetic moment of each sublevel is then given by Equation 13.

$$E_n = E_n^{(0)} + E_n^{(1)}H + E_n^{(2)}H^2 \tag{12}$$

$$\mu_n = -\partial E_n/\partial H = -E_n^{(1)}H - 2E_n^{(2)}H \tag{13}$$

The magnetic moment of a mole of molecules is given by Equation 14 which describes the Boltzmann statistics for $N$ molecules distributed over $n$ states each with energy $E_n$ and magnetic moment $\mu_n$ given by Equation 13.

$$M = \chi \bullet H = N \frac{\sum_n [-E_n^{(1)} - 2E_n^{(2)}H] \exp(-\frac{E_n}{kT})}{\sum_n \exp(-\frac{E_n}{kT})} \tag{14}$$

In the limit that the Zeeman splitting of the sublevels is small compared to $kT$ (i.e., thermal energy, which at 300 K is ~200 cm$^{-1}$) and $\chi$ is independent of $H$, Equation 14 simplifies to the Van Vleck Equation (Equation 15) which is generally used to describe the magnetism of metal complexes.

$$\chi = \frac{N \sum_n [\frac{(E_n^{(1)})^2}{kT} - 2E_n^{(2)}] \exp(-\frac{E_n^{(0)}}{kT})}{\sum_n \exp(-\frac{E_n^{(0)}}{kT})} \tag{15}$$

We now apply the Van Vleck Equation to the coupled binuclear copper proteins. First consider the limit of no exchange coupling in which case the coppers are isolated $S = 1/2$ ions that energetically split in a magnetic field as described in Figure 19A. Here the spin sublevels, $n$, are $M_S = -1/2$ and $+1/2$, the first order Zeeman coefficients are $E^{(1)}_{M_S=-1/2} = (-\frac{1}{2}g\beta)$, $E^{(1)}_{M_S=+1/2} = (+\frac{1}{2}g\beta)$, and $E^{(2)}$ and $E^{(0)}$ are

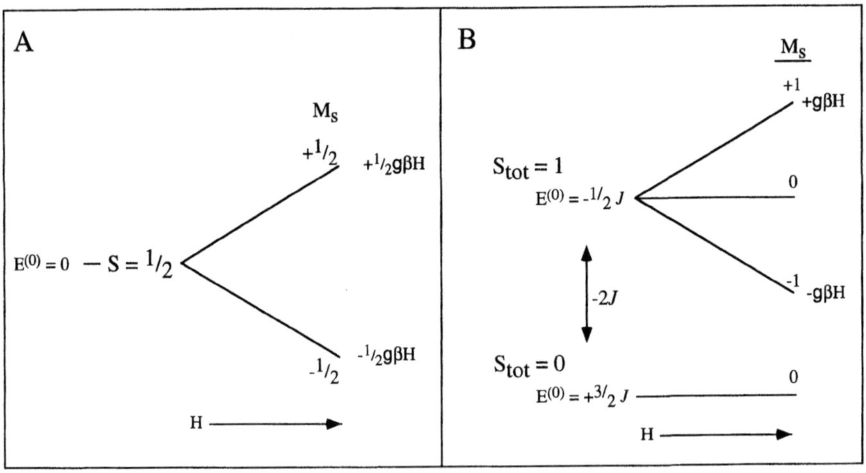


**Figure 19.** Magnetism of coupled binuclear copper systems. A) Without exchange coupling, the isolated $S = 1/2$ copper(II) ion is split in energy by a magnetic field; B) two Cu(II)s with antiferromagnetic exchange coupling split by an external magnetic field.

zero. Substitution into Equation 15 gives Equation 16.

$$\chi_{S=1/2} = \frac{N\left[\frac{(-1/2g\beta)^2}{kT} + \frac{(+1/2g\beta)^2}{kT}\right]}{2} = \frac{N(g\beta)^2}{4kT} \tag{16}$$

For a binuclear copper protein with no exchange coupling, one should observe magnetic susceptibility corresponding to two moles of $S = 1/2$ ions as illustrated in Figure 20.

The general form of Equation 16 for any spin system is given by Equation 17. The susceptibility of a paramagnetic center is proportional to $1/T$ (the Curie law) with a slope which increases with the spin of the ground state, its $g$ value, and the concentration of paramagnetic ion. [Note that the effective magnetic moment $\mu_{\text{eff}}$ is $g\sqrt{S(S+1)}$.]

$$\chi = \frac{N[S(S+1)](g\beta)^2}{3kT} \tag{17}$$

If the two $S = 1/2$ ions are strongly antiferromagnetically coupled and placed in a magnetic field, one gets the energy level diagram shown in Figure 19B. Substitution of these sublevels into Equation 15 [$S = 0$: $E^{(0)} = +(3/2)J$, $E^{(1)} = 0$, $E^{(2)} = 0$; $S = 1$: $E^{(0)} = -(1/2)J$, $E^{(1)}_{M_S=-1} = (-g\beta)$, $E^{(1)}_{M_S=0} = 0$, $E^{(1)}_{M_S=1} = (+g\beta)$,

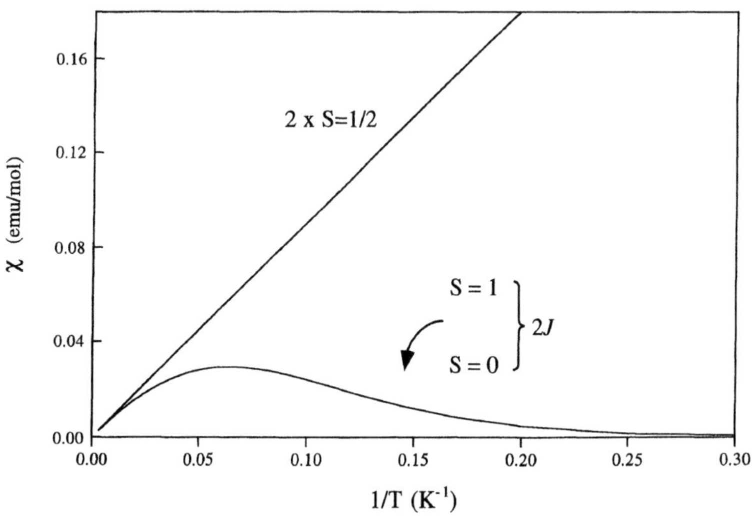


**Figure 20.** Plot of the Bleany Bowers equation (Equation 18) for two $S = 1/2$ ions which are strongly antiferromagnetically coupled. $\chi$ is also plotted for two isolated Cu(II) $S = 1/2$ ions as a reference.

$E^{(2)} = 0]$ gives Equation 18, which is also plotted in Figure 20.

$$\chi = N \frac{\frac{2(g\beta)^2}{kT} \exp\left(\frac{+\frac{1}{2}J}{kT}\right)}{\exp\left(\frac{-\frac{3}{2}J}{kT}\right) + 3\exp\left(\frac{+\frac{1}{2}J}{kT}\right)} = \frac{2N(g\beta)^2}{3kT}\left(\frac{1}{1 + \frac{\exp(-2J/kT)}{3}}\right) \quad (18)$$

This is the Bleany Bowers equation,[36] which shows that at low temperature the susceptibility goes to zero for the antiferromagnetic dimer due to population of only the $S = 0$ sublevel in Figure 19B. At increased temperature the paramagnetic $S = 1$ level at $E = 2J$ is populated, and at higher temperature the behavior of Equation 18 follows that of Equation 16 for two moles of $S = 1/2$ ion, decreasing as $1/T$. For the coupled binuclear copper proteins, the increase in $\chi$ at higher temperature in Figure 20 is not observed experimentally,[26] which places a lower limit on the singlet/triplet splitting in Figure 17 of $-2J > 500\ \text{cm}^{-1}$. This indicates a large covalent interaction through the bridging ligand. This covalent interaction with the bridge can control the two electron redox processes of coupled binuclear copper active sites in proteins.

Note that for metalloprotein systems there is a very large temperature-independent diamagnetic background due to the protein, buffer, and sample container that must be subtracted from the data in order to study the temperature dependent effects described above.[34b]

**2.2.3 *The Cu$_A$ Center*** Historically, the Cu$_A$ center first observed in cytochrome *c* oxidase[37] was difficult to understand in that its X-band EPR spectrum exhibits a low $g$ value and, in particular, little metal hyperfine structure (Figure 21B).[38] A major advance came when S-band EPR spectroscopy (*vide infra*) was applied to this enzyme and Cu hyperfine structure was observed on the Cu$_A$ signal.[39] A similar site is present in nitrous oxide reductase which at S-band shows seven equally spaced Cu hyperfine lines for each $g_i$ value which are particularly well resolved in the second derivative spectrum (Figure 21 C and D).[38] This led to the proposal that this center is binuclear and in a mixed-valence oxidation state with the extra electron strongly delocalized between the two metal centers (i.e., [Cu(1.5) Cu(1.5)]).[40] Such a system would exhibit $2nI + 1 = 7$ hyperfine lines ($n = 2$ Cu atoms, $I_{\text{Cu}} = 3/2$) with a hyperfine coupling of the electron spin to each copper nucleus that is about one half that of a mononuclear cupric center. The $A_{\parallel}^{M}$ value observed for Cu$_A$ in cytochrome *c* oxidase is $\sim 40 \times 10^{-4}\ \text{cm}^{-1}$, which is small enough to be unresolvable at X-band due to broad EPR linewidths but which sharpens at S-band. The proposed binuclear structure of the Cu$_A$ site has been confirmed by recent crystal structures of cytochrome *c* oxidase that show a binuclear copper center (Cu–Cu distance of ~2.5 Å) with bridging thiolate ligands (Figure 1C).[4b,41]

2.2.3.1 *S-Band EPR and g-Strain* EPR linewidths are generally observed to increase with increasing magnetic field and microwave frequency. This phenomenon

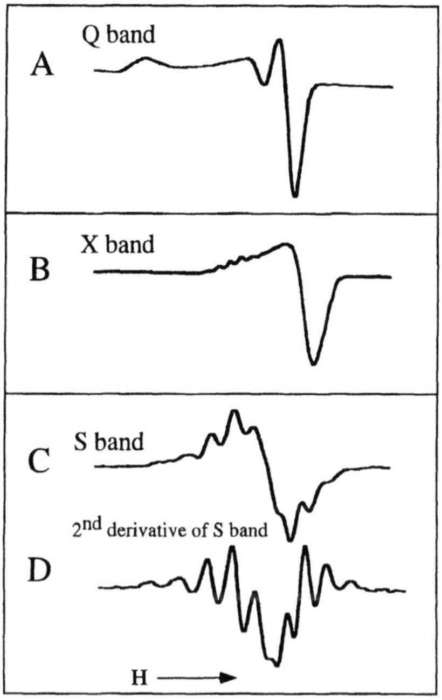


**Figure 21.** EPR spectra of the $Cu_A$ center. A) Q-band EPR spectrum of $Cu_A$ center from nitrous oxide reductase; B) X-band spectrum of $Cu_A$ center from cytochrome *c* oxidase; C) S-band spectrum; and D) its second derivative of $Cu_A$ center from nitrous oxide reductase. Adapted from Reference 38.

is referred to as *g*-strain since a Gaussian distribution of *g* values would show this type of behavior.[42] This *g*-strain is thought to reflect the presence of a distribution of ligand fields around the metal center in a protein, either related to a range of equivalent energy site conformations or induced by the low temperature conditions used to obtain EPR data. From Equation 3, small variations in the ligand environment around a metal center could produce small changes in the ligand field energies $(E_n - E_0)$ and in the covalent mixings of the *d* orbitals with the ligands, $\Psi_L$ in Figure 6D, and this will produce small changes in the *g* values (a gaussian distribution about the mean). This *g*-strain should increase with the deviation of the *g* value from 2.0023 (becoming worse for lower *g* values) and should inversely affect the hyperfine couplings through the effects of the g values on these as described by Equation 4.

These considerations lead to the expression for EPR linewidths given in Equation 19, where $\sigma_{1/2}^R$ is the residual linewidth independent of strain, and $C_{1i}$, $C_{2i}$, $C_{3i}$ $(i = x, y, z)$ are empirical coefficients that give the frequency and $M_I$ dependence of the linewidths.[42b]

$$\sigma_{1/2i} = \left[\left(\sigma_{1/2}^R\right)_i^2 + C_{1i}M_I^2 + C_{2i}\nu^2 + C_{3i}M_I\nu\right]^{1/2} \tag{19}$$

These coefficients can be related to the distribution in ligand field parameters for a specific metal center. From the above discussion the intrinsic residual EPR linewidth associated with hyperfine splittings will be achieved at low frequencies (the $\nu$ giving maximum resolution being dependent on the specific $M_I$ component), and this has provided a major impetus for multifrequency EPR studies in the low energy region (S-band at 2–4 GHz, L-band at 1–2 GHz and C-band at ~4.5 GHz). While high frequency EPR studies (Q-band at 35 GHz, W-band at 95 GHz) have proven to be very useful in resolving rhombic splittings of $g$ values (*vide supra*), these spectra are often very broad due to extensive $g$-strain effects (Figure 21A).

2.2.3.2 *Delocalization in Mixed Valence Complexes* The oxidized $Cu_A$ center formally has one Cu(II) and one Cu(I) in nearly equivalent environments [the major difference being that a long thioether S–Cu bond at $Cu_a$ is replaced by a long carbonyl O–Cu bond at $Cu_b$ (Figure 1C)].[4b,41] The extra electron in such a system can be either localized on $Cu_a$ (with a wavefunction given by $\Psi[Cu_a(I)\ Cu_b(II)]$), localized on $Cu_b$ (wavefunction = $\Psi[Cu_a(II)\ Cu_b(I)]$), or delocalized between the two metal centers as described by Equation 20.

$$\Psi_{\text{ground}} = (1-\alpha^2)^{1/2}\Psi[Cu_a(I)Cu_b(II)] + \alpha\Psi[Cu_a(II)Cu_b(I)] \qquad (20)$$

Robin and Day have provided a classification scheme for mixed valence complexes based on the value of $\alpha^2$, the extent of delocalization.[43] A site with the extra electron completely localized on $Cu_a$, $\alpha^2 = 0$, is called a class I mixed-valence complex, and its properties are the sum of those for the individual oxidized and reduced ions. This

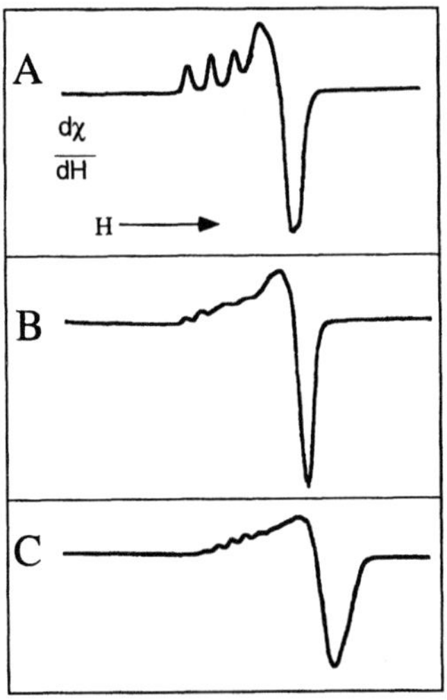


**Figure 22.** EPR spectra of mixed-valence copper sites. A) The $NO_2^-$ derivative of half-met hemocyanin is an example of a Class I complex; B) half-met $Br^-$ hemocyanin is a Class II complex; and C) $Cu_A$ is a Class III complex. Adapted from References 8 and 38.

is exemplified by the half-met $NO_2^-$ derivative of hemocyanin, which shows an EPR spectrum (Figure 22A) having four parallel hyperfine lines due to the fact that the unpaired electron couples to and is localized on only one metal ion in this mixed-valence binuclear copper site. In class II mixed-valence sites partial delocalization of the extra electron between the two metal centers occurs (i.e., $0 < \alpha^2 < 0.5$). In general the spectral properties of class II mixed-valence sites resemble those of the individual ions but are significantly perturbed due to the interaction between metal centers. Figure 22B gives the EPR spectrum of half-met $Br^-$ hemocyanin which shows a complex metal hyperfine splitting pattern in the $g_\parallel$ region. Simulation of this spectrum yields hyperfine parameters of $A_\parallel^{Cu_a} = 92 \times 10^{-4}\ cm^{-1}$ and $A_\parallel^{Cu_b} = 32 \times 10^{-4}\ cm^{-1}$, corresponding to $\alpha^2 \cong 0.30$ (see Reference 8 for details of analysis). The completely delocalized case (i.e., $\alpha^2 = 0.5$) corresponds to the class III mixed-valence limit. These systems exhibit properties strikingly different from those of the individual ions. The $Cu_A$ EPR signal[38] shown in Figure 22C is very broad and exhibits equal hyperfine couplings of the unpaired electron to both copper centers. This indicates $Cu_A$ is in the completely delocalized class III limit.

The extent of electron delocalization is determined by the electronic coupling matrix element, $H_{ab}$, associated with bonding interactions between the valence orbitals on $Cu_a$ and $Cu_b$. The analysis describing this coupling[44] parallels that for exchange interactions between homovalent metal ions as described in Section 2.2.2.1. As seen in Figure 23, the $d_{x^2-y^2}$ orbitals on each copper can directly overlap (depending on metal–metal distance) producing a metal–metal bond, where the splitting between

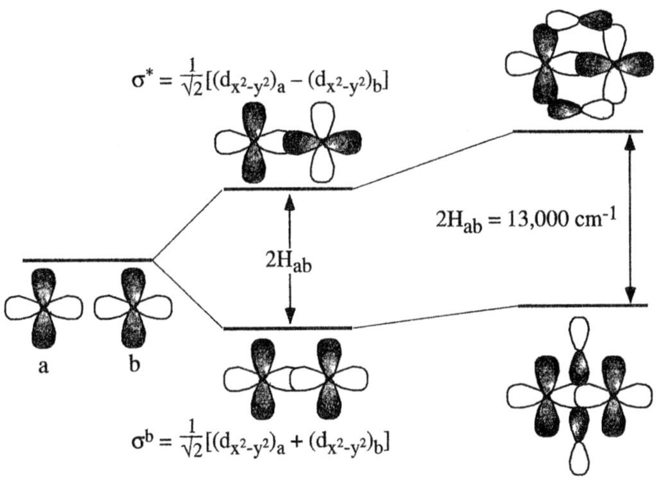


**Figure 23.** Electronic coupling between the valence orbitals on $Cu_a$ and $Cu_b$. Center: a metal–metal bond is formed when the distance between the coppers is small enough that the $d_{x^2-y^2}$ orbitals on each can directly overlap. Right: superexchange pathways can contribute to the electronic coupling when the bridging ligand orbitals interact with the symmetric and antisymmetric combinations of the Cu $d_{x^2-y^2}$ orbitals. The total splitting $= 2H_{ab}$ is 13,000 $cm^{-1}$ for the $Cu_A$ center (*vide infra*). Adapted from Reference 44.

the bonding and the antibonding combinations of $d$ orbitals is $2H_{ab}$. The bridging ligands can also interact with these combinations of $d_{x^2-y^2}$ orbitals and produce a superexchange contribution to this electronic coupling, as seen in the right of Figure 23. If both direct and superexchange contributions are present, they can either be complementary or oppose one another to produce the final energy splitting, $2H_{ab}$. For the $Cu_A$ center with two coppers at a distance of ~2.5 Å, it has been determined[44] that metal–metal direct bonding is present, with a contribution to $2H_{ab}$ estimated to be ~7000 $cm^{-1}$. The superexchange contribution from the bridging thiolates in $Cu_A$ increases the electronic coupling to the experimental value of $2H_{ab} = 13{,}000\ cm^{-1}$.

To this point in the analysis, any non-zero value of $H_{ab}$ between equivalent metal centers would result in an $\alpha^2 = 0.5$ in Equation 20 and produce a class III mixed-valence site. This, however, is often not found experimentally due to vibronic trapping. An oxidized metal center will generally have shorter bond lengths than a reduced metal center and this distortion gives an energy stabilization term which can trap the oxidation on the ligand contracted metal center. Following Piepho, Krausz, and Schatz (PKS) theory (see Chapter 3, Volume II, this series), this is described by the potential energy surfaces for the $\Psi[Cu_a(II)\ Cu_b(I)]$ and $\Psi[Cu_a(I)\ Cu_b(II)]$ states in the $Q_-$ coordinate (normalized antisymmetric combination of breathing modes on the two coppers) given in Figure 24, where the energies of these surfaces are obtained from Equation 21. Here $k_-$ is the force constant in the $Q_-$ mode ($= 4\pi^2c^2m\nu_-^2$ with $\nu_-$ the frequency of the $Q_-$ vibrational mode and m the modal reduced mass) and λ

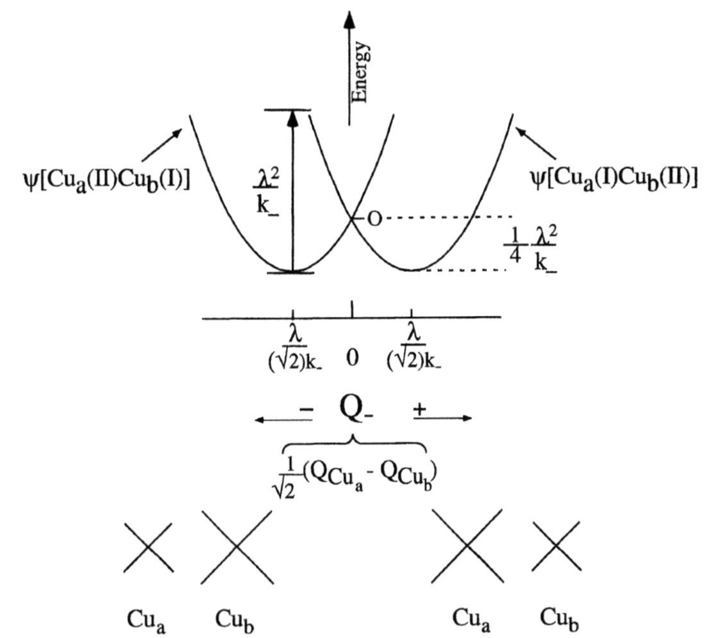


**Figure 24.** Potential energy surfaces in $Q_-$ mode corresponding to oxidation being trapped on $Cu_a$ (left) and $Cu_b$ (right).

is the vibronic coupling parameter, which lowers the energy due to a distortion along the $Q_-$ coordinate.

$$E = 1/2k_-Q_-^2 \pm (\lambda/\sqrt{2})Q_- \quad (21)$$

As indicated in Figure 24, the potential energy surface on the left has the oxidation trapped on $Cu_a$ and that on the right corresponds to the oxidation being trapped on $Cu_b$.

The electronic coupling matrix element, $H_{ab}$, must now be added to Equation 21 and Figure 24 to evaluate the combined effects of electronic coupling ($H_{ab} \neq 0$) and vibronic trapping ($\lambda \neq 0$) on the delocalization of the ground state.[45] This is given by Equation 22 which is plotted in Figure 25 in terms of the dimensionless coordinate $x_- = (k_-/\lambda)Q_-$. In Equation 22 both $\lambda^2/k_-$ and $H_{ab}$ have energy units.

$$E = 1/2\left(\frac{\lambda^2}{k_-}\right)x_-^2 \pm \left[1/2\left(\frac{\lambda^2}{k_-}\right)^2 x_-^2 + H_{ab}^2\right]^{1/2} \quad (22)$$

From Figure 25 it is observed that as $H_{ab}$ increases the two surfaces in Figure 24 interact and split in energy by $2H_{ab}$ at the surface crossing ($Q_- = 0$), and the two minima in the lowest energy potential surface, which correspond to vibronic trapping, become shallow. At a value of $H_{ab} > \lambda^2/2k_-$, there is a single minimum with $Q_- = 0$, and complete delocalization corresponding to the class III limit occurs, even at low temperatures. Note that increasing the magnitude of the vibronic coupling opposes the effect of electronic coupling on the surfaces. The vibronic coupling, $\lambda$, is equal to $k_-(\sqrt{n}\Delta r)$ where $\Delta r$ is the change in ligand–metal bond length

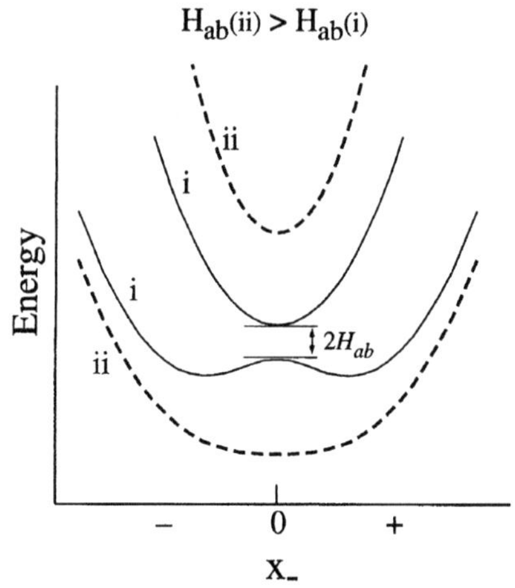


**Figure 25.** Potential energy surfaces in the $x_-$ ($= Q_-k_-/\lambda$) coordinate including the splitting due to the electronic coupling matrix element, $H_{ab}$. Two values of $H_{ab}$ are given ($i$ and $ii$). Adapted from Reference 45e.

on oxidation and $n$ is the number of ligands on one metal center. Thus, increasing the force constant of the $Q_-$ mode, the number of ligands, or the distortion of the ligand–metal bond on oxidation all oppose delocalization. An expression for $\alpha^2$ in terms of $H_{ab}$, $\lambda$, and $k_-$ is given in Equation 23.[45b]

$$\alpha^2 = \frac{H_{ab}^2}{(\lambda^2/k_-)^2}\left[\frac{1}{2} + \left(\frac{1}{4} - \frac{H_{ab}^2}{(\lambda^2/k_-)^2}\right)^{1/2}\right]^{-1} \tag{23}$$

When $H_{ab} \geq \lambda^2/2k_-$, then $\alpha^2 = 0.5$, and the system is class III. $H_{ab} = 0$ corresponds to $\alpha^2 = 0$ (class I), and $0 < H_{ab} < \lambda^2/2k_-$ gives class II mixed-valence behavior.

The fact that the $Cu_A$ center is a class III mixed-valence center indicates that the electronic coupling has overcome the vibronic trapping for this site. This lack of vibronic trapping could play a significant role in lowering the Franck–Condon barrier to electron transfer by the $Cu_A$ center in Biology.

## 3 GROUND STATE SPECTROSCOPIC METHODS: IRON

### 3.1 Normal Spectral Features

**3.1.1** ***EPR: Zero Field Splitting (ZFS)*** Most iron sites in Biology are high-spin as indicated in the one electron $d$ orbital energy diagrams for the ferrous and ferric ground configurations given in Figure 26 A and B. These give $^6A_{1g}$ ($S = 5/2$) and $^5T_{2g}$ ($S = 2$) ground states for the ferric and ferrous centers, respectively, in an octahedral ligand field. For $S > 1/2$, there is a new term in the spin Hamiltonian which will be non-zero when the site symmetry is less than octahedral or tetrahedral. This is called the zero field splitting (ZFS) and is given in Equation 24.

$$\mathcal{H}_{\text{ZFS}} = D\left[S_z^2 - \tfrac{1}{3}S(S+1)\right] + E\left(S_x^2 - S_y^2\right) \tag{24}$$

The $D$ is the axial zero field splitting parameter (for $z \neq x = y$), and the $E$ is rhombic zero field splitting (for $x \neq y$) where the $E/D$ ratio (sometimes defined as $\lambda$) is $\geq 1/3$ (a value $> 1/3$ corresponds to a redefinition of the $z$ axis). These terms split the $(2S+1)M_S$ degeneracy of the ground state in the absence of a magnetic field which will dramatically affect the EPR spectrum producing *fine structure*. This is described below along with the ligand field origin of the zero field splitting parameters.

3.1.1.1 *Fe(III)* $S = 5/2$ Application of Equation 24 with $D > 0$ and $E = 0$ to an $S = 5/2$, $M_S = \pm 5/2, \pm 3/2, \pm 1/2$ ground state will split the $M_S$ degeneracy into three substates of doublets separated in energy by $2D$ and $4D$ as shown in Figure 27A. (A negative $D$ inverts this pattern resulting in the $\pm 5/2$ doublet at lowest energy.) Combining Equation 24 with the Zeeman effect results in the $6 \times 6$ secular determinant for the energy, $W$, given in Equation 25. (Note that for Fe(III) $S = 5/2$

$$
\begin{array}{c|cccccc|}
 & |5/2\rangle & |3/2\rangle & |1/2\rangle & |-1/2\rangle & |-3/2\rangle & |-5/2\rangle \\
\langle 5/2| & \frac{10}{3}D + \frac{5}{2}g\beta H_z - W & \frac{\sqrt{5}}{2}g\beta(H_x - iH_y) & \sqrt{10}E & & & \\
\langle 3/2| & \frac{\sqrt{5}}{2}g\beta(H_x + iH_y) & \frac{-2}{3}D + \frac{3}{2}g\beta H_z - W & \sqrt{2}g\beta(H_x - iH_y) & 3\sqrt{2}E & & \\
\langle 1/2| & \sqrt{10}E & \sqrt{2}g\beta(H_x + iH_y) & \frac{-8}{3}D + \frac{1}{2}g\beta H_z - W & \frac{3}{2}g\beta(H_x - iH_y) & 3\sqrt{2}E & \\
\langle -1/2| & & 3\sqrt{2}E & \frac{3}{2}g\beta(H_x + iH_y) & \frac{-8}{3}D - \frac{1}{2}g\beta H_z - W & \sqrt{2}g\beta(H_x - iH_y) & \sqrt{10}E \\
\langle -3/2| & & & 3\sqrt{2}E & \sqrt{2}g\beta(H_x + iH_y) & \frac{-2}{3}D - \frac{3}{2}g\beta H_z - W & \frac{\sqrt{5}}{2}g\beta(H_x - iH_y) \\
\langle -5/2| & & & & \sqrt{10}E & \frac{\sqrt{5}}{2}g\beta(H_x + iH_y) & \frac{10}{3}D - \frac{5}{2}g\beta H_z - W
\end{array} = 0 \qquad (25)
$$

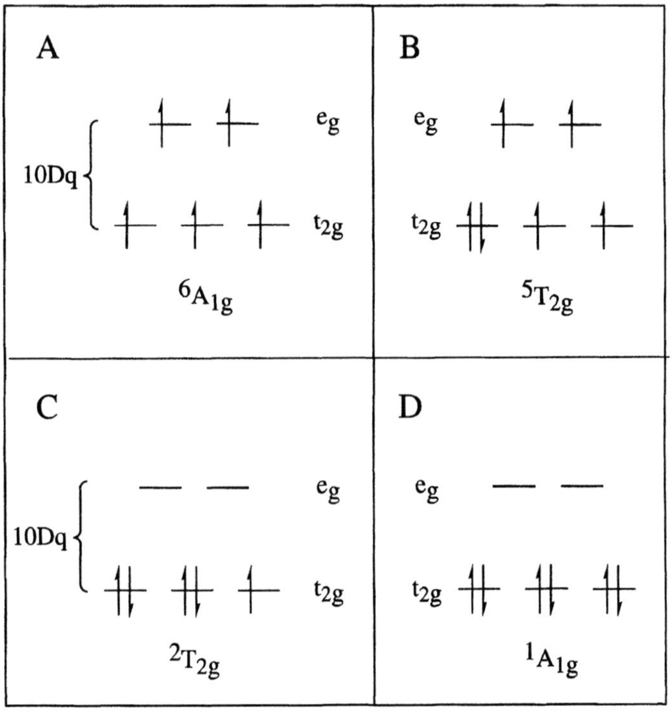


**Figure 26.** One electron *d* orbital energy level diagrams for A) high-spin ferric; B) high-spin ferrous; C) low-spin ferric; and D) low-spin ferrous centers.

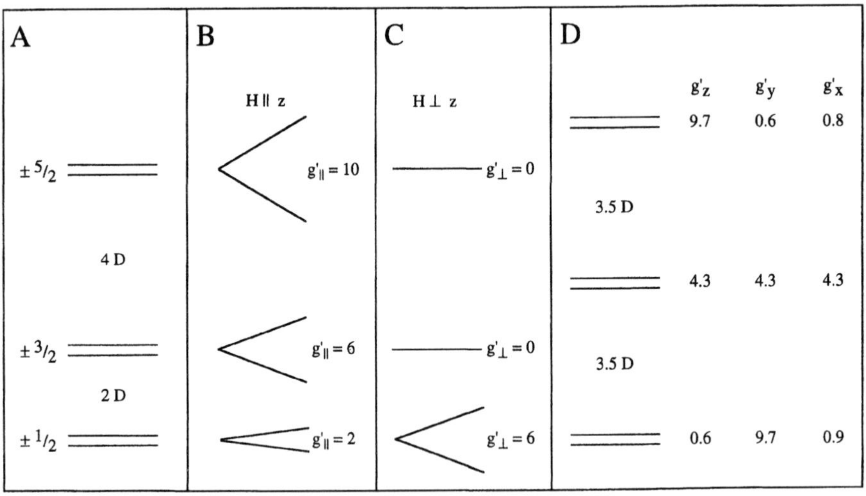


**Figure 27.** Splitting of the $M_S$ levels of a high-spin ferric ($S = 5/2$) center in the case of: A) $D > 0$ and $E = 0$ with no applied field; B) H parallel to the $z$ axis; C) H perpendicular to $z$; and D) large rhombic distortion at zero field with effective $g'$ values for rhombic center listed at right.

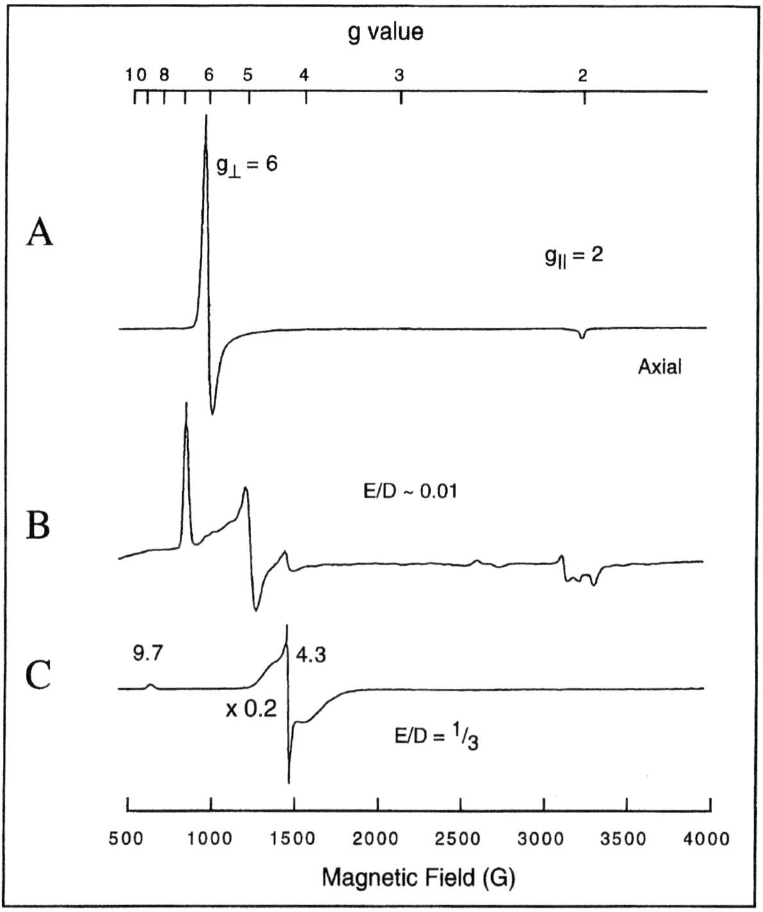


**Figure 28.** Characteristic high-spin ferric EPR spectra. A) Axial symmetry; B) small rhombic distortion; and C) large rhombic distortion. Adapted from Reference 47c.

the $^6A_{1g}$ ground state acquires no orbital angular momentum through spin-orbit mixing with excited states, and all three $g_i$ values are close to 2.0023.) In general, this secular determinant must be solved for all orientations of a ferric site (in a frozen sample) relative to the external magnetic field and a broad complex EPR signal results, its shape dependent on the values of $D$, $E$, and the microwave energy $h\nu$ (see Reference 46). However, in most cases in Bioinorganic Chemistry, the ZFS in Figure 27A is much larger than the microwave energy ($h\nu$ at X-band $\sim$0.3 cm$^{-1}$) and Equation 25 can be separated into the subdeterminants for the $\pm 1/2$, $\pm 3/2$, and $\pm 5/2$ doublets given in Equation 26 (i.e., the energy spacings between doublets are much larger than their off diagonal matrix elements). Diagonalization of the determinants for these doublets gives the energy (W) splitting shown in Figure 27B for $H \parallel$ to the $z$ axis and Figure 27C for $H \perp z$.

$$
\begin{array}{c|cc|c}
 & |+1/2\rangle & |-1/2\rangle & \\
|+1/2\rangle & \frac{-8}{3}D + \frac{1}{2}g\beta H_{\parallel} - W & \frac{3}{2}g\beta H_{\perp} & = 0 \\
|-1/2\rangle & \frac{3}{2}g\beta H_{\perp} & \frac{-8}{3}D - \frac{1}{2}g\beta H_{\parallel} - W &
\end{array}
$$

$$
\begin{array}{c|cc|c}
 & |+3/2\rangle & |-3/2\rangle & \\
|+3/2\rangle & \frac{-2}{3}D + \frac{3}{2}g\beta H_{\parallel} - W & 0 & = 0 \\
|-3/2\rangle & 0 & \frac{-2}{3}D - \frac{3}{2}g\beta H_{\parallel} - W &
\end{array} \qquad (26)
$$

$$
\begin{array}{c|cc|c}
 & |+5/2\rangle & |-5/2\rangle & \\
|+5/2\rangle & \frac{10}{3}D + \frac{5}{2}g\beta H_{\parallel} - W & 0 & = 0 \\
|-5/2\rangle & 0 & \frac{10}{3}D - \frac{5}{2}g\beta H_{\parallel} - W &
\end{array}
$$

Note that each set of doublets behaves as an effective spin $S' = 1/2$ but with unusual (i.e., "effective") $g'$ values ($\pm 1/2$: $g'_{\parallel} = 2$, $g'_{\perp} = 6$; $\pm 3/2$: $g'_{\parallel} = 6$, $g'_{\perp} = 0$; $\pm 5/2$: $g'_{\parallel} = 10$, $g'_{\perp} = 0$). Only the microwave transition between the $\pm 1/2$ levels is allowed by the EPR selection rule ($\Delta M_S = \pm 1$) and this should produce an $S' = 1/2$ type EPR spectrum as described in Section 2.1, but with $g'_{\parallel} = 2.0$ and $g'_{\perp} = 6$ (see Figure 28A). Note that the temperature dependence of the intensity of the EPR signal will reflect the relative Boltzmann population of the $M_S = \pm 1/2$ sublevel in Figure 27A and this allows an estimate of the sign and magnitude of $D$.

If we now allow the rhombic splitting, $E$ in Equation 24, to be greater than 0 but small ($\ll 1/3D$), the $g_{\perp}$ region splits by an amount $g'_x - g'_y = \Delta g' = 48E/D$ (Figure 28B).[47] In the limit of a large rhombic distortion ($E/D = \lambda = 1/3$), the ZFS energy diagram (Figure 27D) has three equally spaced doublets split in energy by an amount 3.5 $D$. The $M_S$ wavefunctions become very mixed due to the rhombic term and transitions within all three doublets in Figure 27D will now contribute to the EPR spectrum with relative weights given by the Boltzmann population of the sublevels. (The temperature dependence of the relative EPR intensities again allow an estimate of $D$.) The first and third doublet in Figure 27D are highly anisotropic with effective $g'$ values ranging from 0.6 to 9.7. Thus they will spread transition probability over a large magnetic field region and contribute little EPR intensity at a given field. Alternatively, the middle doublet is isotropic with all effective $g'$ values equal to 4.3 and thus will appear as a large symmetric derivative shaped feature in the EPR spectrum at a magnetic field associated with this $g'$ value (Figure 28C). Thus the characteristic features of the EPR signals of high-spin Fe(III) are either $g'_{\parallel} = 2$, $g'_{\perp} = 6$ for axial and close to axial sites, or $g' = 4.3$ (intense) for rhombic sites, with the temperature dependence of the intensity of either giving an estimate of the zero field splitting parameters.

Ligand Field Theory provides insight into the information content of the ZFS parameters. Griffith[48] has shown that in less than octahedral or tetrahedral symmetry the first ligand field excited state, the $^4T_1$ (see Section 4.1.1.1) will split into $^4T_1z$ and $^4T_1x, y$ components (in rhombic symmetry the $^4T_1x$ and $^4T_1y$ components are also split). Spin-orbit coupling of the $^6A_{1g}$ ground state to the axial and rhombic split components of the $^4T_1$ excited state [the one electron spin-orbit coupling parameter, $\zeta(\mathrm{Fe}^{+3})$, is 430 cm$^{-1}$] produces the ZFS of the $^6A_{1g}$ ground state (Figure 29). The second order perturbation expressions one obtains are given in Equation 27, which are modified from those of Griffith to allow for the effects of anisotropic covalency through the inclusion of the reduction factors $\kappa_i^2$ ($i = x, y, z$), which are less than 1 due to covalency (see Reference 49 for details).

$$D = \frac{\zeta^2}{10}\left[\frac{2\kappa_z^2}{E(^4T_1z)} - \frac{\kappa_y^2}{E(^4T_1y)} - \frac{\kappa_x^2}{E(^4T_1x)}\right]$$

$$E = \frac{\zeta^2}{10}\left[\frac{\kappa_y^2}{E(^4T_1y)} - \frac{\kappa_x^2}{E(^4T_1x)}\right] \tag{27}$$

One then needs to relate the energy splittings of the $^4T_1$ excited state, $E(^4T_1i)$, and the $\kappa_i^2$s to the energies and covalencies of the $d$ orbitals of the ferric site. These are given in Equation 28 for $E(^4T_1i)$ and Equation 29 for $\kappa_i^2$ in terms of the energy splittings ($\mu, \delta, \alpha$) and covalencies ($\varepsilon^2, \theta^2, \beta^2, \eta^2, \xi^2$) of the one electron $d$ orbitals described in Figure 30. $E(^4T_1$ cubic) is the average energy of the ligand field transition from the $^6A_{1g}$ ground state to the $^4T_{1g}$ excited state (due to the ligand field plus electron repulsion), which is estimated to be in the range of 10,000 cm$^{-1}$ for six coordinate sites. The "+" in Equation 28 applies to tetrahedral complexes, and the

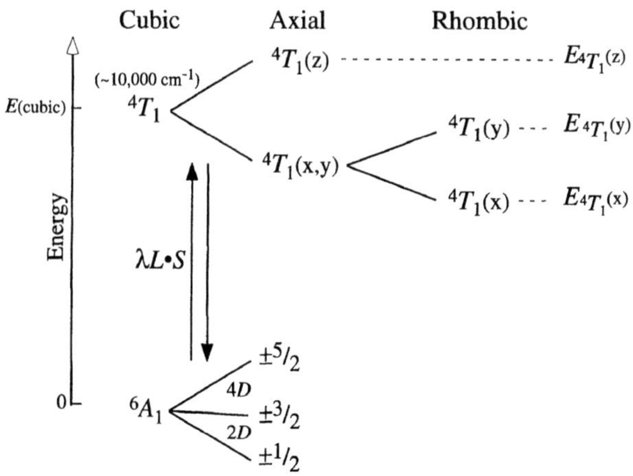


**Figure 29.** Spin-orbit coupling of the $^6A_{1g}$ ground state with the axial and rhombic split components of the $^4T_1$ excited state results in zero field splitting of the ground state.

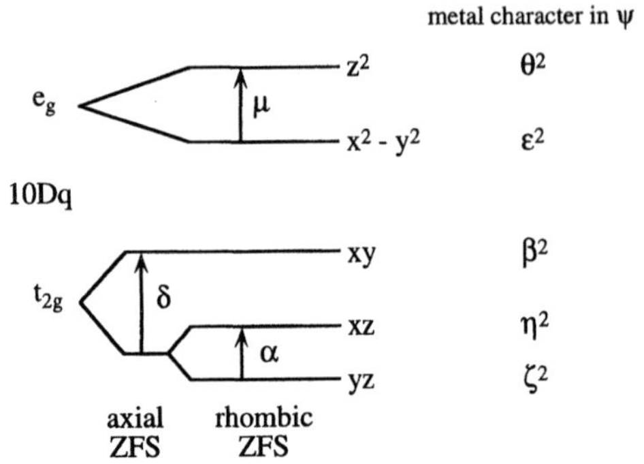


**Figure 30.** Energy splittings of the one electron *d* orbitals and their covalency factors for HS Fe(III) used in the ZFS analysis (Equations 27 to 29).

"−" is appropriate for distorted octahedral and square pyramidal sites.

$$\begin{aligned} E(^4T_1z) &= E(^4T_1\text{ cubic}) + \frac{2}{3}\left(\pm\delta \pm \frac{3}{4}\mu\right) \\ E(^4T_1x) &= E(^4T_1\text{ cubic}) - \frac{1}{3}\left(\pm\delta \pm \frac{3}{4}\mu \pm \frac{3}{2}\alpha\right) \\ E(^4T_1y) &= E(^4T_1\text{ cubic}) - \frac{1}{3}\left[\pm\delta \pm \frac{3}{4}\mu - \left(\pm\frac{3}{2}\alpha\right)\right] \end{aligned} \tag{28}$$

$$\begin{aligned} \kappa_z^2 &= (1-\beta^2)(1-\varepsilon^2) \\ \kappa_x^2 &= \left[\frac{\sqrt{(1-\xi^2)(1-\varepsilon^2)}}{4} + \frac{3\sqrt{(1-\xi^2)(1-\theta^2)}}{4}\right]^2 \\ \kappa_y^2 &= \left[\frac{\sqrt{(1-\eta^2)(1-\varepsilon^2)}}{4} + \frac{3\sqrt{(1-\eta^2)(1-\theta^2)}}{4}\right]^2 \end{aligned} \tag{29}$$

The zero field splitting parameters measured by EPR spectroscopy for high-spin ferric sites directly probe the one electron *d* orbital splittings and their relative covalencies as defined in Figure 30. There are, however, more parameters than experimental observables, so one parameter would have to be dominant for a direct correlation of $D$ and $E$ to the *d* orbital properties in Figure 30; alternatively, ligand field or molecular orbital theory can be used to calculate the parameters in Figure 30 which through Equations 27 to 29 allow an estimate of $D$ and $E$, which can be compared to experiment to quantitatively evaluate the calculation.

3.1.1.2 *Fe(II): $S = 2$* Application of Equation 24 to high-spin Fe(II) $S = 2$, $M_S = \pm 2, \pm 1, 0$, with $E = 0$ and $D$ positive gives the energy splitting in Figure 31A and for $D$ negative the splitting in Figure 31B. If the rhombic splitting $E$ is now allowed to be greater than 0, Equation 24 gives Figure 31C where the $\pm 2$ and $\pm 1$ doublets split in energy even in the absence of a magnetic field. This emphasizes the

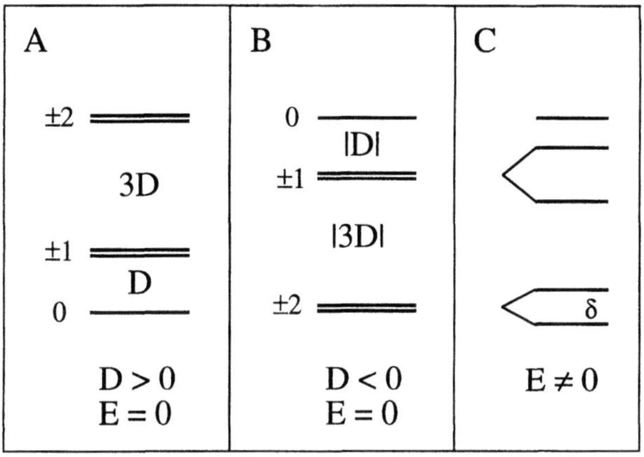


**Figure 31.** Zero field splitting diagrams for a high-spin ferrous center with: A) $D > 0$ and $E = 0$; B) $D < 0$ and $E = 0$; and C) $D < 0$ and $E > 0$.

difference between *Kramers and non-Kramers doublets.* A metal ion with an odd number of electrons has a half integer spin, and all energy levels must be at least 2 fold degenerate in the absence of a magnetic field (Kramers theorem).[50] The magnetic field splits this degeneracy and leads to an EPR transition as described above for Cu(II) $S = 1/2$ and Fe(III) $S = 5/2$. Alternatively, Fe(II) has a $d^6$ configuration and an integer spin $S = 2$ ground state where all degeneracy can be split in rhombic symmetry even in the absence of a magnetic field. The rhombic splitting of the non-Kramers doublet, $\delta$, in Figure 31C can be larger than the microwave energy of the EPR experiment (and usually is for biological ferrous centers), and no transition is observed. Variable temperature, variable field magnetic circular dichroism (VTVH MCD), and magnetic susceptibility provide an alternative approach to experimentally study ground states of high-spin ferrous sites in proteins as described below.

The spin Hamiltonian in Equation 24 producing the energy levels splittings in Figure 31 is, in fact, not rigorously appropriate for the high-spin ferrous center since the ground state in Figure 26B is ${}^5T_{2g}$. The orbital part of the wavefunction, $T_{2g}$, is three fold degenerate corresponding to the fact that the extra electron can be in the $d_{xy}$, $d_{yz}$, or $d_{xz}$ orbital. This corresponds to an effective orbital angular momentum $L' = 1$ which can couple to the spin angular momentum, $S = 2$, to produce the spin-orbit splitting given to the left in Figure 32: 3 fold, 5 fold, and 7 fold degenerate levels separated by $2\lambda$ and $3\lambda$ respectively, where $\lambda(Fe^{+2})$ is the spin-orbit coupling parameter for the ${}^5T_{2g}$ ground state [$\lambda(Fe^{+2}) = \zeta(Fe^{+2})/4 = -100\ cm^{-1}$]. In less than octahedral symmetry, the 3 fold orbital degeneracy of the $T_{2g}$ state is split in energy as described in Figure 33 due to the axial ($\Delta$) and rhombic ($V$) splittings of the $t_{2g}$ set of $d$ orbitals. The effects of the axial and rhombic splittings of the $t_{2g}$ orbitals combined with spin-orbit coupling are shown at the center and right of Figure 32. As can be seen from the expanded scale (Figure 32 right, inset), the energy splittings of the five lowest sublevels is quite different from that associated with the spin Hamiltonian model in Figure 31.[51]

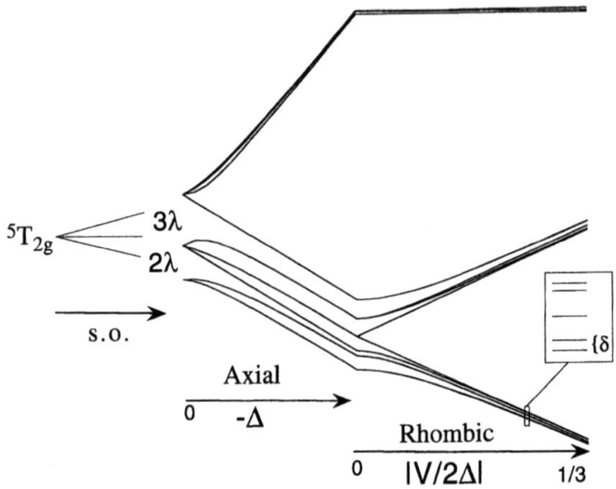


**Figure 32.** Spin orbit coupling plus axial ($D < 0$) and rhombic splitting of the high spin $d^6$ $^5T_{2g}$ ground state ($V$ and $\Delta$ defined in Figure 33). Inset gives an expanded energy scale view of the lowest five energy bands. Adapted from Reference 51b.

The zero field splitting of the lowest energy non-Kramers doublet, $\delta$, at the far right in Figure 32 and its further splitting in a magnetic field, $g_{\parallel}$ ($g_{\perp}$ is zero), can be calculated in terms of the $\Delta$ and $V$ parameters in Figure 33. These are given in Figure 34, which can reciprocally be used to obtain experimental values of $\Delta$ and $V$, the axial and rhombic splittings of the $t_{2g}$ set of $d$ orbitals by the Fe(II) active site, from experimental values of $\delta$ and $g_{\parallel}$. It should be emphasized that the key information content of the ZFS parameters of both the ferric and ferrous sites are the ligand field splitting of the $d$ orbitals, and in the case of ferrous centers the two experimental parameters directly define the energy splitting of the $t_{2g}$ set of $d$ orbitals. These experimental non-Kramers doublet parameters are obtained using saturation magnetization methods as described below.

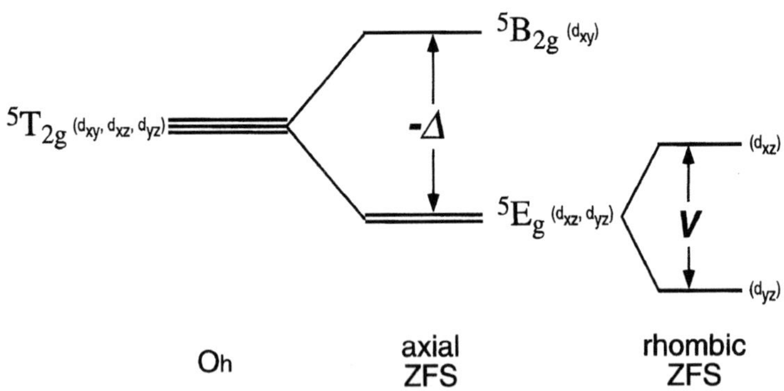


**Figure 33.** Zero-field splitting of the $t_{2g}$ set of $d$ orbitals due to axial ($\Delta$) and rhombic ($V$) distortions of the ligand field at a ferrous site.

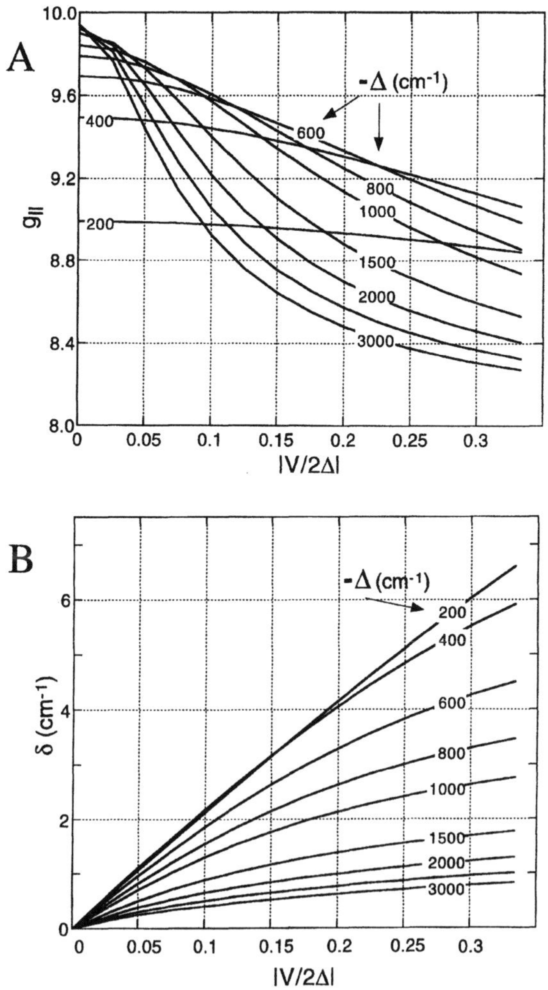


**Figure 34.** Value of $g_{\|}$ and and $\delta$ as a function of the axial ($\Delta$) and rhombic ($V$) splittings of the $t_{2g}$ set of $d$ orbitals, calculated from the full ${}^5T_{2g}$ Hamiltonian using $\lambda = -80\ \text{cm}^{-1}$ $|V/2\Delta| = 0$ at the axial limit and $|V/2\Delta| = 1/3$ at the rhombic limit. Adapted from Reference 51b.

**3.1.2** ***Saturation Magnetization: Magnetic Circular Dichroism (MCD) and Magnetic Susceptibility*** MCD spectroscopy is presented in Section 4.1.3. Here we simply note that the intensity of an MCD feature of a paramagnetic metal center increases with decreasing temperature (as $1/T$). This is known as C-term behavior and is the MCD analog of Curie Law behavior (Equation 17) in magnetic susceptibility. MCD has the potential advantages of higher sensitivity, the ability to discriminate one component in a mixture, and additional information content associated with its dependence on the polarization of the transition being probed (*vide infra*).

Alternatively, the magnitudes of the MCD effect associated with different sublevels are often difficult to calculate and are generally fit as adjustable parameters. For both methods, as one proceeds to lower temperature and higher magnetic fields, the magnetization signal, $M$, starts to deviate from the linear $(1/T)$ limit, and the signal eventually becomes independent of temperature and field and is then saturated.[34,51,52] This corresponds to the limit of Boltzmann population of only the lowest component of the ground state when the Zeeman splitting is very much greater than thermal energy (i.e., $g\beta H/kT$ is large). For Cu(II) $S = 1/2$ with close to isotropic $g$ values, this saturation magnetization behavior is described by Equation 30, which is plotted in Figure 35 ($S = 1/2$ curve). For $S > 1/2$ the appropriate expression is given in Equation 31a where $B_S(x)$ with $x = gS\beta H/kT$ is the Brillouin function given in Equation 31b. Equation 31 simplifies to Equation 30 when $S = 1/2$. The Brillouin curves for $S > 1/2$ are also included in Figure 35 where its shape, on approaching saturation, is dependent on $S$ and can be used to determine the total spin of the ground state.

$$M = \frac{1}{2}Ng\beta \tanh(g\beta H/2kT) \tag{30}$$

$$M = NgS\beta B_S(x) \tag{31a}$$

$$B_S(x) = \frac{2S+1}{2S}\coth\left(\frac{2S+1}{2S}x\right) - \frac{1}{2S}\coth\left(\frac{1}{2S}x\right) \tag{31b}$$

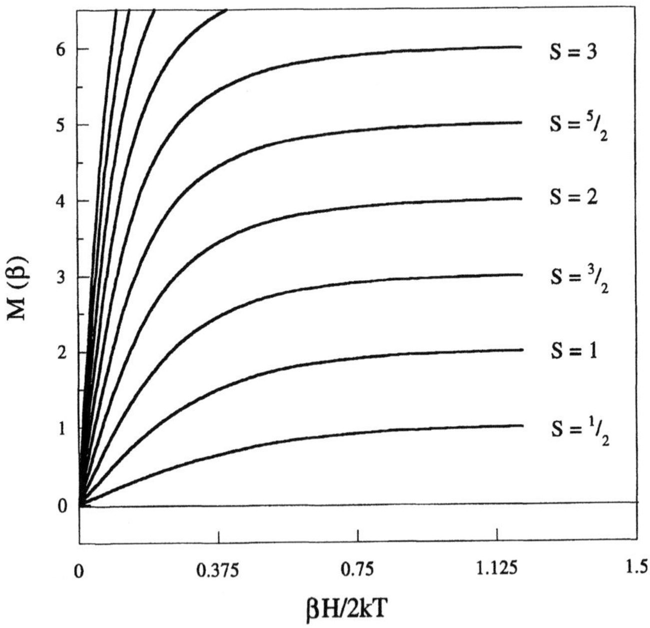


**Figure 35.** Brillouin curves for $S = 1/2$, 1, 3/2, 2, and 5/2, calculated for $g = 2$.

It is important to note that in the above treatment, it is assumed that the $(2S+1)M_S$ components of the ground state are all degenerate in the absence of a magnetic field. In this limit, saturation magnetization curves taken at different fixed temperatures with variation in magnetic field would all superimpose when plotted versus $\beta H/2kT$ in Figure 35. This however, is not the case for ferrous active sites in metalloproteins.[51] As shown in Figure 36A, these experimental curves are in fact nested, and this behavior has been ascribed to the zero field splitting of the $M_S$ components of the ground state.[52c]

Insight into the origin of this nesting is available from Figure 36B, which replots the MCD intensity with decreasing temperature for fixed incremental increases in magnetic field.[51] The MCD intensity at low temperature saturation is found to increase with increasing magnetic field. This behavior is non-linear and eventually levels off at high magnetic field. This indicates that the wavefunction of the lowest component of the ground state is changing in a non-linear way with magnetic field.

The above behavior is analyzed in terms of the non-Kramers doublet model given in Figure 37. At zero and small magnetic fields the $M_S = |+2\rangle$ and $|-2\rangle$ functions are completely mixed due to the rhombic term in the Hamiltonian which has also split the energy of these levels by $\delta$; at this point, there is no MCD C-term intensity or paramagnetic susceptibility which require preferential $|-2\rangle$ or $|+2\rangle$ character. As the field increases, the wavefunctions change in a non-linear manner becoming pure complex $|-2\rangle$ and $|+2\rangle$ and MCD (or magnetic susceptibility) active at high magnetic field. Including effects of Boltzmann population and orientation averaging (over all angles $\theta$ of the magnetic field relative to the molecular $z$ axis) on the non-Kramers doublet model given in Figure 37 for a frozen solution of high-spin ferrous centers gives Equation 32. MCD data sets as in Figure 36 are fit using Equation 32 to obtain experimental estimates of $g_{\parallel}$ and $\delta$ which can then be interpreted in terms of the diagrams in Figure 34 to obtain the experimental axial ($\Delta$) and rhombic ($V$)

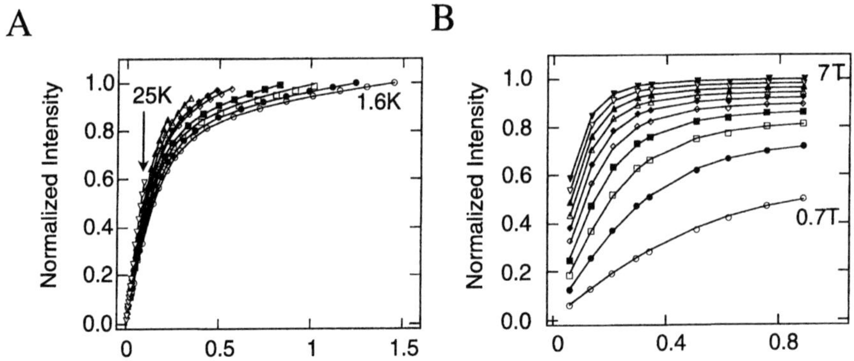


**Figure 36.** Saturation magnetization data (symbols) and fits (lines) using Equation 32 for ferrous bleomycin recorded at 9,900 cm$^{-1}$. A) Nesting behavior shown in plot of intensity versus $\beta H/2kT$; and B) intensity versus temperature for fixed field increments. Taken from Reference 51b.

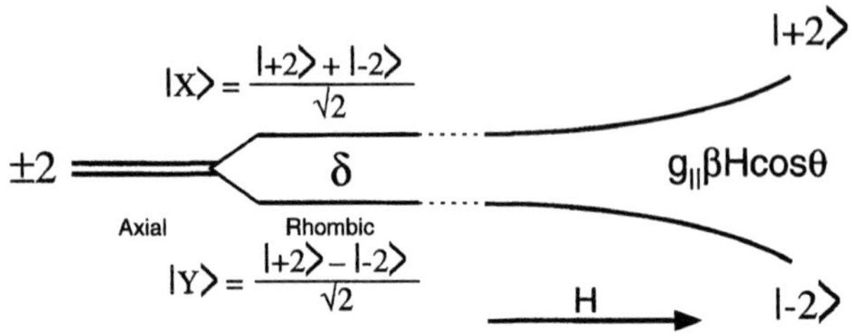


**Figure 37.** Splitting of an $M_S = 2$ non-Kramers doublet due to a zero field rhombic distortion (center) and with increasing applied magnetic field (right). Note that the rhombic splitting $\delta$ mixes the $|+2\rangle$ and $|-2\rangle$ complex wavefunctions which become pure $|+2\rangle$ and $|-2\rangle$ states at the high magnetic field limit.

splittings of the $t_{2g}$ set of $d$ orbitals of the ferrous active site (Figure 33). Additional factors required in this analysis are described in Reference 51b.

$$I_{\text{MCD}} = A_{\text{satlim}} \int_0^{\pi/2} \frac{\cos^2\theta \sin\theta}{\sqrt{\delta^2 + (g_\parallel \beta H \cos\theta)^2}} (g_\parallel \beta H) \times \tanh\left[\frac{\sqrt{\delta^2 + (g_\parallel \beta H \cos\theta)^2}}{2kT}\right] d\theta \tag{32}$$

**3.1.3** ***Magnetic Mössbauer*** A $^{57}$Co Mössbauer source decays by electron capture (half life = 270 days) to form a nuclear excited state of $^{57}$Fe which emits a 14.4 keV $\gamma$-ray upon decay to its ground state. The absorption of this emitted $\gamma$-ray by an $^{57}$Fe center in a sample constitutes the Mössbauer spectrum of the sample. (See Chapter 3, Volume I, this series.) Generally, the nuclear energy levels of the sample will not be in exact resonance with the source, and this difference can be compensated for by Doppler shifting the source $\gamma$-ray energy by moving the source toward (positive) or away (negative) from the sample with a transducer. The velocity required to achieve resonance is given in mm/sec, where 1 mm/sec = $3.87 \times 10^{-4}$ cm$^{-1}$ = 11.6 MHz. In order to have reasonable signal intensity in a Mössbauer experiment, the sample is usually a low temperature solid to reduce the Debye Waller factor and isotopically enriched with $^{57}$Fe (natural abundance = 2.25%).

We first consider the features of a Mössbauer spectrum of a diamagnetic iron compound or a paramagnetic center at a temperature where the spin relaxation is very fast ($\gtrsim$77 K). As shown in Figure 38A, the Mössbauer transition excites the $^{57}$Fe from its $I = 1/2$ ground state to an $I = 3/2$ excited state, the latter having a non-zero quadrupole moment ($I > 1/2$) and thus potentially split into $\pm 1/2$ and $\pm 3/2$ components by the electric field gradient in the vicinity of the nucleus. This quadrupole splitting, $E_Q$, for $I = 3/2$ is given by Equation 33 in terms of the parameters defined earlier. The electric field gradient (EFG) is gen-

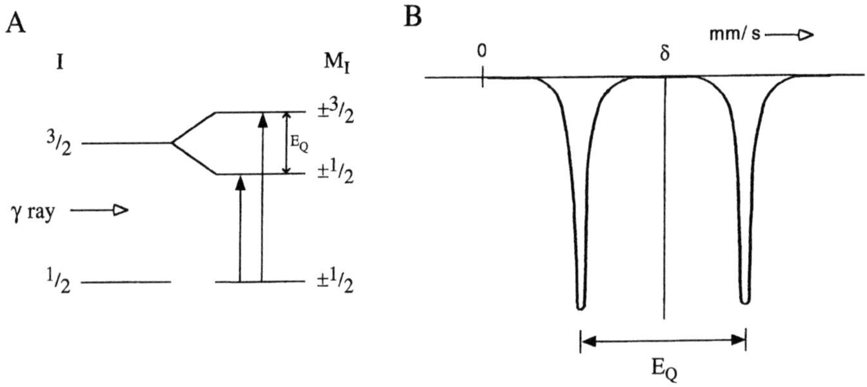


**Figure 38.** The Mössbauer experiment. A) Absorption of a $\gamma$-ray excites $^{57}$Fe from the ground state ($I = 1/2$) to an excited state ($I = 3/2$), the $M_I$ components of which are split due to the electric field gradient at the nucleus; and B) quadrupole doublet spectrum with the isomer shift $\delta$ and the quadrupole splitting $\Delta E_Q$ indicated on spectrum.

erally associated with valence electrons in $d$ orbitals centered on the iron with individual contributions given in Table 7, which assumes equal covalency for all $d$ orbitals.[53]

$$E_Q = \frac{e^2Qq}{2}\sqrt{1 + \eta^2/3} \qquad (33)$$

Note that equal electron populations over the five $d$ orbitals, or the $t_{2g}$ or $e_g$ sets separately, produces no EFG. Alternatively, excess electron density in the $d_{x^2-y^2}$ or $d_{xy}$ orbital will give a positive quadrupole splitting and excess density in the $d_{z^2}$ orbital gives a negative quadrupole splitting, with the $M_I \pm 3/2$ component of the

**Table 7 Values of $q$ and $\eta$ for One Electron in a 3$d$ Orbital[55]**

| | Orbital | $q$ | $\eta$ |
|---|---|---|---|
| $e_g$ | $d_{x^2-y^2}$ | $+4/7\frac{1}{\langle r^3\rangle}$ | 0 |
| | $d_{z^2}$ | $-4/7\frac{1}{\langle r^3\rangle}$ | 0 |
| $t_{2g}$ | $d_{xy}$ | $+4/7\frac{1}{\langle r^3\rangle}$ | 0 |
| | $d_{xz}$ | $-2/7\frac{1}{\langle r^3\rangle}$ | +3* |
| | $d_{yz}$ | $-2/7\frac{1}{\langle r^3\rangle}$ | −3* |

*Note that $q$ and $\eta$ are defined to be along the $z$ axis of the complex. Since the principle axes of the $d_{xz}$ and $d_{yz}$ orbitals are along the $y$ and $x$ directions, respectively, their value of $\eta$ can be $<0$ or $>1$.

$I = 3/2$ excited state in Figure 38A at lowest energy. (Note that the quadrupole moment of this state is positive.) Thus the largest values of $E_Q$ are found for high-spin Fe(II) which has a $t_{2g}^4\ e_g^2$ configuration and low-spin Fe(III) with a $t_{2g}^5$ configuration (Figure 26).

In addition to the quadrupole splitting, the standard Mössbauer spectrum in Figure 38B exhibits a center energy which can be shifted from that of the source (at 0 mm/sec), and this is referred to as the isomer shift, $\delta$. The isomer shift is given by Equation 34, where $|\Psi(0)|_A^2$ is the electron density at the nucleus of the absorber, $|\Psi(0)|_S^2$ is the electron density of the source, and $\partial R/R$ gives the relative change of the nuclear radius in going from the excited to the ground state which is negative for $^{57}$Fe.

$$\delta = \text{const.}(\partial R/R)\left[|\Psi(0)|_A^2 - |\Psi(0)|_S^2\right] \tag{34}$$

Thus, a positive isomer shift indicates that the electron density at the nucleus of the absorber is lower than that of the source, and in general iron compounds with large positive values of $\delta$ have reduced electron density at the nucleus. Electron density at the nucleus is associated with occupied $s$ orbitals, and trends in this value for transition metal complexes have been generally associated with electron occupation of $d$ orbitals which will shield filled inner s-electrons and reduce $|\Psi(0)|_A^2$. Isomer shifts are more positive for Fe(II) than Fe(III) compounds with high-spin (H.S.) larger than low-spin (L.S.) (H.S. Fe(II): $0.6 < \delta < 1.4$; H.S. Fe(III): $0.15 < \delta < 0.5$; L.S. Fe(II): $-0.3 < \delta < 0.5$; L.S. Fe(III): $-0.2 < \delta < 0.2$).[54]

At low temperature, $\lesssim$10 K, where spin relaxation is slow, in applied magnetic fields the $^{57}$Fe ground and excited nuclear states will split in energy due to the nuclear Zeeman effect as shown in Figure 39A. The nuclear Zeeman splitting is given by $-g_n\beta_n I \bullet H$ where $\beta_n$ is the nuclear magneton ($=1/1{,}836\beta$) and $g_n$ is the nuclear $g$ value ($=0.184$ for the $I = 1/2$ ground state and $-0.1$ for the $I = 3/2$ excited state). The magnetic dipole selection rules for the Mössbauer experiment are $\Delta M_I = \pm 1, 0$ which produce the spectrum in Figure 39B for an orientation averaged sample. For a magnetically oriented sample the $\Delta M_I = 0$ transition is forbidden for the magnetic field aligned parallel to the $\gamma$-ray propagation direction, while it maximizes for $H \perp \gamma$-ray (Figure 39C, D).[54b] The combined effects of a magnetic field and the quadrupole splitting are shown in Figure 39E, which indicates that for a positive quadrupole splitting peaks 1 and 6 shift up while 2 through 5 shift to lower velocity. The opposite is the case for a negative EFG, allowing an experimental determination of its sign and thus a determination of the orbital containing the extra electron at the iron site (Table 7).

The magnetic field at the nucleus is an effective field, $\vec{H}_{\text{eff}}$, which is a combination of the external magnetic field, $\vec{H}_{\text{app}}$, and the magnetic field associated with the electron spin which affects the nucleus through the metal hyperfine coupling. The combination of these effects is given by the Hamiltonian in Equation 35, where the internal field associated with the hyperfine coupling, $\vec{H}_{\text{int}}$, is given by $-A\langle\vec{S}\rangle/g_n\beta_n$, where $\langle\vec{S}\rangle$ is the expectation value of the electron spin at low temperature in a mag-

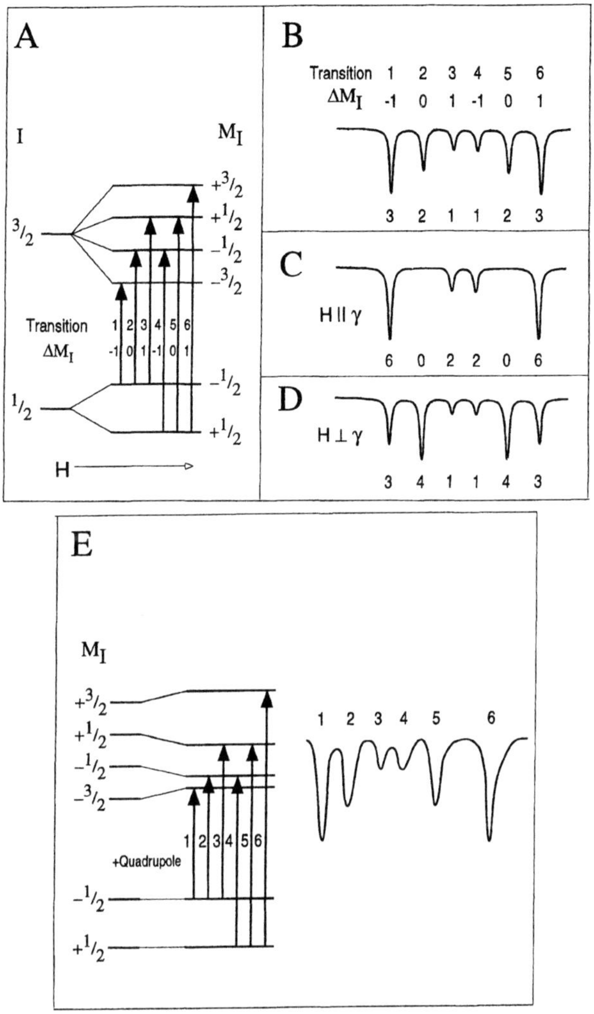


**Figure 39.** Magnetic Mössbauer spectroscopy. A) Splitting of the $^{57}$Fe ground and excited states due to the nuclear Zeeman effect with the allowed $\Delta M_I = \pm 1{,}0$ transitions indicated; B) resulting spectrum for an orientation averaged sample; C) for $H \| \gamma$ the $\Delta M_I = 0$ transition is forbidden; D) for $H \perp \gamma$ it is maximized; and E) the additional effect of a positive quadrupole splitting. Adapted from Reference 54b.

netic field that can be estimated from the effective $g'$ values of the ground state doublets ($\langle S_i \rangle \cong -g'_i/4$ with $i = x, y, z$).

$$\begin{aligned}\mathcal{H} &= -g_N \beta_N \vec{H}_{\text{app}} \bullet \vec{I} + A\vec{S} \bullet \vec{I} = -g_N \beta_N \left( \vec{H}_{\text{app}} - \frac{A\langle \vec{S} \rangle}{g_N \beta_N} \right) \bullet \vec{I} \\ &= -g_N \beta_N (\vec{H}_{\text{app}} + \vec{H}_{\text{int}}) \bullet \vec{I} \qquad (35)\end{aligned}$$

For high-spin Fe(III), the dominant contribution to the metal hyperfine coupling constant $A^M$ is Fermi contact which is isotropic and negative, while for Fe(II) all three contributions to the hyperfine coupling in Equation 4 contribute, and these can be quite anisotropic. The internal magnetic field can be large, $\vec{H}_{\text{int}} = +g_i'/4 \times A_i(\text{cm}^{-1}) \times 2.14 \times 10^8$ gauss, and dominate the external magnetic field.

The experimentally observed Zeeman splitting in the Mössbauer spectrum thus allows an experimental determination of the hyperfine coupling constant for iron sites, and since from Equation 35 the hyperfine field $\vec{H}_{\text{int}}$ and the applied field $\vec{H}_{\text{app}}$ are additive, the effect of changing $\vec{H}_{\text{app}}$ on the Mössbauer spectrum in Figure 39 can give the sign of $A_i$.

Paramagnetic centers can be divided into three classes based on the above considerations.[55] *1) Kramers doublet ground states which have close to isotropic $g'$ values.* These have large internal magnetic fields oriented along the external magnetic field and show magnetic Mössbauer spectra at fairly low applied magnetic fields which are polarized as in Figure 39 C and D. *2) Kramers doublet ground states which have very anisotropic $g'$ values.* These show large internal magnetic fields oriented along the largest $g_i'$ component which is fixed in a specific molecular direction. Thus the internal field of a frozen sample will have all orientations relative to the $\gamma$-ray propagation direction and will show an unpolarized spectrum as in Figure 39B at low magnetic fields. *3) Non-Kramers ground states.* These have rhombic splittings of the $\pm M_S$ components and thus mixed wavefunctions as described in Figure 37. These will only show non-zero values of $\langle S_i \rangle$ at magnetic fields large enough to overcome the rhombic splitting and mixing. Thus, high-spin ferrous centers will show magnetic Mössbauer spectra at fairly large magnetic fields, and since the $g_i'$ values of the $M_S = \pm 2$ non-Kramers doublet are highly anisotropic ($g_{\parallel}' \cong 8,\ g_{\perp}' \cong 0$) these will be orientation averaged as in Figure 39B.

### 3.2 Unique Spectral Features

**3.2.1** ***Heme Sites*** The porphyrin (Figure 2B) is a particularly strong field ligand which, in addition to the high-spin Fe(III) $S = 5/2$ and high-spin Fe(II) $S = 2$ ground states described above, can exhibit the low-spin Fe(III) $S = 1/2$ and low-spin Fe(II) $S = 0$ ground states given in Figure 26C and D. This occurs when the ligand field splitting of the $t_{2g}$ and $e_g$ set of $d$ orbitals, $10\,Dq$, is significantly larger than the spin pairing energy [determined from Tanabe Sugano diagrams to be $Dq/B = 2.8$ for Fe(III) and 2.0 for Fe(II) (*vide infra*)] and is dependent on the ligand field strength of the axial ligands. For axial ligands of intermediate field strength the high-spin and low-spin states can be at comparable energy leading to a spin equilibrium situation.[56] For very weak axial ligands, the ligand field strength in the porphyrin plane can dominate, which will produce a tetragonal splitting of the $d$ orbitals as described in Figure 40A. If the $d_{x^2-y^2}$ orbital is at high enough energy this tetragonal splitting will overcome the spin pairing energy and produce an intermediate spin, Fe(III) $S =$

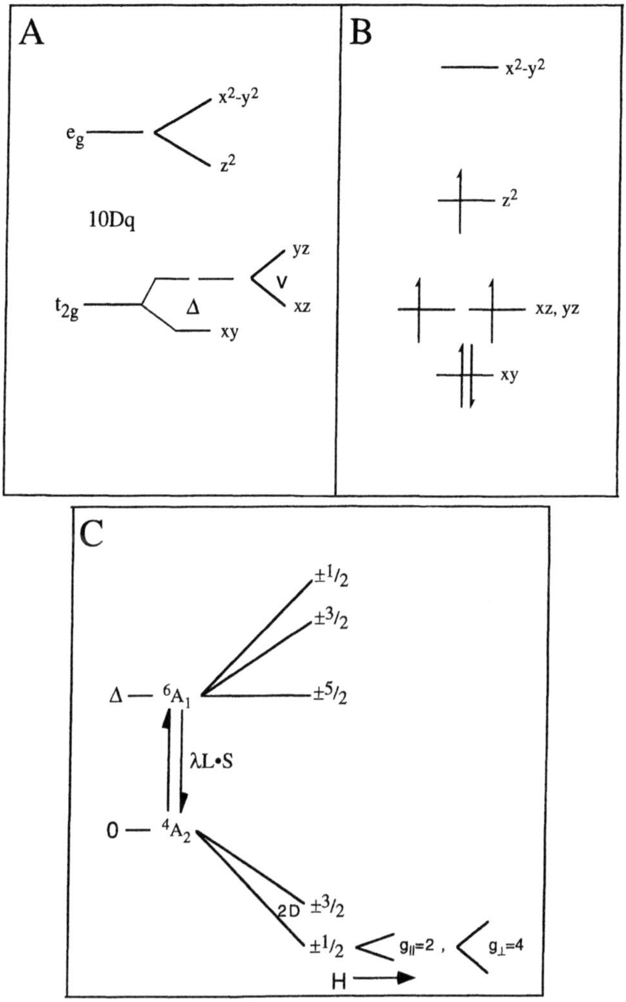


**Figure 40.** Effects of ligand field strength of axial ligands on Fe(III) in heme complexes. A) Weak axial ligation gives tetragonal splitting of *d* orbitals; B) an intermediate spin $S = 3/2$ ground state results when the tetragonal splitting becomes larger than the spin pairing energy; and C) zero field splitting of the intermediate spin $S = 3/2$ ground state due to spin-orbit coupling with the high-spin $^6A_1$ state.

3/2, ground configuration as shown in Figure 40B. For intermediate spin systems the high-spin $S = 5/2$ state can still be close in energy, and while not thermally accessible (i.e., no 3/2 / 5/2 spin equilibrium) it can mix with the ground state to produce a quantum spin-mixed system.[57] The properties of these Fe(III) ground states are described below. The spin states and properties of a number of hemes with variation in axial ligation are summarized in Table 8.[58]

**Table 8 Properties of Ferric Hemes with Varied Axial Ligation**

| Heme | Axial Ligands | Spin State | Spin Parameters | Ref. |
|---|---|---|---|---|
| aquo-metmyoglobin | N(His), $H_2O$ | $S = 5/2$ | $g_\perp = 6$, $g_\parallel = 2$ | 58a |
| cytochrome *c* | N(His), S(Met) | $S = 5/2$ | $g_x = 6.13$, $g_y = 5.80$, $g_z = 2.0$ | 58a |
| $Fe^{III}(TPP)(EtOH)_2$-$BF_4$ | EtOH | $S = 5/2$ | $g_\perp = 6$, $g_\parallel = 2$ | 58b |
| ferricytochrome *c*′ | N(His) | $S = 3/2$–$5/2$ mixed | $g_\perp = 4.77$, $g_\parallel = 1.99$ | 57 |
| $Fe^{III}(TPP)(ClO_4)$ | $ClO_4^-$ | $S = 3/2$–$5/2$ mixed | $g_\perp = 4.75$, $g_\parallel = 2.03$ | 58c |
| $Fe^{III}(TP_{piv}P)(OSO_2CF_3)(H_2O)$ | $OSO_2CF_4$, $H_2O$ | $S = 3/2$–$5/2$ mixed | $g_\perp = 5.7$, $g_\parallel = 2.00$ | 58d |
| Fe(TTP)[C=C(*p*-$C_6H_4)_2$](Cl) | (C = $CAr_2$), Cl | $S = 3/2$ | $g_\perp = 4.1$ (avg), $g_\parallel = 2$ | 58e |
| $Fe^{III}(OEP)(ClO_4)$ | $ClO_4^-$ | $S = 3/2$ | $g_\perp = 5.83$, $g_\parallel = 1.48$ | 58f |
| $Fe^{III}(OEP)(THF_2)$ | THF | $S = 3/2$ | $g_\perp = 4.68$, $g_\parallel = 2.00$ | 58g |
| $Fe^{III}(TPP)(C(CN)_3)$ | $C(CN)_3$ | $S = 3/2$ | $g_\perp = 5.26$, $g_\parallel = 2.00$ | 58h |
| cytochrome *c* (heart) | N(His), S(Met) | $S = 1/2$ | $g_x = 1.24$, $g_y = 2.24$, $g_z = 3.06$ | 58i |
| cytochrome *f* (spinach) | N(His), O(Tyr) | $S = 1/2$ | $g_x = 1.6$, $g_y = 2.07$, $g_z = 3.48$ | 58j |
| cytochrome P-450 | S(Cys), $H_2O$ | $S = 1/2$ | $g_x = 1.93$, $g_y = 2.26$, $g_z = 2.41$ | 58a |
| azide-metmyoglobin | N(His), $N_3^-$ | $S = 1/2$ | $g_x = 1.72$, $g_y = 2.22$, $g_z = 2.80$ | 58a |

3.2.1.1 *Low-Spin Fe(III)* $S = 1/2$ *and Spin Equilibrium* The low-spin Fe(III) configuration in Figure 26C produces a $^2T_{2g}$ ground state. The $T_{2g}$ is three fold orbitally degenerate which produces an orbital angular momentum, $L' = 1$, that spin-orbit couples to the $S = 1/2$ ($\lambda$[L.S.Fe(III)] $= -\xi = -460\,\text{cm}^{-1}$). In addition, the $t_{2g}$ set of $d$ orbitals can split due to the axial ($\Delta$) and rhombic ($V$) symmetry of the heme plus axial ligand site as described at the bottom of Figure 40A, which will split the $T_{2g}$ degeneracy of the ground state. The combined effects of spin-orbit coupling and less than octahedral symmetry on the $^2T_{2g}$ ground state is given by the determinant in Equation 36a, which is written in terms of the $L' = 1$, $M_{L'} = 1, 0, -1$ basis functions, with $\alpha$ indicating $M_S = +1/2$ and $\beta = -1/2$.[46,59] Diagonalization of Equation 36a gives the ground state doublet wavefunctions in Equation 36b.

$$\begin{array}{c|cccccc|}
 & |1\alpha\rangle & |0\beta\rangle & |-1\alpha\rangle & |-1\beta\rangle & |0\alpha\rangle & |1\beta\rangle \\
\langle 1\alpha| & -\frac{\xi}{2} - W & \frac{\sqrt{2}\xi}{2} & \frac{V}{2} & 0 & 0 & 0 \\
\langle 0\beta| & \frac{\sqrt{2}\xi}{2} & \Delta - W & 0 & 0 & 0 & 0 \\
\langle -1\alpha| & \frac{V}{2} & 0 & \frac{\xi}{2} - W & 0 & 0 & 0 \\
\langle -1\beta| & 0 & 0 & 0 & -\frac{\xi}{2} - W & \frac{\sqrt{2}\xi}{2} & \frac{V}{2} \\
\langle 0\alpha| & 0 & 0 & 0 & \frac{\sqrt{2}\xi}{2} & \Delta - W & 0 \\
\langle 1\beta| & 0 & 0 & 0 & \frac{V}{2} & 0 & \frac{\xi}{2} - W
\end{array} = 0 \qquad \text{(36a)}$$

$$\begin{aligned}
\Psi^+ &= A|1\alpha\rangle + B|0\beta\rangle + C|-1\alpha\rangle \\
\Psi^- &= A|-1\beta\rangle - B|0\alpha\rangle + C|1\beta\rangle
\end{aligned} \qquad \text{(36b)}$$

$$\begin{aligned}
g_x &= 2\left[2AC - B^2 + kB(C - A)\sqrt{2}\right] \\
g_y &= 2\left[2AC + B^2 + kB(C + A)\sqrt{2}\right] \\
g_z &= 2\left[A^2 - B^2 + C^2 + k\left(A^2 - C^2\right)\right]
\end{aligned} \qquad \text{(36c)}$$

Evaluation of the Zeeman operator over these gives the expressions for the $g_i$ values in Equation 36c, where $k$ is the Stevens orbital reduction factor ($<1$ due to covalency) and $A^2+B^2+C^2 = 1$. Thus the $g_i$ values from the EPR spectrum of a low-spin ferric site (representative values are given in Figure 41A) provide a direct probe of the axial ($\Delta$) and rhombic ($V$) splittings of the $t_{2g}$ set of $d$ orbitals as defined in Figure 40A. (The experiment does not give the sign of $g_i$, which can be positive or negative, and the experimental designation of the $g_i$ values has the highest to lowest going from $g_z$ to $g_x$, which does not necessarily correlate with the coordinate system used in Equation 36c.)[60a] The axial splitting clearly relates to the $\pi$-bonding interactions of the porphyrin ligand with the $d_{xz,yz}$ vs. $d_{xy}$ orbitals, and the rhombic splitting of $d_{xz}$ and $d_{yz}$ orbitals has been correlated with the anisotropic $\pi$-bonding interaction of the iron with the axial ligands.[61a] (See also Ref. 61b.)

For cases where the axial ligands produce a ligand field strength comparable to the spin pairing energy, both the $^2T_{2g}$ and $^6A_{1g}$ states can be thermally accessible.

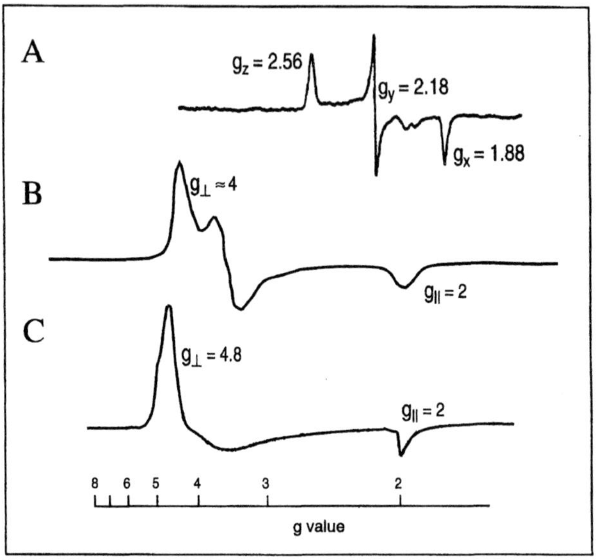


**Figure 41.** Representative EPR spectra for ferric heme systems. A) Low spin ($S = 1/2$); B) intermediate spin ($S = 3/2$); and C) quantum mixed spin ($S = 5/2$ with $S = 3/2$). Adapted from Reference 60a.

Since these differ in spin by $\Delta S = 2$ and spin-orbit coupling can only connect states with $\Delta S = 1$, these can be treated as independent states.[35,56] In this spin equilibrium situation, the features of both states will then contribute to spectra (e.g., EPR) with a relative weight dependent upon their Boltzmann population, which will be temperature dependent.

3.2.1.2 *Intermediate Spin Fe(III) $S = 3/2$ and Quantum Mixing* In the limit of a large tetragonal distortion the electron configuration in Figure 40B is appropriate, which produces an intermediate spin $^4A_2$ $S = 3/2$ Fe(III) ground state which can undergo zero field splitting. Application of Equation 24 (assuming $D > 0$ and $h\nu_{\text{microwave}} \ll D$ with $E = 0$) gives the zero field splitting at the bottom of Figure 40C with the $M_S = \pm 1/2$ Kramers doublet lowest in energy. A magnetic field splits the $\pm 1/2$ doublet by an amount $g'\beta H$ where $g'_\parallel = 2.0$ and $g'_\perp = 4.0$, producing the EPR signal given in Figure 41B.[60b]

The zero field splitting of the $^4A_2$ in Figure 40C derives from spin-orbit coupling in second order with the high-spin $^6A_1$ state (an excited state in the weak axial, tetragonal ligand field), where $D = \zeta^2/5\Delta$ with $\zeta$ the one electron spin-orbit coupling parameter and $\Delta$ the $^4A_2 - {}^6A_1$ energy separation.[62] While the $^6A_1$ state has not been found to be thermally accessible, it can, in some cases, be at low enough energy to significantly "quantum mix" with the $^4A_2$ ground state through spin-orbit coupling.[57] This produces a modified ground state wavefunction which is given by Equation 37a where solution of the determinant in Equation 37b gives the mixing coefficients. The quantum mixed $|\pm 1/2'\rangle$ ground state will have effective $g'_i$ values

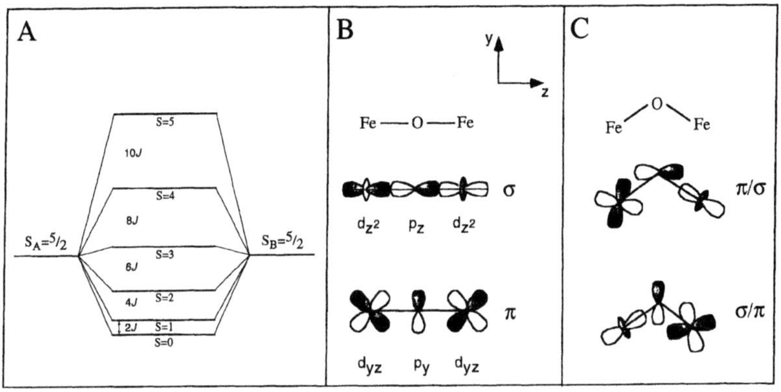


**Figure 42.** Oxo-bridged binuclear ferric systems. A) The exchange ladder for two antiferromagnetically coupled high-spin ferric ($S = 5/2$) ions; B) dominant exchange pathways for a linear oxo-bridged iron dimer; and C) for a bent dimer.

given by the weighted average of those of high-spin and intermediate-spin Fe(III), $g_{\parallel} = 2.0$ and $g_{\perp} = 6a^2 + 4b^2$.

$$|\pm 1/2'\rangle = a|^6A_1 \pm 1/2\rangle + b|^4A_2 \pm 1/2\rangle \quad (37a)$$

$$\begin{array}{c|cc|c} & |^6A_1, \pm 1/2\rangle & |^4A_2, \pm 1/2\rangle & \\ |^6A_1, \pm 1/2\rangle & \Delta - W & -(6/5)^{1/2}\xi & = 0 \\ |^4A_2, \pm 1/2\rangle & -(6/5)^{1/2}\xi & 0 & \end{array} \quad (37b)$$

Therefore, the $g_{\perp}$ value will be greater than 4 and less than 6 in a quantum mixed site (Figure 41C)[60c] and can be used as a quantitative probe of the extent of tetragonality of the heme site.[63]

**3.2.2** ***Oxygen Bridged Binuclear Non-Heme Iron Sites*** From Table 2, a number of enzymes have binuclear non-heme iron sites bridged by oxygen atoms at the binuclear ferric (i.e., met) oxidation level. In all defined cases but methane monooxygenase, this is an oxo bridge with additional carboxylate bridges also present, leading to a bent Fe–O–Fe structure.[64] For methane monooxygenase, the bridge apparently has a higher effective $pK_A$ and is protonated (i.e., hydroxide).[65] The oxo bridged dimers have high-spin ferric ions but with very different ground state properties from those described above in that the bridge results in strong antiferromagnetic coupling between the two $S = 5/2$ centers leading to $S_{total} = 0, 1, 2, 3, 4$, and 5 dimer states. The HDVV Hamiltonian in Equations 8 and 9 gives the exchange ladder for the antiferromagnetically coupled ($J < 0$) dimer in Figure 42. Magnetic susceptibility studies have shown $J$ to be on the order of $-100$ cm$^{-1}$ for met ribonucleotide reductase and met and oxy hemerythrin.[66] The exchange coupling for MMO is much lower, $\sim -7$ cm$^{-1}$, due to the hydroxide bridge.[67] This exchange coupling can be related to individual bonding interactions between half occupied metal $d$ orbitals through the oxo bridge as described by Equation 38, where A and B refer to the Fe ions and $i$ and $j$ to their respective half-occupied $d$ orbitals.

$$J = \frac{1}{25} \sum_{i,j} J_{A_i B_j} \tag{38}$$

Most of the 25 pathways between unpaired electrons on the two irons are negligible, while a few pathways mediate extremely large exchange coupling.[68]

For a linear oxo bridged iron dimer, there are three significant pathways: the $\sigma$-pathway involving $d_{z^2}$ orbitals on each iron and the $p_z$ orbital on the oxo bridge; and two $\pi$-pathways, the $d_{xz_A} - Op_x - d_{xz_B}$ and the $d_{yz_A} - Op_y - d_{yz_B}$. The $\pi$-pathways have been experimentally estimated to contribute ~85% to the total antiferromagnetic coupling and the $\sigma$-pathway contributes ~13%.[68b] On bending the Fe–O–Fe unit to an angle consistent with those of the protein sites, the two $\pi$ pathways are found to decrease their contribution to the exchange coupling (to 72%), while two new mixed $\pi/\sigma$-pathways now significantly contribute (~16%). These $\pi/\sigma$-pathways are shown in Figure 42C. The net effect is that there is little angular variation in $J$ for oxo bridged diferric dimers.[69] This is in contrast to cupric dimers, where there can only be one exchange pathway. On protonation of the bent diferric structure the exchange coupling is greatly reduced, as the bridging ligand charge is lower and the in-plane p $\pi$ orbital on the oxygen is now dominantly involved in bonding to the proton, which significantly reduces its $\pi$-bonding interaction with the iron. Protonation is responsible for the reduced antiferromagnetic coupling in MMO and should contribute significantly to its reactivity relative to the other binuclear non-heme iron enzymes.

On reduction of the diferric protein active sites mentioned above, the bridging oxo group is protonated in the case of deoxy hemerythrin (Hr)[70] and eliminated in the cases of ribonucleotide reductase[71], MMO[67], and $\Delta^9$ desaturase[72], leaving only carboxylate bridges (Figure 2D). Thus the exchange coupling is greatly reduced; for Fe(II) dimers, the $J$ value associated with a $OH^-$ bridge is $\sim -10$ cm$^{-1}$ and carboxylate bridges have even lower exchange couplings which can be weakly ferromagnetic.[73] In addition the zero field splitting on the iron increases from ~1 cm$^{-1}$ for high-spin, non-heme Fe(III) (out-of-state spin-orbit coupling) to ~10 cm$^{-1}$ for high-spin Fe(II) (in-state coupling). These interactions are comparable and must both be treated together to define the ground states of binuclear ferrous sites.

3.2.2.1 *J/D Diagrams* The appropriate Hamiltonian for coupled binuclear ferrous centers which includes exchange coupling, zero field splitting and the Zeeman effect is given in Equation 39a, where the $g_i$ values and zero field splitting parameters are related through ligand field theory (LFT); Equation 39b, and all other parameters have been defined earlier. This Hamiltonian operates on an uncoupled basis set $|S_A M_{S_A}, S_B M_{S_B}\rangle$ yielding a $25 \times 25$ matrix. Diagonalization of this matrix in zero magnetic field with $D_A = D_B$ and $E_A = E_B = 0$ gives the energy level diagrams shown in Figure 43 [A corresponds to antiferromagnetic ($J < 0$) and B to ferromagnetic ($J > 0$) coupling].[73,74]

$$\begin{aligned}\mathcal{H} = {} & -2JS_A \bullet S_B + D_A\left(S_{zA}^2 - \tfrac{1}{3}\left[S_A(S_A+1)\right]\right) \\ & + E_A\left(S_{x_A}^2 - S_{y_A}^2\right) + D_B\left(S_{z_B}^2 - \tfrac{1}{3}\left[S_B(S_B+1)\right]\right) \\ & + E_B\left(S_{x_B}^2 - S_{y_B}^2\right) + g_{zA}\beta H_z S_{zA} + g_{x_A}\beta_x S_{x_A} \\ & + g_{y_A}\beta H_x S_{y_A} + g_{zB}\beta H_z S_{zB} + g_{xB}\beta H_x S_{xB} + g_{y_B}\beta H_y S_{y_B} \end{aligned} \quad (39a)$$

$$D_{\mathrm{Fe(II)}} = \frac{-k^2\lambda}{4}(g_x + g_y - 2g_z), \qquad E_{\mathrm{Fe(II)}} = \frac{-k^2\lambda}{4}(g_y - g_x) \quad (39b)$$

The energy levels that result when there is no exchange coupling are shown at the right of Figure 43A. These states are labeled by the uncoupled basis functions on the two irons ($M_{S_A}$, $M_{S_B}$). Allowing the iron atoms to interact corresponds to $J$ increasing to the left in the diagram. This splits the energy of these levels. The left side of Figure 43A shows the limiting case with large exchange coupling and no axial ZFS. In this limit, two $S = 2$ ions are antiferromagnetically coupled to produce five states having the spin values $S_{\mathrm{total}} = 0, 1, 2, 3, 4$. These levels split into singlets and doublets when $D$ is included, moving toward the middle of the diagram. The singlets and doublets that result are labeled by the dominant contribution in the coupled spin wavefunctions ($S_{\mathrm{total}}$, $M_S$). As the magnitude of $J$ and $D$ become comparable, the splitting and mixing of the levels becomes increasingly more complex. These splittings can be probed experimentally by magnetic susceptibility and MCD. Temperature dependent MCD data on deoxyHr in Figure 44A indicate that the ground state is MCD inactive (diamagnetic) but that a paramagnetic state is populated at higher temperature.[73a] Both the right and left sides of Figure 43A could be used to fit the deoxyHr data. However, in the limit of $D > 0$ and $J = 0$ (right side), the Fe(II) ZFS parameter required would be ~35 $cm^{-1}$ which is larger than the maximum reasonable value of $D \sim 15$ $cm^{-1}$. In the limit of $-J > 0$ and $D = 0$ (left side) the location of the MCD active excited state would give a $-J$ value of ~25 $cm^{-1}$. However, this requires that the ferrous ZFS parameter and the exchange coupling constant be of similar magnitude, and thus both effects must be included in a quantitative fit to the deoxyHr MCD data (center of Figure 43A). This results in $12 < -J < 38$ $cm^{-1}$ for deoxyHr which was early evidence for the presence of a bridging hydroxide.

In contrast, the saturation magnetization MCD data for deoxy$N_3^-$Hr in Figure 44B indicate that the ground state is a MCD active doublet with a $g_{\parallel} \sim 16$.[73a] This requires that the two ferrous centers in deoxy$N_3^-$Hr are weakly ferromagnetically coupled ($J > 0$). Figure 43B treats the ferromagnetic case, allowing for the effects of varying the axial ZFS. Both axes are in units of $J$ so this parameter is also allowed to vary. If $D < 0$ (left side) the $|4, \pm 4\rangle$ non-Kramers doublet is the ground state, which would have a $g_{\parallel} = 16$. Alternatively, if $D > 0$, the ground state is a singlet for any value of $D$ and would not show MCD C-term intensity at lower temperatures. Thus the $g_{\parallel} \sim 16$ observed in deoxy$N_3^-$Hr requires ferromagnetic exchange coupling and negative ZFS between the Fe(II) ions, giving a $M_S = \pm 4$ ground state (Figure 43B bottom left). These diagrams become more complex when $D_A \neq D_B$, $E_A$ and

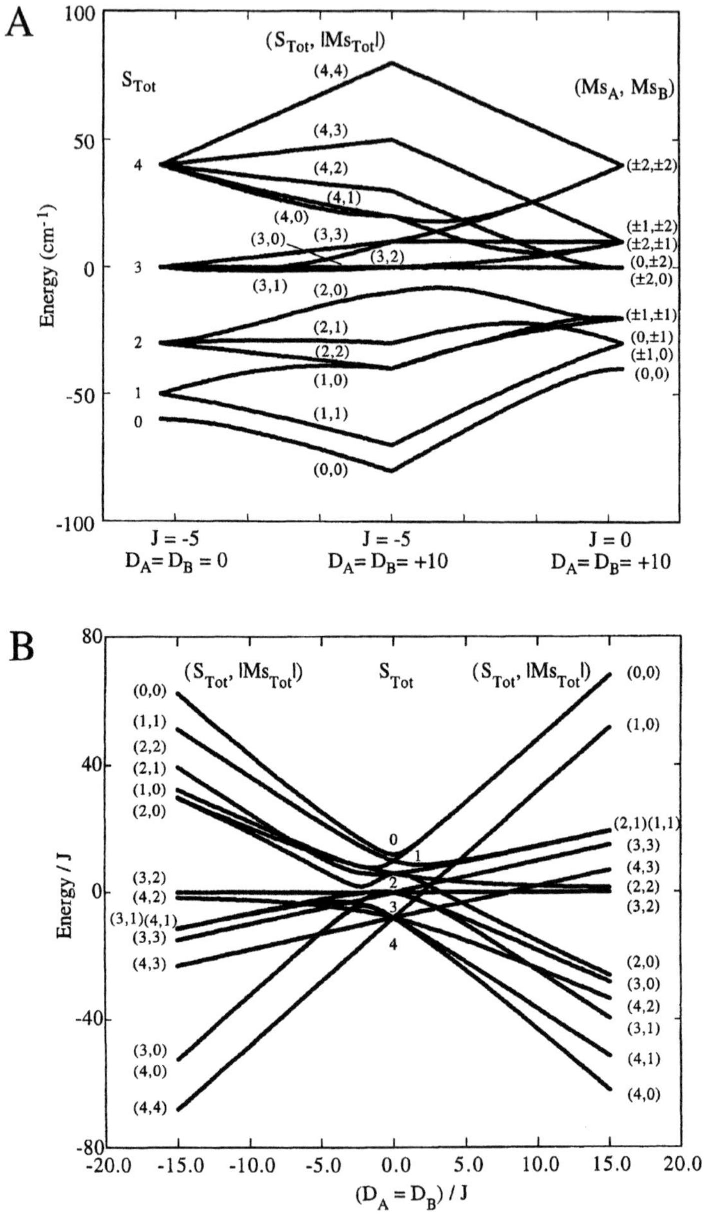


**Figure 43.** Correlation diagrams of the ground state energy levels including single site ZFS and exchange coupling for a binuclear ferrous complex. A) Left: antiferromagnetic exchange coupling with no ZFS; center: both exchange coupling and ZFS; right: limit of positive ZFS only with Fe(II) ions uncoupled; and B) left: ferromagnetic coupling plus negative ZFS; center: ferromagnetic coupling with no ZFS; right: ferromagnetic coupling plus positive ZFS. $S_{\text{total}}$ indicates coupled basis set, $M_{S_A} M_{S_B}$ indicates uncoupled basis set. Adapted from Reference 74.

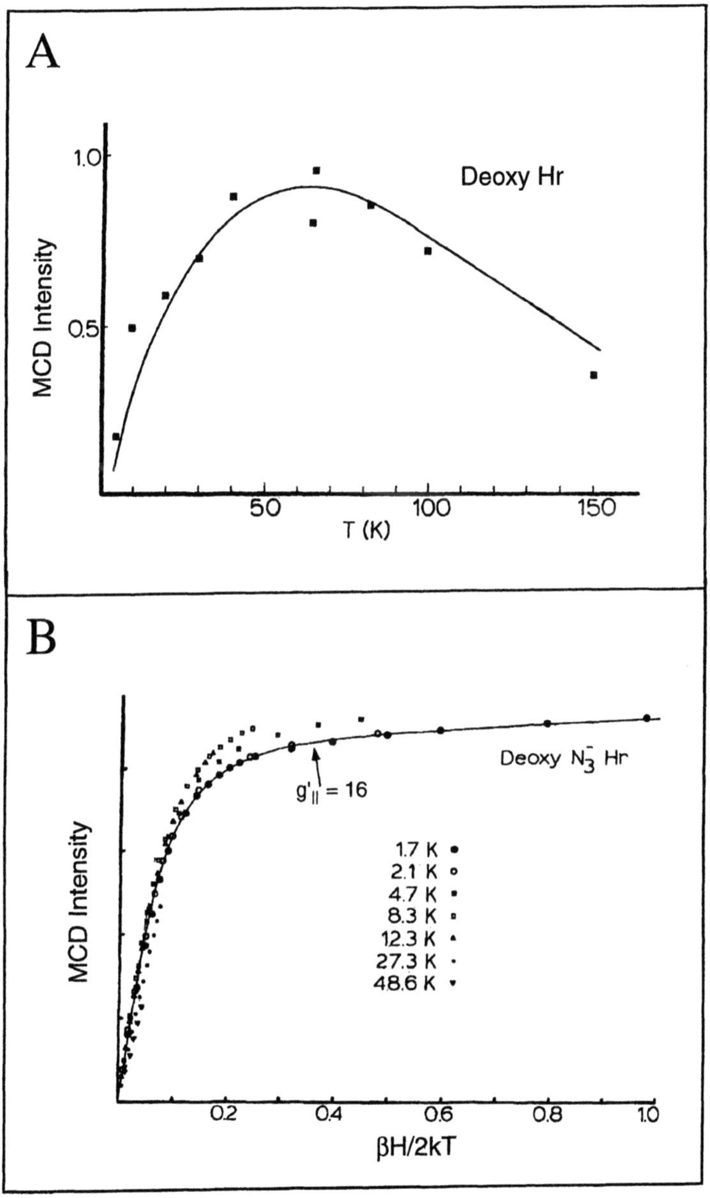


**Figure 44.** Deoxy- and deoxy azide hemerythrin MCD data. A) Temperature dependence of MCD intensity at fixed field for deoxy hemerythrin; and B) saturation magnetization data for deoxy azide hemerythrin taken at 940 nm. A solid line gives fit for low temperature data with Equation 32. Adapted from Reference 73a.

$E_B \neq 0$, and the zero field splitting tensor axis systems can be non-coincident. This limits one's ability to define the ground states of binuclear ferrous active sites; the present estimates for the parameters in Equation 39 for the different binuclear non-heme ferrous enzymes are given in Table 9.[73,75]

There are two interesting points to note. First, in a number of cases [deoxy$N_3^-$Hr, reduced methane monooxygenase (MMO), reduced MMO + component $B$, two $N_3^-$ bound reduced ribonucleotide reductase] the binuclear ferrous site is weakly ferromagnetic. This can be associated with carboxylate and possibly water (i.e., protonated hydroxide) bridges. The origin of this ferromagnetism needs further investigation. Second, the saturation magnetization data in Figure 44B requires that the rhombic ZFS of the $M_S = \pm 4$ non-Kramers doublet, $\delta$ (Figure 45A), is very small. This indicated the possibility of observing a transition within the $M_S = \pm 4$ doublet by EPR spectroscopy.[70a] This transition has now been observed for deoxy$N_3^-$Hr[70a], MMO[76a] and $N_3^-$ bound ribonucleotide reductase.[76b]

3.2.2.2 *Integer Spin EPR* For most mononuclear high-spin Fe(II) centers, the rhombic splitting of the $S = 2 M_S = \pm 2$ non-Kramers doublet, $\delta$ (Figure 37), is on the order of several $cm^{-1}$ ($\cong 3E^2/D$), which is larger than the microwave energy (X-band ~0.3 $cm^{-1}$), and thus this ground state must be studied by alternative methods (e.g., saturation magnetization MCD, see Section 3.1.2). However, in an exchange coupled binuclear ferrous dimer the $S_{total} = 4$, $M_{S_{total}} = \pm 4$ non-Kramers doublet is split by the rhombic term, $E$, in the spin Hamiltonian in Equation 39A only in fourth order, and thus $\delta$ in Figure 45A can be quite small. If $\delta$ is less than 0.3 $cm^{-1}$, a transition between these sublevels can be observed by EPR spectroscopy at X-band frequencies. It should be noted that this is formally a $\Delta M_S = 8$ ($-4 \rightarrow +4$) transition. In a normal EPR experiment the magnetic dipole selection rule is $\Delta M_S = \pm 1$, which allows the $-1/2 \rightarrow +1/2$ transition in Kramers doublets (*vide supra*). This derives from the fact that the electron spin $\vec{S}$ is oriented along the external magnetic field (defining the $z$ axis) and in the normal EPR configuration the oscillating

**Table 9 Ground State Parameters for Binuclear Ferrous Active Sites**

| Binuclear Ferrous Protein Site | $J$ ($cm^{-1}$) | $D_1$ ($cm^{-1}$) | $D_2$ ($cm^{-1}$) | Site Symmetry |
|---|---|---|---|---|
| deoxyHr | $-13 - -38$ | [a] | [a] | |
| deoxy$N_3^-$Hr | $2.2 \pm 0.4$ | $-13.0 \pm 2.0$ | $-13.0 \pm 2.0$ | rhombic |
| reduced RDPR | $-0.44$ | $-10.0 \pm 2.0$ | $+4.0 \pm 1.0$ | rhombic |
| one azide bound reduced RDPR | $-2.5 \pm 1.0$ | $-13.0 \pm 2.0$ | $+4.0 \pm 1.0$ | axial |
| two azide bound reduced RDPR I.[b] | $-2.0 - +2.0$ | $-10.0 \pm 5.0$ | $+4.0 \pm 1.0$ | axial |
| two azide bound reduced RDPR II.[b] | $1.0 \pm 0.5$ | $-10.0 \pm 5.0$ | $-10.0 \pm 5.0$ | |
| reduced MMOH | $0.3 \pm 0.1$ | $-4.0 \pm 3.0$ | $-9.0 \pm 6.0$ | rhombic |
| reduced MMOH-ComponentB | $0.4 \pm 0.1$ | $-6.0 \pm 3.0$ | $-6.0 \pm 3.0$ | rhombic |

[a]Cannot be estimated due to the large magnitude of exchange coupling.
[b] MCD studies have shown that two distinguishable species are formed when a second azide is bound to the one-azide-bound form of reduced RDPR. One of the species has one four-coordinate and one six-coordinate Fe(II) center, while the other has a five-coordinate and a six-coordinate Fe(II) ion.

magnetic field, $\vec{H}_1$, of the microwave radiation is in a transverse orientation (i.e., $\perp$ to the external magnetic field). The expression for EPR intensity in Equation 40 shows that for this configuration, $\vec{H}_1$ projects out the $S_x$ (or $S_y$) operator, where $S_x = 1/2(S_+ + S_-)$. $S_+$ and $S_-$ operating on $M_S$ requires that $M'_S = M'_S \pm 1$ for the EPR intensity integral in Equation 40 to be non-zero. Alternatively, if the oscillating

$$I \propto \left|\langle M'_S | g\beta \vec{H}_1 \bullet \vec{S} | M_s \rangle\right|^2 \tag{40}$$

magnetic field is oriented in a longitudinal configuration (i.e., parallel to the external magnetic field), $\vec{H}_1$ projects out the $S_z$ operator. $S_z$ operating on $M_S$ requires that $M'_S = M_S$ for the integral in Equation 40 to be non-zero (i.e., $\Delta M_S = 0$). From Figure 45A, the $M_S = \pm 4$ wavefunctions are strongly mixed by the rhombic term in the spin Hamiltonian. This leads to a $\Delta M_S = 0$ (i.e., $4 \rightarrow 4$ and $-4 \rightarrow -4$) magnetic dipole allowed component in the formally $\Delta M_S = 8$ transition. As indicated in Figure 45B (from Reference 77) the integer spin EPR signal of deoxy $N_3^-$ hemerythrin increases in intensity for $\vec{H}_1 \| \vec{H}$. It should be noted that since the $g'_i$ values for the $M_S = \pm 4$ doublet are $g'_\| = 16$ and $g'_\perp = 0$, a very anisotropic EPR signal is observed. Also, the $g_\|$ region of the EPR spectrum is very dependent on the value of $\delta$ in Figure 45A, and the resonance condition can be satisfied at lower or even zero magnetic field. Spectral simulations of integer spin EPR signals are described in Reference 78.

**3.2.3** ***Iron-Sulfur Clusters*** As illustrated in Figure 2E, the iron sulfur clusters are commonly found with 2, 3, and 4 iron atoms although higher nuclearity iron and mixed metal clusters are also known. The formal oxidation states of the 2, 3, and 4

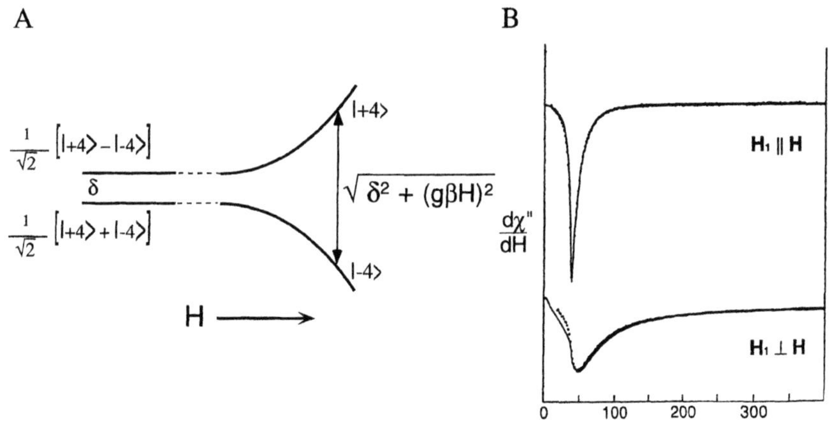


**Figure 45.** Integer spin EPR. A) Rhombic splitting of an $M_S = \pm 4$ non-Kramers doublet at zero field (left) and with increasing applied magnetic field (right); and B) EPR spectra (at X-band, taken from Reference 78) showing the integer spin signal of deoxy azide hemerythrin with microwave field $H_1$ applied parallel (top) and perpendicular (bottom) to the static field $H$. (Dotted line is a simulated spectrum.)

iron sulfur clusters are [2Fe(III)] and [Fe(III)Fe(II)], [3Fe(III)] and [2Fe(III) Fe(II)], [3Fe(III) Fe(II)], and [2Fe(III) 2Fe(II)] and [Fe(III) 3Fe(II)]. Their ground state properties are strongly affected by exchange interactions between different iron centers in the cluster and possible electron delocalization for those in mixed oxidation states. These properties are summarized below.

3.2.3.1 *$Fe_2S_2$ Ferredoxins: Vector Coupling* The oxidized two iron proteins have strong antiferromagnetic coupling ($J$ in Equation 8 $\sim -180\ \text{cm}^{-1}$) between the two Fe(III) $S = 5/2$ ions which produces $S_{\text{total}} = 0, 1, 2, 3, 4, 5$ ground state sublevels split in energy as shown in Figure 42A. All antiferromagnetically coupled homodimers have an $S_{\text{total}} = 0$ ground state and are therefore diamagnetic. Upon reduction, the two iron ferredoxin site has localized Fe(III) $S = 5/2$ and Fe(II) $S = 2$ ions which are also strongly antiferromagnetically coupled ($J \sim -105\ \text{cm}^{-1}$),[79] producing $S_{\text{total}} = 1/2, 3/2, 5/2, 7/2, 9/2$ states split in energy (Equations 8 and 9) as shown on the left side of Figure 46A. For a coupled hetero-dimer with half integer $S_{\text{total}}$ the ground state is paramagnetic ($S_{\text{total}} = 1/2$ for reduced $Fe_2S_2$ ferredoxin) and can be studied in detail by EPR, magnetic Mössbauer, etc. Since in iron sulfur clusters the splitting between $S_{\text{total}}$ states of the dimer is large relative to additional interactions (ZFS, Zeeman, Hyperfine, etc.), good wavefunctions for the $S_{\text{total}}\ M_{S_{\text{total}}}$ states are obtained by *vector coupling* the individual spin angular momenta on $\text{Fe}_A|S_A, M_{S_A}\rangle$ and $\text{Fe}_B|S_B, M_{S_B}\rangle$. The associated coupling coefficients are the Clebsch–Gordon coefficients (or $3j$ symbols where $3j, 6j, 9j$, etc., couple 2, 3, 4, etc., individual angular momenta).[80] Evaluation of the Zeeman and hyperfine operators over these wavefunctions gives the expressions in Equation 41, which relate the properties of a dimer to the properties of the individual metal ions being coupled.[62,81] The lowest energy state for reduced two iron ferredoxin (Figure 46A left) is $S_{\text{total}} = 1/2$.

$$\begin{aligned} g_{S_{\text{total}}} &= C_1 g_A + C_2 g_B \\ A^A_{S_{\text{total}}} &= C_1 A^A \\ A^B_{S_{\text{total}}} &= C_2 A^B \\ &\text{where } C_1 = (1 + C)/2, \quad C_2 = (1 - C)/2 \\ &\text{and } C = \frac{[S_A(S_A + 1) - S_B(S_B + 1)]}{S_{\text{total}}(S_{\text{total}} + 1)} \end{aligned} \tag{41}$$

Substitution of this into Equation 41 (with $S_A = 5/2$, $S_B = 2$) gives $C = 11/3$, $C_1 = 7/3$, $C_2 = -4/3$; thus $g_{S_{\text{total}}=1/2} = (7/3)g_{\text{Fe}^{+3}} - (4/3)g_{\text{Fe}^{+2}}$, $A^{\text{Fe}^{+3}}_{S_{\text{total}}=1/2} = (7/3)A^{\text{Fe}^{+3}}$ and $A^{\text{Fe}^{+2}}_{S_{\text{total}}} = 1/2 = -(4/3)A^{\text{Fe}^{+3}}$. The $g_i$ values for a high-spin Fe(III) should be close to isotropic and equal 2.00 (*vide supra*), while those for a distorted tetrahedral high-spin Fe(II) should be anisotropic with $g_x$ and $g_y > 2.00$ due to spin-orbit coupling of the extra electron in the $d_{z^2}$ orbital (from the sign of $E_Q$ in the Mössbauer spectrum)[82] with higher energy ligand field states. The vector coupling model then predicts dimer $g_x$ and $g_y$ values below 2.00 as is observed experimentally ($g_z = 2.05$, $g_y = 1.96$, and $g_x = 1.90$ for plant ferredoxin).[83] With respect to

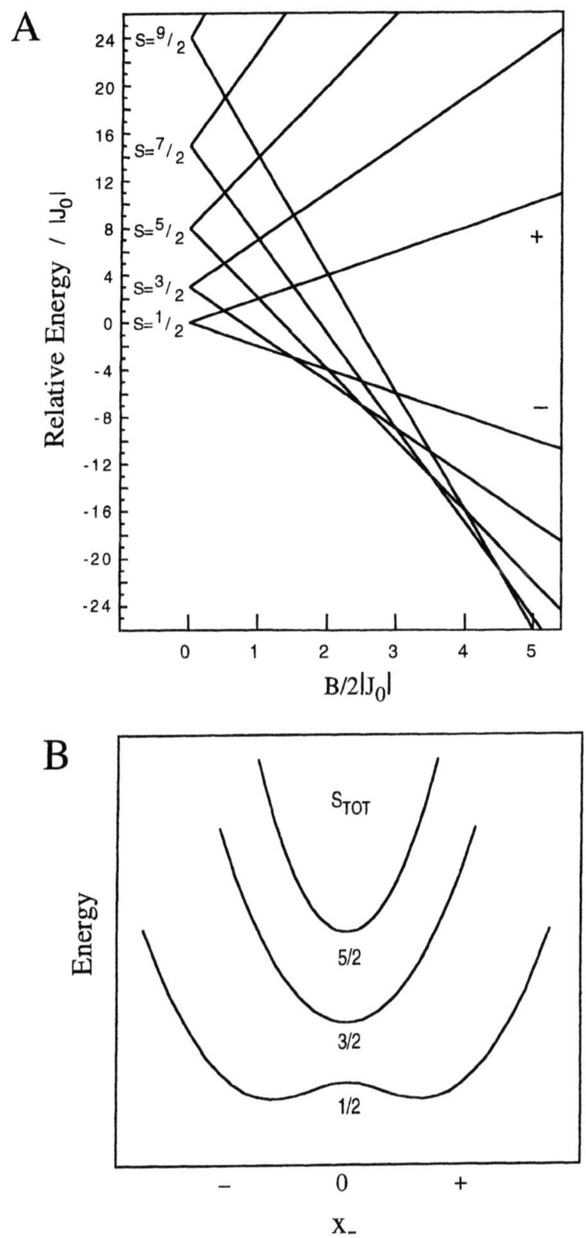


**Figure 46.** Antiferromagnetic coupling in a mixed-valence hetero-dimer with double exchange. A) Spin states produced by coupling of Fe(III) $S = 5/2$ and Fe(II) $S = 2$; right: effects of increasing double exchange ($B/2J$) due to resonance splitting of the + and – states corresponding to symmetric and antisymmetric combinations of fragments with the extra electron on each Fe in the dimer; and B) associated potential energy surfaces for lowest energy spin components (antisymmetric, –, combination).

hyperfine coupling, high-spin Fe(III) has a negative isotropic coupling while high-spin Fe(II) is anisotropic (due to the extra electron in $d_{z^2}$) with negative components. However, as shown above, the hyperfine coupling of the Fe(II) in the $S_{\text{total}} = 1/2$ ground state of the dimer is $-4/3$ that of a monomer ferrous center. As described by Equation 35, the hyperfine coupling produces an internal magnetic field that combines with the applied field to produce a resultant magnetic field at each iron nucleus of the dimer. For high-spin iron monomers the hyperfine field opposes the applied field and results in a decrease in the magnetic Mössbauer splitting with increasing applied magnetic field, as shown in Figure 39B. However, for the Fe(II) center in this exchange coupled site, the hyperfine field is positive and the magnetic Mössbauer splitting increases with increasing $H_{\text{applied}}$.[84] This is a characteristic of antiferromagnetic exchange coupled sites.

3.2.3.2 *$Fe_3(III)S_4$ Clusters: Spin Frustration* From Figure 2E, $Fe_3S_4$ clusters have each pair of irons bridged by two sulfido ligands ($\mu_2$ and $\mu_3$),[85] which should lead to antiferromagnetic coupling between each pair of high-spin Fe(III) $S = 5/2$ ions. The HDVV Hamiltonian for this system is given by Equation 42.

$$\mathcal{H} = -2J_{12}\hat{S}_1 \bullet \hat{S}_2 - 2J_{13}\hat{S}_1 \bullet \hat{S}_3 - 2J_{23}\hat{S}_2 \bullet \hat{S}_3 \tag{42}$$

It is useful to consider the behavior of this system as the $J'$ value between one pair (e.g., $J_{13}$) is allowed to vary relative to $J$ values for the other two pairs ($J_{12} = J_{23} = J$).[62] This is shown schematically in Figure 47. For this system, $Fe_1$ and $Fe_3$ are symmetry related and can be vector coupled to produce states having $S' = 5, 4, 3, 2, 1, 0$. These $S'$ states are further coupled to $Fe_2$ ($S_2 = 5/2$) to produce $S_{\text{total}} = |S_2 + S'|, S_2 + S' - 1 \ldots |S_2 - S'|$ states for each $S'$ intermediate state. Thus for $S' = 0$, $S_{\text{total}} = 5/2$; for $S\prime = 1$, $S_{\text{total}} = 7/2, 5/2, 3/2$; for $S' = 2$, $S_{\text{total}} = 9/2, 7/2, 5/2, 3/2, 1/2$; for $S' = 3$, $S_{\text{total}} = 11/2, 9/2, 7/2, 5/2, 3/2, 1/2$; for $S' = 4$, $S_{\text{total}} = 13/2, 11/2, 9/2, 7/2, 5/2, 3/2$; and for $S' = 5$, $S_{\text{total}} = 15/2, 13/2, 11/2, 9/2, 7/2, 5/2$. Equation 42 then gives the spin Hamiltonian and energy expression in Equation 43 [note that $S^2 = (S_1 + S_2 + S_3)^2$ and $S'^2 = (S_1 + S_3)^2$].

$$\begin{aligned}
\mathcal{H} &= -2J\big(\hat{S}_1 \bullet \hat{S}_2 + \hat{S}_2 \bullet \hat{S}_3 + \hat{S}_1 \bullet \hat{S}_3\big) - 2(J' - J)\hat{S}_1 \bullet \hat{S}_3 \\
&= -J(\hat{S}^2 - \hat{S}_1^2 - \hat{S}_2^2 - \hat{S}_3^2) - (J' - J)\big(\hat{S}'^2 - \hat{S}_1^2 - \hat{S}_3^2\big) \\
E(S, S') &= -J\big[S(S+1) - S_1(S_1+1) - S_2(S_2+1) - S_3(S_3+1)\big] \\
&\quad -(J' - J)\big[S'(S'+1) - S_1(S_1+1) - S_3(S_3+1)\big]
\end{aligned} \tag{43}$$

In the limit of $J' = 0$, substitution of the $|S_{\text{total}}, S'\rangle$ states into Equation 43 reveals that the lowest energy state will have the highest $S'$ and lowest $S$ values (i.e., $S' = 5$, $S_{\text{total}} = 5/2$). (Note that $J < 0$ for antiferromagnetic coupling.) As shown in Figure 47, this corresponds to an antiparallel alignment of the $S = 5/2$ spins between $Fe_1$ and $Fe_2$ and between $Fe_2$ and $Fe_3$ (i.e., $S_{\text{total}} = 5/2$). The net effect of these exchange couplings is that the spins on $Fe_1$ and $Fe_3$ are parallel (i.e., $S' = 5$); however,

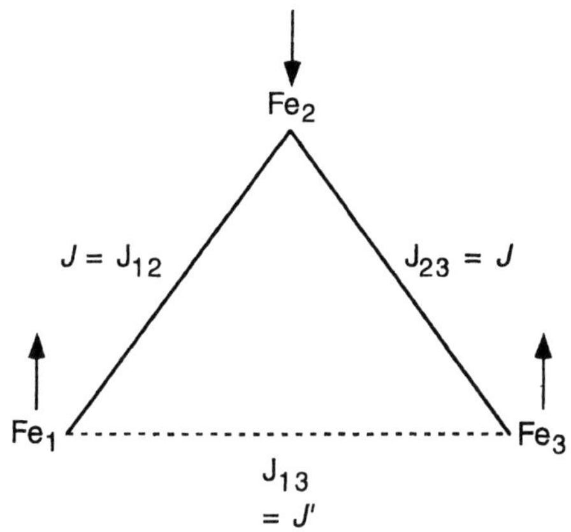


**Figure 47.** Triangular arrangement of three Fe atoms with the exchange coupling between two pairs being equal ($J = J_{12} = J_{23}$). Spin alignment shown for the case of $J' = J_{13} = 0$.

since $J'$ has been set equal to 0 in this limit, there is no actual exchange interaction between $S_1$ and $S_3$.

Now consider the limit where $J' = J$. In this case the energy term dependent on $S'$ in Equation 43 vanishes, and the energy is determined only by $S_{\text{total}}$. Thus the lowest energy state will have the lowest value for $S_{\text{total}}$, namely $1/2$. As indicated above, this corresponds to two degenerate states, one derived from $S' = 2$ and one from $S' = 3$. This degeneracy leads to a phenomenon known as *spin frustration* and derives from the fact that the actual antiferromagnetic coupling between $Fe_1$ and $Fe_3$ associated with bridging ligation, $J'$ in Figure 47, opposes the ferromagnetic alignment of $S_1$ and $S_3$, the latter resulting from antiferromagnetic coupling between pairs $S_1, S_2$ and $S_2, S_3$. In the triangular (i.e., all bridging) arrangement of Figure 47 with all antiferromagnetic exchange interactions close to equal, the anti-parallel alignment of the spin vectors cannot be completely achieved. This leads to intermediate spin values for $S'$ and a ground state that does not simply relate to an antiferromagnetic picture. The $S_{\text{total}} = 1/2$ ground state is observed experimentally for the oxidized $Fe_3S_4$ cluster,[86] indicating that all three pairs of irons have similar bridging sulfide ligation as in the structure shown in Figure 2E.

3.2.3.3 *Reduced $Fe_3S_4$ and $Fe_4S_4$ Clusters: Double Exchange* In reduced $Fe_3S_4^0$ and $Fe_4S_4^{+3,+2,+1}$ clusters, one pair of irons (two in the case of the $Fe_4S_4^{+2}$ cluster) have class III mixed-valence properties; their spin Hamiltonian parameters are intermediate between those of an oxidized and a reduced iron sulfur monomer. These [Fe(II) Fe(III)] mixed-valence pairs have $S_{\text{total}} = 9/2$ ferromagnetic ground states even though the exchange interactions are expected to be strongly antiferromagnetic based on comparison to the $Fe_2S_2$ dimers described above. In the reduced $Fe_3S_4^0$ centers the mixed-valence pairs antiferromagnetically couple to a localized Fe(III) $S = 5/2$ center to produce an $S_{\text{total}} = 2$ ground state, which is observed experimentally.[87] The $Fe_4S_4^{+2}$ cluster is described as two $S = 9/2$ class III mixed-valence dimers that are antiferromagnetically coupled to one another to produce an $S_{total} = 0$ ground state. Similarly, the $Fe_4S_4^{+3}$ cluster has a mixed-valence $S = 9/2$ dimer antiferromagnetically coupled to two $Fe^{+3}$ ions. The latter are an antiferromagnetically coupled pair but are spin frustrated due to the additional exchange interactions with the $S = 9/2$ pair, leading to an intermediate $S = 4$ state for the ferric pair and consequently an $S_{total} = 1/2$ ground state for the cluster. Analysis of the magnetic coupling in reduced $Fe_4S_4^{+1}$ clusters is more complex and leads to $S_{total} = 1/2$, $3/2$, or $7/2$ ground states dependent on specific parameter values.[88] Particularly interesting aspects of these clusters are the interactions within the class III mixed-valence pair that lead to the ferromagnetic $S = 9/2$ ground state, and why the isolated mixed-valence two iron two sulfur center in plant ferredoxin is localized with $S = 1/2$ in contrast to the behavior of this unit in the three and four iron sulfur clusters.

In a mixed-valence system the extra electron can be on $Fe_A$ or on $Fe_B$ and a good wavefunction for the dimer requires taking their + and − combinations. Thus each of the $S_{total}$ states for a mixed-valence dimer shown on the left side of Figure 46A is in fact a pair of states. Electron delocalization between the two irons (i.e., resonance delocalization) leads to a new term in the exchange Hamiltonian, Equations 8 and 9, which is known as double exchange.[89] This is included with exchange coupling in Equation 44 where the ± combination of each $S_{total}$ state is now split in energy by $2B(S + 1)$ where $B$ is the double exchange parameter. This corresponds to the electronic coupling term, $H_{AB}$, in Section 2.2.3.2 applied to exchanged coupled systems. Note that the resonance splitting increases as S increases [i.e., the $E_\pm(1/2)$ states are split by $2B$ while the $E_\pm(9/2)$ states are split by $10B$].

$$E_\pm(S) = -JS(S + 1) \pm B(S + 1/2) \tag{44}$$

This resonance splitting is included in the middle and right of Figure 46A, which shows that when $B/2J > 1.5$ the antiferromagnetic $S = 1/2$ state of the dimer is no longer lowest in energy, and when $B/2J > 4.5$ the ferromagnetic $S = 9/2$ level is the ground state.[45c,87] Thus the double exchange term in Equation 44 associated with resonance delocalization of the extra electron greatly stabilizes the ferromagnetic ground state. The origin of this can be seen from the scheme in Figure 48.[90] For the ferromagnetic ground state on the left, there is no barrier to electron delocalization and this strongly stabilizes this configuration, while for the antiferromagnetic ground state on the right, delocalization of the extra electron requires a spin flip which raises the energy of this process leading to reduced stabilization.

At this point (Equation 44), all the spin states in Figure 46A are delocalized, and vibronic coupling must still be included to allow for the possibility of localization as was done in Section 2.2.3.2.[45c] Thus for a magnetically coupled system, an analog of Equation 22 must include exchange coupling, and the electronic coupling matrix element $H_{AB}$ is given by double exchange whose influence increases with $S_{total}$ (Equation 45 and Figure 46B). For the isolated reduced $Fe_2S_2$ center the ground

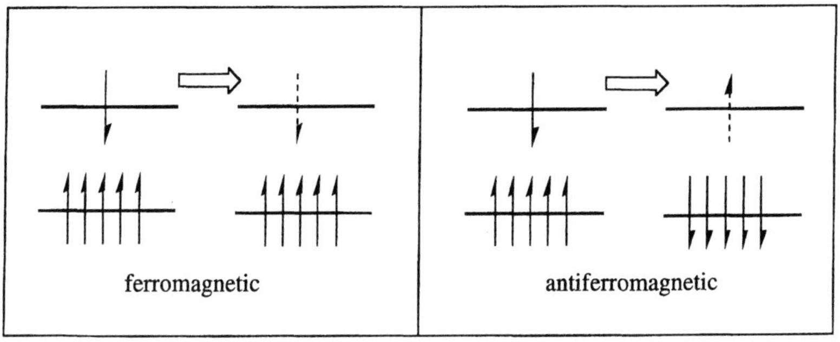


**Figure 48.** Possible ground state configurations for a mixed-valence Fe(II)Fe(III) dimer. The ferromagnetic ground state (left) directly allows for electron delocalization, while in the antiferromagnetic case (right), delocalization requires the electron to flip spin. Adapted from Reference 90.

state is $S = 1/2$ (i.e., $B/2J < 1.5$ in Figure 46A), and, from Figure 46B, vibronic coupling localizes this state as is observed experimentally.

$$E_{\pm} = -JS(S+1) + 1/2\left(\frac{\lambda^2}{k_-}\right)x_-^2 \pm \left[1/2\left(\frac{\lambda^2}{k_-}\right)x_-^2 + B^2(S+1/2)^2\right]^{1/2} \quad (45)$$

From Figure 46B, as $S_{\text{total}}$ increases the double exchange effectively overcomes vibronic trapping and class III delocalization occurs. More complex $B/J$ diagrams have been developed for three and four iron sulfur clusters.[87,88a]

A complex prepared by Wieghardt et al., $[Fe_2(OH)_3(tmtacn)_2]^{2+}$, **1**, is a class III mixed-valence dimer with a ferromagnetic $S = 9/2$ ground state.[91] It is interesting to compare **1** to the $Fe_2S_2^{+1}$ $S = 1/2$ localized cluster to define the factors which may lead to differences in delocalization. The $B$ value for **1** is 1,350 $cm^{-1}$. It has been determined that this resonance delocalization is associated with a direct $d_{z^2} + d_{z^2}$ interaction between the two irons at a distance of 2.509 Å. Importantly, from an analysis of ground and excited state potential energy surfaces, it has been possible to obtain an experimental estimate for the dependence of $B$ on the Fe–Fe distance ($\Delta B/\Delta r_{\text{Fe–Fe}} = \sim 1750$ $cm^{-1}$/Å).[45d,e] This allows an estimate of the $B$ value of the reduced $Fe_2S_2$ center ($r_{\text{Fe–Fe}}$ ~2.73 Å) of 965 $cm^{-1}$. This is still a large resonance stabilization term capable of complete delocalization. The major difference between **1** and $Fe_2S_2^{+1}$ is the magnitude of the antiferromagnetic exchange coupling. For **1**, this is estimated to be $> -70$ $cm^{-1}$, while for $Fe_2S_2^{+1}$ $J$ is approximately $-360$ $cm^{-1}$. This large difference in antiferromagnetic coupling reflects the differences in superexchange pathways associated with hydroxide versus sulfide bridges. (Note that the earlier quoted value of $J$ for the reduced $Fe_2S_2$ center has been modified here to correct for the double exchange contribution.) Thus for **1**, $B/2J \cong 9.6$ producing an $S = 9/2$ ground state (Figure 46) which is delocalized, while for $Fe_2S_2^{+1}$ $B/2J \sim 1.3$,[45d] giving an $S_{\text{total}} = 1/2$ ground state which is localized. Thus the very large antiferromagnetic coupling associated with the $\mu_2$-sulfido bridge leads to localization in the $Fe_2S_2$ center. The $\mu_3$-sulfido bridges in the $Fe_3S_4$ and $Fe_4S_4$ clusters may be significantly perturbed, and a change in these superexchange pathways for antiferromagnetic coupling could contribute significantly to delocalization in the mixed-valence dimer substructure of these clusters.

## 4 EXCITED STATE METHODS—LIGAND FIELD TRANSITIONS

### 4.1 Normal Spectral Features

**4.1.1** ***Electronic Absorption Spectroscopy*** The spectroscopic and magnetic methods considered thus far have dealt with the ground state and sublevels within a few hundred wavenumbers of this which are thermally accessible. We now consider higher energy excitations. In this section, we focus on the near infrared–visible region of the absorption spectrum of transition metal complexes which involves excitation of electrons between molecular orbitals that have dominantly metal-*d* character. These are referred to as $d \rightarrow d$ or ligand field transitions. In Section 5, this is extended to include the ligand-to-metal charge transfer transitions, which involve excitation of an electron from a mostly ligand-centered orbital to a metal-*d* derived molecular orbital and which can dominate the visible-UV spectral region. Finally, Section 6 considers high energy excited states in the X-ray spectral region that involve excitation of core electrons on the ligand or metal to valence orbitals centered on the metal ion.

Electronic absorption spectroscopy probes transitions between the ground state, $\Psi_g$, and the excited states, $\Psi_e$, listed above. On a theoretical level (Figure 49A), the $\Psi_g \rightarrow \Psi_e$ transition is induced by the interaction of a photon with an electron of the transition metal complex through the transition moment operator $\hat{M}$. Experimentally (Figure 49B), this transition produces an absorption band characterized by the energy dependence ($\nu$ in $cm^{-1}$) of its molar extinction coefficient ($\varepsilon$ in $M^{-1}$ $cm^{-1}$). The

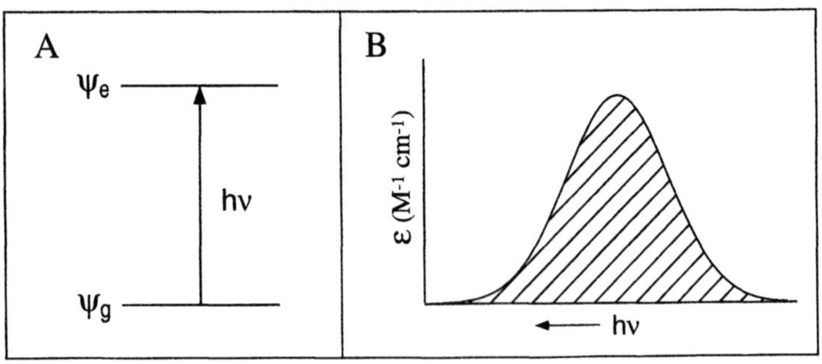


**Figure 49.** An electronic absorption transition. A) The ground-to-excited-state, $\psi_g \rightarrow \psi_e$, transition induced by the interaction of a photon with a bound electron; this transition is governed by the transition moment operator, $\hat{M}$; and B) absorption band resulting from electronic transition.

dimensionless quantity that relates theory and experiment is the oscillator strength, $f$, which is obtained from the integrated intensity under the absorption band through Equation 46a and relates to theory through Equation 46b.

$$f_{\text{exp}} = \frac{10^3 mc^2 \ln 10}{N_A \pi e^2} \int \varepsilon(\nu) d\nu = 4.32 \times 10^{-9} \int \varepsilon(\nu) d\nu \tag{46a}$$

$$f_{\text{theo}} = \frac{8\pi^2 mc}{3he^2} \nu \left[ \int \psi_e^* \hat{M} \psi_g d\tau \right]^2 = 4.702 \times 10^{29} \nu \left[ \int \psi_e^* \hat{M} \psi_g d\tau \right]^2 \tag{46b}$$

In these equations, $m$ is the rest mass of an electron, $c$ is the speed of light, $N_A$ is Avogadro's number, $e$ is the elementary charge, $h$ is Planck's constant, $\nu$ is the frequency in $cm^{-1}$, and $\hat{M}$ is in esu cm. The square of the transition moment integral in Equation 46b gives the dipole strength $D_{g \to e}$ (assuming $\psi_g$ is non-degenerate; for a degenerate ground state, the dipole strength involves a sum over the transitions from the components of the ground state, divided by the ground state degeneracy, c.f. Equation 56), which is then related to the integrated absorption intensity through Equation 47.

$$D_{g \to e} = \frac{3hc10^3 \ln 10}{8\pi^3 N_A} \int \frac{\varepsilon(\nu)}{\nu} d\nu = 91.8 \times 10^{-40} \int \frac{\varepsilon(\nu)}{\nu} d\nu \text{ (in esu}^2\text{ cm}^2\text{)} \tag{47}$$

Thus to observe a transition (i.e., integrated absorption intensity $> 0$) the transition moment integral in Equation 46b must be non-zero. This leads to the *selection rules* for $\Psi_g \to \Psi_e$ transitions. When $|\int \psi_g \hat{M} \psi_e d\tau|$ is greater than zero, the oscillator strength is greater than 0 and a transition is allowed. If the transition moment integral is zero, $f = 0$ and the transition is forbidden.

The form of the transition moment operator, $\hat{M}$, in Figure 49A and Equation 46b is obtained from the interaction of radiation with matter. When the wavelength of light, $\lambda$, is much greater than the radius of an electron's orbital in a molecule, $\hat{M}$ is given by the multipole expansion in Equation 48 where each term is reduced by $\sim 10^{-3}$ relative to the preceding term.[92] For the near IR-visible-UV spectral region $\lambda > 2{,}000$ Å, and the electric dipole term is dominant. The expression for this is given by Equation 49, where $\hat{E}$ is the electric vector of light and $\hat{r}$ operates on the electron coordinates. It should be noted that $\hat{M}_{\text{(electric dipole)}}$ does not operate on the spin part of the electronic wavefunction, thus $\Psi_g$ and $\Psi_e$ must have the same spin function for the $\Psi_g \to \Psi_e$ transition moment integral to be non-zero.

$$\hat{M} = \hat{M}_{\text{(electric dipole)}} + \hat{M}_{\text{(magnetic dipole)}} + \hat{M}_{\text{(electric quadrupole)}} + \cdots \tag{48}$$

$$\hat{M}_{\text{(electric dipole)}} = e\vec{r} \bullet \vec{E} \tag{49}$$

This leads to the spin selection rule $\Delta S = 0$ for spin-allowed transitions. In the ligand field region $\Psi_g$ and $\Psi_e$ involve dominantly metal-centered $d$ orbitals split in energy by the ligand field (*vide infra*). The $d$ orbitals have gerade, $g$, symmetry (i.e., symmetric to inversion or even parity) while the electric dipole operator $\hat{r} = x, y, z$

has ungerade, $u$, symmetry. Thus all $d \rightarrow d$ transitions are parity forbidden (also referred to as *Laporté forbidden*). However, metal sites in proteins usually have low symmetry with no inversion. This results in mixing of higher energy electric dipole allowed charge transfer excited states (*vide infra*) into the $d \rightarrow d$ transitions which then become weakly allowed due to the distorted ligand environment (see Chapter 4, Volume I, this series). Typically values of $\varepsilon \sim$ 10–100 $M^{-1}\,cm^{-1}$ are observed for spin-allowed $d \rightarrow d$ transitions in metalloprotein systems.[93] Note that for a 1 mM protein solution in a 1 mm absorption cell this corresponds to an absorbance of $< 0.01$, which is difficult to observe, particularly since the $d \rightarrow d$ transitions of interest for copper and iron centers are at low energy as given by ligand field theory.

4.1.1.1 *Ligand Field Theory for Cu(II) $d^9$, Fe(II) $d^6$, Fe(III) $d^5$* The ground and ligand field excited states of transition metal complexes can be obtained from the Tanabe–Sugano diagrams which quantify the nature of these states and their energies in terms of the ligand field splitting of the $d$ orbitals in an octahedral environment (10 $Dq$) and the electrostatic repulsive interaction between electrons (given by the Racah parameter $B$, see Chapter 1, Volume I, this series). Cu(II) $d^9$ has a one hole ground configuration which gives a $^2D$ term for the free ion (Figure 50A). Incorporation of the Cu(II) in an octahedral field splits the five fold degenerate $d$-orbitals of the free ion into $t_{2g}$ and $e_g$ sets separated by 10 $Dq$ (Figure 50B). The ground configuration of octahedral $d^9$ complexes is $(t_{2g})^6\ (e_g)^3$, which produces a $^2E_g$ ground state, and the only ligand field excited configuration is $(t_{2g})^5\ (e_g)^4$, which gives a $^2T_{2g}$ excited state at 10 $Dq$. Figure 50C gives the resultant Tanabe–Sugano diagram for the $d^9$ case, which shows the energy splitting of the $^2D$ into $^2E_g$ and $^2T_{2g}$ octahedral states with increasing strength of the ligand field.

For Fe(II) $d^6$ and Fe(III) $d^5$, there are five different octahedral configurations at the following energies: for Fe(II) $d^6$: $(t_{2g})^6$ — 0 $Dq$; $(t_{2g})^5(e_g)^1$ — 10 $Dq$; $(t_{2g})^4(e_g)^2$ — 20 $Dq$; $(t_{2g})^3(e_g)^3$ — 30 $Dq$; and $(t_{2g})^2(e_g)^4$ — 40 $Dq$; for Fe(III) $d^5$: $(t_{2g})^5$ — 0 $Dq$; $(t_{2g})^4(e_g)^1$ — 10 $Dq$; $(t_{2g})^3(e_g)^2$ — 20 $Dq$; $(t_{2g})^2(e_g)^3$ — 30 $Dq$; and $(t_{2g})^1(e_g)^4$ — 40 $Dq$. Each configuration splits into a number of differ-

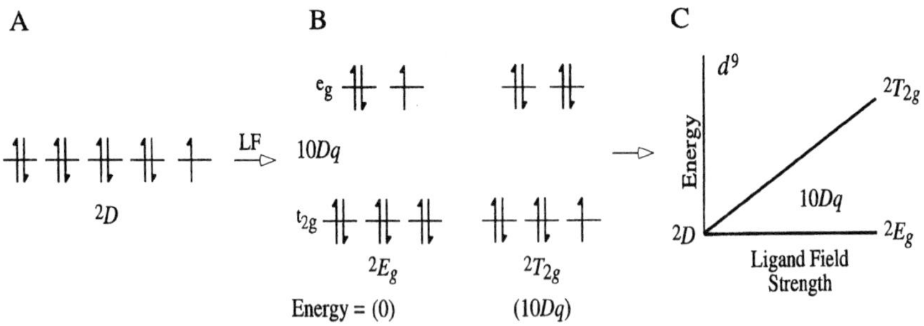


**Figure 50.** Cu(II) $d^9$ ground state. A) $^2D$ free ion ground state configuration; B) splitting of the $d$ orbitals by an octahedral ligand field with the two possible $O_h$–$d^9$ configurations; and C) resultant $d^9$ Tanabe–Sugano diagram.

ent states due to differences in electron repulsion, and all states of the same symmetry and spin can mix through configurational interaction. These effects are included in the Tanabe–Sugano diagrams for the $d^6$ and $d^5$ configurations given in Figure 51. Note that both of these $d^n$ configurations can have two possible ground states, depending on the strength of the ligand field (10 $Dq$) relative to electron–electron repulsion ($B$). The crossover point on the abscissa thus quantifies the spin pairing energy for each metal ion in terms of $Dq/B$. Most iron sites of interest in bioinorganic systems have high-spin ground states (left, weak field side of the Tanabe–Sugano diagrams). From the $d^5$ Tanabe–Sugano diagram, Figure 51B, the high-spin ferric site has a $^6A_1$ ground state, and ligand field excited states are spin quartets and doublets. Thus all ligand field transitions for a high-spin ferric center are spin-forbidden ($\Delta S \neq 0$) and unlikely to be observed in metalloproteins. High-spin Fe(II) has a $^5T_{2g}$ ground state with one spin-allowed $^5T_{2g} \rightarrow {}^5E_g$ transition at an energy of 10 $Dq$. Cu(II) $d^9$ also has one-spin allowed ligand field transition $^2E_g \rightarrow {}^2T_{2g}$ at an energy of 10 $Dq$. These transitions should dominate the ligand field spectroscopy of cupric and ferrous active sites. For divalent metal ions with biologically relevant ligands, 10 $Dq$ is ~10,000–12,000 cm$^{-1}$, thus the spin-allowed $d \rightarrow d$ transitions should contribute in the near infrared spectral region. Both the ground and excited states of Cu(II) and high-spin Fe(II) are degenerate and can split in energy when the symmetry is less than octahedral. These ligand field splittings provide a powerful probe of the geometric and electronic structure of a metalloprotein active site.

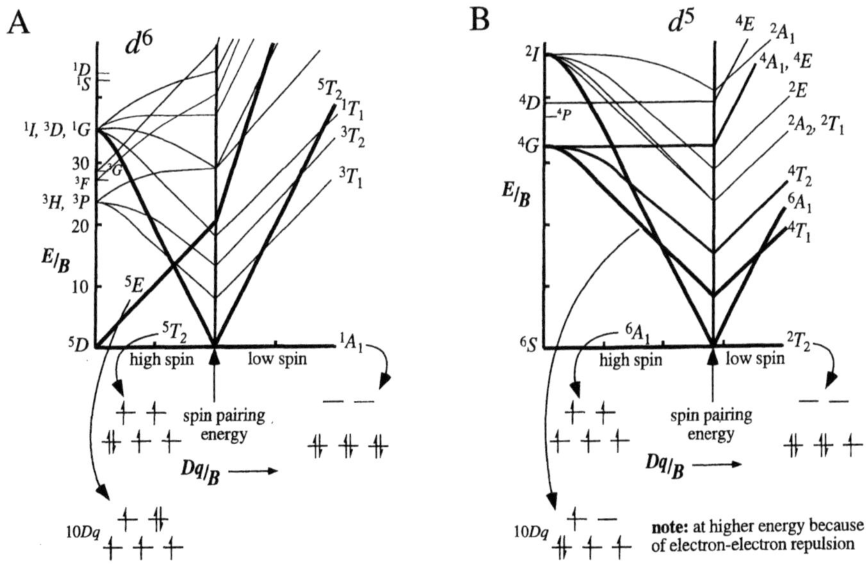


**Figure 51.** Tanabe–Sugano diagrams for $O_h$ states and a few associated configurations for A) $d^6$ Fe(II); and B) $d^5$ Fe(III).

4.1.1.2 *Geometry Dependence of Spin-Allowed Ligand Field Transitions* Ligand field theory quantitates the splitting of the one electron metal $d$ orbitals due to repulsive or antibonding interactions with the ligands. This splitting is dependent upon specific geometry and type of ligand as described in Chapter 1 Volume I and in References 9, 35a, 94. Since for Cu(II) and high-spin Fe(II), the spin-allowed transitions simply involve the promotion of an electron ($d^6$) or hole ($d^9$) among these $d$ orbitals, the ground and excited state energies directly reflect the $d$ orbital energies of these metal sites and can therefore be used to probe their geometric and electronic structures.

Cu(II) $d^9$ The octahedral ground state of Cu(II) complexes, $^2E_g$, has an extra electron in the $e_g$ set of $d$ orbitals that is strongly antibonding with the ligands. This produces a non-spherical charge distribution and results in a symmetry-lowering Jahn–Teller distortion of the metal environment, which splits the orbital degeneracy and lowers the overall energy of the site. The Jahn–Teller distortion generally observed in Cu(II) complexes is a tetragonal elongation of the octahedron as shown in Figure 52B, or in the limit of a larger Jahn–Teller effect a square planar structure as in Figure 52C. The geometric coordinate that has been of interest in copper bioinorganic chemistry is the distortion from a tetragonal (or square pyramidal or square planar)

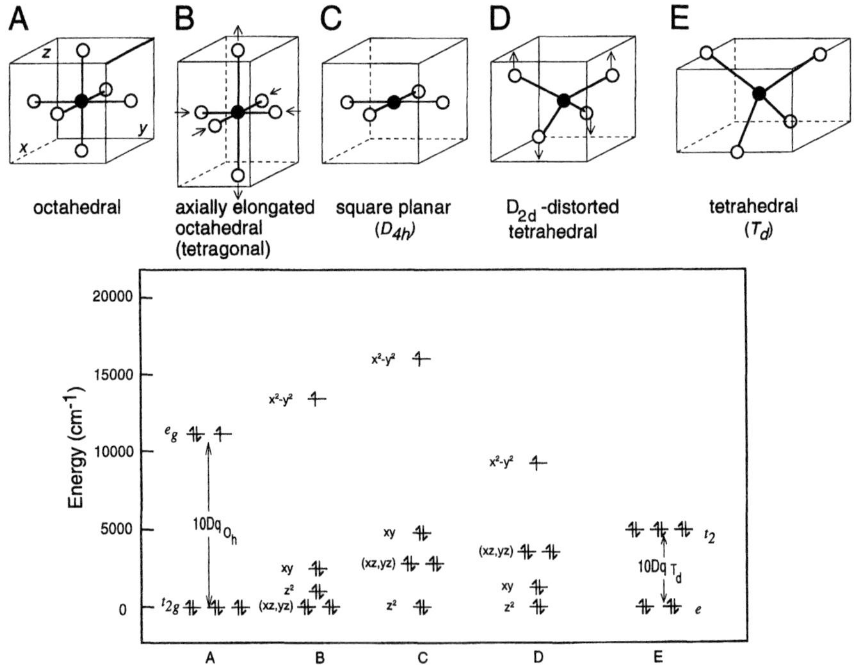


**Figure 52.** Geometry dependence of ligand field transitions for Cu(II): $d$ orbital splittings for A) octahedral limit; B) axial (tetragonal) elongation; C) square planar; D) $D_{2d}$ distorted tetrahedral; and E) tetrahedral limit. Adapted from Reference 95.

structure to a tetrahedral site environment. As shown in Figure 52 C–E, distortion of a Cu(II) complex along a $D_{2d}$ coordinate towards a tetrahedral geometry causes a shift in energy of the ligand field transitions from ~12,000 $cm^{-1}$ to ~5,000 $cm^{-1}$. The experimental transition energies reflect the extent of distortion along the $D_{4h} \rightarrow T_d$ coordinate and thus probe the geometry of the Cu(II) site.[95]

Fe(II) HIGH-SPIN $d^6$ For high-spin Fe(II) the ground state, $^5T_{2g}$, corresponds to a $(t_{2g})^4\ (e_g)^2$ configuration with an extra electron in the $t_{2g}$ set of $d$ orbitals. This provides an asymmetric charge distribution which is only weakly $\pi$-interacting with the ligands. Thus the Jahn–Teller distortion will be limited, and Fe(II) sites are found with a range of six, five, and four coordinate geometries. Calculated energies of the Fe(II) $d$ orbitals in different site geometries are given in Figure 53.[51b] Note that the splitting of the octahedral $^5E_g$ excited state shown in Figure 53 is also very sensitive to the nature of the ligand field. Distorted six coordinate centers (in Figure 53A) generally show two ligand field transitions centered at ~10,000 $cm^{-1}$ and split by ~2,000 $cm^{-1}$. Five coordinate Fe(II) ions also show two transitions but with a splitting >5,000 $cm^{-1}$. For square pyramidal geometry (Figure 53B) these are at

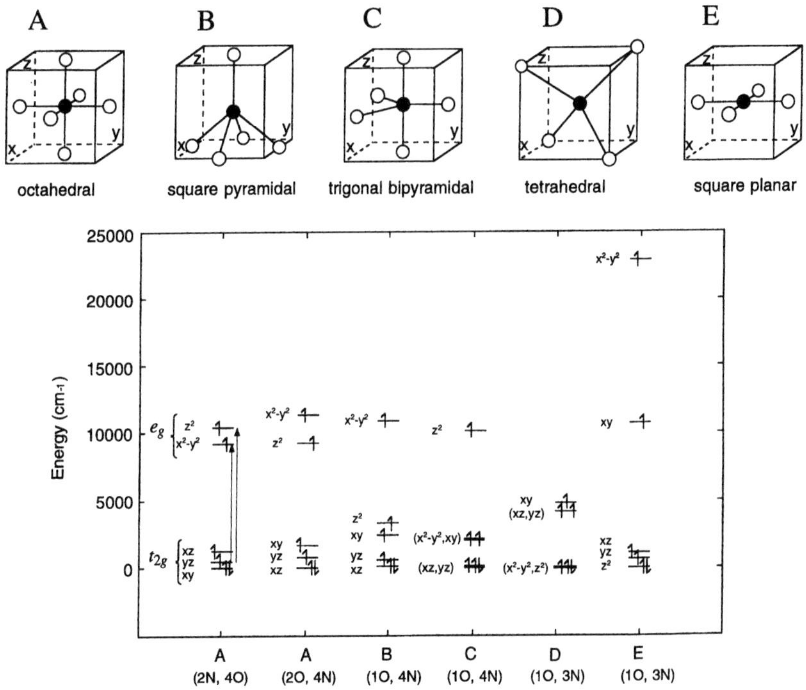


**Figure 53.** Geometry dependence of ligand field transitions for high-spin Fe(II): $d$ orbital splittings for A) octahedral structure; B) square pyramidal; C) trigonal bipyramidal; D) tetrahedral; and E) square planar. Note that two $O_h$ complexes are given with mixed ligation. Adapted from Reference 51b.

$>$10,000 and ~5,000 $cm^{-1}$, while for trigonal bipyramidal geometries these energies are somewhat reduced ($<$10,000 and $<$5,000 $cm^{-1}$ in Figure 53C). Four coordinate tetrahedral Fe(II) sites (Figure 53D) exhibit all ligand field transitions at ~5,000 $cm^{-1}$ (i.e., 10 $Dq_{T_d} = -4/9 \times 10\ Dq_{O_h}$). As seen in Figure 53, the $t_{2g}$ set of $d$ orbitals ($xz$, $yz$, $xy$) also split in energy in a manner reflecting the geometric structure of the Fe(II) site. This has been described in Section 3.1.1.2 where the $^5T_{2g}$ splitting is further correlated to the experimental $\delta$ and $g_{\|}$ values.

In summary, both Cu(II) and high-spin Fe(II) sites exhibit relatively weak spin-allowed ligand field transitions in the 12,000–5,000 $cm^{-1}$ spectral region, and these transitions are sensitive probes of active site geometric and electronic structure. Absorption experiments in this near-IR spectral region are often complicated by intense absorption from the buffer, $H_2O$, and protein vibrations. Ligand field transitions are therefore very difficult to observe by absorption spectroscopy. However, CD and MCD spectroscopies are governed by different selection rules which greatly favor the ligand field transitions and provide powerful probes of this near-IR ligand-field spectral region of metal sites in Biology.

**4.1.2** ***Circular Dichroism (CD) Spectroscopy*** CD spectroscopy[96] measures the difference in absorption of left (L) and right (R) circularly polarized (CP) light ($\Delta\varepsilon = \varepsilon_L - \varepsilon_R$) by a $\Psi_g \rightarrow \Psi_e$ transition. LCP corresponds to light propagating towards an observer with its $\vec{E}$ vector rotating in a counterclockwise direction while RCP involves a clockwise rotation of $\vec{E}$. As shown in Figure 54, transitions in the CD spectrum have a sign as well as a magnitude, and this is often of importance in resolving overlapping bands contributing to an absorption feature.

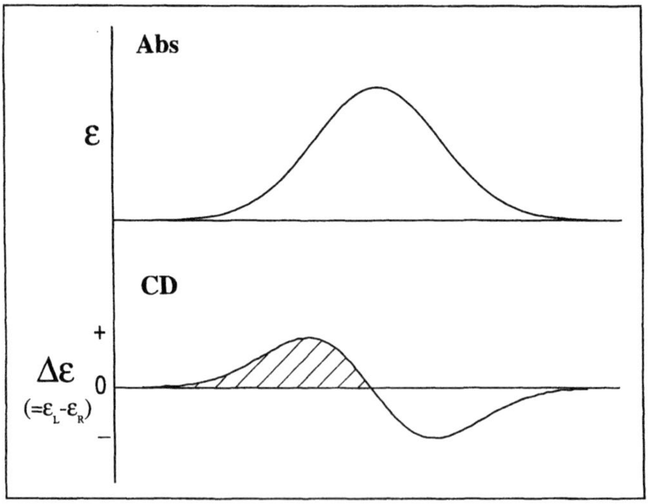


**Figure 54.** Comparison of absorption (top) and CD (bottom) spectra indicates that the sign of the CD transitions can be used to resolve two transitions contributing to an absorption band.

The parameter in CD spectroscopy analogous to the oscillator strength in absorption spectroscopy is the rotational strength $R$. As described by Equation 50a, the experimental rotational strength is proportional to the integrated intensity under a CD peak, while the theoretical rotational strength is given by Equation 50b, in which $\mathrm{Im}(A + Bi) = B$.

$$R_{\mathrm{exp}} = \frac{3hc10^3 \ln 10}{32\pi^3 N_{\mathrm{A}}} \int \frac{\Delta\varepsilon}{\nu} d\nu = (22.9 \times 10^{-40}) \left( \int \frac{\Delta\varepsilon}{\nu} d\nu \right) (\mathrm{esu}^2\,\mathrm{cm}^2) \quad (50a)$$

$$\begin{aligned} R_{\mathrm{theo}} &= \frac{1}{2mc} \mathrm{Im} \int \Psi_g \hat{M}_{(\mathrm{el.\ dip.})} \Psi_e d\tau \bullet \int \Psi_g \hat{M}_{(\mathrm{mag.\ dip.})} \Psi_e d\tau \\ &= (1.8 \times 10^{16}) \mathrm{Im} \int \Psi_g \hat{M}_{(\mathrm{el.\ dip.})} \Psi_e d\tau \bullet \int \Psi_g \hat{M}_{(\mathrm{mag.\ dip.})} \Psi_e d\tau \end{aligned} \quad (50b)$$

Equation 50b shows that for a transition to be observed in the CD spectrum it must be both electric and magnetic dipole allowed and have a non-zero projection of these transition moments onto one another. This expression derives from the fact that CP light excites an electron in a helical motion and thus involves both translation (electric dipole) and rotation (magnetic dipole) operators. In order for there to be a non-zero projection of the electric and magnetic dipole moments for a specific $\Psi_g \rightarrow \Psi_e$ transition, $\hat{M}_{(\mathrm{electric\ dipole})}$ and $\hat{M}_{(\mathrm{magnetic\ dipole})}$ must transform as the same irreducible representation (where $\hat{M}_{(\mathrm{electric\ dipole})}$ transforms as $x$, $y$, $z$ and $\hat{M}_{(\mathrm{magnetic\ dipole})}$ transforms as $R_x$, $R_y$, $R_z$). This can only occur for the molecular point groups $C_n$ and $D_n$ (protein sites have $C_1$ symmetry) which are those associated with optically active molecules.

CD intensity reflects the extent of distortion from the higher symmetry limit (which includes $i$ and/or $\sigma$ symmetry) due to either the ligand orientation at the metal site induced by the protein or the effects of the more distant chiral protein environment on the metal site.[97] These distortions generally induce some electric dipole intensity into a transition and the quantity that is most sensitively reflected in the rotational strength is the magnetic dipole character of the $\Psi_g \rightarrow \Psi_e$ transition. The expression for the magnetic dipole transition moment operator is given in Equation 51, which involves the magnetic vector of light, $\vec{H}$, and the change of orbital angular momentum, $\hat{L}$, in the electronic transition.

$$\hat{M}_{(\mathrm{magnetic\ dipole})} = \beta \hat{L} \bullet \vec{H} \quad (51)$$

The effects of the orbital angular momentum operator on electrons in the $d$ orbitals are given in Table 3. For Cu(II) sites, $\hat{L}_i$ ($i = x, y, z$) transfers the hole from the $d_{x^2-y^2}$ to the $d_{xy}$, $d_{xz}$ or $d_{yz}$ orbital; therefore the associated $d \rightarrow d$ transitions are magnetic dipole allowed. A similar situation exists for $d \rightarrow d$ transitions involving the sixth electron in high-spin Fe(II) centers. Therefore, these $d \rightarrow d$ transitions will have significant rotational strength and appear with reasonable intensity in the CD spectrum, increasing as the electric dipole character mixed into the $\Psi_g \rightarrow \Psi_e$ transition increases. As the electric dipole character also governs the absorption intensity, one often considers the Kuhn anisotropy factor, $g_{g \rightarrow e}$ (not to be confused with the $g$ values of EPR), which is defined as the ratio of CD to absorption intensity (Equation 52). For reasonable estimates of $f$ and $R$, it is found that $g_{g\to e} > 0.01$ for magnetic dipole allowed transitions,[98] and this provides a mechanism for identifying these transitions.

$$g_{g\to e} = \frac{\Delta\varepsilon}{\varepsilon} \tag{52}$$

Note that lower values of the Kuhn anisotropy factor can occur for a magnetic dipole allowed transition if the electric dipole transition moment direction is poorly aligned relative to the magnetic dipole transition moment. Finally, since $\hat{L}$ is a rotational operator, other types of transitions (e.g., charge transfer) that involve excitation of electrons between orbitals that are transformed into one another by a rotation will also have magnetic dipole character and large anisotropy factors. For $d \to d$ transitions of copper(II) and high-spin iron(II) sites, the magnetic dipole character greatly enhances their CD intensity in near IR spectral region relative to that of buffer and protein vibrations and this provides a spectroscopic method for their observation. This is strongly complemented by low-temperature MCD spectroscopy.

**4.1.3** ***Magnetic Circular Dichroism (MCD) Spectroscopy*** MCD spectroscopy[99] combines the CD experiment with a longitudinal magnetic field (i.e., $\vec{H}$ oriented parallel to the direction of propagation of the CP light). The CD spectrum serves as a baseline for the MCD spectrum in that these effects are additive. As described below, MCD probes the Zeeman splittings in the ground and excited states and the magnetic field induced mixing between states. In addition, the selection rules associated with the Zeeman split transitions induced by CP light can provide definitive excited state assignments.

4.1.3.1 *Qualitative Model* Figure 55 illustrates the example of a molecule having an angular momentum $J$ of 1/2 in the ground and excited states. Application of a magnetic field splits the two-fold $M_J = \pm 1/2$ degeneracy of both the ground and excited state by an amount of $g_i \beta H$ [where $g_i$ is the $g$ value of the ground ($g_g$) or excited ($g_e$) state]. The selection rules for absorption of circularly polarized light propagating along the magnetic field direction are $\Delta M = +1$ for left circular polarization (LCP) and $\Delta M = -1$ for right circular polarization (RCP), where $M_{\pm} = \frac{1}{\sqrt{2}}(M_x \pm iM_y)$ and $M_x$ and $M_y$ correspond to light polarized linearly in the $x$ and $y$ directions perpendicular to the magnetic field ($z$). At high temperatures, both components of the ground state are equally populated and there should be two transitions of equal magnitude but opposite sign. The Zeeman splittings will be on the order of 10 cm$^{-1}$ (for $g = 2$ and 7 T magnetic field, the Zeeman splitting is 6.5 cm$^{-1}$) while absorption bandwidths of transitions are on the order of 1,000 cm$^{-1}$ due to excited state distortions (*vide infra*). Thus the two transitions in Figure 55 will mostly cancel in $\Delta\varepsilon = \varepsilon_L - \varepsilon_R$ resulting in a broad, weak, derivative-shaped feature called an $A$-term (Figure 55B). $A$-terms are independent of temperature and depen-

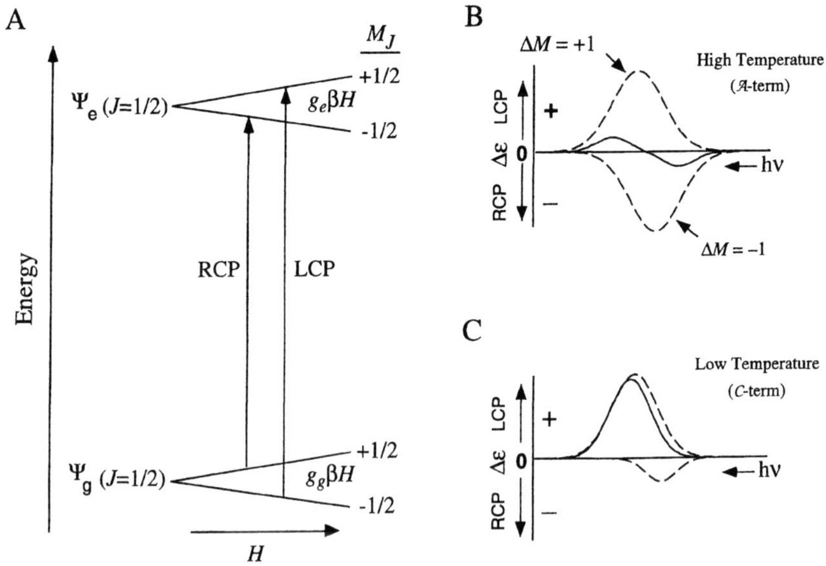


**Figure 55.** MCD mechanism. A) Selection rules for transitions of left and right circularly polarized light between degenerate ground and excited states which have been split by a magnetic field; B) derivative-shaped, temperature-independent A-term and its components; and C) absorption band shaped, temperature-dependent C-term and its components. Adapted from Reference 100.

dent on the $g$ values of the ground and excited states, at least one of which must be degenerate. As the temperature is lowered the Boltzmann population of the higher $M_J$ sublevel shown in Figure 55 is decreased, and the two transitions no longer cancel. This results in a $\Delta\varepsilon$ feature which has the shape of the absorption band and increases in intensity as the temperature is lowered. This is referred to as an MCD C-term. (At low temperature and high magnetic field, the transition saturates as described in Section 3.1.2.) The magnitude of the C-term is dependent on the $g$ value of the ground state, which must be degenerate.

Finally, consider the case of non-degenerate ground and excited states. This is given in Figure 56A, which includes an additional non-degenerate excited state $\Psi_K$ that mixes with $\Psi_e$ due to the applied magnetic field. In Figure 56A, the magnetic field induced mixing of $\Psi_K$ into $\Psi_e$ induces MCD intensity into the $\Psi_g \rightarrow \Psi_e$ transition. This MCD B-term will have the absorption bandshape as with a C-term but is independent of temperature and increases in magnitude as $E_K - E_e$ decreases.

There are several special cases of the above general considerations of A-, B-, and C-terms which can be spectroscopically useful.

1. For B-terms, if only $\Psi_e$ and $\Psi_K$ in Figure 56A interact through the applied magnetic field these will give equal and opposite MCD signals producing a pseudo-A-term.

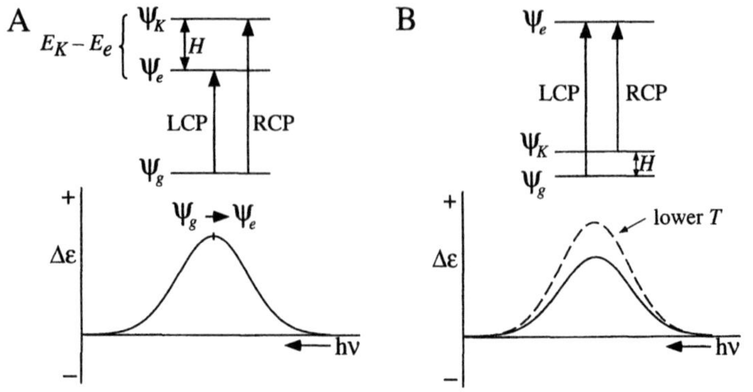


**Figure 56.** Field-induced mixing of a non-degenerate state $\psi_K$ with A) $\psi_e$ yielding absorption band shaped, temperature-independent B-term; and B) $\psi_g$ yielding a temperature-dependent B-term. Adapted from Reference 51b.

2. It is also possible to get equal and opposite C-terms through spin-orbit coupling between two states (*vide infra*), and this produces a pseudo-A-term that is temperature dependent.
3. For B-terms, $\Psi_K$ can also mix with the ground state $\Psi_g$ due to the applied magnetic field (Figure 56B). When $\Psi_K$ is also thermally accessible this results in temperature-dependent B terms, and these can be recognized by their different saturation magnetization behavior. This can occur when $S > 1/2$ through zero field splitting.

4.1.3.2 *Formalism* The generally used expressions for describing MCD spectra are given in Equations 53 and 54, where $\Delta A$ is the field-dependent difference between absorption of left- and right-circularly polarized light, $E$ is $h\nu$, $\alpha$ is the electric permeability, $C$ is the concentration, $l$ is path length, $n$ is the index of refraction, $\beta$ is the Bohr magneton, $H$ is the applied magnetic field (all in cgs units), $f(E)$ is the absorption bandshape, and $\partial f(E)/\partial E$ is its derivative.[99,100]

$$\begin{aligned}\frac{\Delta A}{E} &= \frac{2N_A\pi^3\alpha^2 Cl\log e}{250hcn}\beta H\left[\mathcal{A}_1\left(-\frac{\partial f(E)}{\partial E}\right)+\left(\mathcal{B}_0+\frac{\mathcal{C}_0}{kT}\right)f(E)\right]\\ &= (3.27\times10^{38})\frac{\alpha^2 Cl}{n}\beta H\left[\mathcal{A}_1\left(-\frac{\partial f(E)}{\partial E}\right)+\left(\mathcal{B}_0+\frac{\mathcal{C}_0}{kT}\right)f(E)\right] \qquad (53)\end{aligned}$$

$\mathcal{A}_1$, $\mathcal{B}_0$, and $\mathcal{C}_0/kT$ are the A-, B-, and C-terms, respectively. As seen in Equation 53, all MCD intensity should increase linearly with magnetic field, while C-term intensity is additionally proportional to $1/T$ in the linear limit where $kT > g_g\beta H$. Expressions for the coefficients $\mathcal{A}_1$, $\mathcal{B}_0$, and $\mathcal{C}_0$ are given in Equation 54, where $d_g$ is the degeneracy of $\Psi_g$ and it is assumed that the magnetic field is oriented along the molecular $z$-axis. Equation 54c contains the Zeeman effect on the ground state sublevels and shows that $\mathcal{C}_0$ is thus proportional to $g_g$. Equation 54a contains the difference in Zeeman effects between the ground and excited states. Thus if only the ground state is degenerate $\mathcal{A}_1 = \mathcal{C}_0$, and a difference between the experimental values of $\mathcal{A}_1$ and $\mathcal{C}_0$ can be used to identify excited state degeneracy. The first summation on the right of Equation 54b gives the Zeeman mixing of the intermediate state $\Psi_K$ into the excited state $\Psi_e$ as shown in Figure 56A, and the second summation gives the Zeeman mixing of $\Psi_K$ into the ground state as in Figure 56B $[\mathrm{Re}(a+ib) = a]$.

$$\mathcal{A}_1 = \frac{1}{d_g}\sum \left(\langle e|L_z + 2S_Z|e\rangle - \langle g|L_z + 2S_z|g\rangle\right)\left(|\langle g|m_-|e\rangle|^2 - |\langle g|m_+|e\rangle|^2\right) \tag{54a}$$

$$\mathcal{B}_0 = -\frac{2}{d_g}\mathrm{Re}\sum \left[\begin{array}{l} \displaystyle\sum_{K(K\neq e)} \frac{\langle e|L_z+2S_z|K\rangle}{\Delta E_{Ke}}\left(\langle g|m_-|e\rangle\langle K|m_+|g\rangle \right. \\ \qquad\qquad \left. - \langle g|m_+|e\rangle\langle K|m_-|g\rangle\right) \\ \displaystyle + \sum_{K(K\neq g)} \frac{\langle K|L_z+2S_z|g\rangle}{\Delta E_{Kg}}\left(\langle g|m_-|e\rangle\langle e|m_+|K\rangle \right. \\ \qquad\qquad \left. - \langle g|m_+|e\rangle\langle e|m_-|K\rangle\right) \end{array}\right] \tag{54b}$$

$$\mathcal{C}_0 = -\frac{1}{d_g}\sum \langle g|L_z + 2S_z|g\rangle \times \left(|\langle g|m_-|e\rangle|^2 - |\langle g|m_+|e\rangle|^2\right) \tag{54c}$$

Note that Equations 53 and 54 contain assumptions having to do with the effect of the magnetic field on the bandshape (the rigid shift model), and these can be eliminated through a moment analysis of the bandshape,[101] where the 1 and 0 subscripts on $\mathcal{A}_1$, $\mathcal{B}_0$, and $\mathcal{C}_0$ refer to the associated moments.

4.1.3.3 *Information Content* For paramagnetic metal centers, A- and B-terms are much weaker than C-terms at room temperature, and the C-term will further increase in intensity by approximately two orders of magnitude on going to liquid helium temperatures (from its $1/T$ dependence). Therefore, MCD spectroscopy in Bioinorganic Chemistry usually focuses on C-terms. Generally, ground states are orbitally non-degenerate, and the degeneracy associated with the C-term involves the spin. From the selection rules for the MCD transition ($\Delta M_L = \pm 1$, $\Delta M_S = 0$), it is easy to see that the spin degeneracy can only produce a non-zero C-term through spin-orbit coupling. At this point it is useful to divide protein centers into high and low symmetry classes. High symmetry sites have approximate $C_3$, $C_4$, or $S_4$ axes and are associated with heme and iron sulfur centers. In these cases there can be orbitally doubly degenerate excited states, transitions to which are $x$, $y$ polarized. These have orbital angular momentum which will spin-orbit couple to the electron spin and produce a splitting of the excited state.[49c,52d] The $S = 1/2$ case is given in Figure 57, which shows that a pseudo-A-term will be generated from equal and opposite C-terms with the energy splitting of the C-terms giving the in-state spin-orbit coupling. Pseudo-A-terms allow assignment of orbitally degenerate excited states, quantify the

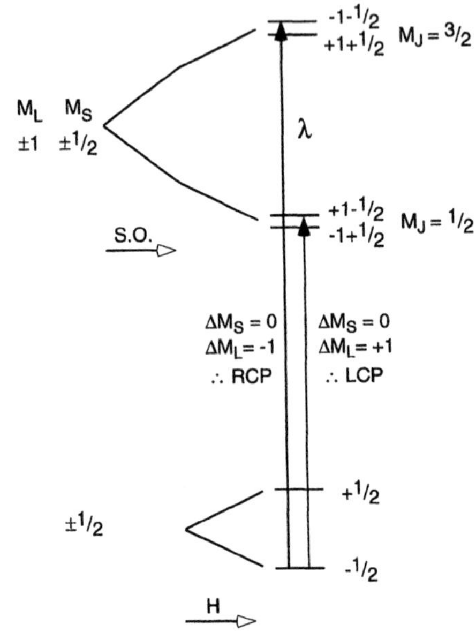


**Figure 57.** Spin-orbit coupling in a two-fold orbitally degenerate excited state with $S = 1/2$ produces C-terms of opposite sign, resulting in a pseudo-A-term, which is temperature-dependent.

spin-orbit coupling and hence the nature of the doubly degenerate orbital involved in the transition, and have a sign (determined by the sign of the higher energy component) which can provide additional insight into the one electron orbitals involved in the transition.[49c] The non-degenerate excited states of high symmetry sites derive C-term intensity through spin-orbit coupling to the doubly degenerate states as described below.

Most metal sites in Bioinorganic Chemistry have low symmetry and no orbital degeneracy. Thus each $\Psi_g \rightarrow \Psi_e$ transition is electric dipole allowed but with a transition moment in only one direction in the molecule. An abbreviated form[52d] of the $C_0$ term expression for a frozen solution of all molecular orientations relative to the external magnetic field is given in Equation 55 where $g_i$ are the components of the ground state $g$ tensor and $M_i$ are the electric dipole transition moments for the $\Psi_g \rightarrow \Psi_e$ transition.

$$C_0 \propto g_z M_x M_y + g_y M_x M_z + g_x M_y M_z \qquad (55)$$

From Equation 55, MCD intensity requires two perpendicular non-zero transition moments. For low symmetry sites with orbitally non-degenerate states this can only occur through spin-orbit coupling. This can mix two excited states with transition moments in different directions ($M_{j \neq i}$ into an $M_i$ polarized transition) which will result in non-zero C-term intensity.[103] Note that coupling between two excited states will produce equal and opposite C-terms resulting in a pseudo-A-term derived from non-degenerate states. More generally spin-orbit mixing among a complete set of excited states should result in equal and opposite C-term intensity distributed over the states which will sum to zero. Any net C-term intensity summed over excited states will result from further spin-orbit coupling into the ground state.[104]

An important point here is that spin-orbit coupling governs the C-term intensity and this is dependent on the magnitude of the spin-orbit coupling parameter which will be much greater in metal-centered than in ligand-centered transitions (for example, $\xi[\text{Cu(II)}] = 830\ \text{cm}^{-1}$, $\xi[(\text{Fe(II)}] = 400\ \text{cm}^{-1}$, $\xi[\text{Fe(III)}] = 460\ \text{cm}^{-1}$, $\xi(\text{O, N}) \sim 60\text{–}70\ \text{cm}^{-1}$, $\xi(\text{S}) \sim 325\ \text{cm}^{-1}$). One would therefore expect the $d \to d$ excited states to generally be more spin-orbit mixed than ligand-based charge transfer states and show larger low-temperature MCD intensity relative to absorption intensity. Defining the electric dipole strength as in Equation 56, the MCD to absorption intensity ratio is often expressed as a $\mathcal{C}_0/\mathcal{D}_0$ ratio which is given experimentally by Equation 57 (with $k/\beta = 1.49 \times TK^{-1}$).

$$\mathcal{D}_0 = \frac{1}{d_g} \sum |\langle g|M|e\rangle|^2 \tag{56}$$

In copper monomers it has been found that $\mathcal{C}_0/\mathcal{D}_0 \sim 0.1$ for $d \to d$ transitions, while much reduced values of $\mathcal{C}_0/\mathcal{D}_0 \sim 0.01$ are observed for charge transfer transitions.[103]

$$\frac{\mathcal{C}_0}{\mathcal{D}_0} = \frac{kT}{\beta H}\left(\frac{\Delta\varepsilon}{\varepsilon}\right) \tag{57}$$

Low-temperature MCD spectroscopy is a powerful method for identifying and studying $d \to d$ transitions. On a quantitative level, the sign and magnitude of the $\mathcal{C}_0/\mathcal{D}_0$ value can be used to make rigorous assignments of bands based on electronic structure calculations which include spin-orbit coupling and thus probe specific bonding interactions as described below.

## 4.2 Unique Spectral Features

**4.2.1** ***Blue Copper Sites*** The excited state spectral features of the blue copper center nicely illustrate the above concepts and provide important insight into the unique electronic structure of these sites. From the room temperature absorption spectrum in Figure 58A, it is not possible to identify electronic transitions in the region below ~12,000 cm$^{-1}$. However, the CD spectrum in this near IR spectral region (Figure 58B) clearly exhibits three new transitions at 11,200, 9,150, and 5,000 cm$^{-1}$. The presence of these low energy transitions was the first indication that the blue copper center has a distorted tetrahedral structure.[105] At low temperature bands sharpen, and from a combination of low temperature absorption, CD, and MCD data (each method governed by a different selection rule), eight electronic transitions are identified, as shown in the gaussian resolved low temperature absorption spectrum in Figure 58C. As seen in Figure 58D bands 5 through 8 are intense in the low temperature

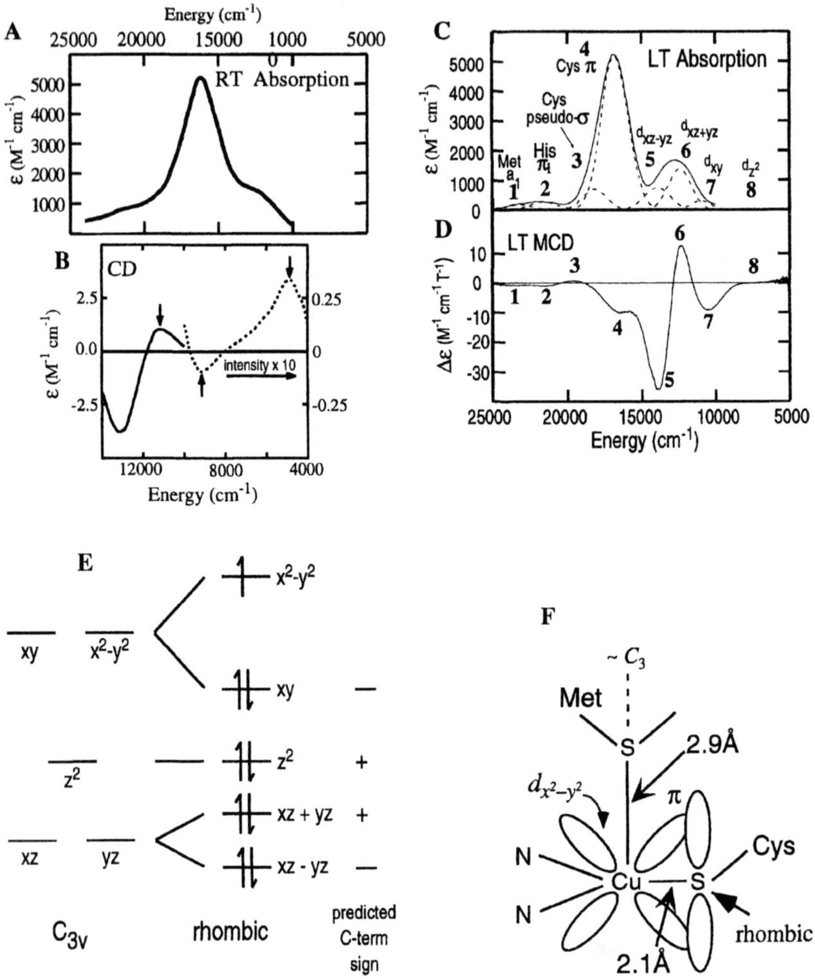


**Figure 58.** Blue copper ligand field transitions and electronic structure. A) Room temperature absorption spectrum of plastocyanin; B) room temperature CD spectrum shows three additional transitions at lower energy; C) low temperature absorption shows eight transitions; D) LT MCD shows eight transitions; E) energy level diagram for a $C_{3V}$ elongated $T_d$ site with rhombic distortion and the expected C-term behavior; and F) strong $\pi$ antibonding interaction between Cys S $p_\pi$ and Cu $d_{x^2-y^2}$ orbitals. Adapted from Reference 2.

MCD relative to the absorption spectrum (i.e., for bands 5 through 8, $\mathcal{C}_0/\mathcal{D}_0 \sim 0.1$; for bands 1 through 4, $\mathcal{C}_0/\mathcal{D}_0 \sim 0.01$), and these are assigned as the ligand field transitions of the blue copper center.[103]

Specific assignments of the bands have been made based on the signs and magnitudes of the $\mathcal{C}_0/\mathcal{D}_0$ ratios using group theory and molecular orbital calculations for a $C_{3V}$ elongated $T_d$ site with significant rhombic distortion. The approximate energy level diagram is shown in Figure 58E. Bands 5 and 6 form a pseudo-A-term comprising positive and negative C-terms associated with the rhombically split set of $d_{xz}/d_{yz}$ orbitals. The positive sign of band 8 indicates that it must be assigned as the $d_{z^2} \rightarrow d_{x^2-y^2}$ transition. The energy of this transition is sensitive to the axial interaction of the thioether with the copper in Figure 58F. Its low energy indicates that this interaction is moderately strong at the Cu–$S_{met}$ distance of 2.9 Å of the blue copper site. Based on its negative sign, band 7 can be assigned as the $d_{xy} \rightarrow d_{x^2-y^2}$ ligand field transition. The high energy of this transition reflects a very large rhombic splitting of the $d_{x^2-y^2}/d_{xy}$ orbitals (~10,000 $cm^{-1}$) which is due to the strong $\pi$-antibonding interaction of the $d_{x^2-y^2}$ orbital with the Cys S $p\pi$ orbital as shown in Figure 58F. This splitting eliminates the Jahn–Teller distorting force that would normally be present for a Cu(II) center in a tetrahedral geometry and lead to site distortion toward a square planar structure.[2,106] Elimination of the Jahn–Teller distorting force results in a low Franck–Condon barrier for electron transfer by the blue copper protein sites. Finally, note that the absorption intensity of band 6 is high for a $d \rightarrow d$ transition. This is due to mixing with a low energy intense charge transfer transition (*vide infra*). This mixing is also observed in a number of other active sites in Bioinorganic Chemistry, for example in the iron-oxo dimer proteins.[68b]

**4.2.2** ***Coupled Binuclear and Trinuclear Copper Sites*** The multicopper oxidases have a trinuclear copper cluster comprised of type II and type III centers (Figure 59A).[3] A combination of CD and low temperature MCD spectroscopies have allowed the ligand field transitions for each center to be differentiated. The mononuclear type II Cu(II) site has a paramagnetic $S = 1/2$ ground state which generates

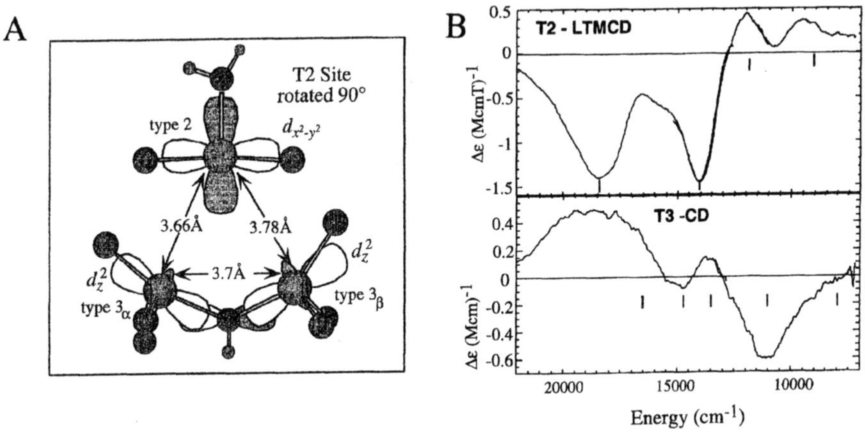


**Figure 59.** Trinuclear copper cluster sites. A) Trinuclear cluster in multicopper oxidases contains one type II Cu and one antiferromagnetically coupled binuclear type III site; B) predominant features in the low-temperature MCD (top) arise from the type II center, while bands in the CD spectrum (bottom) correspond to $d \rightarrow d$ transitions of the type III site. Adapted from Reference 3.

C-terms that dominate the low temperature MCD spectrum in the ligand field region (Figure 59B top). The type III center has two Cu(II) ions strongly antiferromagnetically coupled to produce a $S_{total} = 0$ ground state, which will only exhibit weak B-terms that are essentially not observable in the MCD spectrum. However, experimentally it is observed that the CD spectrum of the trinuclear copper cluster does not change on selective reduction of the type II center. Therefore, the type III $d \rightarrow d$ transitions dominate the CD spectrum (Figure 59B bottom, where the presence of more than four ligand field transitions indicate that the two type III centers are inequivalent).

The $d \rightarrow d$ transitions of each copper can be used in ligand field calculations to correlate to crystallographically determined geometric structures and generate an initial electronic structure description of this active site.[107] As illustrated by the orbitals superimposed on the type II structure shown in Figure 59A, this center has a $d_{x^2-y^2}$ ground state. This three-coordinate site is best described as having a square planar ligand field with an open coordination position oriented toward the type III coppers. Both type III sites are four coordinate and have a $d_{z^2}$ ground state which is oriented along the $OH^-$ bridge. This generates a very good superexchange pathway for antiferromagnetic coupling of the cupric ions. The ligand field for each type III center is best described as having a trigonal bipyramidal geometry with an open coordination position in the equatorial plane directed toward the type II center. This cluster is therefore poised for bridging interactions with exogenous ligands. This appears to play an important role in dioxygen activation for four electron reduction to water by this site.[3]

**4.2.3** ***The Cu$_A$ Site*** The Cu$_A$ center in Figure 1C exhibits an unusual absorption spectrum (Figure 60) for which the assignments are still evolving.[108] The metal $d$ orbital centered transitions appear to contribute to the region below 17,000 $cm^{-1}$. One of these transitions is quite intense, the band at 13,000 $cm^{-1}$ with $\varepsilon \sim$ 1500 $M^{-1}\,cm^{-1}$. This transition is assigned as a Cu–Cu $\sigma \rightarrow \sigma^*$ as shown in Figure 23. The energy of this transition corresponds to $2H_{AB}$, the electronic coupling matrix element for electron delocalization in this class III mixed-valence site (see Section 2.2.3.2), while the high absorption intensity of this transition likely reflects the significant Cu–Cu bonding interaction of this site. Note that in the charge transfer region ($\sim$20,000 $cm^{-1}$), two absorption peaks are observed that exhibit a pseudo-A-term feature in the low temperature MCD spectrum, indicating they have two perpendicular transition moments mixed by spin-orbit coupling.[109] These are assigned as Cys S $\rightarrow$ Cu charge transfer transitions based on resonance Raman spectroscopy[108,110] and may provide insight into the superexchange pathways for electron transfer by this site (*vide infra*).

**4.2.4** ***Iron Sulfur Centers*** While the spin-forbidden $d \rightarrow d$ transitions in high-spin Fe(III) sites are too weak to study in detail in proteins, they have been studied in $Fe(SR)_4^-$ model complexes.[49b,c] As shown in Figure 61, these show a remarkable difference in energy relative to the equivalent ligand field transitions in $FeCl_4^-$ used as a reference.[49a] The shift to lower energy by $\sim$7,000 $cm^{-1}$ in going from Figure 61

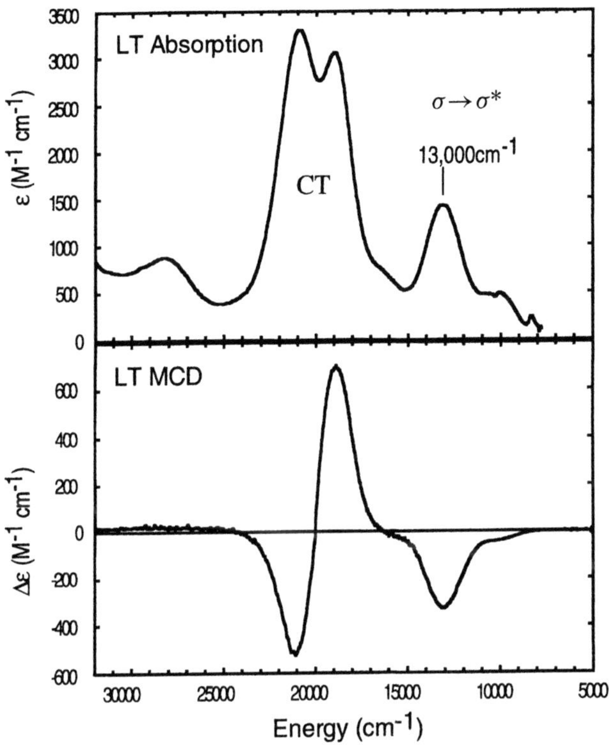


**Figure 60.** Spectroscopy of the $Cu_A$ site. A) Absorption; and B) MCD spectra.

A to B is to be contrasted to, for example, the 1,900 $cm^{-1}$ shift of $d \rightarrow d$ transitions occurring in Cr(III) complexes with thiolate versus chloride ligation. Further, ligand field calculations using the transitions in Figure 61B give an unreasonable covalent reduction in the electron repulsion parameter B to 22 $cm^{-1}$, relative to its free ion value for Fe(III) of 1,100 $cm^{-1}$.

SCF-X$\alpha$-SW calculations[111] combined with photoelectron spectroscopy[109] have shown that this dramatic shift in energy reflects an *inverted* bonding scheme in these high-spin Fe(III) sites attributable to the large exchange effects in $d^5$ ions. Figure 61C illustrates this using the spin unrestricted formalism, which allows the different spins to have different orbital wavefunctions. From Figure 61C, the metal-$d$ levels exhibit a large exchange splitting of the occupied spin-up ($\uparrow$) and unoccupied spin-down ($\downarrow$) orbitals. Alternatively, the exchange splitting of the ligand $\uparrow$ and $\downarrow$ levels is small. For $Fe(SR)_4^-$, the exchange stabilization of the Fe $3d$ is large enough to lower $3d\uparrow$ levels below the filled ligand $p$ valence orbitals. Both $d\uparrow$ and $d\downarrow$ interact with the ligand $p$ orbitals of the same spin resulting in a complex bonding scheme (Figure 61C, center). Importantly, while the LUMO is mostly $3d$ in character as found in normal metal-ligand bonding, the HOMO is mostly ligand in character. Therefore, the lowest energy spin-forbidden "ligand field transitions" are in fact $L(\uparrow)_{Ab} \rightarrow Fe(\downarrow)_{Ab}$

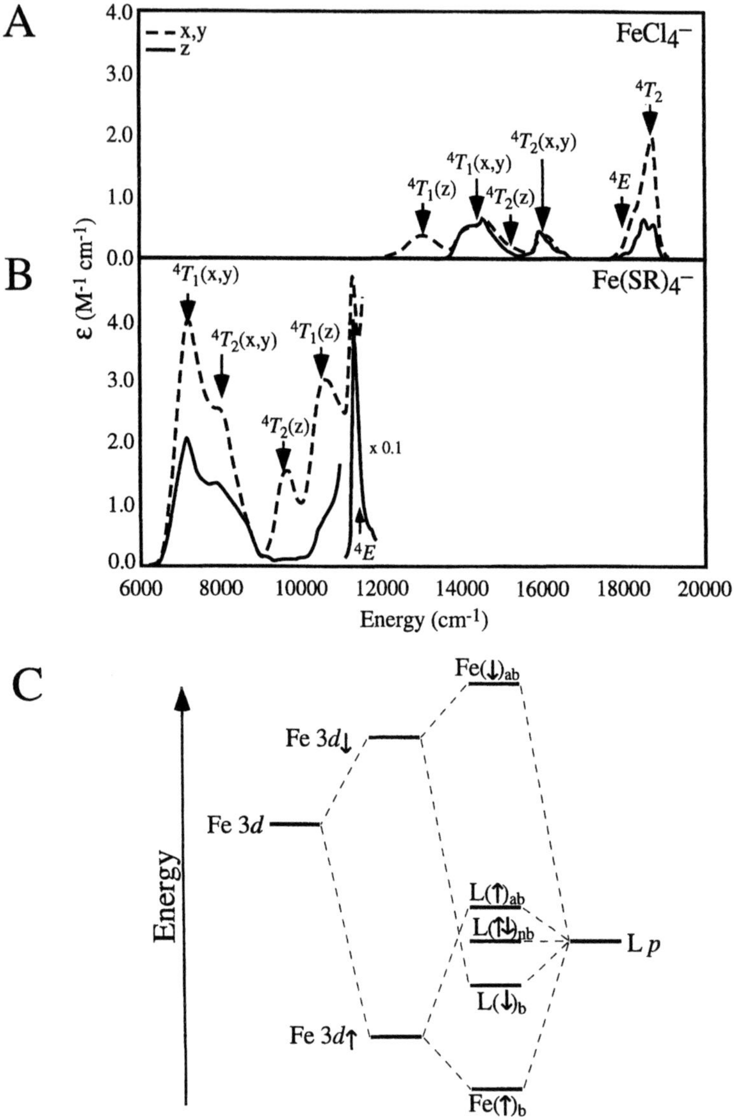


**Figure 61.** Comparison of ligand field transitions in A) $FeCl_4^-$ and B) $Fe(SR)_4^-$ complexes reveals dramatic shift to lower energy in the thiolate complex; C) inverted bonding scheme for high-spin Fe(III) sites. In spin unrestricted formalism the occupied Fe↑ and Fe↓ split in energy due to exchange. In $Fe(SR)_4^-$ the Fe↑ are below the ligand valence orbitals. Bonding results in a HOMO↑ which has mostly ligand character. The LUMO↓ is Fe *d* in character, and the lowest energy transition (seen in the spectra in parts A and B) is $L\uparrow \rightarrow$ Fe $d\downarrow$ charge transfer. For most metal complexes the HOMO is metal $d\uparrow$, and the lowest energy transition is $d\uparrow \rightarrow d\downarrow$ ligand field in nature. Adapted from References 49a,b and 109.

charge transfer in nature.[112b] This provides an explanation for the large shift to lower energy for these transitions in $Fe(III)(SR)_4^-$ relative to $Fe(III)Cl_4^-$, as the thiolate ligand has a lower valence ionization energy. The equivalent ligand field spectroscopy for $Fe(II)(SR)_4^{2-}$ versus $Fe(II)Cl_4^{2-}$ shows only a limited shift in energy of the spin-forbidden $d \rightarrow d$ transitions indicating that a normal bonding description is appropriate.[113] Therefore, there is a large change in electronic structure on oxidation of high-spin iron sites. This is known as *electronic relaxation* and can play a major role in redox processes in iron sulfur centers in Biology.

## 5 EXCITED STATE METHODS—CHARGE TRANSFER EXCITED STATES

### 5.1 Normal Spectral Features

**5.1.1** ***Electronic Absorption Spectroscopy*** In Bioinorganic systems, most ligands (excluding heme) are donors and have filled ligand valence orbitals at deeper binding energy than the metal $d$ orbitals. Photons in the visible/UV spectral region can excite electrons from these filled ligand valence orbitals into the unoccupied or partially occupied metal-$d$ levels, producing ligand-to-metal charge transfer (LMCT) transitions. These can involve a large change in the electric dipole moment. Thus LMCT transitions can be electric dipole allowed, producing intense absorption features at higher energy than the ligand field transitions. The energies and intensities of the LMCT transitions provide a powerful probe of the ligand-metal bond. Further, the CT intensity can be used to quantitate the relative donor strength of a ligand. This information content of LMCT transitions is developed below for the one hole Cu(II) case and then extended to the high-spin iron sites.

5.1.1.1 *Cu(II) Sites* As described above, normal Cu(II) complexes have a tetragonal geometry due to the Jahn–Teller effect. The ligands of interest are generally bound in the equatorial plane, and the half occupied $d_{x^2-y^2}$ orbital on the metal has a lobe oriented along the ligand–metal bond as illustrated in Figure 62A. Note that here we consider simple monatomic anionic ligands (e.g., $Cl^-$, $O^{2-}$) with three valence $p$ orbitals available for bonding. In Section 5.2, these considerations are extended to more complex molecular ligands of relevance to Bioinorganic systems.

CHARGE TRANSFER ENERGIES $E_L$ and $E_M$ in Figure 62B are the valence ionization energies of the ligand and metal ion orbitals involved in bonding. The LMCT transition will shift to lower energy as the ligand becomes easier to oxidize (i.e., lower electronegativity). For example, charge transfer transitions in copper complexes follow the series $CuF_4^{2-}$(>41,500 cm$^{-1}$) > $CuCl_4^{2-}$(22,500 cm$^{-1}$) > $CuBr_4^{2-}$(16,500 cm$^{-1}$).[114] $E_M$ is dependent on the effective nuclear charge on the metal ion, $Z_{eff}$, which is determined by the metal ion charge and number and types of coordinated donor ligands. It is also dependent on the ligand field splitting of the

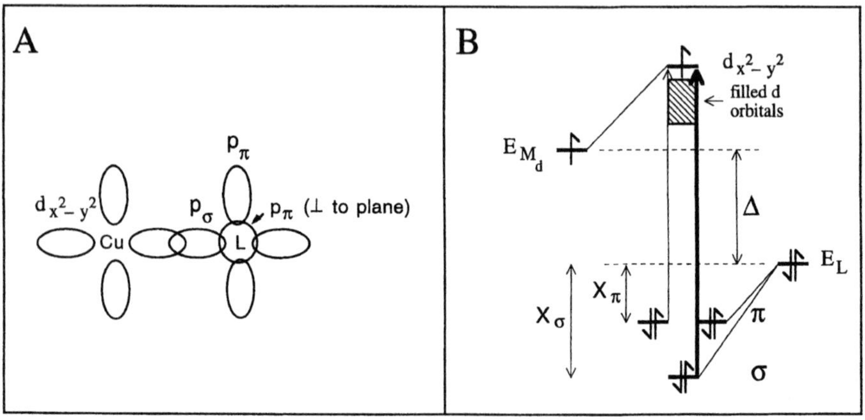


**Figure 62.** Charge transfer spectra of Cu(II) complexes. A) Metal and ligand key orbital interactions in a tetragonal Cu(II) complex; and B) schematic diagram of LMCT transition energies where $E_L$ and $E_M$ are valence ionization energies of the ligand and metal, $X_\sigma$ and $X_\pi$ are stabilization energies due to bonding.

metal *d* orbitals, which derives from antibonding interactions with the ligands (*vide infra*).

Ligand valence orbitals are stabilized in energy by an amount $X$ in Figure 62B due to bonding interactions with the metal valence orbitals $(3d, 4s, 4p)$, where $X$ is defined in Equation 58. $E_M - E_L = \Delta$ is the difference in energy between the metal and ligand valence orbitals before bonding and $\int \phi_M \mathcal{H} \phi_L d\tau$ is the resonance integral.

$$X \approx \frac{-\left[\int \phi_M \mathcal{H} \phi_L d\tau\right]^2}{E_M - E_L} \tag{58}$$

The latter is proportional to the overlap integral of the ligand $\phi_L$ and metal $\phi_M$ valence orbitals, $S_{ML} = \int \phi_M \phi_L d\tau$, which is the key contribution to bonding. In the often used Wolfsberg–Helmholz approximation, $\int \phi_M \mathcal{H} \phi_L d\tau \cong k[(E_M + E_L)/2]S_{ML}$ where $k \sim 1.5$–2.

From Figure 62A, the three valence $p$ orbitals of the ligand split in energy into two sets. The $p_z$ orbital is centered along the ligand–metal bond and has $\sigma$ overlap with the metal; the $p_{x,y}$ levels are doubly degenerate, oriented perpendicular to the ligand–metal bond and have $\pi$-overlap with the metal. Since $\sigma$-overlap is larger than $\pi$-overlap, Equation 58 predicts that $X_{p_\sigma} > X_{p_\pi}$ in Figure 62B. This produces a doubly degenerate ligand $p_\pi \rightarrow$ Cu $d_{x^2-y^2}$ charge transfer transition at lower energy than the ligand $p_\sigma \rightarrow$ Cu $d_{x^2-y^2}$ charge transfer transition.

CHARGE TRANSFER INTENSITIES From Mulliken,[115] in the limit of low metal-ligand orbital overlap, the electric dipole transition moment integral is given by

Equation 59, where $r$ is the ligand–metal bond length and $S_{M'L}$ is the overlap of the donor ($L$) and acceptor ($M'$) orbitals involved in the charge transfer transition.

$$\int \psi_g \vec{r} \psi_e d\tau \approx r S_{M'L} \tag{59}$$

(Note that $M'$ is used to indicate that the charge transfer can be to a metal centered orbital, which is different from that involved in bonding.) From Figure 62 the $p_\sigma$ orbital has significant overlap with the $d_{x^2-y^2}$ orbital while the $p_\pi$ has essentially no overlap. Therefore, the doubly degenerate low energy $\pi$ charge transfer transition should produce a band of very low intensity in the absorption spectrum while the higher energy $\sigma$ LMCT transition should be intense and dominant for normal ligand–metal bonding. Thus the charge transfer intensity distribution is a probe of ligand–metal $\sigma$ and $\pi$ bonding interactions.

The charge transfer intensity also can be used to quantitate the donor strength of specific ligand–metal bonds.[116] Ligand donor strength is defined as the amount of metal character mixed into the filled ligand valence orbitals due to bonding. As the coefficient of metal character in the ligand bonding orbital ($c^2$ in $\psi_L^b$ in Equation 60a) increases, electron density is donated to the metal. The reciprocal metal antibonding orbital is given in Equation 60b.

$$\psi_L^b = \sqrt{1-c^2}(\phi_L) + c(\phi_M) \tag{60a}$$

$$\psi_M^* = \sqrt{1-c^2}(\phi_M) - c(\phi_L) \tag{60b}$$

A transition from the ground to the charge transfer excited state then involves excitation of an electron from $\psi_L^b$ to $\psi_M^*$. This leads to the expansion of the transition moment integral for the $\psi_g \rightarrow \psi_e$ transition in Equation 61. The first term on the right, $\int \phi_M \vec{r} \phi_M d\tau$, is $d \rightarrow d$ in character and thus zero, while calculations show that the last term, $\int \phi_L \vec{r} \phi_L d\tau$, dominates the charge transfer intensity.[117]

$$\begin{aligned}\int \psi_g \vec{r} \psi_e d\tau &= c\sqrt{1-c^2} \int \phi_M \vec{r} \phi_M d\tau + (1-c^2) \int \phi_L \vec{r} \phi_M d\tau \\ &\quad - c^2 \int \phi_M \vec{r} \phi_L d\tau - c\sqrt{1-c^2} \int \phi_L \vec{r} \phi_L d\tau \approx c\vec{r}\end{aligned} \tag{61}$$

This *ligand–ligand term* reflects the ligand character present in the bonding and antibonding molecular orbitals involved in the charge transfer transition. For this integral to be non-zero, the ligand character in $\psi_L^b$ and $\psi_M^*$ must be the same (e.g., $Lp_z$), and when this is the case, the ligand–ligand integral is given approximately by $c \times \vec{r}$ where $\vec{r}$ is a vector along the ligand–metal bond. Thus the charge transfer intensity, which is governed by the transition moment integral squared, is proportional to $c^2$, the covalent mixing of metal character into the ligand donor orbital. From molecular orbital theory, $c$ is given by Equation 62 and reflects the transition moment integral being proportional to the ligand–metal overlap as described by Equation 59.

$$c \approx \frac{-[\int \phi_M \mathcal{H} \phi_M d\tau]}{E_M - E_L} \tag{62}$$

Also $c^2$ and therefore the charge transfer intensity can increase as $E_M - E_L$ decreases (i.e., the energy of the charge transfer transition decreases). *Therefore low energy, intense charge transfer transitions reflect highly covalent ligand–metal bonds (strong donors). This high covalency plays a major role in determining the reactivity of metal sites in Biology.*

5.1.1.2 *Iron Sites* From Figure 62B the energy of a charge transfer transition is dependent on $E_M$, the valence ionization energy of the metal $d$-orbitals, which increases as $Z_{\text{eff}}$ increases. Therefore LMCT transitions of Fe(III) sites are at significantly lower energy than those of Fe(II) sites. From ligand field theory (*vide supra*), high-spin Fe(III) has a $^6A_1$ ground state and no Jahn–Teller distortion. Thus the strong field $z$-direction at an Fe(III) center is usually determined by the most covalent bond (Figure 63A) which should also make the dominant contribution to the LMCT spectrum (note that this figure assumes that one L–Fe bond is dominant). All $d$ orbitals of high-spin Fe(III) are half occupied, and from Figure 63B this leads to a high energy intense $\sigma$ charge transfer transition, $Lp_z(\sigma) \rightarrow Md_{z^2}(\sigma)$, and a lower energy weaker $\pi$ charge transfer, $Lp_{x,y}(\pi) \rightarrow Md_{xz,yz}(\pi)$.[68a] The hole produced in the ligand valence orbital will couple to the unpaired electrons on the metal and results in a number of many electron excited states. However, the repulsion between electrons in metal and ligand centered orbitals is small, and this splitting should not be significant. The LMCT transition produces a metal $d^{n+1}$ final state with ligand

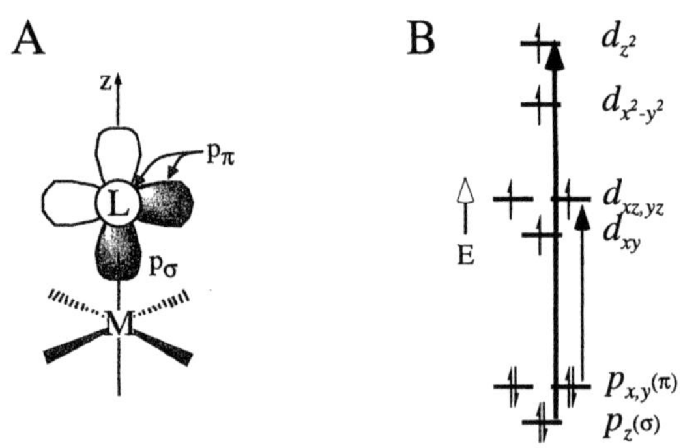


**Figure 63.** Charge transfer spectra for high spin Fe(III) complexes. A) The most covalent ligand bond to the metal determines the strong field $z$ direction (assuming one ligand–metal bond is dominant); and B) the $d$ orbital splittings in the strong tetragonal ligand field shown in A. Each is half-occupied and from overlap has allowed and intense high energy ligand $p\sigma \rightarrow$ metal $d\sigma$ and weaker, lower energy ligand $p\pi \rightarrow$ metal $d\pi$ charge transfer transitions.

field parameters defining the energy splitting of the $d$ orbitals similar to those of Fe(II) (i.e., ~65% those of the high-spin Fe(III) ground state values).[118]

Therefore, while for normal Cu(II) complexes the dominant charge transfer feature probes the sigma bonding interactions of the active site, for high-spin Fe(III) both the $\sigma$ and $\pi$ ligand-to-metal charge transfer transitions can have significant intensity, and these further probe the $\pi/\sigma$ ligand field splitting of the $d$-orbitals of the ferric site (Figure 63B). Once the charge transfer transitions are assigned to specific ligand valence orbitals, these provide significant insight into the properties of the ligand–metal bond. Charge transfer assignments are often elucidated through resonance Raman (rR) spectroscopy.

### 5.1.2 *Resonance Raman Spectroscopy*

5.1.2.1 *The Normal Raman Experiment* In the Raman experiment (see Chapter 7, Volume I, this series), laser photons are scattered from a sample, and one detects the energy lost (Stokes) or gained (anti-Stokes) relative to the intense elastically (Rayleigh) scattered peak. This energy difference ($\nu_L - \nu_{\text{vib}}$ in Figure 64, where $\nu_L$ is the laser energy) is associated with vibrational excitation of the sample. In infrared spectroscopy, the electric dipole operator, $M_{(x,y,z)}$, allows the $0 \rightarrow 1$ vibrational excitation. In Raman spectroscopy the vibrationally scattered intensity is given by Equation 63 where $I_0$ is the laser intensity, $\rho$, $\sigma$ are the $x$, $y$, $z$ coordinates, and $\alpha_{ji}^{\rho\sigma}$ is the $\rho\sigma$ component of the *polarizability tensor* for the $i \rightarrow f$ vibrational transition. Thus the selection rules for Raman scattering are given by $M_{(x^2,xy,xz,y^2,yz,z^2)}$.

$$I = \text{const.} I_0(\nu_L - \nu_{\text{vib}})^4 \left| \sum_{\rho\sigma} \alpha_{ji}^{\rho\sigma} \right|^2 \tag{63}$$

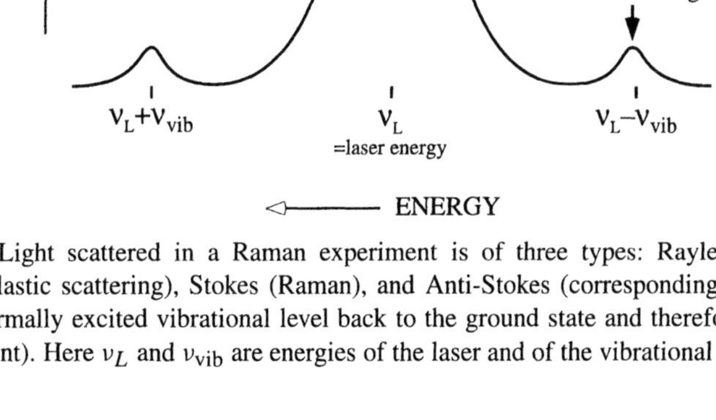


**Figure 64.** Light scattered in a Raman experiment is of three types: Rayleigh (intense peak from elastic scattering), Stokes (Raman), and Anti-Stokes (corresponding to a transition of a thermally excited vibrational level back to the ground state and therefore temperature-dependent). Here $\nu_L$ and $\nu_{\text{vib}}$ are energies of the laser and of the vibrational transition.

Note that in many point groups (including all with inversion symmetry), the vibrational transitions allowed in Raman and infrared spectroscopies are complementary, and application of both techniques is required for rigorous vibrational analysis.

Raman peaks are assigned to specific symmetry vibrations using the *depolarization ratio*, $\rho$.[119] This is given by Equation 64 for a randomly-oriented sample where $I_{\parallel}$ and $I_{\perp}$ are the intensities of light scattered with the $\vec{E}$ vector parallel and perpendicular, respectively, to the polarization direction of the incident laser beam.

$$\rho = \frac{I_{\perp}}{I_{\parallel}} \tag{64}$$

For totally symmetric vibrations of molecules with octahedral or tetrahedral symmetry, $\rho = 0$ and the vibration is polarized ($I_{\parallel} \neq 0$). Totally symmetric vibrations in molecules with less than cubic symmetry have $0 < \rho < 3/4$. Non-totally symmetric Raman allowed vibrations are depolarized with $\rho = 3/4$. In vibrational spectroscopy, the approach generally taken is to use a combination of Raman (with polarizations) and IR spectroscopies, isotopic perturbations, and normal coordinate analysis to describe the molecular normal modes and obtain the interatomic force constants, which provide a direct probe of bond strength.

5.1.2.2 *Resonance Raman Effect* The above experiment requires that the laser excitation energy correspond to a non-absorbing region of the electronic spectrum. As the laser is tuned into an intense absorption band, certain vibrations will be resonance enhanced (i.e., their intensities increase by three to four orders of magnitude). As shown in Figure 65,[120] the excitation energy dependence of the resonance intensity (normalized to an internal standard) generates a resonance Raman excitation profile (RREP). In biological systems resonance Raman spectroscopy provides a direct way to observe vibrations associated with an absorption band, allowing one to focus in on the active site of a large molecule.

The origin of the resonance Raman effect[121] is given by the Kramers-Heisenberg-Dirac dispersion equation for the Raman scattering tensor (Equation 65). $M_{\rho,\sigma}$ is the electric dipole operator in the direction $\rho, \sigma = x, y, z$, $\Psi_i$ is the initial state, $\Psi_f$ the final state and the $\Psi_e$ are intermediate excited states at an energy $\nu_e - \nu_i$ above the initial state.

$$\alpha_{fi}^{\rho\sigma} = \sum_e \left[ \frac{\langle \Psi_f | \hat{M}_\rho | \Psi_e \rangle \langle \Psi_e | \hat{M}_\sigma | \Psi_i \rangle}{\nu_e - \nu_i - \nu_L + i\Gamma_e} + \frac{\langle \Psi_f | \hat{M}_\sigma | \Psi_e \rangle \langle \Psi_e | \hat{M}_\rho | \Psi_i \rangle}{\nu_e - \nu_f + \nu_L + i\Gamma_e} \right] \tag{65}$$

It is important to note from Equation 65 that when the energy of the laser $\nu_L$ equals the transition energy to a specific excited state $\Psi_e (= \nu_e - \nu_i)$ the first term becomes very large. ($\Gamma_e$, the damping factor associated with the $\Psi_e$ excited state lifetime, keeps this term finite.) It is this first term in Equation 65 which is responsible for the resonance enhancement of Raman intensity [where $I \propto (\alpha_{fi}^{\rho\sigma})^2$]. The second term gives a small non-resonance contribution to the Raman intensity. The first term describes resonance enhancement by a specific excited state $\Psi_e$, and thus the resonance

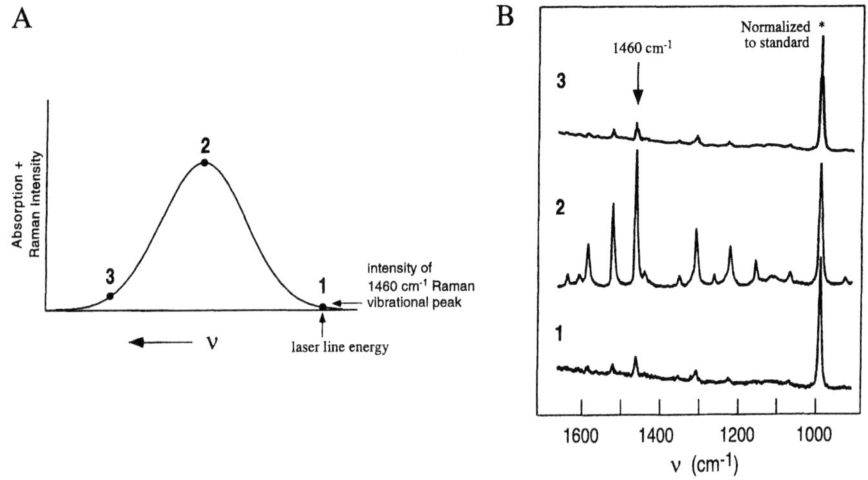


**Figure 65.** Resonance Raman (rR) spectroscopy. A) rR excitation profile shows that Raman signal intensity (dots) is greatly enhanced when the excitation energy corresponds to a maximum in the absorption spectrum; and B) rR spectra obtained at energies 1, 2, and 3. Adapted from Reference 120.

Raman scattering described by this term is dependent on the specific vibrational and electronic properties of $\Psi_e$.

A formalism has been developed by Albrecht et al.[122] which relates the polarizability tensor to the properties of the excited state $\Psi_e$. First, the wavefunctions are expanded in terms of the product (electronic) × (vibrational) functions (i.e., the Born–Oppenheimer approximation). Since both $\Psi_i$ and $\Psi_f$ in Equation 65 are different vibrational levels of the ground electronic state, $\Psi_i = \Psi_g^{\text{elec}}\chi_g^i$, $\Psi_f = \Psi_g^{\text{elec}}\chi_g^f$, $\Psi_e = \Psi_e^{\text{elec}}\chi_e^v$ where $\chi_x^n$ is a vibrational wavefunction of the $x$th electronic state with quantum number $n$. Substitution of these Born–Oppenheimer product functions into Equation 65 gives Equation 66.

$$\alpha_{gf,gi}^{\rho\sigma} = \sum_{ev} \left[ \frac{\langle \Psi_g^{\text{elec}}\chi_g^f | \hat{M}_\rho | \Psi_e^{\text{elec}}\chi_e^v \rangle \langle \Psi_e^{\text{elec}}\chi_e^v | \hat{M}_\sigma | \Psi_g^{\text{elec}}\chi_g^i \rangle}{\nu_{ev} - \nu_{gi} - \nu_L + i\Gamma_{ev}} + \frac{\langle \Psi_g^{\text{elec}}\chi_g^f | \hat{M}_\sigma | \Psi_e^{\text{elec}}\chi_e^v \rangle \langle \Psi_e^{\text{elec}}\chi_e^v | \hat{M}_\rho | \Psi_g^{\text{elec}}\chi_g^i \rangle}{\nu_{ev} - \nu_{gf} + \nu_L + i\Gamma_{ev}} \right] \tag{66}$$

Next, a term is introduced into the electronic Hamiltonian associated with the effects of molecular vibrations. Equation 67 is obtained from a Taylor series expansion of the molecular Hamiltonian in terms of the vibrational normal mode $Q$.

$$\mathcal{H} = \mathcal{H}_0 + \left.\frac{\partial \mathcal{H}}{\partial Q}\right|_0 Q + \cdots \tag{67}$$

The second term on the right enables vibrations to mix electronic wavefunctions.

This Herzberg–Teller coupling mixes the intermediate electronic state $\Psi_S^{\text{elec}}$ into the excited state $\Psi_e^{\text{elec}}$ (Equation 68).

$$\Psi_e^{\text{elec}'} = \Psi_e^{\text{elec}} + \Psi_S^{\text{elec}} \frac{\langle \Psi_S^{\text{elec}} | \left( \frac{\partial \mathcal{H}}{\partial Q} \Big|_0 \right) \Psi_e^{\text{elec}} \rangle Q}{\nu_e - \nu_S} \tag{68}$$

Substitution of this perturbed excited state wavefunction into Equation 66 gives the Albrecht A-, B-, and C-terms of resonance Raman spectroscopy, Equation 69, which are not to be confused with the equivalent nomenclature in MCD spectroscopy. The first two terms dominate and are discussed below.

$$\alpha_{gf,gi}^{\rho\sigma} = \underline{\underline{A + B}} + C + \cdots \tag{69}$$

A-TERM For an allowed transition to an excited state in the absence of Herzberg–Teller coupling, the first (resonance) term in Equation 66 gives Equation 70 where $M_{ge}^{\rho} = \langle \Psi_g^{\text{elec}} | \hat{M}_\rho | \psi_e^{\text{elec}} \rangle$ is the electric dipole transition moment of the resonant absorption transition. The A-term is the most generally observed effect in resonance Raman spectroscopy. As seen in Equation 70, the resonance Raman intensity is proportional to the square of the absorption intensity. Further, the intensity is proportional to the square of the vibrational overlap integral between $\Psi_g$ and $\Psi_e$ states.

$$A = \sum_{\nu} \frac{M_{ge}^{\rho} M_{eg}^{\sigma} \langle \chi_g^f | \chi_e^\nu \rangle \langle \chi_e^\nu | \chi_g^i \rangle}{\nu_{e\nu} - \nu_{gi} - \nu_L + i\Gamma_{e\nu}} \tag{70}$$

Since the Hermite polynomial vibrational functions form an orthonormal set, for these integrals to be non-zero for a Raman transition (i.e., $\chi_g^f \neq \chi_g^i$) there must be an excited state distortion ($\Delta Q$ in Figure 66A). From group theory, a non-degenerate excited state can only distort along totally symmetric normal coordinates. (Degenerate excited states can also distort along Jahn–Teller coordinates.)[10b] From Figure 66B the excited state distortion also leads to the absorption bandshape [i.e., the Franck–Condon progression (indicated by solid lines) in the totally symmetric normal modes] and thus the A-term excitation profile reflects the square of the absorption bandshape. It should also be noted from Figure 66A that there can be non-zero overlap integrals to a number of different vibrational levels of the ground electronic state $\chi_g^f$, and thus one often observes resonance-enhanced *progressions* in totally symmetric modes (and combinations if the excited state distortion is along more than one coordinate). Finally, for totally symmetric modes the depolarization ratio $\rho$ in resonance is $1/3$ for non-degenerate excited states and $1/8$ for degenerate excited states.[121a]

B-TERM For a weakly allowed electronic transition, inclusion of the Herzberg–Teller expansion (Equation 68) into the first (resonance) term in Equation 66 gives

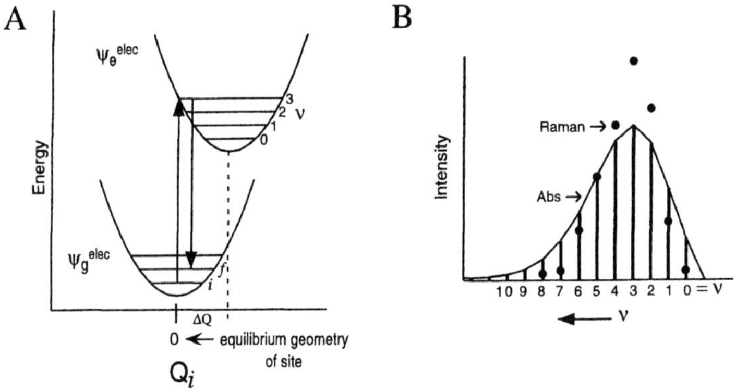


**Figure 66.** Excited state distortions. A) A distortion, $\Delta Q$, and associated vibrational and electronic transitions; and B) absorption bandshape shows Franck–Condon progression (solid lines) and rR A-term intensities (dots) which follow the square of absorption intensities.

the B-term (Equation 71), where $H'_{es} \equiv \langle \Psi_e^{\text{elec}} | (\partial \mathcal{H} / \partial Q|_0) | \Psi_s^{\text{elec}} \rangle$ and $M_{sg}^{\sigma}$ is the allowed electric dipole transition moment for the transition from the ground to intermediate excited state $S$ in $\sigma$ polarization.

$$B = \sum_S \sum_v \left[ \frac{M_{ge}^{\rho} H'_{eS} M_{Sg}^{\sigma} \langle \chi_g^f | \chi_e^v \rangle \langle \chi_e^v | Q | \chi_g^i \rangle}{[\nu_{ev} - \nu_{gi} - \nu_L + i\Gamma_{ev}][\nu_e - \nu_S]} + \frac{M_{gS}^{\rho} H'_{Se} M_{eg}^{\sigma} \langle \chi_g^f | Q | \chi_e^v \rangle \langle \chi_e^v | \chi_g^i \rangle}{[\nu_{ev} - \nu_{gi} - \nu_L + i\Gamma_{ev}][\nu_e - \nu_S]} \right] \tag{71}$$

From group theory $\partial \mathcal{H} / \partial Q$ transforms as $Q$, and for the $H'$ integral to be non-zero, $\Gamma_Q = \Gamma_e \times \Gamma_s$ where $\Gamma_i$ is the symmetry of state $i$. This requires non-totally symmetric normal modes $Q$ which vibronically couple $\Psi_e$ with a more intense state $\Psi_s$ at an energy $\nu_e - \nu_S$.

In normal Raman spectroscopy the polarizability tensor is symmetric (i.e., $\alpha^{\rho\sigma} = \alpha^{\sigma\rho}$), which leads to the depolarization ratios given in Section 5.1.2.1. In resonance, from Equation 71, $\alpha^{\rho\sigma}$ is no longer equal to $\alpha^{\sigma\rho}$, and this can significantly affect the depolarization ratio of non-totally symmetric vibrations.[121a] Raman allowed non-totally symmetric vibrations can have $\rho > 3/4$ (which is called anomalous polarization) and Raman forbidden vibrations can have $\rho \to \infty$ (called inverse polarization). Also note that in Equation 71, the first term has the ground to excited vibrational integral $\langle \chi_g^i | Q | \chi_e^v \rangle$ which can only have a non-zero value for the $0 \to 1$ transition in normal mode $Q$. In the second term, the ground to excited vibrational integral $\langle \chi_e^v | \chi_g^i \rangle$ can only be non-zero for a $0 \to 0$ transition in normal mode $Q$, since the excited state is generally not distorted along the normal coordinates involved in Herzberg–Teller coupling. Therefore B-term profiles peak on the low energy side of an absorption band and can thus be distinguished from A-terms, which also have different depolarization ratios.

EXCITED STATE DISTORTIONS A-terms are generally found to dominate the resonance Raman spectra of many transition metal complexes. From Equation 70, these can be large for intense absorption bands; therefore A-terms are often associated with ligand-to-metal charge transfer transitions and will reflect the nature of the associated excited state distortion. From Figure 67A, ligand-to-metal charge transfer transitions involve excitation of an electron from a bonding to an antibonding molecular orbital. Therefore, the ligand–metal bond gets longer by an amount $\Delta Q$ in the excited state. This leads to the potential energy surfaces of Figure 67B. Note that since ligand $\pi$ bonding interactions with metal ions are weaker than $\sigma$ bonding interactions, $\sigma$ charge transfer transitions should undergo much larger excited state distortions. Representative $\sigma$- and $\pi$-charge transfer resonance Raman profiles are depicted in Figure 67C.

From the above discussion, resonance Raman spectroscopy provides important insight into the properties of an active site. It allows the observation of vibrations of an active site even in the presence of the protein ligand, thus enabling detailed vibrational analysis. Through the nature of the vibrations enhanced it allows electronic transitions to be associated with specific ligand–metal bonds. The behavior of the profiles through the electronic transitions allows specific spectral assignments (e.g., $\pi$ versus $\sigma$ charge transfer in Figure 67C). From a quantitative analysis of the resonance Raman intensities, one can further obtain an experimental estimate of the excited state geometry and thus the change in geometric structure associated with a known change in electronic structure of a metal site.[45e, 123] In Bioinorganic Chemistry, resonance Raman spectroscopy has been used to associate low energy intense LMCT transitions with specific ligand–metal bonds.

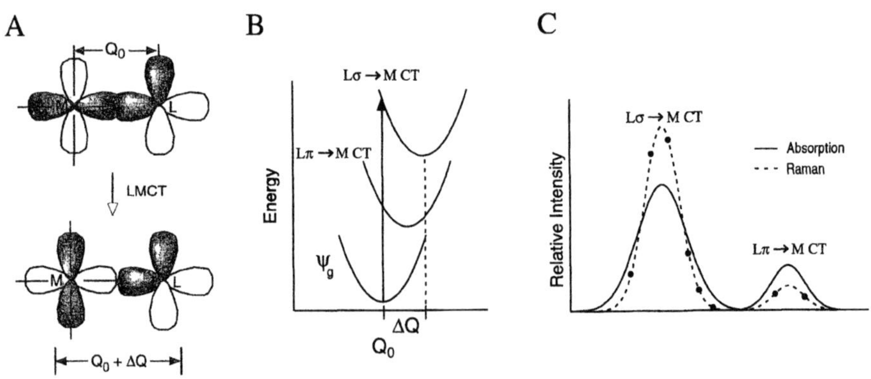


**Figure 67.** LMCT transitions. A) The electron is excited from a bonding orbital to an antibonding orbital, increasing the length of the ligand–metal bond by $\Delta Q$; B) potential energy surfaces representing excited state distortions; note that the $L\sigma \rightarrow M$ CT transition should have a more distorted excited state relative to $L\pi \rightarrow M$ CT; and C) predicted rR excitation profiles for $\sigma$ (more intense and more distorted) and $\pi$ CT transitions.

## 5.2 Unique Spectral Features

Generally when metallobiomolecules are described as having unique spectral features, this refers to their being highly colored, which reflects the presence of low energy intense charge transfer transitions. From the above discussion low-energy intense charge transfer reflects highly covalent ligand–metal bonds (i.e., strong donors) which can make significant electronic structure contributions to the reactivity of active sites in Biology.[124] A combination of absorption, CD, MCD, and resonance Raman spectroscopies have defined the key biological ligands associated with low energy intense charge transfer transitions. These are oxide, sulfide, phenolate, thiolate, and peroxide; their valence orbitals involved in bonding are summarized in Figure 68. Oxide and sulfide ligands have three valence $p$ orbitals that split into $\pi$ and

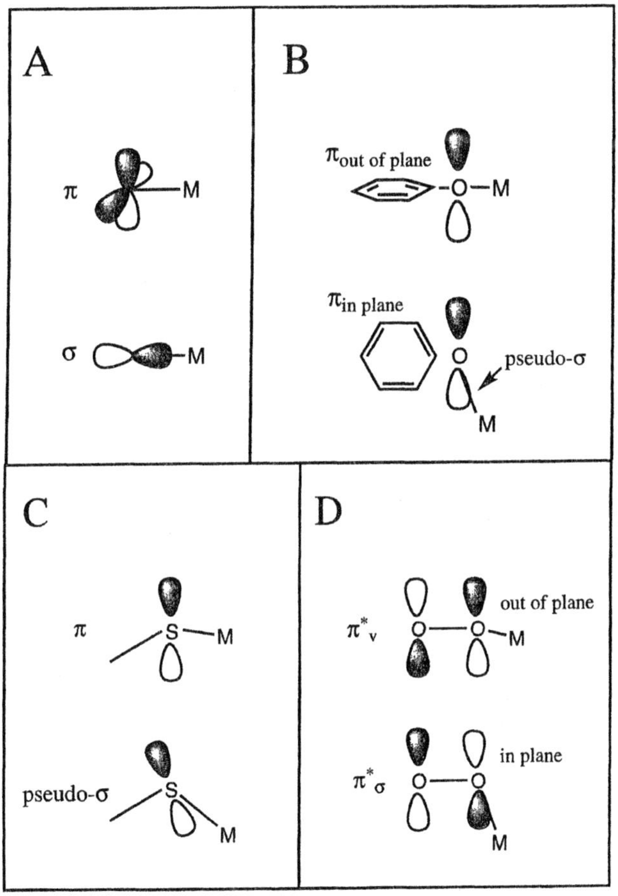


**Figure 68.** Valence orbitals of important biological ligands which exhibit intense low energy LMCT transitions: A) oxide or sulfide; B) phenolate; C) thiolate; and D) peroxide.

$\sigma$ sets (*vide supra*). Phenolate and thiolate have two valence p orbitals perpendicular to the O(S)–carbon bond. As shown in Figure 68B and C, these have $\pi$ and pseudo-$\sigma$ bonding interactions with the metal when the C–O(S)–M bond angle is $<180°$. Here, pseudo-$\sigma$ refers to the fact that the electron density can be tilted relative to the ligand–metal bond. Finally peroxide intermediates, which are often observed in dioxygen binding and activation, have a doubly degenerate $\pi^*$ HOMO which splits into two non-degenerate levels on bonding to a metal ion. The $\pi_\sigma^*$ is directed along the $O_2^{2-}$–M bond and the $\pi_v^*$ is perpendicular or vertical to the ligand–metal bond.

The specific bonding interactions of these ligands with metal active sites are dependent on the type of metal ion, the other bound ligands, the geometry of the site, the ligand orientation at the metal ion imposed by the protein and possible bridging interactions to additional metal centers. Thus each class of active sites is a separate study of geometric and electronic structure and its contributions to function (see Reference 1). Below we briefly consider the blue copper center as a specific case where charge transfer transitions provide an important probe of unique ligand–metal bonding. We further outline the additional effects observed in charge transfer excited states associated with bridging ligand interactions and their relation to ground state superexchange pathways.

**5.2.1** ***The Thiolate–Cu Bond in Blue Copper Centers*** From low temperature absorption and MCD data[103] in Figure 58C and D, band 4 of the blue copper spectrum is intense in absorption and weak in MCD (i.e., $C_0/D_0 < 0.01$). Band 4 can then be assigned as the lowest energy charge transfer transition of the blue copper center. Polarized single crystal electronic absorption spectral data (Figure 69B), combined with the crystal coordinates, indicate that this band is polarized along the Cys S–Cu bond and can be assigned as a Cys $\rightarrow$ Cu(II) charge transfer transition.[125] This assignment is supported by resonance Raman excitation profile data in Figure 69C which show that the Cu–S stretch is resonance enhanced by band 4.[126] Therefore the lowest energy Cys $\pi$ charge transfer transition of the blue copper center is intense while the higher energy Cys $\sigma$ charge transfer (band 3) is weak. As discussed above, normally one observes low energy weak $\pi$ and higher energy intense $\sigma$ charge transfer (Figure 69D). Since charge transfer intensity is proportional to orbital overlap (Equation 59), this $\pi/\sigma$ charge transfer intensity pattern requires that the $d_{x^2-y^2}$ orbital is rotated from its normal orientation with a lobe along the ligand–metal bond by 45° such that there is maximum $\pi$ overlap with the thiolate sulfur as in Figure 69E.[103] This orientation indicates that there is a strong $\pi^*$ interaction of the copper with the thiolate ligand at the blue copper site which is due to the short Cu–S(Cys) bond length of 2.1 Å (Figure 58F).[21] This orientation generates an efficient $\pi$ hole superexchange pathway for rapid electron transfer through the protein.[25]

**5.2.2** ***Charge Transfer Transitions of Bridging Ligands*** As illustrated in Figure 70A and discussed in Section 2.2.2.1, ground state antiferromagnetism involves the second order coupling of the spins on two coppers through the electron pair on the bridging ligand. [Note that Cu(II) dimers are considered here to simplify the problem to one unpaired electron on each metal center.] The bridging ligand-to-metal charge

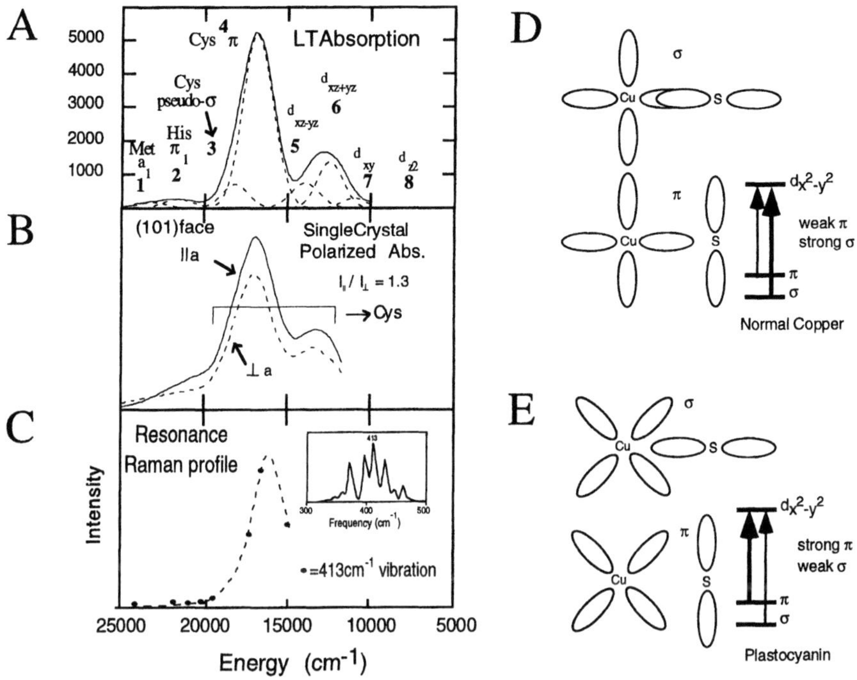


**Figure 69.** Cys $S \rightarrow$ Cu CT transitions in blue copper proteins. A) Absorption spectrum of plastocyanin at low temperature; B) single crystal polarized absorption spectrum of plastocyanin; C) rR excitation profile of the $\nu = 413$ cm$^{-1}$ Raman peak in azurin, with inset showing region of resonance Raman spectrum; D) for normal Cu complexes the $\sigma$ CT transition is more intense than the $\pi$ CT transition due to overlap of a lobe of the Cu $d_{x^2-y^2}$ orbital with the ligand $p_\sigma$, while for E) blue copper complexes the reverse is true due to rotation of the $d_{x^2-y^2}$ orbital by 45°. Adapted from References 2 and 124.

transfer excited state shifts an electron of the pair to $Cu_a$, resulting in direct overlap of an unpaired electron on the bridge with the electron on $Cu_b$. Thus there can be a large antiferromagnetic coupling in LMCT transitions of bridging ligands.[127] It is the purpose of this section to briefly demonstrate that the LMCT transition of a bridging ligand provides the superexchange pathway for ground state antiferromagnetic coupling and that there are large spectral effects in charge transfer transitions of bridging ligands which provide a direct probe of this pathway. This has been elucidated through a valence bond configurational interaction (VBCI) model[128] where the valence bond formalism has been used since the physical reality of the problem involves the coupling of unpaired electrons localized on each Cu(II).

In the VBCI formalism, four configurations are considered with each generating a number of many electron states. The ground configuration in Figure 71 (GS) has one electron on each copper and generates ${}^1\Gamma_+$ and ${}^3\Gamma_-$ zeroth order states at the same energy. The $+$ and $-$ refer to the symmetry with respect to interchange of the two halves of the dimer, and these states are the same as $S$ and $T$ in Figure 18C. The first

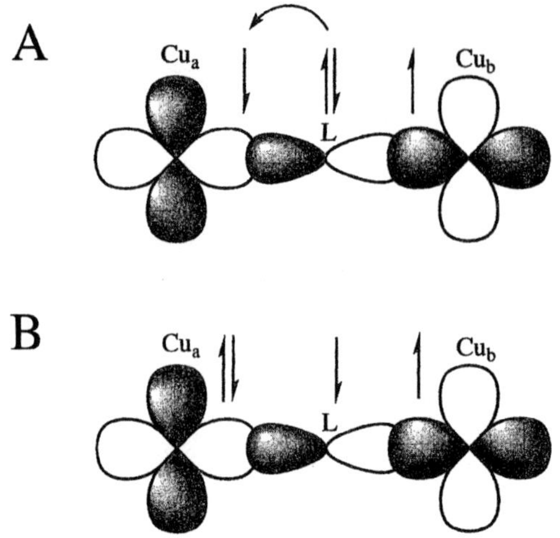


**Figure 70.** LMCT from a bridging ligand. A) Antiferromagnetic alignment of spins on two coupled Cu(II) ions in ground state; and B) the LMCT excited state (excitation of one electron from the bridging ligand to Cu$_a$) has unpaired electron on ligand orbital which directly overlaps with the unpaired electron on Cu$_b$.

excited configuration, CT in Figure 71, involves the ligand-to-metal charge transfer excitation at an energy $\Delta$ above the ground state. This generates four LMCT states, ${}^1\Gamma_+ \; {}^1\Gamma_- \; {}^3\Gamma_+ \; {}^3\Gamma_-$, since the electron can be transferred to Cu$_a$ or Cu$_b$. Next is the MMCT configuration which involves transfer of the electron from Cu$_a$ to Cu$_b$ or Cu$_b$ to Cu$_a$. This produces two states, ${}^1\Gamma_+$, ${}^1\Gamma_-$ at an energy $U$ above the ground state where $U$ is the difference in Coulomb repulsion between the $(\mathrm{Cu}_a)^1$ $(\mathrm{Cu}_b)^1$ and

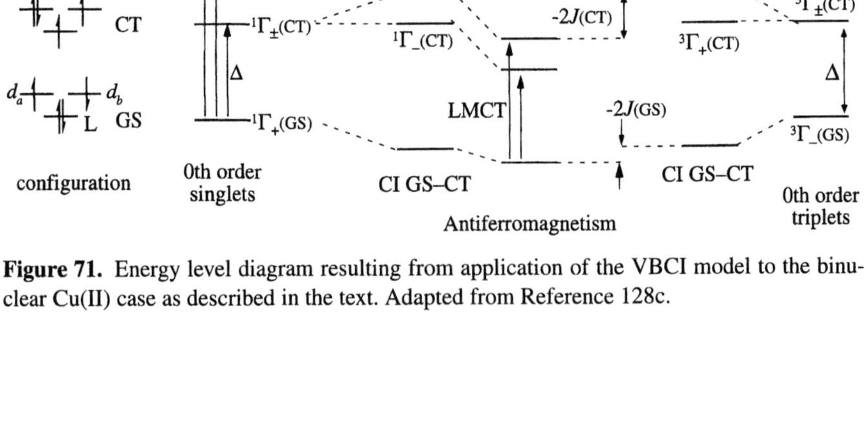


**Figure 71.** Energy level diagram resulting from application of the VBCI model to the binuclear Cu(II) case as described in the text. Adapted from Reference 128c.

$(\mathrm{Cu}_a)^2$ $(\mathrm{Cu}_b)^0$ configurations. Note that there can only be singlet MMCT states. Finally there is also the double CT (DCT) configuration which involves both electrons on the bridge being excited to fill the Cu $d$ orbitals. This produces one singlet state $^1\Gamma_+$ at an energy $E_{\mathrm{DCT}} \gg U$ and can probably be neglected.

These zeroth order singlet and triplet states are given on the left and right sides, respectively, of Figure 71. These can undergo a configurational interaction between states of the same symmetry and spin due to the covalent overlap of the ligand and metal orbitals which is given by $\int \phi_{M_d} \mathcal{H} \phi_L d\tau \equiv h_{dl}$. The appropriate VBCI matrices which describe this interaction are given in Equation 72. Note that if we only include CI between the ground and charge transfer excited states, there is no net antiferromagnetic splitting of the triplet and singlet states of the same configuration (compare CI GS–CT splitting of singlets and triplets).

$$\begin{array}{cc} {}^3\Gamma_-^{GS} & {}^3\Gamma_-^{CT} \end{array} \qquad \begin{array}{cc} {}^1\Gamma_-^{CT} & {}^1\Gamma_-^{MMCT} \end{array}$$

$$\begin{vmatrix} -E & h_{dL} \\ h_{dL} & \Delta - E \end{vmatrix} = 0 \qquad \begin{vmatrix} \Delta - E & h_{dL} \\ h_{dL} & U - E \end{vmatrix} = 0$$

$$\begin{array}{cccc} {}^1\Gamma_+^{GS} & {}^1\Gamma_+^{CT} & {}^1\Gamma_+^{MMCT} & {}^1\Gamma_+^{DCT} \end{array}$$

$$\begin{vmatrix} -E & h_{dL} & 0 & 0 \\ h_{dL} & \Delta - E & h_{dL} & \sqrt{2}h_{dL} \\ 0 & h_{dL} & U - E & 0 \\ 0 & \sqrt{2}h_{dL} & 0 & E_{DCT} - E \end{vmatrix} = 0 \tag{72}$$

However, the MMCT and DCT configurations only generate singlet states, and when these are included in the VBCI matrices in Equation 72 and center of Figure 71, these stabilize the energy of the singlets relative to the triplets leading to net antiferromagnetism. The MMCT coupling into the ground state is the key component of the Anderson theory of superexchange.[129]

There are a number of important points to observe from the resultant VBCI energy diagram in Figure 71.[128] First there is a large stabilization of the charge transfer singlets relative to the triplets. This is excited state antiferromagnetism (ESAF) and indicates that spin allowed singlet charge transfer transitions of bridging ligands can be stabilized to very low energy, resulting in large spectral changes in dimers relative to monomers. Second the VBCI model shows that the ground state antiferromagnetism (GSAF) derives from configurational interaction mixing of the bridging ligand charge transfer excited state into the ground state. Therefore if $\lambda$ is the amount of charge transfer character mixed into the ground state due to covalency, the GSAF coupling $[J(GS)]$ can be related to the charge transfer ESAF $[J(CT)]$ through Equation 73. Since $\lambda^2 \sim 0.1$, ESAF can be orders of magnitude larger than GSAF.

$$-2J^{GS} = \lambda^2(-2J^{CT}) \tag{73}$$

It should further be noted from Figure 71 that there is a configurational interaction splitting of the $\Gamma_+\Gamma_-$ states of the charge transfer excited configuration. It has been shown that this splitting can be related to the HOMO/LUMO splittings of the $\phi_1/\phi_2$ molecular orbitals in Figure 18. Therefore this HOMO/LUMO splitting can be experimentally estimated from the splitting in the LMCT spectrum. Finally, for binuclear metal complexes with $S_a, S_b > 1/2$, one can observe singlet/triplet and $+/-$ splittings for a number of different bridging ligand-to-metal charge transfer excited states, and these can be used to probe the relative contributions of different superexchange pathways to ground state antiferromagnetic coupling. This is described in Reference 68b.

## 6 EXCITED STATE METHODS—X-RAY ABSORPTION SPECTROSCOPY (XAS)

In this section we consider high-energy transitions. Figure 72 presents an extended energy level diagram using $CuCl_4^{2-}$ as an example. In addition to the transitions within the valence orbitals discussed in Sections 4 and 5, there are also high-energy transitions from the core levels. These can be on the ligand or the metal and involve excitation of electrons from the $1s$ ($K$ shell) or $2p$ ($L$ shell) orbitals to either unoccupied bound valence orbitals or into the continuum (involving ionization of the electron with $E > 0$). Figure 73 shows a representative X-ray absorption spectrum which is often considered in four energy regions. The lowest energy *pre-edge* region of the XAS spectrum exhibits discrete peaks which correspond to transitions to bound states. The sharp rising *edge* corresponds to ionization into the continuum and its inflection point can be used to estimate the core binding energy. These provide geometric and electronic structural insight into active sites and will be the focus of this section. The *near-edge* region in Figure 73 contains complex structure known as shape resonances which involve partial localization of the ionized electron due to the potentials of the adjacent atoms and can be analyzed to obtain structural insight through multiple scattering theory. Finally at energies $> \sim 100$ eV above the edge is the extended X-ray absorption fine structure or EXAFS region which provides detailed structural information and is developed in Chapter 9, Volume I, this series.

XAS involves taking an absorption spectrum at energies >500 eV which requires synchrotron radiation. This is produced by electrons at relativistic energies moving in an ultra-high vacuum (UHV $< 10^{-9}$ torr) storage ring in paths bent by a magnetic field. These give off an intense beam of photons which are continuous in energy (particularly important in the X-ray region, Figure 74B) and linearly polarized in the orbital plane (Figure 74A). The absorption of these photons by the sample is either detected directly or more routinely through fluorescence excitation. The information content of metal and ligand K- and L-edges is systematically summarized below.

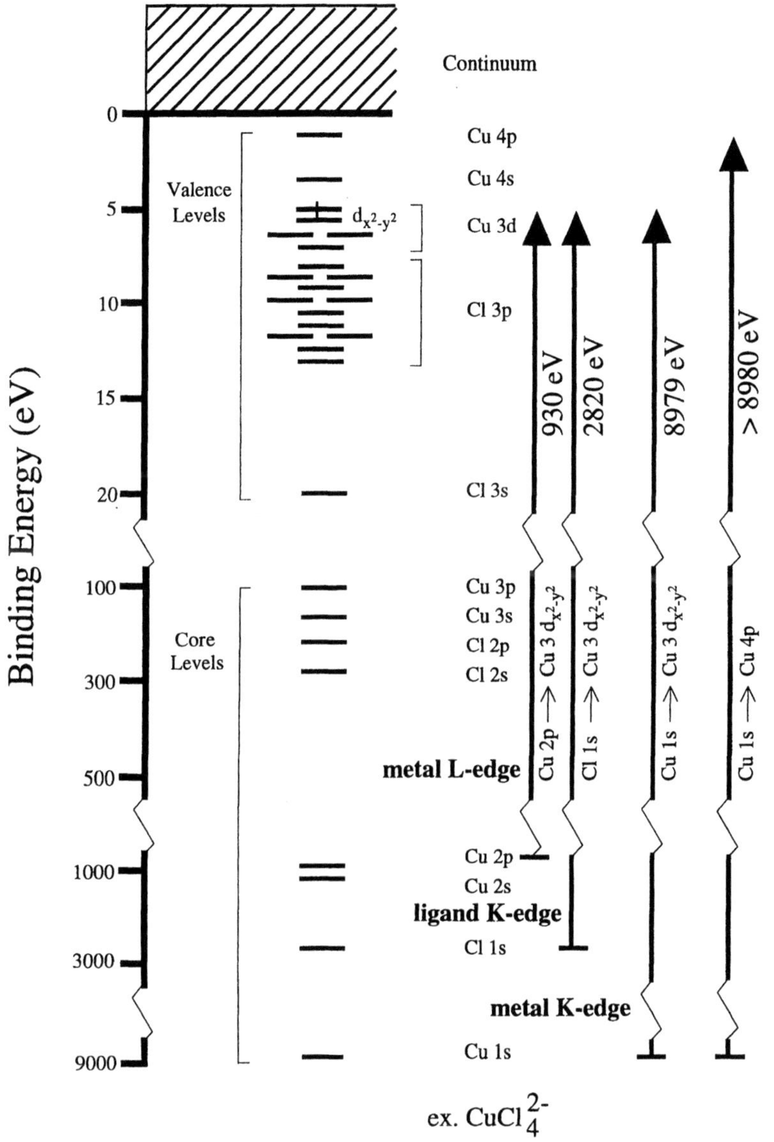


**Figure 72.** Extended energy level diagram of $CuCl_4^{2-}$ showing X-ray absorption transitions of core electrons into valence levels and the continuum.

## 6.1 Metal K-Edges

**6.1.1** ***Cu(I)*** $\boldsymbol{d^{10}}$ This is the most generally studied region in XAS, and XAS is one of the few spectroscopic methods appropriate for the study of closed shell metal ions [see also photoelectron spectroscopy (PES) in References 10b, 106, and Chap-

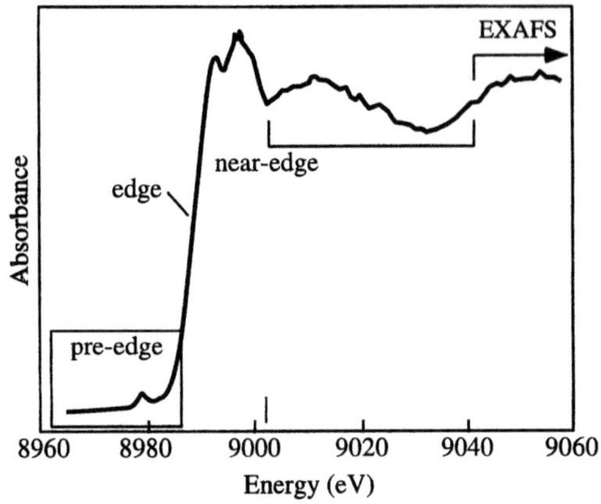


**Figure 73.** Representative XAS spectrum showing the principle energy regions, the pre-edge, edge, near-edge, and the EXAFS. XAS spectrum is of $CuCl_4^{2-}$.

ter 8, Volume I, this series]. From Figure 73, the K-edge of copper complexes is at ~8,990 eV. The energy of the edge corresponds to the threshold of ionization of the Cu 1s electron and is thus dependent on the effective nuclear charge, $Z_{eff}$, on the copper. The edge energy gives the *chemical shift* of the core level and is affected by the metal ion oxidation state and to a more limited extent by the number and types (i.e., covalencies) of the coordinated ligands.

Cu(I) has a $d^{10}$ configuration, thus the lowest energy pre-edge features on the K-edge will be the $1s \rightarrow 4s$ and $4p$ transitions (Figure 72) where only the transition to the $4p$ is electric dipole allowed and thus generates an intense feature in the pre-edge region. Figure 75 shows the pre-edge region of representative members of a series of Cu(I) complexes with different coordination numbers.[130] These spectra illustrate the fact that the pre-edge is very sensitive to geometry and reflects the ligand field splitting of the $4p$ orbitals. From Figure 75A, linear two coordinate Cu(I) sites will have a large repulsive interaction along the z axis destabilizing the $4p_z$ level. This results in low energy $1s \rightarrow 4p_{x,y}$ transitions which produce an intense pre-edge peak in the XAS at 8,984 eV. Addition of a third ligand splits the degeneracy of the $4p_x$ and $4p_y$ levels and thus the XAS of three coordinate Cu(I) shows a weaker double peaked feature in the 8,984 eV region. Finally, four coordinate distorted tetrahedral Cu(I) has all $4p$ orbitals raised in energy by repulsive interactions with the ligands and shows a broad absorption feature at energies >8,985 eV.

**6.1.2** ***Cu(II) $d^9$*** In Cu(II) complexes, there is a hole in the $d_{x^2-y^2}$ orbital which generates a low energy $1s \rightarrow 3d$ transition (Figure 72). From Figure 76B this is observed as a weak peak at 8,979 eV. The $1s \rightarrow 3d$ transition is electric dipole forbidden. The spectrum in Figure 76B is of $D_{4h}CuCl_4^{2-}$, which is centrosymmetric and thus can have no $4p$ mixing into the $3d$ orbitals (*vide infra*), yet it shows some limited absorption intensity in this transition. The origin of this intensity was determined from polarized XAS.[131] Spectra were obtained with the polarization and propagation

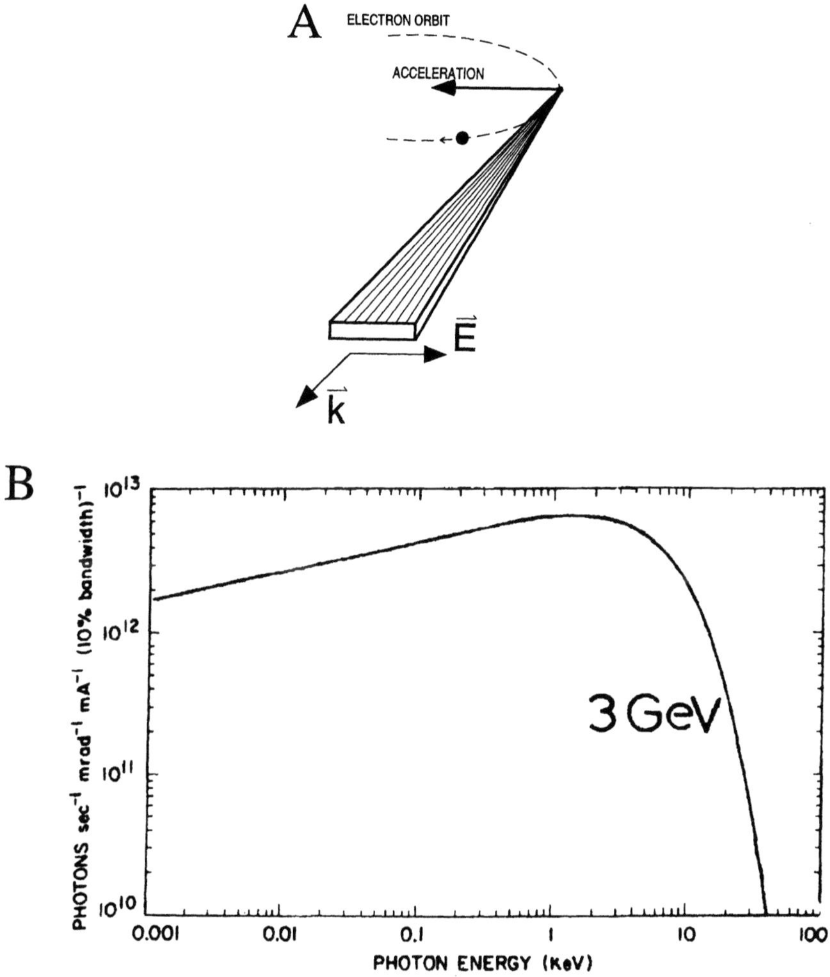


**Figure 74.** Synchrotron radiation. A) Propagation and polarization direction; and B) energy distribution of intensity of X-ray beam from Stanford Synchrotron Radiation Laboratory (SSRL).

directions shown in Figure 76A, and a single crystal of cupric chloride was rotated around its $C_4$ axis (by an angle $\phi$). As shown in the inset in Figure 76B, the 8,979 eV peak is polarized, and the intensity of this peak as a function of $\phi$ is plotted in Figure 76C. Importantly, the intensity maximizes every 90° which demonstrates that it is an electric quadrupole allowed transition. (An electric dipole allowed transition would show a 180° intensity pattern.) The electric quadrupole transition moment integral is given in Equation 74 where $\vec{k}$ is the propagation direction of light, $\vec{p}$ is the electron momentum operator and $\vec{r}$ operates on the electron coordinates. From Equation 74 and Figure 76A where $\vec{E}$ defines the $x$ axis and $\vec{k}$ the $y$ axis, the electric dipole mo-

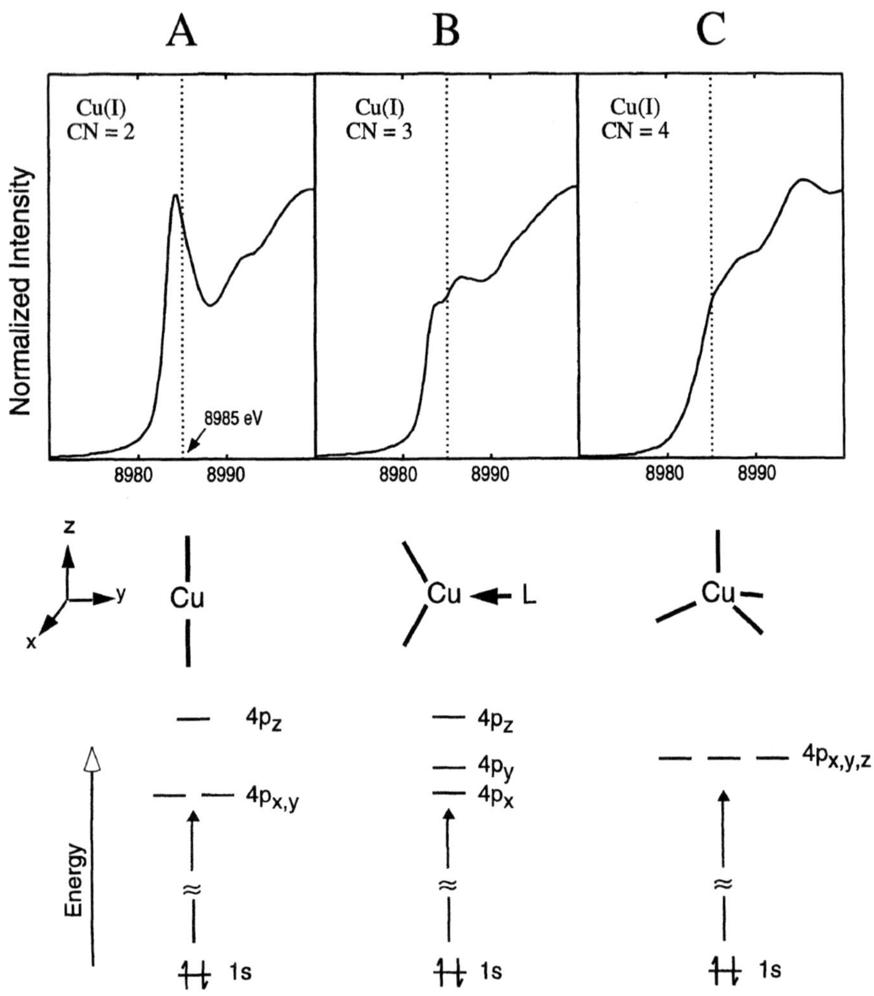


**Figure 75.** Pre-edge features of a series of Cu(I) complexes of varying geometry. A) Linear two-coordinate; B) three-coordinate; and C) four-coordinate distorted tetrahedral. Resulting $4p$ orbital energy splittings by those ligand fields are shown at the bottom with $1s \rightarrow 4p$ transitions indicated. Adapted from Reference 100.

ment transforms as $M_{(xy)}$ and the transition moment is non-zero for $\Psi_e = d_{x^2-y^2}$ when $\vec{k}$ bisects adjacent Cl–Cu bonds ($\phi = 45°$ in Figure 76C).

$$I \propto \langle 1s|(\vec{E} \bullet \vec{p})(\vec{k} \bullet \vec{r})|\Psi_e\rangle \tag{74}$$

The electric quadrupole contribution to absorption intensity is usually very small, but at ~9,000 eV the wavelength of the X-ray is ~1.4 Å and higher terms in the multipole expansion (Equation 48) become important.

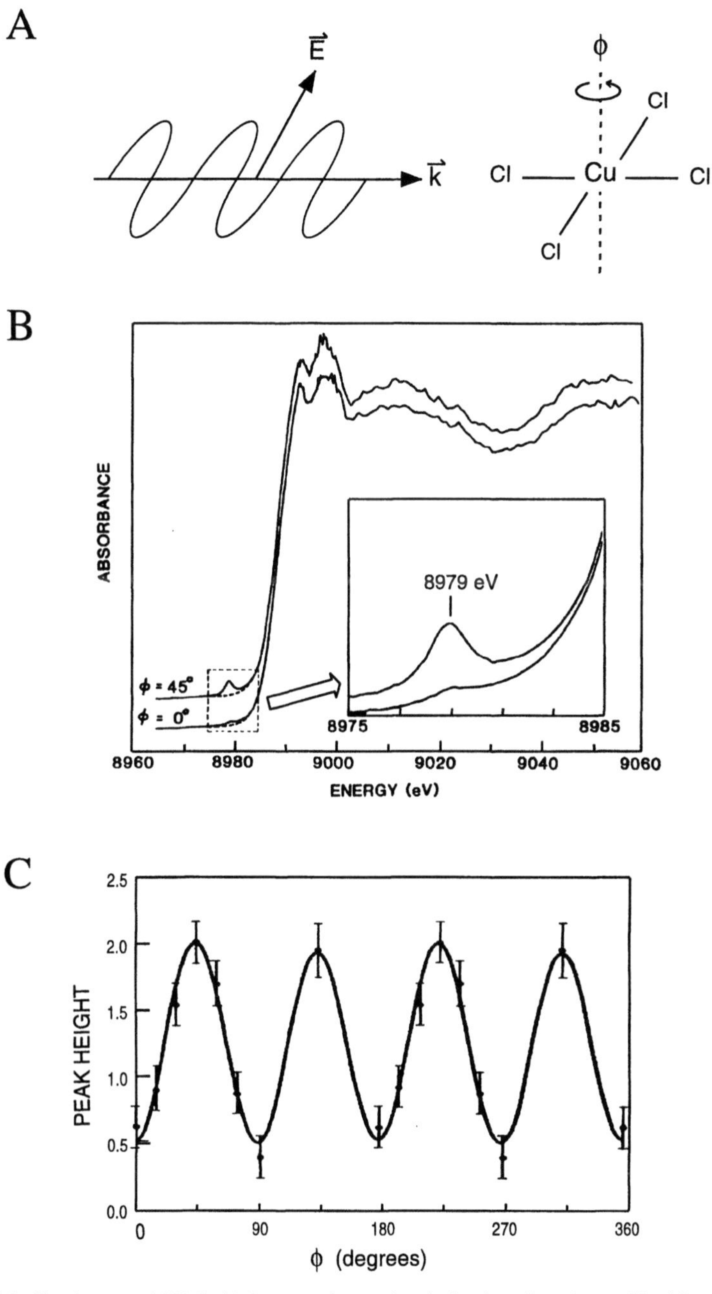


**Figure 76.** Single crystal XAS. A) Propagation and polarization directions of incident X-rays relative to $CuCl_4^{2-}$ single crystal which was rotated about its $C_4$ axis by an angle $\phi$; B) polarized K-edge XAS spectra of $D_{4h}$ $CuCl_4^{2-}$. Inset shows polarization of the pre-edge feature of 8,979 eV; and C) intensity of 8,979 eV peak plotted as a function of rotation angle $\phi$. Adapted from Reference 2.

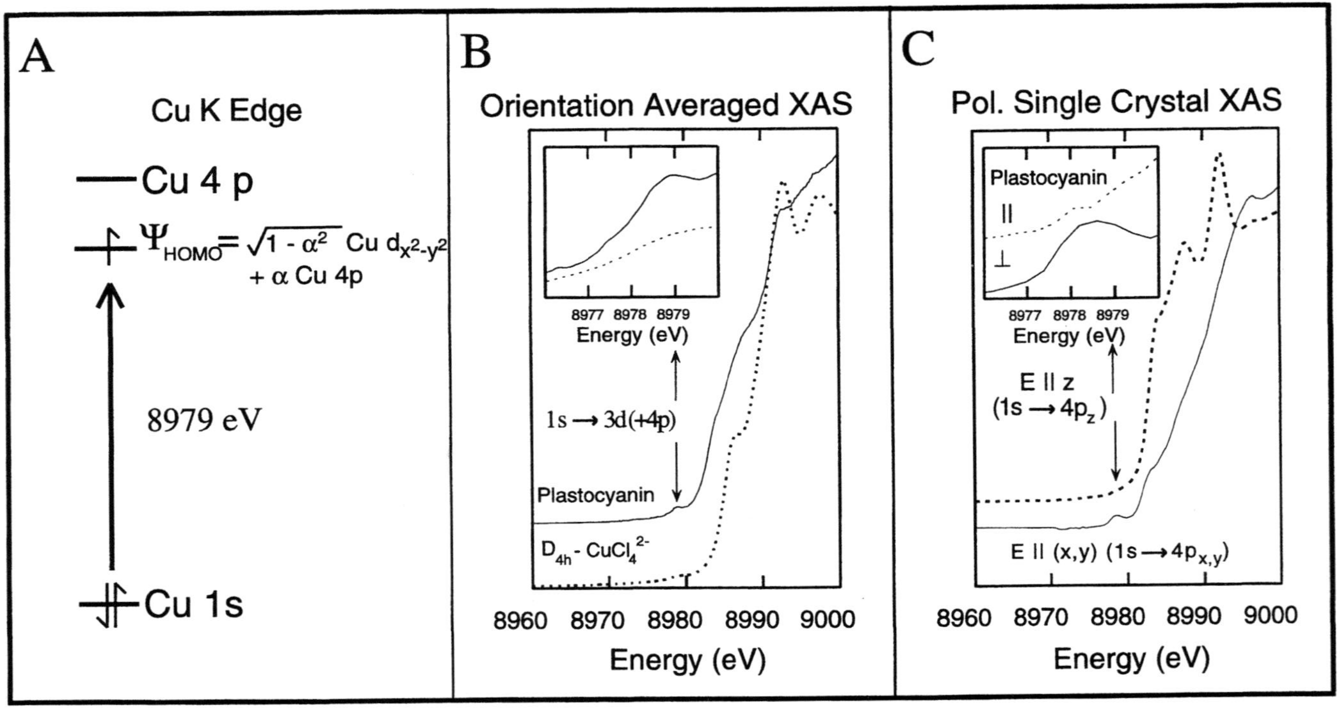


**Figure 77.** K-edge XAS of non-centrosymmetric Cu(II) sites. A) $4p$ mixing into Cu(II) $d_{x^2-y^2}$ HOMO; B) spectrum of plastocyanin compared to that of normal Cu(II), $CuCl_4^{2-}$, insert shows 8,979 eV peak to be much more intense in blue copper; and C) polarization of the 8,979 eV peak in a single crystal XAS experiment. Adapted from Reference 131.

For Cu(II) in non-centrosymmetric sites there can be $4p$ mixing into the $d_{x^2-y^2}$ orbital which will contribute electric dipole character to the $1s \rightarrow$ "$3d$" transition, where "$3d$" now corresponds to the mixed HOMO in Figure 77A. It has been found that the electric dipole intensity for a $1s \rightarrow 4p$ transition is approximately 100 times the electric quadrupole intensity for a $1s \rightarrow 3d$ transition.[23,123] Therefore, a limited amount of $4p$ mixing into the $3d$ level can have a large effect on the 8,979 eV pre-edge feature. Figure 77B compares the pre-edge of the blue copper site in plastocyanin to that of $D_{4h}CuCl_4^{2-}$.[2] The intensity of the 8,979 eV peak for the blue copper site is observed to increase by greater than a factor of 2, indicating $4p$ mixing. It was important to define the nature of the $4p$ orbital mixed into the $3d_{x^2-y^2}$ level to understand the origin of small parallel hyperfine couplings in the EPR spectrum of blue copper centers (see Section 2.2.1). This was accomplished through the analysis[23] of polarized single crystal XAS data.[133] From Figure 77C, the 8,979 eV peak is strongly polarized $\vec{E} \| x, y$, which shows that the $p_x$, $p_y$ orbitals are mixed into the $d_{x^2-y^2}$ orbital and thus $4p_z/d_{x^2-y^2}$ mixing is not the origin of the small $A_\|$ in blue copper sites. The small parallel hyperfine coupling is now understood to relate to a highly covalent thiolate–Cu(II) bond (*vide supra*).[14] The important point here is that a significant increase in the absorption intensity of the $1s \rightarrow$ "$3d$" pre-edge feature is associated with $4p$ mixing which can only occur upon distortion from a centrosymmetric environment.

**6.1.3** ***Fe(II) $d^6$, Fe(III) $d^5$*** The ferrous and ferric $1s \rightarrow 3d$ pre-edge features are at approximately 7,111 and 7,112 eV, respectively. For high-spin iron sites the pre-edge analysis increases in complexity in that more than one $d$ orbital is half-occupied. These $d$ orbitals can split in energy due to the ligand field, and the $1s \rightarrow 3d$ excitation produces $d^{n+1}$ final states which will undergo electron–electron repulsion that can be comparable in energy to the ligand field splitting of the $d$ orbitals.[134] For Fe(II) the high-spin $d^7$ free ion final states are $^4F$ and $^4P$ which split in an octahedral ligand field into $^4A_{2g}$, $^4T_{1g}$, $^4T_{2g}$, $^4T_{2g}$ states. (Note that these further couple to the hole in the $1s$ core, producing quintets and triplets with only the quintets being allowed for transitions from the quintet ground state.) The electric quadrupole intensity associated with the one electron allowed component of these transitions is given in the left side of Figure 78A, which also gives the multiplet splittings of these final states. For high-spin Fe(III) the allowed free ion $d^6$ final state is $^5D$ which splits into $^5T_{2g}$ and $^5E_g$ states separated by $10\,Dq$ in an octahedral ligand field. The quadrupole allowed intensity for these final states is given in the right side of Figure 78A. Distortion to five coordinate square pyramidal or four coordinate tetrahedral changes the ligand field splitting of the $d^{n+1}$ final states. In addition, as described above, $4p$ mixing into the $3d$ orbitals becomes allowed.[135] For tetrahedral geometry the $4p$ set mixes into the $t_2$ set of $d$ orbitals.[134] For square pyramidal it is found that the dominant mixing involves the $4p_z$ into the $d_{z^2}$ orbital, the energy of which is significantly affected by the strength of the axial ligand field. These provide an additional electric dipole contribution to the intensities of pre-edge features in ferrous and ferric sites. From Figure 78 B and C, left and right, it is important to note that this electric dipole mixing is selective for specific final states which have the highest $d_{t_2}$

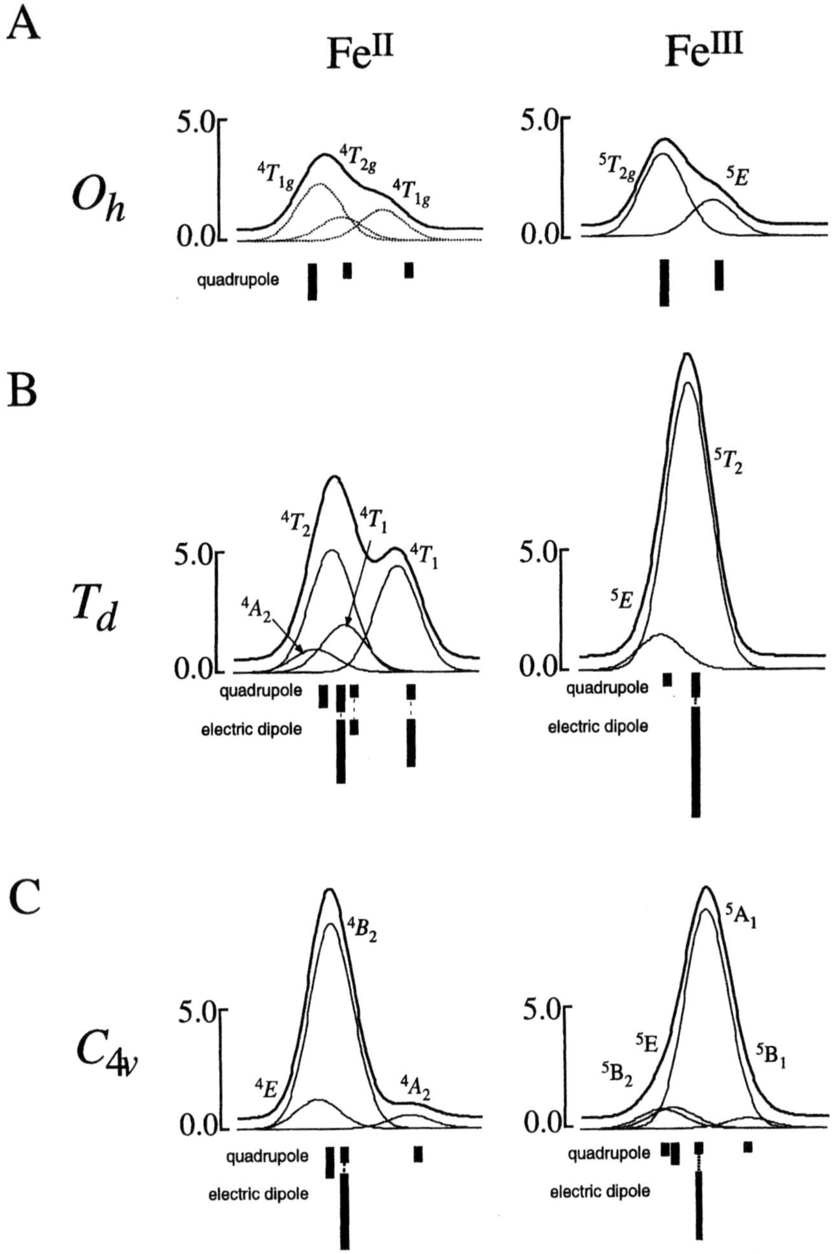


**Figure 78.** Ferrous (left) and ferric (right) pre-edge features in the Fe K-edge including multiplet splittings (light lines) and quadrupole/electric dipole contributions for various geometries (relative contributions of which are shown at the bottom of each spectrum): A) octahedral; B) tetrahedral; and C) square pyramidal.

(for tetrahedral) or $d_{z^2}$ (for square pyramidal) character. Therefore the $1s \rightarrow$ "$3d$" pre-edge feature in Fe(II) and Fe(III) sites will in general be more intense for five and four coordinate geometries and the intensity distribution over energy will be given by the nature of the $4p$ mixing into the $d^{n+1}$ final states which reflects the specific geometric structure.

## 6.2 Metal L-Edges

L-edge absorption features involve transitions from the $(2p)^6$ core to the metal $3d$ and $4s$ levels (Figure 72). Both types of transitions are electric dipole allowed with the $p \rightarrow d$ being much more intense (i.e., the branching ratio between channels is large). Therefore L-edges provide an effective way to study the $d$ levels relative to K-edge XAS where these transitions are electric dipole forbidden. Also the L-edge has a four fold resolution advantage (~0.3 eV) relative to the K-edge (~1.2 eV). However, there is a major difficulty in L-edge spectroscopy in that the transitions are in the 400–1,100 eV range and thus the experiment must be performed in ultra-high vacuum (UHV). The $(2p)^5$ excited core configuration produces a $^2P$ final state. This has orbital as well as spin angular momentum and these spin-orbit couple to produce $^2P_{3/2}$ and $^2P_{1/2}$ final states split by 10–20 eV. The $L_3$ ($^2P_{3/2}$) edge feature is at a lower energy and more intense than the $L_2$ ($^2P_{1/2}$) edge peak.

**6.2.1** ***Cu(I) $d^{10}$, Cu(II) $d^9$*** Cu(I) complexes exhibit a very weak pre-edge feature associated with its $2p \rightarrow 4s$ transition (dotted spectrum in Figure 79B). Alternatively, Cu(II) complexes exhibit an intense $L_3$ edge feature at ~930 eV corresponding to the $(3p)^6(3d)^9 \rightarrow (3p)^5(3d)^{10}$ excited state.[24] The half-occupied $d$ orbital involved in the transition, $\psi^*_{\mathrm{HOMO}}$ in Figure 79A, is a molecular orbital with ligand [S(Cys) for the blue copper site as an example] as well as metal character. Since the $2p$ orbital is localized on the metal, only the $2p \rightarrow 3d$ component of the transition, $(1-\alpha^2)$, can contribute electric dipole intensity to the pre-edge transition. Therefore, the intensity of the L-edge reflects the amount of metal $d$ character in the highest half occupied molecular orbital. Figure 79B compares the $L_3$ edge of the blue copper site to that of $D_{4h}CuCl_4^{2-}$ which is well defined to have 61% Cu $d_{x^2-y^2}$ character in its half occupied molecular orbital. Clearly the blue copper site is more covalent, and from the ratio of intensity of the L-edge in plastocyanin to that of $CuCl_4^{2-}$, the blue copper site is estimated to have only 41% $d_{x^2-y^2}$ character in its redox active orbital.

**6.2.2** ***Fe(II) $d^6$, Fe(III) $d^5$*** For these metal ions there is more than one hole in the $d$ orbitals and thus the number of $(2p)^5(3d)^{n+1}$ final states greatly increases. The energy splittings and intensity distribution over these states is given by a multiplet calculation which includes the ligand field splitting of the $d$ orbitals, the $3d-3d$ and $3d-2p$ electron–electron repulsion, and the $d$ and $p$ spin-orbit coupling.[136] The energy/intensity distribution of the $d^{n+1}$ final states for the oxidized and reduced Fe site in rubredoxin are given at the bottom of Figure 80 A and B.[137] These are lifetime and monochrometer broadened to generate the solid line calculated spectrum in the middle of each figure which also shows the effects of ligand field splittings on these

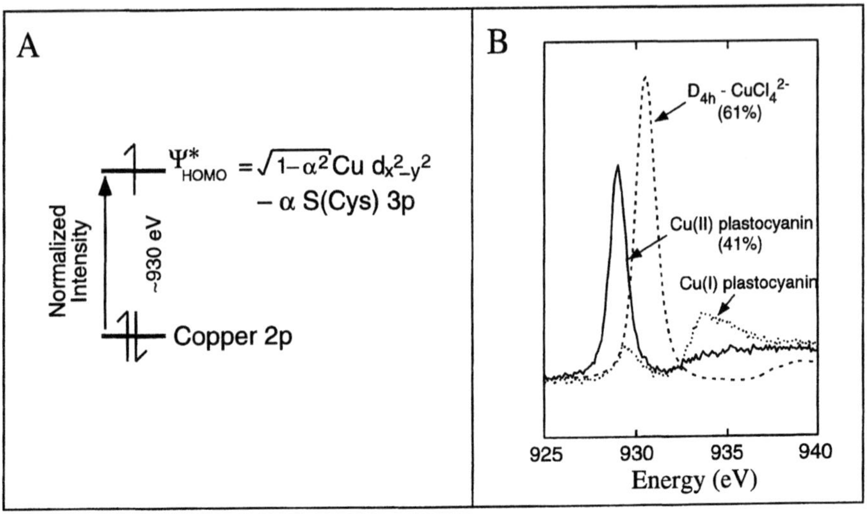


**Figure 79.** L-edge XAS of Cu(I) and Cu(II). A) L-edge transition for Cu(II) involves the half-occupied HOMO with mixed ligand and metal character; and B) comparison of $L_3$ spectra of Cu(I) and Cu(II) forms of plastocyanin and $D_{4h}$ $CuCl_4^{2-}$ showing differing L-edge intensities corresponding to variation in covalency of the Cu(II) sites. Adapted from Reference 133.

multiplet patterns. From Figure 80, the shape of the L-edge is dependent on both the ligand field at the iron site and its oxidation state.

Further insight into these multiplets is starting to be derived from MCD studies at the L-edge (XMCD). As described in Section 4.1.3, paramagnetic metal ions exhibit C-term MCD intensity which increases at low temperature. Figure 81 shows the low temperature XMCD spectrum of ferric rubredoxin.[138] The information content of these type of spectra is now being developed. One interesting approach is application of the XMCD sum rule[139] [Equation 75, where $I_0 = (I_+ + I_-)/2$, $l^n$ is the ground configuration and $c$ denotes a core hole]. This relates the low temperature MCD signal integrated over the total L-edge to the orbital angular momentum of the ground state.

$$\langle L_z \rangle = \frac{2l(l+1)(4l+2-n)}{c(c+1)-l(l+1)-2} \frac{\int_{L_2+L_3}[I_+(\nu) - I_-(\nu)d\nu]}{\int_{L_2+L_3}[I_+(\nu) + I_0(\nu) + I_-(\nu)d\nu]} \qquad (75)$$

This quantity also contributes to the $g$ value deviation from 2.0023 observed in EPR spectroscopy (see Section 2.1.1.2), and it is interesting to compare the numbers obtained by the two methods on the same system.[140] Generally, it will be important to include covalency effects as described above for Cu(II) L-edges into the multiplet calculations for L-edge absorption (Figure 80 A and B, bottom) and XMCD (Figure 81C) spectra for metal ions with more than one hole to estimate the relative covalencies of the partially occupied $d$ orbitals which contribute to the multiplets.

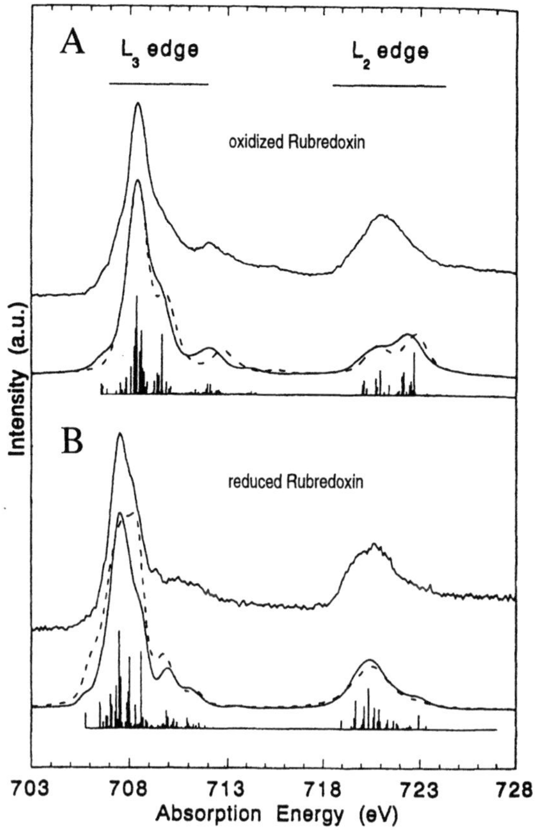


**Figure 80.** $L_2$ and $L_3$ edge XAS of the Fe site of rubredoxin in its A) oxidized and B) reduced forms. Top of each: experimental spectrum. Bottom: results of multiplet calculations (lines) with lifetime broadening and effects of ligand field on spectra (solid spectrum) and without the ligand field added (dotted spectrum). Taken from Reference 135.

## 6.3 Ligand K-Edges

This method provides a most powerful probe of the covalency of ligand–metal bonds.[23,141] For metalloproteins, the ligands of interest have sulfur coordinated to metal ions producing a S K-pre-edge feature at ~2,470 eV (or Cl K-edges with a pre-edge feature at ~2,820 eV). For model complexes the oxygen K-edge at ~530 eV is also accessible in a UHV experiment. The sulfur pre-edge peak corresponds to the $S1s \rightarrow M3d$ transition and thus is only observed for open shell metal ions. Since the $1s$ orbital is localized on the sulfur atom, the only electric dipole allowed component will be $S(1s) \rightarrow S(3p)$. Therefore the intensity of the sulfur pre-edge probes the amount of $S(3p)$ character mixed into the half-occupied or unoccupied metal $3d$ orbitals due to bonding. *Ligand K-edge spectroscopy provides a direct probe of the covalency of the ligand–metal bond.* This parallels superhyperfine structure in

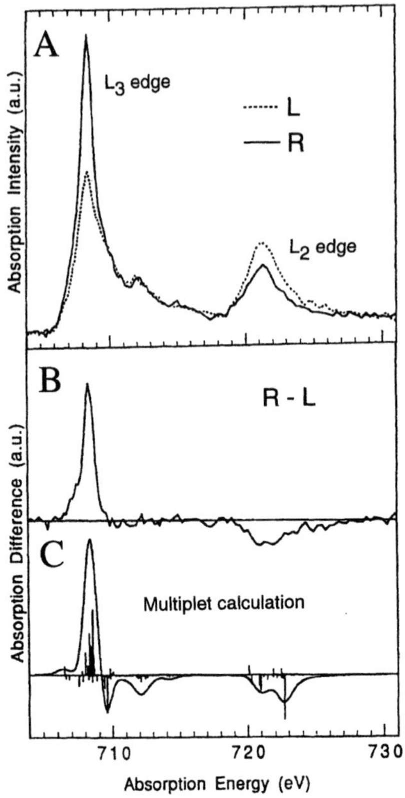


**Figure 81.** Low temperature XMCD spectra of ferric rubredoxin. A) Absorption of left and right circularly polarized soft X-rays; B) difference spectrum; and C) spectrum resulting from multiplet calculations. Taken from Reference 136.

EPR (Section 2.1.1.3) but has the advantages that the site does not have to be EPR detectable, and the edge probes the unoccupied as well as half occupied orbitals.

**6.3.1** ***Ligands Bound to Copper Sites*** The reduced $d^{10}$ blue copper center exhibits no Cys S pre-edge feature. The oxidized site exhibits an extremely intense pre-edge feature at 2,469 eV (Figure 82B).[23] As illustrated in Figure 82A, this corresponds to the $S(1s) \rightarrow$ half occupied $\psi^*_{\text{HOMO}}$ transition with the intensity reflecting $\alpha^2$ the amount of Cys S character mixed into the $d_{x^2-y^2}$ orbital due to bonding at the blue copper site. From Figure 82B, the pre-edge feature in plastocyanin is approximately 2.5 times more intense than that in tet b, a model complex which has a normal thiolate S–Cu bond with ~15% covalency. The intensity ratio in Figure 82B then quantitates the thiolate S character in the redox active orbital of the blue copper site at 38%. This high anisotropic covalency activates cysteine as a very efficient superexchange pathway for rapid electron transfer.[25]

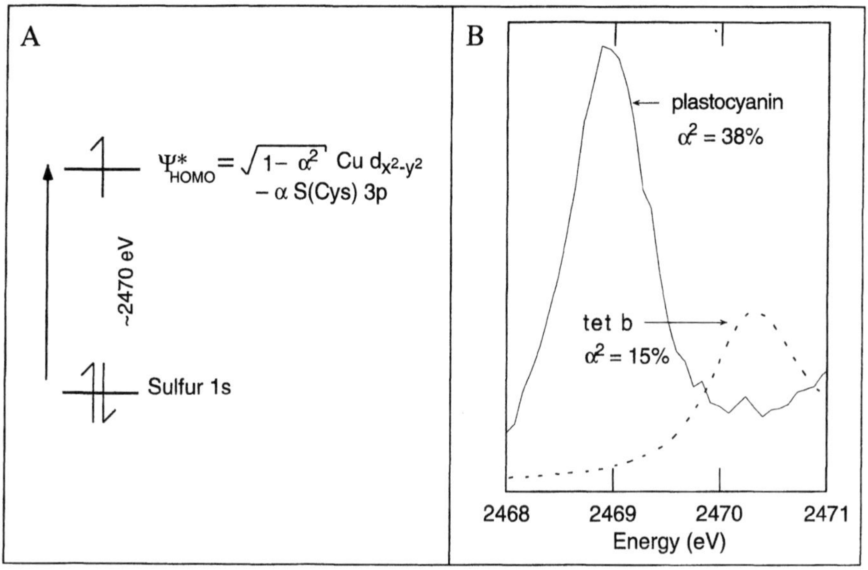


**Figure 82.** Sulfur ligand K-edge XAS. A) Transition from S 1*s* orbital to half-filled Cu(II) HOMO, which has both ligand [S(Cys) for blue copper example] and metal character; and B) pre-edge features of plastocyanin and a Cu(II) model complex, tet b, which has ~15% thiolate character. Adapted from Reference 23.

**6.3.2** ***Ligands Bound to Iron Sites*** Here we consider the Cl K-edge at 2,820 eV of $T_d$–$FeCl_4^-$ $d^5$ as a completed case study[141b] which has now been extended to the S K-edge of the $Fe(SR)_4^-$ site in rubredoxin.[142a] From Figure 83A, high-spin $d^5$ again has more than one, one electron transition possible which generate $d^{n+1}$ final states. These will undergo multiplet splittings due to a combination of electron–electron repulsion and ligand field effects. Further, the $d_e$ orbitals undergo $\pi$ antibonding interactions with the ligands while the $d_{t_2}$ set dominantly undergo $\sigma$ antibonding interactions with the chlorides. This introduces $\alpha_{t_2}^2$ and $\alpha_e^2$ ligand character into the $d$ orbitals which is distributed over the $d^{n+1}$ final states to generate the ligand pre-edge intensity/energy distribution. The analysis has been developed for $d^5$ to $d^9$ metal ions in Reference 141b. As shown in Figure 83B, high-spin Fe(III) exhibits a chloride pre-edge feature at 2,821 eV which is well separated from the edge jump. Contained in this band are two $d^6$ final states, the $^5E$ which corresponds to the $1s \rightarrow e(\pi)$ transition and the $^5T_2$ from the $1s \rightarrow t_2$ ($\sigma$) transition, split by 10 $Dq$ (in the final state). The intensity of the two peaks weighted by orbital degeneracy quantitates the $\pi$ and $\sigma$ covalencies of the ferric site. This methodology is now being extended to the iron sulfur proteins, where it should provide a direct probe of active site covalency and how this is affected by hydrogen bonding and bridging interactions within the clusters.[142]

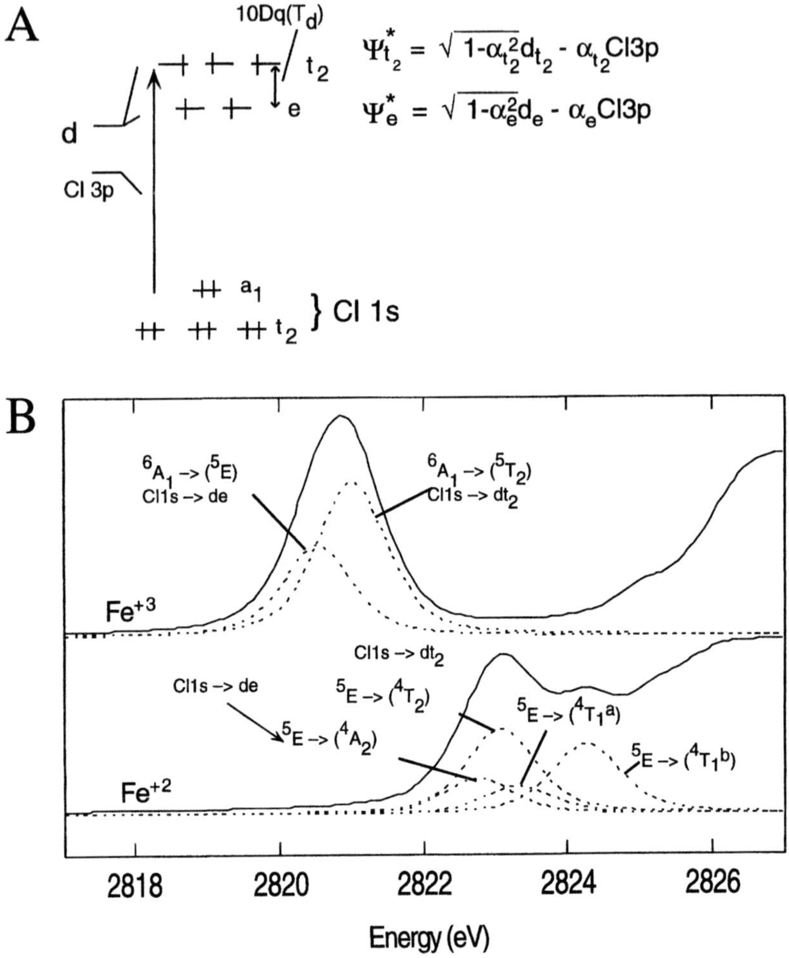


**Figure 83.** Chloride ligand K-edge XAS for $T_d$–$FeCl_4^{2-}$. A) One electron transitions with wavefunctions indicating differences in $\sigma$ and $\pi$ covelencies; and B) pre-edge features and multiplet splittings for ferric (top) and ferrous (bottom) complexes. Adapted from Reference 141b.

## 7 CONCLUDING REMARKS

This review has developed the information content of the different spectroscopic methods in Bioinorganic Chemistry in terms of spectroscopic parameters which in turn have been related to active site geometric and electronic structure. It has further been emphasized that protein sites which exhibit unique spectral features relative to small molecule inorganic complexes of the same metal ion have novel active site electronic and geometric structures which can make key contributions to biological function.

An important goal of spectroscopy in Bioinorganic Chemistry has been to develop effective working models from metalloprotein active sites when direct structural insight is not available. As summarized in Reference 1, the field of Bioinorganic Chemistry is now at the point where many of the basic metalloprotein active site structural motifs are defined at a reasonable resolution through protein crystallography. It is important to recognize that the availability of an active site structure enables an in-depth correlation with spectroscopy to fundamentally understand function. First, spectroscopy combined with electronic structure calculations allow an extension of the resting structure to intermediates to define reaction mechanisms on a molecular level. Further, a correlation of unique spectral features with specific structural features allows the definition of novel electronic structures and an evaluation of their contributions to reactivity. Finally, these novel electronic structures derive from the intimate interactions between the metal ion and the protein ligand which can be evaluated through spectroscopy to determine how the protein activates the metal ion for catalysis.

## ACKNOWLEDGMENTS

This work was supported by the National Science Foundation (grants MCB 93-16768 and CHE 95-28250) and the National Institute of Health (grants GM 40392 and DK 31450). E.I.S. expresses his sincere appreciation to all the students and collaborators listed as co-authors in the literature cited for their commitment and contributions to this research.

## REFERENCES

1. Holm, R.H.; Kennepohl, P.; Solomon, E.I. *Chem. Rev.* **1996**, *96*, 2239–2314.
2. Solomon, E.I.; Penfield, K.W.; Gewirth, A.A.; Lowery, M.D.; Shadle, S.E.; Guckert, J.A.; LaCroix, L.B. *Inorg. Chim. Acta* **1996**, *243*, 67–78.
3. Solomon, E.I.; Sundaram, U.M.; Machonkin, T.E. *Chem. Rev.* **1996**, *96*, 2563–2605.
4. (a) Neese, F.; Zumft, W.G.; Anthroline, W.E.; Kroneck, P.M.H. *J. Am. Chem. Soc.* **1996**, *118*, 8692–8699. (b) Wilmanns, M.; Lappalainen, P.; Kelly, M.; Sauer-Eriksson, E.; Saraste, M. *Proc. Natl. Acad. Sci. USA* **1995**, *92*, 11955–11959.
5. Ohlendorf, D.H.; Orville, A.M.; Lipscomb, J.D. *J. Mol. Biol.* **1994**, *244*, 586–608.
6. Waller, B.J.; Lipscomb, J.D. *Chem. Rev.* **1996**, *96*, 2625–2657.
7. Cammack, R. In *Iron–Sulfur Proteins*; Sykes, A.G., Ed.; Academic Press, Inc.: New York, **1992**; Vol. 38, p. 487.
8. Westmoreland, T.D.; Wilcox, D.E.; Baldwin, M.J.; Mims, W.B.; Solomon, E.I. *J. Am. Chem. Soc.* **1989**, *111*, 6106–6123.
9. Ballhausen, C.J. *Introduction to Ligand Field Theory*; McGraw-Hill Book Company, Inc.: New York, **1962**.
10. (a) Chow, C.; Chang, K.; Willett, R.D. *J. Chem. Phys.* **1973**, *59*, 2629–2640. (b) Solomon, E.I. *Comments Inorg. Chem.* **1984**, *3*, 225–320.

11. Gerloch, M.; Miller, J.R. *Prog. Inorg. Chem.* **1968**, *10*, 1–47.

12. McGarvey, B.R. In *Transition Metal Chemistry*; Carlin, R. L., Ed.; Marcel-Dekker, **1967**; Vol. 3, pp. 89–201.

13. (a) Abragam, A.; Pryce, M.H.L. *Proc. Roy. Soc. A.* **1951**, *206*, 164–172. (b) Abragam, A.; Pryce, M.H.L. *Proc. Roy. Soc. (London) A.* **1951**, *205*, 135–152.

14. Penfield, K.W.; Gewirth, A.A.; Solomon, E.I. *J. Am. Chem. Soc.* **1985**, *107*, 4519–4529.

15. Didziulis, S.V.; Cohen, S.L.; Gewirth, A.A.; Solomon, E.I. *J. Am. Chem. Soc.* **1988**, *110*, 250–268.

16. Goodman, B.A.; Raynor, J.B. *Advances in Inorganic Chemistry Radiochemistry* **1970**, *13*, 135–361.

17. (a) Feher, G.; Isaacson, R.A.; Scholes, C.P.; Nagel, R. *Annals New York Academy of Sciences* **1973**, 86–101. (b) Hoffman, B.M.; DeRose, V.J.; Doan, P.E.; Gurbiel, R.J.; Houseman, A.L.P.; Telser, J. In *Biological Magnetic Resonance*; Berliner, L.J., Reuben, J., Eds.; Plenum Press: New York, **1993**; Vol. 13, pp. 151–218. (c) Schweiger, A. *Struct. Bonding* **1982**, *51*, 1–122.

18. (a) Pilbrow, J.R. *Transition Ion Electron Paramagnetic Resonance*; Clarendon Press: Oxford, **1990**, pp. 440–480. (b) Mims, W.B.; Peisach, J. In *Biological Magnetic Resonance*; Berliner, L.J., Reuben, J., Eds.; Plenum Press: New York, **1981**; Vol. 3, pp. 213–263. (c) Mims, W.B.; Peisach, J. In *Advanced EPR Applications in Biology and Biochemistry*; Hoff, A.J., Ed.; Elsevier: New York, **1989**; pp. 1–57. (d) Mims, W.B.; Peisach, J. *Biochemistry* **1976**, *15*, 3863–3869. (e) Britt, R.D. *Curr.Opinion Struct. Biol.* **1993**, *3*, 774–779. (f) Thomann, H.; Mims, W.B. In *Pulsed Magnetic Resonance: NMR, ESR, and Optics*; Bagguley, D.M.S., Ed.; Clarendon Press: Oxford, **1992**; pp. 362–389.

19. (a) Edmonds, D.T. *Phys. Reports* **1977**, *29*, 233–290. (b) Hunt, M.J.; Mackay, A.L.; Edmonds, D.T. *Chem. Phys. Lett.* **1975**, *34*, 473–475.

20. (a) Blumberg, W.E.; Peisach, J. *Biochim. Biophys. Acta* **1966**, *126*, 269–273. (b) Malkin, R.; Malmström, B.G. *Advances in Enzymology* **1970**, *33*, 177. (c) Vänngård T. In *Biological Applications of Electron Spin Resonance*; Bolton, J.R., Schwartz, H.M., Borg, D.C., Eds.; Wiley-Interscience: New York, **1972**; Vol. 3, pp. 411.

21. Guss, J.M.; Bartunik, H.D.; Freeman, H.C. *Acta Cryst.* **1992**, *B48*, 790–811.

22. (a) Bates, C.A.; Moore, W.S.; Standley, K.J.; Stevens, K.W.H. *Proc. Phys. Soc.* **1962**, *79*, 73–83. (b) Roberts, J.E.; Brown, T.G.; Hoffman, B.M.; Peisach, J. *J. Am. Chem. Soc.* **1980**, *102*, 825–829.

23. Shadle, S.E.; Penner-Hahn, J.E.; Schugar, H.J.; Hedman, B.; Hodgson, K.O.; Solomon, E.I. *J. Am. Chem. Soc.* **1993**, *115*, 767–776.

24. George, S.J.; Lowery, M.D.; Solomon, E.I.; Cramer, S.P. *J. Am. Chem. Soc.* **1993**, *115*, 2968–2969.

25. Lowery, M.D.; Guckert, J.A.; Gebhard, M.S.; Solomon, E.I. *J. Am. Chem. Soc.* **1993**, *115*, 3012–3013.

26. (a) Solomon, E.I.; Dooley, D.M.; Wang, R.-H.; Gray, H.B.; Cerdonio, M.; Mogno, F.; Romani, G.L. *J. Am. Chem. Soc.* **1976**, *98*, 1029–1031. (b) Dooley, D.M.; Scott, R.A.; Ellinghaus, J.; Solomon, E.I.; Gray, H.B. *Proc. Natl. Acad. Sci. USA* **1978**, *75*, 3019. (c) Cole, J.L.; Clark, P.A.; Solomon, E.I. *J. Am. Chem. Soc.* **1990**, *112*, 9534.

27. Magnus, K.A.; Hazes, B.; Ton-That, H.; Bonaventura, C.; Bonaventura, J.; Hol, W.G.J. *Proteins: Structure, Function, and Genetics* **1994**, *19*, 302–309.

28. (a) Messerschmidt, A.; Rossi, A.; Ladenstein, R.; Huber, R.; Bolognesi, M.; Gatti, G.; Marchesini, A.; Petruzzelli, R.; Finazzi-Agró, A. *J. Mol. Biol.* **1989**, *206*, 513–529. (b) Messerschmidt, A.; Ladenstein, R.; Huber, R.; Bolognesi, M.; Avigliano, L.; Petruzzelli, R.; Rossi, A.; Finazzi-Agró, A. *J. Mol. Biol.* **1992**, *224*, 179–205.

29. Smith, T.D.; Pilbrow, J.R. *Coord. Chem. Rev.* **1974**, *13*, 173–278.

30. Goodenough, J.B. *Magnetism and the Chemical Bond*; John Wiley & Sons: New York, **1963**; Vol. I.

31. van Vleck, J.H. *The Theory of Electric and Magnetic Susceptibilities*; Oxford University Press: London, **1965**.

32. Hay, P. J.; Thibeault, J.C.; Hoffman, R. *J. Am. Chem. Soc.* **1975**, *97*, 4884–4899.

33. McKee, V.; Zvagulis, M.; Reed, C.A. *Inorg. Chem.* **1985**, *24*, 2914–2919.

34. (a) Day, E.P. *Methods Enzymol.* **1993**, *227*, 437–463. (b) Day, E.P.; Kent, T.A.; Lindahl, P.A.; Münck, E.; Orme-Johnson, W.H.; Roder, H.; Roy, A. *Biophys. J.* **1987**, *52*, 837–853. (c) Butzlaff, Ch.; Trautwein, A.X.; Winkler, H. *Methods Enzymol.* **1993**, *227*, 412.

35. (a) Figgis, B.N. *Introduction to Ligand Fields*; Interscience Publishers: New York, **1966**, pp. 248–292. (b) Figgis, B. N.; Lewis, J. *Prog. Inorg. Chem.* **1964**, *6*, 57–239. (c) Mabs, F.E.; Machin, D.J. *Magnetism and Transition Metal Complexes*; Halsted Press: New York, **1973**.

36. Bleaney, B.; Bowers, K.D. *Proceedings of the Royal Society of London Series A.* **1952**, *A214*, 451–465.

37. (a) Beinert, H.; Griffiths, D.E.; Wharton, D.C.; Sands, R.H. *J. Biol. Chem.* **1962**, *237*, 2337–2346. (b) Beinert, H.; Palmer, G. *J. Biol. Chem.* **1964**, *239*, 1221–1227.

38. Antholine, W.E.; Kastrau, D.H.W.; Steffens, G.C.M.; Buse, G.; Zumft, W.G.; Kroneck, P.M.H. *Eur. J. Biochem.* **1992**, *209*, 875–881.

39. (a) Froncisz, W.; Scholes, C.P.; Hyde, J.S.; Wei, Y.-H.; King, T.E.; Shaw, R.W.; Beinert, H. *J. Biol. Chem.* **1979**, *254*, 7482–7484. (b) von Wachenfeldt, C.; de Vries, S.; van Der Oost, J. *FEBS Lett.* **1994**, *340*, 109–113.

40. (a) Kroneck, P.M.H.; Antholine, W.E.; Reister, J.; Zumft, W.G. *FEBS Lett.* **1988**, *242*, 70–74. (b) Kroneck, P.M.H.; Antholine, W.E.; Kastrau, D.H.W.; Buse, G.; Steffens, G.M.; Zumft, W.G. *FEBS Lett.* **1990**, *268*, 274–276.

41. (a) Tsukihara, T.; Aoyama, H.; Yamashita, E.; Tomizaki, T.; Yamaguchi, H.; Shinzawa-Itoh, K.; Nakashima, R.; Yaono, R.; Yoshikawa, S. *Science* **1996**, *272*, 1136–1144. (b) Tsukihara, T.; Aoyama, H.; Yamashita, E.; Tomizaki, T.; Yamaguchi, H.; Shinzawa-Itoh, K.; Nakashima, R.; Yaono, R.; Yoshikawa, S. *Science* **1995**, *269*, 1069–1074. (c) Iwata, S.; Ostermeier, C.; Ludwig, B.; Michel, H. *Nature* **1995**, *376*, 660–669.

42. (a) Hyde, J.S.; Froncisz, W. *Ann. Rev. Biophys. Bioeng.* **1982**, *11*, 391–417. (b) Froncisz, W.; Hyde, J.S. *J. Chem. Phys.* **1980**, *73*, 3123–3131. (c) Hagen, W.R.; Hearshen, D.O.; Sands, R.H.; Dunham, W.R. *J. Magn. Reson.* **1985**, *61*, 220–232. (d) Pilbrow, J.R. *Transition Ion Paramagnetic Resonance*; Clarendon Press: Oxford, **1990**, pp. 44–55.

43. Robin, M.B.; Day, P. *Adv. Inorg. Chem. Radiochem.* **1967**, *10*, 247–403.

44. Williams, K.R.; Gamelin, D.R.; LaCroix, L.B.; Houser, R.P.; Tolman, W.B.; Mulder, T.C.; deVries, S.; Hedman, B.; Hodgson, K.O.; Solomon, E.I. *J. Am. Chem. Soc.* **1997**, *119*, 613–614.

45. (a) Hush, N.S. In *Mixed-Valence Compounds Theory and Applications in Chemistry, Physics, Geology, and Biology*; Brown, D.B., Ed.; D. Reidel Publishing Company: Boston, **1980**, pp. 151–188. (b) Wong, K.Y.; Schatz, P.N. *Prog. Inorg. Chem.* **1981**, *28*, 369–449. (c) Blondin, G.; Girerd, J.-J. *Chem. Rev.* **1990**, *90*, 1359–1376. (d) Gamelin, D.R.; Bominaar, E.L.; Kirk, M.L.; Weighardt, K.; Solomon, E.I. *J. Am. Chem. Soc.* **1996**, *118*, 8085–8097. (e) Gamelin, D.R.; Bominaar, E.L.; Mathonière, C.; Kirk, M.L.; Weighardt, K.; Girerd, J.-J.; Solomon, E.I. *Inorg. Chem.* **1996**, *35*, 4323–4335.

46. (a) Gibson, J.F. In *ESR and NMR of Paramagnetic Species in Biological and Related Systems*; Bertini, I., Drago, R.S., Eds.; D. Reidel Publishing Company: Boston, **1979**, pp. 225–253. (b) Aasa, R.; Malmström, B.G.; Saltman, P.; Vänngård, T. *Biochim. Biophys. Acta* **1963**, *75*, 203–222.

47. (a) Blumberg, W.E.; Peisach, J. In *Electron Spin Resonance and Nuclear Magnetic Resonance in Biology and Medicine and Magnetic Resonance in Biological Systems*; New York Academy of Sciences: New York, **1973**; Vol. 222, pp. 539–560. (b) Blumberg W.E. In *Magnetic Resonance in Biological Systems*; Ehrenberg, A., Malmström, B. G., Vänngård, T., Eds.; Pergamon Press: New York, **1967**; pp. 119–133. (c) Cammack, R.; Cooper, C.E. *Methods Enzymol.* **1993**, *227*, 353–384.

48. (a) Griffith, J.S. In *Quantum Aspects of Polypeptides and Polynucleotides*; Weissbluth, M., Ed.; Interscience Publishers, 1964; Vol. Bipolymers Symposia No. 1, **1964**, pp. 35–46. (b) Griffith, J.S. *Mol. Phys.* **1965**, *8*, 213–224.

49. (a) Deaton, J.C.; Gebhard, M.S.; Koch, S.A.; Millar, M.; Solomon, E.I. *J. Am. Chem. Soc.* **1988**, *110*, 6241–6243. (b) Deaton, J.C.; Gebhard, M.S.; Solomon, E.I. *Inorg. Chem.* **1989**, *28*, 877–889. (c) Gebhard, M.S.; Deaton, J.C.; Koch, S.A.; Millar, M.; Solomon, E.I. *J. Am. Chem. Soc.* **1990**, *112*, 2217–2231. (d) Zhang, Y.; Gan, Q.-F.; Pavel, E.G.; Sigal, E.; Solomon, E.I. *J. Am. Chem. Soc.* **1995**, *117*, 7422–7427. (e) Zhang, Y.; Gebhard, M.S.; Solomon, E.I. *J. Am. Chem. Soc.* **1991**, *113*, 5162–5175.

50. Kramers, H.A. *Proceedings of the Section of Sciences, Koninklifke Akad. van Wetenschappen, Amsterdam* **1930**, *XXXIII*, 962–965.

51. (a) Whittaker, J.W.; Solomon, E.I. *J. Am. Chem. Soc.* **1988**, *110*, 5329–5339. (b) Solomon, E.I.; Pavel, E.G.; Loeb, K.E.; Campochiaro, C. *Coord. Chem. Rev.* **1995**, *144*, 369–460.

52. (a) Stephens, P.J. *Adv. Chem. Phys.* **1976**, *35*, 197–264. (b) Schatz, P.N.; Mowery, R.L.; Krausz, E.R. *Mol. Phys.* **1978**, *35*, 1537–1557. (c) Thompson, A.J.; Johnson, M.K. *Biochem. J.* **1980**, *191*, 411–420. (d) Thompson, A.J.; Cheesman, M.R.; George, S.J. *Methods Enzymol.* **1993**, *226*, 199–232.

53. Debrunner, P.G. In *Spectroscopic Approaches to Biomolecular Conformation*; Urry, D.W., Ed.; AMA Press: Chicago, **1967**, pp. 209–262.

54. (a) Huynh, B.H. Presented at the 209th National Meeting of the American Chemical Society, Anneheim, CA, **April 1995**. (b) Gütlich, P.; Link, R.; Trautwein, A. *Mössbauer Spectroscopy and Transition Metal Chemistry*; Springer-Verlag: New York, **1978**.

55. (a) Oosterhuis, W.T. *Struct. Bonding* **1974**, *20*, 59–99. (b) Münck, E.; Champion, P.M. *J. de Physique* **1974**, *35*, C6–33. (c) Münck, E. *Methods Enzymol.* **1978**, *54*, 346–379.

56. Ewald, A H.; Martin, R.L.; Ross, I.G.; White, A.H. *Proc. Roy. Soc.* **1964**, *A280*, 235.

57. Maltempo, M.M.; Moss, T.H. *Quarterly Reviews of Biophysics* **1976**, *9*, 181–215.

58. (a) Peisach, J.; Blumberg, W.E.; Ogawa, S.; Rachmilewitz, E.A.; Oltzik, R. *J. Biol. Chem.* **1971**, *246*, 3342–3355. (b) Gans, P.; Buisson, G.; Duie, E.; Regnard, J.R.; Marchon, J.C. *J. Chem. Soc., Chem. Commun.* **1979**, 393. (c) Kastner, M.E.; Scheidt, W.R.;

Mashiko, T.; Reed, C.A. *J. Am. Chem. Soc.* **1978**, *100*, 666. (d) Gismelseed, A.; Bominaar, E.L.; Bill, E.; Trautwein, A.X.; Winkler, H.; Nasri, H.; Doppelt, P.; Mandon, D.; Fischer, J.; Weiss, R. *Inorg. Chem.* **1990**, *29*, 2741–2749. (e) Mansuy, D.; Morgenstern-Badarau, I.; Lange, M.; Gans, P. *Inorg. Chem.* **1982**, *21*, 1427–1430. (f) Dolphin, D.H.; Sams, J.R.; Tsin, T. B. *Inorg. Chem.* **1977**, *16*, 711. (g) Masuda, H.; Taga, T.; Osaki, K.; Sugimoto, H.; Yoshida, Z.; Ogoshi, H. *Bull. Chem. Soc. Jpn.* **1982**, *55*, 3891–3895. (h) Summerville, D.A.; Cohen, I.A.; Hatano, K.; Scheidt, W. R. *Inorg. Chem.* **1978**, *17*, 2906–2910. (i) Salmeen, I.; Palmer, G. *J. Chem. Phys.* **1968**, *48*, 2049–2052. (j) Siedow, J.N.; Vickery, L.E.; Palmer, G. *Arch. Biochem. Biophys.* **1980**, *203*, 101–107.

59. Griffith, J.S. *The Theory of Transition-Metal Ions*; Cambridge University Press: Cambridge, UK, **1961**.
60. (a) Blumberg, W.E.; Peisach, J. In *Probes of Structure and Function of Macromolecules and Membranes*; Chance, B., Yonetani,T., Mildvan, A.S., Eds.; Academic Press: New York, **1971**; Vol. 2, pp. 215–227. (b) Niarchos, D.; Kostikas, A.; Simopoulos, A.; Coucouvanis, D.; Piltingsrud, D.; Coffman, R.E. *J. Chem. Phys.* **1978**, *69*, 4411–4418. (c) Maltempo, M.M. *Biochim. Biophys. Acta* **1974**, *342*, 290–305.
61. (a) Palmer, G. In *Iron Porphyrins, Part II*; Lever, A.B.P., Gray, H.B. Eds.; Addison-Wesley Publishing Company: Reading, Massachusetts, **1983**, pp. 43–88. (b) Taylor, C.P.S. *Biochim. Biophys. Acta* **1977**, *491*, 137–149.
62. Kahn, O. *Molecular Magnetism*; VCH Publishers, Inc.: New York, **1993**.
63. Reed, C.A.; Guiset, F. *J. Am. Chem. Soc.* **1996**, *118*, 3281–3282.
64. (a) Holmes, M.A.; Stenkamp, R.E. *J. Mol. Biol.* **1991**, *220*, 723–737. (b) Nordlund, P.; Eklund, H. *J. Mol. Biol.* **1993**, *232*, 123–164.
65. (a) Rosenzweig, A.C.; Takahara, P.M.; Frederick, C.A.; Lippard, S.J.; Nordlund, P. *Nature* **1993**, *366*, 537–543. (b) Rosenzweig, A.C.; Nordlund, P.; Takahara, P.M.; Frederick, C.A.; Lippard, S.J. *Chemistry & Biology* **1995**, *2*, 409–418. (c) Elango, N.; Radhakrishman, R.; Froland, W.A.; Waller, B.J.; Earhart, C.A.; Lipscomb, J.D.; Ohlendorf, D.H. *Protein Sci.* **1997**, *6*, 556–568.
66. (a) Dawson, J.W.; Gray, H.B.; Hoenig, H.E.; Rossman, G.R.; Schredder, J.M.; Wang, R.-H. *Biochemistry* **1972**, *11*, 461–465. (b) Petersson, L.; Gräslund, A.; Ehrenberg, A.; Sjöberg, B.-M.; Reichard, P. *J. Biol. Chem.* **1980**, *255*, 6706–6712.
67. Lee, S.-K.; Fox, B.G.; Froland, W.A.; Lipscomb, J.D.; Münck, E. *J. Am. Chem. Soc.* **1993**, *115*, 6450–6451.
68. (a) Reem, R.C.; McCormick, J.M.; Richardson, D.E.; Devlin, F.J.; Stephens, P.J.; Musselman, R.L.; Solomon, E.I. *J. Am. Chem. Soc.* **1989**, *111*, 4688–4704. (b) Brown, C.A.; Remar, G.J.; Musselman, R.L.; Solomon, E.I. *Inorg. Chem.* **1995**, *34*, 688–717. (c) Hotzelmann, R.; Weighardt, K.; Flörke, U.; Haupt, H.J.; Weatherburn, D.C.; Bonvoison, J.; Blondin, G.; Girerd, J.-J. *J. Am. Chem. Soc.* **1992**, *114*, 1681–1696.
69. Gorun, S.M.; Lippard, S.J. *Inorg. Chem.* **1991**, *30*, 1625–1630.
70. (a) Reem, R.C.; Solomon, E.I. *J. Am. Chem. Soc.* **1984**, *106*, 8323. (b) Holmes, M.A.; Trong, I.L.; Turley, S.; Sieker, L.C.; Stenkamp, R.E. *J. Mol. Biol.* **1991**, *218*, 583–593.
71. Logan, D.T.; Su, X.-D.; Åberg, A.; Regnström, K.; Hajdu, J.; Eklund, H.; Nordlund, P. *Structure* **1996**, *4*, 1053–1064.
72. Lindqvist, Y.; Huang, W.; Schneider, G.; Shanklin, J. *The EMBO Journal* **1996**, *15*, 4081–4092.
73. (a) Reem, R.C.; Solomon, E.I. *J. Am. Chem. Soc.* **1987**, *109*, 1216–1226. (b) Hartman, J.R.; Rardin, R.L.; Chaudhuri, P.; Pohl, K.; Weighardt, K.; Nuber, B.; Weiss, J.; Papaefthymiou, G.C.; Frankel, R.B.; Lippard, S.J. *J. Am. Chem. Soc.* **1987**, *109*, 7387.

74. Pulver, S.; Froland, W.A.; Fox, B.G.; Lipscomb, J.D.; Solomon, E.I. *J. Am. Chem. Soc.* **1993**, *115*, 12409–12422.

75. (a) Pulver, S.C.; Thong, W.H.; Bollinger, J.M.; Stubbe, J.; Solomon, E.I. *J. Am. Chem. Soc.* **1995**, *117*, 12664. (b) Pulver, S.C.; Froland, W.A.; Lipscomb, J.D.; Solomon, E.I. *J. Am. Chem. Soc.* **1997**, *119*, 387–395.

76. (a) Fox, B.G.; Hendrich, M.P.; Surerus, K.K.; Andersson, K.K.; Froland, W.A.; Lipscomb, J.D.; Münck, E. *J. Am. Chem. Soc.* **1993**, *115*, 3688–3701. (b) Elgren, T.E.; Hendrich, M.P.; Que, L., Jr. *J. Am. Chem. Soc.* **1993**, *115*, 9291–9292.

77. Hendrich, M.P.; Pearce, L.L.; Que, L., Jr.; Chasteen, N.D.; Day, E.P. *J. Am. Chem. Soc.* **1991**, *113*, 3039–3044.

78. (a) Hendrich, M.P.; Debrunner, P.G. *Biophys. J.* **1989**, *56*, 489. (b) Münck, E.; Surerus, K.K.; Hendrich, M.P. *Methods Enzymol.* **1993**, *227*, 463–479.

79. (a) Petersson, L.; Cammack, R.; Rao, K.K. *Biochim. Biophys. Acta* **1980**, *622*, 18–24. (b) Palmer, G. In *Iron- Sulfur Proteins*; Lovenberg, W. Ed.; Academic Press: New York, **1973**, pp. 285–325.

80. Condon, E.U.; Shortley, G.H. In *The Theory of Atomic Spectra*; Cambridge University Press: New York, **1935**, pp. 75–78.

81. Bencini, A.; Gatteschi, D. *EPR of Exchange Coupled Systems*; Springer-Verlag: New York, **1990**.

82. Rao, K.K.; Evans, M.C.W.; Cammack, R.; Hall, D.O.; Thompson, C.L.; Jackson, P.J.; Johnson, C.E. *Biochem. J.* **1972**, *129*, 1063–1070.

83. Sands, R.H.; Dunham, W.R. *Quarterly Reviews of Biophysics* **1975**, *7*, 443–504.

84. Dunham, W.R.; Bearden, A.J.; Salmeen, I.T.; Palmer, G.; Sands, R.H.; Orme-Johnson, W.H.; Beinert, H. *Biochim. Biophys. Acta* **1971**, *253*, 134–152.

85. Hofmeister, A.E.; Buckel, W. *Eur. J. Biochem.* **1992**, *206*, 547.

86. Münck, E. In *Iron-Sulfur Proteins*; Spiro, T. G., Ed.; John Wiley & Sons: New York, **1982**; Vol. 4, pp. 147–175.

87. Münck, E.; Papaefthymiou, V.; Surerus, K.K.; Girerd, J.-J. In *Metal Clusters in Proteins*; Que, L., Jr., Ed.; American Chemical Society: Washington, D.C., **1988**; Vol. 372, pp. 302–325.

88. (a) Noodleman, L.; Peng, C.Y.; Case, D. A.; Mouesca, J.-M. *Coord. Chem. Rev.* **1995**, *144*, 199–244. (b) Noodleman, L.; Case, D.A. *Adv. Inorg. Chem.* **1992**, *38*, 423–470.

89. (a) Zener, C. *Phys. Rev.* **1951**, *82*, 403–405. (b) Anderson, P.W.; Hasegawa, D.A. *Phys. Rev.* **1955**, *100*, 675.

90. Noodleman, L.; Baerends, E.J. *J. Am. Chem. Soc.* **1984**, *106*, 2316–2327.

91. Drüeke, S.; Chaudhuri, P.; Pohl, K.; Wieghardt, K.; Ding, X.-Q.; Bill, E.; Sawaryn, A.; Trautwein, A.X.; Winkler, H.; Gurman, S.J. *J. Chem. Soc., Chem. Commun.* **1989**, 59.

92. Davydov, A.S. *Quantum Mechanics*; NEO Press: Ann Arbor, Michigan, **1966**.

93. Lever, A.B.P. *Inorganic Electronic Spectroscopy*; Elsevier: Amsterdam, **1984**.

94. (a) Sugano, S.; Tanabe, Y.; Kamimura, H. *Multiplets of Transition-Metal Ions in Crystals*; Academic Press: New York, **1970**; Vol. 33. (b) Schläfer, H.L.; Gliemann, G. *Basic Principles of Ligand Filed Theory*; Wiley-Interscience: New York, **1969**. (c) McClure, D.S. *Electronic Spectra of Molecules and Ions in Crystals*; Academic Press: New York, **1959**.

95. Solomon, E.I.; Lowery, M.D.; LaCroix, L.B.; Root, D.E. *Methods Enzymol.* **1993**, *226*, 1–33.
96. Gillard, R.D. In *Physical Methods in Advanced Inorganic Chemistry*; Hill, H.A.O., Day, P., Eds.; Interscience: New York, **1968**, pp. 167–213.
97. (a) Liehr, A.D. *J. Phys. Chem.* **1964**, *68*, 665. (b) Liehr, A. *J. Phys. Chem.* **1964**, *68*, 3629. (c) Richardson, F.S. *Chem. Rev.* **1979**, *79*, 17. (d) Liehr, A.D. In *Transition Metal Chemistry*; Carlin, R.L. Ed.; Marcel-Dekker: **1966**; Vol. 2.
98. Mason, S.F. *Proc. Chem. Soc.* **1962**, 137.
99. Piepho, S.; Schatz, P.N. *Group Theory in Spectroscopy With Applications to Magnetic Circular Dichroism*; John Wiley & Sons: New York, **1983**.
100. Buckingham, A.D.; Stephens, P.J. *Ann. Rev. Phys. Chem.* **1966**, *17*, 399–432.
101. Henry, C.H.; Schnatterly, S.E.; Slichter, C.P. *Phys. Rev.* **1965**, *A137*, 583.
102. Solomon, E.I. In *Metal Clusters in Proteins*; Que, L., Jr., Ed.; American Chemical Society: Washington, D.C., **1988**; Vol. 372, pp. 116–150.
103. Gewirth, A.A.; Solomon, E.I. *J. Am. Chem. Soc.* **1988**, *110*, 3811–3819.
104. Gerstman, B.S.; Brill, A.S. *J. Chem. Phys.* **1985**, *82*, 1212–1230.
105. Solomon, E.I.; Hare, J.W.; Gray, H.B. *Proc. Natl. Acad. Sci. USA* **1976**, *73*, 1389–1393.
106. Guckert, J.A.; Lowery, M.D.; Solomon, E.I. *J. Am. Chem. Soc.* **1995**, *117*, 2817–2844.
107. Sundaram, U.M.; Zhang, H.H.; Hedman, B.; Hodgson, K.O.; Solomon, E.I. *J. Amer. Chem. Soc.* **1977**, *119*, 12525–12540.
108. Gamelin, D.R.; Randal, D.W.; Hay, M.T.; Houser, R.P.; Mulder, T.C.; Canters, G.W.; de Vries, S.; Tolman, W.B.; Lu, Y.; Solomon, E.I. *J. Amer. Chem. Soc.* **1998**, *120*, 5246–5263.
109. Farrar, J.A.; Neese, F.; Lappalainen, P.; Kroneck, P.M.H.; Saraste, M.; Zumft, W.G.; Thompson, A.J. *J. Am. Chem. Soc.* **1996**, *118*, 11501–11514.
110. (a) Andrew, C.R.; Fraczkiewicz, R.; Czernuszewicz, R.S.; Lappalainen, P.; Saraste, M.; Sanders-Loehr, J. *J. Am. Chem. Soc.* **1996**, *118*, 10436–10445. (b) Andrew, C.R.; Han, J.; de Vries, S.; van der Oost, J.; Averill, B.; Loehr, T.M.; Sanders-Loehr, J. *J. Am. Chem. Soc.* **1994**, *116*, 10805–10806. (c) Andrew, C.R.; Lappalainen, P.; Saraste, M.; Hay, M.T.; Lu, Y.; Dennison, C.; Canters, G.W.; Fee, J.A.; Slutter, C.; Nakamura, N.; Sanders-Loehr, J. *J. Am. Chem. Soc.* **1995**, *117*, 10759–10760. (d) Wallace-Williams, S.E.; James, C.A.; de Vries, S.; Saraste, M.; Lappalainen, P.; van der Oost, J.; Fabian, M.; Palmer, G.; Woodruff, W.H. *J. Am. Chem. Soc.* **1996**, *118*, 3986–3987.
111. Jackels, S.C.; Norman, J.G., Jr. *J. Am. Chem. Soc.* **1975**, *97*, 3833.
112. (a) Butcher, K.D.; Didziulis, S.V.; Briat, B.; Solomon, E.I. *J. Am. Chem. Soc.* **1990**, *112*, 2231–2242. (b) Butcher, K.D.; Gebhard, M.S.; Solomon, E.I. *Inorg. Chem.* **1990**, *29*, 2067–2074.
113. Gebhard, M.S.; Koch, S.A.; Millar, M.; Devlin, F.J.; Stephens, P.J.; Solomon, E.I. *J. Am. Chem. Soc.* **1991**, *113*, 1640.
114. (a) Laiho, R. *Phys. Status Solidi B* **1975**, *69*, 579. (b) Bird, B.D.; Day, P. *J. Chem. Phys.* **1968**, *49*, 392.
115. Mulliken, R.S.; Reike, C.A.; Orloff, D.; Orloff, H. *J. Chem. Phys.* **1949**, *17*, 1248.
116. Baldwin, M.J.; Root, D.E.; Pate, J.E.; Fujisawa, K.; Kitajima, N.; Solomon, E.I. *J. Am. Chem. Soc.* **1992**, *114*, 10421–10431.
117. Ros, P.; Schuit, G.C.A. *Theo. Chim. Acta* **1966**, *4*, 1.

118. Karpishin, T.B.; Gebhard, M.S.; Solomon, E.I.; Raymond, K.N. *J. Am. Chem. Soc.* **1991**, *113*, 2977–2984.
119. Koningstein, J.A. *Introduction to the Theory of the Raman Effect*; D. Reidel Publishing Company: Dordrecht, Holland, **1972**.
120. Strommen, D.P.; Nakamoto, K. *J. Chem. Ed.* **1977**, *54*, 474–477.
121. (a) Clark, R.J.H.; Stewart, B. *Struct. Bonding* **1979**, *36*, 1–80. (b) Spiro, T.G.; Czernuszewicz, R.S. *Methods Enzymol.* **1995**, *246*, 416–460.
122. Tang, J.; Albrecht, A.C. In *Raman Spectroscopy*; Szymanski, H.A., Ed.; Plenum: New York, **1970**; Vol. 2, pp. 33–68.
123. (a) Warshel, A. *Ann. Rev. Biophys. Bioeng.* **1977**, *6*, 273–300. (b) Warshel, A.; Dauber, P. *J. Chem. Phys.* **1977**, *66*, 5477–5488. (c) Blazej, D.C.; Peticolas, W.L. *Proc. Natl. Acad. Sci. USA* **1977**, *74*, 2639–2643. (d) Myers, A.B.; Mathies, R.A. In *Biological Applications of Raman Spectroscopy*; Spiro, T.G., Ed.; John Wiley & Sons: New York, **1987**; Vol. 2, pp. 1–58.
124. Solomon, E.I.; Lowery, M.D. *Science* **1993**, *259*, 1575–1581.
125. Penfield, K.W.; Gay, R.R.; Himmelwright, R.S.; Eickman, N.C.; Norris, V.A.; Freeman, H.C.; Solomon, E.I. *J. Am. Chem. Soc.* **1981**, *103*, 4382–4388.
126. Andrew, C.R.; Sanders-Loehr, J. *Acc. Chem. Res.* **1996**, *29*, 365–372 (and references therein).
127. Desjardins, S.R.; Wilcox, D.E.; Musselman, R.L.; Solomon, E.I. *Inorg. Chem.* **1987**, *26*, 288–300.
128. (a) Tuczek, F.; Solomon, E.I. *Inorg. Chem.* **1991**, *31*, 944–953. (b) Tuczek, F.; Solomon, E.I. *J. Am. Chem. Soc.* **1994**, *116*, 6916–6924. (c) Solomon, E.I.; Tuczek, F.; Root, D.E.; Brown, C.A. *Chem. Rev.* **1994**, *94*, 827–856.
129. Anderson, P.W. In *Magnetism*; Rado, G.T., Suhl, H., Eds.; Academic Press: New York, **1963**; Vol. I., pp. 25–83.
130. Kau, L.-S.; Spira-Solomon, D.J.; Penner-Hahn, J.E.; Hodgson, K.O.; Solomon, E.I. *J. Am. Chem. Soc.* **1987**, *109*, 6433–6442.
131. Hahn, J.E.; Scott, R.A.; Hodgson, K.O.; Doniach, S.; Desjardins, S.R.; Solomon, E.I. *Chem. Phys. Lett.* **1982**, *88*, 595–598.
132. Bair, R.A.; Goddard, W.A., III *Phys. Rev. B* **1980**, *22*, 2767–2776.
133. Scott, R.A.; Hahn, J.E.; Doniach, S.; Freeman, H.C.; Hodgson, K.O. *J. Am. Chem. Soc.* **1982**, *104*, 5364–5369.
134. Westre, T.E.; Kennepohl, P.; DeWitt, J.G.; Hedman, B.; Hodgson, K.O.; Solomon, E.I. *J. Am. Chem. Soc.* **1997**, *119*, 6297–6314.
135. Roe, A.L.; Schneider, D.J.; Mayer, R.J.; Pryz, J.W.; Widom, J.; Que, L., Jr. *J. Am. Chem. Soc.* **1984**, *106*, 1676–1681.
136. deGroot, F.M.F.; Fuggle, J.C.; Thole, B.T.; Sawatsky, G.A. *Phys. Rev. B* **1990**, *42*, 5459.
137. George, S.J.; van Elp, J.; Chen, J.; Ma, Y.; Chen, C.T.; Park, J.-B.; Adams, M.W.W.; Searle, B.G.; de Groot, F.M.F.; Fuggle, J.C.; Cramer, S.P. *J. Am. Chem. Soc.* **1992**, *114*, 4426–4427.
138. van Elp, J.; George, S.J.; Chen, J.; Peng, G.; Chen, C.T.; Tjeng, L.H.; Meigs, G.; Lin, H.-J.; Zhou, Z.H.; Adams, M.W.W.; Searle, B.G.; Cramer, S.P. *Proc. Natl. Acad. Sci. USA* **1993**, *90*, 9664–9667.

139. (a) Thole, B.T.; Carra, P.; Sette, F.; van der Laan, G. *Phys. Rev. Lett.* **1992**, *68*, 1943–1946. (b) Carra, P.; Thole, B.T.; Altarelli, M.; Wang, X. *Phys. Rev. Lett.* **1993**, *70*, 694–697.

140. Wang, H.; Bryant, C.; Randall, D.W.; Lacroix, L.B.; Solomon, E.I.; LeGros, M.; Cramer, S.P. *J. Phys. Chem. B* **1998**, *102*, 8347–8349.

141. (a) Hedman, B.; Hodgson, K.O.; Solomon, E.I. *J. Am. Chem. Soc.* **1990**, *112*, 1643–1645. (b) Shadle, S.E.; Hedman, B; Hodgson, K.O.; Solomon, E.I. *Inorg. Chem.* **1994**, *33*, 4235–4244. (c) Shadle, S.E.; Hedman, B; Hodgson, K.O.; Solomon, E.I. *J. Am. Chem. Soc.* **1995**, *117*, 2259–2272.

142. (a) Rose, K.; Shadle, S.E; Eidsness, M.K.; Kurtz, D.M., Jr.; Scott, R.A.; Hedman, B.; Hodgson, K.O.; Solomon, E.I. *J. Amer. Chem. Soc.* **1998**, *120*, 10743–10747. (b) Rose, K.; Shadle, S.E.; Glaser, T.; de Vries, S.; Cherepanov, A.; Canters, G.W.; Hedman, B.; Hodgson, K.O.; Solomon, E.I. *J. Amer. Chem. Soc.* **1999** (*in press*). (c) Glaser, T.; Hedman, B.; Hodgson, K.O.; Solomon, E.I., results to be published.

# 2 Electron-Transfer Reaction Rate Theory

DAVID E. RICHARDSON

Department of Chemistry
University of Florida
Gainesville, FL 32611-7200, USA
E-mail: der@chem.ufl.edu



*Inorganic Electronic Structure and Spectroscopy, Volume II: Applications and Case Studies.*
Edited by E. I. Solomon and A. B. P. Lever.
ISBN 0-471-32682-8. 

## 1 INTRODUCTION

Electron-transfer rate theory is the single most successful kinetic theory for predicting the rates of solution reactions from physical and spectroscopic data for the reactants and the solvent. A number of excellent review articles[1–7] on theoretical approaches to electron-transfer phenomena in solution have been published in the 40 years since Marcus[8,9] and Hush[10] advanced the first truly quantitative treatments[11,12] for predicting kinetic barriers in redox reactions.

It is not the intent of this chapter to repeat the advanced presentations that have appeared elsewhere. Instead, some of the fundamental concepts will be explained in more detail than generally found in the specialized literature, and a different point of entry to the material will be used. In particular, aspects of gas-phase electron-transfer reactions that have a close relation to the solution reactions more familiar to inorganic and organic chemists will be presented first. From there, the role of solvent is introduced and its impact on electron-transfer rates is considered. This separation of the molecular and solvent contributions to reaction barriers is a central feature of current theories for solution reactions. The discussion of the origin of the solvent barrier to electron transfer will be built around the viewpoint that the differential solvation energy of reactants and transition states leads to the barrier. The references given are not exhaustive and only provide an entry to the vast literature on electron transfer. The important contributions of many workers in the field cannot be covered here, but the literature cited can be used as a starting point for delving into more theoretical and experimental details.

Cannon, recognizing the relationships between the two fields, summarized both gas-phase and solution electron-transfer reactions in his thorough 1980 monograph.[13] Until the mid-80's, however, there was little experimental work that bridged gas-phase and solution electron-transfer systems involving larger polyatomic reactants. Direct comparisons between rates of electron-transfer reactions in solution and the gas phase are now available for both organic and metal complex reactants,[14] and the conceptual approach used by Cannon and here is therefore on firmer ground experimentally. This perspective provides helpful insights into the nature of electron-transfer phenomena.

The discussion will focus initially on bimolecular reactions. The next section will consider gas-phase collisions and possible barriers to electron transfer between colliding reactants. Following that, the solvent will be introduced and its impact on collisional and electron-transfer barriers will be worked out.

In the relatively recent history of electron-transfer reactions, experiments have been devised that permanently link redox partners together, thus removing the complexities of bimolecular kinetics. Such intramolecular electron-transfer studies turn out to be very relevant to biological electron-transfer, where redox sites are often at firmly fixed distances from one another. Some introductory material on these topics is therefore provided.

## 2 GAS-PHASE ELECTRON TRANSFER

The development of solution electron-transfer chemistry was paralleled by increased experimental and theoretical research on the nature of gas-phase electron-transfer reactions (also referred to as *charge-transfer* reactions in the literature).[15,16] However, the overlap between these developments was sparse, perhaps because the gas-phase reactions were more commonly investigated by the physics community in the early research. Quantum chemists have now considered gas-phase electron-transfer reactions in detail using a variety of computational methods.[17] Not surprisingly, many of the theoretical principles that apply in the gas-phase also find use in solution models, but the complexity of the liquid environment necessarily leads to much less detailed quantum mechanical treatments than can be accorded reactions involving only a few nuclei.

The initial focus here will be on bimolecular reactions that produce two products following a one-electron transfer, i.e.

$$A^n + B^m \rightarrow A^{n-1} + B^{m+1} \tag{1}$$

where $n$ and $m$ are the charges on the reactants.

### 2.1 Collision Energetics

Let us first consider the characteristic types of collisions that might occur in a gas-phase reaction and their corresponding potential energy surfaces. Later, the extent to which these surfaces are modified by solvation of the reactants will be described.

In the case of both reactants being neutral, the collision is governed by weak London forces along with dipolar forces if the reactants are polar. Surface $n/n$ in Figure 1 is a plot of the common Lennard–Jones potential[18] for $N_2 + N_2$. Neutral–neutral bimolecular reactions are rarely considered in the context of electron transfer. The most widely studied gas-phase electron-transfer reactions are ion–molecule reactions, usually between a monocation or monoanion and a neutral molecule. The principle attractive forces are between the ion and an induced dipole of the polarizable neutral and (if present) the neutral's permanent dipole. In Figure 1, the corresponding curve $(0/+)$ is estimated for the $N_2 + N_2^+$ collision.

Ion–ion collisions involve recombination reactions between oppositely charged reactants (e.g., $Cl^- + Cl^+ \rightarrow Cl_2$) or reactions between like-charged reactants and are not often considered in the gas-phase electron-transfer literature. In the former case,

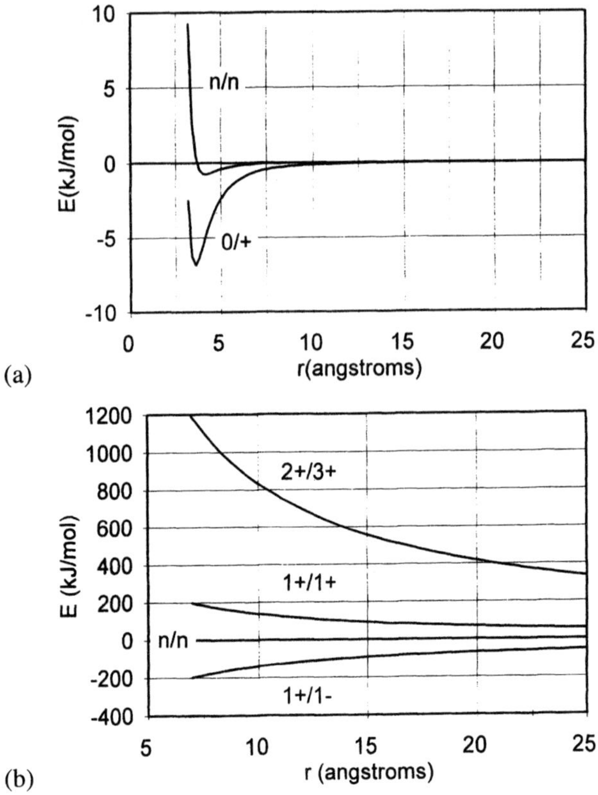


**Figure 1.** Potential energy plots for gas-phase collisions. The $r$ axis is the separation between the reactants. (a) Neutral–neutral ("$n/n$") and ion/neutral ("0/+") interactions. Both plots use the Lennard–Jones potential for $N_2$ to calculate Van der Waals forces. (b) Electrostatic energies for various ion–ion combinations. Note the difference between the energy scales for (a) and (b).

neutrals are formed, and in the latter case, ion repulsion (estimated by Equation 2) makes collisions highly improbable.

$$E = \frac{Z_A Z_B e^2}{4\pi \varepsilon_0 r} \tag{2}$$

In Equation 2, $Z$ is the integer charge on the ion A or B, $\varepsilon_0$ is the vacuum permitivity, and $r$ is the ion separation. The $1+/1+$ surface in Figure 1 shows how significant the electrostatic repulsion can be in vacuum. For more highly charged reactants, such as 2+ or 3+, the repulsion increases with increasing charge (Figure 1). In solution such reactions are common thanks to the solvent influence, as discussed later.

Collision rates for ion–neutral[19] reactions are readily estimated and provide an upper limit for the rate of any subsequent electron-transfer reaction. Not surprisingly, collision rates (or *cross sections*) for such reactions are greater than those of neutral–

neutral reactions because an ion can capture a neutral at a greater distance by virtue of the stronger ion–molecule attraction.

## 2.2 Evolution from Reactants to Products in a Gas-Phase Collision

Most chemists think of reactions as occurring on a potential energy surface, with the sizes of barriers between reactant and product states determining the rates of the reactions. Such surfaces arise from solutions to the time-independent version of the Schrödinger equation, from which the energies of stationary states along the reaction coordinate are calculated. These calculations can be used to find the potential energy surfaces that control nuclear motions under the Born–Oppenheimer approximation. However, quantum calculations based on the time-dependent Schrödinger equation reveal important features of electron transfer between small reactants. These calculations go beyond the usual time-independent form of the wave functions and show how electronic distributions evolve during a reactive collision.

Consider the exchange reaction between H (an electron and proton "A") and $H^+$ (proton "B") and separated by distance $R$. The system $H_2^+$ has well-known, exact stationary state wave functions for fixed values of $R$. A simple LCAO molecular orbital calculation based on the time-independent Schrödinger equation can estimate the total energy of this system, and both the exact solution and the LCAO model imply that the electron is delocalized over both nuclei at any distance between A and B. When $R$ is large (and basis orbital overlap is negligible and the overlap integral $S_{AB}$ is very small), the ground state wave function in the LCAO model is $(1/\sqrt{2})(\phi_A + \phi_B)$, where $\phi_A$ and $\phi_B$ are the basis wavefunctions ($1s$ atomic orbitals on nucleus A or B in the most simple model). While it is true that the probability of finding the electron on each center is $(1/\sqrt{2})^2 = 0.5$ *if averaged over an infinite amount of time*, the experimental observation is that the H atom remains intact for long periods at large distances from $H^+$. When H and $H^+$ collide (or pass close to one another), there is a certain probability that the system will evolve with net transfer of the electron from A to the B proton. This reaction probability, expressed as a cross-section, depends on the relative velocities of the reactants and the distance of closest approach (the *impact parameter*). The estimation of the probability of electron-transfer requires the introduction of the time dependent Schrödinger equation, $\boldsymbol{H}\boldsymbol{\Psi} = (ih/2\pi)\partial\boldsymbol{\Psi}/\partial t$.

Electron-transfer cross-sections can be estimated from a stationary state description of a general $M^+$+ M electron-transfer reaction if the time-dependent Hamiltonian is approximated via perturbation theory as $\boldsymbol{H}_s + \boldsymbol{H}_t$, where $\boldsymbol{H}_s$ is the stationary state Hamiltonian and $\boldsymbol{H}_t$ is the time dependent operator. With this addition, the total wave function of the system becomes an evolving, time-dependent combination of the wavefunctions $\psi_i$ used to solve the stationary state equation $\boldsymbol{H}_s\psi = \boldsymbol{E}\psi$. Either a valence-bond or molecular orbital approximation can be used to provide the $\psi_i$ functions.

Consider again the case of the reaction of A + $e^-$ (an H atom) colliding with proton B, initially at an infinite separation. The time-dependent wave function can be approximated as $\boldsymbol{\Psi} = c_1\psi_A + c_2\psi_B$, where $\psi_A$ is a H $1s$ wave function centered on proton A and $\psi_B$ is a H $1s$ wave function centered on proton B. At $t = 0$,

$c_1 = 1$ and $c_2 = 0$. Details of the calculations are given in the literature.[20] As the ion and neutral approach, the electronic wave function begins to evolve with time and "transfers" from proton A to proton B, and $c_2 > 0$. During the time the two nuclear positions cross paths close by in space, the wave function location can oscillate many times between proton A and proton B. If, after the oscillation stops, proton B has the electron associated with it as A and B separate, an electron-transfer reaction has occurred. Figure 2 shows the oscillation of the wave function with time during a collision, expressed as a Löwdin population of the original H atom as a function of time. The picture of a time-dependent oscillation between two possible electronic states is a critical concept when the influence of solvation is described later in this chapter.

Gas-phase electron-transfer reactions continue to generate intense interest in the theory community. The analysis of electron-transfer probability by time-dependent theoretical methods[21] provides details not available in models based on stationary-state wave functions. Indeed, there is no requirement that such reactions be viewed as occurring on the traditional diabatic potential surfaces for nuclear motions of reactants and products. In this chapter, however, the focus will be on traditional potential energy surfaces that arise from solutions to the time-independent Schrödinger equation. In this viewpoint, chemical reactions take place through the necessary pas-

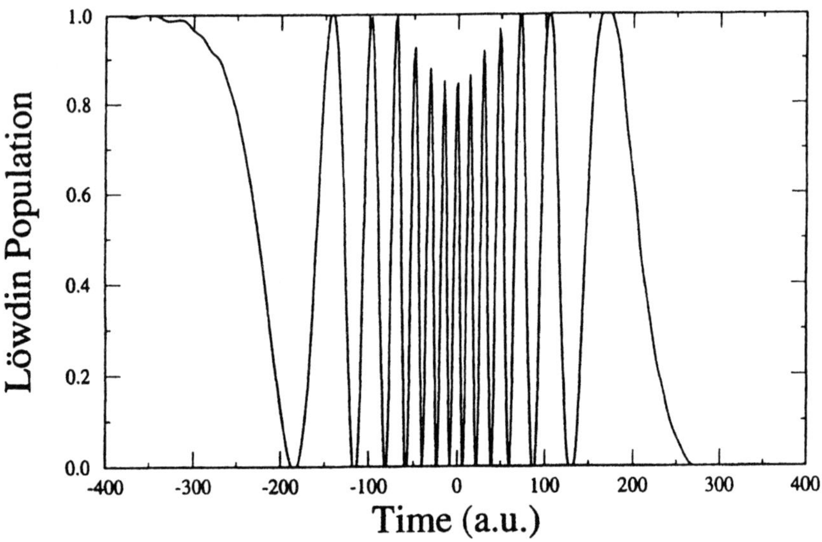


**Figure 2.** Löwdin population of the target 1 s orbital vs. time (in atomic units) for a $H/H^+$ collision at 10 eV relative translational energy. The reactants closest approach for this calculation is 1.2 Bohr radii or about 0.64 Å. As the ion and neutral approach, the wavefunction oscillates between two possible quantum states. The oscillation frequency around the collision (closest approach) at $t = 0$ units is $\sim 10^{15}\ s^{-1}$ (1 a.u. $= 2.419 \times 10^{-17}$ s). At long times after the point of closest approach, the transfer probability is 100% for the particular conditions considered here. Reprinted from Runge, K. and Micha, D. *Phys. Rev. A* **1996**, *53*, 1388.

sage through a transition state with a definable activated complex structure. A brief description of a recent, non-traditional theoretical approach will be covered in Section 4.1 on intramolecular electron transfer.

## 2.3 Electron Transfer Potential Energy Surfaces

Figure 3 shows a potential surface for an ion-neutral electron-transfer reaction in the gas phase.[22] Note that these double well potentials are *not* equivalent to the double well diagrams often seen in the electron-transfer literature. Instead, they are essentially a combination of the Marcus double well (described in more detail below) and the reactant and product association/dissociation surfaces. Reactants first collide to form the *precursor complex* $A^+$—B. For electron transfer to occur, a barrier must be traversed for electron transfer, forming the *successor complex* A—$B^+$. Dissociation of the successor complex leads to products. A more complete picture is shown in the 3-dimensional surface of Figure 4.

For simple atomic gas-phase reactions, such as $H + H^+$, a barrier due to nuclear rearrangement is absent since the individual reactants do not have associated vibrational degrees of freedom. However, these reactions can be slowed by poor energy matching of the initial and final state wavefunctions for the system.[23] The evolution of the reacting system from precursor to successor state will become increasingly unlikely as the energy difference between their electronic states increases.[24] When the reactants are more complex polyatomics, electronic state energy matching becomes more probable since more electronic levels are present, but a true nuclear barrier develops due to structural differences between the reactants and products. Although not called such in the physics literature, these are the *inner-sphere reorganization en-*

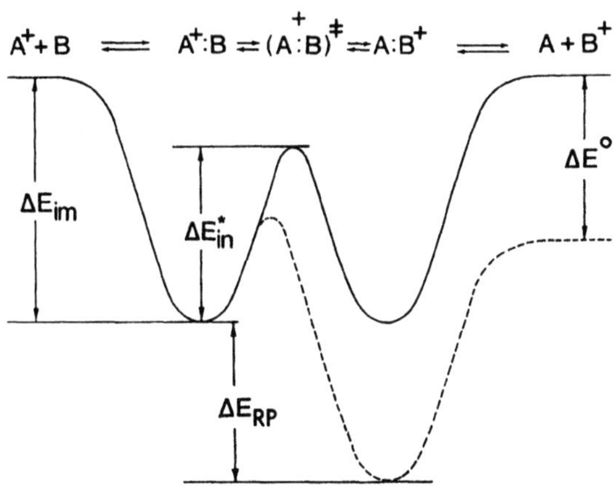


**Figure 3.** A generalized potential energy surface for polyatomic gas-phase electron transfer reactions of an ion with a neutral. The initial lowering of potential energy arises from the ion-neutral attractive forces. The central barrier is the result of requirements for reorganization of bond lengths and angles as described in the text. The solid curve illustrates the symmetrical surface for the self-exchange case ($A = B$).

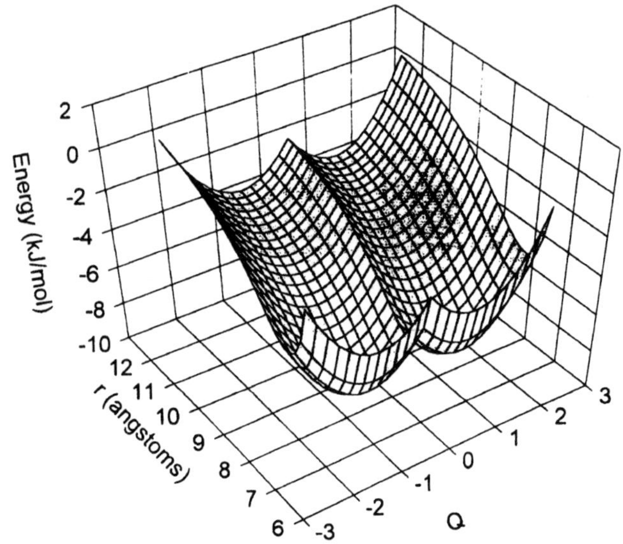


**Figure 4.** A three dimensional plot illustrating the full potential surface for polyatomic gas-phase electron transfer reactions of an ion with a neutral molecule. The reacting system enters at the top of one of the channels. As the distance of separation $r$ decreases, electronic overlap becomes significant and ET can occur for those reactants with sufficient energy to pass over the central barrier.

*ergy* barriers encountered in Marcus theory. The origin of these inner-sphere nuclear barriers will now be examined in more detail.

## 2.4 Self-Exchange Reactions

Consider the $N_2/N_2^+$ *self-exchange* reaction, where a net electron transfer from $N_2$ to $N_2^+$ occurs. (The reaction can be followed by using a mass spectrometer and isotopic labeling of one of the reactants.) Self-exchange reactions are ergoneutral (i.e., have no driving force since $\Delta G^\circ = 0$). Since the bond lengths of $N_2$ and $N_2^+$ are somewhat different, electron transfer within the precursor complex with equilibrium N–N distances will produce products in which the bond distances are not at their equilibrium values. Thus, electron transfer without altering the bond lengths would require energy input (which can be provided by a photon—see Chapter 3 on mixed valency and intervalence transitions in this volume). In other words, a Franck–Condon barrier to electron transfer is expected, as pointed out by Libby in 1952.[25] The classical value of the resulting Franck–Condon barrier, $\lambda_{\text{in}}$, can be estimated by calculating the internal energy needed to deform the ground state reactants to a point where the bond distances are that of the other reactant. The general result is

$$\lambda_{\text{in}} = \tfrac{1}{2} \sum f_i (\Delta d_0)^2 \tag{3}$$

where, for the example, $f_i$ are the N–N force constants for the neutral and cation, and $\Delta d_0$ is the difference between the bond lengths in $N_2$ and $N_2^+$. If it is assumed that the two force constants are equal to some average value, then $\lambda_{in} = f_{avg}(\Delta d_0)^2$ for the $N_2/N_2^+$ example.

The inner reorganizational barrier to a thermal electron-transfer reaction is obtained easily by noting that the requirement for electron exchange is that the initial and final states must have the same energy, which is clearly not the case if electron transfer occurs without a modification of the bond distances in the reactants. It is easily shown that for a self-exchange reaction, the condition is met when the bond distances change by $\Delta d_0/2$ and the energy barrier is given by Equation 4 after pulling out the resulting $(1/2)^2$ term from Equation 3.

$$\Delta G^*_{in} = \frac{\lambda_{in}}{4} \tag{4}$$

Equation 4 is written as the *free energy* of the rearrangement of the bond distances rather that the energy change alone ($\Delta E^*_{in}$). This will be useful later on when kinetic equations for solution reactions are developed based on the magnitude of barriers. Note that Equation 4 does not include any attractive or repulsive energies for bringing the reactants to a certain separation from infinite separation as it only is concerned with intramolecular rearrangements.

In Figure 5, the crossing of two parabolic harmonic oscillator potential surfaces indicates the point of intersection which corresponds to the activated complex. The normal coordinate $Q_{in}$ is given by Equation 5, where $q_A$ and $q_B$ are the vibrational coordinates for the reactants A and B.

$$Q_{in} = \frac{1}{\sqrt{2}}(q_A - q_B) \tag{5}$$

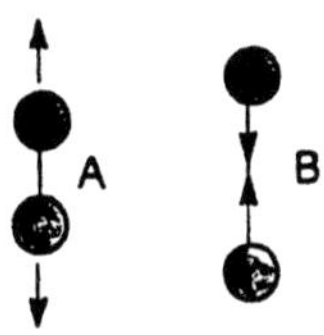


The $Q_{in}$ coordinate is essentially the lowest energy path from reactants to products in a multidimensional space and is the asymmetric combination of bond stretches (illustrated) that can form the activated complex. The energy cost for moving from the minimum on the reactant curve to the intersection is given by Equation 4. Note that the symmetrical combination $q_A + q_B$ will *not* lead to an activated complex since both bond lengths increase and decrease in phase. From Equations 3 and 4, larger barriers are predicted as the force constant and the $\Delta d_0$ values increase.

The statement that $\Delta E^*_{in} = \Delta G^*_{in}$ requires that the partition functions for the ground and transition states are equivalent and no pressure-volume work is done. The latter requirement is met as an intramolecular distortion does essentially no work on the surroundings and allows us to write $\Delta E^*_{in} = \Delta H^*_{in}$. The former requirement means that the vibrational frequencies in the ground and transition states are the same, which seems reasonable. However, some caution is needed here. In transition state theory, the partition function of the activated complex has the reaction coordi-

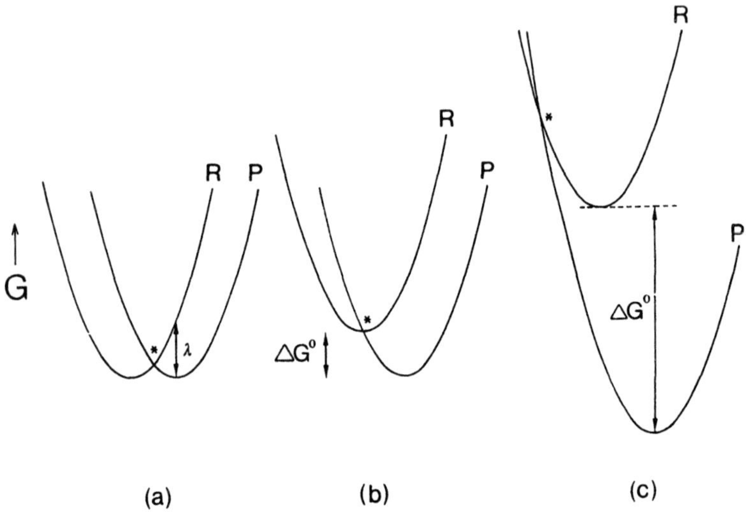


**Figure 5.** Parabolic potential surfaces depicting reactant (R) and product (P) complex energies as a function of $Q_{\text{in}}$. (a) A ergoneutral reaction ($\Delta G^\circ = 0$) with barrier $\Delta G^*$. The Franck–Condon barrier $\lambda$ is indicated. The barrier indicated by * has a height $\lambda/4$. (b) For this case, the reaction conversion is exoergic ($\Delta G^\circ < 0$). Note that the barrier separating the minima is reduced for the exoergic reaction compared to (a). If $-\Delta G^\circ = \lambda_{\text{in}}$, the barrier would be zero as shown. (c) In this highly exoergic reaction, the reactant surface has become "embedded" in the product surface, and the classical barrier (*) is large again. Thus, the classical prediction is that the rate will decrease at such high driving forces.

nate $Q_{\text{in}}$ removed from consideration, but $\Delta G^*_{\text{in}}$ is *not* equivalent to $\Delta G^\ddagger_{\text{in}}$ of transition state theory (as will be shown in Section 3.7). Instead, $\Delta G^*_{\text{in}}$ represents the *minimum* free energy cost of producing a nuclear configuration where the reactant state and product state have the same energy. Thus, it is best to note that $\Delta E^*_{\text{in}} = \Delta G^*_{\text{in}}$ basically requires that the entropy change for formation of that state is zero.

For more complex reactants, many more vibrational degrees of freedom are present, but often one normal vibrational mode will serve to interconvert the bond distances and angles that are changed in the reactant-to-product conversion. Consider, for example, a $Cp_2Fe^{0/+}$ self exchange (Cp = cyclopentadienyl), where the most significant structural change is in the Fe–Cp distance in going from $Cp_2Fe^+$ to $Cp_2Fe$ and vice versa. Here, Equation 4 can be used by substituting the force constants estimated for the symmetric Fe–Cp stretches and the mass of Cp. The value of $\Delta d_0$ is known from x-ray crystallography and an inner-sphere barrier is estimated as follows.

***Example Calculation*** The Fe–C distance changes by 0.033 Å between $Cp_2Fe^+$ (2.097 Å) and $Cp_2Fe$ (2.064 Å). Assuming the Cp moves as a unit, the Fe–(Cp centroid) distance change is estimated by noting that the center of the Cp ring is 1.219 Å from the C atoms and that $(d_{\text{Fe–Cp}})^2 + (1.219)^2 = (d_{\text{Fe–C}})^2$. Solving, the Cp distance increases by $\Delta d_0 = 0.040$ Å when $Cp_2Fe$ is oxidized. To get the average force constant for the Fe–Cp interaction, the following relation for the symmetric Cp–Fe–Cp stretch can be used, where $\nu$ is the vibrational frequency observed in $cm^{-1}$ and the mass, $m_L$, of the Cp ligand is in amu.

$$f = 5.89 \times 10^{-7} \nu^2 m_L$$

This equation can be used to find the normal mode force constant for any totally symmetric vibrational mode involving two or more equivalent metal–ligand bonds in a $ML_n$ complex. From $\nu = 303\ cm^{-1}$ for $Cp_2Fe$, $\nu = 310\ cm^{-1}$ for $Cp_2Fe^+$, and $m_L = 65$, $f = 3.6$ mdyn/Å on average. From Equations 3 and 4, $\Delta G^*_{in} = 0.0029$ mdyn Å $= 1.7$ kJ/mol.

A rather small inner-sphere barrier is expected for many reactants since the energetic cost of reorganizing the structures to the activated complex is low for the usual small geometry changes that typically accompany addition or removal of an electron from a molecule. As a result, the overall rates of gas-phase electron-transfer reactions tend to be high and approach the collisional rate. In other words, every precursor complex formed rapidly converts to the successor complex and can then dissociate to products. For a self-exchange reaction, the successor complex will of course "back electron transfer" to form the precursor complex just as rapidly and dissociate to reactants with equal probability. Thus, for self exchange reactions the maximum observed rate is half the collision rate (and the efficiency $k_{obs}/k_{coll} = 0.5$).

When significant electronic rearrangement occurs upon electron transfer, the geometries can change markedly and large inner-sphere barriers can be produced. Such reactions can have predicted efficiencies $\ll 1$, and several experimental examples have been reported for transition metal, organic, and main group self exchange reactions in the gas phase. Examples include $SF_6^{0/-}$ and $Cp_2Mn^{0/+}$ self-exchange reactions.[26,27]

One feature of gas-phase reactions that distinguishes them from solution reactions is that the energy released in the formation of the precursor complex must remain in the complex since no third bodies are immediately present to take away the excess energy. As a result, the internal energy distribution is not Boltzmann as it would be in solution but instead is higher and narrower. Such a complex is said to be *chemically activated* or *collisionally activated*. The latter term usually refers to collisions of neutrals with accelerated ions as opposed to thermal ions. Given that all this internal energy is available in the precursor complex, it is surprising to most chemists familiar with solution reactions that an electron-transfer barrier *below* the reactant energy (as in the case shown in Figure 3) can still slow a reaction. The explanation involves application of statistical rate theory and the details are beyond the scope of this chapter, but articles on statistical rate theory for gas-phase electron-transfer reactions have appeared.[28,29] A brief summary is given here.

In essence, the electron-transfer transition state can be disfavored relative to the dissociation of the precursor complex back to reactants, and therefore only a fraction of the collisions that form the precursor complex lead to products even though the energy barrier to the electron transfer state is lower. The explanation for this situation comes from the explicit recognition that entropies of activation can also affect the rates of reactions. Thus, a free energy barrier $\Delta G^*$ to a reaction can arise from both the usual potential energy considerations associated with bond deformation (appearing in $\Delta H^*$) and entropies (in $\Delta S^*$). A "tight" transition state is so called by analogy to an oscillator with a high force constant, where the quantum mechanical result is that the energy spacing between allowed levels is large (leading to a low entropy $S$). For a "loose" transition state in which vibration modes (e.g., A—B stretching modes) are converted into rotational modes (since A and B are separated), the spacing between levels is less, leading to a high density of available states and a relatively high $S$. In deducing the probability that the precursor complex will go through a tight vs. loose transition state, the density of available states for the two activated complexes is paramount.

As shown below, the breakdown of a precursor complex P having a particular energy along the path with the higher activation barrier (path 1) can be favored relative to lower barrier path 2 by the availability of more states for the transition state of path 1.

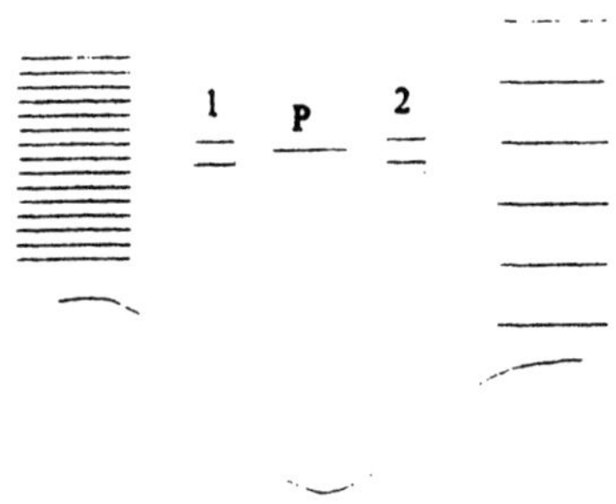


Such a situation can apply in Figure 3, where the statistical component can make the electron transfer less likely than unproductive dissociation of the precursor complex. The role of entropy changes in solution reactions will be addressed later in the chapter. Interestingly, Marcus also contributed to the most popular version of statistical rate theory, Rice–Ramsperger–Kassel–Marcus (or RRKM) theory.[30,31]

## 2.5 Reactions with $\Delta G° \neq 0$ and the Inverted Region

When the driving force for an electron-transfer reaction is non-zero, the surface related to the inner-sphere barrier must also change (Figure 5). As the reaction becomes more exoergic, it is seen that the classical barrier to electron transfer must decrease for the forward reaction (assuming that the bond distance changes and force constants are unchanged). When $-\Delta G = \lambda$, the barrier disappears (Figure 5b). In addition, conversion of the successor (S) complex back to the precursor (P) complex will now involve a larger energy barrier and will be less probable than the P $\rightarrow$ S process. Hence, most P complexes will dissociate to products and the maximum efficiency will approach unity with increasing driving force. Conversely, for increasingly endoergic reactions, the rate will be reduced as the formation of product becomes less probable. Figure 6 show a plot of log $k_{et}$ vs. driving force to illustrate these ideas.

In 1960, Marcus made the intriguing prediction (for solution reactions) that when the driving force of an electron-transfer reaction becomes sufficiently large the rate of the reaction will decrease.[32] The classical explanation is seen in curve (c) of Figure 5 for $-\Delta G > \lambda$, where the crossing of the reactant and product surfaces is higher in energy than that of curve (b) and the classical activation energy is therefore increasing as the curves separate vertically. (Equations defining the relationship between thermodynamics of a redox reaction and the activation barriers will be covered later in this chapter.) More detailed quantum mechanical theories also predict inversion, although with less rapid fall off in the *inverted region* than predicted by the classical model.[33]

It was 28 years after Marcus' prediction before definitive experimental confirmation of the inverted region was found for liquid-phase reactions, and these were for *intramolecular* electron transfer.[34] In these reactions, the electron donor and acceptor were separated by a rigid organic spacer and the acceptor end was varied chemically to change the free energy change of the reaction. Bimolecular inverted behavior is harder to observe since the top of the rate profile can be cut off by the limits of the collision rate.

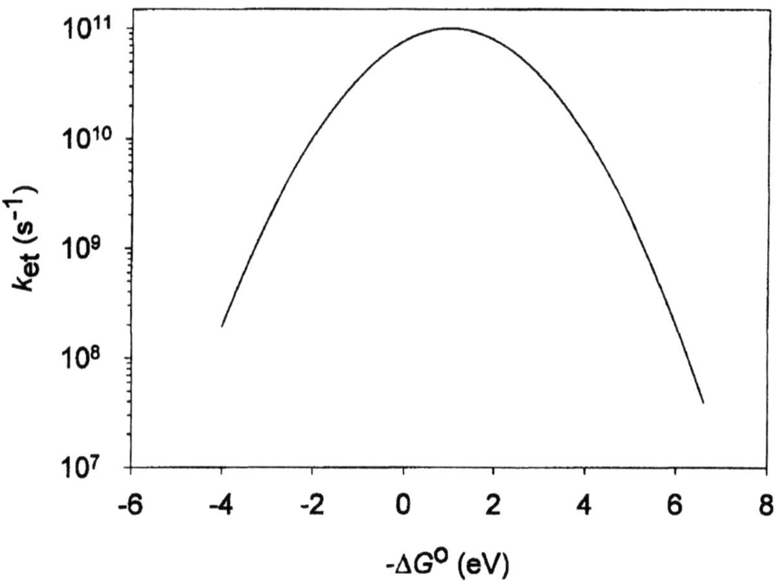


**Figure 6.** Plot of $k_{et}$ vs. $-\Delta G^\circ$ for the classical model. For this example calculation, $\lambda = 0.5$ eV and the prefactor is set at $10^{11}$ s$^{-1}$ (see Equation 27 and associated discussion for details). The unimolecular electron-transfer reaction is assumed to be adiabatic (Section 2.6). The maximum rate corresponds to case (b) in Figure 5 and occurs when $-\Delta G^\circ = \lambda$. Detailed quantum mechanical models also predict inversion, although the decrease in the inverted region is usually less steep.

Interestingly, the inverted region for bimolecular reactions was well known to those working in gas-phase electron transfer well before 1988,[35] although it did not go by that name. An example of bimolecular gas-phase electron-transfer inverted behavior is shown in Figure 7 for the reaction $A^+ + B \rightarrow A + B^+$, with $B = NH_3$. When the oxidizing agent $A^+$ is a cation and the neutral B is a polyatomic molecule, then the vibrational manifold of interest is that observed in the photoelectron (PE) spectrum of B. The lowest energy band in the photoelectron spectrum of B will approximate the Franck–Condon factors connecting the neutral B and the $B^+$ ion ground state. If the Franck–Condon factor connecting the B and $B^+$ state is "good" at a particular energy (i.e., the PE spectrum has intensity at that energy), then $A^+$ can abstract an electron efficiently if the $A \rightarrow A^+$ ionization energy matches the energy of the $B \rightarrow B^+$ ionization. If the $A \rightarrow A^+$ ionization energy is outside the $B \rightarrow B^+$ Franck–Condon manifold, the reaction is less probable. Thus, the plot in Figure 7 shows the electron-transfer reaction rates for various cations superimposed on the

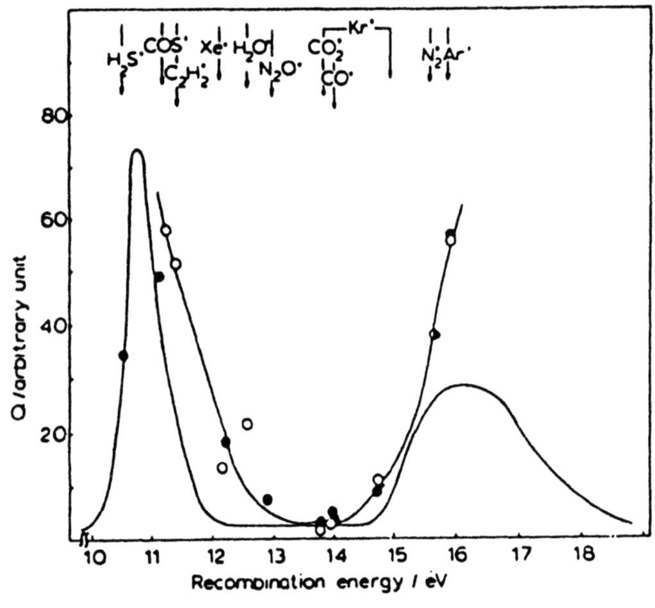


**Figure 7.** An example of inverted region behavior in bimolecular gas-phase electron transfer reactions. The rate constants for the reactions $A^+ + NH_3 \rightarrow A + NH_3^+$ are shown for the various $A^+$ species at the top of the figure. The data are superimposed on the photoelectron spectrum of $NH_3$ (ionization energy = 10.2 eV), and the $A^+$ positions indicate the ionization energy for A. Thus, the driving force for the reaction is given by IE(X) − 10.2 eV. When the driving force is <1.5 eV, the Franck–Condon factors for the $NH_3 \rightarrow NH_3^+$ conversion are high, as indicated by the photoelectron band. In the region of driving forces 1.5 ∼ 4 eV, poor Franck–Condon factors lead to low rates. When the driving force is >5 eV, $NH_3^+$ can be formed readily in an excited state (corresponding to the removal of a more strongly bound electron from $NH_3$), as indicated by the onset of the second photoelectron band and higher electron transfer rates. Reprinted from Jówko, A; Foryś, M; Jonsson, B. et al., *Int. J. Mass Spectrom. Ion Phys.* **1979**, *29*, 249.

PE spectrum of ammonia. This plot demonstrates an inverted region; that is, at large driving forces (higher $A \rightarrow A^+$ ionization energies) the reaction rate decreases. If higher lying excited electronic states are available (i.e., a higher binding energy PE peak as shown in Figure 7), then the efficiency can go up again.

## 2.6 Adiabaticity *vs.* Nonadiabaticity

For small reactants (e.g., atoms, diatomics), the overlap of atomic or molecular wavefunctions between the reactants is generally substantial in the precursor complex. Therefore, if energy matching conditions are met as outlined above, electron transfer can proceed rapidly and the transition probability approaches unity. For larger reactants, it is not clear whether electronic overlap in a precursor complex formed by a collision will always be sufficient to make the electron-transfer transition probability unity and the reaction *adiabatic*. In an adiabatic reaction, the reaction is viewed as proceeding on a potential surface such as those in Figures 3 and 4, where arrival of the reacting system at the transition state will always lead to the successor complex. In other cases where the overlap of wavefunctions on the two reactants is poor, reactions may become *nonadiabatic* even in an intimate collision. In this case, only a fraction of transition states formed lead to product since the level of electronic coupling required for adiabaticity is not met. Consider a precursor complex for the $N_2/N_2^+$ reaction where the ion and neutral are separated by a large distance. Even if the requisite nuclear reorganization takes place, the transfer probability is low because the rate of wavefunction evolution to the product state is low (as predicted by the time-dependent Schrödinger equation for localized wavefunctions with low overlap).

The radial properties of wavefunctions lead to the approximation of Equation 6 for electronic interaction between the precursor (P) and successor (S) states, $H_{ps}$, which drops off exponentially from that at the close contact interaction ($H_{ps}^0$) at separation distance $r_0$.

$$H_{ps} = H_{ps}^0 \exp\left[-\frac{b(r - r_0)}{2}\right] \tag{6}$$

The value of the coefficient $b$ determines the rate of drop off as $r$ increases and depends on the detailed nature of the wavefunctions. This issue will resurface when quantum approaches for electron transfer in solution are considered later in this chapter. Unlike the situation in gas-phase electron transfer, where most simple reactions would be adiabatic, nonadiabatic reactions are more common in complex systems such as biological electron transfer over long distances.

## 2.7 Constructing Potential Energy Surfaces

Constructing approximate classical potential surfaces for nuclear motion in gas-phase electron-transfer reactions is straightforward since the reaction coordinate is $Q_{in}$ (Equation 5) and the potential energy is determined by a harmonic oscillator function (Equation 3).

Considering the inner-sphere reorganizational barrier alone, one can readily arrive at a potential surface that takes into account the electronic interaction $H_{ps}$, which leads to two surfaces, a lower and an upper (Equations 7 and 8, respectively). Expressions of this type are derived in Chapter 3. The lower surface is the reaction surface and forms the classic double well potential energy diagram. In adiabatic reactions, the system remains only on this lower surface.

$$E_L = h\nu\left[\frac{Q_{in}{}^2}{2} - \left(\left(\Lambda_{in}Q_{in} + \frac{\Delta E}{h\nu}\right)^2 + \frac{H_{ps}^2}{(h\nu)^2}\right)^{1/2}\right] \tag{7}$$

$$E_U = h\nu\left[\frac{Q_{in}{}^2}{2} + \left(\left(\Lambda_{in}Q_{in} + \frac{\Delta E}{h\nu}\right)^2 + \frac{H_{ps}^2}{(h\nu)^2}\right)^{1/2}\right] \tag{8}$$

In Equations 7 and 8, $h\nu$ is the energy of the inner-sphere vibrational mode with frequency $\nu$, $\Delta E$ is the energy difference between the precursor and successor complexes, $H_{ps}$ is the electronic interaction energy between the initial (precursor) and final (successor) states, and $\Lambda_{in}$ is the *vibronic coupling parameter*. This dimensionless coupling parameter has been used by Schatz and coworkers in the PKS model for mixed-valence compounds (see Chapter 3).* The parameter is readily obtained from Equation 9, where $n$ is the number of bonds deformed and $f_{avg}$ is the average force constant for those bonds.

$$\Lambda_{in} = \left(\frac{nf_{avg}}{4h\nu}\right)^{1/2} \Delta d_0 \tag{9}$$

Large values of $\Lambda_{in}$ are associated with large bond distance differences between the oxidized and reduced reactants and high force constants. In the PKS model, potential energy surfaces are expressed in units of $h\nu$ and dimensionless parameters for electronic coupling and the energy difference $\Delta E$. The barrier on the lower surface is approximately $(h\nu\Lambda^2/2) - H_{ps}$ when $H_{ps}/h\nu \ll \Lambda$. In these types of plots, it should be noted that, although potential energy is shown, they are equivalent to plots of free energy in light of the discussion surrounding Equation 4.

## 3 SOLUTION ELECTRON TRANSFER

### 3.1 Solvation Effects on Electron-Transfer Rates and Collision Energetics

Transferring a gas-phase molecule or ion to solution has a dramatic effect on its redox reactivity. Electron-transfer reactions between ions and neutrals are generally

* Note that the PKS publications use $\lambda$ for the vibronic coupling parameter instead of $\Lambda$ used here. The notation for Franck–Condon reorganization energies in much of the electron-transfer literature is $\lambda$, as used elsewhere in this chapter. The standard PKS notation was therefore altered here to avoid confusion. In addition, $\varepsilon = H_{ps}/h\nu$ in PKS terminology and should not be confused with the permitivity of a dielectric medium.

slower in solution than in the gas phase. For example, the second order rate constant near room temperature for the $Cp_2Fe^{0/+}$ self-exchange reaction is a factor of $\sim 10^4$ slower in polar solvents than in the gas phase.[14] It will be shown how this lower rate in solution arises from the difference between the solvent interactions for the reactants and the electron-transfer transition state. On the other hand, reaction rates for cation–cation or anion–anion ion redox reactions become vastly higher in polar solvents. For example, the $[Ru(NH_3)_6]^{2+/3+}$ self-exchange reaction would be essentially unobservable in the gas phase due to the 2+/3+ repulsion (Equation 2), but the reaction is facile in aqueous solution, with $k = 3 \times 10^3\ M^{-1}s^{-1}$ at 298 K.[36] The self-exchange reaction $[Fe(CN)_6]^{3-/4-}$ can also be cited, where an even larger electrostatic barrier would be encountered in the gas phase, but a rapid reaction is observed in aqueous solution ($k \simeq 10^3\ M^{-1}s^{-1}$ at 277 K).[37]

The rate acceleration in solution compared to the gas phase for two ionic reactants (cation–cation or anion–anion) arises from the reduced repulsive energy between the reactants, as calculated to a first approximation by Equation 10, where $Z$ is the integer charge of an ion, $r$ is the ion–ion separation, and ionic strength is negligible.

$$E = \frac{Z_A Z_B e^2}{4\pi \varepsilon r} \tag{10}$$

Some care in applying Equation 10 is required. As shown in SI units, the absolute permitivity $\varepsilon$ of the solvent medium is used rather than the solvent dielectric constant $D$, which is the ratio of the solvent permitivity to the vacuum permitivity ($\varepsilon/\varepsilon_0$, often called the *relative permitivity* of the medium) but is sometimes also given by the symbol $\varepsilon$ in the literature. If a calculation is done for water (static dielectric constant $D_s = \varepsilon/\varepsilon_0 = 78.5$ at 298 K), the denominator of Equation 10 would be $4\pi r(78.5\varepsilon_0)$. Electrostatic repulsions are greatly diminished when like-charged reactants are immersed in a high dielectric solvent such as water or acetonitrile ($D_s = 37$ at 298 K).

A collision between a 2+ and a 3+ ion in the gas phase ($\varepsilon = \varepsilon_0$) with a separation of $\sim$7 Å would have repulsive electrostatic energy of about 280 kcal mol$^{-1}$, and the rate constant of a 2+/3+ gas-phase electron-transfer reaction would thus be incredibly small since an intimate collision would be unlikely due to electrostatic repulsion. From Equation 10, the 2+/3+ repulsion at $\sim$7 Å would be reduced to only 3.6 kcal mol$^{-1}$ in water. Since $\Delta G^\circ = -2.303RT \log K_{eq}$ and $2.303RT =$ 1.4 kcal mol$^{-1}$ at 298 K, the 2+/3+ precursor complex should form in water with an equilibrium constant on the order of $10^{-3}$ to $10^{-2}$.

For ion-neutral reactions, ion-solvent interactions will essentially balance ion-dipolar forces between reactants that are so important in accelerating the gas-phase reactions. Therefore the significant potential well associated with formation of the precursor complex in the gas phase (Figure 3) will be eliminated to a large extent in solution.

Summarizing, one major effect of solvation is to greatly attenuate the electrostatic repulsions and attractions that accompany collisions of charged reactants with other charged or neutral reactants. Another effect is the appearance of a new kinetic barrier that will slow electron transfer by increasing the activation energy for the precursor to successor conversion.

### 3.2 Rate Laws for Bimolecular Electron Transfer

For an *irreversible* bimolecular reaction where diffusion of reactants together and dissociation of the formed precursor complex are rapid, the following mechanism is operative.

$$\mathrm{A + B} \rightleftharpoons \mathrm{A{-}B}, \quad K_\mathrm{p} \tag{11}$$

$$\mathrm{A{-}B} \rightarrow \mathrm{A^+{-}B^-}, \quad k_\mathrm{et} \tag{12}$$

$$\mathrm{A^+{-}B^-} \rightarrow \mathrm{A^+ + B^-}, \quad \textit{fast}$$

and

$$R_\mathrm{et} = -d[\mathrm{A}]/dt = k_\mathrm{et}K_\mathrm{p}[\mathrm{A}][\mathrm{B}]/(1 + K_\mathrm{p}) \tag{13}$$

The reactants A and B can be charged, as in Equation 1, but charges are left off for simplicity. When $K_\mathrm{p} \ll 1$, as would be expected for like-charged ionic reactants,

$$R_\mathrm{et} = k_\mathrm{et}K_\mathrm{p}[\mathrm{A}][\mathrm{B}] = k_\mathrm{obs}[\mathrm{A}][\mathrm{B}] \tag{14}$$

where $k_\mathrm{obs}$ is the observed second-order rate constant. (Note that the rate for a self-exchange reaction would be half that in Equation 14 since all the steps are reversible.) Therefore, to calculate a bimolecular electron-transfer rate, methods are needed for estimating the value of $K_\mathrm{p}$ and the effect of solvent on the rate constant for electron transfer in the precursor *supercomplex*, $k_\mathrm{et}$, for Equation 12. Such methods are covered in the sections that follow.

### 3.3 The Reaction Model and the Born Equation

A typical 2+/3+ self-exchange reaction (e.g., $[\mathrm{Ru(NH_3)_6}]^{2+/3+}$) will be used to illustrate the basic treatment of $K_\mathrm{p}$ and the solvent barrier to electron transfer. In this chapter, the reactants will be treated as spheres with radii $a$ and center-to-center separation $r$. It is possible to use more sophisticated shapes to define the reactants, of course, but the spherical approximation is a useful starting point for developing the basic theory. The M-donor atom distances, which are critical in the determination of the inner-sphere barrier, are given by $d$. Outside of the spheres, the solvent is modeled as a *dielectric continuum.*

In discussing the solvation energetics of ions in solution, the Born equation (Equation 15), which arises from the treatment of solvent as a dielectric continuum, can be used as a first approximation.

$$\Delta G_\mathrm{Born} = -\frac{(Ze)^2}{4\pi\varepsilon_0(2a)}\left(1 - \frac{1}{D_\mathrm{s}}\right) \tag{15}$$

The Born energy of a 3+ complex is larger than that of a 2+ ion by a factor of $3^2/2^2 = 2.25$. Note that the Born equation has been occasionally misinterpreted in the literature as the solvation free energy change when a ion is transferred from vacuum to solvent [the "single" ion solvation energy $A^Z(g) \rightarrow A^Z(soln)$]. It is correctly the free energy released when a charge $Ze$ is discharged from a sphere of radius $a$ in the gas phase and a sphere of the same size is recharged in solution. In other words, the energy of cavity formation in the solvent to accommodate the sphere and the work of crossing the solvent/vacuum interface are not included. For purposes of comparison with experiment, the Born energy can be compared to the deduced free energy change for the reaction $A^Z(g) + A(soln) \rightarrow A^Z(soln) + A(g)$. This free energy change can be obtained from the difference between ionization free energies (or electron attachment free energies) for the $A/A^Z$ redox couple in the gas phase and solution.[38]

### 3.4 The Work Term and $K_p$

Sutin[4] has compared a number of quantitative methods for determining effective equilibrium constants $K_p$ for formation of precursor complexes from charged reactants (Equation 11). It is important to recognize that electron transfer can occur at any separation between the reactant ions but is, of course, most likely in an intimate collision complex where substantial electronic interaction can occur. For like-charged reactants, an intimate collision is disfavored on electrostatic grounds, so a balance of these factors must be considered. In general, the overall rate constant of bimolecular electron transfer is the summed total of all possible microscopic rate constants $k_{et}(r)$ for precursor complexes at various separations $r$ weighted by the probability of formation of each complex. Equation 16 expresses this summation as an integration over all values of $r$.

$$k_{obs} = \int_0^{\infty} 4000\pi L r^2 e^{-(w(r)/RT)} k_{et}(r) dr \tag{16}$$

In Equation 16, $w(r)$ is the work required to bring the ions from infinite separation to a separation $r$ (in meters for Equations 16–18), and $L$ is the Avogadro constant. If electron transfer is only significant at near a close contact distance, Equation 16 is approximated by Equation 17, where $r$ is now the average distance, $\delta r$ is the width of the reactive zone, and $k_{et}(r)$ is the rate constant at the average reactive separation.

$$k_{obs} \approx 4000\pi L r^2 \delta r\, e^{-(w(r)/RT)} k_{et}(r) \tag{17}$$

Noting from Equation 14 that $k_{obs} = k_{et} K_p$ leads to the commonly employed Equation 18 for estimating $K_p$,

$$K_p(M^{-1}) = 4000\pi L r^2 \delta r\, e^{-w/RT} \tag{18}$$

where, for reactants of equal radii and a reactive separation of $r = a_A + a_B$,

$$w = \frac{Z_A Z_B e^2}{4\pi\varepsilon_0 r D_s(1 + \beta r\sqrt{\mu})}$$

$$\beta(m^{-1}) = \frac{5.02 \times 10^{11}}{(D_s T)^{1/2}}$$

and $\mu$, the ionic strength, corrections have now been explicitly included in Equation 10 to calculate the work $w$. In the above equations, $r$ must be expressed in meters, and $\mu$ is taken as dimensionless. For water at 298 K, $\beta = 3.29 \times 10^9\ \mathrm{m}^{-1} = 0.329\ \text{Å}^{-1}$. (Note that much of the literature on electron transfer has electrostatic equations written expressly for cgs units rather than SI units, which are used here. With the given constants and $r$ in meters, Equation 18 estimates $K_p$ with units of $\mathrm{M}^{-1}$, i.e., $\mathrm{mol}^{-1}\ \mathrm{dm}^3$. Ionic strength is also derived from molarity for use in the above equation).

As shown in examples by Sutin,[4] the above equations and various other approaches give $K_p$ values for the *outer-sphere* complex $[Fe(H_2O)_6]^{2+}/[Fe(H_2O)_6]^{3+}$ in the range of $3 \times 10^{-3}$ to $8 \times 10^{-2}\ \mathrm{M}^{-1}$, in accord with the simple electrostatic estimate above.

***Example Calculation*** Here we will estimate a $K_p$ value using the above equations for a $2+/3+$ precursor complex in water with an ionic strength $\mu = 0.50$ at 298 K. For this example, the sphere radii $a$ are both set at 3.5 Å and the reactive zone width $\delta r = 0.80$ Å. Thus, $r = 7.0 \times 10^{-10}$ m. We find $\beta = 3.29 \times 10^9\ \mathrm{m}^{-1}$ and then $w = 9.58 \times 10^{-21}$ J or 5.77 kJ mol$^{-1}$. From Equation 18, $K_p = 0.029\ \mathrm{M}^{-1}$.

## 3.5 The Classical Solvent Reorganization Energy

The deceleration of electron-transfer reactions for ion–neutral reactions in going from the gas phase to solution arises from the second major effect of solvation, i.e., an increase in electron-transfer activation energy due to nonequilibrium polarization of the solvent in the transition state. It is this solvent barrier, also known as the *outer-sphere reorganization energy*, that received the greatest attention in the early days of modern electron-transfer theory, particularly by Marcus and Hush. The same barrier is also present in ion–ion reactions, so solvation can make bimolecular electron transfer more probable by reducing repulsions but always will slow down the "unimolecular" precursor-to-successor electron-transfer step within the supercomplex formed by a collision.

A qualitative view of solvation in electron-transfer reactions is provided by the potential surface in Figure 8, which shows that the additional free energy barrier to electron transfer due to solvation can be regarded as originating in the difference in solvation free energy of the transition state and reactants. Calculating the value of this difference requires the choice of some appropriate model for the transition state, which by its nature is not amenable to the usual methods of structural analysis. Indeed, it is at first hard to see why the solvation energy of the transition state should be all that different from that of the precursor supercomplex. After all, the transition state is still a charged ion of the same approximate structure as the precursor supercomplex. Why then is solvation energy of the transition state less than that of the precursor? The answer to this question leads to the solvent Franck–Condon barrier, which arises because the solvation of the activated complex must be equivalent for both the reactant *and* product electronic states in electron transfer. As will be shown,

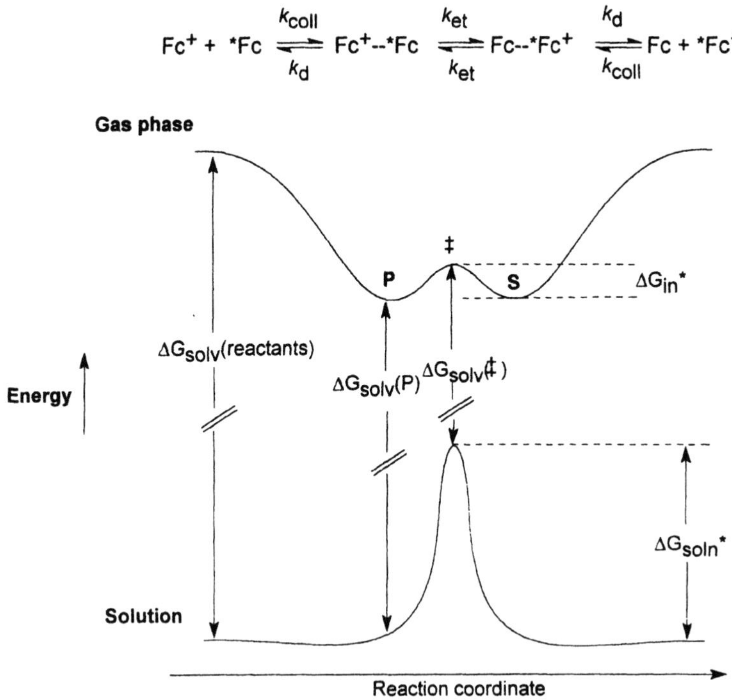


**Figure 8.** Reaction free energy surfaces illustrating the relationship between a gas-phase ion-molecule self-exchange ET reaction and its solution counterpart. A reaction mechanism for the example ferrocene/ferrocenium self-exchange reaction is shown at top for reference. The stability of the precursor (P) and successor (S) complexes relative to separated reactants and products is greatly reduced (or absent) in solution because the interaction energies of the ion with the neutral reactant and solvent molecules are comparable. The solvation energy of the transition state for ET is lower than that of the P and S complexes for the reasons described in the text. Thus, the overall activation energy for ET in solution is higher because of the decreased stability of the P complex relative to the reactants and the lower solvation of the activated complex at the transition state. Note that the reaction coordinate changes from the gas phase to solution since the solution coordinate is a complex combination of inner-sphere and solvent reorganization.

this condition means that the solvation energy of the transition state cannot be as large as it is for the precursor and successor complexes (i.e., the solvent molecules cannot orient and polarize around the activated complex to achieve the lowest possible potential energy). Therefore, the precursor complex has a greater solvation energy than the activated complex, and an additional barrier due to solvent is observed.

The details of these solvent effects will now be quantified by using a dielectric continuum model for the solvent.

Returning to the example of a 2+/3+ self-exchange reaction, it is noted that the Born energy of a 3+ ion is over twice as large as that of the 2+ ion. As described above for inner-sphere barriers in gas-phase reactions, there must be a Franck–Condon barrier due to the solvent since instantaneous transfer of an electron from the 2+ to the 3+ ion results in a state with non-equilibrium solvation (i.e., non-equilibrium polarization of the dielectric medium in the dielectric continuum model). Thus, some *fluctuation* in the solvent polarization must occur so that the total energy of either charge distribution in the precursor supercomplex (2+/3+ or 3+/2+) is the same. Only in this case has a transition state been formed. The electron-transfer event in the activated complex can be thought of as an oscillating, time-dependent probability for the charge state to be one of two equal-energy states, 2+/3+ or 3+/2+ (recall the simple $H + H^+$ reaction described earlier). If the successor complex dissociates with a charge distribution different from that of the precursor complex, electron transfer has occurred in that collision.

It is now clear why the solvation free energy of the transition state is lower than that of the precursor complex (Figure 8). Since the solvation of the transition state has been rearranged to some intermediate polarization around the 2+ and 3+ ions, the full free energy of solvation (i.e., that estimated via Equation 15) cannot be realized.

Our task is then to estimate the energy required to form the necessary non-equilibrium polarization. In molecular terms, the solvent molecules must redistribute to some "average" positions that produce the same energy for the 2+/3+ *and* 3+/2+ states. Treating the solvent as a dielectric continuum (much easier mathematically!) means the polarization of the continuum must change to a level that solvates the two possible charge distributions equivalently. However, the frequency dependence of the dielectric constant of any real liquid must be accommodated in the calculation. The static dielectric constant $D_s$ applies to a *stationary* electric field and is therefore used in the Born equation (Equation 15) to get the solvation energy of the precursor complex, which has a fixed charge distribution. However, $D_s$ does not apply for *oscillating* electric fields as found in the transition state.

As an electric field begins to oscillate, the solvent permitivity begins to decrease and approach first the optical dielectric permitivity $\varepsilon_{op}$ and ultimately the vacuum permitivity value ($\varepsilon_0$). This phenomenon is shown in Figure 9, where polarizability of a liquid medium is plotted vs. electric field frequency. At the molecular level, as the frequency of the oscillating field gets higher it will eventually shift faster than the molecules of the solvent can rearrange their positions (*orientation* polarization with a time scale $\sim 10^{11}$ Hz—on the order of rotational or tumbling frequencies). At the frequencies of molecular vibrations ($\sim 10^{13}$ Hz), molecular distortions that change the dipole moment cannot keep up and the *distortion* polarization is lost. At frequencies in the visible region (*optical* frequencies, $\sim 10^{15}$ Hz) only the electronic distributions can be polarized and follow the field shifts exactly. At that point the appropriate dielectric constant for the medium is much lower than the static constant $D_s$ and is given by $D_{op}$.

The transition state in an electron-transfer reaction presents just such an oscillating field to the solvent, and the frequency of oscillation is fast enough that the orientation and distortion polarization of the solvent can't follow it. Thus, the solvent must fluctuate through normal thermal motion to the necessary averaged orientation before electron transfer can take place. Otherwise, electron transfer produces a state in which the solvent orientation is in non-equilibrium configuration and corresponds to a vertical Franck–Condon transition.

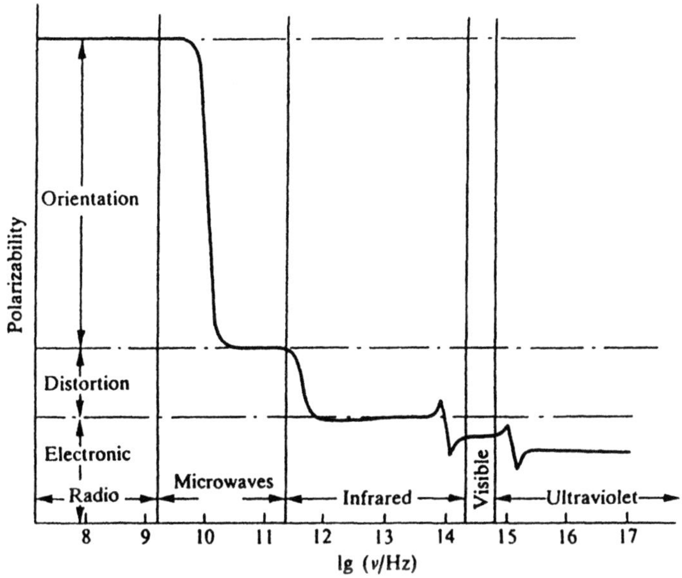


**Figure 9.** A plot of the polarizability of a typical liquid as a function the frequency of an applied electric field. At low frequency the polarizability (and dielectric constant) approach that determined for a static electric field. As the field oscillation rate increases, first the orientation then the distortion components of the polarizability are lost. At very high frequencies the permitivity of the liquid will approach the vacuum permitivity. Reprinted from Atkins, P. *Physical Chemistry, 4th Ed.*; W. H. Freeman: New York, 1990.

Formally, the Born equation (Equation 15) can be split for an ion of charge $Z$ into a sum of the electronic contribution to solvation and the remaining contribution that is realized when the solvent molecular distortion/orientation responds to the charge, respectively (Equation 19),

$$\Delta G_{\text{Born}} = -\frac{(Ze)^2}{4\pi\varepsilon_0(2a)}\left(1 - \frac{1}{D_{\text{op}}}\right) - \frac{(Ze)^2}{4\pi\varepsilon_0(2a)}\left(\frac{1}{D_{\text{op}}} - \frac{1}{D_{\text{s}}}\right) \tag{19}$$

where the dielectric constant at the field oscillation frequency $v$ is equal to $D_{\text{op}}$. In the electron-transfer literature, the square of the refractive index ($\eta^2$) at the sodium-D line (589 nm) is usually used, and the optical dielectric constant is therefore given as $D_{\text{op}} = \eta^2$.

The Born solvation energy of the transition state, $\Delta G^*_{\text{Born}}$, is needed so that the barrier presented by the solvent can be calculated from the difference between it and the solvation energy of the reactants in the precursor complex ($\Delta G^*_{\text{Born}} - \Delta G^{\text{p}}_{\text{Born}}$). This difference in Born free energies yields a good approximation for the difference in solvation free energies for the transition state and precursor complex. For a precursor complex AB and transition state AB* with charge z, the term ($\Delta G^*_{\text{Born}} - \Delta G^{\text{p}}_{\text{Born}}$)

is the free energy change for the hypothetical reaction

$$\begin{aligned}&AB^{*z}(g) + AB^{*}(soln) + AB(g) + AB^{z}(soln)\\ &\quad\rightleftharpoons AB^{*z}(soln) + AB^{*}(g) + AB(soln) + AB^{z}(g)\end{aligned}$$

which is equivalent to the desired free energy change $\Delta G^{*}_{\text{solv}} - \Delta G^{\text{p}}_{\text{solv}}$ for the reaction

$$AB^{*z}(g) + AB^{z}(soln) \rightleftharpoons AB^{*z}(soln) + AB^{z}(g)$$

if the solvation free energies for the hypothetical neutrals AB and AB* are assumed to be equal.

Since the electronic polarization equally solvates both the reactants and transition state (assuming a negligible change in electronic distribution), the first term arising from Equation 19 will drop out in the difference. Thus, from the second term in Equation 19 the underlying origin of the term $(1/D_{\text{op}} - 1/D_{\text{s}})$, often encountered in the electron-transfer literature, is revealed. It describes that part of the solvation energy that differs between the reactant and transition states because of the slow response of solvent orientation to a rapid precursor $\rightleftharpoons$ successor interconversion.

A derivation for the barrier arising from solvation for a rather simple though instructive model system is given here. The model consists of spherical reactants A and B of charges $Z$ and $(Z - 1)$, respectively, and radii $a$, sufficiently separated that the electrostatic repulsion between the ions is negligible (as might be found for remote, charged redox sites bridged by a protein or polypeptide, for example). Charge exchange between these two ions will produce charges $(Z - 1)$ and $(Z)$ on A and B, respectively. For any reacting pair, there is a certain probability that A and B will have one of the two charges; indeed, if A = B, as will be assumed here, the probability is 1/2 and each reactant has charge $Z$ 50% of the time and $(Z - 1)$ 50% of the time. Thus, for *each* of the two ions in the precursor supercomplex $[\text{A—B}]^{Z+Z-1}$, the *average* Born solvation energy is given by Equation 20.

$$\Delta G^{\text{p}}_{\text{Born}} = -\frac{\left[\frac{Z^2}{2} + \frac{(Z-1)^2}{2}\right]e^2}{4\pi\varepsilon_0(2a)}\left(1 - \frac{1}{D_{\text{op}}}\right) - \frac{\left[\frac{Z^2}{2} + \frac{(Z-1)^2}{2}\right]e^2}{4\pi\varepsilon_0(2a)}\left(\frac{1}{D_{\text{op}}} - \frac{1}{D_{\text{s}}}\right) \quad (20)$$

In the transition state, the electronic polarization of the solvent follows the electron oscillation and still solvates the average charge as above. However, the charge on A or B seen by the slowly changing orientation and distortion polarization is $(Z - 1/2)$, and the solvation energy of *each* ion in the transition state is therefore given by Equation 21.

$$\Delta G^{*}_{\text{Born}} = -\frac{\left[\frac{Z^2}{2} + \frac{(Z-1)^2}{2}\right]e^2}{4\pi\varepsilon_0(2a)}\left(1 - \frac{1}{D_{\text{op}}}\right) - \frac{(Z - 0.5)^2 e^2}{4\pi\varepsilon_0(2a)}\left(\frac{1}{D_{\text{op}}} - \frac{1}{D_{\text{s}}}\right) \quad (21)$$

The difference between the solvation energy for an ion of radius $a$ and the solvation energy of that ion in the hypothetical transition state is then the difference between Equations 20 and 21, which is shown in Equation 22.

$$\Delta\Delta G_{\mathrm{solv}} = \Delta G^{*}_{\mathrm{Born}} - \Delta G^{\mathrm{p}}_{\mathrm{Born}} = \left(\frac{1}{4}\right)\frac{e^2}{4\pi\varepsilon_0(2a)}\left(\frac{1}{D_{\mathrm{op}}} - \frac{1}{D_{\mathrm{s}}}\right) \tag{22}$$

The $\Delta\Delta G_{\mathrm{solv}}$ then is the outer-sphere barrier presented by the solvation of *each* ion to the electron-transfer reaction, and for *two* ions in the present model the outer-sphere barrier is *twice* that of Equation 22. Therefore, when $\eta^2$ is substituted for $D_{\mathrm{op}}$,

$$\Delta G^{*}_{\mathrm{out}} = \frac{\lambda_{\mathrm{out}}}{4} \approx \left(\frac{1}{4}\right)\frac{e^2}{4\pi\varepsilon_0}\left(\frac{1}{a}\right)\left(\frac{1}{\eta^2} - \frac{1}{D_{\mathrm{s}}}\right) \tag{23}$$

Equation 23 introduces the symbol for the outer-sphere Franck–Condon energy $\lambda_{\mathrm{out}}$, which, as in the inner-sphere case (Equations 1 and 2), is four times the size of the barrier. The quantity $\lambda_{\mathrm{out}}$ is the energy input required to change from the precursor to the successor state without altering the solvation shell (except for the electronic contribution to the polarization).

The repulsive interaction between the ions must be accounted for in the more usual case of close contact between the reactants. The repulsive energy in the precursor complex is given by Equation 10. Once again, however, only the electronic portion of the dielectric medium will remain in phase with the oscillating charge of the transition state. The charge repulsion for the remaining orientation polarization will be that for two ions of charge $(Z - 1/2)$ in the model. After accounting for this difference in the total solvation energies for the precursor and transition states, the result is the famous Equation 24, where the two reactant ions can have different radii.

$$\lambda_{\mathrm{out}} = \frac{e^2}{4\pi\varepsilon_0}\left(\frac{1}{2a_{\mathrm{A}}} + \frac{1}{2a_{\mathrm{B}}} - \frac{1}{r}\right)\left(\frac{1}{\eta^2} - \frac{1}{D_{\mathrm{s}}}\right) \tag{24}$$

Equation 24 assumes that the distance $r$ is sufficiently large that the polarization around each ion center is that for a sphere. At the usual collision distance, $r \simeq a_{\mathrm{A}} + a_{\mathrm{B}}$, this assumption is unreasonable and significantly more work must be done to find an expression for $\lambda_{\mathrm{out}}$. However, the order of magnitude for $\lambda_{\mathrm{out}}$ obtained from Equation 24 is generally close to experimentally deduced values. Note that Equation 24 reduces to Equation 23 when $a_{\mathrm{A}} = a_{\mathrm{B}} = a$ and $r \gg a$. The outer-sphere reorganization barrier leads to a contribution of $\lambda_{\mathrm{out}}/4$ to the activation barrier for thermal electron transfer.

***Example Calculation*** The outer-sphere barrier for the $Cp_2Fe^{0/+}$ self exchange in acetonitrile is obtained as follows: The constants of Equation 24 are evaluated as:

$$\frac{e^2}{4\pi\varepsilon_0} = \frac{(1.602 \times 10^{-19}\mathrm{C})^2}{4\pi(8.854 \times 10^{-12}\mathrm{C^2m^{-1}J^{-1}})} = 2.307 \times 10^{-28}\ \mathrm{Jm}$$

Setting $a_A = a_B = 3.8 \times 10^{-10}$ m, $r = 2a$ (i.e., touching spheres), and for acetonitrile at 293 K $\eta^2 = (1.344)^2$ and $D_s = 37.5$, then $\lambda_{out} = 1.60 \times 10^{-19}$ J. On a molar basis, i.e., multiplied by the Avogadro's constant one obtains $\lambda_{out} = 96.3$ kJ mol$^{-1}$. Thus, the activation barrier due to solvent for this self-exchange reaction is $\lambda_{out}/4 = 24$ kJ mol$^{-1}$. The vast majority of the kinetic barrier for the $Cp_2Fe^{0/+}$ self exchange arises from solvent reorganization, so it is not surprising that the rate in solution is several orders of magnitude lower than the gas-phase rate.

As noted, more general equations can be obtained following the same general assumptions but relaxing the restrictions concerning shape, charge, etc. For example, it is possible to derive $\lambda_{out}$ values for model systems such as charged reactants in an ellipsoid surrounded by the dielectric.[13]

## 3.6 Free Energy Barriers and Rate Constants for Electron Transfer

When the inner-sphere barrier is included, the total classical Franck–Condon barrier to thermal electron transfer for a *self-exchange* reaction is

$$\Delta G^* = \lambda/4 = (\lambda_{out} + \lambda_{in})/4 \tag{25}$$

When a free energy change for the reaction is added as required for exoergic or endoergic reactions, Equation 26 results for intersecting parabolic surfaces.[7]

$$\Delta G^* = \frac{(\lambda + \Delta G^\circ)^2}{4\lambda} \tag{26}$$

Equation 26 reduces to Equation 25 when the free energy difference between the precursor and successor complexes $\Delta G^\circ = 0$. In typical cases where inner-sphere reorganization is small (such as octahedral low spin $d^5/d^6$ transition metal complex couples), $\lambda$ is dominated by the $\lambda_{out}$ term. However, the role of inner-sphere configuration changes has been clearly demonstrated in studies where the contribution of $\lambda_{in}$ increases with experimentally determined metal–ligand bond distance changes, $\Delta d_0$.[39]

The classical electron-transfer rate constant (for the unimolecular P $\rightarrow$ S conversion, not the bimolecular rate constant) is

$$k_{et} = \kappa \nu_n \exp\left(\frac{-\Delta G^*}{RT}\right) \tag{27}$$

where $\kappa$ is the unitless *transmission coefficient* ($\leqslant 1$) and $\nu_n$ (in s$^{-1}$) is the frequency at which the nuclear fluctuations occur that form the transition state. The exponential term is the fraction of the precursor complexes with minimum free energy content $\Delta G^*$ above the equilibrium reactant energy and therefore determines what fraction can pass on to the successor state.

The transmission coefficient $\kappa = 1$ for an adiabatic reaction, as every time the system reaches the transition state the successor complex is formed. In nonadiabatic reactions, $\kappa < 1$. Later in this chapter, an expression for $\kappa$ is given that arises from a quantum mechanical approach to the problem.

The predicted value of $k_{obs}$ in Equation 14 for a bimolecular solution reaction is given by

$$k_{obs} = \kappa Z_{soln} \exp\left(\frac{-\Delta G^{*\prime}}{RT}\right) \tag{28}$$

where $Z_{soln}$ is an effective collision frequency for the reactants (usually $\sim 10^{11} M^{-1}$ $s^{-1}$) and $\Delta G^{*\prime}$ is the *overall* free energy barrier due to both Franck–Condon barriers *and* precursor complex formation. From the equations above,

$$\Delta G^{*\prime} = w + \Delta G^{*} = w + \frac{\lambda}{4}(1 + \Delta G^{\circ}/\lambda)^2 \tag{29}$$

The term $w$ is the work term for formation of the precursor complex from the reactants (Equation 18). The overall free energy change for a bimolecular reaction, $\Delta G^{\circ}_{reacn}$, is typically obtained from the differences between the electrode potentials for the relevant redox couples. $\Delta G^{\circ}$ in Equation 29 is only equal to $\Delta G^{\circ}_{reacn}$ if the work of bringing the reactants together is equal to the work of bringing the products together at the same separation.

## 3.7 Activation Parameters

The fundamental equation for the rate of a reaction according to transition state theory[40] is given in Equation 30, where $L$ is Avogadro's number, $h$ is Planck's constant, and the equilibrium constant for formation of the activated complex is given by $K^{\ddagger}$:

$$k = \left(\frac{RT}{Lh}\right)K^{\ddagger} = \left(\frac{RT}{Lh}\right)e^{-\Delta G^{\ddagger}/RT} \tag{30}$$

It is common to analyze the temperature dependence of the rates of reactions by the Eyring analysis, plotting $\ln(k/T)$ vs. $1/T$ to obtain the activation parameters $\Delta H^{\ddagger}$ and $\Delta S^{\ddagger}$ for the transition state model from Equation 31.

$$\frac{k}{T} = \left(\frac{R}{Lh}\right)e^{\Delta S^{\ddagger}/R}e^{-\Delta H^{\ddagger}/RT} \tag{31}$$

The enthalpy and entropy of activation, $\Delta H^{\ddagger}$ and $\Delta S^{\ddagger}$, are *not* equivalent to the analogous components of the overall reaction free energy barrier $\Delta G^{*\prime}(= \Delta H^{*\prime} - T\Delta S^{*\prime})$. Equation 32 is obtained by equating the right hand side of Equation 30 to that of Equation 28, setting $\kappa = 1$ for an adiabatic reaction, and taking the natural logarithm of both sides.

$$\Delta G^{\ddagger} = \Delta G^{*\prime} - RT\ln(LhZ_{soln}/RT) \tag{32}$$

Now, recognizing from the relation $\Delta G = -RT \ln K$ that

$$-RT \ln K^{\ddagger} = \Delta H^{\ddagger} - T\Delta S^{\ddagger} \tag{33}$$

and differentiating with respect to temperature, Equation 34 is obtained.

$$\frac{d \ln K^{\ddagger}}{dT} = -\frac{d\left(\frac{\Delta H^{\ddagger}}{RT}\right)}{dT} \tag{34}$$

After completing the derivative on the right side and rearranging, Equation 35 results:

$$\Delta H^{\ddagger} = RT^2 \frac{d \ln K^{\ddagger}}{dT} \tag{35}$$

Substituting $K^{\ddagger} = kLh/RT$ from Equation 30, Equation 36 is obtained:

$$\Delta H^{\ddagger} = RT^2 \frac{d \ln k}{dT} - RT \tag{36}$$

To obtain the relationship between the enthalpy of activation and $\Delta H^{*\prime}$, Equation 37 is usually assumed to apply with the collision theory result that $Z_{\text{soln}}$ is proportional to $(T)^{1/2}$. In Equation 37, the term $B$ is a constant that is independent of temperature, and the free energy change $\Delta G^{*\prime}$ of Equation 28 has been replaced by $\Delta H^{*\prime} - T\Delta S^{*\prime}$.

$$k = BT^{1/2} \exp(-\Delta H^{*\prime}/RT) \exp(\Delta S^{*\prime}/R) \tag{37}$$

Since from Equation 37, $\ln k = -\Delta H^{*}/RT + \Delta S^{*\prime}/R + \ln T/2 + \ln B$, substituting $\ln k$ into Equation 36 leads to Equation 38.

$$\Delta H^{\ddagger} = \Delta H^{*\prime} - RT/2 \tag{38}$$

Equation 38 states that the observed activation enthalpy is less than the actual minimum energy barrier on the potential surface for the reaction that would be calculated from the properties of the reactants and solvent.

From Equations 38 and 32, we can obtain an expression for $\Delta S^{\ddagger}$ (Equation 40).

$$\Delta S^{\ddagger} = \frac{\Delta H^{\ddagger} - \Delta G^{\ddagger}}{T} = \frac{\Delta H^{*\prime} - \Delta G^{*\prime}}{T} - R/2 + R \ln(LhZ_{\text{soln}}/RT) \tag{39}$$

$$\Delta S^{\ddagger} = \Delta S^{*\prime} - R/2 + R \ln(LhZ_{\text{soln}}/RT) \tag{40}$$

Although thermodynamic parameters for comparison to theory should be determined from experiment by applying the relations in Equations 32, 38, and 40, the usually small corrections are not always made in the literature. Normally, however, experi-

**Table 1 Outer-Sphere Electron Exchange Reactions**[a]

| Couple | $k_{obs}$ ($M^{-1}\ s^{-1}$) | $\Delta d$ (nm) | $r$ (nm) | $k_{calcd}$ ($M^{-1}\ s^{-1}$) |
|---|---|---|---|---|
| $Fe(H_2O)_6^{2+/3+}$ | 4.2 | 0.014 | 0.65 | 3 |
| $Ru(H_2O)_6^{2+/3+}$ | $\sim 10^2$ | 0.009 | 0.65 | $4 \times 10^3$ |
| $Ru(NH_3)_6^{2+/3+}$ | $3 \times 10^3$ | 0.004 | 0.67 | $1 \times 10^5$ |
| $Ru(bpy)_3^{2+/3+}$ | $4 \times 10^8$ | 0.000 | 1.36 | $1 \times 10^9$ |
| $Co(NH_3)_6^{2+/3+}$ | $\geqslant 10^{-7}$ | 0.022 | 0.66 | $4 \times 10^{-6}$ |
| $Co(bpy)_3^{2+/3+}$ | 18 | 0.019 | 1.36 | 20 |

[a] Adapted from Reference 1, which can be consulted for more details.

mental and theoretical rate constants are compared directly to assess the quality of the theory (as in Table 1).

### 3.8 Cross Reactions and the Cross Relation

From Equations 26 and 29, it is apparent that changing the driving force of a reaction will alter the rate in predictable ways, assuming $\lambda$ is constant. It is also of interest to ask if the overall $\lambda$ value can be calculated for a case where the $\lambda$ values for the individual self-exchange reactions for the donor and acceptor are known. If so, then rates for a large number of *cross reactions* could be estimated without doing the kinetic experiment. The so-called *cross relation* arises from treatments outlined above.

In the derivation of the cross relation, it is assumed that the Franck–Condon free energy barrier $\lambda_{12}$ for a reaction involving two different redox couples is the average of the barriers for the component self-exchange reactions. For example, consider the two self-exchange reactions of Equations 41 and 42, where $R_i$ and $O_i$ are the reduced and oxidized forms of the couple 1 and 2.

$$R_1 + O_1 \rightarrow O_1 + R_1 \ k_{11} \tag{41}$$

$$R_2 + O_2 \rightarrow O_2 + R_2 \ k_{22} \tag{42}$$

If the reactions have Franck–Condon barriers $\lambda_{11}$ and $\lambda_{22}$, respectively, the cross reaction (Equation 43)

$$R_1 + O_2 \rightarrow O_1 + R_2 \ k_{12} \tag{43}$$

is assumed to have a barrier $\lambda_{12} = 0.5(\lambda_{11} + \lambda_{22})$. After accounting for the influence of a nonzero $\Delta G^\circ$ (Equation 26), the cross relation is obtained for the cross reaction rate constant $k_{12}$:

$$k_{12} = (k_{11} k_{22} K_{12} \gamma_{12})^{1/2} \tag{44}$$

where $K_{12} = \exp[-\Delta G^\circ/RT]$ and $\gamma_{12}$ is defined by

$$\ln \gamma_{12} = \frac{\frac{1}{4}(\ln K_{12})^2}{\ln(k_{11}k_{22}/Z_{soln}^2)} \tag{45}$$

In Equation 45, $Z_{soln}$ is the effective collision frequency of the reactants (*not* the charge). For a modest driving force, $\gamma_{12} \simeq 1$. Note Equation 44 only applies if the work terms for all the various reactions are the same (i.e., all reactant couples are similar in charge and size) or if $k_{11}$ and $k_{22}$ are corrected to account for the terms (more general expressions are given in Reference 1).

Although typically of high predictive accuracy for metal complexes, the cross relation (Equation 44) has been less successful for redox reactions of main group compounds. Problems with its use appear to arise from the estimates of self-exchange rate constants for main-group couples (e.g., $O_2^{0/-}$, $ClO_2^{0/-}$, $N_3^{0/-}$) that were derived from application of Equation 44 to cross reactions with metal complex couples.[42] The large difference in radii for such couples invalidate central assumptions in the original derivation of Equation 44, and appropriate corrections yield much improved agreement between theory and experiment.

It is worthwhile to explore the thermodynamics of cross reactions in more detail to assess the separate roles of enthalpy and entropy changes on predicted rates. Let $\Delta X_{11}^*$ and $\Delta X_{22}^*$ be the thermodynamic quantities ($X = G,\ H$, or $S$) for the intrinsic barriers to self-exchange reactions 1 and 2 (these are the total barriers including both Franck–Condon contributions and work terms for formation of the precursor, but the work terms are assumed to be the same for the couples in what follows). For the cross reaction, the quantities $\Delta X_{12}^*$ are given by Equations 46–48.

$$\Delta G_{12}^* = \frac{\Delta G_{11}^* + \Delta G_{22}^*}{2} + \frac{\Delta G_{12}^\circ}{2}(1+\alpha) \tag{46}$$

$$\Delta H_{12}^* = \left(\frac{\Delta H_{11}^* + \Delta H_{22}^*}{2}\right)(1-4\alpha^2) + \frac{\Delta H_{12}^\circ}{2}(1+2\alpha) \tag{47}$$

$$\Delta S_{12}^* = \left(\frac{\Delta S_{11}^* + \Delta S_{22}^*}{2}\right)(1-4\alpha^2) + \frac{\Delta S_{12}^\circ}{2}(1+2\alpha) \tag{48}$$

where $\alpha$ is defined by

$$\alpha = \frac{\Delta G_{12}^\circ}{4(\Delta G_{11}^* + \Delta G_{22}^*)} \tag{49}$$

### 3.9 Entropy Effects in Cross Reactions

In the section on gas-phase reactions, it was noted that adiabatic electron-transfer reactions with negative activation barriers for overall reaction (e.g., Figure 3) can be relatively inefficient under certain circumstances. In the context of Equation 28, this result makes little sense as adiabatic solution reactions with no *free energy* barrier would be predicted to proceed with rate constant $Z_{soln}$. The explanation offered for inefficient gas-phase reactions arises from the differences in densities of states between the nonproductive dissociation of the precursor complex and the electron transfer step. In essence, a high entropy transition state for dissociation can be favored over a low entropy electron-transfer transition state. Entropic barriers can be substantial in solution electron-transfer reactions as well and will be manifested most often in cross reactions with large negative entropy changes $\Delta S^\circ_{12}$.[42] A reaction may have a low or absent potential energy barrier, which is associated with a low enthalpic barrier, but the entropic barriers can contribute to the free energy barrier used in Equation 28.

As an example, consider a cross reaction with a very low free energy change for the cross reaction. Substituting $\Delta G^\circ_{12} \sim 0$ into Equations 46–48, it is readily seen that Equations 50–52 result.

$$\Delta G^*_{12} \simeq 0.5(\Delta G^*_{11} + \Delta G^*_{22}) \tag{50}$$
$$\Delta H^*_{12} \simeq 0.5(\Delta H^*_{11} + \Delta H^*_{22} + \Delta H^\circ_{12}) \tag{51}$$
$$\Delta S^*_{12} \simeq 0.5(\Delta S^*_{11} + \Delta S^*_{22} + \Delta S^\circ_{12}) \tag{52}$$

In other words, about half of the overall enthalpy and entropy changes will appear in the cross reaction enthalpic and entropic barriers, respectively, while the free energy barrier (Equation 50) will be approximately the average of the component self-exchange barriers. Under these conditions $\Delta H^\circ_{12} \simeq T\Delta S^\circ_{12}$, and reactions with large negative entropies will have large negative enthalpy changes $\Delta H^\circ_{12}$. This negative $\Delta H^\circ_{12}$ offsets the enthalpic barriers and $\Delta H^*_{12}$ (Equation 51) can be small or even negative as a result. In these cases, the rate of electron transfer can be slowed primarily by the entropy barrier since $\Delta S^*_{12}$ (Equation 52) will tend to be large and negative.

Large entropy changes are rather common in cross reactions involving transition metal complexes as a result of high possible charges on the reactants and low inner-sphere vibrational frequencies. High ionic charge will affect the ordering of solvent molecules around the complex, and changes in electronic occupancy in the metal *d* orbitals upon change in oxidation state can lead to large changes in metal–ligand frequencies thereby affecting vibrational entropies. As an example, consider the cross reaction

$$\mathrm{Ru(NH_3)_4bpy^{2+} + Fe(CN)_4bpy^- \rightarrow Ru(NH_3)_4bpy^{3+} + Fe(CN)_4bpy^{2-}} \tag{53}$$

which has a $\Delta G^\circ_{12} \simeq 0$ in aqueous solution. From the reaction entropies of the individual couples, $\Delta S^\circ_{12} = -170$ J/mol K can be estimated and $\Delta H^\circ_{12} = -52$ kJ/mol at 298 K.[43] The negative entropy change can be traced primarily to the increase in charges on the products compared to the reactants and the resulting increased ordering of the polar water molecules around the complex ions. From Equation 52, the entropy change therefore contributes about $-T\Delta S^\circ_{12}/2$ or 25 kJ/mol to the activation barrier, and the enthalpic barrier $\Delta H^*_{12}$ (Equation 51) is expected to be small.

Thus, the same entropy considerations that accounted for "slow" gas-phase reactions with small or negative enthalpic barriers can also become important in certain solution reactions, where the passage from precursor complex to the successor complex is disfavored by entropy changes manifested in the transition state. The density of states in the electron transfer transition state for a reaction such as Equation 53 is lowered by the smaller entropy (and lower density of states) in the successor complex, but, unlike gas-phase reactions, much of this reduced entropy arises from the solvation component of the barrier.

Table 1 compares experimental and calculated rate constants for several self-exchange reactions.[1] In addition to inner- and outer-reorganization barriers and ion–ion interactions, corrections for nuclear tunneling have been applied in the theoretical estimates (which effectively increase the calculated rate constants less than a factor of 10 in the cases shown). Calculated rate constants are often within an order of magnitude of the observed, covering a range of ~15 orders of magnitude. Such agreement is a strong validation of the general ideas advanced thus far and suggests that the reactions shown are essentially adiabatic with transmission coefficients near unity.

Some specific cases deserve comment and illustrate general principles described earlier. For the M = Ru series with L = $H_2O$, $NH_3$, and bpy (2,2′-bipyridine), the observed increase in electron transfer rate constants is attributable to declining inner-sphere and outer-sphere barriers as $\Delta d$ decreases and $r$ increases. Comparisons for M = Ru and M = Co with identical ligand sets illustrate the role of inner-sphere barriers, since $\Delta d$ values for M = Co are uniformly larger than for M = Ru due to the conversion from low-spin cobalt(III) to high-spin cobalt(II). Another consequence of the spin change following the one-electron reduction of cobalt(III) complexes is a large, positive contribution to the reaction entropy, primarily as a result of the substantial decrease in skeletal vibrational frequencies in the reduced state.[44]

## 3.10 Potential Energy Surfaces for Solvent Reorganization

Equations for the potential energy surfaces associated with the inner-sphere reorganization energy were given earlier. Less obvious is how to produce a surface for the outer-sphere barriers, since the thermal fluctuation of a dielectric is not readily transformed into a reaction coordinate. However, a molecular view of solvent suggests that the orientational fluctuations embodied in the Franck–Condon solvent barrier are associated with vibrations of the solvent molecules about their positions (called *librations* in a quasi-crystalline model for the solvent). If it is assumed that the overall change in free energy of solvation as the reaction proceeds to the transition state is quadratic and the highly dispersed phonon frequencies can be condensed into a single frequency $\nu_{solv}$ (around 1–50 cm$^{-1}$), then it is apparent that parabolic wells might be used to represent free energy along the reaction coordinate just as in the inner-sphere case. Thus, the energetics of the solvent fluctuations can be obtained from the classical expression $E = \frac{1}{2}kQ_{out}^2$, where $k$ is the effective solvation force constant. The reaction coordinate $Q_{out}$ is implied to be a representation of motion of solvent nuclei toward appropriate bond lengths (and angles) for the transition state, by analogy with the inner-sphere reaction coordinate $Q_{in}$. This is obviously fiction since the chaotic motion of solvent molecules (each undergoing non-harmonic motion) is not "concerted", and, more properly, the reaction coordinate is represented as the difference in energy between the reactant and products.[7]

A solvent vibronic coupling parameter is readily obtained and is analogous to $\Lambda_{in}$ derived in Equation 54. Since the classical barrier in the symmetrical two parabola case is $\Delta E^* = h\nu\Lambda^2/2 = \lambda/4$ (assuming the impact of electronic interaction on the surface is negligible *and* the energy change is equal to the free energy change $\lambda/4$), Equation 54 is obtained by setting $a_A = a_B = a$ in Equation 24. Note that equating the energy change to the free energy change is essentially the same as neglecting entropy changes, and, as shown earlier, this assumption is not always valid.

$$\Lambda_{out} = \left[\frac{2C}{h\nu_{solv}}\left(\frac{1}{a} - \frac{1}{r}\right)\right]^{1/2} \tag{54}$$

$$C = \frac{e^2}{4\pi\varepsilon_0}\left(\frac{1}{\eta^2} - \frac{1}{D_s}\right) \tag{55}$$

Combining and defining $\rho = \nu_{solv}/\nu$, where $\nu$ is the characteristic inner-sphere frequency, Equation 56 is obtained for the lower potential surface

$$E_L = h\nu\left(\frac{Q_{in}^2 + \rho Q_{out}^2}{2} - \left[\left(\Lambda_{in}Q_{in} + \rho\Lambda_{out}Q_{out} + \frac{\Delta E}{h\nu}\right)^2 + \frac{H_{ps}^2}{(h\nu)^2}\right]^{1/2}\right) \tag{56}$$

Note that $\rho < 1$ in all real-world examples since bound ligands have much tighter binding than solvent molecules in the bulk, which interact with one another through weak intermolecular forces. Considering the origins of $\Lambda_{in}$ and $\Lambda_{out}$, it is apparent that energy surface of Equation 56 is most properly a free energy surface.

Figure 10 shows a plot of the free energy of the reaction system as a 3-D surface with $Q_{in}$ and $Q_{out}$ as axes. The resulting surface shows that the transition state is reached through fluctuations along both inner and outer reaction coordinates. Note that the $Q_{out}$ scale has been greatly compressed in the figure since the solvent vibronic coupling parameters are high. The treatment of an electron-transfer reaction on such a surface derived from a "fast" vibrational mode (such as a M–L vibration at $\sim 10^{13}\ s^{-1}$) and a "slow" vibrational modes (such as solvent librations at $\sim 10^{11}\ s^{-1}$) has been described in detail in the literature.[45]

The usual 2-D double well diagram for a reaction in solution can be derived by slicing the 3-D free energy surface along the straight line path from P to S. The reaction coordinate is then a combination of the $Q_{out}$ and $Q_{in}$ modes and the transition state barrier is composed of contributions from the solvent and inner-sphere vibrations. The classical reaction path on this type of surface is actually not the straight line from precursor to successor but is instead a curved path defined by the steepest descent from the transition state to the P and S minima.

From Equation 54, the vibronic coupling parameter $\Lambda_{out}$ gets smaller as the reactants approach and $r$ gets smaller. A plot of the variation of the lower surface vs.

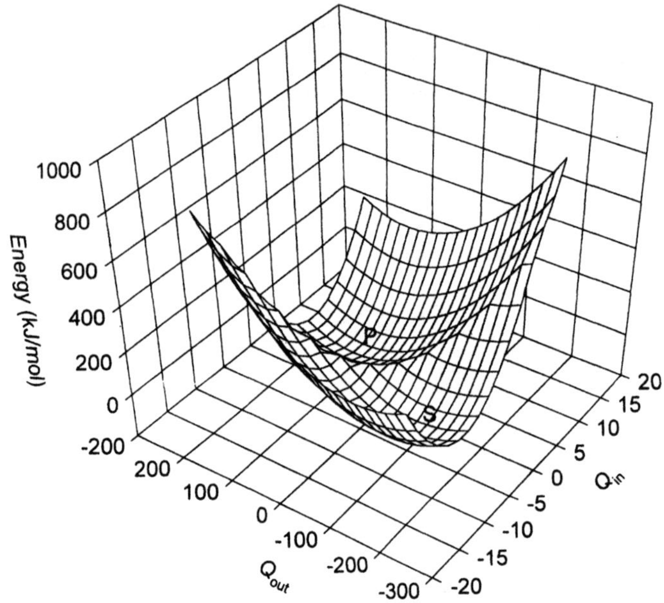


**Figure 10.** Three dimensional classical potential surface for a precursor (P) to successor (S) interconversion (lower surface given by Equation 30). The three dimensional surface is shown as a function of the reaction coordinates $Q_{\text{in}}$ and $Q_{\text{out}}$. This type of surface is obtained by treating the complex motions of the solvent molecules as a single harmonic oscillator of low frequency.

$Q_{\text{out}}$ and $r$ is shown in Figure 11. The plot has been normalized so the minimum energy of the surface is zero at all distances. As $r$ gets smaller, the barrier presented by the solvent is lower. Thus, reducing the distance between reactants is doubly effective in increasing the rate of electron transfer; the solvent barrier is lowered and the electronic coupling increases (Equation 6).

### 3.11 Nonadiabatic Reactions and Quantum Methods

According to Equation 27, the rate constant for electron transfer will be lowered if the transmission coefficient $\kappa$ is less than unity. Such a situation can arise when the electronic orbital overlap between reactants is poor. In the classical model employed so far, $\kappa < 1$ means that an activated complex is formed but the evolution of the system to the successor complex has low probability. In essence, the required solvent and nuclear fluctuations have occurred, overcoming the Franck–Condon barriers, but the reaction stays on a reactant-like surface. Such reactions are said to be *nonadiabatic*. In order for the reaction to be adiabatic, the electronic coupling between the reactants has to be sufficiently large that the electronic states can rapidly oscillate between the two possible electron distributions on the time scale of a collision. In

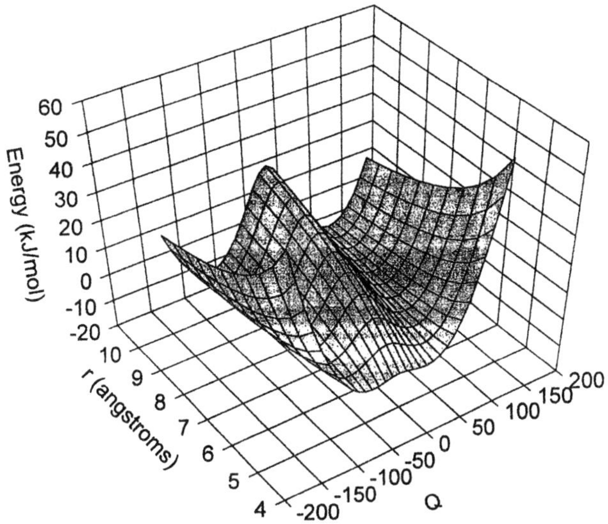


**Figure 11.** Free energy surfaces illustrating the variation in reaction surfaces as a function of the distance of separation $r$ and the outer sphere reaction coordinate $Q_{\text{out}}$ for a symmetrical electron transfer reaction. For an adiabatic process, the reaction is confined to the lower surface as it moves from one side of $Q_{\text{out}} = 0$ to the other. At large separation $r$, the outer sphere reorganizational barrier is higher and the electronic coupling is weak (Equation 6). At smaller $r$ values, the barrier is lower due not only to the lower $\lambda_{\text{out}}$ but also because of electronic delocalization, which lowers the central barrier (Equation 30). The plot has been normalized so the minima are at zero on the energy scale for all values of $r$.

such a case, the reaction can follow the surface in Figure 10, smoothly converting from the precursor to the successor state. The general effect of electronic interaction on potential surfaces is illustrated in Figure 12.

The classical model described so far does an admirable job of calculating inner-sphere and outer-sphere activation barriers to electron transfer and yields good estimates for rate constants of presumably adiabatic reactions. However, it is totally inadequate to predict the rates of nonadiabatic reactions. The classical model also does not predict temperature effects correctly; certain reactions are known to become temperature independent at low temperatures, demonstrating the reactions can proceed by nuclear tunneling without thermal activation. In addition, the classical model underestimates the rates of reactions in the inverted region, where the details of quantum states are critical for imbedded potential surfaces. Thus, much of the theoretical work on solution electron transfer has focused on the quantum approach.[46–49] Some key results will be briefly summarized here.

The time-dependent evolution of the nonadiabatic reacting system is generally assumed to follow the Fermi "Golden Rule", Equation 57. This expression arises from a time-dependent perturbation theory analysis of the conversion of a given electronic-

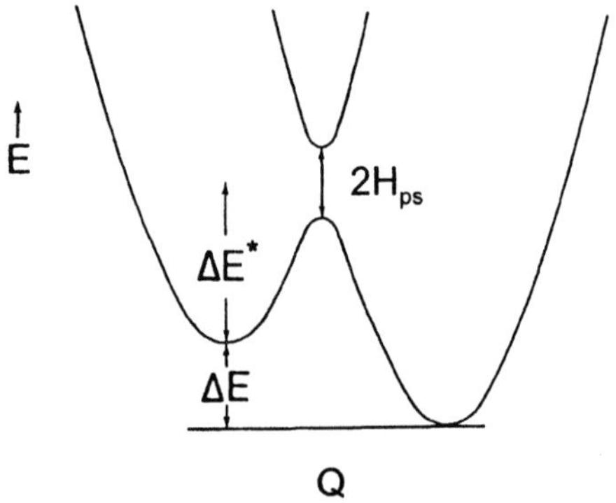


**Figure 12.** Typical potential surface for an exoergic reaction showing the effect of electronic coupling $H_{ps}$ on the activation barrier for electron transfer. The classical barrier $\Delta E^*$ is lowered by the electronic interaction. If the coupling is small, a nonadiabatic reaction results since every passage of the system to the transition state does not lead to products. For large electronic coupling, the barrier is reduced and can even disappear.

vibrational state of the precursor to a different state of the successor.

$$k_{et} = \frac{4\pi^2}{h} H_{ps}^2 (\Omega_{FC}) \tag{57}$$

The electronic coupling of the initial and final states (i.e., precursor and successor states) is quantified by $H_{ps}^2$. Obviously, as the electronic overlap approaches zero then the reaction rate also approaches zero. $\Omega_{FC}$ is the thermally averaged Franck–Condon factor linking the precursor and successor *vibrational* states. It is composed of contributions from all of the precursor P vibrational levels weighted according to their thermal populations, with each P level going to one or more successor levels having the same total energy. In Figure 13, precursor vibrational wavefunctions are shown with the corresponding wavefunctions of the successor state. The lower level (vibrational quantum number $n = 0$) has little overlap with the product function, so $\Omega_{FC}$ is small and the rate for that level is low. On the other hand, at higher values of $n$, which are near the classical transition state energy (the curve crossing at $n = 7$), the overlap is high and the rate via Equation 57 is much higher. The total rate is determined by a weighted sum over all of the contributing levels, and at higher temperatures the population of the higher $n$ levels is large enough so those states dominate the conversion. *Note that Equation 57 is for electron transfer in a reacting complex and does not yield bimolecular rate constants.*

Equation 57 is applicable for reactions with relatively low $H_{ps}$ values (i.e., cases where the electronic communication between the reacting centers is weak), and appropriate expressions for the $\Omega_{FC}$ term can then be developed. In one approach, the low frequency solvent modes are treated as classical oscillators (since $h\nu \ll kT$) while the inner-shell vibrations are handled quantum mechanically. The resulting

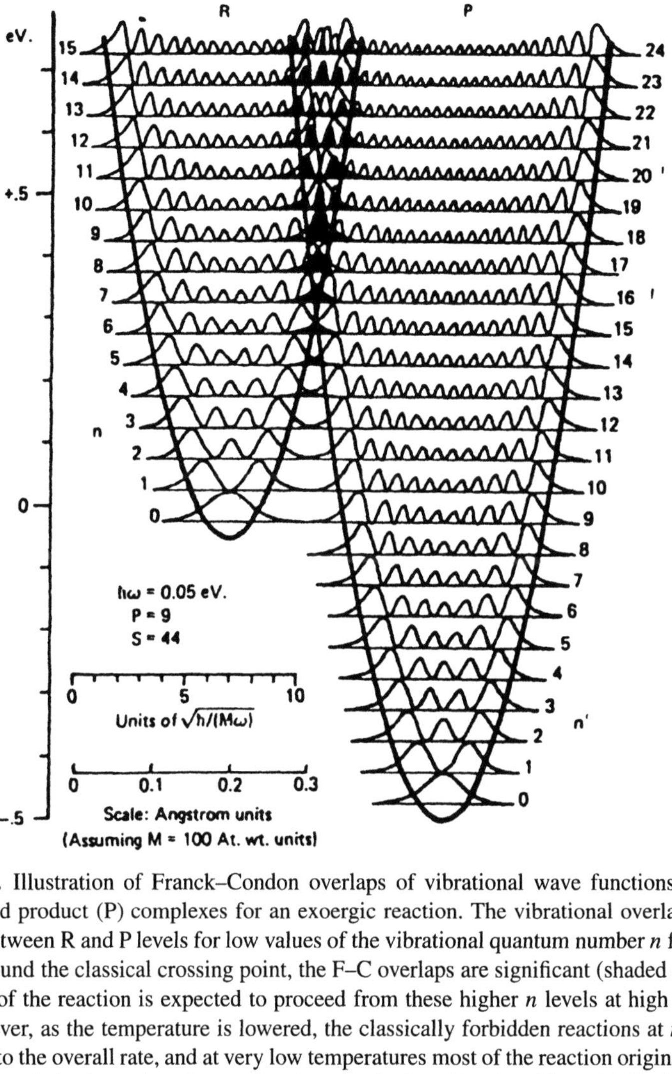


**Figure 13.** Illustration of Franck–Condon overlaps of vibrational wave functions in reactant (R) and product (P) complexes for an exoergic reaction. The vibrational overlap factors are poor between R and P levels for low values of the vibrational quantum number $n$ for the reactant. Around the classical crossing point, the F–C overlaps are significant (shaded overlaps) and much of the reaction is expected to proceed from these higher $n$ levels at high temperature. However, as the temperature is lowered, the classically forbidden reactions at $n < 7$ do contribute to the overall rate, and at very low temperatures most of the reaction originates from the $n = 1$ level. At that point, the reaction is activationless. Reprinted from DeVault, D., Ed., *Quantum Mechanical Tunneling in Biological Systems*; Cambridge University Press: Cambridge, 1989.

expression for $k_{et}$ is given by

$$k_{et} = \frac{4\pi^2}{h} H_{ps}^2 \sum_{\upsilon_r \upsilon_p} (4\pi\lambda_{out} RT)^{-1/2} e^{-(\Delta G^0 + \Delta\varepsilon + \lambda_{out})^2 / 4\lambda_{out} RT} (\Omega_{FC})_\upsilon \, p(\upsilon_r) \quad (58)$$

In Equation 58, $\upsilon_r$ and $\upsilon_p$ are the vibrational quantum numbers of the reactant and product states, $p(\upsilon_r)$ is the Boltzmann probability of the system being in the state

$\upsilon_r$, $\Delta\varepsilon$ is the difference in energy between the reactant and product vibrational energy levels, and the Franck–Condon factors are only those for the inner-shell vibrational modes. In the high temperature limit, even the inner-shell vibrations behave classically and Equation 59 results.

$$k_{et} = \frac{4\pi^2}{h} H_{ps}^2 (4\pi\lambda RT)^{-1/2} e^{-(\Delta G^0+\lambda)^2/4\lambda RT} \tag{59}$$

Comparing Equation's 59 with the classical Equation 27, $\kappa_{el}$ is interpreted as

$$\kappa_{el} = \frac{2\pi H_{ps}^2}{h\nu_n}\left(\frac{\pi}{\lambda RT}\right)^{1/2}. \tag{60}$$

The magnitude of the electronic coupling term $H_{ps}$ has been the subject of quantum mechanical studies. Conceptually, the *superexchange* model is a convenient starting point for such calculations. Since the *d* orbitals that serve as donor and acceptor orbitals have poor overlap, it is supposed that the ligand based orbital states provide a pathway for electronic interaction. In some cases, the transferring electron is delocalized from the donor into an unoccupied orbital on a ligand, thereby increasing its overlap with an accepting orbital on the other center—this is commonly called the *electron transfer* superexchange pathway. In other cases, an electron is delocalized from a ligand to the acceptor orbital in a process called *hole transfer*. The resulting hole in the ligand orbital manifold is filled by the donor metal electron to complete the transfer. For metal complex reactants, the energies and transition probabilities of the corresponding charge transfer transitions (metal to ligand, MLCT, and ligand to metal, LMCT) can be used to ascertain the importance of these two pathways, which are defined schematically in Figure 14. Note that the superexchange mechanism does *not* imply that the ligand-based states are true intermediates in the electron transfer process, but instead provides a method for estimating the overall matrix element $H_{if}$ between the initial and final states of the process. Equation 61 shows the perturbation theory result for the total interaction matrix element.[50]

$$H_{if} = \sum_j \frac{H_{0,CTj}^{ii} H_{CTj,CTj}^{if} H_{CTj,0}^{ff}}{(\Delta E_{0,CTj}^{ii})^2} \tag{61}$$

where the off-diagonal matrix elements are

$$H^{if} = \int \Psi_i^* H \Psi_f d\tau \tag{62}$$

and the diagonal matrix elements are

$$H^{ii} = \int \Psi_i^* H \Psi_i d\tau \tag{63}$$

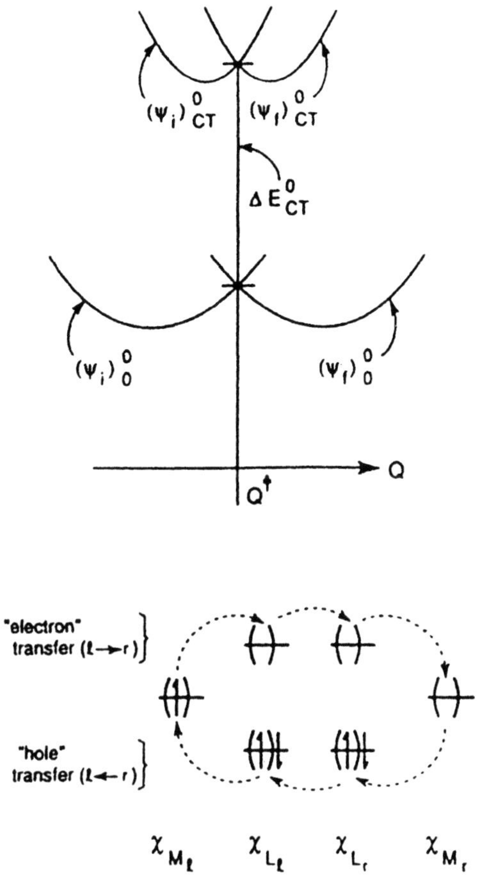


**Figure 14.** Illustration of electron vs. hole transfer in the ligand-mediated superexchange mechanism of electronic coupling. The top diagram shows the initial state $i$ and final state $f$ potential surfaces along with charge transfer (CT) excited states. Mixing of CT excited state character with the ground state provides coupling of the donor and acceptor orbitals via ligand orbitals, which have substantial overlap. In the "electron transfer" mechanism, empty ligand orbitals provide a pathway for coupling and MLCT states are involved. In "hole transfer", filled ligand orbitals provide coupling via LMCT states. Reprinted from Newton, M.D.; Ohta, K; Zhong, E. *J. Phys. Chem.* **1991**, *95*, 2317.

Equation 61 essentially states that it is the CT states $j$ that connect the initial and final zero order states to each other since the direct interaction matrix element $(H^{\text{if}}_{0,0})$ is unimportant. The perturbation theory expression in Equation 61 also indicates that the interaction depends inversely on the energy between the ground and CT states, $\Delta E_{0,\text{CT}}$. In some cases, the CT state energies can be assessed experimentally as spectroscopic transition energies. If the MLCT states are low lying, then the electron transfer mechanism can dominate since $\Delta E$ is small for those terms. Likewise, hole transfer can dominate if the LMCT states are low in energy. A computationally tractable approach for transition metal complexes uses semi-empirical molecular orbital theory to find the interaction energies.

As an example, for the prototype ferrocene-ferrocenium self-exchange reaction, the calculated value of $H_{if}$ depends on the relative orientations of the reactants and ranges from 11–135 $cm^{-1}$ when the metallocenes are in close contact,[51] placing the reaction at the estimated crossover (usually estimated to be $\sim$100 $cm^{-1}$) from nonadiabatic to adiabatic behavior.

In addition, the magnitude of $H_{if}$ can be estimated from the intensities of intervalence transfer bands in mixed-valence systems (Chapter 3).[52]

## 4 INTRAMOLECULAR ELECTRON TRANSFER

Although the electron-transfer theory described so far was developed for bimolecular reactions, it is clear that when donors and acceptors are at fixed distances one can use a subset of the overall model above. Essentially, expressions given above that estimate the unimolecular rate constant $k_{et}$ for the collisional supercomplex are used for intramolecular rate calculations. Both inner-sphere and outer-sphere barriers are present, although in really complex systems such as protein–protein reactions it can be unrewarding to model the nature of the medium as a dielectric continuum. Obviously, one need not be concerned about the precursor complex formation constant, since 100% of the reactants are in the precursor configuration. The effect of weak electronic coupling becomes more important since reactants may be at rather large distances from one another and the electronic overlap between the initial and final states can be quite weak (i.e., $H_{ps}$ in Equation 57 is small). Thus, many such reactions are in the nonadiabatic regime ($\kappa_{el} \ll 1$).

The great advantage of studying intramolecular electron transfer arises from the removal of collision kinetics and thermodynamics from the kinetic modeling problem. In addition, it also becomes possible in some systems to study the Franck–Condon reorganization energy directly by measuring the intervalence transition energy spectroscopically (see Chapter 3).

Modeling of complex intramolecular systems has usually been accomplished with nonadiabatic theory. Once again, the issue of the magnitude of the $H_{ps}$ becomes relevant, and superexchange theory can provide insight. Here the important states are clearly those that involve the bridging ligand between the donor and acceptor sites. For simple ligands, the theoretical approaches are rather easily established,[53] but for complex media (e.g., proteins), the appropriate methods are less obvious.

### 4.1 Advanced Time-Dependent Methods

Intramolecular electron-transfer reactions for gas-phase species have not received the attention in the quantum theoretical community that bimolecular reactions have attracted. There has been some initial progress for systems studied by the time-dependent electron-nuclear dynamics (END) theory.[21] The advantage of this and related methods is that it is not necessary to break the transfer into two separate steps (geometry change and electronic transition), but instead the whole electron-nuclear system is considered in the calculations. Furthermore, there is no conceptual reason to separate the calculation into inner-sphere and outer-sphere (solvent) subsystems, although all work thus far has been strictly for gas-phase systems. Some initial results have been reported for the prototype system $[Li–H–Li]^+$. For a linear structure, the two minimum energy states are $Li^+$—HLi and LiH—$Li^+$. The reaction coordinate in a traditional potential surface analysis would be the motion of the bridging H from one Li center to the other. To apply the END theory to the interconversion of these two states (a formal electron transfer), the Li–H bond distances are distorted and the evolution of the system is calculated over time. Figure 15 presents the time-dependent change in the Mulliken populations for $\alpha$ and $\beta$ electrons (spins $1/2$ and $-1/2$) on one of the Li centers. The loss of one of the $\alpha$ electrons as the system evolves corresponds to electron transfer of an $\alpha$ electron to the other Li center. The rate of electron transfer for the system, based on a rather simple model for a macroscopic sampling of reactants, is $\sim 10^{13}$ $s^{-1}$ at 300 K. This example shows how changes in nuclear positions coupled to electron transfer can lead to "slow" electron transfer without using the traditional potential surface methods. Such detailed quantum calculations for more complex species and solvated systems are (currently) daunting computationally but could introduce a new level of sophistication to the theory of electron-transfer reactions.

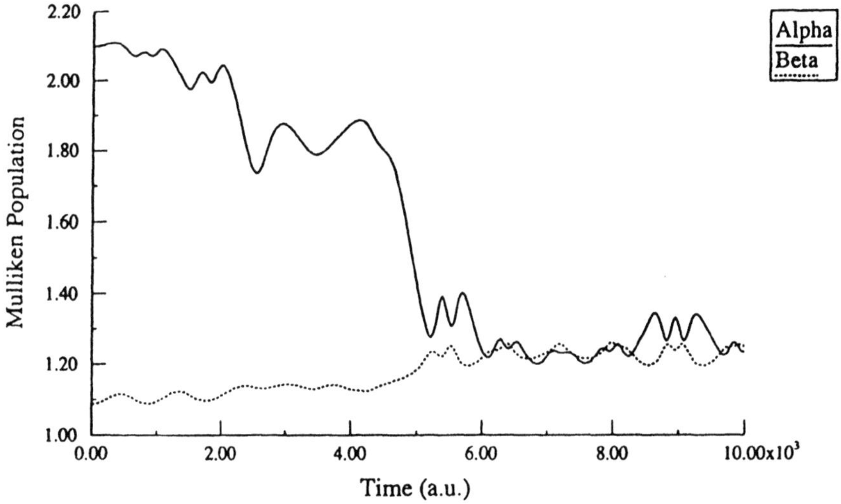


**Figure 15.** Change in the Mulliken populations on the "donor" Li in the model $Li–H–Li^+$ binuclear electron transfer system. An alpha spin electron transfers to the acceptor Li center in about 5,000 atomic time units or $\sim 10^{-13}$ s. The oscillations in the evolution of the spin populations arises from direct, simultaneous evaluation of the combined electron/nuclear wavefunctions. The reaction was initiated at $t = 0$ with a 10% displacement of the equilibrium Li–H distances. Reprinted from Deumens, E.; Diz, A.; Longo, R.; Ohrn, Y. *Rev. Mod. Phys.* **1994**, *66*, 917.

## 5 BIOLOGICAL ELECTRON TRANSFER

Biological electron transfer is currently an area of intense interest among theorists and experimentalists.[1,54] Electron transfer is a central feature of the some of the most critical aspects of biochemistry, especially those involved in energy conversion and storage. Examples include oxidative phosphorylation in mitochondria of eukaryotes and photosynthesis in chloroplasts of plant cells. A remarkable feature of biological electron transfer is the coupling of the electron transfer event to other energy-requiring processes such as proton pumping across membranes.[55] In simple homogeneous chemical electron-transfer reactions, any excess free energy of the reaction can be released as heat ($\Delta H$) or loss of entropy ($T\Delta S$).

A key question to be answered for biological electron transfer is how the protein matrix that is usually present around and between the donor and acceptor sites acts to control the rate and specificity of the redox process. Substantial progress on these issues has been made by the use of various synthetic models composed of modified redox proteins such as cytochrome *c* and plastocyanin.[56] Central theoretical issues include the "pathway" of electron transfer through proteins when long peptide chains connect donors and acceptors. Such pathways can be modeled by variants of the superexchange model described earlier in this chapter. In addition to the important electronic coupling of the redox active metal center to the protein chain, the many different types of linkages in the protein bridge provide electronic interactions of varying efficiency. Presently, a thorough quantum mechanical evaluation of the possible pathways has not been accomplished, but one can use a parameterization approach in which polypeptide structural features such as covalent links, hydrogen bonds, and through-space links are assigned a coupling efficiency.[57] The overall efficiency of coupling by any particular route through the protein can then be evaluated as a product of individual links.

## REFERENCES


1. Marcus, R. A.; Sutin, N. *Biochim. Biophys. Acta* **1985**, *811*, 265.
2. Marcus, R. A. *Electrochim. Acta* **1968**, *13*, 995.
3. Newton, M. D.; Sutin, N. *Annu. Rev. Phys. Chem.* **1984**, *35*, 437.
4. Sutin, N. *Prog. Inorg. Chem.* **1983**, *30*, 441.
5. Sutin, N. *Acc. Chem. Res.* **1982**, *15*, 275.
6. Marcus, R. A. *Rev. Mod. Phys* **1993**, *65*, 599.
7. Barbara, P. F.; Meyer, T. J.; Ratner, M. A. *J. Phys. Chem.* **1996**, *100*, 13148.
8. Marcus, R. A. *J. Photochem. Photobiol.* **1994**, *82*, 1–3.
9. Sutin, N. *J. Phys. Chem.* **1986**, *90*, 3465.
10. Hush, N. S. *Trans. Faraday Soc.* **1961**, *57*, 557.
11. Marcus, R. A. *J. Chem. Phys.* **1956**, *24*, 966.
12. Hush, N. S. *J. Chem. Phys.* **1958**, *28*, 962.
13. Cannon, R. D. *Electron Transfer Reactions*; Butterworth: London, 1980.

14. Richardson, D. E.; Christ, C. S.; Sharpe, P.; Eyler, J. R. *J. Am Chem. Soc.* **1986**, *109*, 3894.
15. Bates, D. R.; McCarroll, R. *Adv. Phys.* **1962**, *11*, 39.
16. Bates, D. R.; Reid, R. H. G.; *Proc. Royal Soc. London A* **1969**, *310*, 1.
17. McCarroll, R. In *Atomic and Molecular Collision Theory*, Gianturco, F. A., Ed; Plenum Press: New York, 1980; Ch. 2.
18. See any physical chemistry textbook, such as Atkins, P. *Physical Chemistry*, $4^{th}$ Ed.; Freeman: New York, 1990.
19. Su, T.; Bowers, M. T. In *Gas Phase Ion Chemistry*, Vol. 1, Bowers, M. T., Ed.; Academic Press: New York, 1979.
20. Gurnee, E. F.; Magee, J. L. *J. Chem. Phys.* **1957**, *26*, 1237.
21. Deumens, E.; Diz, A.; Longo, R.; Öhrn, Y. *Rev. Mod. Phys.* **1994**, *66*, 917.
22. Richardson, D. E. *J. Phys. Chem.* **1986**, *90*, 3697.
23. Bowers, M. T.; Elleman, D. D. *Chem. Phys. Lett.* **1972**, *16*, 486.
24. Bates, D. R.; Lynn, N. *Proc. Royal Soc. 253*, 141.
25. Libby, W. F. *J. Phys. Chem.* **1952**, *56*, 863.
26. Grimsrud, E. P.; Chowdhury, S.; Kebarle, P. *J. Chem. Phys.* **1985**, *83*, 1059.
27. Eyler, J. R.; Richardson, D. E. *J. Am. Chem. Soc.* **1985**, *107*, 6130.
28. Richardson, D. E.; Eyler, J. R. *Chem. Phys.* **1993**, *176*, 439.
29. Han, C.; Wilbur, J. L.; Brauman, J. I. *J. Am. Chem. Soc.* **1992**, *114*, 887.
30. Forst, W. *Theory of Unimolecular Reactions*; Academic Press: New York, 1973.
31. Robinson, P. J.; Holbrook, K. A.; *Unimolecular Reactions*; Wiley-Interscience: New York, 1972.
32. Marcus, R. A. *Disc. Faraday Soc.* **1960**, *29*, 21.
33. Siders, P. Marcus, R. A. *J. Am. Chem. Soc.* **1981**, *103*, 741.
34. Miller, J. R.; Calcaterra, L. T.; Closs, G. L. *J. Am. Chem. Soc.* **1984**, *106*, 3047.
35. Laudenslager, J. B.; Huntress, W. T.; Bowers, M. T. *J. Chem. Phys.* **1974**, *61*, 4600.
36. Brown, G; Sutin, N. *J. Am. Chem. Soc.* **1979**, *101*, 883.
37. Wahl, A. C. *J. Am. Chem. Soc.* **1954**, *76*, 4054.
38. Richardson, D. E. *Inorg. Chem.* **1990**, *29*, 2779.
39. Brunschwig, B. S.; Creutz, C.; Macartney, D. H.; Sham, T.-K.; Sutin, N. *Faraday Disc Chem. Soc.* **1982**, *74*, 113.
40. Laidler, K. J. *Chemical Kinetics*; Harper and Row: New York, 1987.
41. Awad, H. H.; Stanbury, D. M. *J. Am. Chem. Soc.* **1993**, *115*, 3636.
42. Marcus, R. A.; Sutin, N. *Inorg. Chem.* **1975**, *14*, 213.
43. Yee, E. L.; Weaver, M. J. *Inorg. Chem.* **1980**, *19*, 1077.
44. Richardson, D. E.; Sharpe, P. *Inorg. Chem.* **1993**, *32*, 1809.
45. Brunshwig, B. S.; Sutin, N. *J. Am. Chem. Soc.* **1989**, *111*, 7454.
46. Dogonadze, R. R.; Kuznetsov, A. M.; Levich, V. G. *Electrochim. Acta* **1968**, *13*, 1025.
47. Levich, V. G. *Adv. Electrochem. Electrochem. Eng.* **1966**, *4*, 249.
48. Jortner, J. *J. Chem. Phys.* **1976**, *64*, 4860.
49. Ulstrup, J. *Charge Transfer Processes in Consensed Media. Lecture notes in Chemistry*; Springer-Verlag: New York, 1979; Vol. 10.
50. Newton, M. D. *Chem. Rev.* **1991**, *91*, 767.

51. Newton, M. D.; Ohta, K.; Zhong, E. *J. Phys. Chem.* **1991**, *95*, 2317.
52. Hush, N. S. *Prog. Inorg. Chem.* **1967**, *8*, 391.
53. Richardson, D. E.; Taube, H. *J. Am. Chem. Soc.* **1983**, *105*, 40.
54. Bowler, B. E.; Raphael, A. L.; Gray, H. B. *Prog. Inorg. Chem.* **1990**, *38*, 259.
55. Fraústo da Silva, J. J. R.; Williams, R. J. P. *The Biological Chemistry of the Elements*; Clarendon Press: Oxford, 1991; Ch. 5.
56. Gray, H. B.; Winkler, J. R. *Annu. Rev. Biochem.* **1996**, *65*, 537–561.
57. Beratan, D. N.; Onuchic, J. N.; Hopfield, J. J. *J. Chem. Phys.* **1987**, *86*, 4488.

# 3 Mixed Valence

PAUL N. SCHATZ
Chemistry Department
University of Virginia
Charlottesville, VA 22901, USA
E-mail: pns@virginia.edu



*Inorganic Electronic Structure and Spectroscopy, Volume II: Applications and Case Studies.*
Edited by E. I. Solomon and A. B. P. Lever.
ISBN 0-471-32682-8. 

## 1 INTRODUCTION

My aim in this article is to present a concise, self-contained account of one general method of treating mixed-valence systems, based on a model (PKS[1,2] and its MO and two-mode extensions) which takes explicit account of the vibronic coupling inherent in such systems. For simplicity, I discuss only "dimeric" (two-center) systems. And I restrict the discussion to the work of past collaborators and myself in order to achieve a unified, self-consistent account within the allotted space. In doing this, I do not mean to imply that the work of others is unimportant. I note particularly the pioneering work of M. J. Ondrechen, who considered both the role of the bridging ligand in coupling the dimer centers and the contribution of additional modes to the vibronic coupling, and who also introduced the use of a molecular orbital basis.

We shall start by considering a symmetric mixed-valence molecule in which two "monomer" centers differ by one in oxidation number. In the absence of interaction between the two centers, the result is in fact a very simple Jahn–Teller system. When the centers are allowed to interact, the electronic degeneracy is lifted, but the essential Jahn–Teller nature of the system remains. The essence of the PKS model is the treatment of such a system. In the course of solution, we get a very illuminating example of the breakdown of the Born–Oppenheimer appoximation and the significance of Jahn–Teller interactions, in particular why and when the Born–Oppenheimer approximation fails and what to do about it. Because of the relative simplicity of the system, it is possible to present an essentially exact solution with minimal mathematics and easily pictured (two-dimensional) potential energy surfaces in the simplest case. The first part of this article closely follows a review[3] which described the original PKS model in detail.

Following this initial formulation, a variety of extensions of the model are presented and illustrated. Most important is the introduction of a second vibrational coordinate (Section 3), because in its absence, the model is only applicable to strongly localized systems. Other important extensions treat the unsymmetrical dimer and the two-electron case, i.e., two centers which differ by two in oxidation number. But in all cases, the crucial need is to deal with vibronic interactions which invalidate the usual Born–Oppenheimer view that each state can be approximated as the product of a single electronic and vibrational function.

## 2 ORIGINAL PKS MODEL

We use as a prototypical model system the celebrated Creutz–Taube (CT-ion), $\mu$-pyrazine bis(pentaammineruthenium) (5+):

and we stress at the outset that though we repeatedly use this ion to illustrate and compare different approaches, no inference should be drawn that the various models are in any way restricted either to Ru ions as mixed-valence centers or the pyrazine ring as the electronic coupling agent between them. The mixed-valence nature of the CT-ion is evident from the fact that the formal oxidation states of the two Ru atoms are different—2+ and 3+ or 3+ and 2+, respectively. Thus we picture a (dimeric) mixed-valence system as consisting of two "monomers," designated A and B, which are associated with formal oxidation states M and N, respectively. In the CT-ion, for example, the subunits would be the Ru atoms and surrounding ligands with M $= 3+$, N $= 2+$. We designate the electronic Hamiltonion operators associated with these two moieties as $H_{el}^{A}$ and $H_{el}^{B}$ respectively, and if the corresponding wave functions are $\psi_{M}^{A}$ and $\psi_{N}^{B}$, then:

$$\begin{aligned} H_{el}^{A}\psi_{M}^{A} &= W_{M}^{A}\psi_{M}^{A} \\ H_{el}^{B}\psi_{N}^{B} &= W_{N}^{B}\psi_{N}^{B} \end{aligned} \tag{1}$$

We assume for the moment that all nuclei are held fixed so that the nuclear kinetic energy ($T_n$) is zero. If there is no interaction (coupling) between these subunits, the Schrödinger equation for the composite system $H_{el}^{A} + H_{el}^{B}$ is:

$$\begin{aligned} \left(H_{el}^{A} + H_{el}^{B}\right)\psi_{M}^{A}\psi_{N}^{B} &\equiv \left(H_{el}^{A} + H_{el}^{B}\right)\psi_{a} \\ &= \left(W_{M}^{A} + W_{N}^{B}\right)\psi_{a} \equiv W_{a}\psi_{a} \end{aligned} \tag{2}$$

Let us suppose that moieties B and A are, respectively, oxidizable and reducible. Then we can anticipate another state of the system,

$$\psi_{b} = \psi_{N'}^{A}\psi_{M'}^{B} \tag{3}$$

where $n$ electrons have been transferred from $B$ to $A$ ($M' = N + n$, $N' = M - n$). Then an equation directly analogous to Equation 2 applies to Equation 3 (substituting $M'$ for $N$, $N'$ for $M$, and $b$ for $a$). Let us now imagine that an electronic interaction, symbolized by $V^{AB}$, occurs between $A$ and $B$. We can readily show that $V^{AB}$ couples (mixes) $\psi_a$ and $\psi_b$. Thus writing

$$\psi_k = c_a\psi_a + c_b\psi_b \tag{4}$$

use of the linear variation method yields the secular equations

$$\sum_{j=a,b} c_j(H_{ij} - S_{ij}W_k) = 0; \quad (i = a, b);\ (k = 1, 2) \tag{5}$$

and the $W_k$ are the two roots of the secular determinant

$$|H_{ij} - S_{ij}W_k| = 0; \quad (i, j = a, b);\ (k = 1, 2) \tag{6}$$

where

$$H_{ij} = \langle\psi_i|H_{el}|\psi_j\rangle, \qquad S_{ij} = \langle\psi_i|\psi_j\rangle, \qquad H_{el} = H_{el}^A + H_{el}^B + V^{AB} \tag{7}$$

Using Equation 2 and its analogue, we obtain

$$H_{aa} = W_a + V_{aa}, \qquad H_{bb} = W_b + V_{bb}, \qquad H_{ab} = V_{ab} \tag{8}$$

where $V_{ij} \equiv \langle\psi_i|V^{AB}|\psi_j\rangle$ with $i, j = a, b$, and we have assumed that centers $A$ and $B$ are so far apart that

$$\langle\psi_i|\psi_j\rangle = \delta_{ij}, \quad i, j = a, b \tag{9}$$

Using the definition

$$E_l \equiv W_b + V_{bb} - W_a - V_{aa} \tag{10}$$

and defining our energy zero by $W_a + V_{aa} = 0$, Equation 6 becomes

$$\begin{vmatrix} -W_k & V_{ab} \\ V_{ab} & E_l - W_k \end{vmatrix} = 0 \tag{11}$$

The lower root of Equation 11, corresponding to the ground state, is

$$W_1 = \frac{1}{2}\left[E_l - \sqrt{E_l^2 + 4V_{ab}^2}\right] \tag{12}$$

and the corresponding coefficients in Equation 4 are given by

$$\frac{c_b}{c_a} = \frac{W_1}{V_{ab}} = \frac{1}{2}\left[R - \sqrt{R^2 + 4}\right] \tag{13}$$

with $c_a^2 + c_b^2 = 1$ where

$$R \equiv \frac{E_l}{V_{ab}} \tag{14}$$

We see from Equation 13 that $V_{ab} \neq 0$ indeed couples $\psi_a$ and $\psi_b$, that is, produces a finite ratio, $c_b/c_a$. The ratio, $c_b^2/c_a^2$, determines the fraction of $\psi_b$ in the ground state (Equation 4), and thus directly measures the degree of *valence mixing or valence delocalization* in the ground state. The quotient $c_b/c_a$ in turn is determined by $R$ (Equation 13), which is directly proportional to $E_l$, the energy separation of the two "valence states," and inversely proportional to the interaction ($V_{ab}$) that couples them—Equation 14.

## 2.1 The Robin and Day (R&D) Classification Scheme

The R&D classification scheme,[4,5] though simple, has significant utility. Besides marking the first systematic classification of mixed-valence systems, the scheme at least roughly correlates, within each of three classes, a variety of physical properties including spectroscopic, electrical, and magnetic. We show later that the inclusion of a vibronic coupling parameter along with $E_l$ and the electronic coupling ($V_{ab}$) permits a more precise interpretation of the R&D scheme over the entire range of valence delocalization. The R&D scheme focuses on the quantity $R$, and more particularly on $E_l$. Mixed-valence compounds are divided into three classes (I, II, III) depending on the size of $E_l$. If $E_l$ is very large (i.e., $R^{-1} \ll 1$), then appropriately expanding Equation 13 gives:

$$\frac{c_b^2}{c_a^2} \approx \left(\frac{V_{ab}}{E_l}\right)^2 + \cdots \tag{15}$$

R&D define Class I compounds to be those in which the ions of differing valences are in sites of very different symmetry and/or ligand field strength (very large $E_l$) so that by Equation 15, $c_b^2 \approx 0$. In the other extreme, if the sites become very similar so that $E_l$ is very small (i.e., $R \ll 1$), expanding Equation 13 gives:

$$\frac{c_b^2}{c_a^2} \approx 1 - \frac{E_l}{V_{ab}} + \frac{1}{2}\left(\frac{E_l}{V_{ab}}\right)^2 + \cdots \tag{16}$$

Then in the limit of identical sites ($E_l = 0$), $c_b^2 = c_a^2$. R&D characterize compounds in this category as Class III. However, as we shall see later (Section 2.6), *this description is incomplete because vibronic coupling has been neglected. Though the two sites may be identical, the system can nevertheless vary from completely (valence) delocalized to completely (valence) localized, depending on the magnitude of the vibronic coupling*. With appropriate modification and in the spirit of the R&D scheme, Class III systems are those in which there is strong valence delocalization. That is, there is a single stable nuclear configuration corresponding to intermediate oxidation states on the two centers. Class II compounds are then intermediate between Class I and Class III. There is some delocalization, but different, distinct oxidation states are still identifiable on the two centers.

## 2.2 Solution of the Static Problem in the Symmetrical Case

Let us return to Equation 1 *et seq.* and initially assume that centers A and B are equivalent (the symmetrical $A \equiv B$ case). We later assign a precise meaning to this statement. For the moment, it is simply assumed that $\psi_a$ and $\psi_b$ (Equations 2 and 3) are electronically degenerate ($W_a = W_b$) with $M' = M$, $N' = N$. The Creutz–Taube ion, as mentioned earlier, is an example of such a system with $M = 3+$ and $N = 2+$. We also assume that there is no interaction between the two subunits, that is, $V^{AB} = 0$. Our zeroth-order electronic wave functions are thus

$$\psi_a = \psi_M^A \psi_N^B, \qquad \psi_b = \psi_N^A \psi_M^B \tag{17}$$

We now allow nuclear motion and seek to associate vibrational wave functions with $\psi_a$ and $\psi_b$. The first step is to work out vibrational functions for the subunits $A$ and $B$ by expanding the potential energy in a power series in the normal coordinates, keeping terms through quadratic (the harmonic approximation).[6,7] Solution of the vibrational wave equation yields eigenfunctions in the form of products of harmonic oscillator wave functions.[6] If it is assumed, however, that the subunits have the same point group symmetry in both oxidation states ($M$ and $N$)—an assumption made throughout this treatment—then it is only necessary to consider totally symmetric normal coordinates. This is explicitly demonstrated in Section 2.7. Thus assuming that there is just one totally symmetric normal coordinate associated with each subunit, designated respectively $Q_A$ and $Q_B$, the vibrational potential energy of subunit $A$ in oxidation state $M$ can be written

$$W_M^A = W_M^0 + l_M^A Q_A + \frac{1}{2} k_M^A Q_A^2 \tag{18}$$

with analogous expressions for $W_N^A$, $W_M^B$, and $W_N^B$, where $W_M^0 \equiv W_M^A(Q_A = 0) = W_M^B(Q_B = 0)$. Equation 18 is a parabola with curvature $k_M^A$; its minimum occurs at $(-l_M^A/k_M^A)$ with the value, $W_M^0 - \frac{1}{2}(l_M^A)^2/k_M^A$. We write the Hamiltonian operator for subunit $A$,

$$H^A = H_{el}^A + T_n^A \tag{19}$$

where $T_n^A$ is the nuclear kinetic energy operator, and internuclear repulsion terms are included in $H_{el}^A$. Thus the electronic Schrödinger equation for the composite system in which subunit $A$ is in oxidation state $M$ and subunit $B$ is in oxidation state $N$ is given by:

$$\left(H_{el}^{A} + H_{el}^{B}\right)\psi_a = \left(W_M^A + W_N^B\right)\psi_a \equiv W_a\psi_a \tag{20}$$

Equations exactly analogous to 19 and 20 apply if the interchanges $A \rightleftharpoons B$, and $a \to b$ are made. We define the zero of energy by $W_N^0 + W_M^0 = 0$, and arbitrarily assume that the lower oxidation state ($N$) is lower in energy. We choose origins by setting $Q_A$ and $Q_B$ to zero at the minima of $W_N^A$ and $W_N^B$, respectively, so $l_N^A = l_N^B = 0$. Setting $l_M^A = l_M^B \equiv l$, $k_M^A = k_M^B \equiv k_M$ and $k_N^A = k_N^B \equiv k_N$ and using Equation 18 and its analogues, we get:

$$\begin{aligned} W_a &= lQ_A + \frac{1}{2}k_M Q_A^2 + \frac{1}{2}k_N Q_B^2 \\ W_b &= lQ_B + \frac{1}{2}k_M Q_B^2 + \frac{1}{2}k_N Q_A^2 \end{aligned} \tag{21}$$

Let us introduce the new coordinates:

$$Q_{\pm} = \frac{1}{\sqrt{2}}(Q_A \pm Q_B) \tag{22}$$

Equation 21 becomes

$$\begin{aligned} W_a = {} & \frac{1}{\sqrt{2}}lQ_- + \frac{1}{4}(k_N + k_M)Q_-^2 + \frac{1}{2}(k_M - k_N)Q_+Q_- \\ & + \frac{1}{\sqrt{2}}lQ_+ + \frac{1}{4}(k_N + k_M)Q_+^2 \end{aligned} \tag{23}$$

and $W_b$ is obtained if $(-Q_-)$ is substituted for $Q_-$ in Equation 23. *We now make the approximation $k_M = k_N$*. This often reasonable assumption greatly simplifies the model, both conceptually and computationally. Thus with equal force constants in both oxidation states, we obtain (setting $k_M = k_N \equiv k$):

$$\begin{aligned} W_a &= \frac{1}{\sqrt{2}}lQ_- + \frac{1}{2}kQ_-^2 + \frac{1}{\sqrt{2}}lQ_+ + \frac{1}{2}kQ_+^2 \\ W_b &= \frac{-1}{\sqrt{2}}lQ_- + \frac{1}{2}kQ_-^2 + \frac{1}{\sqrt{2}}lQ_+ + \frac{1}{2}kQ_+^2 \end{aligned} \tag{24}$$

Since the cross-term in $Q_+Q_-$ has vanished, the problem is separable with respect to $Q_+$ and $Q_-$. In $Q_+$ space, $W_a$ and $W_b$ give identical parabolas with minima at $Q_+ = (-l/k\sqrt{2})$. We show later that $Q_+$ does not enter the calculation of spectroscopic or other properties of interest to us (when $k_M = k_N$). *The crucial coordinate is* $Q_-$. Defining the dimensionless variables:

$$q \equiv 2\pi(\nu_-/h)^{1/2}Q_-, \qquad \lambda \equiv \left(8\pi^2 h\nu_-^3\right)^{-1/2} l \tag{25}$$

where $\nu_- = (2\pi)^{-1}\sqrt{k}$, is the fundamental vibrational frequency associated with normal coordinate $Q_-$(or $q$), Equation 24 *in q space becomes*:

$$W_a/h\nu_- = \lambda q + \frac{1}{2}q^2, \qquad W_b/h\nu_- = -\lambda q + \frac{1}{2}q^2 \tag{26}$$

The physical significance of $q$ (or $Q_-$) can be understood by reference to Figure 1 using the Creutz–Taube ion as an example. The lines from centers $A$ and $B$ show the Ru–N bond distances, assuming that the $RuN_6$ moiety is octahedral. Observe from our earlier discussion that $\lambda$ is directly proportional to the *difference*, in the two oxidation states, of the equilibrium displacement along the totally symmetric normal coordinate. It is thus proportional to the difference in the Ru–N equilibrium distances ($\Delta r$) in the 2+ and 3+ oxidation states in the Creutz–Taube ion example. It can be shown[3] for octahedral subunits that $\Delta r = \lambda(h\nu_-/3m_L k)^{1/2}$ where $m_L$ is the mass of

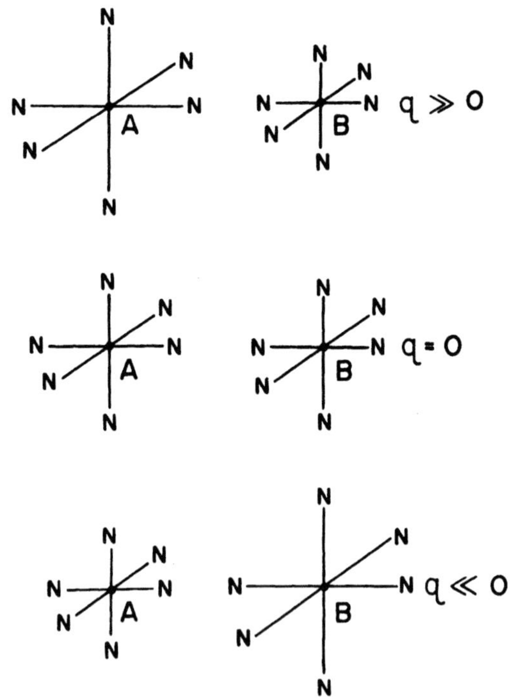


**Figure 1.** Crucial vibrational coordinate $q$ for two coupled octahedral centers. Adapted from Reference 3.

the coordinated ligand (17 amu for the Creutz–Taube ion). Since we have permitted no interaction between subunits $A$ and $B$, the vibronic (vibrational-electronic) wave functions for electronic states $\psi_a$ and $\psi_b$ are:

$$\phi_a = \psi_a \chi_{n_a}, \qquad \phi_b = \psi_b \chi_{n_b} \tag{27}$$

where the $\chi$ are defined by:

$$\begin{aligned} \left[T_n(q) + W_a(q)\right]\chi_{n_a}(q) &= E_{n_a}\chi_{n_a}(q) \\ \left[T_n(q) + W_b(q)\right]\chi_{n_b}(q) &= E_{n_b}\chi_{n_b}(q) \end{aligned} \tag{28}$$

with $T_n(q)$ the kinetic energy operator associated with coordinate $q$. The terms $W_a(q)$ and $W_b(q)$ are referred to as potential surfaces (or potential energy surfaces) because they describe the potential energy of the system as a function of nuclear coordinates—in this case, the single normal coordinate $q$. Equations 28 are the familiar result of the Born–Oppenheimer approximation whereby nuclear motions take place on only a single potential surface ($W_a$ or $W_b$), the form of which is determined by solution of the electronic Schrödinger equation (Equation 20 and its analogue for $\psi_b$). The harmonic approximation (Equation 18) requires that solutions of the Schrödinger equation in the vicinity of the equilibrium configuration produce energies which depend quadratically on nuclear coordinates.

Since $W_a$ and $W_b$ are harmonic potentials according to Equation 26 (i.e., are parabolas), $\chi_{n_a}$ and $\chi_{n_b}$ are harmonic oscillator functions centered at $-\lambda$ and $\lambda$, respectively, and the vibronic eigenvalues are

$$\frac{E_{n_a,n_b}}{h\nu_-} = -\frac{\lambda^2}{2} + \left(n_{a,b} + \frac{1}{2}\right) \tag{29}$$

with $n_a$ and $n_b = 0, 1, 2, \ldots$. These vibrational levels are shown in Figure 2.

Let us suppose that the system is in state $\psi_a (= \psi_M^A \psi_N^B)$ and it is desired to compute the probability (or intensity) of an optical transition to state $\psi_b (= \psi_N^A \psi_M^B)$. On the basis of Figure 2, this transition corresponds classically (in the Creutz–Taube ion example) to an electron hop from $Ru_B^{2+}$ to $Ru_A^{3+}$ (an intervalence transition) producing a final state consisting of $Ru_B^{3+}$ and $Ru_A^{2+}$ after absorption of a photon. We show in Section 2.7 that in fact this transition occurs with zero intensity if there is no electronic coupling between $A$ and $B$. The transition is shown by the vertical (Franck–Condon) arrow in Figure 2.

It is important to note that the final electronic state ($\psi_b$) is vibrationally excited ($n_b = 9$ in the diagram). This is due physically to the rapidity of the electronic motion (transition) compared to the vibrational motion: Thus immediately *after* the transition, $Ru_A^{2+}$ finds itself in the equilibrium environment (Ru–N distances) of a $Ru^{3+}$, and vice versa for $Ru_B^{3+}$. It is seen that the degree of vibrational excitation is proportional to $\lambda$, and this is physically reasonable because we know from earlier

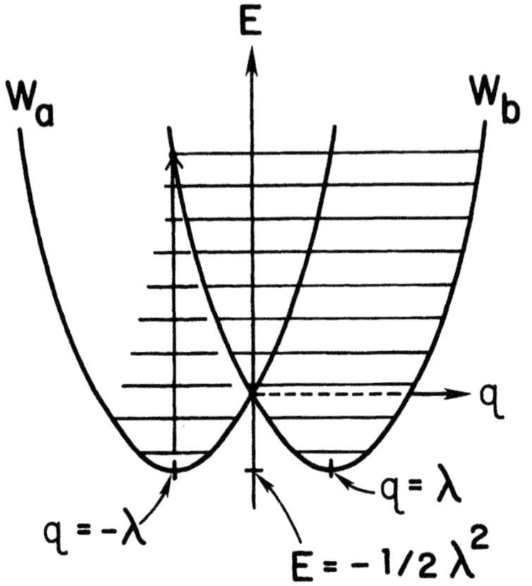


**Figure 2.** Potential energy surfaces and associated vibrational levels for a symmetrical mixed-valence dimer in the absence of electronic coupling ($\varepsilon = 0$). Adapted from Reference 3.

discussion that $\lambda$ is a direct measure of the difference in equilibrium displacement along the totally symmetric normal coordinate in the two oxidation states. We describe the vibrational change produced by the electronic transition (i.e., we recognize that the electronic and vibrational excitations are coupled) by saying that the system is *vibronically coupled.* Since the magnitude of this coupling is proportional to $\lambda$, we refer to the latter as the vibronic (or electron–phonon) coupling parameter. Note that as $\lambda \to 0$, the two curves (potential surfaces) in Figure 2 move into coincidence, and the transition energy goes to zero. Note also that when $\lambda \neq 0$, we have (at $q = 0$) a Jahn–Teller-like situation; that is, the two electronic states ($\psi_a$, $\psi_b$) are degenerate and two equivalent, *lower energy* configurations can be reached by moving distances $\pm\lambda$ along $q$. As is customary in Jahn–Teller systems, *we make symmetry classifications at the nuclear configuration corresponding to the electronic degeneracy—not at a minimum energy configuration.* Consequently the symmetrical ($A \equiv B$) case applies if, *when coordinate $q$ is held fixed at the value $q = 0$*, a symmetry operation exists which permits the interchange of the atoms in subunits $A$ and $B$.

The system becomes much more interesting if we let the subunits interact. We follow the earlier treatment starting just below Equation 3. We hold the nuclei fixed at $q = 0$ with all other normal coordinates also at zero (equilibrium) and indicate this by appending superscript zeros:

$$V_{ij}^0 = \langle \psi_i | V^{AB} | \psi_j \rangle^0, \quad i, j = a, b \tag{30}$$

Note in the symmetrical case that $V_{aa}^0 = V_{bb}^0$. We define a new zero of energy by $V_{aa}^0 = 0$ (not $W_a + V_{aa} = 0$ as was done below Equation 10, because $W_a$ is now a function of $q$). Defining a parameter $\varepsilon$ by the relation

$$(h\nu_-)\varepsilon \equiv V_{ab}^0 \tag{31}$$

and substituting the definitions above and Equation 26 into Equations 7 and 8 using Equation 9, then Equation 6 becomes

$$\begin{vmatrix} \lambda q + \frac{1}{2}q^2 - W_k & \varepsilon \\ \varepsilon & -\lambda q + \frac{1}{2}q^2 - W_k \end{vmatrix} = 0 \tag{32}$$

with $W_k$ in units of $h\nu_-$. The roots of Equation 32 are

$$W_{1,2} = \frac{q^2}{2} \mp \sqrt{\varepsilon^2 + \lambda^2 q^2} \tag{33}$$

and the corresponding electronic wave functions obtained from Equation 5 are

$$\begin{aligned} \psi_1 &= -\frac{1}{N\sqrt{2}}\left[\left(\varepsilon + \lambda q - \sqrt{\varepsilon^2 + \lambda^2 q^2}\right)\psi_a + \left(\varepsilon - \lambda q - \sqrt{\varepsilon^2 + \lambda^2 q^2}\right)\psi_b\right] \\ \psi_2 &= \frac{1}{N\sqrt{2}}\left[-\left(\varepsilon - \lambda q - \sqrt{\varepsilon^2 + \lambda^2 q^2}\right)\psi_a + \left(\varepsilon + \lambda q - \sqrt{\varepsilon^2 + \lambda^2 q^2}\right)\psi_b\right] \end{aligned} \tag{34}$$

where $N = [\lambda^2 q^2 + (\varepsilon - \sqrt{\varepsilon^2 + \lambda^2 q^2})^2]^{1/2}$. Keep in mind in this and subsequent equations that we are now working in a one-dimensional ($q$) space. Thus, for example, we use Equation 26 for $W_a$ and $W_b$, not Equation 24; that is, the coordinate $Q_+$ has been dropped.

We call $\varepsilon$ the electronic coupling parameter because it is defined by the electronic functions $\psi_a$ and $\psi_b$ (at $q = 0$) and couples them. A basic assumption of the model is that $\varepsilon$ is small compared to electronic binding energies within the subunits. If this is not the case, it would be necessary to include additional terms in Equation 4 corresponding to excited states of $\psi_M$ and $\psi_N$; that is, the first-order perturbation treatment implied by Equation 4 would be inadequate. Put another way, potential surfaces in addition to those of Equation 33 would play an important role in the mixed-valence properties.

Equation 34 clearly exhibits the coupling (mixing) of $\psi_a$ and $\psi_b$ when $\varepsilon$ is "turned on". In addition, it is seen that in general $\psi_1$ and $\psi_2$ *are explicit and sensitive functions of* $q$. Physically this is so because when $\varepsilon$ is turned on, the mixing of $\psi_a$ and $\psi_b$ depends on both $\varepsilon$ and the energy difference $\Delta W = |W_a - W_b|$, and in fact by Equation 15 is directly proportional to the former and inversely proportional to the latter for large $\Delta W$. In turn, $\Delta W$ is generally a strong function of $q$ *because the potential surfaces* $W_a$ *and* $W_b$ *are not parallel* (Figure 2).

Inspection of Equation 34 shows that the $q$ dependence of $\psi_1$ and $\psi_2$ vanishes only in two limiting cases, namely, $\varepsilon = 0$ ($\lambda \neq 0$) or $\lambda = 0$ ($\varepsilon \neq 0$). This result follows in the former case because there is no perturbation mixing $\psi_a$ and $\psi_b$, and Equation 34 reduces to $\psi_1 = \psi_b$, $\psi_2 = \psi_a$. It happens in the latter case ($\lambda = 0$, $\varepsilon \neq 0$) because *the two potential surfaces are parallel*—in fact by Equation 33 they are identical parabolas vertically displaced by $2|\varepsilon|$. With a little rearranging, Equation 34 can be shown in that case to reduce to $\psi_1 = (1/\sqrt{2})(\psi_a + \psi_b)$, $\psi_2 = (1/\sqrt{2})(\psi_a - \psi_b)$. In anticipation of our later discussion, note that these two limiting cases correspond respectively to the valence bond (VB) and molecular orbital (MO) limits.

The interesting cases arise *when* $\varepsilon$ *and* $\lambda$ *are simultaneously nonzero*. Figure 3 shows the potential surfaces (Equation 33) for a variety of cases including the unsymmetrical ($A \not\equiv B$) case to be discussed later. It follows from Equation 33 that the height of the barrier on the lower potential surface is given by:

$$\frac{E_a}{h\nu_-} = \frac{1}{2}\lambda^2 - |\varepsilon| + \frac{\varepsilon^2}{2\lambda^2}; \qquad |\varepsilon| \leqslant \lambda^2 \tag{35}$$

$E_a$ is often referred to as the activation energy for thermal electron transfer.

The basic difference between the earlier treatment (Equations 11–13) and the present one (Equations 32–34) is the inclusion in the latter of vibronic coupling ($\lambda$). This new feature emerges as a direct consequence of examining electronic energy changes as a function of nuclear coordinates. Indeed, one can confirm that the two treatments are identical for $q = 0$, remembering that we are presently considering only the symmetrical case so that $E_l = 0$. It is clear immediately that the properties of a symmetrical mixed-valence system depend strongly on the values of both the electronic ($\varepsilon$) and vibronic ($\lambda$) coupling. If $E_a$ is large ($\lambda^2$ large and $|\varepsilon|/\lambda^2 \ll 1$), the system is strongly localized (Class I). As $|\varepsilon|/\lambda^2$ gets larger we move to the Class II case, and the transition to Class III occurs in the vicinity of $|\varepsilon| \to \lambda^2$ (when $E_a \to 0$). These matters are discussed in some detail (Section 2.6) after solution of the dynamic problem.

We now show that an exact solution of the dynamic problem can be accomplished[1] using a method developed many years ago by Fulton and Gouterman to treat the excited states of symmetrical dimers.[8,9]

## 2.3 Solution of the Dynamic Problem in the Symmetrical Case

A complete set of vibronic energies and eigenfunctions, the ground vibronic manifold, is associated with the potential surfaces (Figure 3). It is clearly of great interest to determine these, since by quantum mechanical principles one can then calculate, within the context of the model, observables associated with this ground vibronic manifold; this includes the spectroscopic properties of the intervalence band, a distinctive feature of mixed-valence systems. If we confine ourselves to the two potential surfaces of Equation 33, the most general vibronic function is a $q$-dependent linear combination of $\psi_1$ and $\psi_2$ (Equation 34). We therefore write our *vibronic* wave

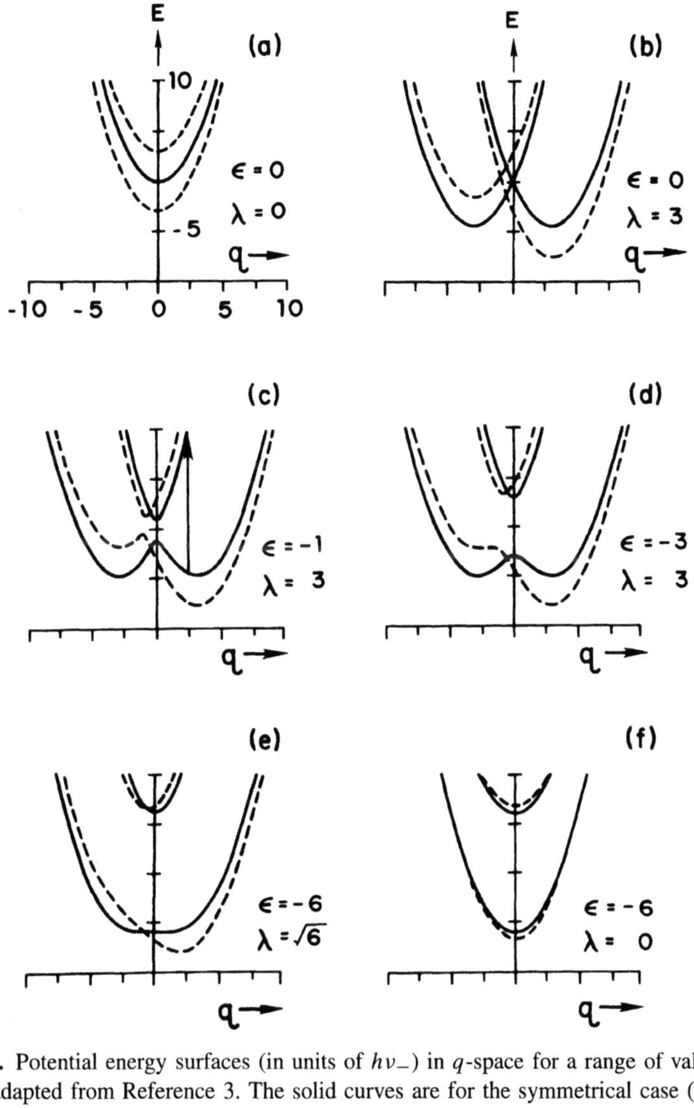


**Figure 3.** Potential energy surfaces (in units of $h\nu_-$) in $q$-space for a range of values of $\varepsilon$ and $\lambda$—adapted from Reference 3. The solid curves are for the symmetrical case ($A \equiv B$) using Equation 33, and the dashed curves are for the unsymmetrical case ($W = 3$) obtained replacing $\lambda q$ by $\lambda q + W$ in Equation 33—see Section 2.4. The vertical arrow in (c) shows a possible Franck–Condon intervalence transition.

functions in the form

$$\Psi_k(r, q) = \psi_1(r, q)\chi_{1,k}(q) + \psi_2(r, q)\chi_{2,k}(q) \tag{36}$$

where $k = 1, 2, \ldots, \infty$, and $r$ symbolizes all the electronic coordinates implicit in the functions $\psi_M$ and $\psi_N$ (Equation 1). However, whereas in Equations 5 and 6 (therefore Equations 32–34) we used only the electronic Hamiltonian (since we were holding the nuclei fixed), we must now use the dynamic Hamiltonian ($H$); that is, we must include the nuclear kinetic energy operator ($T_n$) because we are allowing nuclear motion. Putting it another way, we are now solving *the dynamic problem.* Recalling again that we are working in $q$ space, we write,

$$H = H_{el}(r, q) + T_n(q) \tag{37}$$

where $T_n(q)$ is the kinetic energy operator for vibrational coordinate $q$, and by definition, $H_{el}$ is diagonal in $\psi_1$ and $\psi_2$. If we multiply the vibronic Schrödinger equation, $(H_{el} + T_n)\Psi_k = E_k\Psi_k$, by $\psi_1^*$ and $\psi_2^*$ in turn and integrate over electronic coordinates (i.e., take scalar products), the result is

$$\sum_{j=1,2} \left[(W_i - E_k)\delta_{ij} + \langle\psi_i|T_n(q)|\psi_j\rangle\right]\chi_{j,k} = 0; \quad i = 1, 2;\ k = 1, 2, \ldots, \infty \tag{38}$$

where $\langle\psi_i|T_n(q)|\psi_j\rangle\chi_{j,k} \equiv \int \psi_i^*(r, q)T_n(q)\psi_j(r, q)\chi_{j,k}(q)d\tau_{el}$, and the $E_k$ are the vibronic eigenvalues. If $\psi_i$ and $\psi_j$ are *not* strong functions of $q$ (i.e., if *either* $\lambda$ or $\varepsilon = 0$), then

$$T_n(q)\psi_j\chi_{j,k} \approx \psi_j T_n(q)\chi_{j,k} \tag{39}$$

In that case (since $\langle\psi_i|\psi_j\rangle = \delta_{ij}$), Equations 38 and 36 become, respectively,

$$\left[W_i(q) + T_n(q)\right]\chi_{i,k}(q) = E_k\chi_{i,k}(q); \quad i = 1, 2;\ k = 1, 2, \ldots, \infty \tag{40}$$

$$\Psi_k^{(i)}(r, q) = \psi_i(r, q)\chi_{i,k}(q); \quad i = 1, 2;\ k = 1, 2, \ldots, \infty \tag{41}$$

*Equations 40 and 41 are the familiar results of the Born–Oppenheimer approximation.* In particular, $\psi_i$ is assumed to be at most a slowly varying function of $q$ so that Equation 39 is a good approximation. In the crude adiabatic approximation, $\psi_i$ would be set equal to its value at $q = 0$. The vibrational function $\chi_{i,k}$ is determined *only* by the $W_i$ potential surface. *Physically, the nuclear motion is confined to that surface alone. Thus we see that the Born–Oppenheimer approximation results if the nuclear kinetic energy operator is assumed to commute with the electronic functions* $\psi_j$ (see Equation 39). Furthermore, since the surface is harmonic under these circumstances ($\varepsilon$ or $\lambda = 0$), the $\chi_{i,k}$ are simply harmonic oscillator functions, and the dynamic problem has been solved.

The situation changes profoundly if the electronic functions depend strongly on $q$ ($\varepsilon$ and $\lambda$ both $\neq 0$), since Equation 39 no longer applies. Thus $T_n(q)$ must be applied to $\psi_j\chi_{j,k}$ (i.e., this product must be differentiated twice with respect to $q$) *before* the integration over electronic coordinates can be performed in $\langle\psi_i|T_n(q)|\psi_j\rangle$ (Equation 38). Under these circumstances, this matrix element is *not* diagonal in $\psi_i$ and $\psi_j$, and so both $\chi_{1,k}$ and $\chi_{2,k}$ appear in *both* ($i = 1, 2$) equations (Equation 38). Thus

$\chi_{1,k}$ and $\chi_{2,k}$ are determined by *both* potential surfaces $W_1(q)$ and $W_2(q)$; that is, the vibronic functions $\psi_1\chi_{1,k}$ and $\psi_2\chi_{2,k}$ are coupled (mixed) by the nuclear motion, and so the Born–Oppenheimer approximation, Equation 41, fails and must be replaced by Equation 36. Physically, this result means that *the nuclear motion cannot be confined to a single potential surface.* Aside from the complicated nature of the $\chi_{i,k}$ in this case—they are not in general harmonic oscillator functions—$\psi_1$ and $\psi_2$ are not convenient in spectroscopic applications because their $q$ dependence invalidates the Franck–Condon approximation ($\langle\psi_i\chi_i|\mathbf{m}_{el}|\psi_j\chi_j\rangle \approx \langle\psi_i|\mathbf{m}_{el}|\psi_j\rangle^0\langle\chi_i|\chi_j\rangle$), where $\mathbf{m}_{el}$ is the electronic part of the electric dipole moment operator, and the right superscript zero indicates that the nuclear coordinates are held fixed at equilibrium in evaluating the matrix element.

In such circumstances, it is much more convenient to use electronic wave functions that have no explicit $q$ dependence, and so we return to the functions $\psi_a$ and $\psi_b$. To take advantage of the interchange symmetry ($A \equiv B$), we use the functions $\psi_\pm$ defined by:

$$\psi_\pm = \frac{1}{\sqrt{2}}(\psi_a \pm \psi_b) \tag{42}$$

If we substitute $\psi_1$ and $\psi_2$ from Equation 34 into Equation 36, the latter can be rearranged to the form

$$\Phi_\nu(r,q) = \psi_+(r)\chi_{+,\nu}(q) + \psi_-(r)\chi_{-,\nu}(q) \tag{43}$$

where the $\chi_+$ and $\chi_-$ are ($q$-dependent) linear combinations of the $\chi_1$ and $\chi_2$; we use $\Phi_\nu$ in place of $\Psi_k$ to emphasize the change in electronic basis. Proceeding as before (Equations 36–38), Equation 43 leads to the equations:

$$\sum_{j=+,-} \left[\langle\psi_i|H_{el} + T_n|\psi_j\rangle - \langle\psi_i|\psi_j\rangle E_\nu\right]\chi_{j,\nu} = 0; \quad i = +,-;\ \nu = 1,2,\ldots,\infty \tag{44}$$

From the definition of $\psi_\pm$, it follows that $\langle\psi_i|\psi_j\rangle = \delta_{ij}$, *but* $\langle\psi_i|H_{el}|\psi_j\rangle$ *is not diagonal.* We assume that $\psi_a$ and $\psi_b$ are at most slowly varying functions of $q$. In particular, the subunit electronic states ($\psi_M^A, \psi_N^A, \psi_M^B, \psi_N^B$) may *not* show significant Jahn–Teller coupling. We may then assume that $T_n(q)$ commutes with $\psi_a$ and $\psi_b$ (therefore with $\psi_\pm$), and it follows immediately that $\langle\psi_i|T_n|\psi_j\rangle\chi_{j,\nu} = \delta_{ij}T_n\chi_{j,\nu}$. The matrix elements of $\langle\psi_i|H_{el}|\psi_j\rangle$ $(i,j = a,b)$ are displayed in Equation 32, so it is an elementary matter to compute the corresponding matrix elements for $i,j = +,-$, namely

$$V = \begin{pmatrix} \frac{1}{2}q^2 + \varepsilon & \lambda q \\ \lambda q & \frac{1}{2}q^2 - \varepsilon \end{pmatrix} \tag{45}$$

$V$ is referred to as the electronic potential energy (or vibronic) matrix. Its roots map out the potential (energy) surfaces. Equation 44 then becomes

$$
\begin{aligned}
&\left[T_n(q) + \frac{1}{2}q^2 + \varepsilon - E_\nu\right]\chi_{+,\nu}(q) + \lambda q\,\chi_{-,\nu}(q) = 0 \\
&\lambda q\,\chi_{+,\nu}(q) + \left[T_n(q) + \frac{1}{2}q^2 - \varepsilon - E_\nu\right]\chi_{-,\nu}(q) = 0; \qquad \nu = 1, 2, \ldots, \infty
\end{aligned}
\tag{46}
$$

with $E_\nu$ in units of $h\nu_-$.

If this pair of *coupled equations* can be solved for each value of $\nu$, that is, if $\chi_{+,\nu}$, $\chi_{-,\nu}$ (and $E_\nu$) can be determined, then Equation 43 gives the complete set of vibronic eigenfunctions of the ground manifold. As Fulton and Gouterman have shown,[8] this is easily accomplished if $\chi_{+,\nu}$ and $\chi_{-,\nu}$ are expanded in the complete orthonormal set of harmonic oscillator functions $\chi_n(q)$, $n = 0, 1, 2, \ldots, \infty$; that is, *harmonic oscillator functions in coordinate q are used as the vibrational basis.* Thus

$$
\chi_{+,\nu} = \sum_{n=0}^{\infty} c_{\nu,n}\chi_n; \qquad \chi_{-,\nu} = \sum_{n=0}^{\infty} c'_{\nu,n}\chi_n \tag{47}
$$

and Equation 43 assumes the form

$$
\Phi_\nu(r, q) = \psi_+ \sum_{n=0}^{\infty} c_{\nu,n}\chi_n + \psi_- \sum_{n=0}^{\infty} c'_{\nu,n}\chi_n \tag{48}
$$

Because of the symmetry ($A \equiv B$), group theoretic principles permit us to classify the $\Phi_\nu$ in accord with their behavior under interchange; $\Phi_\nu^-$ and $\Phi_\nu^+$ designate, respectively, vibronic functions that do and do not change sign when "monomers" are interchanged. Note again that this classification is with respect to the nuclear configuration $q = 0$. Since $\psi_+ \to \psi_+$, $\psi_- \to (-\psi_-)$ and $q \to (-q)$ [thus $\chi_n(-q) \to (-1)^n\chi_n(q)$] under an interchange operation, it follows that Equation 48 can be written

$$
\begin{aligned}
\Phi_\nu^+ &= \psi_+ \sum_{n=0,2,4,\ldots}^{\infty} r_{\nu n}\chi_n + \psi_- \sum_{n=1,3,5,\ldots}^{\infty} r_{\nu n}\chi_n \\
\Phi_\nu^- &= \psi_+ \sum_{n=1,3,5,\ldots}^{\infty} s_{\nu n}\chi_n + \psi_- \sum_{n=0,2,4,\ldots}^{\infty} s_{\nu n}\chi_n
\end{aligned}
\tag{49}
$$

Our task is to calculate the coefficients $r_{\nu n}$ and $s_{\nu n}$ for each value of $\nu$. Let us first do this for the $r_{\nu n}$. We note that

$$\begin{aligned} \left(T_n + \frac{1}{2}q^2\right)\chi_n &= \left(n + \frac{1}{2}\right)\chi_n \\ \langle \chi_n | q | \chi_{n+1} \rangle &= \langle \chi_{n+1} | q | \chi_n \rangle = \left(\frac{n+1}{2}\right)^{1/2} \end{aligned} \tag{50}$$

That is, the $\chi_n$ are eigenfunctions of $(T_n + \frac{1}{2}q^2)$ with eigenvalues $(n + \frac{1}{2})$ in units of $h\nu_-$; this of course is why they are such a convenient basis. The second of Equations 50 is a well-known property of harmonic oscillator functions. Considering the first of Equations 49 and substituting Equation 47 into Equation 46 gives

$$\begin{aligned} \left[T_n(q) + \frac{1}{2}q^2 + \varepsilon - E_\nu^+\right] \sum_{n \text{ even}} r_{\nu n}\chi_n + \lambda q \sum_{n \text{ odd}} r_{\nu n}\chi_n &= 0 \\ \lambda q \sum_{n \text{ even}} r_{\nu n}\chi_n + \left[T_n(q) + \frac{1}{2}q^2 - \varepsilon - E_\nu^+\right] \sum_{n \text{ odd}} r_{\nu n}\chi_n &= 0 \end{aligned} \tag{51}$$

We now multiply Equations 51 by each $\chi_n$ in turn $(n = 0, 1, 2, \ldots, \infty)$ and integrate over $q$ (i.e., we take scalar products). Using Equation 50, the following set of secular equations is obtained:

$$\sum_{n=0}^{\infty} r_{\nu n}(H_{mn} - \delta_{mn}E_\nu^+) = 0; \quad m = 0, 1, 2, \ldots, \infty; \ \nu = 0, 1, 2, \ldots, \infty$$

with

$$H_{mn} = \lambda\left[\left(\frac{m}{2}\right)^{1/2}\delta_{m,n+1} + \left(\frac{m+1}{2}\right)^{1/2}\delta_{m,n-1}\right] + \left(m + \frac{1}{2} + (-1)^m\varepsilon\right)\delta_{mn} \tag{52}$$

The corresponding secular determinant yields the eigenvalues $E_\nu^+$. Proceeding in exactly the same way for the second of Equations 49, the analogous set of equations for the $s_{\nu n}$ is obtained by making the substitutions $r_{\nu n} \to s_{\nu n}$, $\varepsilon \to -\varepsilon$, $E_\nu^+ \to E_\nu^-$ in Equation 52.

Thus the solution to the dynamic problem has been reduced to a very simple matrix diagonalization. It is true that this matrix is infinite, but drastic truncation is possible without significant loss of accuracy. For example, to calculate the intervalence band contour (see later), a basis of 40 quanta ($n = 0, 1, \ldots, 39$ in Equation 52) is often entirely satisfactory. This requires on the order of seconds on a modern PC. Furthermore, if there is any doubt about the numerical accuracy, the basis size can be increased to see if significant changes occur.

An excellent, concise discussion of electronic potential surfaces and associated nuclear motions may be found in Stephens.[10]

### 2.4 Solution of the Dynamic Problem in the Unsymmetrical Case ($A \not\equiv B$)

The symmetrical case is relatively uncommon in practice. It is more usual for the two centers to contain different atoms, or for identical atoms to be located in different environments. For example, $[(NC)_5\text{–Ru–C–N–Fe–}(CN)_5]^{6-}$ and $[(NC)_5\text{–Fe–C–N–Fe–}(CN)_5]^{6-}$ illustrate these two cases. Such systems can be treated by the same basic vibronic coupling model if some additional assumptions are made. We outline this case.

Equations analogous to 1, 2, and 17–19 still apply, but now $W_M^A(Q_A = 0) \equiv W_M^{\circ A} \neq W_M^{\circ B}$ and $W_N^{\circ A} \neq W_N^{\circ B}$. In addition, although we again define coordinate origins by $l_N^A = 0 = l_N^B$, it is no longer a symmetry requirement that $l_M^A = l_M^B$, $k_M^A = k_M^B$, or $k_N^A = k_N^B$. These equalities are the *additional* assumptions required in the unsymmetrical case to allow the degree of freedom in coordinate $Q_+$ to again be ignored. With these assumptions, $W_a$ and $W_b$ in Equation 21 *et seq.* contain the additional constant terms, $(W_M^{\circ A} + W_N^{\circ B})$ and $(W_N^{\circ A} + W_M^{\circ B})$, respectively, and it is again assumed that $k_M = k_N \equiv k$. Finally, following Equation 30, it is no longer true that $V_{aa}^0 = V_{bb}^0$. Proceeding exactly as before, one finds that Equations 31–34, 36–48, and 50 continue to apply, provided only that the zero of energy is now defined by

$$W_M^{\circ A} + W_N^{\circ B} + W_M^{\circ B} + W_N^{\circ A} + V_{aa}^0 + V_{bb}^0 = 0 \tag{53}$$

*and* the term $\lambda q$ is replaced everywhere by the term $\lambda q + W$, where

$$(h\nu_-)W \equiv \frac{1}{2}\left[\left(W_M^{\circ A} + W_N^{\circ B}\right) - \left(W_N^{\circ A} + W_M^{\circ B}\right) + V_{aa}^0 - V_{bb}^0\right] \tag{54}$$

An analytic expression analogous to Equation 35 cannot be written when $W \neq 0$.

Thus $W$ is the one additional parameter introduced in the unsymmetrical case; it goes to zero when $A \equiv B$. Since the interchange symmetry has been lost, Equation 49 is replaced by

$$\Phi_\nu = \sum_{n=0}^{\infty}(\psi_+ r_{\nu n}\chi_n + \psi_- r'_{\nu n}\chi_n) \tag{55}$$

The secular equations corresponding to Equation 55 are then found to be:

$$\sum_{n=0}^{\infty} r_{\nu n}(H_{mn} - \delta_{mn}E_\nu) + \sum_{n=0}^{\infty} r'_{\nu n}H''_{mn} = 0$$

$$\sum_{n=0}^{\infty} r'_{\nu n}(H'_{mn} - \delta_{mn}E_\nu) + \sum_{n=0}^{\infty} r_{\nu n}H''_{mn} = 0$$

$$H_{mn} = \left(m + \frac{1}{2} + \varepsilon\right)\delta_{mn}; \qquad H'_{mn} = \left(m + \frac{1}{2} - \varepsilon\right)\delta_{mn} \tag{56}$$

$$H''_{mn} = \lambda\left[\left(\frac{m}{2}\right)^{1/2}\delta_{m,n+1} + \left(\frac{m+1}{2}\right)^{1/2}\delta_{m,n-1}\right] + W\delta_{m,n}$$

$$m = 0, 1, 2, \ldots; \quad \nu = 0, 1, 2, \ldots$$

Thus it is necessary to diagonalize a single block twice the size of the two separate blocks corresponding to $\Phi_\nu^+$ and $\Phi_\nu^-$ of the symmetrical case.

Potential surfaces for the $W \neq 0$ case are included in Figure 3. As expected, the interchange symmetry is lost, and we find later that a system is strongly localized if $(\lambda^2 + W)$ is large *and* $|\varepsilon| \ll (\lambda^2 + W)$. It is clear that if the centers $A$ and $B$ differ greatly, there may be no justification for assuming that bond length changes between oxidation states are the same for different centers ($l_M^A = l_M^B$) and/or that force constants are the same for the same oxidation state on different centers ($k_M^A = k_M^B$ and $k_N^A = k_N^B$). Thus considerable caution is required in applying the model to unsymmetrical mixed-valence systems.

## 2.5 Localized Limit; Intervalence Band and the Hush Formulas

Let us consider the potential surfaces described by Equation 33 (the symmetrical case) for a strongly localized system ($|\varepsilon| \ll \lambda^2$). A typical case is shown in Figure 3c, solid lines. Our task is to calculate the absorption band contour that results when transitions occur "between" (careful!) the lower and upper surfaces. This contour is the so-called mixed-valence or intervalence band, which is a distinctive feature of mixed-valence systems. In view of the treatment in Section 2.3, the method for accomplishing this task is perfectly straightforward. Since the complete set of eigenfunctions and eigenvalues is available for any values of $\varepsilon$ and $\lambda$, via Equations 49 and 52 (or via Equations 55 and 56 if $W \neq 0$), one can clearly calculate the energy and intensity (dipole strength) of each vibronic line (e.g., $\Phi_\nu^\pm \to \Phi_{\nu'}^\mp$). The totality of such lines, each weighted by an appropriate Boltzmann factor, comprises the calculated intervalence band. And we shall later (Section 2.7) calculate intervalence band contours in precisely this manner. Since our eigenfunctions are in a numerical rather than analytical form, the procedure must be accomplished numerically (i.e., by computer).

However, for a strongly localized system, it is possible to derive an approximate, analytical formula for the intervalence band contour. As a prologue, we emphasize again that the picture of an intervalence transition implied by the vertical arrow in Figure 3c is subject to serious misinterpretation. In particular, *one cannot in general regard the transition as occurring between the two potential surfaces*. The reason is that each vibronic eigenfunction (left-hand side of Equation 36) is a sum containing a vibronic contribution from *each* surface (right-hand side of Equation 36), and hence the nuclear motion in any given vibronic state is not confined to a single surface. Thus an inherent conceptual difficulty arises when the Born–Oppenheimer approximation (Equation 41) fails. As a consequence, caution must be exercised in drawing conclusions about intervalence band contours simply from pictures of the potential surfaces. Nevertheless, if the system is sufficiently strongly localized ($|\varepsilon| \ll \lambda^2 + W$), the band contour *can* be calculated analytically, and transitions *can* be pictured as occurring between the two surfaces in accord with the Franck–Condon principle. Space limitations preclude a detailed presentation here, but the method uses a semiclassical treatment of the Franck–Condon principle formulated many years ago by Lax[11] and later used by Toyozawa and Inoue[12] to discuss bandshapes in Jahn–Teller systems. Our application to mixed-valence systems is described in detail elsewhere,[2,3] where among other things, we give an explicit and unified derivation of the following Hush Formulas:[13,14]

$$\frac{E_{op}}{E_a} = 4 \tag{57}$$

$$\frac{(h\Delta\nu_{1/2})^2}{h\nu_{max}} = 16kT\ln 2 = 2{,}312\ \text{cm}^{-1} \quad \text{at 300 K} \tag{58}$$

$$(h\Delta\nu_{1/2})^2 = 16kT\ln 2\,(h\nu_{max} - 2h\nu_- W) \tag{59}$$

$$\begin{aligned} \alpha_d^2 &= 4.335 \times 10^{34}\frac{\langle\alpha\rangle_0}{R^2} = 3.983 \times 10^{-4}\frac{\langle\varepsilon\rangle_0}{R^2} \\ &= 4.24 \times 10^{-4}\frac{\varepsilon_{max}\Delta\nu_{1/2}}{\nu_{max}R^2} \end{aligned} \tag{60}$$

In Equation 57, which applies only in the symmetrical case ($A \equiv B$), $E_{op}$ ($op$ = optical) is the energy of the vertical excitation from a minimum in the lower potential surface ($W_1$) to the upper surface ($W_2$), and $E_a$ ($a$ = activation) is the height of the energy barrier in a thermal (radiationless) electron transfer. In Equations 58 and 59, $\Delta\nu_{1/2}$ is the full width of the intervalence band at half-maximum, and $\nu_{max}$ is the frequency of the band maximum. Equations 58 and 59 apply, respectively, in the symmetrical and unsymmetrical cases, and $W = E_l/h\nu_-$ (Equation 10), where $\nu_-$ is the fundamental frequency of the coupling vibrational mode (Section 2.2). In Equation 22, $\alpha_d^2$ is the degree of (valence) delocalization, $\langle\alpha\rangle_0$ and $\langle\varepsilon\rangle_0$ are the zeroth moments (band areas) of the intervalence band, dipole moments are expressed in debye units, and $R$ is the distance (in Å) between the centers $A$ and $B$. This equation applies for a single electron transfer ($n = 1$ in Equation 3), and the last equality in Equation 60 assumes a Gaussian bandshape. $\varepsilon$ is the molar extinction coefficient.

The assumptions required in this derivation are as follows. First the vibronic states must be adequately described by simple Born–Oppenheimer products (Equation 41). Second, it is assumed that the vibrational motion can be treated semiclassically. Specifically, for a given vibronic transition it is assumed that the entire contribution to the vibrational overlap integral is at the classical turning points (i.e., the values of $q$ at which $W_2$ and $W_1$ [Equation 40] are equal to $E_{k'}$ and $E_k$, respectively). Finally, it is assumed that a continuum of vibrational energy levels exists on each of the potential surfaces, that is, that the spacing between successive vibrational levels is infinitesimal. It is clear that the first of these assumptions is favored by large vibronic coupling because Equation 41 is increasingly valid as $\lambda^2$ becomes $\gg |\varepsilon|$. The second and third assumptions are favored by high temperature because an increasing number of vibrational levels are populated and the harmonic oscillator functions approach classical behavior for large quantum number.

The Hush formulas are still often used, and a detailed analysis of their range of validity has been presented elsewhere.[2,3] The general conclusion is that these formulas have a wider range of validity than might be expected considering the severity of the assumptions required for their derivation.

## 2.6 Refinement of the Robin and Day (R&D) Classification Scheme

To refine the R&D[4,5] classification scheme, we explicitly include vibronic coupling via the parameter $\lambda$. This parameter thus joins $\varepsilon$ and $W$, which together with a value for $\nu_-$ determine the complete ground vibronic manifold of the system. We recall that the parameters $\varepsilon$, $\lambda$ and $W$, respectively, measure the electronic coupling between the two subunits, the difference in equilibrium displacements of the $a_{1g}$ subunit normal coordinate in the two oxidation states, and the difference in zero point energy of the two valence states $\psi_a$ and $\psi_b$. The terms $\varepsilon$ and $W$ are directly proportional, respectively, to the R&D parameters $V_{ab}$ and $E_l$, although R&D also associate the latter parameter with differences in subunit site symmetry.

Classes I, II, and III correspond respectively to the cases $(\lambda^2+W)$ large and $\gg |\varepsilon|$, $\lambda^2 + W \geqslant |\varepsilon|$, and $\lambda^2 + W < |\varepsilon|$. The qualitative rationale for this scheme is clear from Figure 3 and the generalization (Section 2.4) of Equation 33 which replaces $\lambda^2 q^2$ by $(\lambda q + W)^2$. The relative sizes of $(\lambda^2 + W)$ and $|\varepsilon|$ define the shape of the lower potential surface and in particular determine the depth of the potential minima (or minimum). The deeper such a minimum, the more probable is the corresponding nuclear configuration. If a deep minimum corresponds to a nuclear configuration in which the two different oxidation states are approximately at their "equilibrium" configurations, the valences will be strongly localized. If a single minimum occurs at $q \approx 0$, the electronic distribution corresponds to a situation in which both centers have "intermediate" oxidation states (i.e., there is strong valence delocalization). The requirement "$(\lambda^2+W)$ large" in regard to Class I systems is included to ensure a deep minimum in the potential surface even if $|\varepsilon|$ is very small.

These qualitative ideas can be most easily visualized in the case $W = 0$. Let us use the Creutz–Taube ion (CT-ion) as an example. If $\lambda^2$ is large and $\gg |\varepsilon|$, there are two deep minima in the potential surface (Figure 3c) and the system is a superposition of the configurations $Ru_A^{2+}Ru_B^{3+}$ and $Ru_A^{3+}Ru_B^{2+}$ (sometimes abbreviated, 2–3 $\longleftrightarrow$ 3–2). Distinct $Ru^{2+}$ and $Ru^{3+}$ ions are present. As $|\varepsilon|$ gets larger, the barrier in the lower potential surface shrinks, disappearing entirely when $|\varepsilon| = \lambda^2$. The presence of distinct 2+ and 3+ oxidation states becomes progressively less pronounced and ultimately is replaced ($\lambda^2 \leqslant |\varepsilon|$) by a single intermediate situation (sometimes abbreviated, $2\frac{1}{2} \longleftrightarrow 2\frac{1}{2}$). Obviously these changes occur in a continuous fashion, and sharp distinctions cannot be made at the borderline between classes. We show later (Section 4.3), however, that a *quantitative* measure of the state of valence delocalization can be obtained for specified values of $\varepsilon$, $\lambda$, and $W$ from the corresponding eigenfunctions; these ideas are illustrated in our subsequent discussion of the Creutz–Taube ion. We emphasize that even when $W = 0$, a system can vary from strongly localized to strongly delocalized depending on the relative values of $\varepsilon$ and $\lambda$. Note that the refinements in the R&D scheme discussed here apply only for two centers having the same local symmetry. We do not include cluster systems nor the case of two centers with different site symmetries.

### 2.7 Spectroscopy of the Intervalence Band; Simple PKS Fit of the Creutz–Taube Ion (CT-Ion) as an Example

We here illustrate the model by fitting the observed intervalence band (IVB) of the Creutz–Taube ion shown by the dashed curve in the bottom panel of Figure 4. We emphasize immediately *that the simple PKS model contains a fundamental omission which makes it unacceptable for systems with a significant degree of delocalization.* However, the model offers a simple, clean framework for presenting the basic spectroscopy. The refinements necessary for a minimally acceptable model, i.e., addition of a second mode, are then easily appended later (Section 3).

The CT-ion is an example of the symmetrical ($A \equiv B$) case, so (interchange) symmetry requirements allow only $\pm \rightarrow \mp$ transitions; the dipole strength (intensity) of a vibronic line, arising from a collection of randomly oriented molecules, is given by[7]

$$D(\nu' \rightarrow \nu) = \sum_{\gamma=x,y,z} \frac{(N_{\nu'} - N_{\nu})}{N} \left| \langle \Phi^{+}_{\nu'} | m_{\gamma} | \Phi^{-}_{\nu} \rangle \right|^2 \tag{61}$$

where $N_{\nu'}/N$ is the fractional population in state $\nu'$ given by $\exp(-E_{\nu'}/kT)/\sum_{\nu'} \exp(-E_{\nu'}/kT)$; $m_{\gamma}$ is the $\gamma$th component of the molecule-fixed electric dipole operator, $\mathbf{m} = \sum_i e_i \mathbf{r}_i$. We first rewrite Equation 49 in the form of Equation 43:

$$\begin{aligned} \Phi^{+}_{\nu'} &= \psi_{+}\chi^{+}_{+,\nu'} + \psi_{-}\chi^{+}_{-,\nu'} \\ \Phi^{-}_{\nu} &= \psi_{+}\chi^{-}_{+,\nu} + \psi_{-}\chi^{-}_{-,\nu} \end{aligned} \tag{62}$$

The electric dipole operator in Equation 61 may be written

$$m_{\gamma} = m^{e}_{\gamma} + m^{n}_{\gamma} \tag{63}$$

where $e$ and $n$ denote the electronic and nuclear contributions, respectively. Denoting $\langle \Phi^{+}_{\nu'} | m_{\gamma} | \Phi^{-}_{\nu} \rangle \equiv TD_{\gamma} = \langle \Phi^{+}_{\nu'} | m^{e}_{\gamma} + m^{n}_{\gamma} | \Phi^{-}_{\nu} \rangle = TD_{\gamma}(e) + TD_{\gamma}(n)$, we examine the various contributions. Using Equation 62,

$$TD_{\gamma}(e) = \langle \psi_{+}\chi^{+}_{+,\nu'} + \psi_{-}\chi^{+}_{-,\nu'} | m^{e}_{\gamma} | \psi_{+}\chi^{-}_{+,\nu} + \psi_{-}\chi^{-}_{-,\nu} \rangle \tag{64}$$

We now make the Franck–Condon approximation; that is, we assume that since $\psi_{+}$ and $\psi_{-}$ are slowly varying functions of $q$, the integrals over electronic wave functions can be evaluated holding the nuclei fixed at equilibrium (including $q = 0$). Then the vibrational functions can be factored out and noting that (see Equation 49)

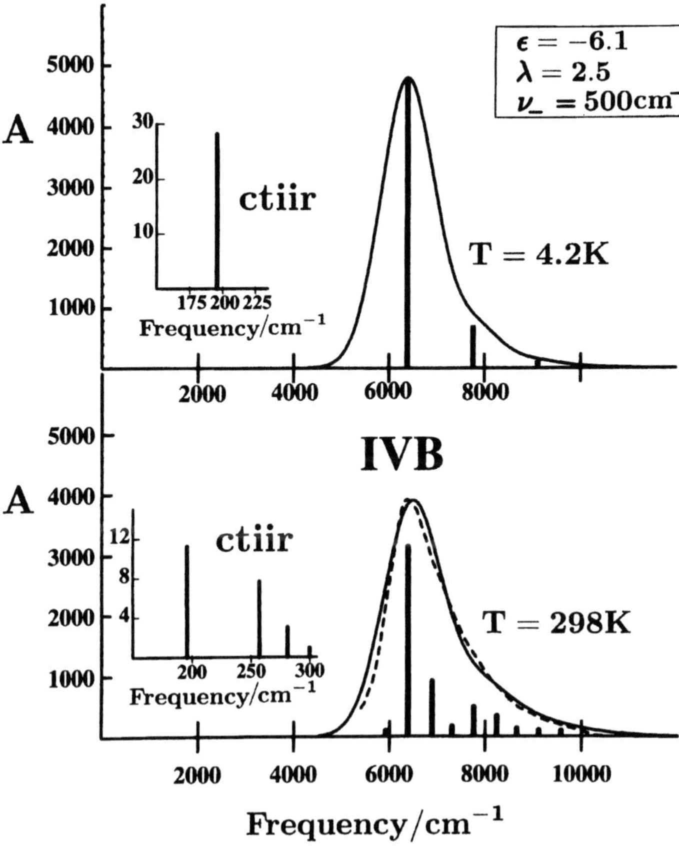


**Figure 4.** One-mode theoretical fit (solid curve) of the Creutz–Taube ion spectrum (dashed curve) at room temperature (bottom panel) using the parameters shown in the upper right box. Stick heights are directly proportional to absorbance and are numerically equal to the product of dipole strength and transition energy ($\text{cm}^{-1}$) with $\mathcal{M} \equiv \langle \psi_+ | m_z | \psi_- \rangle$ set to unity. The IVB contour is obtained by convoluting each stick with a Gaussian whose area is directly proportional to stick height and with *ad hoc* half-width of 781 $\text{cm}^{-1}$ at 1/e of maximum. The left inset shows the calculated charge-transfer-induced infrared (ctiir) stick spectrum. The upper panel shows the theoretical spectrum at 4.2 K using the same parameters.

$\langle \chi^+_{+,\nu'} | \chi^-_{+,\nu} \rangle = \langle \chi^+_{-,\nu'} | \chi^-_{-,\nu} \rangle = 0$, Equation 64 becomes

$$TD_\gamma(e) = \langle \psi_+ | m^e_\gamma | \psi_- \rangle^0 \langle \chi^+_{+,\nu'} | \chi^-_{-,\nu} \rangle + \langle \psi_- | m^e_\gamma | \psi_+ \rangle^0 \langle \chi^+_{-,\nu'} | \chi^-_{+,\nu} \rangle \tag{65}$$

where superscript zero indicates that the matrix element is evaluated at $q = 0$. Note that this treatment would not work using the strongly $q$-dependent functions $\psi_1$ and $\psi_2$ (Equation 34), since the Franck–Condon approximation would be unjustified. At this point we can see clearly why the vibrational motion in normal coordinate $Q_+$, and indeed also in all non-totally symmetric normal coordinates (except $q$), plays no significant role in the model (recall discussion following Equations 17 and 24). In the harmonic approximation, inclusion of $Q_+$ and all non-totally symmetric coordinates (except $q$) would multiply each of the two terms in Equation 65 by Franck–Condon overlap factors of the form $\langle \chi_{i',n'}(Q_i)|\chi_{i,n}(Q_i)\rangle$ for each such normal coordinate. But with the assumption of equal force constants in both oxidation states ($k_M = k_N$), the harmonic oscillator functions $\chi(Q_+)$ for example, on potential surface $W_a$ and $W_b$ are members of the same set of orthonormal functions, since the two surfaces are identical in $Q_+$-space (Equation 24). Thus via Equation 42 and using the orthonormality of these vibrational functions, it follows that $\langle \chi_{n'}(Q_+)|\chi_n(Q_+)\rangle = 0$ unless $n' = n$. Thus no change can occur in the vibrational quantum number associated with $Q_+$ in the course of the transition. Precisely the same argument applies for all the nontotally symmetric normal coordinates (except $q$) because we have assumed throughout that the subunits have the same point group symmetry (and force constants) in both oxidation states. [If the $TD_\gamma(n)$ analogue of Equation 64 is written, it is evident that $TD_\gamma(n) = 0$ in the Franck– Condon approximation.]

Working through the algebra,[1–3] we obtain,

$$D(\nu' \to \nu) = \left(\frac{N_{\nu'} - N_\nu}{N}\right) S^2_{\nu'\nu} |\langle \psi_+|m^e_z|\psi_-\rangle^0|^2 \tag{66}$$

$$S_{\nu'\nu} = \sum_n r_{\nu' n} s_{\nu n} \tag{67}$$

where $|\langle \psi_+|m^e_z|\psi_-\rangle^0|^2 = e^2R^2/4$ with $R$ the distance between subunit centers, $e$ the electronic charge, and the $r_{\nu n}$, $s_{\nu n}$ defined by Equation 49 and determined by Equation 52. *We see that the transition is z-polarized.* To fit the intervalence band (IVB), we assumed $\nu_- = 500\ \text{cm}^{-1}$ based on the totally symmetric frequency in $Ru(NH_3)_6^{3+}$.[1] We may now proceed by a trial and error fitting procedure, guessing values of $\varepsilon$ and $\lambda$ and solving the secular equations (Equation 52). The IVB is the totality of lines $\nu' \to \nu$, Equation 61, where each line is convoluted with a normalized gaussian [$f_{aj}(\mathcal{E})$], usually of the same *ad hoc* width. Thus the absorption coefficient $\alpha$ is:

$$\alpha(\mathcal{E}) = \sum_{a,j} D(a \to j) f_{aj}(\mathcal{E}) \tag{68}$$

with $\mathcal{E} = h\nu$ the photon energy. Note that the molar extinction coefficient $\varepsilon$ (the usual experimental quantity—and not to be confused with the PKS $\varepsilon$) is related to $\alpha$ by

$$\varepsilon(\mathcal{E}) = \alpha(\mathcal{E})\mathcal{E} \times \text{constant} \tag{69}$$

where the explicit value of the constant is given elsewhere.[15] Thus in convoluting the dipole strength ($D_{aj}$) with $f_{aj}$, it is necessary to multiply by the transition energy $E_{aj}$. This was not done in the original treatment[1,3] of the Creutz–Taube ion but has been done in Figure 4 (solid curves). This correction has a relatively small effect on the parameters of the IVB fit but greatly reduces the *relative* intensity of the far infrared transitions (ctiir) discussed below. In practice, we chose the constant in Equation 69 to scale the calculated spectrum to experiment since we make no attempt to calculate an absolute value of the (z-polarized) transition moment. (The expression given for this quantity just below Equation 67 is very crude at best.) We do not dwell on the fit in Figure 4 because, as mentioned earlier, the simple PKS model has a fundamental omission which makes it inapplicable to more or less strongly delocalized systems, such as the Creutz–Taube ion.

### 2.8 Low Energy (ctiir) Transitions

If both $|\varepsilon|$ and $\lambda^2$ are greater than unity, one can divide intervalence transitions into two types: the first occurs "between" the two surfaces and gives rise to the conventional IVB. The second occurs between levels "on" the lower potential surface and is labelled ctiir in Figure 4.

Toward the localized limit, these transitions occur at very low frequency and are akin to the inversion doubling transition observed in $NH_3$ around 1 $cm^{-1}$. We originally referred to such transitions as "tunneling transitions." Such transitions have not been observed in localized systems to this writer's knowledge. Toward the delocalized limit, however, such transitions have been observed in quite spectacular fashion in mixed-valence linear chain systems, first by Rice and coworkers[16,17] in $(TCNQ)_2^-$ and later by Collman et al.[18] in mixed-valence OsOEP polymers. We refer to these as charge-transfer induced infrared (ctiir) transitions. This nomenclature emphasizes that these differ from conventional (vibrationally) allowed infrared transitions *in that the intensity is borrowed from the intervalence band.* Specifically, they are a direct consequence of the mixing of vibronic functions between the two surfaces and thus directly reflect the breakdown of the Born–Oppenheimer approximation. These transitions therefore vanish in the limit of zero vibronic coupling ($\lambda = 0$) or zero electronic coupling ($\varepsilon = 0$). Perturbational expressions for these transitions have been given in both limits.[19]

## 3 THE PIEPHO MO APPROACH

### 3.1 Introduction

It will be noted that the original PKS electronic basis (Equation 17), which is "exact" in the localized limit (substitute $\varepsilon = 0$, $\lambda \neq 0$ in Equation 34), i.e., satisfies the Born–Oppenheimer approximation, is a valence-bond-type basis. It is also clear in the delocalized limit (substitute $\lambda = 0$, $\varepsilon \neq 0$ in Equation 34) that the electronic functions become Equation 42:

$$\psi_{\pm} = \frac{1}{\sqrt{2}}(\psi_a \pm \psi_b) \tag{70}$$

Again the Born–Oppenheimer approximation is satisfied, and $\psi_\pm$ in turn forms a MO-type basis. For example, if the highest occupied orbital on each center were nondegenerate and the mixed-valence compound had a single hole (as is the case for the Creutz–Taube ion), then $\psi_+$ would represent a doubly occupied bonding MO and $\psi_-$ a singly occupied antibonding MO. In this limit, the intervalence band (IVB) transition corresponds to the promotion of an electron from the bonding ($\psi_+$) to antibonding ($\psi_-$) MO.

In solving the dynamic problem (Equation 37), we have chosen the Equation 70 electronic basis because then it is possible to use a vibrational basis of undisplaced harmonic oscillators which have *very* convenient convergence and selection-rule properties. However, it is also possible to use the Equation 17 electronic basis. In that case, the appropriate vibrational basis is displaced harmonic oscillators, and literature discussions of this case have been given.[8,9,22] Of course the same results for all observables must be obtained using either basis.

The crucial point in the Piepho approach[20,21] is the recognition that while $Q_-$ (Equation 22) is an essential coordinate and that $Q_+$ decouples from the problem to a good approximation, *another coordinate is essential even in the simplest mixed-valence system*, namely a coordinate which measures the distance between the two interacting centers. This is seen most clearly in the MO limit (Equation 70) since the IVB (actually a single line in this case) arises from a transition from a bonding to antibonding MO. But of course the $A - B$ equilibrium distance must change in such a case which means that there is significant vibronic coupling in a coordinate which measures the $A - B$ distance. ***Thus a minimum of two vibrational coordinates is an absolute essential*** to give a meaningful, unified discussion of mixed-valence dimer systems over the range from strongly localized to strongly delocalized!

### 3.2 Summary of Piepho Approach—The Inclusion of $Q_{AB}$

As before we consider a symmetrical mixed-valence "dimer" containing two "monomeric" units ($A \equiv B$), and we consider only the HOMO of each monomer unit. The two "monomers" interact to give rise to a bonding (irrep $b$) and antibonding (irrep $a$) MO separated in energy by $\Delta$ and occupied by three electrons. Thus our electronic basis states are

$$\phi_1 = |b^2 a\rangle, \qquad \phi_2 = |ba^2\rangle \tag{71}$$

with (neglecting overlap),

$$|b\rangle = \frac{1}{\sqrt{2}}\big[a_1(A) + a_1(B)\big], \qquad |a\rangle = \frac{1}{\sqrt{2}}\big[a_1(A) - a_1(B)\big] \tag{72}$$

where $a_1(A)$ designates a HOMO on $A$ which (arbitrarily) has totally symmetric local symmetry. *We stress that the detailed form of Equation 72 is irrelevant. We only subsequently require that* |b⟩ *and* |a⟩ *respectively do not and do change sign under an interchange symmetry operation.*

Following Piepho, *we hereafter relabel the original PKS coordinate* $q$ *(or* $Q_-$*) with a subscript* 2 and introduce a new coordinate with subscript 1.

$$\begin{aligned} Q_+^{molec} &\equiv Q_1 = Q_{AB} \\ Q_- \equiv Q_2 &= \frac{1}{\sqrt{2}}(Q_A - Q_B) \end{aligned} \tag{73}$$

$Q_A$ and $Q_B$ are totally symmetric breathing coordinates of the respective monomers, and $Q_{AB}$ measures the distance *between* the monomers. For a strongly localized system, vibronic interaction is dominated by $Q_2$ since an optical (vertical) excitation moves an electron from one center to the other thus exciting $Q_2$, whereas for a strongly delocalized system the vibronic interaction is dominated by $Q_1$, since an optical excitation moves an electron from a bonding to antibonding orbital thus exciting $Q_1$.

The total molecular Hamiltonian of the system is written,

$$H_T(r, Q) = H_{el}(r, Q) + T_N(Q) \tag{74}$$

with

$$H_{el}(r, Q) = T_e(r) + V_{ee}(r) + V(r, Q) \tag{75}$$

where $r$ and $Q$ respectively designate electronic and mass-weighted nuclear coordinates. $T_N(Q)$ designates the nuclear kinetic energy operator and $H_{el}$ includes all other terms. Our zero-order electronic functions $[\phi_j(r, Q_0) \equiv \phi_j^0]$ satisfy the equation,

$$H_{el}(r, Q_0)\phi_j^0 = W_j^0(Q_0)\phi_j^0; \quad j = 1, 2 \tag{76}$$

at the ground state nuclear configuration $Q_0 = 0$. Our goal, as before, is to solve the dynamic Schrödinger equation,

$$H_T \Psi_k = E_k \Psi_k \tag{77}$$

in the presence of strong vibronic interactions. This is done by expanding $H_{el}$ in a Taylor series in the nuclear coordinates and expressing $\Psi_k$ in the non-adiabatic form,

$$\Psi_k(r, Q) = \phi_1^0 \chi_{1k}(Q) + \phi_2^0 \chi_{2k}(Q) \tag{78}$$

so that the Born–Oppenheimer approximation is not made.

Equation 78 in essence is identical in form to Equation 43, but $\chi_{j,k}(Q)$ is a function of two coordinates: $Q_1$ and $Q_2$. We have here explicitly recognized the MOs $\phi_1$ and $\phi_2$, and we have used the Piepho notation, $\Psi_k$. Proceding then to the analogue of Equation 44, we obtain

$$\sum_{j'=1}^{2} \left[(T_N + W_j^0 - E_k)\delta_{jj'} + W_{jj'}(Q)\right]\chi_{j'k}(Q) = 0; \tag{79}$$

where

$$W_{jj}(Q) = \sum_{\alpha=1}^{2} \left[l_\alpha^{(j)} Q_\alpha + \frac{1}{2} k_\alpha^{(j)} Q_\alpha^2\right] \tag{80}$$

$$W_{jj'}(Q) = \sum_{\alpha=1}^{2} l_\alpha^{(j,j')} Q_\alpha, \quad (j, j' = 1, 2);\ (k = 1, 2, \ldots, \infty) \tag{81}$$

and

$$\begin{aligned} l_\alpha^{(j)} &= \langle \phi_j^0 | (\partial V/\partial Q_\alpha)_0 | \phi_j^0 \rangle \\ l_\alpha^{(j,j')} &= \langle \phi_j^0 | (\partial V/\partial Q_\alpha)_0 | \phi_{j'}^0 \rangle \\ k_\alpha^{(j)} &= \langle \phi_j^0 | (\partial^2 V/\partial Q_\alpha^2)_0 | \phi_j^0 \rangle \end{aligned} \tag{82}$$

Transforming to the dimensionless variables

$$\begin{aligned} q_\alpha &\equiv (2\pi/h)(h\nu_\alpha)^{1/2} Q_\alpha \\ \lambda_\alpha &\equiv h(4\pi^2 h\nu_\alpha)^{-1/2}\left(\sqrt{2}\, h\nu_\alpha\right)^{-1} l_\alpha \end{aligned} \tag{83}$$

we assume $k_\alpha^{(1)} = k_\alpha^{(2)}$ and writting $\nu_\alpha = (1/2\pi)\sqrt{k_\alpha}$, we obtain

$$\begin{aligned} W_{jj}(q) &= \sum_{\alpha=1}^{2} h\nu_\alpha\left[\sqrt{2}\,\lambda_\alpha^{(j)} q_\alpha + (1/2) q_\alpha^2\right] \\ W_{jj'}(q) &= \sqrt{2} \sum_{\alpha=1}^{2} h\nu_\alpha \lambda_\alpha^{(jj')} q_\alpha \end{aligned} \tag{84}$$

Note that $W_{22}$ and $W_{12}$ contain no linear terms in $q_2$ and $q_1$ respectively:

$$\langle \phi_i^0 | (\partial V/\partial q_2)_0 | \phi_i^0 \rangle = 0 \quad (i = 1, 2) \tag{85}$$

$$\langle \phi_i^0 | (\partial V/\partial q_1)_0 | \phi_j^0 \rangle = 0 \quad (i \neq j = 1, 2) \tag{86}$$

This follows since $\phi_i^0 \otimes \phi_i^0$ has even interchange symmetry and $(\partial V/\partial q_2)_0$ which transforms like $q_2$ is correspondingly odd, whereas $\phi_i^0 \otimes \phi_j^0$ $(i \neq j)$ is odd and $(\partial V/\partial q_1)_0$ which transforms like $q_1$ is even. Furthermore, $l_1^{(1)}$ can be set to zero by choosing $Q_1 = 0$ at the potential minimum. Thus we end up with two vibronic coupling constants: $\lambda_1^{(2)} \equiv \lambda_1$ and $\lambda_2^{(1,2)} \equiv \lambda_2$. The electronic potential energy matrix is

$$\begin{pmatrix} W_{11} & W_{12} \\ W_{12} & W_{22} + \Delta \end{pmatrix} = \begin{bmatrix} \left(\frac{1}{2}q_1^2 h\nu_1 + \frac{1}{2}q_2^2 h\nu_2\right) & \sqrt{2}\,\lambda_2 q_2 h\nu_2 \\ \sqrt{2}\,\lambda_2 q_2 h\nu_2 & \begin{pmatrix} \frac{1}{2}q_1^2 h\nu_1 + \frac{1}{2}q_2^2 h\nu_2 + \\ \sqrt{2}\,\lambda_1 q_1 h\nu_1 + \Delta \end{pmatrix} \end{bmatrix} \tag{87}$$

Comparing this equation with Equation 45, recalling that the latter is in units of $h\nu_- \equiv h\nu_2$, we find that $\lambda_2 = (1/\sqrt{2})\lambda_{\text{PKS}}$, where $\lambda_{\text{PKS}}$ denotes the single vibronic coupling constant that appears in the simple PKS treatment—the $\lambda$ of Equation 25 or 45—and the electronic coupling is $\Delta = 2|\varepsilon|h\nu_2$.

Equation 84 is now substituted into Equation 79, the $\chi_{jk}(Q)$ are expanded in complete sets of harmonic oscillator functions in $q_1$ and $q_2$:

$$\chi_{jk}(Q) = \sum_{n_1,n_2=0}^{\infty} c_{jn_1n_2}^{k}\,\chi_{n_1}(q_1)\chi_{n_2}(q_2) \tag{88}$$

and the expansion coefficients are determined as before (Equation 51, 52) by taking scalar products of Equation 79. Using the Piepho notation[20] $[\Phi_{jn_1n_2} \equiv \phi_j^0 \chi_{n_1}(q_1)\chi_{n_2}(q_2)]$, the non-zero matrix elements of the resulting secular equations are

$$\begin{aligned} \langle \Phi_{1n_1n_2}|H_T|\Phi_{1n_1n_2}\rangle &= \left(n_1 + \frac{1}{2}\right)h\nu_1 + \left(n_2 + \frac{1}{2}\right)h\nu_2 \\ \langle \Phi_{2n_1n_2}|H_T|\Phi_{2n_1n_2}\rangle &= \Delta + \left(n_1 + \frac{1}{2}\right)h\nu_1 + \left(n_2 + \frac{1}{2}\right)h\nu_2 \end{aligned} \tag{89}$$

$$\begin{aligned} &\langle \Phi_{2n_1n_2}|H_T|\Phi_{1n_1n_2'}\rangle = \langle \Phi_{1n_1n_2}|H_T|\Phi_{2n_1n_2'}\rangle \\ &= \sqrt{2}\,\lambda_2 h\nu_2 \left\{ \left(\frac{n_2+1}{2}\right)^{1/2} \delta_{n_2',n_2+1} + \left(\frac{n_2'+1}{2}\right)^{1/2} \delta_{n_2,n_2'+1} \right\} \\ &\langle \Phi_{2n_1n_2}|H_T|\Phi_{2n_1'n_2}\rangle \\ &= \sqrt{2}\,\lambda_1 h\nu_1 \left\{ \left(\frac{n_1+1}{2}\right)^{1/2} \delta_{n_1',n_1+1} + \left(\frac{n_1'+1}{2}\right)^{1/2} \delta_{n_1,n_1'+1} \right\} \end{aligned} \tag{90}$$

The basis in Equation 88 is truncated at sufficient size to give acceptable accuracy, transition dipoles are calculated and the resulting stick spectra are convoluted with gaussians of a single *ad hoc* linewidth to produce the final intervalence band contour, exactly as is done in the original PKS model (Section 2.7).

Piepho presents[20] typical contours for a range of values of $\lambda_1$ and $\lambda_2$. In general, strong coupling to $q_1$ (large $\lambda_1$) favors a symmetrical band while strong coupling to $q_2$ (large $\lambda_2$) produces a strongly asymmetric band. (Piepho also explicitly considers a third coordinate in her treatment, namely $Q_+$, and shows why and under what circumstances this coordinate decouples from the problem. For simplicity, we have omitted $Q_+$ right from the start in our summary and subsequent extensions.)

In Figure 5 we show a "best fit" of the CT-ion IVB using the Piepho two-mode model. We note at the outset that finding such a "best" fit is quite subjective, since the CT-ion IVB is largely featureless, and we now have two additional parameters as compared to the one-mode case. What we have preferred to do is start with the Piepho parameters for the Creutz–Taube ion (see below) and vary our lesser number of them as little as possible to get a reasonable fit. The fit in fact is seen to be quite good

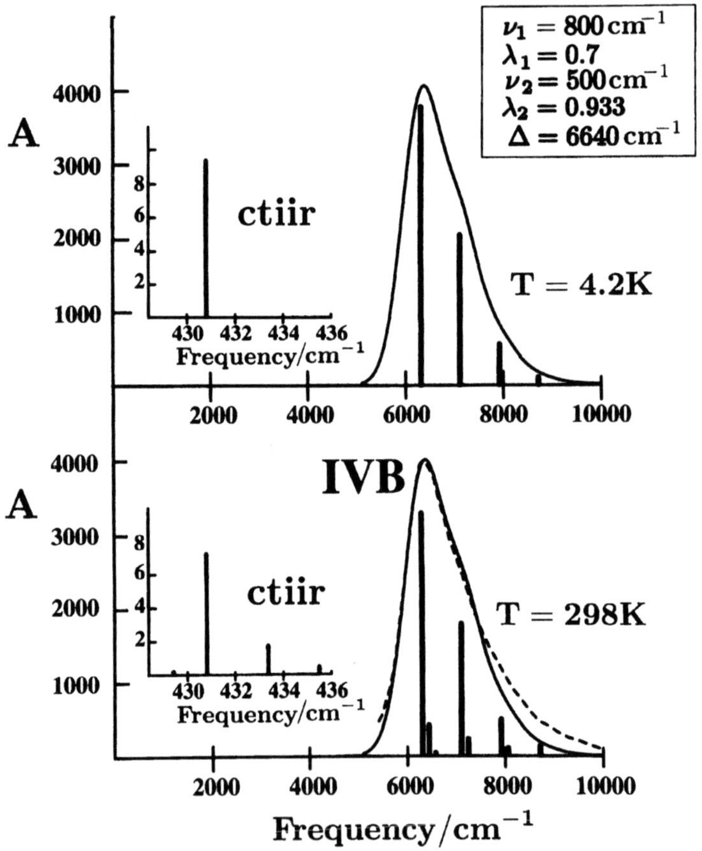


**Figure 5.** Two-mode theoretical fit of the Creutz–Taube ion spectrum. Display and arrangement are as in Figure 4 except that the convolution half-width is 550 cm$^{-1}$.

except on the high energy tail where there is reason to think that some of the intensity arises from a mechanism not included in our model.[21] One notes immediately that $\lambda_{PKS}(=\sqrt{2}\,\lambda_2)$ is considerably smaller than in the original one-mode PKS treatment (Section 2.7) suggesting a more strongly delocalized system. Quantitative statements on this score are made in Section 4.

### 3.3 The Electronic Coupling Mechanism in the Creutz–Taube Ion

In our discussion so far, we have designated the electronic coupling between subunits by the symbol $\varepsilon$ (Equation 31), but we have given no physical picture of this important parameter which actually encompasses much of the chemistry of the system. In the valence-bond basis of the original PKS treatment, $\varepsilon$ is seen to be the resonance stabilization that occurs when the two equivalent structures $\psi_a$ and $\psi_b$ are permitted to interact (see Equation 30). Equivalently, from the MO viewpoint we have shown below Equation 87 that $\varepsilon$ is the energy of the bonding MO formed when the HOMOs of the two subunits interact. Either point of view is apt to be physically illuminating only when the simplest kind of through-space interaction is relevant, as for example in the benzene dimer cation with the two rings face-to-face. In the case of the CT-ion, as one example, the coupling of the two Ru centers clearly must be through the pyrazine ring, and thus $\varepsilon$ must be replaced by parameters which describe MO formation between appropriate orbitals on the Ru atoms and the pyrazine ring.

The reader is referred to the excellent, detailed treatment of the Creutz–Taube ion presented in the second MO paper of Piepho.[21] Spin-orbit and low symmetry perturbations ($D_{2h}$ Ru site symmetry) as well as specific MO formation between the pyrazine ring and Ru orbitals are all explicitly treated. Furthermore, parameters are determined by fitting both the IVB and experimental g-values available from esr measurements. It is shown that MO formation between Ru 4d orbitals and both $\pi$ and $\pi^*$ pyrazine orbitals are expected to be important. Reference is also given to the extensive work of Ondrechen, and a comparison of their respective treatments of the CT-ion is presented. The analysis supports the conclusion that the CT-ion is a delocalized mixed-valence system. This analysis also emphasizes the point that all relevant experimental data must be incorporated into a single unified treatment if one is to obtain a convincing picture of this much-studied ion with a rather nondescript IVB.

## 4 VALENCE TRAPPING AND DELOCALIZATION IN MIXED-VALENCE SYSTEMS

### 4.1 Introduction

One of the immediate questions that arises in mixed-valence systems is the following: Is the system localized or delocalized? Since the PKS-type approach yields the entire ground vibronic manifold once the various parameters are determined, it should be possible to make well-defined statements. Our starting point is the vibronic wave function of each thermally accessible level of the system.

### 4.2 The Probability Distribution in Configuration Space, $P(q_2)$

We start by squaring the vibronic wave function (Equation 78) noting that all quantities are real:

$$|\Psi_k(r,Q)|^2 = \phi_1^{0^2}(r)\chi_{1k}^2(Q) + \phi_2^{0^2}(r)\chi_{2k}^2(Q) + 2\phi_1^0(r)\phi_2^0(r)\chi_{1k}(Q)\chi_{2k}(Q) \quad (91)$$

We then define a quantity $P(q_2)$ by

$$P(q_2) \equiv \int |\Psi_k(r,Q)|^2 d\tau_{el} dq_1 \quad (92)$$

that is, we integrate $|\Psi_k(r,Q)|^2$ over all coordinates except $q_2$ (reiterating that $q_2$ now denotes the $q$ of Equation 25). Substituting for the $\chi_{i,k}(Q)$ from Equation 88, noting the orthonormalities

$$\langle \chi_{n_1}(q_1)|\chi_{n_1'}(q_1)\rangle = \delta_{n_1 n_1'}; \qquad \langle \phi_i^0(r)|\phi_j^0(r)\rangle = \delta_{ij} \quad (93)$$

$$P_k(q_2) = \sum_{n_1,n_2,n_2'=0}^{\infty} c_{1n_1n_2}^k c_{1n_1n_2'}^k \chi_{n_2}(q_2)\chi_{n_2'}(q_2) + \sum_{n_1,n_2,n_2'=0}^{\infty} c_{2n_1n_2}^k c_{2n_1n_2'}^k \chi_{n_2}(q_2)\chi_{n_2'}(q_2) \quad (94)$$

where the $c_{jn_1n_2}^k$ are defined in Equation 88. The $\chi_{n_2}(q_2)$ are the familiar harmonic oscillator functions given by $\chi_{n_2}(q_2) = (2^{n_2}n_2!\sqrt{\pi})^{-1/2}e^{-q_2^2/2}H_{n_2}(q_2)$ where the $H_{n_2}(q_2)$ are the Hermite polynomials.

$P_k(q_2)$ is thus the probability distribution in $q_2$-space of vibronic state $k$. If more than one vibronic level is populated, then

$$P(q_2) = \frac{\sum_k P_k(q_2)e^{-E_k/kt}}{\sum_k e^{-E_k/kt}} \quad (95)$$

Recalling the form of $q_2(\equiv q)$, Figure 1, we see the basis for a qualitative distinction between a localized and delocalized system: If $P(q_2)$ shows a single maximum at $q_2 = 0$, the system is delocalized, and if $P(q_2)$ shows two maxima, the system is localized. Three cases are illustrated in Figure 6 for the simple PKS model (no $q_1$ coordinate), namely a strongly localized (R&D Class I), a strongly delocalized (R&D

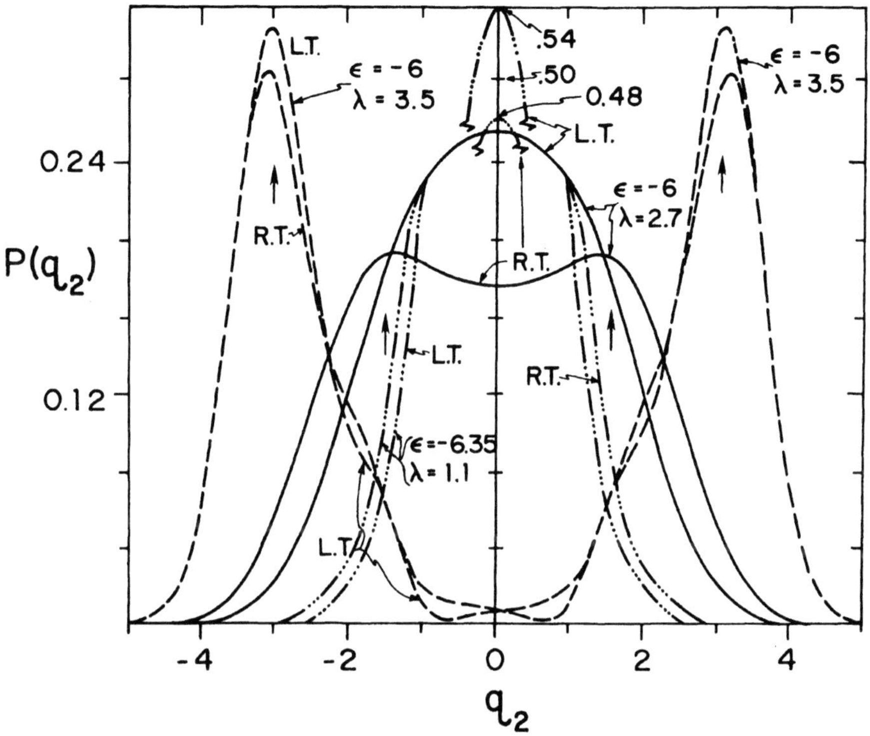


**Figure 6.** Calculated probability distribution, $P(q_2)$, at 298 K (R. T.) and 4.2 K (L. T.) for the indicated parameters in the symmetrical case (adapted from Reference 3). The calculated minima in the lower potential surface are shown by the vertical arrows for $\lambda = 2.7$ ($q_2 = \pm 1.53$) and $\lambda = 3.5$ ($q_2 = \pm 3.05$). Calculated degrees of delocalization ($\beta_d^2$) for the three cases shown are 0.269, 0.781, 0.986 and 0.282, 0.695, 0.982 at 4.2 K and 298 K, respectively.

Class III), and a borderline (R&D Class II) case. The potential surfaces corresponding to these three cases are shown in Figure 7.

One interesting feature of these $P(q_2)$ is the fact that the systems tend to become more localized as the temperature rises. This is seen most clearly in the borderline (Class II) case where the system shows a single maximum at low temperature and two maxima at room temperature. This presumably reflects the fact that (classically speaking) as the temperature rises, the amplitude of $q_2$ increases, and since an excited harmonic oscillator spends most of its time at the turning points (which represent localized configurations), the degree of delocalization decreases with increasing temperature. In Figure 8, $P(q_2)$ is compared for "best fits" of the Creutz–Taube ion IVB using simple PKS and the Piepho two-mode model, i.e., the theoretical parameters used in Figures 4 and 5, respectively.

The quantity $\lambda_\alpha^2 h\nu_\alpha$ should be a reasonable measure of the contribution of mode $q_\alpha$ to the vibronic coupling. On this basis, we note that modes $q_1$ and $q_2$ make compa-

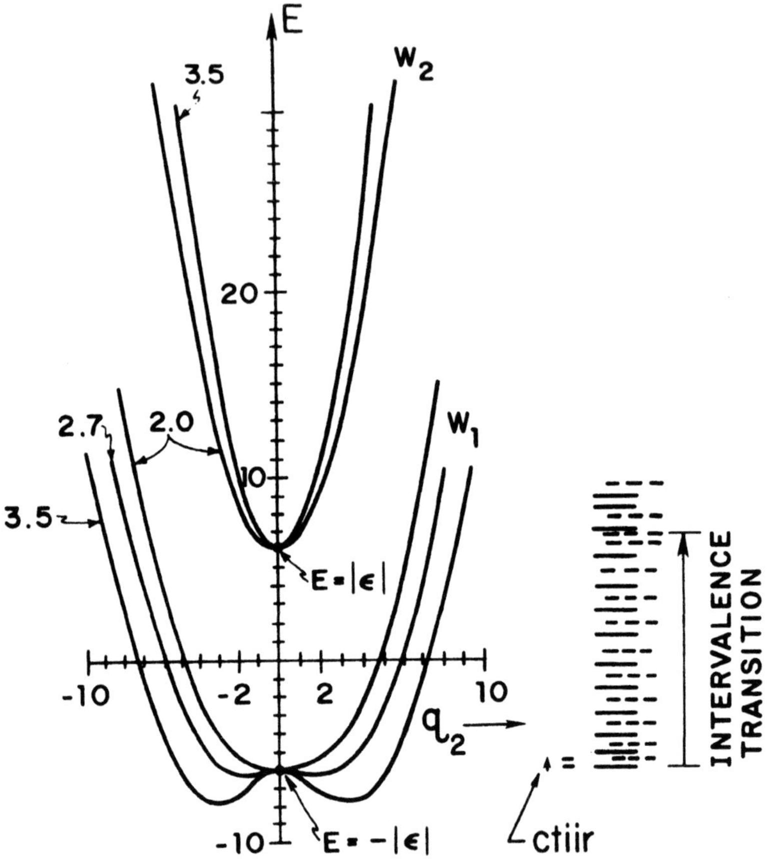


**Figure 7.** Potential energy surfaces in $q_2$-space using Equation 33 in units of $h\nu_-$ for $\varepsilon = -6.0$, and $\lambda = 2.0$, 2.7, and 3.5 (adapted from Reference 3). ($W_2$ is omitted for $\lambda = 2.7$.) Lower-lying vibronic levels are shown to the right for $\lambda = 2.7$ with the solid and dashed lines representing levels of even and odd interchange symmetry respectively. The long and short vertical arrows show respectively the strongest intervalence and the strongest charge-transfer-induced-infrared (ctiir) transitions.

rable contributions in the two-mode case whereas in the one-mode case, the vibronic coupling is (unreasonably) constrained to a single asymmetric mode. It is therefore not surprising that the two-mode treatment suggests a significantly more delocalized structure (Figure 8). In the next section, we put a quantitative value on this property, but we once again emphasize that our treatment is more in the way of an illustration of the formalism. An acceptable treatment of the Creutz–Taube ion must take account of spin-orbit, low symmetry effects, and the esr g-values, and thus a treatment such as the one given by Piepho[21] is required for a convincing description.

**Finally, we stress that a necessary condition for the use of the term mixed-valence is the existence of non-negligible vibronic coupling in mode $q_2$, i.e., $\lambda_2$**

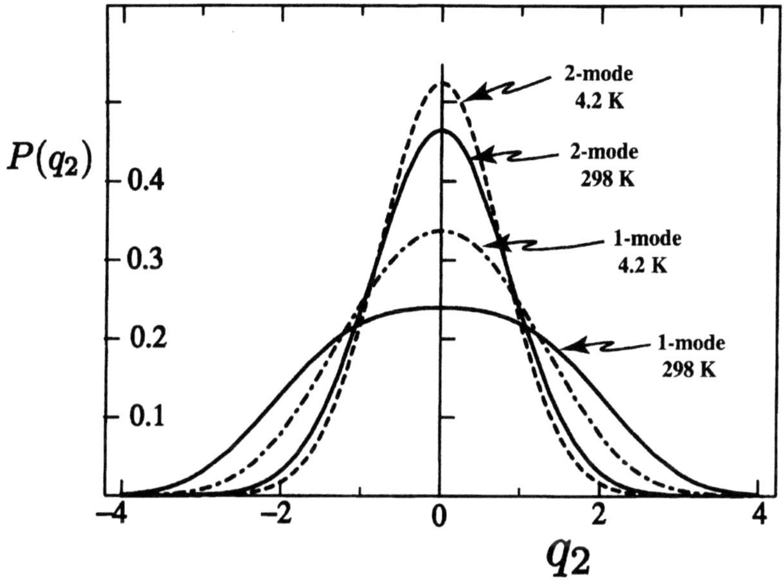


**Figure 8.** Calculated probability distributions, $P(q_2)$, at 298 K and 4.2 K for the one- and two-mode fits of Figures 4 and 5, respectively. Calculated degrees of delocalization ($\beta_d^2$) for the two cases shown are 0.857 (one mode), 0.980 (two mode) and 0.789 (one mode), 0.974 (two mode) at 4.2 K and 298 K, respectively.

**non-negligible.** Otherwise, the conventional MO description of an odd-electron system is applicable.

## 4.3 The Degree of Delocalization ($\beta^2$)

$P(q_2)$ is an incomplete description of delocalization because we have integrated over the electronic coordinates. Thus we may instead integrate Equation 91 over vibrational coordinates. In completing this process, we note the following:

$$\sum_{n_1,n_2} \left[\left(c^k_{1n_1n_2}\right)^2 + \left(c^k_{2n_1n_2}\right)^2\right] = 1; \quad \sum_{n_1,n_2} c^k_{1n_1n_2} c^k_{2n_1n_2} = 0 \tag{96}$$

The first of these comes directly from the normalization requirement: $\int |\Psi_k(r, Q)|^2 \times d\tau_{el} dq_1 dq_2 = 1$. To obtain the second, we note that a symmetrical ($A \equiv B$) system must show interchange symmetry, and it is evident from Equations 71 and 72 that $\phi_1$ and $\phi_2$ must have opposite interchange symmetries, i.e., one is odd and the other even. It then immediately follows from Equation 78 that the same is true for $\chi_{1,k}(Q)$ and $\chi_{2,k}(Q)$, and they are therefore orthogonal. The second of Equation 96 is the statement of that orthogonality. Therefore, integrating Equation 91 over $q_1$ and $q_2$, using Equations 96, 88, 71, and 72, we obtain after some rearranging

$$P_k(el) \equiv \int |\Psi_k(r, Q)|^2 dq_1 dq_2$$
$$= \frac{3}{2}[a_{1u}^2(A) + a_{1u}^2(B)] + a_{1u}(A)a_{1u}(B) \sum_{n_1,n_2} [(c_{1n_1n_2}^k)^2 - (c_{2n_1n_2}^k)^2] \quad (97)$$

We can reasonably regard the coefficient of $a_{1u}(A)a_{1u}(B)$ as a measure of delocalization, and we therefore *define* the degree of delocalization in vibronic state $k$ by

$$\beta_d^2(k) \equiv \left( \sum_{n_1,n_2} [(c_{1n_1n_2}^k)^2 - (c_{2n_1n_2}^k)^2] \right)^2 \quad (98)$$

and as above (Equation 95), we take a Boltzmann average over populated levels. Note that Equation 98 is the direct two-mode analogue of the degree of delocalization defined in the original PKS model.[2,3]

As an illustration, $\beta_d^2 = 0.269, 0.786, 0.986$ and $0.282, 0.695, 0.982$ at 4.2 K and 298 K, respectively, for the three cases shown in Figure 6; and $\beta_d^2 = 0.857$ (one mode), 0.980 (two mode) and 0.789 (one mode), 0.974 (two mode) at 4.2 K and 298 K, respectively, for the one- and two-mode fits shown in Figures 4, 5, and 8. Once again (except in the strongly localized case), raising the temperature favors localization. Also, on the basis of $\beta_d^2$, the simple two-mode treatment suggests a much more delocalized Creutz–Taube ion than the one-mode treatment.

The quantity $\beta_d^2$ has not been much used but may be a useful measure of delocalization in a series of related systems. We have also defined a quantity $\gamma_{dis}$ which measures the disymmetry of a system when $W \neq 0$ (see Reference 3, page 421); its analogue could readily be defined in the two-mode case.

Finally, as we leave the CT-ion for the last time, we reiterate that although this fascinating ion has been repeatedly used in our various comparisons, the models presented are quite general and do not depend on the detailed features of any specific system.

## 5 TWO-CENTER TWO-ELECTRON ONE-MODE SYSTEMS IN THE SYMMETRICAL CASE

### 5.1 Introduction

Though our original formulation of the PKS model (Equation 1, *et seq.*) allows the mixed-valence oxidation states to differ by more than one electron, we have so far only treated the one-electron (or one-hole) case, i.e., $n = 1$ in Equations 1–3. We now explicitly consider the case $n = 2$. We find that an additional electronic parameter (the on-site repulsion, $U$) and *three* potential surfaces are an inevitable consequence.

## 5.2 Simple Treatment[22]

We return to Equation 1 *et seq.* and now assume that $n = 2$ so that $N = M - 2$. Then Equation 2 reads

$$\begin{aligned}(H_{el}^A + H_{el}^B)\psi_M^A\psi_{M-2}^B &\equiv (H_{el}^A + H_{el}^B)\psi_a \\ &= (W_M^A + W_{M-2}^B)\psi_a \equiv W_a\psi_a \end{aligned} \tag{99}$$

Clearly an analogous equation applies for $\psi_c = \psi_{M-2}^A\psi_M^B$ if the replacements $A \rightleftharpoons B$ $a \to c$ are made in Equation 99. However, another state ($\psi_b$) must also exist in which only one electron is transferred, namely

$$\begin{aligned}(H_{el}^A + H_{el}^B)\psi_{M-1}^A\psi_{M-1}^B &\equiv (H_{el}^A + H_{el}^B)\psi_b \\ &= (W_{M-1}^A + W_{M-1}^B)\psi_b \equiv W_b\psi_b \end{aligned} \tag{100}$$

Proceeding as before, we make the harmonic approximation and assume that the subunits have the same point group symmetry in all three oxidation states. We here restrict ourselves to a single totally symmetric normal coordinate on each moiety ($Q_A$ and $Q_B$). The potential energy of subunit $j$ in oxidation state $i$ can then be written

$$W_i^j = W_i^{0j} + l_i^j Q_j + \frac{1}{2}k_i^j Q_j^2; \quad j = A, B; \; i = M, M-1, M-2 \tag{101}$$

where $W_i^{0j} \equiv W_i^j(Q_j = 0)$. We assume that centers $A$ and $B$ are equivalent (the symmetrical case). The hexachloroantimonate (III, V) compounds and Wolffram's Red Salt (Section 5.4) are examples of such systems with $M = 5$ and 4, respectively. We define the zero of energy by $W_M^{0A} + W_{M-2}^{0B} = W_{M-2}^{0A} + W_M^{0B} = 0$ and define $2\overline{WW} \equiv W_{M-1}^{0A} + W_{M-1}^{0B} - (W_M^{0A} + W_{M-2}^{0B})$. As before we introduce the sum and difference coordinates:

$$Q_\pm = \frac{1}{\sqrt{2}}(Q_A \pm Q_B) \tag{102}$$

We now make the crucial assumptions (which also apply for $A \to B$)

$$\begin{aligned} k_M^A &= k_{M-1}^A = k_{M-2}^A \equiv k \\ l_{M-1}^A &= \frac{1}{2}(l_M^A + l_{M-2}^A) \equiv l \end{aligned} \tag{103}$$

With Equation 103, $W_a$, $W_b$, and $W_c$ give identical parabolas in $Q_+$ space. Thus the $Q_+$ coordinate decouples from the problem and may be dropped as before. We choose origins by setting $Q_A$ and $Q_B$ to zero at the minima of $W_{M-2}^A$ and $W_{M-2}^B$, respectively. We define the dimensionless variables as

$$
\begin{aligned}
q &\equiv 2\pi(\nu_-/h)^{1/2}Q_- \\
\lambda &\equiv \left(8\pi^2 h\nu_-^3\right)^{-1/2} l \\
2WW &\equiv 2\overline{WW}/h\nu_- + \lambda^2/2
\end{aligned}
\tag{104}
$$

where $\nu_- = (2\pi)^{-1}k^{1/2}$ is the vibrational frequency associated with the crucial coordinate $Q_-$. We obtain (after changing the energy zero by $\lambda^2/2$)

$$
\begin{aligned}
W_a/h\nu_- &= \frac{1}{2}(q+\lambda)^2 \\
W_b/h\nu_- &= 2WW + \frac{1}{2}q^2 \\
W_c/h\nu_- &= \frac{1}{2}(q-\lambda)^2
\end{aligned}
\tag{105}
$$

We now permit interaction ($V^{AB}$) between centers $A$ and $B$, so $H_{el} = H_{el}^A + H_{el}^B + V^{AB}$. We hold the nuclei at $q = 0$ and define

$$
V_{ij}^0 \equiv \left\langle \psi_i \middle| V^{AB} \middle| \psi_j \right\rangle^0; \quad i, j = a, b, c \tag{106}
$$

In the symmetrical case, $V_{aa}^0 = V_{cc}^0$. We again define the parameter $\varepsilon$ by the relation

$$
\varepsilon \equiv V_{ab}^0/h\nu_- \tag{107}
$$

Changing the zero of energy by $V_{aa}^0/h\nu_-$ and using the above definitions and Equations 99 and 100, we obtain the electronic potential energy matrix in the $\psi_a, \psi_b, \psi_c$ electronic basis:

$$
\begin{bmatrix}
\frac{1}{2}(q+\lambda)^2 & \varepsilon & 0 \\
\varepsilon & \frac{1}{2}q^2 + 2WW & \varepsilon \\
0 & \varepsilon & \frac{1}{2}(q-\lambda)^2
\end{bmatrix}
\tag{108}
$$

The constant term $(V_{bb}^0 - V_{aa}^0)/h\nu_-$ has now been absorbed into $2WW$. [We had also introduced an *intersite* electronic parameter ($\varepsilon'$) in our original treatment.[22] However, as noted later,[23] such a parameter is inconsistent with our model.]

Comparing Equation 108 with the original PKS treatment—for example, Equation 32 or 45— we note the appearance of an additional electronic parameter ($2WW$) and a $3 \times 3$ electronic matrix. The latter mandates the existence of three potential surfaces. A more rigorous treatment[22] which starts with the full molecular Hamiltonian and takes account of the Pauli principle leads again to Equation 108, but provides an important interpretation of $WW$, namely

$$
2WW = \lambda^2/2 - U \tag{109}
$$

$$U = \left\langle \psi_a \middle| \frac{e^2}{r_{12}} \middle| \psi_a \right\rangle = \left\langle \psi_c \middle| \frac{e^2}{r_{12}} \middle| \psi_c \right\rangle \tag{110}$$

$U$ is the on-site coulomb repulsion integral (because both electrons are on the same center in $\psi_a$ or $\psi_c$). *Thus WW is a direct measure of the competition between the on-site interelectronic coulomb repulsion and the vibronic coupling* (electron–phonon coupling in solid-state language). As we shall see ahead, if $2WW$ is positive, the $(M-2, M \leftrightarrow M, M-2)$ state is more stable than the $(M-1, M-1)$ state. That is, it is as if putting two electrons on the same center results in an attractive interaction, *the so-called "negative U" effect.*[24]

The next step is the solution of the dynamic problem, $(H_{el} + T_n)\Psi_k = E_k\Psi_k$. The procedure is exactly analogous to that used earlier (Equation 37 *et seq.*), and we refer the reader to the literature[22] for the detailed treatment which also includes the unsymmetrical case and a perturbational solution in the localized limit. We here confine our discussion to the potential surfaces.

### 5.3 Potential Energy Surfaces

As has been stressed before, the adiabatic potential surfaces offer physical insight into the mixed-valence problem (or in general into the "phase diagram" of the two-site two-electron problem), but they can be used in only a qualitative fashion to predict absorption profiles and other physical observables. The potential energy surfaces are obtained by diagonalization of Equation 108; this also leads to the electronic eigenfunctions associated with these surfaces. These are eigenfunctions of the electronic Hamiltonian and thus apply only if the kinetic energy operators are omitted (i.e., in the "static" limit); they *do not commute* with $T_N$ which is why we do not use them as basis functions in the dynamic treatment.[22] However, in special cases where either $\varepsilon = 0$ or $\lambda = 0$, they do not depend on the vibrational coordinate $q$. Then we can separate the nuclear (vibrational motion) from the electronic problem using the Born–Oppenheimer approximation.

If we start by assuming that $\varepsilon = 0$, Equations 108 and 109 give the adiabatic potentials (Figure 9):

$$\begin{aligned} W_a/h\nu_- &= \frac{1}{2}(q+\lambda)^2 \\ W_b/h\nu_- &= \frac{1}{2}q^2 + \frac{1}{2}\lambda^2 - U \\ W_c/h\nu_- &= \frac{1}{2}(q-\lambda)^2 \end{aligned} \tag{111}$$

The energy term $(1/2)\lambda^2$ represents the stabilization due to electron–phonon coupling of the mixed-valence states (or the double occupancy states) relative to the single-valence state (or the one-by-one occupancy state). The parameter $2WW[\equiv (1/2)\lambda^2 - U]$ directly measures their relative stabilities. If $WW > 0$, the mixed-valence state is stabilized at the expense of the single-valence one, or disproportion-

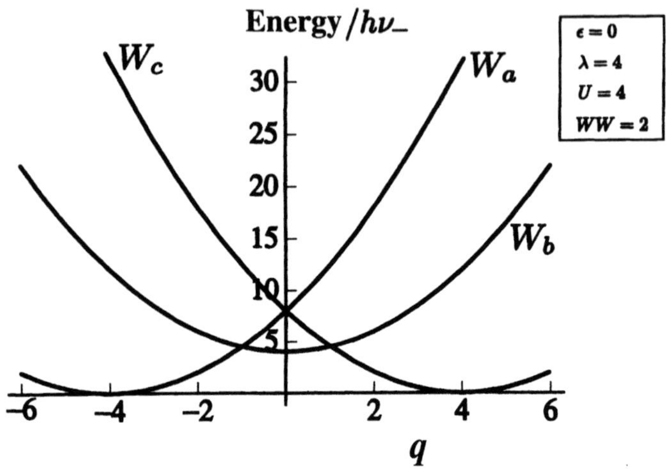


**Figure 9.** Potential energy surfaces for the two-site two-electron case in units of $h\nu_-$ according to Equation 108 for the parameters shown in the upper right box.

ation of the type

$$(e^1) + (e^1) \rightarrow (e^2) + (e^0) \tag{112}$$

takes place with the electron-pair localized on only one site—the "negative-$U$" effect mentioned above.[24] If $WW < 0$, the "delocalized" (single-valence) state is more stable. Furthermore, a potential barrier exists between the minima of $W_a$ (or $W_c$) and $W_b$ for values of $WW$ greater than $(-1/4)\lambda^2$. A second limiting case is $\lambda = 0$ (or $2W = -U$). Then from Equation 108 we obtain the adiabatic potentials

$$\begin{aligned} W_a/h\nu_- &= \frac{1}{2}(q^2 - U) - \frac{1}{2}\left(U^2 + 8\varepsilon^2\right)^{1/2} \\ W_b/h\nu_- &= \frac{1}{2}q^2 \\ W_c/h\nu_- &= \frac{1}{2}(q^2 - U) + \frac{1}{2}\left(U^2 + 8\varepsilon^2\right)^{1/2} \end{aligned} \tag{113}$$

The ground state tends to the valence bond or molecular orbital state of the hydrogen molecule as $|\varepsilon|/U \rightarrow 0$ or $\infty$, respectively.[25]

In general, the surfaces are no longer harmonic when both $\varepsilon$ and $\lambda$ are nonzero. A variety of cases are presented and discussed in Figure 1 *et seq.* of Prassides et al.,[22] and an illuminating discussion of the "phase diagram" of these systems has been given by Toyozawa.[25]

## 5.4 Wolffram's Red Salt (WRS) as an Example

The original stimulus for the study of quasi-one-dimensional mixed-valence compounds was the observation of a spectacular resonance Raman (RR) spectrum[26,27] for Wolffram's Red Salt ($[Pt^{II}(EtNH_2)_4][Pt^{IV}(EtNH_2)_4Cl_2]Cl_4\cdot 4H_2O$). This compound consists of linear chains of halogen bridged alternating Pt(II) and Pt(IV) subunits, and excitation into its intervalence band produces essentially a single long harmonic progression ($\approx$ 16 quanta) in a Pt(IV)–Cl stretching mode of frequency 307.4 $cm^{-1}$ at 4.2 K.[28]

We model WRS as a mixed-valence dimer, $AB$, and associate formal oxidation states $M$ and $M-2$ with subunits $A$ and $B$, respectively: This is equivalent to con-

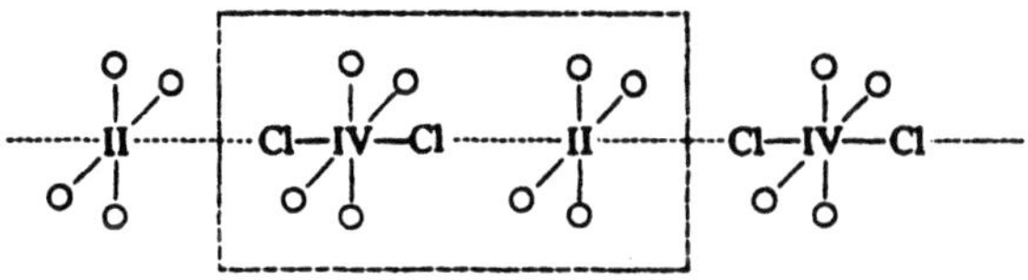


centrating our analysis on one "unit cell" along the WRS chain—the dashed box. The problem can be reduced to the standard PKS form, Equation 108, if we use the following definitions for the vibrational coordinate q and the vibronic (electron-phonon) coupling constant λ:

$$q = 2\pi(2M_{Cl})^{1/2}(\nu_-/h)^{1/2}x \tag{114}$$

$$\lambda = 2\pi(2M_{Cl})^{1/2}(\nu_-/h)^{1/2}|x_0^A| \tag{115}$$

Here $M_{Cl}$ is the mass of a chlorine atom, $\nu_-$ is the Cl–$Pt^{IV}$–Cl symmetric stretching frequency, $x$ designates the position of the Cl atoms with respect to an origin halfway between $A$ and $B$, and $|x_0^A|$ is the equilibrium displacement for $Pt_A$ in the ground state in the absence of other interactions. We focus our attention on a single pair of electrons in the $d_{z^2}$ orbital.[22] This orbital may be doubly occupied on $Pt_B$ and empty on $Pt_A$ giving rise to the state $\psi_a = \psi_M{}^A\psi_{M-2}^B$. There is clearly an equivalent state $\psi_c = \psi_{M-2}^A\psi_M^B$, and of course the (higher energy) state $\psi_b = \psi_{M-1}^A\psi_{M-1}^B$ in which the $d_{z^2}$ orbital of each of the two Pt atoms of the unit cell is singly occupied.

The solution of the dynamic problem and the fitting of the various spectra are discussed in detail in Reference 23. The results are summarized in Figures 10 and 11. Figure 10 shows the potential surfaces which arise when Equation 108 is diagonalized for the best fit parameters given in the caption. Figure 11 shows a comparison of experimental and theoretical spectra, the latter calculated using the vibronic functions resulting from solution of the dynamic problem. It is concluded that this simple model which focuses on a single coordinate in a single unit cell is able to account for the most important features of the optical spectra of WRS.[23]

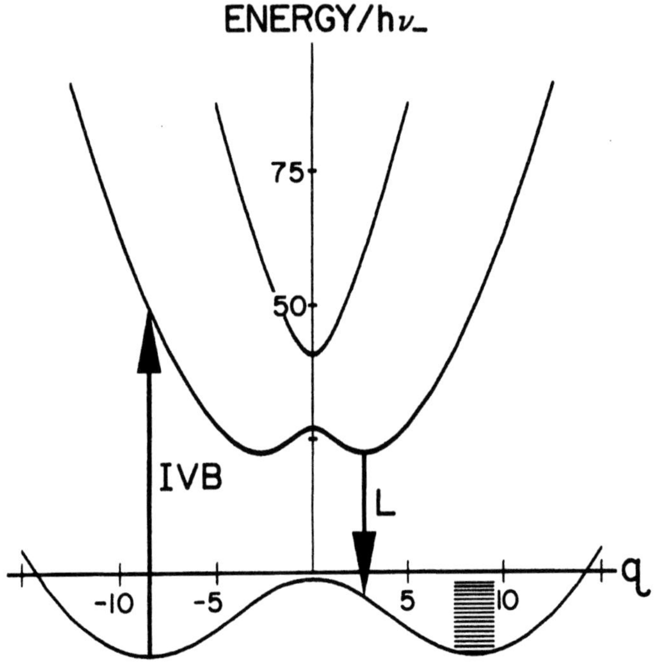


**Figure 10.** Potential energy surfaces for the two-site two-electron case according to Equations 108 and 109 for the parameters $\varepsilon = -14.0$, $\lambda = 8.9$, and $U = 14.5$, all in units of $h\nu_{-} = 327.7\ \mathrm{cm}^{-1}$ (adapted from Reference 23). The vertical arrows symbolize the excitation (IVB) and luminescence (L) processes. The location of the first 16 vibronic levels is shown by the parallel horizontal lines in the right hand well. Each of these is actually a doublet ($\pm$ interchange symmetry), but the splittings are unresolved to the scale of the drawing. The zero point energy is 0.47, and the height of the barrier on the lowest surface is 14.30. For convenience of display, the energy zero is $(1/4)\lambda^2 - (1/2)U$ lower than in Equation 108.

## 6 COMPARISON AND EQUIVALENCE OF APPROACHES

It was pointed out earlier that the simple (original) PKS model has a fundamental omission in that a single-mode treatment is unable to give a consistent description of mixed-valence systems, even at the simplest level. This omission can be remedied[29] by expanding the electronic coupling (associated with the parameter $\varepsilon$) in a power series, in which case it can be shown that the PKS model becomes precisely equivalent to the Piepho MO approach. We briefly summarize this procedure here.

### 6.1 Extension of the Simple PKS Model to Two Modes: Symmetrical and Unsymmetrical Cases

Starting with Equation 30, we remove the implicit assumption that the electronic interaction between subunits ($V^{AB}$) has no dependence on distance. Consequently, we introduce the coordinate $Q_1$ which measures the distance between subunits (Equa-

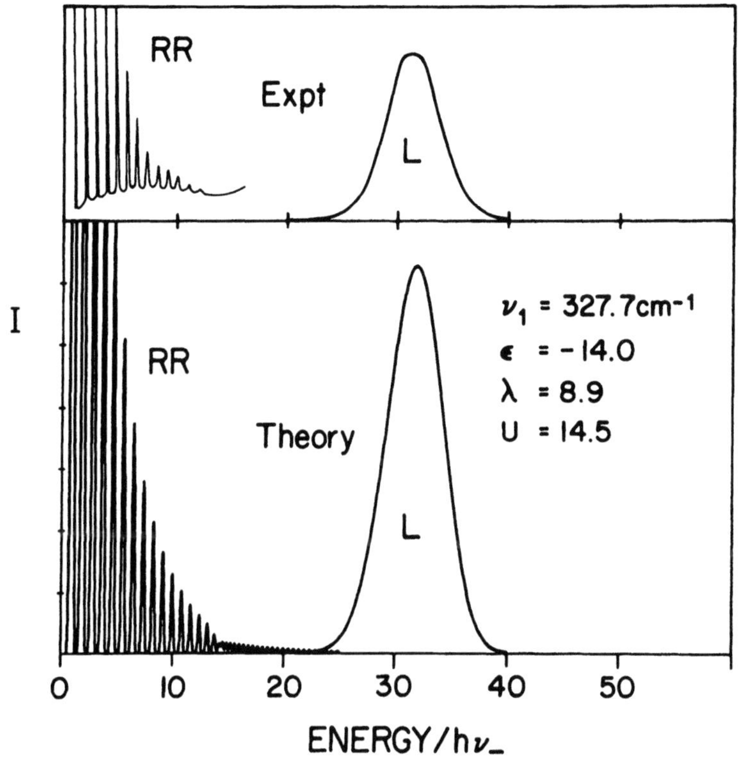


**Figure 11.** Comparison of experimental[28] resonance Raman (RR) and luminescence (L) bands at 4.2 K with the theoretical simulation based on the parameters used in Figure 10 (adapted from Reference 23).

tion 73), and we expand $V^{AB}$ in a Taylor's series around the equilibrium configuration $Q_1 = 0$:

$$V^{AB} = V_0^{AB} + \left(\frac{\partial V^{AB}}{\partial Q_1}\right)_0 Q_1 + \frac{1}{2}\left(\frac{\partial^2 V^{AB}}{\partial Q_1^2}\right)_0 Q_1^2 + \cdots \tag{116}$$

Then in place of Equation 45, we obtain

$$V = \begin{pmatrix} \begin{bmatrix} \frac{1}{2}q_2^2 h\nu_2 + \frac{1}{2}(q_1')^2 h\nu_1 + \\ V_{ab}^{0'} \end{bmatrix} & \sqrt{2}\,\lambda_2 q_2 h\nu_2 + W' \\ \sqrt{2}\,\lambda_2 q_2 h\nu_2 + W' & \begin{bmatrix} \frac{1}{2}q_2^2 h\nu_2 + \frac{1}{2}(q_1')^2 h\nu_1 - \\ 2\sqrt{2}\,\lambda_1^{(a,b)} q_1' h\nu_1 - V_{ab}^{0'} \end{bmatrix} \end{pmatrix} \tag{117}$$

with the $q_i$ and $\lambda_i$ defined as in Equation 83, $V_{ab}^0$ defined by Equation 30, and recalling that $q_2 \equiv q$ is the original PKS coordinate, Equation 25. We have at the same time included the unsymmetrical case by following the procedure outlined in

Section 2.4 with $W' \equiv (h\nu_-)W \equiv (h\nu_2)W$, and $W$ is defined in Equation 54. The primes on $q_1$ and $V_{ab}^0$ denote new choices of coordinate and energy zeros respectively. In the course of this derivation, we have assumed equal force constants for mode $q_1$ in the basis states $\psi_a$ and $\psi_b$ just as we have done previously for mode $q_2$ (Equation 23 *et seq*).

## 6.2 Equivalence of the Two Approaches in the Symmetrical Case

If we subtract $V_{ab}^{0'}$ from the diagonal of Equation 117 and compare with the treatment in Section 3.2, Equation 87, we find that the two models have identical form and indeed give identical electronic potential energy matrices if we make the identifications, $q_1' \to q_1$, $-2\lambda_1^{(a,b)} \to \lambda_1$, $-2V_{ab}^{0'} \to \Delta$, where $\Delta$ is the separation between the bonding and anti-bonding MOs ($|b\rangle$ and $|a\rangle$ of Equation 72). Thus the two approaches are formally equivalent.

We note that extension of the two-electron PKS treatment (Section 5.2) to the case of two modes is immediate using Equation 116. The complete expressions are given in Reference 29. Likewise, it is straightforward to extend the simple Piepho MO model to the two-electron case. The explicit details are also given in Reference 29. Again the PKS and MO treatments are found to be exactly equivalent with appropriate correlations of the parameters. Thus equivalence of the two models applies for both one- and two-electron dimer systems.

## 6.3 Extension of the Piepho MO Approach to the Unsymmetrical Case ($A \not\equiv B$)[30]

We return to Section 3.2, but now we turn on a small asymmetry factor, so that the two centers are no longer identical. The effect, to first order, is to mix the original MOs ($|b\rangle$, $|a\rangle$), i.e., configuration interaction occurs to form new MOs ($\gamma$, $\gamma^*$) and new electronic states ($\phi_1'$, $\phi_2'$):

$$\phi_1^{0'} = |\gamma^2\gamma^*\rangle; \qquad \phi_2^{0'} = |\gamma\gamma^{*2}\rangle \tag{118}$$

$$|\gamma\rangle = c_1|b\rangle + c_2|a\rangle, \quad |\gamma^*\rangle = -c_2|b\rangle + c_1|a\rangle, \quad c_1^2 + c_2^2 = 1; \ |c_1| \gg |c_2| \tag{119}$$

Using the new Equation 118 basis, we write Equations 80 and 81 in their most general form:

$$\begin{aligned} W_{ii}(q) &= (1/2)h\nu_1 {q_1}^2 + \sqrt{2}\lambda_1^{(i)} h\nu_1 q_1 + (1/2)h\nu_2 q_2^2 \\ &\quad + \sqrt{2}\lambda_2^{(i)} h\nu_2 q_2; \quad i = 1, 2 \\ W_{12}(q) &= W_{21}(q) = \sqrt{2}\left[h\nu_1 \lambda_1^{(1,2)} q_1 + h\nu_2 \lambda_2^{(1,2)} q_2\right] \end{aligned} \tag{120}$$

with

$$\left.\begin{aligned}\sqrt{2}\,h\nu_\alpha\lambda_\alpha^{(i)} &= \langle\phi_i^{0'}|(\partial V/\partial q_\alpha)_0|\phi_i^{0'}\rangle\\ \sqrt{2}\,h\nu_\alpha\lambda_\alpha^{(1,2)} &= \langle\phi_1^{0'}|(\partial V/\partial q_\alpha)_0|\phi_2^{0'}\rangle\end{aligned}\right\}\quad \alpha=1,2;\ i=1,2 \tag{121}$$

We substitute Equations 118 and 119 into Equation 120 making use of Equations 85 and 86. We eliminate $\lambda_1^{(1)}$ by a new choice of origin for $q_1$ since Equation 86 continues to apply under these circumstances, but the origin for $q_2$ is not changed. Following Bersuker[31] and Piepho[20], we note that $(\partial V/\partial q_\alpha)_0$ is a one-electron operator so that

$$(\partial V/\partial q_\alpha)_0 = \sum_{i=1}^{3}\left[\partial v(i)/\partial q_\alpha\right]_0 \tag{122}$$

The $[\partial v(i)/\partial q_\alpha]_0$ are the so-called orbital vibronic constants (OVC) of Bersuker.[31] Using Equations 72, 73, 118, 119, and 122, and expanding the matrix elements to one-electron terms, the result is

$$\begin{aligned}h\nu_2\lambda_2^{(1)} = -h\nu_2\lambda_2^{(2)} &= c_1c_2\big[\langle a_1(A)|\left[\partial v(i)/\partial q_A\right]_0|a_1(A)\rangle\\ &\quad - \langle a_1(A)|\left[\partial v(i)/\partial q_B\right]_0|a_1(A)\rangle\big]\\ h\nu_1\lambda_1^{(2)} &= \sqrt{2}(c_2^2-c_1^2)\langle a_1(A)|\left[\partial v(i)/\partial q_1\right]_0|a_1(B)\rangle\\ h\nu_1\lambda_1^{(1,2)} &= \sqrt{2}\,c_1c_2\langle a_1(A)|\left[\partial v(i)/\partial q_1\right]_0|a_1(B)\rangle\\ h\nu_2\lambda_2^{(1,2)} &= (1/2)(c_2^2-c_1^2)\big[\langle a_1(A)|\left[\partial v(i)/\partial q_A\right]_0|a_1(A)\rangle\\ &\quad - \langle a_1(A)|\left[\partial v(i)/\partial q_B\right]_0|a_1(A)\rangle\big]\end{aligned} \tag{123}$$

Inspecting Equation 123, we note that

$$\frac{\lambda_2^{(1)}}{2\lambda_2^{(1,2)}} = \frac{\lambda_1^{(1,2)}}{\lambda_1^{(2)}} \equiv R \tag{124}$$

so Equation 120 becomes

$$\begin{aligned}W_{11} &= (1/2)q_1^2h\nu_1 + (1/2)q_2^2h\nu_2 + 2\sqrt{2}R\lambda_2^{(1,2)}q_2h\nu_2\\ W_{12} &= \sqrt{2}R\lambda_1^{(2)}q_1h\nu_1 + \sqrt{2}\,\lambda_2^{(1,2)}q_2h\nu_2\\ W_{22} &= (1/2)q_1^2h\nu_1 + (1/2)q_2^2h\nu_2 + \sqrt{2}\,\lambda_1^{(2)}q_1h\nu_1 - 2\sqrt{2}R\lambda_2^{(1,2)}q_2h\nu_2\end{aligned} \tag{125}$$

When $R \to 0$, these equations reduce to the symmetrical case, Equation 87 $[\lambda_2^{(1,2)} \equiv \lambda_2,\ \lambda_1^{(2)} \equiv \lambda_1]$,[20] as expected. Thus we see that the effect of the asymmetry is to introduce one additional parameter ($R$) which has the simultaneous effect of coupling $q_2$ into the diagonal terms and $q_1$ into the off-diagonal terms of the potential energy matrix $W$. (Note that in Reference 30, $\Delta$—the energy separation of the the original bonding and antibonding MOs—was inadvertently included in $W_{22}$, when in fact it is already implicit in the $W_j^\circ$ terms of Equation 79. It properly appears in the electronic potential energy matrix, Equation 87.) Solution of the vibronic Schrödinger equation (Equation 77) is now obtained by diagonalizing the appropriate dynamic matrix. The elements of this matrix for the symmetric case ($A \equiv B$) have already been given in Equations 89 and 90, so we simply list the additional terms which arise when $A \not\equiv B$ [again using the Piepho notation $\Phi_{jn_1n_2} \equiv \phi_j^0 \chi_{n_1}(q_1)\chi_{n_2}(q_2)$].

$$\begin{aligned} \langle \Phi_{1n_1n_2} | H_T | \Phi_{1n_1n_2'} \rangle &= -\langle \Phi_{2n_1n_2} | H_T | \Phi_{2n_1n_2'} \rangle \\ &= 2\sqrt{2}R\lambda_2 h\nu_2 \left\{ \left( \frac{n_2+1}{2} \right)^{1/2} \delta_{n_2',n_2+1} + \left( \frac{n_2'+1}{2} \right)^{1/2} \delta_{n_2,n_2'+1} \right\} \\ \langle \Phi_{2n_1n_2} | H_T | \Phi_{1n_1'n_2} \rangle \\ &= \sqrt{2}R\lambda_1 h\nu_1 \left\{ \left( \frac{n_1+1}{2} \right)^{1/2} \delta_{n_1',n_1+1} + \left( \frac{n_1'+1}{2} \right)^{1/2} \delta_{n_1,n_1'+1} \right\} \end{aligned} \tag{126}$$

Thus the elements in Equation 126 are to be added to those in Equations 89 and 90.

This treatment has been applied to the IVB in Lupc$_2$[30] and to the IVB in the oxidized photosynthetic bacterial reaction center (P$^+$) of Rhodobacter sphaeroides.[32] The fit of the IVB of P$^+$ and the resulting potential surfaces are shown in Figures 12 and 13. The IVB of P$^+$ is of considerable interest both inherently because it is involved in the photosynthetic electron transfer process, and because it occurs well into the vibrational infrared region with significant resolved structure, and with vibronic coupling apparently larger than the electronic coupling.[32] The treatment in Reference 32 suggests that P$^+$ is a strongly delocalized system.

## 6.4 Comparison of the MO and PKS Two-Mode Cases for Unsymmetrical Systems

To compare these two cases, it is only necessary to compare the corresponding electronic potential energy matrices. The PKS matrix is given by Equation 117 and the elements of the corresponding MO matrix are given by Equation 125 provided only that $W_1^\circ$ and $W_2^\circ$ are respectively added to $W_{11}$ and $W_{22}$ (see Equation 79). Subtracting $V_{ab}^{0'}$ from the diagonal in Equation 117 and $W_1^\circ$ from the diagonal elements of $W_{11}$ and $W_{22}$ as augmented above, we may make the identification, $-2V_{ab}^{0'} \rightarrow (W_2^\circ - W_1^\circ) \equiv \Delta$. Nevertheless, it is apparent that the potential energy matrices for the PKS and MO case are not equivalent, nor can they be made so by new definitions of coordinate and/or energy zeros.

To get an equivalence of the two methods, we return to Section 6.3. Rather than assume a configuration interaction (Equation 119), we use unperturbed MOs (i.e,

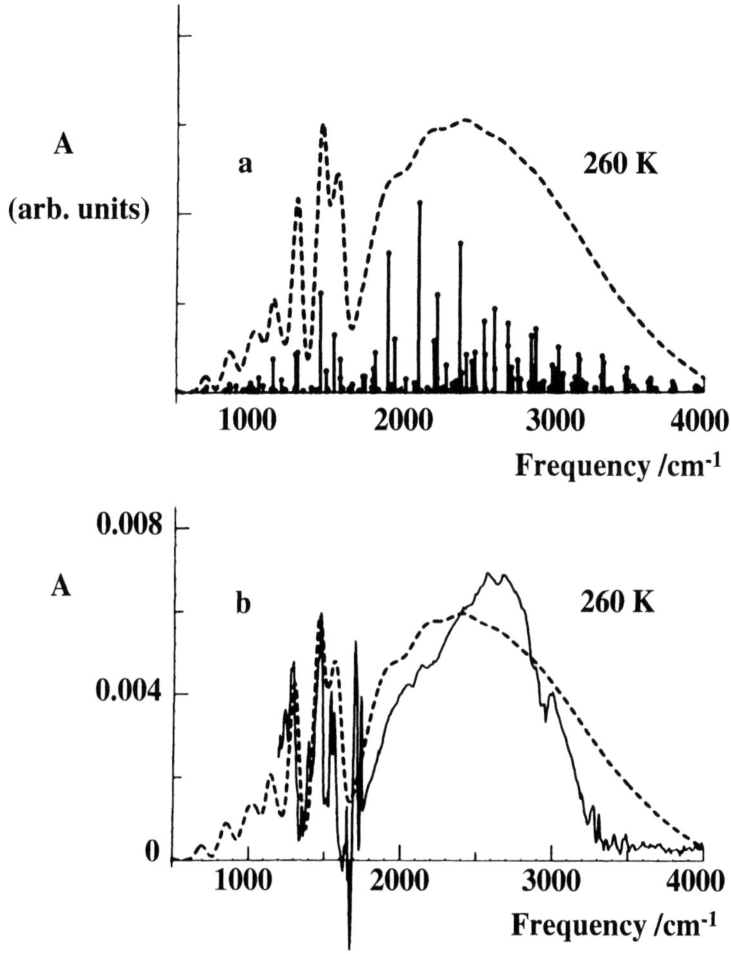


**Figure 12.** Intervalence band of *Rhodobacter sphaeroides*. Panel (a) shows the theoretical simulation, and panel (b) compares the simulation with experiment (solid curve). Each stick in panel (a) is convoluted with a Gaussian whose area is directly proportional to stick height which in turn is directly proportional to absorbance. Gaussian half-widths (at $1/e$ of maximum) above and below 1,630 $cm^{-1}$ are 145 and 45 $cm^{-1}$, respectively (adapted from Reference 32).

$c_2 = 0$ in Equation 119), and we represent the asymmetry as an off-diagonal matrix element ($W'$) in this unperturbed basis. Then repeating the treatment in Section 3.2, one finds that both methods give equivalent electronic potential energy matrices.

The obvious question arises: Which of the two MO methods (for the unsymmetrical case) is to be preferred? In the first method (Equation 125), configuration interaction (CI) is allowed to occur such that the low symmetry perturbation is diagonal in the new CI basis, and the original equilibrium configuration relaxes to a new equilibrium configuration according to the first equality in Equation 123. In the second

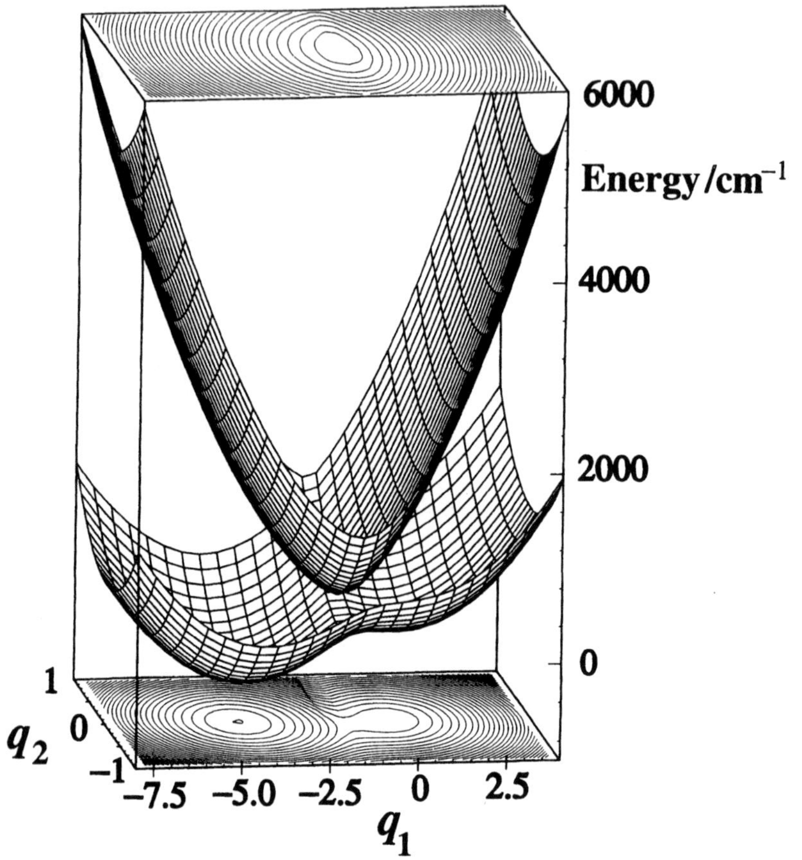


**Figure 13.** Calculated potential energy surfaces using Equation 125 and the parameters[32] which produced the simulation in Figure 12. Contour plots of the upper and lower surfaces are projected respectively at the top and bottom. The two minima in the lower surface have energies $-527$ and $-1.45\ cm^{-1}$ at $q_1 = -4.39$, $q_2 = -0.020$ and $q_1 = -0.015$, $q_2 = -0.044$, respectively. The single minimum on the upper surface has energy $323\ cm^{-1}$ at $q_1 = -1.49$, $q_2 = 0.13$ (adapted from Reference 32).

method, the lowering of symmetry is simply expressed as an off-diagonal matrix element in the unperturbed electronic basis. The first method seems clearly preferable since an improved, i.e., first-order wave function is used. It would be interesting to investigate how much the two methods differ in fitting experimental intervalence bands.

### 6.5 Why the MO Approach is Better

Although we have demonstrated equivalencies in the PKS (valence bond) and MO approaches, there are clear reasons for preferring the MO approach. First, the MO approach (and language) is used in the vast majority of spectroscopic applications. Second, and more specifically, the MO approach permits the use of the Bersuker orbital vibronic constants (OVC). One such application has been illustrated in Section 6.3, where important relationships were established among various vibronic coupling parameters (Equation 124). Piepho has discussed and illustrated the application of the OVCs in much greater detail.[20,21,33] It is clear that the use of the OVCs permits powerful insights and correlations.

## 7 APPLYING THE MODELS TO A NEW SYSTEM

An obvious question which arises is: How do I apply all this to a new system? Assuming that one knows the nature of the system—one- or two-electron, symmetrical or non-symmetrical—the question translates to determining the parameters of the applicable model. This is so because it has been shown in this chapter that such parameters permit exact numerical solutions which provide eigenfunctions and eigenvalues from which observables of interest can be calculated. Since the parameters in the various models are connected to the IVB (and ctiir), they are usually obtained by fitting the observed IVB (and possibly other available data) starting with initial guesses. If the system is "one electron," i.e., if the two subunits differ by one in oxidation number, computer programs are available that, for example, permit one to simulate the spectrum and calculate $P(q_2)$. Download from:

**www.virginia.edu/~chem/emeritus/schatz/pks**

The question then is: What starting parameters should be used? In the one-electron case, seven parameters are required: $\nu_1$, $\nu_2$, $\lambda_1$, $\lambda_2$,$\Delta$, $R$, $\delta$. $R$ is zero in the symmetrical case, and $\delta$—the *ad hoc* linewidth—is usually assigned toward the end of the fitting process. One usually starts from available experimental data. For example, $\nu_1$and $\nu_2$ can often be estimated from infrared and Raman vibrational data plus some intuition about what vibrational motions are apt to be excited by an intervalence transition. $\lambda_2$ can sometimes be estimated from known bond length differences between oxidation states (see discussion just above Equation 27). $\Delta$ can often be estimated from the known electronic spectrum of the system or from semiempirical or *ab initio* calculations. Every case is different. The reader is referred to Reference 21 for an illuminating example of this parameter estimation process.

Having the initial parameters, one can obtain some preliminary feel for the system by plotting its potential energy surfaces. This is easily done by diagonalizing the relevant $2\times2$ (one-electron case) or $3\times3$ (two-electron case) potential energy matrix, all of which are explicitly given in this article. Inspection of such surfaces may show immediately whether the chosen parameters have produced a localized or delocalized structure. In addition, simple Franck–Condon arguments should give one some idea of the expected spectrum though it must be kept in mind that such arguments become increasingly erroneous as one moves away from either the localized or delocalized limit.

Once initial parameters have been chosen, the computer programs mentioned above may be used to refine the parameters through an iterative least squares or simplex procedure which hopefully converges to a "best fit" of the experimental spectrum and other possibly available data. This is likely to be a somewhat subjective procedure since intervalence bands tend to be relatively unstructured, and a range of parameters is apt to give comparably good fits. Thus a generous margin of error must usually be anticipated for at least some of the "best fit" parameters. It is clearly important to bring all available information to bear in the initial estimate of parameters, because if these estimates are poor, it is entirely possible to achieve a local minimum ("best fit") in parameter space which is far off the mark.

Similar remarks apply for the two-electron case (Section 5), which has been formulated as a one-mode model. There are five parameters in the symmetrical case: $\varepsilon$, $\lambda$, $\nu_-$, $U$, and $\delta$. An iterative fit in such a case is illustrated in Reference 23. Computer programs for this case are not included in the programs which can be downloaded, but they could be readily developed from those supplied by substituting the matrix elements for the two-electron case which are given in Reference 22.

## 8 CONCLUDING REMARKS

I have tried in this review to give a systematic and consistent account of the treatment of mixed-valence dimer systems. Although the mathematical details can at times seem complex, the conceptual basis of all the treatments is the same. A ground state electronic basis is chosen containing one or two electrons (or holes) located on two nuclear centers between which an electronic interaction occurs. A harmonic nuclear potential function is constructed focussing on a small number of essential vibrational coordinates associated with these centers, namely those coordinates whose equilibrium values would be expected to change significantly if electronic excitation occurs *within the ground state electronic basis.* The resulting ("coupled") potential energy surfaces are referred to as the solution of the static problem. Nuclear motion is then allowed, i.e., the nuclear kinetic energy operator ($T_n$) is turned on, and the dynamic Schrödinger equation, viz., $(H_{el}+T_n)\Psi = E\Psi$ is to be solved. The absolutely crucial point is that the assumed form of solution must be non-adiabatic, i.e., it must consist of a *sum* of Born–Oppenheimer Products ($\psi_{el}\chi_{vib}$) from the two (one-electron case) or three (two-electron case) potental surfaces resulting from solution of the static problem. Physically, this means that the nuclear motion cannot be confined to a single surface but executes some complicated trajectory over all (two or three) of the coupled surfaces. Conversely, if adiabatic functions (single Born–Oppenheimer products) give acceptable solutions, then the essential mixed-valence nature of the problem is absent and, for example, a conventional molecular orbital treatment is appropriate.

The mathematical complications arise because the dynamic problem involves a set of coupled equations which generally must be solved numerically. We do this, following Fulton and Gouterman by employing a vibrational basis of (usually) undisplaced harmonic oscillators. But these numerical details should not be allowed to obscure the essentially simple steps, which are the same in every case, and which carry over to more complicated systems, for example, three-center mixed-valence cases. From another point of view, our treatment of mixed-valence systems is closely analogous to the treatment of Jahn–Teller systems where again the heart of the problem is solution of a set of coupled equations arising in the dynamic problem because of the necessarily nonadiabtic form of the solutions. In Section 2.2, it was pointed out that Figure 2 displays a Jahn–Teller-like situation since a vibronic double degeneracy occurs at $q = 0$. However, this system is of trivial interest until electronic coupling is allowed, in which case the vibronic double degeneracy vanishes (Figure 3). However, provided the electronic coupling is not too large, nonadiabatic solutions are required, and we are solving a psudo-Jahn–Teller problem which requires the same mathematical machinery as a true Jahn–Teller problem, where the vibronic degeneracy is not removed by electronic interactions.

## ACKNOWLEDGMENTS

I would like to acknowledge the extensive contributions of my coworkers, particularly Professor Susan B. Piepho, Drs. K. Y. Wong, E. R. Krausz, K. Prassides, and Z. Gasyna. I would particularly like to emphasize the crucial insights of Professor Piepho who first explicitly took account of the inherently nonadiabatic, dynamic nature of the problem and thus established the basis of the PKS model. This work has been supported from the start by the National Science Foundation, most recently under grant NSF CHE9523697.

## REFERENCES

Computer programs are available for the one electron case. See p. 223.

1. Piepho, S.B.; Krausz, E.R.; Schatz, P.N. *J. Am. Chem. Soc.* **1978**, x *100*, 2996–3005.
2. Wong, K.Y.; Schatz, P.N.; Piepho, S.B. *J. Am. Chem. Soc.* **1979**, *101*, 2793.
3. Wong, K.Y.; Schatz, P.N. *Prog. Inorg. Chem.* **1981**, *28*, 369–449.
4. Robin, M. B.; Day, P. *Adv. Inorg. Radiochem.* **1967**, *10*, 247.
5. Day, P. *Endeavour* **1970**, *29*, 45.
6. Wilson, Jr., E. B.; Decius, J. C.; Cross, P. C. *Molecular Vibrations*. McGraw Hill; New York, 1965.
7. Piepho, S. B.; Schatz, P. N. *Group Theory in Spectroscopy with Applications to Magnetic Circular Dichroism Spectroscopy*. Wiley: New York, 1983.
8. Fulton, R. L.; Gouterman, M. *J. Chem. Phys.* **1961**, *35*, 1059.
9. Fulton, R. L.; Gouterman, M. *J. Chem. Phys.* **1964**, *41*, 2280.
10. Stephens, P. J. *Adv. Chem. Phys.* **1976**, *35*, 197.
11. Lax, M. *J. Chem. Phys.* **1952**, *20*, 1752.
12. Toyozawa, Y.; Inoue, M. *J. Phys. Soc. Jap.* **1966**, *21*, 1663.
13. Allen, G. C.; Hush, N.S. *Prog. Inorg. Chem.* **1967**, *8*, 357.
14. Hush, N.S. *Prog. Inorg. Chem.* **1967**, *8*, 391.
15. See Reference 3; footnote, p. 409.

16. Rice, M.J. *Solid State Commun.* **1979**, *31*, 93.
17. Rice, M.J.; Yartsev, V.M.; Jacobsen, C.S. *Phys. Rev. B: Condens. Matter* **1980**, *21*, 3437.
18. Collman, J.P.; McDevitt, J.T.; Leidner, C.R.; Yee, T.G.; Torrance, J.B.; Little, W.A. *J. Am. Chem. Soc.* **1987**, *109*, 4606-4614.
19. Prassides, K.; Schatz, P.N. *J. Phys. Chem.* **1989**, *93*, 83-89.
20. Piepho, S. B. *J. Am. Chem. Soc.* **1988**, *110*, 6319–6326. [Note three typos: (1) The elastic restoring terms were inadvertantly omitted from Equation 23, *cp.* with Equation 87 of present article; (2) the $\sqrt{2}$ in the first of Equation 20 goes only with the first term in the square brackets; and (3) the parentheses in Equation 47 are raised to the one-half power.]
21. Piepho, S. B. *J. Am. Chem. Soc.* **1990**, *112*, 4197–4206.
22. Prassides, K.; Schatz, P.N.; Wong, K.Y.; Day, P. *J. Phys. Chem.* **1986**, *90*, 5588.
23. Prassides, K.; Schatz, P.N. *Chem. Phys. Lett.* **1991**, *178*, 227–234.
24. Anderson, P. W. *Phys. Rev. Lett.* **1975**, *34*, 953.
25. Toyozawa, Y. *J. Phys. Soc. Jap.* **1981**, *50*, 1861.
26. Clark, R. J. H.; Franks, M. L.; Trumble, W. R. *Chem. Phys. Lett.* **1976**, *41*, 287.
27. Clark, R. J. H.; Turtle, P. C. *Inorg. Chem.* **1978**, *17*, 2526.
28. Tanino, H.; Kobayashi, K. *J. Phys. Soc. Jap.* **1983**, *52*, 1446.
29. Schatz, P. N. *Vibronic Coupling Models of Mixed Valency: Relation of the PKS and MO Models for One- and Two-Electron Systems*, in Mixed Valency Systems: Applications in Chemistry, Physics and Biology, K. Prassides ed.; Kluwer: Dordrecht, **1991**, pp. 7–28.
30. Gasyna, Z.; Schatz, P. N.; Boyle, M. E. *J. Phys. Chem.* **1995**, *99*, 10159. [Note typo: The second of Equation 2 in this reference should contain a minus sign.]
31. Bersuker, I.B. *The Jahn–Teller Effect and Vibronic Interactions in Modern Chemistry.* Plenum: New York, 1984.
32. Gasyna, Z.; Schatz, P.N. *J. Phys. Chem.* **1996**, *100*, 1445.
33. Piepho, S. B. *Vibronic Coupling Models for Mixed-Valence Line Shapes: Going Beyond the PKS Approach*, in Mixed Valency Systems: Applications in Chemistry, Physics and Biology, K. Prassides ed.; Kluwer: Dordrecht, **1991**, pp. 329–334.

# 4 Electrochemistry, Charge Transfer Spectroscopy, and Electronic Structure

A. B. P. LEVER AND ELAINE S. DODSWORTH
Department of Chemistry
York University
Toronto, Ontario M3J 1P3, Canada
E-mail: blever@yorku.ca
dod@yorku.ca



*Inorganic Electronic Structure and Spectroscopy, Volume II: Applications and Case Studies.*
Edited by E. I. Solomon and A. B. P. Lever.
ISBN 0-471-32682-8. 

## 1 GENERAL INTRODUCTION

Electrochemical techniques can provide fundamental information about the free energies required to add an electron, or electrons, to the lower unoccupied molecular orbitals or remove an electron, or electrons, from the higher occupied molecular orbitals of a chemical species thereby providing a map of the frontier levels of the species.

Charge transfer spectroscopy refers to that aspect of electronic spectroscopy dealing with the excitation of an electron generally from one component of the molecule to another, usually metal to ligand (MLCT), ligand to metal (LMCT), ligand to a different ligand (LLCT), or in polynuclear species, metal to metal (MMCT). Additionally one may characterize metal to solvent transitions and outer sphere charge transfer (OSCT) referring, for example, to an electron jumping from a donor to an acceptor in an ion-pair (see Section 4.3). The orbitals involved in these various transitions are commonly also accessible to electrochemical techniques, and thus these two areas of chemistry have grown together and interacted in a synergistic fashion.

The combined study of electrochemistry and optical spectroscopy has proven to be a powerful means of elucidating electronic structure through analysing the relationships between the observables of these two techniques. Information not readily available from either technique alone can be extracted by consideration of both techniques. We explore these relationships in some depth in this chapter.

In developing the information content of an electrochemical experiment, various ligand additivity parameter strategies have been developed by Bursten,[1] Pletcher,[2] Chatt,[3] Treichel[4] and Lever.[5] These schemes have permitted us both to predict the electrochemical potentials of chemical species from knowledge of the metal ion and ligands involved, and to extract electronic structural information from the magnitudes of the various parameters. We show here how they can also provide additional insight into the combined optical/electrochemical analysis.

## 2 INTRODUCTION TO ELECTROCHEMISTRY

It is not intended to provide a detailed introduction to inorganic electrochemistry which is a topic covered by many excellent textbooks.[6–8] However, some basic ideas, pertinent to the remainder of this chapter, are presented.

Various electrochemical methods can be used to provide, for any given complex, a set of observable redox potentials as exemplified below. Generally, a chemical species is dissolved in a solvent and a three electrode cell consisting of working, reference and counter electrodes is set up. A supporting electrolyte is added to carry the bulk of the current in this cell. Ideally neither the electrolyte nor the solvent should enter into any chemical interaction with the chemical species being studied, in any of its redox states. Using a potentiostat, the polarization of the counter electrode sets up a potential at the working electrode which is measured against the reference electrode. No current flows through the reference electrode which establishes the zero of the experiment. Common reference electrodes include the saturated calomel electrode (SCE) [$Hg_2Cl_2$/Hg/KCl sat] whose potential, in water, lies at +0.24 V vs the normal hydrogen electrode (NHE) at 25°C, and the silver chloride electrode [AgCl/Ag/KCl sat] lying at 0.197 vs NHE at 25 °C.[8] Thus a working electrode polarized at 0 V vs calomel, will have a potential of 0.24 V vs NHE.

While water is commonly used as a solvent in electrochemical studies, it does have several disadvantages. Reduction to hydrogen gas occurs readily, especially at a platinum working electrode, limiting the extent to which the negative potential regime can be explored. Solvolysis and protonation equilibria may also occur such that the actual species being oxidized or reduced in solution may be uncertain. It is also true that many metal complexes are insoluble in water preventing their study in this medium.[9]

Thus in recent years, a great deal of attention has been paid to studying metal complexes dissolved in non-aqueous solvents.[10] A range of organic solvent soluble electrolytes have been developed to carry the current in these experiments; commonly these are tetraethyl or tetrabutylammonium salts. While it can be very convenient to use an organic solvent in which the complex of interest dissolves unchanged, it does introduce a further set of problems. The most serious is the problem of comparing electrochemical data recorded in one organic solvent with data collected in another organic solvent or in water. One cannot simply compare the potentials recorded versus the reference since an unknown potential, termed the junction potential,[8] is set up at the interface between the solvent and the reference electrode. The common procedure to resolve this problem is to add a calibrating internal standard and relate the potentials of the solute of interest to this standard. The most common standard is ferrocene which exhibits a fairly well behaved one-electron ferrocenium/ferrocene ($Fc^+$/Fc) couple. There has been much discussion as to whether this potential, and those of related species, are independent of solvent and whether ferrocene is a useful standard.[11–17] The consensus is that the $Fc^+$/Fc couple is probably not independent of solvent. Various laboratories have reported values for this couple versus calomel as a function of solvent; some values are shown in Table 1. Some laboratories simply report their data versus the $Fc^+$/Fc couple, while others correct their data by

**Table 1 Ferrocene Calibration as an Internal Reference**[19]

| Solvent | $E_{1/2}$ vs NHE (V) |
|---|---|
| Acetonitrile | +0.665 |
| Dichlorobenzene, Dichloroethane | +0.73 |
| Dimethylformamide | +0.64 |
| Pyridine, Dimethylsulfoxide | +0.74 |
| Dichloromethane | +0.69 |
| Dimethylacetamide | +0.71 |

Data collected from References 19–24.

the corrections shown in Table 1 and report potentials versus calomel (or NHE). Researchers studying ruthenium complexes often use the potential of the $[Ru^{III/II}]$ couple of the $[Ru(bpy)_3]^{2+}$ ion as an internal standard. Its potential is +1.554 V vs NHE in acetonitrile.[5]

The actual recorded $E_{1/2}$ will contain a differential solvation energy contribution reflecting the relative solvation energy of the species in each oxidation state. For many systems, the recorded $E_{1/2}$ in one organic solvent is very similar to that in another, suggesting that this differential term is small. However there are certainly examples where this is not true; solvatochromism, for example, where specific solvent interactions, such as hydrogen bonding, may occur.[12,25–27a] Data collected in water may differ significantly from the same complex and couple recorded in an organic solvent as a consequence of protonation equilibria.

Other potential problems with organic solvents relate to the question of purity and dryness. Clearly the solvent should not be contaminated with potentially reactive species, such as peroxides, and water should be excluded to as great a degree as possible. Traces of water, aside from possibly reacting with the solute in one or more of its oxidation states, will usually restrict the usable window for electrochemical study. Purification and drying procedures for common organic solvents are well documented, e.g., a useful "trick" is to add a gram or so of activated alumina to the organic solvent in the electrochemical cell during the electrochemical experiment. The alumina serves to adsorb trace water in the solvent. In some cases however, it may also adsorb the solute, in which case passing the solvent down a column of activated alumina prior to use is an alternative procedure.

The nature of the working electrode may also play a role in the electrochemistry. Platinum is commonly used because of its chemical inertness and because electron transfer between the species being oxidized or reduced and the platinum surface is usually rapid. Other electrodes, especially glassy carbon, may be used for special situations. For example, in water a rather more negative potential is required, at given pH, to reduce protons to hydrogen gas on glassy carbon than on platinum, permitting a wider potential range to be studied.

The potential at which a molecular species can be oxidized or reduced is, of course, dependent upon its electronic structure. A chemical species can be oxidized when, in contact with the working electrode, it becomes thermodynamically downhill to transfer an electron from the chemical species to the electrode. Thus, if electrons travel from the chemical species to the electrode when the latter is polarized at 0 V vs SCE, then the oxidation potential of the chemical species must be more negative than 0 V vs SCE. Figure 1 shows how electrons will move from a chemical species to a working electrode, or from a working electrode to a chemical species, depending upon the relative magnitudes of the redox potentials of the chemical species and the polarization of the electrode. A chemical species has valence molecular orbitals occupied as in the center of the diagram. The solid levels on the two electrodes depict the energies of the electrons therein, with increasing negative potential up the diagram. Electrode A has a moderately positive potential and can oxidize the chemical species, i.e., an electron can "fall" from the chemical species down into the electrode. Electrode A cannot reduce the chemical species. Electrode B has a more negative potential and cannot oxidize the chemical species. It can however reduce the chemical species, i.e., an electron can "fall" from the electrode into the half-empty orbital level.

A potentiostat allows one to scan the potential of the working electrode over a potential range to explore the relative energies of the chemical species involved.

The most common potentiodynamic technique is cyclic voltammetry (CV) (Figure 2) wherein the working electrode is scanned over a potential range to explore the potentials at which redox events can occur, and the current flow, when the bulk species is oxidized or reduced, is recorded. In an electrochemically reversible process (*vide infra*), we observe two waves for each redox event:

$$\text{ox} + ne^- \Longleftrightarrow \text{red} \tag{1}$$

corresponding (Equation 1) to the reduction of the oxidized species (ox) to (red) and its re-oxidation to (ox).

Consider the CV shown in Figure 2. There will be a potential for the electrode such that no chemical reaction will occur, neither oxidation nor reduction of the chemical species, i.e., no current flow to the working electrode. This nil current flow occurs at the open-circuit potential (ocp) which is that potential which the cell would

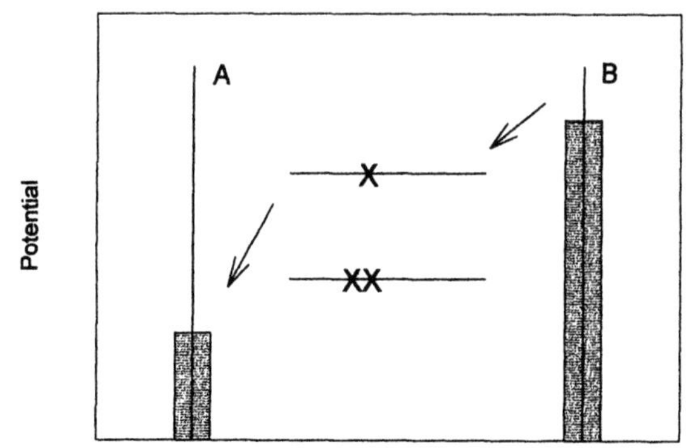


**Figure 1.** A cartoon showing relative potentials of two electrodes A and B (more reducing) with respect to orbital levels of a solute. See text for explanation.

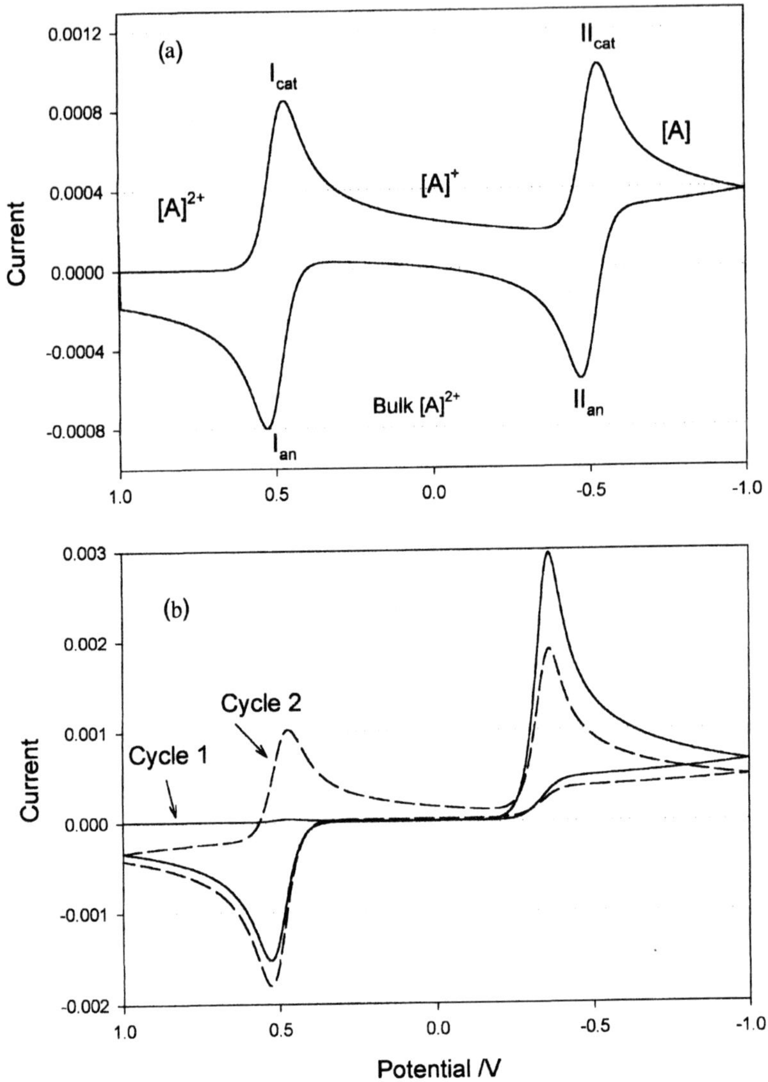


**Figure 2.** (a) A typical cyclic voltammogram (CV) showing the behaviour of two reversible couples. (b) A cyclic voltammogram of an irreversible process. See text for details. The CV shows the initial scan (cycle 1) and the subsequent scan (cycle 2). For bulk $[A]^{2+}$, the open circuit potential (ocr) lies positive of wave $\mathbf{I}_{cat}$.

register when measured by a high impedance voltmeter. It lies between the first oxidation potential and the first reduction potential.

As the potential is scanned negatively we reach a potential where it is now thermodynamically possible to reduce the species in solution (call it $[A]^{2+}$), i.e., to add an electron to the LUMO of the molecule. This is called a Faradaic process and the current at $\mathbf{I}_{cat}$ may be referred to as Faradaic current. The remaining current, and that which we observe in the region where there is no electroactivity, is referred to as charging or capacitive current, being the current necessary to charge the electrode–solution interface, i.e., the layer of electrolyte in close proximity to the electrode. Hence at wave $\mathbf{I}_{cat}$, we observe the process:

$$[A]^{2+} + e^- \Longleftrightarrow [A]^+ \tag{2}$$

All the chemical species $[A]^{2+}$ diffusing to the electrode is reduced in the vicinity of wave $\mathbf{I}_{cat}$.

Indeed as the potential moves more negative of the reduction peak, the concentration of the bulk $[A]^{2+}$ at the electrode approaches zero. Since the reduction current is proportional to the concentration gradient of $[A]^{2+}$, the Faradaic current declines to zero. This accounts for the peak shape of the wave, rather than a plateau of constant current. If the scan were stopped here and the system allowed to reach equilibrium, the current would decay to zero and there would only be $[A]^+$ in the immediate vicinity of the electrode. As we scan more negatively, species diffusing from the bulk will be reduced at the electrode to $[A]^+$ and then will pick up an additional electron at wave $\mathbf{II}_{cat}$, according to:

$$[A]^+ + e^- \Longleftrightarrow [A] \tag{3}$$

The potential at which this occurs depends upon the electronic structures of $[A]^+$ and [A] and not upon the electronic structure of the starting material $[A]^{2+}$. At the most negative point in the voltammogram, the ramp is reversed (switching potential) and the potential is swept in a positive direction. At the switching potential, the species local to the electrode is [A] rather than $[A]^{2+}$ and so re-oxidation will occur at $\mathbf{II}_{an}$, being the reverse of Equation 3, in the lower part of the voltammogram, and a further re-oxidation, the reverse of Equation 2 occurs at $\mathbf{I}_{an}$. The cycle can be repeated (Figure 2a).

Similarly if the solution is scanned positively, $[A]^{3+}$ is generated normally[27b] by removing an electron from the HOMO, and any subsequent oxidation process will be carried out on $[A]^{3+}$ and not $[A]^{2+}$. This explains why, if we suppose Figure 1 provides the energy levels of $[A]^{2+}$ the CV only provides some information about the HOMO and LUMO of $[A]^{2+}$, and one should be cautious about drawing conclusions about any other energy levels of $[A]^{2+}$.[27b]

It is also good practice to record the first cycle voltammogram, then to allow many cycles to reach equilibrium and record the final voltammogram. These two voltammograms will often look identical, but there are certainly conditions where they do not. If they are identical, then the electrochemistry is straightforward with a series of electrochemically reversible waves which are probably not associated with any irreversible chemical processes.

As an example of where they may differ, consider a species $[A]^{2+}$ being reduced to $[A]^+$ at, say, −0.5 V, with no electroactivity at 0 V. If this reaction is fully reversible, then we would see a pair of waves centered at −0.5 V when scanning from, say, 0 to −1.0 V and returning. The first scan and equilibrium scan would look essentially identical. Now suppose that $[A]^+$ is slowly irreversibly converted, by a following chemical reaction, to $[B]^+$, which we suppose can be re-oxidized reversibly to $[B]^{2+}$ at +0.5 V. The first scan from 1 V shows no wave between +1 and 0 V since there is no $[B]^+$ in the system. However, following the first scan, $[A]^+$ is formed, which generates $[B]^+$, and on the second and subsequent scans a new reversible wave occurs at +0.5 V. Thus the equilibrium and first voltammograms would differ (Figure 2b).

Note that the peak potentials in the cathodic and anodic scans (Figure 2a) do not lie at the same potential. For an $n$-electron process, at room temperature, the cathodic peak occurs $28.50/n$ mV more negative than the half-wave potential (see Equation 4) while the anodic peak occurs $28.50/n$ mV more positive than the half-wave potential. Thus the peak-to-peak separation for an electrochemically reversible process is $59/n$ mV at room temperature (25 °C).[8] Note, in passing, that for species adsorbed on an electrode surface, a reversible process will exhibit identical potentials for both the anodic and cathodic processes.

## 2.1 Redox Potentials

The average potential of the pair of waves, such as $\mathbf{I}_{\text{cat}}$ and $\mathbf{I}_{\text{an}}$, is known as the *half-wave potential*, $E_{1/2}$, and is related to the formal potential $E^{0\prime}$, by

$$E_{1/2} = E^{0\prime} + (RT/n\mathcal{F}) \ln \left[D_{\text{red}}^{1/2}/D_{\text{ox}}^{1/2}\right] \tag{4}$$

where $D_{\text{red}}$ and $D_{\text{ox}}$ are the diffusion coefficients of the reduced and oxidized species respectively, $n$ is the number of electrons involved and $\mathcal{F}$ is the Faraday, namely 96,485 C mol$^{-1}$.[8]

The formal potential, $E^{0\prime}$, is related to the standard potential, $E^0$, by

$$E^{0\prime} = E^0 + (RT/n\mathcal{F}) \ln[\gamma_{\text{ox}}/\gamma_{\text{red}}] \tag{5}$$

where $\gamma_{\text{ox}}$ and $\gamma_{\text{red}}$ are the activity coefficients of the oxidized and reduced species. The formal potential will vary with the ionic strength of the solution, since this affects the activity coefficients. Finally, the standard potential is related to the Gibbs free energy of the redox event, $\Delta G^0$, by

$$\Delta G^0 = -n\mathcal{F}E^0 \tag{6}$$

From these equations, if we assume that the activities of the oxidant and reductant are equal at the mid point potential between the oxidation and reduction waves and further that the diffusion coefficients of the oxidized and reduced species are similar, then we can approximate the $E_{1/2}$ as equal to the thermodynamic $E^0$ value. This is commonly assumed to be the case. This thermodynamic information, as we shall develop, can be used to obtain information about the electronic structure of the species, and about possible charge transfer transitions.

To establish the thermodynamic quantity more clearly, consider a reaction, with the following potential and free energy:

$$\begin{aligned} &[\mathrm{Fe^{III}L_6}]^{3+} + e^- \Longleftrightarrow [\mathrm{Fe^{II}L_6}]^{2+} \\ &E_{1/2}[\mathrm{Fe^{III/II}L_6}] = -(\Delta G^0/\mathcal{F})(\mathrm{Fe^{III/II}L_6}) \end{aligned} \quad (7)$$

and consider also the following three processes:

$$\mathrm{Fe^{3+}} + e^- \Longleftrightarrow \mathrm{Fe^{2+}} \qquad E_{1/2}[\mathrm{Fe^{III/II}}] = -(\Delta G^0/\mathcal{F})(\mathrm{Fe^{III/II}}) \quad (8)$$

$$\mathrm{Fe^{3+}} + 6\mathrm{L} \Longleftrightarrow [\mathrm{Fe^{III}L_6}]^{3+} \qquad \Delta G^0 = -RT \ln \beta_6(\mathrm{Fe^{III}L_6}) \quad (9)$$

$$\mathrm{Fe^{2+}} + 6\mathrm{L} \Longleftrightarrow [\mathrm{Fe^{II}L_6}]^{2+} \qquad \Delta G^0 = -RT \ln \beta_6(\mathrm{Fe^{II}L_6}) \quad (10)$$

where $\beta_6$ is the overall stability constant for binding 6 ligands L to the metal center. Subtracting Equation 8 from 7 and Equation 9 from 10 leads to:

$$\begin{aligned} &[\mathrm{Fe^{II}L_6}]^{2+} + \mathrm{Fe^{3+}} \Longleftrightarrow [\mathrm{Fe^{III}L_6}]^{3+} + \mathrm{Fe^{2+}} \\ &\Delta G^0 = RT[\ln \beta_6(\mathrm{Fe^{II}L_6}) - \ln \beta_6(\mathrm{Fe^{III}L_6})] \end{aligned} \quad (11)$$

and:

$$E_{1/2}[\mathrm{Fe^{III/II}L_6}] - E_{1/2}[\mathrm{Fe^{III/II}}] = (RT/\mathcal{F})(\ln[\beta_6(\mathrm{Fe^{II}L_6})/\beta_6(\mathrm{Fe^{III}L_6})]) \quad (12)$$

Thus the observed redox potential for, say, oxidizing $[ML_6]^{2+}$ to $[ML_6]^{3+}$ is equal thermodynamically to the ratio of the stability constants for binding L to $M^{2+}$ and to $M^{3+}$ offset by the free energy of the $M^{III}/M^{II}$ couple, usually for the aquo ion.

**2.1.1** ***Redox Processes*** Before employing this thermodynamic information, it is necessary to ensure that the process being investigated is electrochemically reversible. Processes may be irreversible for a variety of reasons, most commonly a chemical reaction preceding or following the electrochemical process, or an especially slow rate for electron transfer across the solution/electrode interface. For example, if reduction of $[\mathrm{Ru^{II}Cl_2(bpy)_2}]$ should be followed by rapid loss of a chloride ion, and perhaps its replacement by solvent, then the potential at which this combined process occurs will not necessarily correspond with the potential for Equation 13a:

$$[\mathrm{Ru^{II}Cl_2(bpy)_2}] + e^- \Longrightarrow [\mathrm{Ru^{II}Cl_2(bpy)(bpy^-)}]^- \quad (13a)$$

$$[\mathrm{Ru^{II}Cl_2(bpy)(bpy^-)}]^- + \mathrm{sol} \Longleftrightarrow [\mathrm{Ru^{II}Cl(sol)(bpy)(bpy^-)}] + \mathrm{Cl^-} \quad (13b)$$

It is therefore vitally important that, when electrochemical data are collected for use in various thermodynamic arguments, the electrochemical reversibility of the process is assured. These criteria are not explored in depth here, but basically, in a CV experiment, the currents associated with the two redox waves (Equation 1,

Figure 2) should be identical, the wave separation as noted above should be 59 mV for a one-electron process at 25 °C and both anodic and cathodic currents (i) should increase linearly with the square root of the scan rate ($\nu$), i.e., $i \alpha \nu^{1/2}$.[8]

Varying the scan rate is the most common procedure for evaluating the possible existence of coupled chemical reactions. These chemical processes will have a rate constant ($k$) associated with them. For example, if, in the above example Equations 13, the loss of chloride ion is fairly slow, it may be feasible to scan the system rapidly enough (i.e., reduce $[Ru^{II}Cl_2(bpy)_2]$ and then re-oxidize it rapidly) that there is no time for loss of chloride ion to occur. Under such a fast scan situation, the measured potential might then correspond to the process shown in Equation 13a (see Reference 28 for a detailed analysis of this specific situation). Scan rates are commonly varied from a few $mVs^{-1}$ to 5–10 $Vs^{-1}$ but can, under special conditions, specifically using an ultramicroelectrode,[29–32] be increased to the order of $10^6$ $Vs^{-1}$. Other strategies to obtain thermodynamic information involve cooling the solution to slow down or stop coupled chemical reactions, and changing the solvent, to a poorer donor, if solvolysis is occurring. In an appropriate solvent, it is not difficult to collect data at −40 °C or at lower temperatures.

In assessing the identities of redox waves, it is also critical to evaluate the number of electrons, n, involved. This can be done by coulometry, simply by measuring the charge required to totally oxidize or reduce a known amount of complex. Alternatively, it is useful to undertake a potentiodynamic scan using a microelectrode[33] or rotating disk electrode (RDE).[8] In both these experiments, the current is directly proportional to the number of electrons involved, while in the CV experiment $i$ is proportional to $n^{3/2}$.

$$
\begin{aligned}
i_p &= (2.69 \times 10^5) n^{3/2} A D_0^{1/2} \nu^{1/2} C_0^* && \text{[Linear potential sweep current (CV)]} \\
i_l &= 0.620 n F A D_0^{2/3} \omega^{1/2} \upsilon^{-1/6} C_0^* && \text{(Levich Equation, RDE current)} \\
i_p &= 4 n F D C_0^* r && \text{(Plateau current, microelectrode)} \qquad (14)
\end{aligned}
$$

where current, $i$, is in $A$, $n$ is the number of electrons, $C_0^*$ is the concentration of bulk active species in mol $cm^{-3}$, $D$ is diffusion coefficient in $cm^2$ $s^{-1}$, the scan rate $\nu$ is in $Vs^{-1}$, the kinematic viscosity, $\upsilon$, is in cm $s^{-1}$, $A$ is the area of the electrode in $cm^2$, $F$ is the Faraday constant, $r$ is the radius of the microelectrode in cm and $\omega$ is the rotation rate in $s^{-1}$.

The peak current in a CV experiment is a composite of the charging and Faradaic current and this, together with the wave shape characteristics make it somewhat difficult to extract an accurate value for $n$. In the microelectrode or RDE experiments, the wave shape is sigmoidal and the current is readily evaluated from the rising portion of the curve. The relative currents, and relative $n$, for several redox waves in a long range potential scan, are immediately evaluated from the relative heights of these sigmoidal scans. An absolute measurement of $n$ is achieved either by evaluating the terms in the current expression for a microelectrode or RDE, or by adding a known amount of a calibrant of known $n$, and comparing wave heights (subject to any difference in diffusion coefficient).

These redox processes may, in principle, be events localized at the metal, or at a ligand, or may involve orbitals spanning the whole molecule. Thus a complex such as $[Ru^{II}Cl_2(bpy)_2]$ could, in principle, exhibit processes involving reduction or oxidation at the ruthenium center, e.g., to form species containing $Ru^{I}$ or $Ru^{III}$ or $Ru^{IV}$ etc., or reduction or oxidation may occur at the bipyridine ligand, involving, commonly, $\pi^*$ or $\pi$ (or possibly $\sigma$) ligand orbitals, and finally, oxidation of the chloride ligand is also possible.

Of the many processes which are feasible for a molecule such as $[Ru^{II}Cl_2(bpy)_2]$, subject to kinetic pathways being available, processes will occur in order of their themodynamic energies. Thus, for example, in the parent molecule, the oxidation of $Ru^{II}$ to $Ru^{III}$ may require less energy than the oxidation of bipyridine to ($bpy^+$). Assuming that the oxidation of bound chloride requires even more energy, then the first oxidation process to be observed will be

$$[Ru^{II}Cl_2(bpy)_2] \Longrightarrow [Ru^{III}Cl_2(bpy)_2]^+ + e^- \tag{15}$$

Thus, the alternate oxidation process:

$$[Ru^{II}Cl_2(bpy)_2] \Longrightarrow [Ru^{II}Cl_2(bpy)(bpy^+)]^+ + e^- \tag{16}$$

will not be observed experimentally, though, as we shall explore, its magnitude may be accessible theoretically or by using spectroscopic information. Similarly one might envisage at least two reduction processes, namely Equation 13a above, or:

$$[Ru^{II}Cl_2(bpy)_2] + e^- \Longrightarrow [Ru^{I}Cl_2(bpy)_2]^- \tag{17}$$

The process requiring the least energy to achieve (most positive redox potential) will be that observed. In a molecular orbital sense, the distinction between Equations 15 and 16 depends upon whether the HOMO is localized on ruthenium or bipyridine, while the distinction between Equations 13a and 17 depends upon whether the LUMO is localized on bipyridine or ruthenium. This is a useful working argument but is only strictly valid in the situation where there is no electronic relaxation following the redox process and the redox product has the same nuclear structure as the reactant, i.e., there are no significant changes in bond distances or angles following the redox reaction. The assumption that there is no electronic relaxation (which would necessarily mean no nuclear rearrangements) aside from the number of electrons, is known as Koopmans' theorem. Exceptions to Koopmans' theorem are quite common so that a decision as to the localization of a redox process, based solely on the localization of the HOMO or LUMO, should be treated with caution (see Section 2.2.1).

In this instance, reduction occurs localized at the bipyridine ligand so that Equation 17 is unobservable. Note that even if Equation 17 might theoretically occur only, say, 100 mV more negatively than Equation 13a, then Equation 17 will usually not be observed; all reduction will proceed through the thermodynamically preferred pathway (Equation 13a). The only exception might be when a kinetic complication inhibits or slows the thermodynamically preferred pathway. This discussion assumes a LUMO localized entirely on ruthenium or on bipyridine, but one of the facets which we would wish to analyze by optical and electrochemical means, is the consequence of significant mixing between metal and ligand orbitals, such that, in this example, the LUMO is a mixed ruthenium $d$ and ligand $\pi^*$ orbital.

The site at which a redox process occurs is commonly determined experimentally by controlled potential electrolysis. Thus, the species is exhaustively oxidized or reduced by the appropriate number of electrons and the properties (UV/visible spectrum, infrared, EPR or NMR, X-ray absorption spectroscopy (XAS), etc.) of the resulting oxidized or reduced species are probed to locate the site of localization of the redox processes. Spectroelectrochemistry is the name accorded to the most common of these techniques; a thin layer cell with a transparent electrode (e.g., $SnO_2$ or a gold minigrid) sits inside a UV/vis or IR or Raman spectrometer,[34] and potential is applied to totally oxidize or reduce the bulk solution. The resulting spectroscopic changes, perhaps over several redox processes (e.g.,[35,36]) can be followed and interpreted in terms of the localization of the redox site. Similarly, the electrochemical experiment could be carried out *in situ*, inside an EPR cavity, to obtain definitive information as to the localization of the added electron (or hole).[37,38] These experiments require that the products of oxidation or reduction be stable for the length of time required to run the experiment (e.g., minutes to hours).

## 2.2 Electrochemical Potentials and Ionization Potentials

We discuss here the relationships between the ionization potential, IP, and the observed redox potential. First, however, recognize that terms included within the ionization potential will vary depending upon the method of generation. Using gas phase pulsed high pressure mass spectrometry or ion cyclotron resonance mass spectrometry methods, electron attachment energies, $M_g^n + e^- \rightarrow M_g^{n-1}$ are obtained.[39–41] The inverse of this attachment energy (Joules) is then the ionization potential ($\times\mathcal{F}$) of the $M_g^{n-1}$ species. An adiabatic IP is obtained, i.e., the difference in energy between the thermally equilibrated starting species, R, and the thermally equilibrated oxidized product O. We term this adiabatic conversion of R to O, $IP_g^A$. The corresponding solution phase quantity, $IP_s^A$, is, of course, the observable redox potential (Equation 18). The subscripts $g$ and $s$ refer to gas and solution phase respectively.

The photoelectron emission experiment, on the other hand, yields a vertical ionization potential, $IP_g^V$, wherein the oxidized product, O, is obtained with the nuclear coordinates of the reductant, R, i.e., it is not thermally equilibrated. The corresponding solution quantity, $IP_s^V$, can be obtained experimentally by photoelectron threshold emission[42,43] and has recently been shown to be derivable from the outer sphere charge transfer spectra of ion pairs[44] (see Section 4.3).

The redox potential may be written:

$$IP_s^A = E^0[\mathrm{O/R}] = (1/\mathcal{F})\left[\left(\Delta G_{\mathrm{O}}^0\right)_s - \left(\Delta G_{\mathrm{R}}^0\right)_s\right] + C \tag{18}$$

In these expressions, $C$ is a constant potential depending upon the reference electrode. The various ionization potentials may be written:

$$\mathrm{IP}_g^A = (1/\mathcal{F})\left[\left(\Delta G_\mathrm{O}^0\right)_g - \left(\Delta G_\mathrm{R}^0\right)_g\right] \tag{19}$$

$$\mathrm{IP}_g^V = (1/\mathcal{F})\left[\chi_i + \left(\Delta G_\mathrm{O}^0\right)_g - \left(\Delta G_\mathrm{R}^0\right)_g\right] \tag{20}$$

$$\mathrm{IP}_s^V = (1/\mathcal{F})\left[\chi_i + \chi_0 + \left(\Delta G_\mathrm{O}^0\right)_s - \left(\Delta G_\mathrm{R}^0\right)_s\right] \tag{21}$$

where $\chi_i$ and $\chi_0$ are the inner- and outer-sphere reorganization energies; in this case, specific to the oxidized species, $(\Delta G_\mathrm{O}^0)_g$ is the gas phase Gibbs free energy of formation of the thermally equilibrated oxidized species and $(\Delta G_\mathrm{R}^0)_g$ is the gas phase Gibbs free energy of the initial species R, and the '$s$' subscripted terms are the corresponding solution phase quantities (Figure 3). The reorganization energies are commonly met when using the Marcus relationships in kinetic studies (see discussion following Equation 59).

It is now possible to relate the oxidation potential to the various ionization energies via:

$$E^0[\mathrm{O/R}] = \mathrm{IP}_g^A - 4.44 + (1/\mathcal{F})\left[\Delta G_s^{0\,\mathrm{O}} - \Delta G_s^{0\,\mathrm{R}}\right] \tag{22}$$

$$E^0[\mathrm{O/R}] = \mathrm{IP}_g^V - 4.44 + (1/\mathcal{F})\left[\Delta G_s^{0\,\mathrm{O}} - \Delta G_s^{0\,\mathrm{R}} - \chi_i\right] \tag{23}$$

$$E^0[\mathrm{O/R}] = \mathrm{IP}_s^V - 4.44 - (1/\mathcal{F})[\chi_i + \chi_0] \tag{24}$$

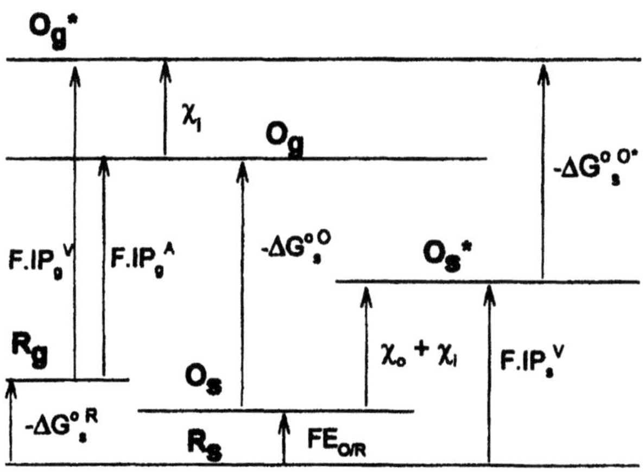


**Figure 3.** A free energy cycle showing relationships between ionization potentials, free energies of solvation and reorganization energies. The bolded letters represent the species involved. $\mathbf{R}_g$ and $\mathbf{R}_s$ are the gas phase and solvent equilibrated reduced species **R**. $\mathbf{O}_g$ and $\mathbf{O}_s$ are the oxidized species **O** in the ground state gas and solvent equilibrated forms. $\mathbf{O}_g^*$ and $\mathbf{O}_s^*$ are the vibrationally excited **O** species with the nuclear coordinates of the **R** species, in the gas phase and solvent sphere appropriate for **R**, not **O**, respectively. The various energies are discussed in the text and defined according to the energy differences shown in the figure. $\mathcal{F} \cdot \mathrm{IP}$ is potential multiplied by the Faraday to convert potential to energy units.

where $\Delta G_s^{0\,\mathrm{R}}$ and $\Delta G_s^{0\,\mathrm{O}}$ are the Gibbs free energies of solvation of the reduced, R, and oxidized, O, species. The $(1/\mathcal{F})$ factor can be omitted if the free energies are expressed in V. The value 4.44 relates the vacuum scale electrostatic zero potential energy to the NHE scale, via:[45–47]

$$H_{\mathrm{aq}}^{+} + e^{-} \Longleftrightarrow \tfrac{1}{2}H_2, \quad \Delta G^0 = -429\ \mathrm{KJ\ mol^{-1}}, \quad E^0 = 4.44\ \mathrm{V} \tag{25}$$

The solvation energy terms can be estimated using the Born equation which is used to derive the change in free energy, $\Delta G_{\mathrm{el}}$, when charge, $z$, on an sphere of radius, $r$, is moved from vacuum to a solution of dielectric constant, $D_s$ [this is rarely used now]:

$$\Delta G_{\mathrm{el}} = \left(Nz^2e^2/2r\right)\left(1 - \tfrac{1}{D_s}\right) \tag{26}$$

Assuming that the effective radii, $r$, of the oxidized and reduced species are similar $\Delta G_{\mathrm{el}} \approx (\Delta G_{\mathrm{O}}^0)_s - (\Delta G_{\mathrm{R}}^0)_s = \Delta\Delta G_{\mathrm{solv}}^0$.[45] However such a procedure is not very accurate and more recent studies in calculating hydration energies, take account of the influence of both first and second water coordination spheres using density functional theory.[48] The development of theoretical approaches to the inclusion of solvent effects is an on-going and important area of research.[49–51]

The correlation of ionization potentials with redox potentials has been probed most commonly in organic chemistry. The aromatic hydrocarbons and related molecules provide a convenient group to test relationships between ionization potentials and oxidation potentials. For a series of alkyl benzenes, a good linear relationship exists between the experimental (gas phase) vertical ionization potentials and the oxidation potentials, but the slope is 0.71 rather than unity.[52] However a study of azoalkanes did lead to a slope close to unity.[53] Indeed there have been many reports of linear correlations between ionization potentials and redox potentials.[54–65]

Holm, Kennepohl and Solomon have discussed[66] the relative redox potentials of copper proteins in terms of Equation 22. The adiabatic ionization energy is determined by the energy of the redox orbital, by the change in effective nuclear charge, $Z'_{\mathrm{eff}}$, following the redox process, and by the relaxation of electrons in all the other orbitals whose energies will respond to a change in the net oxidation state of the species. Changes in $Z'_{\mathrm{eff}}$, are determined by the donor strength of the ligands coordinated to the metal center and their influence can be parametrized as discussed in Section 3. Blue copper proteins contain a distorted tetrahedral copper center with a $CuN_2SS'$ chromophore [the nitrogen sites are histidines (imidazole residues) and the two sulfur ligands are cysteine and methionine with the latter at an extended distance] and they have $E_{1/2}[\mathrm{Cu^{II/I}}]$ values some 500 mV more positive than the much more common square planar copper(II) complexes. The redox orbital is destabilized in a square planar complex to a much greater degree than in a tetrahedral species as follows from the application of simple ideas of crystal field theory and the influence of ligands on the $d$-orbital splitting energies as a function of stereochemistry. This difference in redox orbital energy modifies the $\mathrm{IP}_g^A$ value and thereby influences the redox potential in the direction indicated. Compared with a normal length Cu–

N(imidazole) bond, the long Cu–S(methionine) bond in these blue copper proteins has the effect of decreasing the $Z'_{\text{eff}}$ at the copper site and therefore causes a further positive shift of the redox potential.

The question of entropy changes involved in a redox process are not addressed here. Intramolecular entropy differences between the oxidized and reduced species are contained within the $\text{IP}_g^A$ term, while solvational entropy changes are contained within $\Delta\Delta G_s$. The magnitudes of these changes have been addressed in some detail.[67] Some other subtle features are also not addressed here, e.g., the experimental ionization energies are determined at temperatures above 0 K so that the difference in heat capacity of the oxidized and reduced species should really be taken account of. Some small relativistic corrections have also been ignored.

**2.2.1** ***Electrochemical Potentials and HOMO and LUMO Energies*** It is commonly assumed that oxidation and reduction potentials correlate with HOMO and LUMO energies, respectively. Certainly for homologous series of complexes, this is generally true. However it is clearly incomplete since the redox potential reflects the relative free energies of both the oxidized and reduced species, while the HOMO and LUMO energies belong only to one species. If we were to assume that the product after oxidation or reduction retains the same nuclear geometry and electronic structure, and retains the same solvation energy, then the HOMO and LUMO energies, if derived from a sufficiently high level analysis, would, with an appropriate reference, correspond precisely with the observed oxidation and reduction potentials. Clearly this is not the case, so caution should be exercised in relating molecular orbital energies with redox potentials.

Koopmans' theorem indeed proposes that the ionization potential of a species is the negative of the HOMO orbital energy, and similarly the electron affinity of a species is the negative of the LUMO energy. This is a "frozen orbital" approach in that the orbitals of the product do not relax. If we permit electron relaxation, then the product oxidized and reduced species must become more stable and thus Koopmans' approach produces too positive an ionization potential and too negative an electron affinity. Simple treatments such as Hartree–Fock do not take care of electron correlation effects so that the proper inclusion of these would also lead to a further correction to the ionization energies and electron affinities. Semi-empirical methods such as ZINDO partially compensate for this. Correlation effects are greater for the greater number of electrons, so that their omission tends to cancel the error due to omitting electron relaxation for ionization potentials, but add to the error for electron affinities.[68]

In some early contributions, Treichel and co-workers[4] and McCleverty and co-workers[69] noted a linear relationship between the number of isocyanide ligands in a series of complexes $[\text{M(CO)}_{6-n}(\text{CNMe})_n]^{m+}$ (M = Mn, $m = 1$; Cr, $m = 0$) and their metal centered oxidation potentials, respectively. This led naturally to the idea of a simple additive effect by the isocyanide ligands upon the HOMO energy[70] and hence the redox potential. Sarapu and Fenske[71] then demonstrated, in the series $[\text{Mn(CO)}_{6-n}(\text{CNMe})_n]^+$, that a linear relationship exists between the HOMO energy

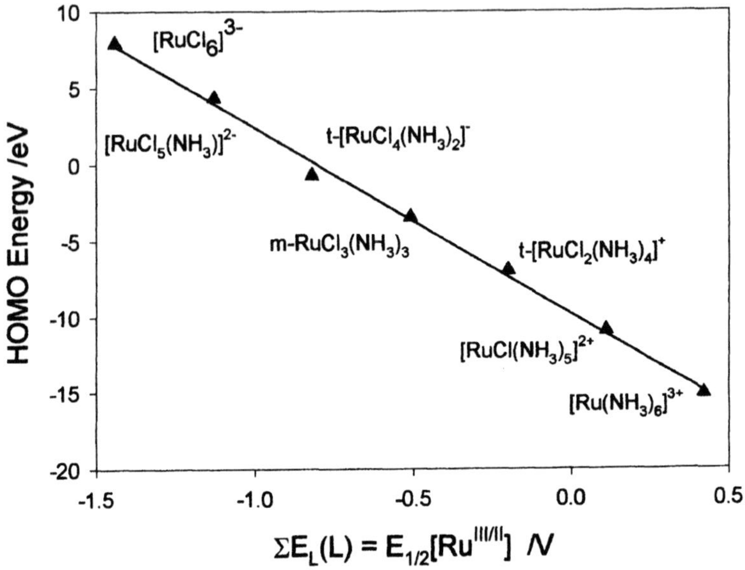


**Figure 4.** A plot of the HOMO energy of the $Ru^{II}$ species, calculated using ZINDO/S and the $Ru^{III/II}$ redox potential calculated using $E_L$(L) parameters (see Section 3.4) for a generic series of ruthenium complexes. [A. B. P. Lever, unpublished observations.]

as calculated by the Hartree–Fock method, and the observed metal centered oxidation potential is represented generally by:

$$E_{1/2} = k\varepsilon(\text{HOMO}) + C \tag{27}$$

In the absence of major nuclear re-arrangements occurring subsequent to oxidation or reduction, we can expect that the oxidation or reduction potentials of a series of related molecules would correlate linearly with the HOMO or LUMO energies, respectively, as derived by Extended Hückel (for HOMO only), Hartree–Fock or higher level calculations. This has been shown to be the case for many organic systems, and some inorganic or organometallic species, as outlined, for example, in[1,71–82] (Figure 4[80]). These ideas are further explored below within the purview of ligand additivity.

**2.2.2** ***Optical Transitions and the HOMO-LUMO Gap*** Related to the above ideas it is often assumed, erroneously, in the literature that the energy difference between the HOMO and the LUMO corresponds with the optical energy of the HOMO → LUMO electronic transition. Writing the HOMO and LUMO orbital energies as $\varepsilon_{\text{HOMO}}$ and $\varepsilon_{\text{LUMO}}$, the optical transition energy between ground state, $\psi_g$, and excited state, $\psi_e$, for a closed shell ground state, in the absence of configurational interaction, is given by:*

$$^1\psi_e \leftarrow {}^1\psi_g = \varepsilon_{\mathrm{LUMO}} - \varepsilon_{\mathrm{HOMO}} - J_{\mathrm{HL}} + 2K_{\mathrm{HL}} \tag{28}$$

and similarly for a spin doublet-doublet transition (configuration $a^2b \rightarrow ab^2$):[84]

$$^2\psi_e \leftarrow {}^2\psi_g = \varepsilon_{\mathrm{LUMO}} - \varepsilon_{\mathrm{HOMO}} - J_{\mathrm{HL}} + 0.5J_{\mathrm{LL}} + 0.5K_{\mathrm{HL}} \tag{29}$$

where $J_{\mathrm{HL}}$, $J_{\mathrm{LL}}$ and $K_{\mathrm{HL}}$ are the coulombic and exchange inter-electronic interaction integrals appropriate to the HOMO and LUMO as labeled H and L. Depending on the system, these terms can be quite large, often of the order of eV, and $J > 2K$ so that the optical transition energy is usually significantly less than the HOMO-LUMO gap energy.

## 3 LIGAND ADDITIVITY

These developments led to a series of different ligand additivity models which are now discussed.

### 3.1 Pickett and Pletcher Model

Pickett and Pletcher[2] proposed a model where the change in oxidation potential for a series $[\mathrm{M(CO)}_{6-n}\mathrm{L}_n]^{m+}$ is a linear function of ligand L but is not dependent on $n$. Further, the effect of changing charge, $m+$, for a given L is to add a constant $Q_{\mathrm{PL}}$, viz.:

$$E_{1/2} = A + n(\delta E^0/\delta n)_{\mathrm{L}} + Q_{\mathrm{PL}} \tag{30}$$

where the magnitude of the constant $A$ depends upon the solvent and reference electrode. The magnitude of $Q_{\mathrm{PL}}$ was found to be fairly constant for a series of isoelectronic pairs of complexes and was set empirically at 1.48 V. The so-called "ligand inductive parameter," measures the shift to more negative potentials as CO is replaced by a more electron rich ligand.†[85] Table 2 lists some of these ligand inductive

* Note the distinction between one-electron energies, $h_i$, and orbital energies, $\varepsilon_i$. The one-electron energy of an electron in an orbital is its kinetic energy and the energy of its attraction to the nucleus but excludes the energy of its interaction with all other electrons in the atom (molecule). Thus, in an $N$-electron atom, an electron in orbital "a," has a one-electron energy $h_a$. The orbital energy, $\varepsilon_a$, is given by $h_a$ plus the electrostatic interaction of this electron with the other $N-1$ electrons in the atom (molecule), namely:

$$\varepsilon_a = h_a + \sum_{b \neq a}(J_{ab} - K_{ab})$$

For more details, see Reference 83.

† The datum for the bromide ion is probably incorrect since it predicts a shift to higher potential when CO is substituted by $\mathrm{Br}^-$. The oxidation process was likely misinterpreted.

**Table 2 Ligand Inductive Parameters (LIP)[2]**

| L | LIP | L | LIP |
|---|---|---|---|
| $NH_3$ | −0.80 | $Ph_3P$ | −0.36 |
| $CH_3CN$ | −0.54 | $P(OPh)_3$ | −0.19 |
| $CH_3NC$ | −0.44 | CO | 0.00 |
| p-Tol–NC | −0.42 | Br | +0.42 |
| p-ClPh–NC | −0.37 | | |

parameter values. This model does not distinguish possible differences between the redox potentials of isomers such as *cis*- and *trans*-$ML_4X_2$.

### 3.2 Bursten's Model

Based on an idea initially proposed by Bond and co-workers,[87] Bursten and co-workers[1,77,78,86] developed a sophisticated approach to using ligand additivity to correlate a range of calculated and experimental data including molecular orbital energies, photoelectron spectroscopy ionization energies and electrochemical potentials. A primary strength of this model is that it could distinguish the behavior of different isomers.

The fundamental idea is that the energies of the $d\pi$ orbitals of a substituted carbonyl species, $M(CO)_{6-n}L_n$, could be defined by[86]

$$\varepsilon_i = a + bn + cx \tag{31}$$

where $\varepsilon_i$ is the energy of the $i$th-$d\pi$ orbital, $x$ is the number of CO $\pi$-orbitals with which this $i$th-$d\pi$ orbital interacts, and $a$, $b$ and $c$ are empirically derived parameters. The distinction between stereoisomers may be appreciated as follows. In the *trans*-$ML_4(CO)_2$ species (CO groups along the $z$ axis), simple Angular Overlap Model theory[88,89] would predict the energy order $(xz, yz) < xy$ (HOMO), assuming CO is a better $\pi$-acceptor than ligand L. In this situation, the $(xz, yz)$ orbitals each overlap a $\pi$-orbital on two CO groups, while $xy$ interacts with no CO $\pi$-orbitals. Thus, for this stereochemistry, $x$ is zero, in Equation 31 for the HOMO.

More generally, for a complex $ML_nL'_{6-n}$ Equation 32 can be written:[86]

$$\varepsilon_i = a_M^0 + nb_M^L + (6-n)b_M^{L'} + x_i c_M^L + (4-x_i)c_M^{L'} \tag{32}$$

which corresponds with Equation 31 when:

$$\begin{aligned} a &= a_M^0 + 6b_M^{L'} + x_i c_M^L + 4c_M^{L'} \\ b &= b_M^L - b_M^{L'} \\ c &= c_M^L - c_M^{L'} \end{aligned} \tag{33}$$

**Table 3 The $d\pi$ Orbital Order and Number of Interacting L* $\pi$-Orbitals as a Function of Stereochemistry, and the $E_{1/2}$ Values for $d^6$ and $d^1$ Species; L Is a Better $\pi$-Acceptor than L′**

| Stereoisomer | Octahedral-Based Geometry | $d\pi$ Order (No. of L $\pi$-Orbitals) | $E_{1/2}d^6$ | $E_{1/2}d^1$ |
|---|---|---|---|---|
| $ML'_6$ | | $(xz, yz, xy)(0)$ | A | A |
| $ML'_5L$ | L along $z$ | $(xz, yz)(1) < (xy)(0)$ | A + B | A + B + C |
| *trans*-$ML'_4L_2$ | L along $z$ | $(xz, yz)(2) < (xy)(0)$ | A + 2B | A + 2B + 2C |
| *cis*-$ML'_4L_2$ | L's in $xy$ plane | $(xy)(2) < (xz, yz)(1)$ | A + 2B + C | A + 2B + 2C |
| mer-$ML'_3L_3$ | L's in $xz$ plane | $(xz)(3) < (yz)(2) < (xy)(1)$ | A + 3B + C | A + 3B + 3C |
| fac-$ML'_3L_3$ | L's along $x$, $y$, $z$ | $(xz, yz, xy)(2)$ | A + 3B + 2C | A + 3B + 2C |
| *trans*-$ML'_2L_4$ | L's along $z$ | $(xy)(4) < (xz, yz)(2)$ | A + 4B + 2C | A + 4B + 4C |
| *cis*-$ML'_2L_4$ | L's in $xy$ plane | $(xz, yz)(3) < (xy)(2)$ | A + 4B + 2C | A + 4B + 3C |
| $ML'L_5$ | L′ along $z$ | $(xy)(4) < (xz, yz)(3)$ | A + 5B + 3C | A + 5B + 4C |
| $ML_6$ | | $(xz, yz, xy)(4)$ | A + 6B + 4C | A + 6B + 4C |

*For example, L = CO, L′ = most other ligands. Data for two right-hand columns come from Equation 34.[1]

In this model, $a_M^0$ is a function of the metal ion concerned, and $b_{\mathrm{M}}^{\mathrm{L}}$ and $b_{\mathrm{M}}^{\mathrm{L}'}$ refer to the overall binding effect of the ligands L and L′, which is somewhat akin to the spherical term in crystal field theory in its effect on $d$ orbital energies. The differential $\pi$-interaction on the $d\pi$ orbitals is taken account of through the $(c_{\mathrm{M}}^{\mathrm{L}} - c_{\mathrm{M}}^{\mathrm{L}'})$ term. The term in $x_i$ provides the means to distinguish the energies of each $d\pi$ orbital as a function of the stereochemistry of the species using the number of interacting L and L′ $\pi$-orbitals in a manner parallel to that shown in Table 3 (also see Table XI of Reference 1). Bursten notes[1] that in this model the average $d\pi$ MO energies depends only on "$n$" in the sense that no allowance is made for the effect of a given ligand to change with stereochemistry. Thus synergistic effects where, for example the behaviour of L *trans* to L′ is different from that of L *trans* to another L, are not recognized by the model and would be averaged out when considering a homologous series of complexes. The model also assumes that each of the $d\pi$ orbitals can be recognized and behaves independently, for the purpose of assigning the number of interacting ligand $\pi$-orbitals, even though mixing between these $d$ orbitals may occur in some symmetries.

This model, using the Fenske–Hall method for calculating MO energies, appears to work well for a range of $d^1$ and $d^6$ species for which it has been tested.[77,78]

Given the relationship between $E_{1/2}$ and $\varepsilon_i$, it is obvious that an equation similar to Equation 31, and based upon Equation 27, is readily derived for the metal centered oxidation potential of $ML_nL'_{6-n}$, namely:[1]

$$E_{1/2} = A + Bn + Cx_{\mathrm{HOMO}} \tag{34}$$

where:

$$A = A_{\mathrm{M}}^{0} + 6B_{\mathrm{M}}^{\mathrm{L}'} + 4C_{\mathrm{M}}^{\mathrm{L}'}$$
$$B = B_{\mathrm{M}}^{\mathrm{L}} - B_{\mathrm{M}}^{\mathrm{L}'}$$
$$C = C_{\mathrm{M}}^{\mathrm{L}} - C_{\mathrm{M}}^{\mathrm{L}'} \qquad (35)$$

where $x_{\mathrm{HOMO}}$ is the number of ligand $\pi$-orbitals with which the highest energy $d\pi$ orbitals interacts. Clearly these upper case $A$, $B$, and $C$ have significance closely aligned with their lower case analogs. The predicted $E_{1/2}$ values for the set of substituted $\mathrm{ML}_n\mathrm{L}'_{6-n}$ species is shown in Table 3.

Not unexpectedly, these equations work well for a range of substituted organometallic species[1,77–79] and indeed Equation 27 is valid for simple organic hydrocarbons as demonstrated 40 years ago.[74]

### 3.3 The $P_{\mathrm{L}}$ Model

In a variation of the "ligand inductive parameter," and based on a series of papers detailing ligand additivity behaviour,[93,94] Chatt, Leigh, Pickett, and co-workers[95,96] introduced a new parameter, termed $P_{\mathrm{L}}$, and defined by:

$$P_{\mathrm{L}} = E_{1/2}^{\mathrm{ox}}[\mathrm{Cr(CO)_5L}] - E_{1/2}^{\mathrm{ox}}[\mathrm{Cr(CO)_6}] \qquad (36)$$

i.e., defined as the change in redox potential when one carbonyl group is replaced by the ligand L in $\mathrm{Cr(CO)_6}$. Its sign is such that the more electron releasing the ligand (or a substituent thereof), the more negative is $P_{\mathrm{L}}$. It is also a measure of the net $\sigma$-donor minus $\pi$-acceptor ability of the ligand. Suppose one now plots data for a series of different complexes $E_{1/2}[\mathrm{M(VWXYZ)L}]$ (where ligands VWXYZ are fixed, such that the fragment M[VWXYZ] represents a fixed core often called $\mathrm{M}_s$)[96] and varying L against the $E_{1/2}^{\mathrm{ox}}[\mathrm{Cr(CO)_5L}]$, one obtains, in general, a series of parallel lines, suggesting that the potential for a general complex $\mathrm{M}_s\mathrm{L}$ is given by:

$$E_{1/2}^{\mathrm{ox}}[\mathrm{M}_s\mathrm{L}] = E_s + \beta_{\mathrm{L}} P_{\mathrm{L}} \qquad (37)$$

The parameter $E_s$ is a measure of the electron richness at the site $\mathrm{M}_s$, while $\beta_{\mathrm{L}}$ reflects how, in a relative sense, information is transferred from the ligand L to the HOMO of the metal ion; it may be thought of as a measure of the polarizability of the site. While the theory is referenced to the $\mathrm{Cr(CO)_5L}$ complex, it is clear that $P_{\mathrm{L}}$ values for ligands not involved in binding with $\mathrm{Cr(CO)_6}$ can be obtained by extrapolating values from a series of $\mathrm{M}_s\mathrm{L}$ species where the parameters in Equation 37 are known. Table 4 contains a representative set of $P_{\mathrm{L}}$ data.

Where the necessary parameter values exist, this model can obviously be used to predict $E_{1/2}^{\mathrm{ox}}$ values of new complexes. It has also been shown to correlate well with other properties, principally infrared stretching frequencies such as $\nu(\mathrm{N_2})$ and $\nu(\mathrm{NO})$

**Table 4 A Collection of $P_L$ Values[96,98,100–102]**

| Ligand | $P_L$ | Ligand | $P_L$ |
|---|---|---|---|
| $NO^+$ | +1.40 | Py | −0.59 |
| CO | 0 | $NH_3$ | −0.77 |
| $N_2$ | −0.07 | $CF_3CO_2^-$ | −0.78 |
| $P(OPh)_3$ | −0.18 | o-$CN(H)C_6H_4C(PR_3)$[b] | −0.83[c] |
| o-$R_3PCH_2C_6H_4NC(1+)$ | −0.32[a] | $NCNC(NH_2)_2$ | −0.86 |
| 2,6-$Cl_2PhNC$ | −0.33 | NCNHCN | −0.88 |
| $PPh_3$ | −0.35 | $NCS^-$ | −0.88 |
| 4-ClPhNC | −0.37 (−0.11)[d] | o-$N{=}C(H)C(PR_3)(C_6H_4)$[e] | −0.96[f] |
| PhNC | −0.38 (−0.12)[d] | $CN^-$ | −1.00 |
| 2-MePhNC | −0.38 | $NCO^-$ | −1.16 |
| 4-MePhNC | −0.39 (−0.13)[d] | $NCNC(NH)NH_2^-$ | −1.22 |
| PhCN | −0.40 | $H^-$ | −1.22 |
| 4-MeOPhNC | −0.40 (−0.14)[d] | $NCNCN^-$ | −1.14 |
| $CH_3NC$ | −0.43 (−0.17)[d] | $I^-$ | −1.15 |
| $^t$BuNC | −0.44[g] (−0.18)[d] | $Br^-$ | −1.17 |
| Protonated indoles | −0.53[h] | $Cl^-$ | −1.19 |
| Protonated carbenes | −0.55[i] | $N_3^-$ | −1.26 |
| $CH_3CN$ | −0.58 | $OH^-$ | −1.55 |

[a]Average of range, (−0.28) – (−0.36); [b]carbene; [c]average of (−0.79) – (−0.87); [d]bent isocyanide; [e]Indole[100b]; [f]average of −(0.92) to (−1.00); [g]$P_L$ = −0.09 when bound to $TcH(dppe)_2$, see text; [h]average of (−0.44) – (−0.62); [i]average of (−0.51) – (−0.60). For additional data, see Reference 104.

in general species such as $M(Y)(L)X_4$ with Y = $N_2$ or NO. Such correlations probably arise through the stretching frequency being sensitive to the electron richness of the site as L is varied. Where rate constants show such dependence, they may also correlate with $P_L$. Chatt has also discussed[96] how the variation in $M_s$ and $\beta_L$ values might be used to identify sites which would preferentially bind $N_2$ or hydride ion, in other words, the ability to predict binding behaviour from a combination of electron richness and polarizability.

Table 5 shows that both the electron richness, $E_s$, and polarizability, $\beta_L$, values can vary widely. The $TcH(dppe)_2$ center, for example, is evidently extremely polarizable. When bound to isocyanide ligands, the redox potentials then observed are consistent with the $P_L$ parameter for isocyanide ligands being significantly less negative than that shown in Table 4. This may be associated with the non-innocent character of the isocyanide ligand where the M-CNR group can be bent or linear and whose $\pi$-acceptor strength can obviously vary with stereochemistry (vide infra[97]). Pombeiro has discussed at some length[85,98] the probability of observing bent isocyanide ligands when they are bound in electron rich sites such as $[ReCl(CNR)(dppe)_2]$ and some data for bent isocyanides are shown in Table 4.

While the $P_L$ model is generally used in systems where one ligand is being changed and Equation 37 is used for a variety of different $ML_5$ cores. Datta has shown[99] that it can be extended to at least two different ligands (X, Y) in a system $ML_4XY$ where redox potentials were shown to linearly correlate with $P_L(X) + P_L(Y)$.

**Table 5 A Collection of $M_s$ and $\beta_L$ Values[85,96,100–108] for Species MY(L)$X_4$**

| $M_s$ [$MYX_4$][a] | $E_s$(V vs SCE) | $\beta_L$ | $P_L$(Y) |
|---|---|---|---|
| $W(CO)_5$ | 1.52 | 0.90 | 0 |
| $Cr(CO)_5$ | 1.50 | 1 | 0 |
| $Mo(CO)_5$ | 1.44 | 0.86 | 0 |
| $[Re(CO)(dppe)_2]^+$ | 1.42 | 0.62 | 0 |
| $[Tc(^tBu\text{-}NC)(dppe)_2]^+$ | 1.34 | 0.99 | −0.44 |
| $[Re(CNMe)(dppe)_2]^+$ | 1.19 | 0.93 | −0.43 |
| $[Fe(H)(dppe)_2]^+$ | 1.04 | 1.0 | −1.22 |
| $[Mo(NO)(dppe)_2]^+$ | 0.91 | 0.51 | 1.40 |
| $[Mn(CO)(dppm)_2]^+$ | 0.86 | 0.75 | 0 |
| $Re(CN)(dppe)_2$ | 0.78 | 3.7 | −1.0 |
| $ReCl(dppe)_2$ | 0.68 | 3.4 | −1.19 |
| $TcH(dppe)_2$ | 0.34 | 4.0 | −1.22 |
| $Mo(CO)(dppe)_2$ | −0.11 | 0.72 | 0.00 |
| $Mo(N_2)(dppe)_2$ | −0.13 | 0.84 | −0.07 |
| $Mo(PhCN)(dppe)_2$ | −0.40 | 0.82 | −0.40 |
| $[Mo(N_3)(dppe)_2]^-$ | −1.00 | 1.0 | −1.26 |

[a]dppe = $Ph_2PCH_2CH_2PPh_2$; dppm = $Ph_2PCH_2PPh_2$.

The use of the parameters to predict redox potentials for a range of complexes requires knowledge of the $M_s$ and $\beta_L$ parameters. Some of these are shown in Table 5.

This theory has not been developed to discriminate between different isomers, as Bursten's model has, and an evident weakness is that new values of $E_s$ and $\beta_L$ must be derived for any new group of complexes, though relationships between similar groups of compounds have been developed.[85] Perhaps for these reasons, with the exception of studies from the Pombeiro laboratory[85,98,100–110] it is now rarely utilised.

### 3.4 Lever's Electrochemical Parameter Series

In 1990, Lever introduced the ligand electrochemical parameter $E_L(L)$ series.[5] This has proven to be a very useful methodology for solving problems in electrochemistry and the related spectroscopy and has been extensively employed.[111] We discuss here aspects relevant to this chapter. The parameters are based upon the reversible $E_{1/2}[Ru^{III}/Ru^{II}]$ couple, an idea that had been explored in a preliminary fashion by Datta[90], such that for a general complex Ru(UVWXYZ). The observed $E_{1/2}[Ru^{III}/Ru^{II}]$ potential versus NHE is given by

$$E_{1/2}\left[Ru^{III}/Ru^{II}\right] = E_L(U) + E_L(V) + E_L(W) + E_L(X) + E_L(Y) + E_L(Z) = \sum E_L(L) \quad (38)$$

By statistical analysis of the $E_{1/2}[Ru^{III}/Ru^{II}]$ redox couples for hundreds of ruthenium complexes recorded in organic solvents (usually acetonitrile), the individual $E_L(L)$ values were extracted[5] (Table 6). $E_L(L)$ parameter values range from

**Table 6 A Selection of $E_L$(L) Values[a]**

| Ligand | $E_L$(L) |
|---|---|
| 1,10-Phenanthroline (phen) | 0.26 |
| 1,10-phenanthroline, 4,7-$Me_2$ | 0.23 |
| 1,1,1,5,5,5-hexafluoro-2,4-pentanedionato (hfac) (1−) | 0.17 |
| 1,1,1,-trifluoro-2,4-pentanedionato (tfac) (1−) | 0.03 |
| 1,1-bis(diphenylphosphino)methane | 0.43 |
| 1,2-bis(diphenylarsino)ethane ($Ph_2AsCCAsPh_2$) | 0.44 |
| 1,2-bis(diphenylphosphino)ethane ($Ph_2PCCPPh_2$) | 0.31 |
| 1,2-bis(ethylthio)ethane | 0.32 |
| 1,2-bis(methylthio)ethane | 0.33 |
| 1,2-diamino-2-methyl-propane | 0.11 |
| 1,2-bis(4-pyridyl)ethane | 0.26 |
| 2,2′-Bipyridine | 0.26 |
| 2,2,6,6,-tetramethyl-3,5-heptanedionato (dpm) (1−) | −0.13 |
| 2,2-bipyridine, 5,5-dimethyl | 0.23 |
| 2,4-pentanedionato (1−) (acac) | −0.08 |
| 2-methylthioquinoline | 0.30 |
| 3,3′-biisoquinoline (biq) | 0.24 |
| 3-phenyl-2,4-pentanedionato (1−) (3-Phacac) | −0.09 |
| 8-hydroxyquinolinato (1−) | −0.09 |
| 8-methylthioquinoline | 0.30 |
| Acetonitrile | 0.34 |
| Ammonia | 0.07 |
| Azide (1−) | −0.30 |
| Benzoisonitrile | 0.41 |
| Benzonitrile | 0.37 |
| Benzylamine | 0.14 |
| Benzylisonitrile | 0.56 |
| Bipyrazinium(2,2′) (1+) | 0.75 |
| Bipyrazine(2,2′) | 0.36 |
| Bipyridazine | 0.30 |
| Bipyrimidine | 0.31 |
| Bipyrimidine(4,4-$Me_2$) | 0.24 |
| Biquinoline (biq) | 0.29 |
| Bis(4-pyridyl)acetylene | 0.27 |
| Bi-benzimidazole | 0.17 |
| Bi-imidazole | 0.13 |
| Bromide (1−) | −0.22 |
| Butylamine | 0.13 |
| Butyronitrile | 0.35 |
| Carbon monoxide (CO) | 0.99 |
| Chloride (1−) | −0.24 |
| Cyanate (1−) | −0.25 |
| Cyanide (1−) | 0.02 |
| Cyclohexylisonitrile | 0.32 |
| *c*-1,2-bis(diphenylphosphino)ethene ($Ph_2PCH{=}CHPPh_2$) | 0.49 |
| Diazabutadiene, di-R-alkyl (DAB) | 0.14 |
| Diethyldithiocarbamato (1−) | −0.08 |
| Diethylsulphide | 0.35 |
| Dimethyldithiocarbamato (1−) | −0.12 |
| Dimethylphenylphosphine (MMP) | 0.34 |

**Table 6** (*Continued*)

| Ligand | $E_L$(L) |
|---|---|
| Dimethylphosphine | 0.34 |
| Dimethylsulfide | 0.31 |
| Dimethylsulfoxide (DMSO) | 0.47 |
| Dinitrogen | 0.68 |
| Di-2-pyridylamine | 0.18 |
| Ethane thiolate (1−) | −0.56 |
| Ethyl nitrite | 0.70 |
| Ethylenediamine | 0.06 |
| Ethylnitrile | 0.33 |
| Ethylxanthato (1−) | −0.02 |
| Fluoride (1−) | −0.42 |
| Formate (1−) | −0.30 |
| Glycine (1−) | −0.05 |
| Hydroxide (1−) | −0.59 |
| Imidazole | 0.12 |
| Iodide (1−) | −0.24 |
| Isonicotinamide | 0.26 |
| Isopropylnitrite | 0.68 |
| Isopropylamine | 0.05 |
| Isopropylisonitrile | 0.36 |
| Maleonitriledithiolate (2−) | −0.33 |
| Methyl nitrite | 0.72 |
| Methyldiphenylphosphine (MPP) | 0.37 |
| Methylisonitrile | 0.37 |
| Methylphenylsulfide | 0.33 |
| *m*-Toluonitrile | 0.34 |
| Nitrate (1−) | −0.11 |
| Nitrite (1−) | 0.02 |
| *n*-Butylisonitrile | 0.45 |
| *N*-methylpyrazinium (NMePyz(1+)) | 0.79 |
| *N*-methylpyridinium, 4-cyano (1+) | −0.17 |
| Oxalate (2−) | −0.17 |
| *o*-Phenylenebisdiphenylphosphine | 0.45 |
| *o*-phenylenebis(dimethylarsine) (diars) | 0.34 |
| *o*-Toluonitrile | 0.34 |
| Pentafluorothiolato (1−) | −0.33 |
| Perchlorate (1−) | 0.06 |
| Phenanthrolinedione | 0.28 |
| Phenylazopyridine | 0.41 |
| Polyvinylimidazole | 0.11 |
| Pyrazine | 0.33 |
| Pyrazine,2,3-(di-2-pyridyl) | 0.32 |
| Pyrazole | 0.20 |
| Pyrazole (1−) | −0.24 |
| Pyridine | 0.25 |
| Pyrimidine | 0.29 |
| *p*-Chlorothiophenolate (1−) | −0.43 |
| *p*-Methylthiophenolate (1−) | −0.48 |
| *p*-toluenethiolate (1−) | −0.48 |

**Table 6** *(Continued)*

| Ligand | $E_L$(L) |
|---|---|
| Salicylaldehyde (1−) | −0.04 |
| Selenocyanate (1−) | −0.23 |
| Taphen | 0.37 |
| Terpyridine | 0.25 |
| Thiocyanate (1−) | −0.06 |
| Thiophenolato (1−) | −0.50 |
| Thiourea | −0.13 |
| Triazole,1,2,4 | 0.18 |
| Triethylphosphine | 0.34 |
| Trifluoroacetate (1−) | −0.15 |
| Trifluorosulfonate (1−) | 0.13 |
| Trimethylphosphine | 0.33 |
| Trimethylphosphite | 0.42 |
| Triphenylarsine | 0.38 |
| Triphenylphosphine | 0.39 |
| Triphenylphosphite | 0.58 |
| Triphenylstibine | 0.38 |
| Tri-*n*-Butylphosphine | 0.29 |
| Tri-*n*-propylphosphine | 0.34 |
| trans-1,2-diamminocyclohexane | 0.09 |
| trans-1,2-bis(4-pyridyl)ethene (BPE) | 0.26 |
| tert-Butylisonitrile | 0.36 |
| tert-butylthiolate (1−) | −0.55 |
| Vinyl imidazole | 0.13 |
| Water | 0.04 |

[a]Selected from the more complete Table 1 in Reference 5. The reliability of a given datum depends upon from how many complexes it was derived. The original publication[5] should be consulted. An up-to-date listing of $E_L$(L) can be found on the internet at
http://www.science.yorku.ca/chem/profs/blever/elparam98.htm

about −0.59 for $OH^-$ to 0 for saturated amines, 0.2–0.3 for unsaturated amines, 0.3–0.5 for phosphines and phosphites, and 0.99 V for CO. In general, the values decrease with increasing basicity and increase with increasing Lewis acid or acceptor character. Relationships have been developed between Hammett substituent constants [$\rho(R)$] and the $E_L$ parameters such that $E_L$(L) parameters can now be calculated for an extremely large number of variously substituted ligands[112] (Table 7), viz. ($R_p$ is the so-called reaction parameter):

$$E_L(RL) = 2.303(RT/nF)R_p\rho(R) + E_L(L_H) = f'\rho(R) + E_L(L_H) \qquad (39)$$

where $L_H$ is the parent ligand.

The Lever ligand electrochemical parameter theory is quite general, and certainly not restricted to ruthenium, thus for any other metal couple, or for a different redox couple of ruthenium, we can write:

$$E_{1/2}[M^{n+1}/M^{n+}] = S_M \sum E_L(L) + I_M \qquad (40)$$

**Table 7 Hammett Reaction Parameters for Deriving $E_L(L)$ Parameters of Substituted Ligands (Equation 39)**

| Ligand | $f^a$ | $E_L(L_H)$ | $R_p{}^b$ | Correlation Coeff. $R$ | No. of Points |
|---|---|---|---|---|---|
| phosphines – $\sigma^{*(c)}$ | 0.17 | 0.35 | 2.9 | 0.95 | 10 |
| pyridines – $\sum\sigma^{(d)}$ | 0.13 | 0.24 | 2.2 | 0.95 | 17 |
| bipyridines – $\sum\sigma^{(d)}$ | 0.07 | 0.25 | 1.2 | 0.99 | 16 |
| diketones – $\sigma^+$ | 0.12 | 0.01 | 2.0 | 0.98 | 13 |
| direct – $\sigma_p^{(e)}$ | 0.62 | −0.37 | 10.5 | 0.95 | 20 |
| benzoquinonediimines – $\sum\sigma^{(d)}$ | 0.17 | 0.26 | 2.9 | 0.98 | 6 |

[a]slope of Equation 39; [b]$R_p$ = reaction parameter = slope/0.059; [c]Taft parameters; [d]Hammett $\sigma_m$ and $\sigma_p$ parameters; [e]substituents directly as ligands.

The potentials of other reversible couples can be derived from the corresponding ruthenium complexes scaled by the factor $S_M$ and offset by $I_M$.

This procedure has been shown to work extremely well for a wide range of Werner type classical and organometallic species[5] (Table 8), including sandwich organometallic species,[113] and half-sandwiches[114,115] though the latter two require an $E_{1/2}[Fe^{III}/Fe^{II}]$ standard rather than an $E_{1/2}[Ru^{III}/Ru^{II}]$ standard. As discussed above, redox potentials are thermodynamic measurements and reflect the ratios of the stability constants for binding of the ligands to the metal ions in the two oxidation states. $S_M$ for a given couple reflects how the ratio for that couple compares with the ratio for the $E_{1/2}[Ru^{III}/Ru^{II}]$ couple. The slope is dependent on spin states so that for example the $S_M$ value for high spin $E_{1/2}[Fe^{III}/Fe^{II}]$ is different from that for low spin $E_{1/2}[Fe^{III}/Fe^{II}]$, etc. The $I_M$ parameter can be thought of as being a composite of contributions from the difference in free energy of solvation for the pair of ions concerned (differential solvation energy), from the gas phase redox potential for the free ions involved, and trivially from the reference electrode used. The differential solvation energy term also appears to influence $S_M$ since, for example, the slopes for the $Ru^{III/II}$ couples in organic solvents and in water differ appreciably. This field is still young and there is much to learn in extracting useful chemical information from $S_M$ and $I_M$ values.

Because of differential solvation contributions one might have anticipated that $[ML_6]^{n+}$ and $[ML_6]^{m-}$ species with various $n$ and $m$ might have different $S_M$ and $I_M$ values for the same metal and redox system. Fortunately this does not appear to be true for organic solvents where, generally the overall charge on the complex is not a factor. However, there are exceptions where specific solvation effects do occur, as alluded to above, e.g.[12,25,26,117–121] In water, however, solvation contributions are much larger and, in general, $S_M$ and $I_M$ may depend upon $n(m)$ for a given system.[5] There are also examples of significant solvatochromic behavior with redox potentials depending critically on the organic solvent in which the measurement is made. These cannot be handled by the $E_L(L)$ model without accounting for this solvent interaction.

**Table 8 $S_M$ and $I_M$ Data[a] for Metal Centered Redox Couples in Water and in Organic Solvents[5]**

| Redox Couple | $S_M$ | $I_M$ | Correlation Coeff. $R$ | No. | Solvent |
|---|---|---|---|---|---|
| Cr(III)/Cr(II) (LS) | 1.18(0.06) | −1.72(0.15) | 0.98 | 18 | organic |
| Cr(III)/Cr(II) (LS) | 0.575(0.04) | −1.12(0.05) | 0.98 | 8 | water |
| Cr(III)/Cr(II) (HS) | 0.84(0.05) | −1.18(0.09) | 0.98 | 14 | organic |
| Cr(I)/Cr(O) | 0.52(0.02) | −1.75(0.12) | 0.97 | 39 | organic |
| Fe(III)/Fe(II) (LS) | 0.68(0.02) | 0.24(0.04) | 0.99 | 24 | water |
| Fe(III)/Fe(II) (LS) | 1.10(0.05) | −0.43(0.12) | 0.99 | 14 | organic |
| Fe(III)/Fe(II) (HS) | 0.89(0.04) | −0.25(0.04) | 0.99 | 8 | organic |
| Fe(III)/Fe(II)[5CN] | 1.61(0.12) | b | 0.99 | 5 | organic |
| Mn(III)/Mn(II)[5CN] | 0.38(0.12) | b | 0.85 | 6 | organic |
| Mn(II)/Mn(I) | 0.81(0.02) | −1.74(0.08) | 0.99 | 28 | organic |
| Mo(I)/Mo(O) | 0.74(0.03) | −2.25(0.10) | 0.99 | 24 | organic |
| Nb(V)/Nb(IV) | 0.76(0.02) | 1.24(0.02) | 0.999 | 3 | organic |
| Nb(IV)/Nb(III) | 0.75(0.01) | −0.12(0.01) | 0.999 | 3 | organic |
| Os(V)/Os(IV)[c] | 1.00(0.02) | 2.98(0.02) | 1 | 4 | organic |
| Os(IV)/Os(III)[d] | 0.86(0.03) | 1.02(0.08) | 0.98 | 22 | organic |
| Os(III)/Os(II) | 1.01(0.02) | −0.40(0.11) | 0.98 | 80 | organic |
| Os(III)/Os(II) | 1.61(0.05) | −1.30(0.11) | 0.99 | 18 | water |
| Re(IV)/Re(III) | 0.86(0.04) | 0.51(0.13) | 0.98 | 20 | organic |
| Re(III)/Re(II) | 1.17(0.06) | −0.88(0.17) | 0.97 | 22 | organic |
| Re(II)/Re(I)[e] | 0.76(0.03) | −0.95(0.13) | 0.96 | 59 | organic |
| Re(II)/Re(I)[e] | 0.27(0.07) | −1.43(0.11) | 0.76 | 12 | organic |
| Ru(III)/Ru(II) | 1.14(0.04) | −0.35(0.09) | 0.97 | 44 | water |
| Ru(IV)/Ru(III) | 0.93(0.04) | 1.65(0.07) | 0.99 | 17 | organic |
| Ta(V)/Ta(IV) | 0.79(0.05) | 0.66(0.07) | 0.985 | 5 | organic |
| Tc(IV)/Tc(III) | 1.00(0.04) | 0.65(0.09) | 0.92 | 12 | organic |
| Tc(III)/Tc(II) | 1.28(0.08) | −0.89(0.11) | 0.86 | 22 | organic |
| Tc(II)/Tc(I) | 1.42(0.12) | −2.09(0.10) | 0.94 | 14 | organic |

[a]Numbers in parentheses here and in later tables are standard deviations from regression line; LS and HS are low spin and high spin, respectively; 5CN is 5 coordinate species; [b]Indeterminate due to use of a macrocycle of unknown $E_L$(L) value; [c]from Reference 116; [d]some data tend to deviate badly from this line and may require additional parameterization to deal with synergistic interactions between ligands, e.g., $[OsBr_5(CO)]^-$ $Os^{IV/III}$ obs. +1.36 V, calc. 0.91 V vs. NHE; [e]there are two Re(II)/Re(I) lines dependent on $\sum E_L(L)$.[97] See text.

The model is remarkably successful in predicting electrochemical potentials with surprising accuracy (usually better than 50 mV). Certain conclusions can be drawn from such success. It is an additive model so that synergistic interactions between ligands (i.e., where binding ligand X to M, influences significantly the M–Y bond, and therefore might change the $E_L$(Y) value) are usually not a significant factor even in organometallic complexes[5] or sandwich complexes.[113] However such synergistic interactions must surely be present in strongly coupled metal complexes and consequently variation of the $E_L$(L) parameter could be used as a probe to explore this possibility.

Heath and Humphrey[122a] noted significant synergistic interactions in the series $[OsBr_{6-x}(CO)_x]^{n-}$ where insertion of the first CO group into $[OsBr_6]^{3-}$ generates a species, $[OsBr_5(CO)]^{2-}$, with predicted $E_{1/2}[Os^{IV/III}]$ and $E_{1/2}[Os^{III/II}]$ potentials that vary markedly from those predicted by the use of standard $E_L(L)$ parameters. In particular this species is dramatically more difficult to oxidise [from $Os^{II} \rightarrow Os^{III}$ (about 0.3 V greater) or $Os^{III} \rightarrow Os^{IV}$ (about 0.4 V greater)] than predicted by the additivity theory. Curiously, in the series $[OsBr_{6-x}(CO)_x]^{n-}$ and $[OsBr_{6-x}(CH_3CN)_x]^{n-}$, the Br $\rightarrow$ $Os^{III}$ LMCT band correlates fairly well with the $E_{1/2}[Os^{III/II}]$ potential (see Section 4.1.1) without any anomaly at $[OsBr_5(CO)]^{2-}$. The species $[RuCl_5(CO)]^{2-}$ also appears anomalously easy to reduce ($E_{1/2}[Ru^{III/II}]$).[123] There are certainly some interesting features to these carbonyl series which are as yet poorly understood. For example, the more heavily CO substituted members of these series have potentials much closer to those predicted by the $E_L(L)$ theory and indeed the deviation appears, if anything, to be in the opposite sense, i.e., they are easier to oxidize than predicted.[122b]

One can also expect some breakdown of the simple $E_L(L)$ theory in species where the $E_L(L)$ values of non-innocent (or redox active) ligands, such as quinones, may be tunable by other ligands bound to the metal center. Some evidence, as yet very limited, suggests that tuning is possible.[124] The $E_L(L)$ theory can in fact be used to search for such behavior.[125]

There are certainly examples in the literature of complexes in which *cis* and *trans* or *mer* and *fac* isomers have significantly different redox potentials, signalling that at least one of the pair will not fit the simple equation. Procedures exist[5] to correct for stereochemistry in such cases, being based on analyses carried out earlier by Bursten; these are discussed above.[1] They add some complexity to the model and are not further discussed here (but see[108]).

Pombeiro has explored the relationship which exists between the $E_L(L)$ and $P_L$ parameters and derived values for both sets of parameters for a wide range of organometallic species. His group has especially studied variously substituted carbenes, vinylidenes and isocyanides, including the problem of variation in isocyanide $E_L(L)$ value[97] possibly due to the geometry of the isocyanide linkage.[96,98,100,102–110] Carbenes are reported to have $E_L(L)$ values near 1.2, i.e., greater than a carbonyl. Vinylidenes fall in the range $E_L(L) = 0.5–0.62$.[104]

Rhenium complexes offer some unusual behavior. While the $Re^{IV/III}$ and $Re^{III/II}$ couples behave normally, the $Re^{II/I}$ couple seems to afford two $\sum E_L(L)$ versus $E_{1/2}$ regression lines, the so-called "upper line" and "lower line"[97] with different $S_M$ and $I_M$ values (Equation 40). Complexes with $\sum E_L(L) >$ ca. 2.0 fall on the "upper line," and those with $\sum E_L(L) <$ ca. 2.0 fall on the "lower line." The reasons for this are not clear. It is possible that the "true" regression line is curved. The problem may be better understood when more complexes are analyzed with $\sum E_L(L)$ in the range 1.0–2.0. Some evidence for curvature has been presented.[104]

Finally, most, but not all, systems studied to date[5] involve electron loss or gain at $d\pi^*$ orbitals rather than $d\sigma^*$ orbitals. The reason is simply that the former commonly give rise to reversible electrochemical behavior while the latter often do not. Future studies will focus more on $d\sigma^*$ systems carried out under conditions where a reversible couple is more likely to be observed, e.g., at low temperatures.

In the sections which follow, we explore possible applications of the $E_L(L)$ parameter theory showing how its predictive power can 1) provide a diagnostic of a correct assignment of electrochemical potential or optical transition energy, 2) through a regression analysis expose information concerning the bonding characteristics of a complex and 3) when it fails to agree with experiment, alert the researcher to a possibly unusual or important observation. The development is guided by the availability of experimental data, and possible correlations for which experimental data do not yet exist, will only be briefly touched upon.

The $P_L$ parameter was based upon additivity relationships within carbonyl complexes while $E_L(L)$ is based upon the $Ru^{III/II}$ potential. However, the $E_L(L)$ parameter can be used to define the additivity relationships within the carbonyl species, and hence one can expect the two parameters to be related. Indeed they are via:

$$E_L(L) = 0.82(0.015)P_L + 0.70(0.02) \quad [24 \text{ species}, \ R = 0.996] \qquad (41)$$

The regression in Equation 41 is exceptionally good with a deviation between observed and calculated of 0.04 V or less for all 24 species taken from Table 4 [where $E_L(L)$ data were independently available].

Four species were poorly behaved and not included in the correlation. These were CO ($\Delta$ 0.28), $[CF_3CO_2]^-$($\Delta$−0.21), $CN^-$ ($\Delta$ 0.14), $CH_3CN$ ($\Delta$ 0.11) where the deviation (from the current line) cited is experimental $E_L(L)$ − calculated $E_L(L)$. It is rather curious that the carbonyl ligand is poorly behaved in this correlation. Datta[91] has also explored correlation of $E_L(L)$ and $P_L$ with frontier orbital energies.

**3.4.1** ***Reduction Potentials*** These $E_L(L)$ parameters may also be employed to predict reduction potentials localized on the ligand. Thus for some reducible fragment $M^n$(LL) in a complex, $M^n$(LL)WXYZ:

$$E_{1/2}\left[LL^{0/-}(M^{n+})\right] = S_L \sum E_L(WXYZ) + I_L \qquad (42)$$

where $E_{1/2}[LL^{0/-}(M^{n+})]$ signifies the potential for reduction of $LL^0$ to $LL^-$ when bound to $M^{n+}$.

$S_L$ is a very interesting parameter.[126] Clearly, if $S_L$ is 0, the reduction potential at the M(LL) site is independent of the other ligands WXYZ, i.e., there is no information transfer from these ligands to the M(LL) site. If $S_L$ is large, then information concerning the binding of the other ligands is transmitted to the M(LL) site through the metal center. Thus $S_L$ conveys information about coupling of the WXYZ ligands to the LL ligand. Technically, this should be stated as *relative* binding of the WXYZ ligands to LL in the LL and $LL^-$ oxidation states (Table 9).

$I_L$ reflects a number of factors including the extent to which the metal perturbs the ligand reduction orbital. Thus, in the extreme situation where binding the metal center to the ligand has no effect upon its energy levels, $S_L = 0$, and $I_L$ is the reduction potential of the free ligand. Thus both $S_L$ and $I_L$ contain information about the extent of mixing between metal orbitals and the ligand reduction orbital. Methods

**Table 9 Summary of $S_L$ and $I_L$ Values[a] for Reducible M(LL) Fragments[126] vs. NHE (Equation 41)**

| Fragment | $S_L$ | $I_L$ | No. | Correlation Coeff. *R* |
|---|---|---|---|---|
| $Mo^0$(bpy) | 0.06(0.02) | −1.56(0.04) | 4 | 0.900 |
| $Mo(NO)_2$ | 0.43(0.01) | −0.35(0.02) | 9 | 0.996 |
| Os(NO) | 0.5(0.06) | −0.58(0.03) | 3 | 0.996 |
| $Os^{II}$(bpy) | 0.27(0.01) | −1.38(0.02) | 14 | 0.985 |
| $Re^I$(bpy) | 0.27(0.04) | −1.76(0.03) | 7 | 0.943 |
| $Re^I$(dmbpy) | 0.49(0.11) | −2.55(0.04) | 40 | 0.960 |
| $Ru^{II}$(NO) | 0.62(0.04) | −0.06(0.03) | 7 | 0.987 |
| $Ru^{II}$(bpy) | 0.25(0.01) | −1.40(0.03) | 30 | 0.964 |
| $Ru^{II}$(bpym) | 0.31(0.04) | −1.08(0.03) | 6 | 0.975 |
| $Ru^{II}$(biq)[b] | 0.60(0.11) | −1.28(0.01) | 6 | 0.94 |
| $Ru^{II}$(bpz)[c] | 0.41(0.05) | −1.05(0.02) | 9 | 0.96 |
| $Ru^{II}$(terpy) | 0.31(0.055) | −1.26(0.04) | 6 | 0.94 |
| Pt(ppy) (d) | 0.55(0.07) | −1.63(0.09) | 6 | 0.97 |

[a]See Equation 41; [b]$[Ru^{II}(bpy)(biq)(LL)]^{2+}$ data set;[127] [c]Data from;[128] [d]Abbreviations: ppy = 2-phenylpyridine;[129] bpz = 2,2′-bipyrazine; bpym = 2,2′-bipyrimidine; dmbpy = 4,4′-dimethyl-2,2′-bipyridine; biq = bi-isoquinoline; terpy = 2,2′,2″-terpyridine.

to extract this information in a quantitative manner are under development.[130] We may also expect that $S_L$ may no longer be a constant if the extent of mixing between the metal and the ligand which is reduced can be significantly tuned by the other coordinated ligands.

**3.4.2** ***Reduction Potentials Versus Oxidation Potentials*** It is now evident, from the electrochemical parameter treatment discussed above, that, within homologous series of complexes, $E_{ox}$ and $E_{red}$ must linearly correlate, as noted experimentally, for example, by Meyer and Rillema.[35,128,131–133] Thus, rearrangement of Equations 40 and 42 generates the equation for this linear correlation:

$$E_{1/2}\left[LL^{0/-}(M^{II})\right] = (S_L/S_M)E_{1/2}\left[M^{III/II}(L)\right] + \left[I_L - (S_L/S_M)I_M - S_L d \sum E_L(LL)\right] \quad (43)$$

In Equation 43, LL is the ligand being reduced, *d* is the denticity of that ligand and the term in square brackets is a constant for the given reducible ligand, LL. Table 10 collects some of these data. Note that the slopes do indeed equal $(S_L/S_M)$ to within experimental error.

For completion, consider the oxidation potential of ligand L in a $M^{III}$L(VWXYZ) complex and suppose that this potential, $E_{1/2}[L^{+/0}(M^{III})]$, (Equation 44) will track with the $E_L$(VWXYZ) through an equation similar to Equation 42, i.e.,

$$E_{1/2}\left[L^{+/0}(M^{III})\right] = S_{L+} \sum E_L(VWXYZ) + I_{L+} \quad (44)$$

**Table 10 Examples of Various Electrochemical Correlations and Relevant Equation Numbers**

| Complexes | Correlation | No. | Correlation Coeff. $R$ | Eqn. # | Ref. |
|---|---|---|---|---|---|
| Ru(bpz)(WXYZ) | $E[(bpz)^{0/-}(Ru)] = 0.41(0.03)E[Ru^{III/II}(bpz)] - 1.36(0.02)$ | 9 | 0.98 | 42 | 128 |
| Ru(bpym)(WXYZ) | $E[(bpym)^{0/-}(Ru)] = 0.32(0.03)E[Ru^{III/II}(bpym)] - 1.31(0.02)$ | 6 | 0.98 | 42 | 128 |
| $[Ru(bpy)_2(LL)]^{2+}$ [a] | $E[(LL)^{0/-}(Ru)] = 1.17(0.2)E[LL^{0/-}](\text{free lig}) + 1.01(0.06)$ | 5 | 0.96 | 48 | 83 |
| $[Ru(LL)_3]^{2+}$ | $E[(LL)^{0/-}(Ru)] = 1.03(0.06)E[LL^{0/-}](\text{free lig}) - 0.79(0.04)$ | 11 | 0.98 | 48 | 134 |
| $Re(LL)(CO)_3Cl$ | $E[(LL)^{0/-}(Re)] = 0.91(0.07)E[LL^{0/-}](\text{free lig}) + 0.685(0.03)$ | 5 | 0.99 | 48 | 135 |
| $[Ru(R_2\text{-bpy})_3]^{2+}$ | $E[(R_2\text{-bpy})^{0/-}(Ru)] = 0.95(0.03)E[LL^{0/-}](\text{free lig}) + 0.62(0.03)$ | 6 | 0.997 | 48 | 136 |
| $MoO(R_i^{cat})(S_2CNEt_2)_2$ [b] | $E[(R_i\text{cat})^{-/2-}(Mo)] = 1.04(0.05)E[R_i\text{cat}^{-/2-}](\text{free lig}) + 0.72(0.03)$ | 5 | 0.997 | 48 | 137 |
| $[Ru(R_2\text{-bpy})_3]^{2+}$ | $E[(R_2\text{-bpy})^{0/-}(Ru)] = 8.19(0.26)\ E_L(R_2\text{-bpy}) - 3.07(0.02)$ | 4 | 0.999 | 46 | 136 |
| $[Os(R_2\text{-bpy})_2(CO)Cl]^+$ | $E[(R_2\text{-bpy})^{0/-}(Os)] = 1.44(0.48)E[Os^{III/II}] - 3.07(0.10)$ | 6 | 0.83 | 52 | 138 |
| $[Os(R_i\text{-phen})_2(dppm)]^{2+}$ | $E[(R_i\text{-phen})^{0/-}(Os)] = 1.55(0.22)E[Os^{III/II}] - 3.44(0.02)$ | 4 | 0.98 | 52 | 127 |
| $[Os(bpy)_2XY]^{m+}$ | $E[(bpy)^{0/-}(Os)] = 0.18(0.02)E[Os^{III/II}] - 1.47(0.03)$ | 11 | 0.93 | 42 | 127 |

[a] Original dataset had some extra species which also fit the regression but which were structurally different from $[Ru(bpy)_2(LL)]^{2+}$; free ligand potentials were adjusted slightly to agree with other literature values; [b] cat $\rightarrow$ sq oxidation process.

This is a reasonable supposition, but not yet proven. Then, by analogy to Equation 43, one may write:

$$E_{1/2}\left[\mathrm{L}^{+/0}(\mathrm{M}^{\mathrm{III}})\right] = (S_{\mathrm{L}+}/S_{\mathrm{M}})E_{1/2}\left[\mathrm{M}^{\mathrm{III/II}}(\mathrm{L})\right] + \left[I_{\mathrm{L}+} - (S_{\mathrm{L}+}/S_{\mathrm{M}})I_{\mathrm{M}} - S_{\mathrm{L}+}d\sum E_{\mathrm{L}}(\mathrm{L})\right] \quad (45)$$

**3.4.3** ***Redox Processes of Homologous Series of Ligand Fragments*** Studies with substituted ligands (homologous series of ligands) reveal more subtle changes in electrochemical behaviour (or optical spectra, see Section 4) which can be analyzed as a function of substituent to reveal details of the metal–ligand binding properties.

3.4.3.1 *Reduction* Consider a homologous series of complexes such as $M(RL)X_5$ and $M(RLL)X_4$ maintaining X constant and varying R. We can write:

$$E_{1/2}\left[\mathrm{RL}^{0/-}(\mathrm{M})\right] = df\rho(R) + E_{1/2}\left[\mathrm{L}_{\mathrm{H}^{0/-}}(\mathrm{M})\right] \quad (46a)$$

$$= (df/f')E_{\mathrm{L}}(\mathrm{RL}) + C \text{ (from Equation 39)} \quad (46b)$$

where $E_{1/2}[\mathrm{L}_{\mathrm{H}^{0/-}}(\mathrm{M})]$ is the reduction potential, specific to the parent ligand metal complex, $ML_H$, in the complex $[M(L_H)X_5]^{n-}$ (or $[M(LL_H)X_4]^{n+}$) and will be a constant here, $f$ is a reaction parameter indicating the sensitivity of the function to the variation in Hammett substituent parameter, and $\rho$(R) is an appropriate set of Hammett substituent constants.

Thus, there is much literature precedent[101,138–159] for the observation that redox potentials in homologous series of complexes are linearly dependent upon a Hammett parameter $\rho$(R) (Table 11). These include data for $[M(RLL)_2X_2]^{n+}$ and $[M(RLL)_3]^{n+}$ species. These generally obey Equation 46 though in principle they should not. One can argue as to whether reduction is localized at one RLL ligand in a multi-RLL complex or whether it is not, but in any event, the constant terms in Equation 46 will not be precisely constant in these multi-RLL (or multi-RL) species. Evidently they are frequently sufficiently "constant" within the error of the experiment.

Note in passing, the fact that similar correlations exist with metal centered reduction processes as a function of ligand substituent, e.g.[160,161]

We expect that the free ligand reduction potential, $E_{1/2}[\mathrm{RL}^{0/-}]$, will also correlate with the Hammett substituent constant via:

$$E_{1/2}[\mathrm{RL}^{0/-}] = f^0\rho(\mathrm{R}) + E_{1/2}\left[\mathrm{L}_{\mathrm{H}^{0/-}}\right] \quad (47)$$

Examples with substituted bipyridines and phthalocyanines are shown in Table 11. The reduction potential at a metal complexed ligand should linearly track the free ligand reduction potential, if other ligands are held constant. Thus in a general series $[M(RL)X_5]^{n-}$ or $[M(RLL)X_4]^{n-}$ with constant M and X and varying RL

**Table 11 Redox Potentials vs Hammett Substituents for R–L Species [eV] Data vs. NHE**

| Complex Series | Redox Process Equation[a] | No. | Correlation Coeff. $R$ | Ref. |
|---|---|---|---|---|
| $[Co(Rpho)_3]^+$ [b] | $E[Co^{III/II}] = 0.09(0.01)\sigma_p - 0.34(0.03)$ | 6 | 0.96 | 151 |
| $[Co(Reto)_3]^+$ [c] | $E[Co^{III/II}] = 0.17(0.01)\sigma_p - 0.26(0.03)$ | 6 | 0.986 | 151 |
| $[Cu(Reto)_3]^+$ [c] | $E[Cu^{III/II}] = 0.11(0.01)\sigma_p - 0.51(0.02)$ | 6 | 0.98 | 151 |
| $[Fe(Rpho)_3]^+$ [b] | $E[Fe^{III/II}] = 0.11(0.01)\sigma_p - 0.18(0.03)$ | 6 | 0.98 | 151 |
| $[Fe(Reto)_3]^+$ [c] | $E[Fe^{III/II}] = 0.16(0.01)\sigma_p - 0.46(0.02)$ | 6 | 0.993 | 151 |
| $[Fe(CN)_5(4R\text{-}Py)]^{3-}$ | $E[Fe^{III/II}] = 0.11(0.01)\sigma_p + 0.47(0.01)$ | 9 | 0.98 | 140,142 |
| $[Fe(CN)_5(3R\text{-}Py)]^{3-}$ | $E[Fe^{III/II}] = 0.12(0.02)\sigma_p + 0.48(0.01)$ | 11 | 0.93 | 140,142 |
| $Mo(CO)_4(4,4R_2\text{-bpy})$ | $E[bpy^{0/-}] = 0.33(0.04)\sigma_p - 0.15(0.08)$ | 4 | 0.99 | 162 |
| $Mo(N_2\text{-}C_6H_4\text{-}X)(S_2CNMe_2)_3$ | $E[Mo^{V/IV}] = 0.31(0.02)\sigma_p + 0.75(0.02)$ | 13 | 0.98 | 93 |
| $Pt(R_2\text{-bpy})(tdt)$ [d] | $E[bpy^{0/-}] = 0.36(0.05)\sigma_p - 1.27(0.06)$ | 6 | 0.97 | 163 |
| Ru($p$R-TPP)(CO) | $E[Ru^{III/II}(TPP^+)] = 0.041(0.01)\sigma_p + 1.46(0.02)$ | 6 | 0.92 | 156 |
| Ru($p$R-TPP)(CO) | $E[TPP^{+/0}\ (Ru^{II})] = 0.060(0.01)\sigma_p + 1.045(0.02)$ | 6 | 0.95 | 156 |
| Ru($p$R-TPP)(CO) | $E[TPP^{+/0}\ (Ru^{II})] = 0.060(0.01)\sigma_p - 1.06(0.01)$ | 6 | 0.97 | 156 |
| $[Ru(R_2\text{-bpy})_3]^{2+}$ | $E[R_2\text{-bpy}^{0/-}] = 0.16(0.01)\sigma_p - 1.07(0.06)$ | 8 | 0.98 | 164 |
| $[Ru(bpy)_2(Me,X\text{-}L^-)]^+$ [e] | $E[Ru^{III/II}] = 0.085(0.003)\sigma_p + 0.59(0.002)$ | 5 | 0.998 | 165 |
| $[Ru(bpy)_2(H,X\text{-}L^-)]^+$ [f] | $E[Ru^{III/II}] = 0.10(0.006)\sigma_p + 0.61(0.005)$ | 5 | 0.995 | 165 |
| $[Ru(pap)_2(Reto)]^+$ [c,g] | $E[Ru^{III/II}] = 0.19(0.03)\sigma_p + 1.04(0.02)$ | 6 | 0.996 | 155 |
| $Ru(EtN(O){=}N\text{-}NH\text{-}C_6H_4\text{-}p\text{-}X)_3$ | $E[Ru^{III/II}] = 0.18(0.015)\sigma_p - 0.93(0.03)$ | 5 | 0.99 | 152 |
| $Ru(EtN(O){=}N\text{-}NH\text{-}C_6H_4\text{-}p\text{-}X)_3$ | $E[Ru^{IV/III}] = 0.13(0.007)\sigma_p + 0.54(0.015)$ | 5 | 0.996 | 152 |
| $Ru(PhN(O){=}N\text{-}NH\text{-}C_6H_4\text{-}p\text{-}X)_3$ | $E[Ru^{III/II}] = 0.10(0.02)\sigma_p - 0.55(0.03)$ | 4 | 0.965 | 152 |
| $Ru(PhN(O){=}N\text{-}NH\text{-}C_6H_4\text{-}p\text{-}X)_3$ | $E[Ru^{IV/III}] = 0.11(0.003)\sigma_p + 0.75(0.003)$ | 4 | 0.999 | 152 |
| $Ru(diketonates)_3$ [h] | $E[Ru^{III/II}] = 0.38(0.025)\sigma_p - 0.207(0.10)$ | 9 | 0.98 | 157 |
| $(R_i\text{-bpy})_2Os(CO)Cl$ | $E[Os^{III/II}] = 0.18(0.02)\sigma_p + 1.40(0.05)$ | 7 | 0.97 | 138 |
| $(R_i\text{-bpy})_2Os(CO)Cl$ | $E[bpy^{0/-}] = 0.18(0.02)\sigma_p - 1.04(0.07)$ | 7 | 0.96 | 138 |
| $H_2(R_iPc)$ [i] | $E[Pc^{2-/3-}] = 0.10(0.007)\sigma_p - 0.65(0.04)$ | 5 | 0.992 | 166 |
| $Zn(R_iPc)$ | $E[Pc^{1-/2-}] = 0.10(0.008)\sigma_p + 0.99(0.03)$ | 8 | 0.979 | 166 |

**Table 11 (*Continued*)**

| Complex Series | Redox Process Equation[a] | No. | Correlation Coeff. $R$ | Ref. |
|---|---|---|---|---|
| $Zn(R_iPc)$ | $E[Pc^{2-/3-}] = 0.11(0.01)\sigma_p - 0.61(0.09)$ | 7 | 0.972 | 166 |
| $Co^{II}(R_iPc)$ | $E[Pc^{1-/2-}] = 0.14(0.01)\sigma_p + 1.16(0.10)$ | 15 | 0.96 | 166 |
| $(CN)_2Co^{III}(R_iPc)$ | $E[Pc^{1-/2-}] = 0.13(0.01)\sigma_p + 1.12(0.05)$ | 6 | 0.99 | 166 |
| $(CN)_2Co^{III}(R_iPc)$ | $E[Co^{III}/Co^{I}] = 0.11(0.007)\sigma_p - 0.58(0.03)$ | 4 | 0.995 | 166 |
| $VO(R_iPc)$ | $E[Pc^{2-/3-}] = 0.24(0.01)\sigma_p - 0.28(0.04)$ | 3 | 0.993 | 167 |
| $VO(R_iPc)$ | $E[Pc^{3-}/Pc^{4-}] = 0.25(0.01)\sigma_p - 0.62(0.02)$ | 3 | 0.998 | 167 |
| Emission | | | | |
| $Pt(R_2\text{-bpy})(tdt)$ | $h\nu[\pi^* R_2\text{-bpy} \rightarrow Pt(d)] = -0.26(0.03)\sigma_p + 1.82(0.03)$ | 5 | 0.98 | 163 |
| $Re(R_4\text{-bpy})(CO)_3(EtPy)$ | $h\nu[\pi^* R_4\text{-bpy} \rightarrow Re(d)] = -0.24(0.02)\sigma_p + 2.06(0.04)$ | 8 | 0.98 | 168 |
| Absorption | | | | |
| $(R_4\text{-bpy})Re(CO)_3(EtPy)$ | $h\nu[Re^{I}(d) \rightarrow \pi^* R_4bpy] = -0.24(0.02)\sigma_p + 3.47(0.05)$ | 8 | 0.97 | 168 |
| $(R_2\text{-bpy})Mo(CO)_4$ | $h\nu[Mo^{0}(d) \rightarrow \pi^* R_2\text{-bpy}] = -0.25(0.02)\sigma_p + 2.56(0.06)$ | 9 | 0.98 | 162 |
| $[Re(bpy)(CO)_3(R\text{-}Py)]^+$ | $h\nu[Re^{I}(d) \rightarrow \pi^* bpy] = 0.34(0.03)\sigma_p + 3.46(0.04)$ | 11 | 0.96 | 169 |

[a]Equations are based on the total Hammett substituent contribution to the molecule; for example, $3\sigma_p$(R) is used for a species $M(R\text{–}L)_3$; the intercept should correspond with the redox potential of the R = H species; [b]Rpho = $PhN(O)NNC_6H_4$-*p*-R, triazene-1-oxides; [c]Reto = $EtN(O)NNC_6H_4$-*p*-R, triazene-1-oxides; [d]tdt = toluenedithiolate; [e]$MeN(O)C(O)C_6H_4$-*p*-X, hydroxamates; [f]$HN(O)C(O)C_6H_4$-*p*-X, hydroxamates; [g]pap = phenylazo-2-pyridine; [h]$R_1.COCHCO.R_2$; [i]$R_iPc$ is a substituted phthalocyanine, with 4 to 16 substituents; also see substituted ferrocene species in Reference 104.

(or RLL) combining Equations 46 and 47:

$$E_{1/2}[\mathrm{RL}^{0/-}(\mathrm{M})] = (df/f^0)\{E_{1/2}[\mathrm{RL}^{0/-}] - E_{1/2}[\mathrm{L}_{\mathrm{H}^{0/-}}]\} + E_{1/2}[\mathrm{L}_{\mathrm{H}^{0/-}}(\mathrm{M})] \quad (48)$$

This is considered valid also for $[M(RLL)_2XY]^{n+}$ and $[M(RLL)_3]^{n+}$ if one RLL ligand is considered reduced while the others are treated as spectator ligands, XY.

Experimentally, this analysis is true for a variety of different ligands in homologous series and with $(f/f^0)$ close to unity (Table 10).[83,134,137,148,150,170] It should not be true for unrelated ligands since the right-hand terms (in Equation 48) will vary. However, included in Table 10 are datasets for $Re(LL)(CO)_3Cl$ and $[Ru(LL)_3]^{2+}$, where LL is a range of *different* diimine ligands. In this case, Equation 48 appears to be valid statistically, even though $L_H$ itself is varying. This probably arises because the various diimine ligands are similar enough that the various "constant" terms compensate in some fashion and leave the factor a constant within normal experimental limits. However this is surprising and should not be taken for granted in considering other sets of similar ligands. There is, however, no especial reason to believe it to be fundamentally true, even for a homologous series, since it must surely depend upon the extent to which the $MX_5$ fragment perturbs/mixes with the ligand $\pi^*$ orbital. It may well be a useful correlation when this perturbation is small, i.e., when mixing is not very important. Deviations might be expected where extensive mixing occurs (*vide infra*).

Compounds such as $[M(LL)X_3(RL)]^{n+}$ are complementary to those just described. There are few systematic attempts to investigate M(LL) reduction potentials as a function of RL and X. The relevant equation (from Equations 39 and 42) would simply be [for given X]:

$$E_{1/2}[\mathrm{LL}^{0/-}(\mathrm{M})] = S_\mathrm{L} f'\sigma(\mathrm{R}) + C \quad \text{(vary RL)} \quad (49)$$

The function in Equation 49 is independent of the X ligand, provided X is kept constant. However, if X does mediate and modify the coupling between RL and LL, the slope factors will not be constant. This allows for an interesting study of how modifying X (and RL) tunes the intra-ligand interactions in the molecule. Such studies are virtually unknown but the complexes $[Re(bpy)(CO)_3(R\text{-}Py)]^+$ fall into this class.[169]

3.4.3.2 *Oxidation* Table 11 contains data relating the metal centered oxidation potentials of $M(RL)_xX_y$ species to Hammett substituents. Beginning with a modification of Equation 39 and using Equation 46, the required relationship [assuming X is monodentate]:

$$\begin{aligned} E[\mathrm{M}^{n+1/n}(\mathrm{RL})] &= S_\mathrm{M} y E_\mathrm{L}(\mathrm{X}_i) + dS_\mathrm{M} x E_\mathrm{L}(RL) + I_\mathrm{M} &(50)\\ &= S_\mathrm{M} \sum E_\mathrm{L}(\mathrm{X}_i) + dS_\mathrm{M} x f'\sigma(\mathrm{R}) + dS_\mathrm{M} x E_\mathrm{L}(\mathrm{L_H}) + I_\mathrm{M} &\\ &= dS_\mathrm{M} x f'\sigma(\mathrm{R}) + C &(51)\end{aligned}$$

$S_M$ is usually close to unity, $d$ will usually be 1 or 2 and $f'$ is a small number (Table 7), so the slope here is expected to be small (Table 10).

For completion, oxidation and reduction potentials in a homologous series $[M(RL)_xX_y]^{n+}$ will correlate via (Equations 46 and 51) (Table 10):

$$E_{1/2}\left[RL^{0/-}(M)\right] = \left[f/(S_M x f')\right]E\left[M^{n+1/n}(RL)\right] + C. \tag{52}$$

**3.4.4 *The σ and π Contributions to $E_L$(L) Parameters*** Evaluation of the $E_L$(L) parameter of a newly synthesized ligand provides some insight to its electronic characteristics. Thus, generally speaking, strong Lewis donor or negatively charged ligands have small or negative $E_L$(L) values while acceptor ligands, in particular positively charged ligands, have more positive $E_L$(L) values. Preliminary studies[172] have explored factorizing $E_L$(L) into a $\sigma$- and $\pi$-donor component and a $\pi$-acceptor component which are intrinsic to the ligand and independent of any metal to which the ligand might be bound. When considering ligands which lacked $\pi$-acceptor character, i.e., purely donor ligands such as the chloride ion, the $E_L$(L) value was shown to correlate linearly with the molecular electrostatic potential, $V_{MEP}$,[171] of the ligand evaluated at the bonding site. $V_{MEP}$ is a continuous function describing the electric field arising from a molecule, generated by virtue of its charge distributions. It is calculated by determining the electrostatic interaction between a point probe of unit positive charge and the total electron density at a set of grid loci, $r_i$, in the vicinity of the molecule according to Equation 53. Here it provides a measure of the combined $\sigma$- and $\pi$-donating ability of the ligand.

$$V_{MEP}(r_j) = \sum_A \left[Z_A/(|r_j - R_A)|\right] - \int \left[\sigma(r)/(|r_j - r|)\right] dr \tag{53}$$

where $Z_A$ and $R_A$ are the atomic charge and position of atom A, and $\sigma(r)$ is the total electron density function of the molecule. $V_{MEP}$ is a measure, primarily, of electron availability, and, for example, correlates with a nucleophile's basicity when evaluated at the site of an acidic proton.[173]

Using this function, $E_L$(L) values of non-$\pi$-accepting ligands can be reproduced by:

$$\begin{aligned} E_L(L) &= a_0 + a_1 V_{MEP} \\ &= 0.246 + 0.056 V_{MEP} \text{ (correlation coeff.} = 0.96 \text{ for 13 ligands)} \end{aligned} \tag{54}$$

Representative data are shown in Table 12. The $\pi$-acceptor capability was modelled using second order perturbation theory. Thus, the back-bonding ability of a ligand is related to the energy of interaction, $\Delta E$, of the metal center with all the unoccupied ligand orbitals of appropriate symmetry. The approximate energy of interaction between filled orbitals on atom A with unoccupied orbitals on atom B, can be written

**Table 12 Experimental $E_L$(L) Values Compared with Calculated Values Derived from Ligand Molecular Electrostatic Potentials ($V_{MEP}$) and Back-Bonding Capability ($V_b$)[172]**

| Ligand | $E_L$(L) Expt. | $V_{MEP}$ | $V_b$ | $E_L$(L) calc. |
|---|---|---|---|---|
| | | Non-$\pi$-backbonding ligands | | |
| $F^-$ | −0.42 | −11.1 | 0 | −0.38 |
| $H^-$ | −0.3 | −9.97 | 0 | −0.33 |
| $[HCO_2]^-$ | −0.3 | −9.28 | 0.035 | −0.25 |
| $NCO^-$ | −0.25 | −8.85 | 0.034 | −0.23 |
| $Cl^-$ | −0.24 | −8.2 | 0 | −0.24 |
| $Br^-$ | −0.22 | −7.03 | 0 | −0.18 |
| $[ox]^{2-}$ | −0.17 | −8.41 | 0.042 | −0.19 |
| $[CF_3CO_2]^-$ | −0.15 | −8.33 | 0.042 | −0.19 |
| $\underline{N}CS^-$ | −0.06 | −6.38 | 0.085 | −0.04 |
| $H_2O$ | 0.04 | −2.45 | 0.001 | 0.05 |
| $iPrNH_2$ | 0.05 | −3.28 | 0.056 | 0.08 |
| $NH_3$ | 0.07 | −3.09 | 0.034 | 0.06 |
| $PhCH_2NH_2$ | 0.14 | −3.01 | 0.047 | 0.08 |
| | | $\pi$-backbonding ligands | | |
| $CN^-$ | 0.02 | −8.59 | 0.165 | −0.04 |
| py | 0.25 | −2.86 | 0.134 | 0.21 |
| pyz | 0.33 | −2.57 | 0.308 | 0.44 |
| 3,5-$Cl_2$py | 0.33 | −2.37 | 0.164 | 0.27 |
| $CH_3CN$ | 0.34 | −2.41 | 0.228 | 0.35 |
| PhCN | 0.37 | −2.3 | 0.238 | 0.37 |
| $P(pTolP)_3$ | 0.37 | −1.57 | 0.192 | 0.34 |
| $PPh_3$ | 0.39 | −1.5 | 0.195 | 0.35 |
| $P(OMe)_3$ | 0.42 | −0.59 | 0.358 | 0.61 |
| p-$(CN)_2$Ph | 0.49 | −2.09 | 0.262 | 0.41 |
| CO | 0.99 | −1.51 | 0.633 | 0.92 |

(assuming interaction is weak):

$$\Delta E = \sum_i^{\text{occ}} \sum_j^{\text{unocc}} \left[ 2\left( \sum_\mu \sum_u c_{\mu i} c_{uj} \beta_{\mu u} \right)^2 / (|\varepsilon_i - \varepsilon_j|) \right] \tag{55}$$

where $\varepsilon_i$ and $\varepsilon_j$ are the MO eigenvalues for the respective fragments, $c_{\mu i}$ and $c_{uj}$ are the coefficients of orbital $\theta_\mu$ in molecular orbital ion fragment A and $\theta_u$ in molecular orbital $j$ on fragment B, and $\beta_{\mu u}$ is the resonance integral between $\theta_\mu$ and $\theta_u$.

An arbitrary probe atom, the chloride ion, was used to act as a generic $\pi$-donor metal atom. The chloride ion was chosen because of the accuracy in determining its atomic properties using the AM1 level of theory. The use of this arbitrary probe is analogous to the use of a point positive charge in probing the $V_{MEP}$. Since the ligand properties being evaluated are independent of the metal center, any choice of probe atom could have been made and a different choice should only scale the output values. Using an M–L bond lying along the $z$ axis, only $p_x$ and $p_y$ orbitals on the ligand will have $\pi$-symmetry and be capable, in principle, of accepting $\pi$-electron density from the metal. The perturbation, $V_b$, can now be written:

$$V_b = \sum_i^{\text{unocc}} \left[ \left( \sum_\mu c_{\mu i} \beta_{\mu p x} \right)^2 + \left( \sum_\mu c_{\mu i} \beta_{\mu p y} \right)^2 / (|\varepsilon_i - \varepsilon_p|) \right] \tag{56}$$

where the coefficients of the probe orbital are assumed to be unity (see text following Equation 55 for meaning of terms). The outer summation is expressed over all unoccupied valence orbitals of the ligand (of appropriate symmetry). $\varepsilon_i$ are the energies of the ligand $p_x$ and $p_y$ orbitals and $\varepsilon_p$ is the energy of the probe (chloride) orbital. The quantity $V_b$ then correlates with the $\pi$-accepting capability of a ligand such that the total $E_L(L)$ value can be expressed as (all data in eV):

$$\begin{aligned} E_L(L) &= a_0 + a_1 V_{MEP} + a_2 V_b \\ &= 0.175 + 0.050 V_{MEP} + 1.290 V_b \\ &\quad \text{(regression coeff. } = 0.98 \text{ for 24 ligands)} \end{aligned} \tag{57}$$

In Table 12, it can be seen that good $\pi$-accepting ligands have relatively large $V_b$ values, and poor or non-$\pi$-accepting ligands have very small values of $V_b$, as anticipated. Clearly, using this method, the $E_L(L)$ of a newly synthesized ligand can be evaluated and factorized into its bonding components without necessarily carrying out any electrochemical measurement. It would surely be prudent, however, since this model has not yet been thoroughly tested, to ensure that the predicted $E_L(L)$ value agrees with an experimental electrochemical value. Future studies will seek to correlate the $V_{MEP}$ and $V_b$ values with other chemical and physical properties of the ligands.

**3.4.5** ***Future Developments*** The $S_L$ parameter is a measure of inter-ligand communication mediated by the metal center (Section 3.4.1). This concept of communication could be extended to consider chains of compounds and the electrochemical consequences of varying one component of the chain upon another as measured by changes in $S_L$. This information could be combined with the optical spectroscopic features of charge transfer between the two communicating components and with high level computation to broaden our understanding of electron transport along molecular paths. For example, $S_L$ should correlate with metal to bridge electronic coupling matrix elements.[174] It should also prove possible to relate these data to the kinetics of electron transport over molecular chains.

The $S_M$ parameter is a little more complex as it does contain a measure of the binding of ligands to two oxidation states, relative to their binding to Ru(II) and Ru(III). Given that the $E_L(L)$ parameters form a set independent of the metal center, we apparently face some dilemmas with respect to our ideas on hard and soft behavior. It is a consequence of the theory that, with $S_M$ tending to be near unity, and especially not changing in a dramatic fashion from one redox couple to another, the relative binding of soft ligands to a hard Lewis acid couple such as Nb(V)/Nb(IV) is very similar to the relative binding to a soft Lewis acid couple such as Mo(I)/Mo(0). The consequences of this have yet to be properly understood.

The factorization of the $E_L(L)$ parameter into $\sigma$- and $\pi$-components promises us the possibility of an empirical method to deduce $E_L(L)$ parameters by theory. Perhaps more importantly, it permits these properties to be distinguished numerically and should lead to an assessment of their individual contributions to the chemical reactivity of the ligand both free and bound to a metal center.

## 4 CHARGE TRANSFER SPECTROSCOPY

In addition to the long established *d–d* spectroscopy of transition metal ions[175–178] a variety of charge transfer (CT) transitions[179] can take place. These involve excitations from one atom to a different atom in the same molecule. Thus in a general species such as $M(L)L'X_2$ where, say, L and L′ are different unsaturated polyatomic species and X is a monoatomic species, we can theoretically anticipate in the near infrared, visible and near UV regions, the possibility of transitions of the following types: excitation from M $\rightarrow$ L and M $\rightarrow$ L′ (metal to ligand charge transfer, MLCT), X $\rightarrow$ M, L $\rightarrow$ M and L′ $\rightarrow$ M (ligand to metal charge transfer, LMCT), X $\rightarrow$ L, X $\rightarrow$ L′ and L $\rightarrow$ L′ (ligand to ligand charge transfer, LLCT), where additional distinctions can be drawn between excitation of $\sigma$ or $\pi$ electrons on X, L and L′. The acceptor orbitals on the unsaturated ligands will commonly be $\pi^*$ orbitals.

These are generally electronically allowed transitions with molar absorption coefficients ($\varepsilon$) ranging from about $1 \times 10^2$ to $4 \times 10^4$ (or more) L mol$^{-1}$cm$^{-1}$. The low intensity limit refers to CT transitions with poor overlap between the donor and acceptor orbital locations but which are otherwise electronically allowed. The corresponding spin-forbidden transitions are rarely observed in absorption but can often be identified in luminescence experiments.

The early definitive work in this field was stimulated by Jørgensen[180] who recorded the charge transfer spectra of many low spin transition metal hexahalides $[MX_6]^{n-}$. He noted various patterns of behavior which laid the basis of much of what we know today. Fundamentally, he noted the existence of four regions of absorption corresponding to transitions from the $\sigma$- and $\pi$-MOs localized on the six halides, to vacancies in the $t_{2g}(\pi)$ and $e_g(\sigma^*)$ levels of the metal. Transitions between $\sigma$ and $\pi$ levels were shown to be weak due to poor overlap, while the set of transitions terminating in the $t_{2g}(\pi)$ sub-shell disappeared for $d^6$ and higher d configurations since there was no longer a vacancy to excite to in this subshell.[175,177]

At high resolution, e.g., in a low temperature single crystal, far more than 4 electronic transitions may be observed. For example, the set of six halides has 12 lone pairs with $\pi$-symmetry transforming as $t_{1u} + t_{2u} + t_{1g} + t_{2g}$ in $O_h$ symmetry (see Chapter 1, Volume 1). As an example, consider a low spin $d^5$ hexahalide, with $^2T_{2g}$ ground state which would have allowed transitions to excited states of $A_{2u}$, $E_u$, $T_{1u}$ and $T_{2u}$ symmetry given that the dipole moment operator transforms as $T_{1u}$ in $O_h$ symmetry.[177–182] Transitions from the $t_{1u}$ and $t_{2u}$ $\pi$-MOs to the hole in the $t_{2g}$ shell, generate $A_{1u} + A_{2u} + 2E_u + 2T_{1u} + 2T_{2u}$. Thus the single transition in the Jørgensen analysis as $\pi - X_6 \rightarrow d(t_{2g})$ actually contains seven allowed transitions in the $d^5$ case. These will lie close together but may be partially or wholly resolved under appropriate high resolution conditions. We do not pursue this argument in detail here since under the usual routine conditions of room temperature solution spectroscopy, these various transitions would be observed as one broad absorption band. An additional complication of the heavier ligands, such as bromide and iodide, is the possible, and indeed usual, splitting of charge transfer bands into several components due to spin-orbit coupling.[175–178,180] Heavy metal atoms, such as osmium, also frequently give rise to transitions which are split by spin-orbit coupling.[175,177]

Of more direct relevance to this chapter is the notion that there are, of course, five $d$-orbitals, and, depending upon their occupation, more than one $d \rightarrow$ L MLCT transition. Similarly there can be more than one L $\rightarrow d$ LMCT transition. The latter case has been barely explored from the viewpoint of electrochemistry since only rarely has the ligand oxidation potential been systematically studied. The former situation is much more common. In analyzing relationships which may exist, in a specific complex or series of complexes, between optical and electrochemical data, the possible charge transfer transitions and their assignments, must first be assessed.

As an example where the procedure is typical of any problem which may be relevant, consider the $M(NH_3)_4(DAB)$ where DAB = diazabutadiene is a simple model for a bidentate $\pi$-acceptor ligand. Further, assume the $NH_3$ groups to be axially symmetric leading to the point group $C_{2v}$ for this species. We wish to know how many CT bands to expect and their relative intensities and assignments. Assuming the framework shown in Figure 5, the $t_{2g}$ orbitals of the $O_h$ group would transform as $(xy)a_2 + (yz)b_2 + (x^2 - z^2)a_1$, where the real $d$ orbital labels in the framework of Figure 5, are also indicated.

The lowest lying $\pi^*$ orbital on the DAB ligand has $b_2$ symmetry and can obviously combine in- and out-of-phase with the $yz$ orbital to provide the mechanism for $\pi$-back donation. The $(x^2 - z^2)a_1$ orbital lies in the M-DAB plane and has pseudo $\sigma$-symmetry[‡] with respect to DAB. The $(xy)a_2$ orbital has sideways $\delta$-symmetry with respect to the DAB ligand and can interact with a higher lying $a_2$ DAB($\pi^*$) orbital (LUMO + 1). Given the very different interactions possible for these three orbitals, we can expect that they will have different energies, but their order and their energy differences cannot be reliably predicted without calculation. In a very early paper, Magnuson and Taube[183] proposed that in a situation like this, when only one $d$ orbital ($b_2$ in this case) is significantly involved in $\pi$-back donation, then this $d$ orbital is stabilized by this bonding interaction to lie below the other two orbitals assumed to be only weakly interacting with the ligands. Many authors have assumed this to be the case in subsequent studies of parallel systems. However it is certainly an over simplification. Depending on the nature of the $\pi$-acceptor ligand, there will be lower-lying filled $\pi$-orbitals which may have similar energies to the $d(t_{2g})$ orbitals. These will raise the energy of the $d$-orbitals of the same symmetry as the $\pi$-levels

‡ This orbital has $\sigma$-symmetry in the sense that it lies in the $xz$ plane, perpendicular to the $\pi$-orbitals of the diimine ligand. However it does not lie along the M–N $\sigma$-bonding axis, hence the term pseudo-$\sigma$.

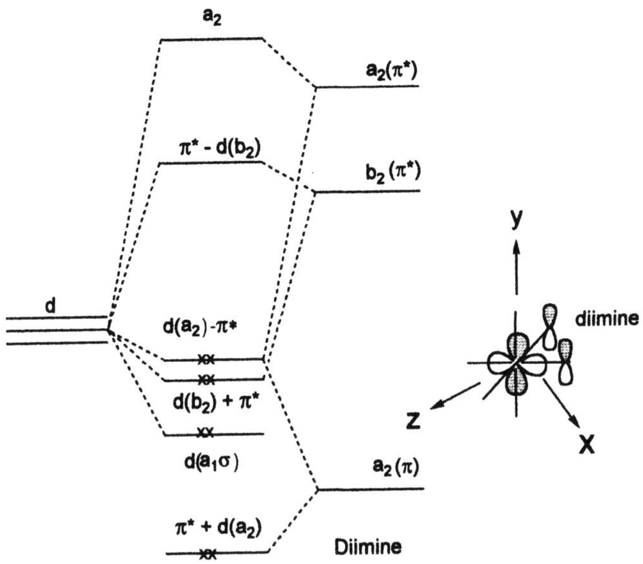


**Figure 5.** A qualitative MO diagram for a $d^6$ $ML_4$ (diimine) species (e.g., diimine = diazabutadiene or *o*-benzoquinonediimine). The $z$ axis bisects the diimine N=C–C=N bond and the diimine plane is $xz$.

concerned such that the order of the split $t_{2g}$ levels will vary from one system to another and the strongly $\pi^*$-interacting $d$-orbital may even be the HOMO.[133,184] For example in this specific DAB case, HOMO-3 is a bonding combination composed of a DAB $a_2\pi$-orbital mixed with the $a_2$ metal $d$-orbital, while the HOMO is the corresponding anti-bonding combination. HOMO-6 is another filled $\pi$-orbital of $b_2$ symmetry.

From Figure 5, one might conclude that transitions from the "$t_{2g}$" subshell to the LUMO $\pi^*$ level, if allowed, would all cluster together and be separated approximately by the extent of splitting of the "$t_{2g}$" subshell. This is not necessarily the case. Consideration of Equation 28 shows that the transition energy is determined not just by the difference in orbital energies but also by the $J$ and $K$ electron interaction contributions. Differences in configurational interaction with other states are also a factor. In the situation described here, there will be an intense transition, $\pi d(b_2) \rightarrow \pi^*(b_2)$, described in more detail as $[\pi d(b_2) + \pi^*(\text{LUMO, DAB})]$ to the LUMO $[\pi^*(\text{LUMO, DAB}) - \pi d(b_2)]$ where the extent of mixing between the $\pi d(b_2)$ and $\pi^*(b_2)$ orbitals will vary from one system to another. The transition will be intense because these two orbitals will overlap well in space. The transition from the $\pi d(a_2) \rightarrow \pi^*(b_2)$ orbital, on the other hand, will be very much weaker because of much poorer overlap, while the remaining possible transition, $\sigma d(a_1) \rightarrow \pi^*(b_2)$, will have vanishing intensity due to orthogonality. When the electronic transition occurs between two orbitals which are extensively mixed, electroabsorption spectroscopy experiments[185] have revealed that there is very little net charge transfer. Formally an electron has moved from the metal to the ligand, but because of the extensive metal $d$-ligand $\pi^*$ mixing, the charge does not move far from the metal center, and the $J_{d,L}$ and $K_{d,L}$ terms will be relatively larger. In particular, $K_{d,L}$ will be much larger than the corresponding term in the $\pi d(a_2) \rightarrow \pi^*(b_2)$ [or $\sigma d(a_1) \rightarrow \pi^*(b_2)$] transition, both of which have a much greater degree of charge transfer character. This effect, via Equation 28, causes the the $\pi d(a_2) \rightarrow \pi^*(b_2)$ to lie at much lower energy than $\pi d(b_2) \rightarrow \pi^*(b_2)$, even though the splitting of the "$t_{2g}$" set may be small.[133,184]

Since these transitions all involve the formal oxidation of one site and the formal reduction of another, their energies reflect the ease of such redox processes. If complexes are formed between a readily oxidizable metal and a readily reducible ligand, or vice-versa, then CT transitions between these components are expected at relatively low energies.

At a very simple level of theory a relationship has long been recognized between optical charge transfer energies and the formal redox potentials for metal oxidation (or reduction) and ligand reduction (or oxidation). We plan here to investigate the theoretical basis for such correlations and to illustrate this theory with a range of relevant examples.

## 4.1 Theoretical Basis; Absorption [MLCT Transitions]

In the development which follows, we identify the various contributions to the optical or electrochemical energies in some detail, recognizing, however, that not all are derivable by current experiment or theory. However some combinations of these parameters can be derived experimentally, and therefore some test of the theory can be carried out; no doubt in due course all will be derivable by theory.

Consider a diamagnetic metal complex, represented by $M^{II}L$, in which charge transfer can occur from M to L (an acceptor ligand), giving an excited singlet state[§] which can be formally described as $^1[M^{III}L^-]^*$ where the star indicates a Franck–Condon (FC) state, i.e., a vibrationally excited electronic state having the same nuclear coordinates as the ground state species. In the equations which follow, we designate the spin only where there is a possible ambiguity with respect to spin.

$$M^{II}L \longrightarrow {}^1[M^{III}L^-]^* = h\nu(\text{MLCT}) \tag{58}$$

This FC state can relax to the vibrationally equilibrated (relaxed) excited state, $[M^{III}L^-]$, termed as a VEqES and discussed elsewhere in this volume.[186] A relationship between these two energy states (for complexes in fluid solution), written as Gibbs free energies, is:

$$\Delta G^{0\,1}[M^{III}L^-]^* = \Delta G^{0\,1}[M^{III}L^-] + \chi_i + \chi_o \tag{59}$$

§ We use singlet state as a simple example but the theory is generally applicable to any allowed MLCT transition between states of the same spin.

where $\chi_i$ is the so-called inner-sphere reorganization energy and represents the vibrational excitation of the excited electronic state, and $\chi_0$ is the outer-sphere or solvent reorganization energy.[89] The first term, $\chi_i$, is the vibrational excitation of the FC state. The second term, $\chi_0$, arises because the solvent sheath around the excited state at the moment of electronic excitation corresponds on average to the equilibrium environment of the ground state. For many complexes which do not interact strongly with the solvent, $\chi_0$ will be small, e.g., <500 cm$^{-1}$.

Now consider the difference between the two related half-cell redox potentials:

$$\mathrm{M^{II}L} \longrightarrow [\mathrm{M^{III}L}]^+ + e^- \qquad -E_{1/2}[\mathrm{M^{III/II}(L)}] \tag{60}$$

$$[\mathrm{M^{III}L}]^+ + e^- \longrightarrow {}^1[\mathrm{M^{III}L^-}] \qquad E_{1/2}{}^1[\mathrm{L^{0/-}(M^{III})}] \tag{61}$$

A convention is followed whereby placing the electrons on the left corresponds with $+E$(process), while placing them on the right corresponds with $-E$(process); $E$(process) itself may be positive or negative according to the reaction concerned. Adding Equations 60 and 61 yields:

$$\begin{aligned} &\mathrm{M^{II}L} \longrightarrow {}^1[\mathrm{M^{III}L^-}] \\ &E_{1/2}{}^1[\mathrm{L^{0/-}(M^{III})}] - E_{1/2}[\mathrm{M^{III/II}(L)}] = -\Delta E'(\mathrm{Redox}) \end{aligned} \tag{62}$$

Equation 62 apparently corresponds with an excitation from the ground state to the VEqES.[186] Before further development, recall that the free energy is related to an electrochemical potential via $\Delta G^0 = -nFE$ and that the free energy, $\Delta G^0$, associated with Equation 62 will be positive (i.e., uphill, since energy must be input to excite the molecule). Optical transition energies are usually written as positive quantities so we define $\Delta E'$(Redox) as shown in Equation 62 so that now the several equations may be combined to write (all data in eV):

$$h\nu(\mathrm{MLCT}) = \Delta E'(\mathrm{Redox}) + \chi_0 + \chi_i \tag{63}$$

Technically, we should write $F\Delta E'$(Redox) but taking care to appreciate the sign of this quantity and writing it in eV, it can be simplified to $\Delta E'$(Redox). Here several implicit assumptions are being made. Since the charge transfer energy is an enthalpy whereas the redox potentials are Gibbs free energies, it is assumed here that there is no significant entropy difference between the ground and excited state species.[186,188] Further, it is necessary that the optical transition have a Gaussian band shape.[26] It is also necessary that the orbitals involved in the optical process are the same as those in the redox process. The solvation energies of the two equilibrated species, ground and excited electronic state, are each included in the optical and redox expressions, so that they cancel and do not appear explicitly.

Clearly, this analysis deals only with two states in isolation. It is likely that configurational interaction will mix ground and/or excited state with other excited states. Any error introduced by this is probably offset somewhat because the equilibrium redox potentials will contain the same admixtures of higher excited states. Nevertheless, some caution should be exercised in interpreting the various equations developed here.

To continue, a different pair of redox potentials can be written to yield the same result:

$$M^{II}L + e^- \longrightarrow [M^{II}L^-]^- \qquad E_{1/2}[L^{0/-}(M^{II})] \tag{64}$$

$$[M^{II}L^-]^- \longrightarrow {}^1[M^{III}L^-] + e^- \qquad -E_{1/2}{}^1[M^{III/II}(L^-)] \tag{65}$$

By adding Equations 64 and 65, Equation 62 is again obtained. The difference between these two redox potentials is also precisely equal to $-\Delta E'$(Redox). However, for each of these pairs of potentials one of the pair is a potential that usually cannot be observed in an electrochemical experiment, i.e., Equations 61 and 65. In other words, we cannot electrochemically reduce $[M^{III}L]^+$ to $M^{III}L^-$ because in most systems, the metal would be reduced first with subsequent reduction of the ligand to form $[M^{II}L^-]^-$ not $M^{III}L^-$. Later we discuss means to evaluate these particular potentials, which are not available from the electrochemistry experiment.

An early analysis using the specific potentials discussed above (Equation 63) dealt with phthalocyanine (Pc) CT spectra;[189] both LMCT and MLCT transition energies were predicted from redox potential data. In this Pc case the unobservable ligand reduction potentials, Equation 61, were estimated quite accurately from the reduction potentials of main group $M^{III}$Pc species since the Pc ring-based potentials are largely independent of the central metal, and depend only on its oxidation state and size, provided that the metal lies in the plane of the ring. This treatment was extremely successful for Pc complexes because of their large size. Thus the transfer of an electron to or from the Pc ring has very little effect upon its nuclear geometry so that $\chi_i$ is very small. Further the (organic) solvent effects on these complexes are generally small ($\chi_0$ small), as are the entropy differences between the ground and excited states.

However, as will be discussed in depth below, many authors, from as early as the 1960s (very early examples discussed in References 190,193) reported correlations between $E_{op}$(CT) and the electrochemically observable $\Delta E$(Redox). Indeed much of the basis for the development presented here is based upon the early pioneering work of Antonin Vlcek,[194–196] especially the 1968 review[194] which was truly a paper ahead of its time. In our example:

$$\Delta E(\text{Redox}) = E_{1/2}[M^{III/II}(L)] - E_{1/2}[L^{0/-}(M^{II})] \tag{66}$$

noting that the second term on the right refers to ligand reduction/oxidation attached to $M^{II}$, not $M^{III}$ as in Equation 61. In this case, as developed below, the various solvation energies do not cancel from the optical and redox terms and must be explicitly considered, but their total sum is often quite small so that the measured $\Delta E$(Redox) often gives a good approximation of the CT band energy, typically to within about 0.3 eV.

In order to understand the significance of this relationship Equation 66, consider a thermodynamic cycle as shown in Figure 6, from which the following equation can

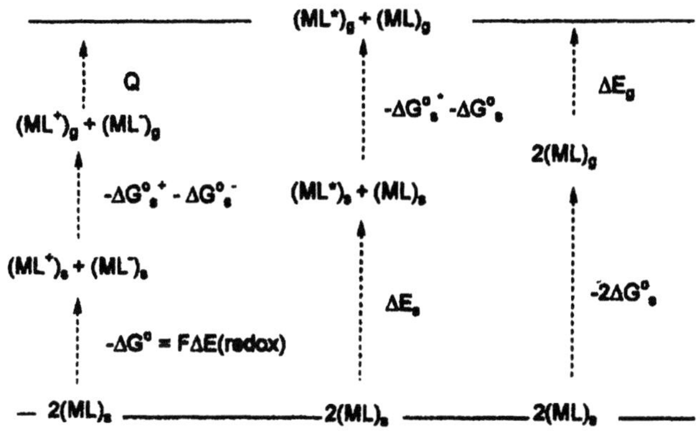


**Figure 6.** A free energy diagram relating equilibrated electrochemical and spectroscopic parameters. ML, ML*, $ML^+$ and $ML^-$ are the neutral, excited, oxidized and reduced species respectively. $\Delta G_s^0$, $\Delta G_s^{0*}$, $\Delta G_s^{0+}$ and $\Delta G_s^{0-}$ are the solvation free energies for the corresponding species. The *s* and *g* subscripts refer to solvent and gas phase respectively. $\Delta E$(Redox) is defined in the text. $\Delta E_s$ and $\Delta E_g$ are the solvent and gas phase excitation to the equilibrated excited state.

be obtained (data in eV:[35])

$$h\nu(\text{MLCT}) = \left[\chi_i + \Delta E(\text{Redox}) + \Delta\Delta G_s^0 + Q\right] + \chi_0 + \Delta_{\text{sol}}^0 \qquad (67)$$

This refers to a transition to the spin singlet state and assumes the HOMO and LUMO of the complex are involved in both the optical transition and the redox process. It also assumes that the optical transition is pure, i.e., that there is no significant configuration interaction between the CT excited state and other states of the same symmetry, nor should there be mixing with other states due to spin-orbit coupling. The entropy difference between the ground and excited states is also assumed to be negligible in the MLCT transition. Table 13 displays some exemplar data. See Reference 133 for possible important deviations from Equation 67 when there is significant metal–ligand coupling.

Equation 67 requires unit slope if $h\nu$(MLCT) is plotted against $\Delta E$(Redox). The linear regression data shown in Table 13 have slopes close to unity. In the case of the archetypical $Ru(bpy)_2XY$ data set initially studied[35] the best line has a slope of $0.87 \pm 0.05$. The population standard deviation of calculated less observed $h\nu$(MLCT) is 0.09 eV. If the slope is arbitrarily set at unity, then this specific data set which should be valid for all (bipyridine)ruthenium species where the bipyridine ligand is the most easily reducible ligand, satisfies (data in eV):

$$h\nu(\text{Ru}\, d\pi \longrightarrow \text{bpy}\pi^*) = \Delta E(\text{Redox}) + 0.21 \qquad (68)$$

Using Equation 68, the population standard deviation of calculated less observed $h\nu(\text{Ru}\ d \rightarrow \text{bpy}\ \pi^*)$ is 0.10 eV. Thus within experimental error, one may assume unit slope.

**Table 13 Summary of Regression Data—MLCT and LMCT Transitions vs Electrochemical Data**

| Complex[a] | Equation | No. | $R$ | Ref. |
|---|---|---|---|---|
| | MLCT Transitions vs. Electrochemical Data | | | |
| $Fe(II)(phen)_2(CN\text{-}R)_2$ | $h\nu[\text{Fe } d \rightarrow \pi^*(1) \text{ phen}] = 0.94\ (0.01) E_{1/2}[\text{Fe}^{III/II}] + 1.35(0.005)$ | 5 | 0.999 | 190 |
| | $h\nu[\text{Fe } d \rightarrow \pi^*(2) \text{ phen}] = 1.03\ (0.03) E_{1/2}[\text{Fe}^{III/II}] + 1.58(0.02)$ | 5 | 0.997 | 190 |
| $Fe(II)(dmgBF_2)_2X_2$ | $h\nu[\text{Fe } d \rightarrow \pi^* \text{ dmgBF}_2] = 0.87\ (0.05) E_{1/2}[\text{Fe}^{III/II}] + 1.75(0.03)$ | 4 | 0.996 | 191 |
| $[Fe(II)(CN)_5(4\text{-}R\text{-}Py)]^{3-}$ | $h\nu[\text{Fe } d \rightarrow \pi^* \text{ R-Py}] = -11.00(0.87) E_{1/2}[\text{Fe}^{III/II}] + 8.39(0.08)$ | 8 | 0.98 | 141,142 |
| $[Ru(II)(bpy)(biq)(LL)]^{2+}$ | $h\nu[\text{Ru } d \rightarrow \pi^* \text{ bpy}] = 0.44\ (0.09) E_{1/2}[\text{Ru}^{III/II}] + 1.77(0.03)$ | 7 | 0.91 | 192 |
| $[Ru(II)(bpy)(biq)(LL)]^{2+}$ | $h\nu[\text{Ru } d \rightarrow \pi^* \text{ bpy}] = 0.97(0.24) \Delta E(\text{Redox}) + 0.17(0.02)$ | 6 | 0.89 | 192 |
| $[Ru(II)(NH_3)_5(R\text{-}Py)]^{2+}$ | $h\nu[\text{Ru } d \rightarrow \pi^* \text{ R-Py}] = -6.70(0.64) E_{1/2}[\text{Ru}^{III/II}] + 5.07(0.08)$ | 8 | 0.97 | 148,193 |
| $[RuCl_{6-x}(PhCN)_x]^{n-}$ | $h\nu[\text{Ru } d \rightarrow \pi^* \text{ PhCN}] = 0.67(0.03) E_{1/2}[\text{Ru}^{III/II}] + 2.94(0.06)$ | 5 | 0.995 | 197 |
| $[Ru(II)(bpy)_2XY]^{n+}$ | $h\nu[\text{Ru } d \rightarrow \pi^* \text{ bpy}] = 0.87(0.05) \Delta E(\text{Redox}) + 0.54(0.1)$ | 30 | 0.95 | 35 |
| $Re(I)(R\text{-}bpy)(CO)_3Cl$[b] | $h\nu[\text{Re } d \rightarrow \pi^* \text{ bpy}] = -0.55(0.08) E_{1/2}[(\text{R-bpy})^{0/-}] + 2.67(0.08)$ | 11 | 0.94 | 198 |
| $Re(I)(R\text{-}bpy)(CO)_3Cl$[f] | $h\nu[\pi^* \text{ bpy} \rightarrow \text{Re } d] = -0.50(0.08) E_{1/2}[(\text{R-bpy})^{0/-}] + 1.77(0.08)$ | 8 | 0.93 | 198 |
| $Os(II)(bpy)_2XY$ | $h\nu[\text{Os } d \rightarrow \pi^* \text{ bpy}] = 1.21(0.09) \Delta E(\text{Redox}) - 0.50(0.1)$ | 12 | 0.97 | 127,199 |
| | LMCT Transitions vs Electrochemical Data | | | |
| $(NH_3)_5Ru(III)(R_2pycd)$ | $h\nu[\pi \text{ pycd} \rightarrow \text{Ru } t_{2g}] = 1.05(0.07) \Delta E(\text{Redox}) + 0.41(0.03)$ | 16 | 0.967 | 200 |
| $t\text{-}[Ir(IV)L_2Cl_4]$ | $h\nu[\pi \text{ Cl} \rightarrow \text{Ir}^{IV}\ t_{2g}] = -0.49(0.11) E_{1/2}[\text{Ir}^{IV/III}] + 2.84(0.02)$ | 7 | 0.89 | 201 |
| $t\text{-}[Ir(IV)L_2Cl_4]$[c] | $h\nu[\pi \text{ L} \rightarrow \text{Ir}^{IV}\ t_{2g}] = -0.46(0.03) E_{1/2}[\text{Ir}^{IV/III}] + 2.35(0.006)$ | 6 | 0.98 | 201 |
| $[OsBr_{6-x}L_x]^{n-}$ [d] | $h\nu[\pi \text{ Br} \rightarrow \text{Os}^{III}\ t_{2g}] = -0.49(0.05) E_{1/2}[\text{Os}^{III/II}] + 2.57(0.11)$ | 6 | 0.98 | 122 |
| $[RuCl_{6-x}(PhCN)_x]^{n-}$ | $h\nu[\pi \text{ Cl} \rightarrow \text{Ru}^{III}\ t_{2g}] = -0.34(0.04) E_{1/2}[\text{Ru}^{III/II}] + 3.00(0.09)$ | 6 | 0.98 | 197 |
| $[MCl_6]^{2-}$ [e] | $h\nu[\pi \text{ Cl} \rightarrow \text{M}^{IV}\ t_{2g}] = -1.28(0.09) E_{1/2}[\text{M}^{IV/III}] + 3.19(0.15)$ | 8 | 0.98 | 197 |
| $[L(Im)Ru(III)(NH_3)_4]^{n+}$ (In water) | $h\nu[\text{Im} \rightarrow \text{Ru}^{III}\ t_{2g}] = -1.17(0.08) E_{1/2}[\text{Ru}^{III/II}] + 3.00(0.01)$ | 4 | 0.995 | 202 |

[a]Some data outlying the main group of complexes were excluded; "No.:" is number of complexes in the correlation. [b]various different R-bpy ligands being reduced; [c]tentative assignment (differs from that proposed by authors), see text; [d]L is $CH_3CN$ and/or CO; potential data are corrected to NHE using the data in Table 1; [e]references includes complexes of Re, Rh, Ru, Os, Tc, Mo, W, and Ir; [f]emission.

The term in square brackets Equation 67 is independent of solvent because the solvation energies are contained both in the $\Delta\Delta G_s^0$ and the $\Delta E$(Redox) terms. $\Delta\Delta G_s^0$ is the sum of the solvation energies contained in the two redox processes, using our example:

$$\Delta\Delta G_s^0 = 2\Delta G_s^0[\mathrm{M^{II}L}] - \Delta G_s^0[\mathrm{M^{II}L^-}]^- - \Delta G_s^0[\mathrm{M^{III}L}]^+ \tag{69}$$

$\Delta_{\mathrm{sol}}^0$ is the difference in solvation energy between the equilibrated (vibrationally and solvationally) excited state, VEqES, and the ground state:

$$\Delta_{\mathrm{sol}}^0 = \Delta G_s^{0\,1}[\mathrm{M^{III}L^-}] - \Delta G_s^0[\mathrm{M^{II}L}] \tag{70}$$

$Q$ is the energy involved in transferring an electron from $[\mathrm{M^{II}L^-}]_g^-$ to $[\mathrm{M^{III}L}]_g^+$ yielding $[\mathrm{M^{II}L}]_g$ and $[\mathrm{M^{III}L^-}]_g$ (VEqES) which involves both one-electron orbital energy differences and a significant contribution from $J_{d,L}$, the coulombic repulsion energy between an electron in the ligand-based LUMO and one in the (mainly metal-based) HOMO. This can be examined using Figure 6, which depicts the energies of these various states such that:

$$Q = 2K_{d,L} - J_{d,L} + \sum E \tag{71}$$

where $\sum E$ is the net sum of changes in the HOMO and LUMO one-electron energies between the two pairs of species.¶ Without a calculation, this is difficult to assess. Since $J$ is usually substantially larger than $2\,K$, the interaction terms will make $Q$ negative but this may be partially offset by $\sum E$. It is useful also to relate $Q$ to the energy of redox self-quenching of thermally equilibrated excited state singlet ${}^1[\mathrm{M^{III}L^-}]_s^*$, in solution, as initially discussed by Meyer and co-workers, i.e., the energy of the process:

$$\begin{gathered} {}^1[\mathrm{M^{III}L^-}]_s^* + [\mathrm{M^{II}L}]_s \longrightarrow [\mathrm{M^{III}L}]_s^+ + [\mathrm{M^{II}L^-}]_s^- \\ \Delta G^0 = -F\{E_{1/2}[\mathrm{L^{0/-}(M^{II})}] - E_{1/2}{}^1[\mathrm{L^{0/-}(M^{III})}]\} = D^S \end{gathered} \tag{72}$$

This experimental observable is related to $Q$ by:

$$D^S = -[Q + \Delta\Delta G_s^0 + \Delta_{\mathrm{sol}}^0] \tag{73}$$

The discussion which follows assumes the involvement of low spin $d^6$, diamagnetic $\mathrm{Ru^{II}}$, and low spin $d^5$, $S = 1/2$, paramagnetic $\mathrm{Ru^{III}}$ species since most authors have focused on these systems. The principles will be general for all metal ions, but some details of the equations may vary where other spin states need to be considered.

Other authors, in particular Meyer and DeArmond along with their groups, have investigated the relationships between optical CT transitions and redox potentials.[12,26,203] They generally use different terms, mainly to break down $Q$ or to

¶ More accurately the net change in the energies of all orbitals in each species should be considered.

approximate it. In both cases their equations were also written specifically for $Ru^{II}$ complexes. Meyer includes the $T\Delta S$ term explicitly, and uses first order perturbation theory to approximate $Q$ in terms of the change in the extent of mixing of the donor and acceptor orbitals. However, this approach is of limited usefulness as the various terms are not generally known quantities. Thus Meyer's equation[12] is:

$$h\nu(\text{MLCT}) = \chi_i + \chi_0 + \left[1/(1 - \Delta\alpha)\right]\left[\Delta E(\text{Redox}) + e^2/D_s d + T\Delta S\right] \quad (74)$$

The quantity $e^2/D_s d$ is an electrostatic term which assumes that the complex can be treated as two spheres separated by distance $d$ in a medium of dielectric constant $D_s$; $\Delta\alpha$ is the change in extent of mixing of the electron donor and acceptor orbitals before and after charge transfer. The sum $[-\Delta\alpha/(1 - \Delta\alpha)]\Delta E(\text{Redox}) + [1/(1 - \Delta\alpha)]e^2/D_s d]$ is approximately equivalent to $Q$ in Equation 67. The $1/(1 - \Delta\alpha)$ term makes it clear that the slope of this correlation may not be exactly unity because, in effect, $Q$ is dependent to some extent on $\Delta E(\text{Redox})$ as alluded to above. Comparison with Equation 68 shows that this treatment neglects most of the solvation energy terms, $\Delta\Delta G_s^0$ and $\Delta_{\text{sol}}^0$, and only includes $\chi_0$, the solvent reorganization energy associated with the electronic transition. The $e^2/D_s d$ and $T\Delta S$ terms may be relatively small.[12]

DeArmond and co-workers have also reported equations relating both triplet (emission) and singlet (absorption) electronic transition energies, in ruthenium complexes, to $\Delta E(\text{Redox})$.[203] For the singlet state, analogous to Equation 68 above, these authors have essentially broken down $Q$ into $\alpha$, $2K$, $X_1$ and $X_0$. $\alpha$ is the difference in coulomb repulsion energy of an $L^-$ electron with a $Ru^{II}$ or a $Ru^{III}$ core, $K$ is the exchange energy, $X_1$ is the charge transfer resonance energy of the excited singlet state and $X_0$ is the sum of the charge transfer interaction energies for the triplet state. The last two terms both relate to the changes in orbital energies of the two relaxed excited states due to the attraction between the formally negatively charged ligand and the $Ru^{III}$ core. The solvation energy difference is given as $2S$, which is almost equivalent to $\Delta\Delta G_s^0$ in Equation 69, except that it refers to the triplet state (i.e., it is not exactly the correction required) and $\Delta_{\text{sol}}^0$ is neglected. DeArmond noted[203] that the two relevant potentials, of reducing L bound to $M^{II}$ or to $M^{III}$, are connected by:

$$-F\left\{E_{1/2}{}^3\left[\text{L}^{0/-}(\text{M}^{\text{III}})\right] - E_{1/2}\left[\text{L}^{0/-}(\text{M}^{\text{II}})\right]\right\} = -(\alpha - 2S) = -D^T \quad (75)$$

where $D^T$ is similar to $D^S$ in Equation 73 but refers to the thermally equilibrated triplet excited state (also see Section 4.3).[204] Consideration of these potentials shows that $D$ must be positive.

### 4.1.1 *Dependence of MLCT Energy upon Oxidation or Reduction Potentials*

Given the connection between oxidation and reduction potentials via the $E_L(L)$ parameters, as discussed in Section 3.4, it is obvious that MLCT energies should correlate both with metal centered oxidation potentials and ligand centered reduction potentials. Such information is of value in circumstances when only one of these is experimentally available so that a $\Delta E(\text{Redox})$ correlation is not obtainable. The slope of such a correlation is related to the $S_M$ and $S_L$ values and so provides a mechanism to obtain such data when the relevant electrochemical data may not be available. Table 13 displays a collection of such data.

For example, in a general species M(LL)(WXYZ), the observed $h\nu(M \rightarrow LL)$ MLCT band should correlate with $E_{1/2}[M^{n+1/n}(LL)]$ as (from Equations 38 and 42)

$$h\nu(M \longrightarrow LL) = (S_M - S_L)E_{1/2}\left[M^{n+1/n}(LL)\right] + C \qquad (76)$$

Since $(S_M - S_L)$ may be a small number, a fairly large dataset is desirable to deduce a reliable value of $S_M$ or $S_L$. These correlations can also be used to derive an oxidation (or reduction) potential which may not be otherwise measurable. For example, in a species $M^{II}(LL)(bpy)X_2$, (LL) may be oxidized to $(LL^+)$ before $M^{II}$ is oxidized to $M^{III}$. Yet it is often of interest, in mapping energy levels, to know the $E_{1/2}[M^{III/II}(LL)]$ potential rather than $E_{1/2}[M^{III/II}(LL^+)]$. Thus, for example,[35] (data in eV):

$$E_{1/2}\left[M^{III/II}(bpy)\right] = 0.65[h\nu(M \longrightarrow bpy)] + 2.0 \qquad (77)$$

Other similar equations may be derived from the data in Table 13. In a general complex M(LL)(WXYZ), as $\sum E_L(WXYZ)$ becomes more positive, the metal becomes more difficult to oxidize, $E_{1/2}[M^{n+1/n}(LL)]$ increases and $h\nu[M \rightarrow LL]$ will blue-shift. Thus the slope of the correlation between $h\nu[M \rightarrow LL]$ with $E_{1/2}[M^{n+1/n}(LL)]$, varying WXYZ, is positive (see Table 13). At first sight then, it is counter-intuitive that in a series M(RL)(WXYZ), varying RL, the corresponding slope is negative (Table 13). As the metal becomes more difficult to oxidize the MLCT transition $h\nu[M \rightarrow RL]$ red shifts. This result can be appreciated by combining Equations 52 and 67 [treating all the terms in Equation 67 as constant except $\Delta E$(Redox)] to yield:

$$h\nu[M \longrightarrow RL] = E_{1/2}\left[M^{n+1/n}(RL)\right]\left[1 - (f/S_M x f')\right] + C \qquad (78)$$

Reference to the data in Table 10 shows that the term $(f/S_M x f')$ commonly exceeds unity, and thus the slope in Equation 78 is negative.

## 4.2 LMCT Transitions

The theory for an LMCT transition is fundamentally the same as for MLCT. Commonly such transitions are seen with anionic ligands, such as $Cl^-$ or $S^{2-}$, but for simplicity here, we treat the ground state ligand as the neutral L. Thus for a transition to the equilibrated excited state (VEqES):

$$h\nu\left[M^{III}L \longrightarrow M^{II}L^+\right] \qquad (79)$$

$$M^{III}L + e^- \longrightarrow M^{II}L \qquad E_{1/2}\left[M^{III/II}(L)\right] \qquad (80)$$

$$M^{II}L \longrightarrow M^{II}L^+ + e^- \qquad -E_{1/2}\left[L^{+/0}(M^{II})\right] \qquad (81)$$

and summing:

$$\mathrm{M^{III}L} \longrightarrow \mathrm{M^{II}L^{+}} \quad E_{1/2}[\mathrm{M^{III/II}(L)}] - E_{1/2}[\mathrm{L^{+/0}(M^{II})}] \tag{82}$$

Thus, for the Franck–Condon state (eV):

$$\begin{aligned} h\nu(\mathrm{LMCT}) &= -F\{E_{1/2}[\mathrm{L^{+/0}(M^{II})}] - E_{1/2}[\mathrm{M^{III/II}(L)}]\} + \chi_i + \chi_0 \\ &= -F\{E_{1/2}{}^{1}[\mathrm{L^{+/0}(M^{III})}] - E_{1/2}[\mathrm{M^{III/II}(L)}]\} - D^S + \chi_i + \chi_0 \quad (83) \\ &= \Delta E''(\mathrm{Redox}) - D^S + \chi_i + \chi_0 \quad (84) \end{aligned}$$

where $\Delta E''$(Redox) is defined as in Equation 83. Usually, in this situation, one is studying a higher oxidation state complex, such as a [$\mathrm{M^{III}L}$] species so that $\Delta E''$(Redox) is the difference in potential between reducing the species to [$\mathrm{M^{II}L}$] and oxidizing it to a ligand radical [$\mathrm{M^{III}L^{+}}$] species. However, ligand oxidation, $E_{1/2}{}^{1}[\mathrm{L^{+/0}(M^{III})}]$, is commonly not observable or is irreversible and usually one relates the LMCT transition to the observable reduction potential, $E_{1/2}[\mathrm{M^{III/II}(L)}]$ in this example. There is one example listed in Table 13 and indeed the slope is essentially unity. Combining Equation 45 with Equation 83 leads to a simple equation:

$$h\nu(\mathrm{LMCT}) = [(S_{\mathrm{L+}}/S_{\mathrm{M}}) - 1]E_{1/2}[\mathrm{M}^{n+1/n}(\mathrm{L})] + C \tag{85}$$

With $S_{\mathrm{M}}$ close to unity and $S_{\mathrm{L+}}$ not directly available but likely to be similar to $S_{\mathrm{L}}$ and hence less than unity, the slope here will be negative. Indeed a negative slope is anticipated since as the oxidation potential of the metal increases, LMCT transition should move to lower energy. This analysis indeed provides a mechanism to evaluate $S_{\mathrm{L+}}$. Relationships between LMCT data and various observable redox potentials are shown in Table 13.

### 4.3 OSCT Transitions

Outer sphere charge transfer (OSCT), sometimes called ion-pair charge transfer (IPCT), bands are those transitions which are often observed as a consequence of outer sphere electron transfer from the anion to the cation. If the former has strong reducing properties and the cation is an oxidant, these transitions can occur at quite low energies, e.g., at 14,000 $\mathrm{cm^{-1}}$ for the ion pair $[\mathrm{Ru(NH_3)_6}]^{3+}[\mathrm{Fe(CN)_6}]^{4-}$ (see[205]) and otherwise can be observed through the visible and ultraviolet region.[206] The basic principles outlined above for charge transfer transitions also clearly apply to OSCT transitions except that here the electronic transition occurs between two different molecules (ions). It must therefore be possible to write the OSCT energy in terms of the oxidation potential of the anion, $E[\mathrm{D/D^+}]$, and reduction potential of the cation, $E[\mathrm{A/A^-}]$, namely:[44]

$$h\nu(\mathrm{OSCT}) = F\{E_{1/2}[\mathrm{D^+/D}] - E_{1/2}[\mathrm{A/A^-}]\} + (\chi_0 + \chi_i)^D + (\chi_0 + \chi_i)^A - W \tag{86}$$

where the several reorganization energy terms are distinguished by $D$ (donor) or $A$ (acceptor) superscripts, and $W$ is a (differential, between ground and excited state) work term associated with bringing the two charged species together to form an ion pair. This term is very small, of the order of $10^{-1}$ eV.[207] Linear correlations between OSCT transition energies and the difference in oxidation potentials defined in Equation 86 have been reported.[44,208]

Recently, the photoelectron emission threshold energies for some simple anions[42,43] have been shown to correlate[44] with OSCT transitions from these anions to various cations. Since these photoelectron emission threshold energies are related to solution phase vertical ionization potentials, $IP_s^v$, it is now feasible to derive solution phase vertical ionization potentials for any anion and solution phase vertical electron affinities ($EA_s^v$) of any cation by consideration of their OSCT spectra.[44] This procedure also provides a new method to derive reorganization energies.[44]

## 4.4 Emission

The fundamental approach to emission spectra is, of course, almost the same as to absorption. However, there are a few caveats. Experimentally, one must ensure that the spectrum is corrected for variation of the phototube response with wavelength. Any data reported for which this is not explicitly stated, should, for correlation purposes, be disregarded as potentially very inaccurate particularly for long wavelength (red) emission where most phototubes have poor response. Emission spectra frequently show vibrational structure corresponding to transitions to various vibrational levels of the ground state. The highest energy emission should correspond with the 0–0′ energy gap, expressed as $E_{00}$, i.e., the energy separation between the 0th vibrational level of the excited state and that of the ground state. This will not necessarily be a peak in the emission spectrum, or correspond with the most intense feature; often it will be a shoulder, and it may not even be identified by the authors of the work concerned. However, it is this $E_{00}$ energy which is the most useful parameter to correlate with an electrochemical observable since $E_{00}$ will correspond, subject to solvation effects noted below, to the energy difference between ground and excited states.

Since the excited state is vibrationally relaxed (VEqES), and we consider the $E_{00}$ transition, the inner reorganization contribution to this energy, $\chi_i$, vanishes. However the emission terminates on the ground state with a solvation environment appropriate to the excited state, and therefore there will be a contribution to the solvent reorganization energy, $\chi_0$. This will decrease the emission energy since the ground state energy is raised. Considering the emission process:

$$^3[M^{III}L^-]_s^* \longrightarrow [M^{II}L]_s \tag{87}$$

Then in fluid solution, $E_{00}$ being *negative*, because emission is spontaneous:

$$E_{00} = \Delta G^{0\,3}[M^{III}L^-]_s^* - \Delta G^0[M^{II}L]_s + \chi_0 \tag{88}$$

where $\Delta G^0[\mathrm{M^{II}L}]_s$ and $\Delta G^{0\,3}[\mathrm{M^{III}L^-}]_s^*$ are the Gibbs free energies of formation of the vibrationally and solvent relaxed ground and excited states, respectively. Emission energies are frequently reported for samples at room temperature and at low, frozen solvent, temperature (commonly 77 K). These relative energies are invariably different because, at low temperatures, the excited state is frozen into the solvation environment of the ground state. This will increase the emission energy, namely in frozen solvent:

$$E_{00} = \Delta G^{0\,3}\Delta G^0[\mathrm{M^{II}L}]_s - [\mathrm{M^{III}L^-}]_s^* - \chi_0 \quad \text{(frozen solution)} \tag{89}$$

where we assume that the magnitudes of the $\chi_0$ correction are the same. This is probably roughly true. Then the difference between the emission energies $E_{00}$ in fluid and frozen solution would be $2\chi_0$.[186]

Emission spectra can be related to redox processes via pairs of equations related to Equations 64 and 65. Thus, $E_{00}$ from the MLCT spin-triplet state can be expressed as the sum of two potentials, via two alternate pairs of potentials, viz:

$$^3[\mathrm{M^{III}L^-}] \longrightarrow [\mathrm{M^{III}L}]^+ + e^- \qquad -E_{1/2}{}^3[\mathrm{L^{0/-}(M^{III})}] \tag{90}$$

$$[\mathrm{M^{III}L}]^+ + e^- \longrightarrow [\mathrm{M^{II}L}] \qquad E_{1/2}[\mathrm{M^{III/II}(L)}] \tag{91}$$

which sum to Equation 87, or:

$$^3[\mathrm{M^{III}L^-}] + e^- \longrightarrow [\mathrm{M^{II}L^-}]^- \qquad E_{1/2}{}^3[\mathrm{M^{III/II}(L^-)}] \tag{92}$$

$$[\mathrm{M^{II}L^-}]^- \longrightarrow \mathrm{M^{II}L} + e^- \qquad -E_{1/2}[\mathrm{L^{0/-}(M^{II})}] \tag{93}$$

which also sum to Equation 87. In fluid solution, the $E_{00}$ energy can then be written:

$$E_{00} = -F\{E_{1/2}[\mathrm{M^{III/II}(L)}] - E_{1/2}{}^3[\mathrm{L^{0/-}(M^{III})}]\} + \chi_0 \tag{94}$$

From the second pair of potentials, one can similarly write:

$$E_{00} = -F\{E_{1/2}{}^3[\mathrm{M^{III/II}(L^-)}] - E_{1/2}[\mathrm{L^{0/-}(M^{II})}]\} + \chi_0 \tag{95}$$

But (see Equation 75):

$$\begin{aligned} &-F\{E_{1/2}[\mathrm{L^{0/-}(M^{II})}] - E_{1/2}{}^3[\mathrm{L^{0/-}(M^{III})}]\} \\ &\quad = -F\{E_{1/2}{}^3[\mathrm{M^{III/II}(L^-)}] - E_{1/2}[\mathrm{M^{III/II}(L)}]\} = -D^T \end{aligned} \tag{96}$$

Then Equation 94 can be re-written using the observable $\Delta E(\text{Redox})$ (in eV) for fluid phase, as:

$$E_{00} = -[\Delta E(\text{Redox}) - \chi_0 - D^T] \tag{97}$$

Both $\chi_0$ and $D^T$ [$= (\alpha - 2S)$, Equation 75] are positive quantities, so the derived $E_{00}$ should always be less than $\Delta E$(Redox) when other boundary conditions, such as equivalent orbitals being involved, are met. In terms of the parameters used by Dodsworth and Lever,[35] the $E_{00}$ can also be expressed as:

$$E_{00} = -\left[\Delta E(\text{Redox}) + \Delta\Delta G_s + Q - 2K_{d,L} + \Delta(\text{sol}) \pm \chi_0\right] \qquad (98)$$

where $K_{d,L}$, is the exchange energy, the difference in energy between the spin singlet and spin triplet excited states being $2K_{d,L}$, and the $\pm\chi_0$ term reflects the choice between fluid and frozen solvent (see Equations 88 and 89). Equation 98 follows from Equations 73 and 99:

$$D^S = D^T + 2K_{d,L} \qquad (99)$$

A final comment about the quantities $D^S$ and $D^T$ which are the Gibbs free energy differences due to the replacement of $M^{II}$ by $M^{III}$, or of L by reduced ligand $L^-$, in both halves of the redox couples. These definitions includes all energy changes, due to structural, solvation, resonance and entropy changes.[204]

For a large number of ruthenium(II) complexes $D^T$ is approximately constant at $0.48 \pm 0.09$ eV (where 0.09 eV is equal to one standard deviation),[204] close to DeArmond's estimate of 0.5 eV, based on one complex.[203] For a series of platinum diimine toluenedithiolate complexes,[163] an average value of $D^T = 0.17 \pm 0.06$ eV can be extracted.

Some relevant emission energy versus electrochemical data are collected in Table 14. Because of the problem of ensuring one is recording phototube-corrected $0''$–$0'$ data, such information must be treated with caution. There is room here for more definitive data to be collected.

### 4.5 Charge Transfer Spectra and Hammett Substituent Constants

It is obvious from our earlier discussions that the energies of charge transfer spectra, MLCT or LMCT, in homologous series of complexes where one or more substituent is varied, will correlate both with Hammett substituent constants and, via equations such as Equation 47 with $E_L$(L) parameters. Some data are collected in Table 11. MLCT transitions to RLL species show negative coefficients on the Hammett substituent constant since there is a red shift in the MLCT transition as the substituent becomes a better electron acceptor. Conversely, in systems of the type M(LL)(RL) where an MLCT transition to LL is monitored as a function of RL, the coefficient is positive as the metal becomes a poorer donor with increasing electron acceptor character of $R$. Appropriate equations can be derived by combining Equations 46 and 51 with Equations 67 or 97; linear dependence upon $\sigma(R)$ follows, but analysis of the significance of the slope is complex.

**Table 14 Summary of Regression Data—MLCT Emission vs. Redox Potentials**

| $h\nu$(MLCT) emission vs $\Delta E$(Redox) | | No. | $R$ | Ref. |
|---|---|---|---|---|
| Ru(II)(bpy)(biq)(LL) | $h\nu[\pi^*\ \text{bpy} \rightarrow \text{Ru}\ d] = 0.49(0.03)\Delta E(\text{Redox}) + 1.06(0.01)$ | 7 | 0.98 | 127,209 |
| $Os(II)(bpy)_nL_m$ | $h\nu[\pi^*\ \text{bpy} \rightarrow \text{Os}\ d] = 0.62(0.02)\ \Delta E(\text{Redox}) + 0.34(0.01)$ | 6 | 0.997 | 127,210 |
| $Os(II)(LL)_nL_m$ [a] | $h\nu[\pi^*\ \text{LL} \rightarrow \text{Os}\ d] = 0.62(0.04)\Delta E(\text{Redox}) + 0.37(0.04)$ | 25 | 0.96 | 127,210 |
| $Ru(II)(bpm)_nL_{6-2n}$ | $h\nu[\pi^*\ \text{bpm} \rightarrow \text{Ru}\ d] = 1.04(0.03)\Delta E(\text{Redox}) - 0.75(0.01)$ | 5 | 0.999 | 128 |
| $ReL(CO)_3(44'\text{-}Me_2\text{-bpy})$ | $h\nu[\pi^*\ \text{bpy} \rightarrow \text{Re}\ d] = 0.96(0.12)\Delta E(\text{Redox}) - 0.63(0.02)$ | 4 | 0.98 | 211 |
| $[Os(II)(CO)Cl(R_2\text{-bpy})_2]^+$ | $h\nu[\pi^*\ \text{bpy} \rightarrow \text{Os}\ d] = 0.81(0.10)\Delta E(\text{Redox}) - 0.02(0.02)$ | 7 | 0.96 | 138 |
| $h\nu$(MLCT) vs $E_{red}$ $(LL)M^{n+}/(LL^-)M^{n+}$ | | | | |
| $LRe(CO)_3(4,4'\text{-}Me_2bpy)$ [b] | $h\nu[\pi^*\ \text{Me}_2\text{bpy} \rightarrow d\ \text{Re}^{I}] = 0.63(0.18)E_{1/2}[(\text{Me}_2\text{-bpy})^{0/-}] + 2.82(0.04)$ | 4 | 0.93 | 211 |
| $Re(I)(R\text{-bpy})(CO)_3Cl$ | $h\nu[\pi^*\ \text{bpy} \rightarrow \text{Re}\ d] = -0.50(0.08)E_{1/2}[\text{R-bpy})^{0/-}] + 1.77(0.08)$ | 8 | 0.93 | 198 |

[a] A range of substituted bipyridines and phenanthrolines; [b] vary L. Numbers in parentheses are standard deviations.

## 5 CONCLUSIONS

Electrochemistry and optical spectroscopy clearly play pivotal roles in elucidating the nature of the chemical bond in a coordination complex. Electrochemical potentials, via the various parameter models, provide information about the bonding characteristics of the ligand and the way in which it influences the metal binding site. They are probes of electron richness at the metal or ligand site which is a property of considerable interest with respect to chemical reactivity. A pertinent example would be the reactivity of $H_2$ when bound at sites of varying richness, as modeled by the Lever parameter.[212] Charge transfer electronic spectroscopy paints a picture of the relative frontier orbital energies in metal and ligand. When combined with electrochemical information, it gives a synergism allowing a more detailed picture of the energy levels within the molecule to emerge.

It is now possible to predict the electrochemical and optical properties of a wide range of species through use of the correlations discussed here. Clearly, they can also be used to confirm spectroscopic and electrochemical assignments. However, they have much more fundamental significance, providing greater insight into the nature of the metal–ligand bond. For example, apparent deviations from Equation 67 can be expected if the orbital from which excitation occurs to generate a strong charge transfer band differs from that from which redox occurs. Considering just the $d(t_{2g})$ set, this deviation will be noticed if the $d(t_{2g})$-orbitals are markedly split by low symmetry distortion and/or strong covalent binding, e.g., see References 124,133. The combination of molecular computation, especially Density Functional Theory or ZINDO, with electrochemical and optical data can be expected to develop rapidly in the future leading to a more fundamental understanding of metal–ligand binding. Obviously such studies can also involve the many other physical techniques discussed in these volumes.

## ACKNOWLEDGEMENTS

The authors are indebted to the Natural Science and Engineering Research Council for financial support during the development of the ideas discussed in this chapter. We are also indebted to Professor Antonin A. Vlcek for his collaboration over the years, and for his in-depth reading of a draft of this manuscript, to Jon Veinot for collecting some of the data upon which the manuscript is based, and to Serge Gorelsky, Prof. Armando Pombeiro and Dr. Brett Yeomans for their careful reading of the manuscript.

## REFERENCES

1. B. E. Bursten and M. R. Green, *Progr. Inorg. Chem.* **36**, 393 (1988).
2. C. J. Pickett and D. Pletcher, *J. Organomet. Chem.* **102**, 327 (1975).
3. J. Chatt, *Coord. Chem. Rev.* **43**, 337 (1982)

4. P. M. Treichel, G. E. Dureen and H. J. Mueh, *J. Organomet. Chem.* **44**, 339 (1972).

5. A. B. P. Lever, *Inorg. Chem.* **29**, 1271 (1990).

6. J. B. Headridge, "Electrochemical Techniques for Inorganic Chemists," Academic Press, NY, 1969.

7. A. Weissberger and B. W. Rossiter, *Physical Methods in Chemistry*, Vol. 1, Part IIB, Electrochemical Methods (1971).

8. A. J. Bard and L. R. Faulkner, *Electrochemical Methods, Fundamentals and Applications*, John Wiley & Sons, New York (1980).

9. R. L. Bretz and H. D. Abruna, *J. Electroanal. Chem.* **388**, 123 (1995); A. L. B. Marques, J. J. Zhang, A. B. P. Lever and W. J. Pietro, *J. Electroanal. Chem.* **392**, 43 (1995); E. P. Marques, J. J. Zhang, Y. H. Tse, R. A. Metcalfe, W. J. Pietro and A. B. P. Lever, *J. Electroanal. Chem.* **395**, 133 (1995); Y. Sato and F. Mizutani, *Denki Kagaku* **63**, 1173 (1995).

10. M. Lamrani, G. Mousset and P. Pouillen, *Bull. Soc. Chim. Fr.* **129**, 209 (1992); J. Hanzlik, L. Pospisil, A. A. Vlcek and M. Krejcik, *J. Electroanal. Chem.* **331**, 83178 (1992); B. S. Kim and S. M. Park, *J. Electrochem. Soc.* **142**, 26 (1995); P. Sharpe, N.G. Alameddin and D. E. Richardson, *J. Am. Chem. Soc.* **116**, 11098 (1994).

11. R. N. Bagchi, A. M. Bond, R. Colton, D. L. Luscombe and J. E. Moir, *J. Am. Chem. Soc.* **108**, 3352 (1986).

12. J. C. Curtis, B. P. Sullivan and T. J. Meyer, *Inorg. Chem.* **22**, 224 (1983).

13. S. M. Drew, R. M. Wightman and C. A. Amatore, *J. Electroanal. Chem.* **317**, 117 (1991).

14. R. R. Gagne, C. A. Koval and G. C. Lisensky, *Inorg. Chem.* **19**, 2854 (1980).

15. P. A. Lay, N. S. Mcalpine, J. T. Hupp, M. J. Weaver and A. M. Sargeson, *Inorg. Chem.* **29**, 4322 (1990).

16. D. A. Bohling, J. F. Evans and K. R. Mann, *Inorg. Chem.* **21**, 3546 (1982).

17. J. Daschbach, D. Blackwood, J. W. Pons and S. Pons, *J. Electroanal. Chem.* **237**, 269 (1987).

18. T. Iguro, N. Ikeda and T. Ohno, *Inorg. Chim. Acta* **226**, 203 (1994).

19. A. B. P. Lever, E. Milaeva and G. Speier, *The Phthalocyanines, Properties and Applications*, Vol. 3. VCH, New York (1993) p. 1.

20. C. C. Leznoff, S. Greenberg, S. M. Marcuccio, P. C. Minor, P. Seymour and A. B. P. Lever, *Inorg. Chim. Acta* **89**, L35 (1984).

21. T. M. Mezza, N. R. Armstrong, G. W. Ritter II, J. P. Iafelice and M. E. Kenney, *J. Electroanal. Chem.* **137**, 227 (1982).

22. D. W. DeWulf, J. K. Lelend, B. L. Wheeler, A. J. Bard, D. A. Batzel, D. R. Dininny and M. E. Kenney, *Inorg. Chem.* **26**, 266 (1987).

23. A. B. P. Lever and J. P. Wilshire, *Inorg. Chem.* **17**, 1145 (1978).

24. R. O. Loutfy and Y. C. Cheng, *J. Chem. Phys.* **73**, 2902 (1980).

25. See Section 3.4 for references.

26. C. J. Timpson, C. A. Bignozzi, B. P. Sullivan, E. M. Kober and T. J. Meyer, *J. Phys. Chem.* **100**, 2915 (1996).

27. (a) L. Karki and J. T. Hupp, *J. Am. Chem. Soc.* **119**, 4070 (1997). (b) Normally, oxidation occurs from the HOMO orbital, but in some circumstances, it may occur from a lower lying orbital such as HOMO-1. Then HOMO-1 becomes the singly occupied HOMO, termed SOMO, of the oxidized species. Similarly if there is re-structuring of the reduced species, it may be true that reduction does not occur at the LUMO of the parent but at a higher energy orbital which then becomes the SOMO of the reduced species.
28. B. P. Sullivan, D. Conrad and T. J. Meyer, *Inorg. Chem.* **24**, 3640 (1985).
29. C. D. Baer, *Coord. Chem. Rev.* **93**, 1 (1989).
30. A. M. Bond, *Analyst* **119**, R1-R21 (1994).
31. R. J. Forster, *Chem. Soc. Rev.* **23**, 289 (1994).
32. F. Magno and I. Lavagnini, *Anal. Chim. Acta* **305**, 96 (1995).
33. C. Amatore, in "Physical Electrochemistry," Marcel Dekker, New York (1995).
34. M. Krejcik, M. Danek and F. Hartl, *J. Electroanal. Chem.* **317**, 179 (1991).
35. E. S. Dodsworth and A. B. P. Lever, *Chem. Phys. Lett.* **124**, 152 (1986).
36. M. Haga, E. S. Dodsworth and A. B. P. Lever, *Inorg. Chem.* 25, 4474 (1986).
37. A. C. Samuels and M. K. DeArmond, *Inorg. Chem.* **34**, 5548 (1995).
38. D. A. Fiedler, M. Koppenol and A. M. Bond, *J. Electrochem. Soc.* **142**, 862 (1995).
39. P. Kebarle and S. Chowdhury, *Chem. Rev.* **87**, 513 (1987).
40. S. G. Lias, J. E. Bartmess, J. F. Liebman, J. I. Holmes, R. D. Levin and W. G. E. Mallard, *Gas-Phase Ion and Neutral Thermochemistry*, American Institute of Physics, New York (1988).
41. M. F. Ryan, D. E. Richardson, D. L. Lichtenberger and N. E. Gruhn, *Organomet.* **13**, 1190 (1994).
42. K. Von Burg and P. Delahay, *Chem. Phys. Lett.* **78**, 287 (1981).
43. N. Takahashi, K. Sakai, H. Tanida and I. Watanabe, *Chem. Phys. Lett.* **246**, 183 (1995).
44. S. I. Gorelsky, V. Yu. Kotov and A. B. P. Lever, *Inorg. Chem.* **37**, 4574 (1998).
45. D. E. Richardson, *Inorg. Chem.* **29**, 3213 (1990).
46. This is commonly rounded to 4.5 eV.
47. H. Reiss and A. Heller, *J. Phys. Chem.* **89**, 4207 (1985).
48. J. Li, C. L. Fisher, J. L. Chen, D. Bashford and L. Noodleman, *Inorg. Chem.* **35**, 4694 (1996).
49. C. J. Cramer and D. G. Truhlar, *J. Comput. Chem.* **13**, 1089 (1992); D. J. Giesen, C. J. Cramer and D. G. Truhlar, *J. Phys. Chem.* **99**, 7137 (1995).
50. J. Li, L. Noodleman and D. A. Case, Ch. 11, in *Inorganic Electronic Structure and Spectroscopy, Vol. I, Methodology*, E. I. Solomon and A. B. P. Lever, Eds., John Wiley & Sons, Inc.: New York (1999).
51. C. H. Martin and M. Zerner, Ch. 10, in *Inorganic Electronic Structure and Spectroscopy, Vol. I, Methodology*, E. I. Solomon and A. B. P. Lever, Eds., John Wiley & Sons, Inc.: New York (1999).
52. J. O. Howell, J. M. Goncalves, C. Amatore, L. Klasine, M. Wightman and J. K. Kochi, *J. Am. Chem. Soc.* **106**, 3968 (1984).
53. W. M. Nau, W. Adam, D. Klapstein, C. Sahin and H. Walter, *J. Org. Chem.* **62**, 5128 (1997).
54. E. S. Pysh and N. C. Yang, *J. Am. Chem. Soc.* **85**, 2124 (1963).

55. L. L. Miller, G. D. Nordblom and E. A. Mayeda, *J. Org. Chem.* **37**, 916 (1972).
56. P. G. Gassman and R. Yamaguchi, *J. Am. Chem. Soc.* **101**, 1308 (1979).
57. S. F. Nelsen, S. C. Blackstock, P. A. Petillo, I. Agmon and M. Kaftory, *J. Am. Chem. Soc.* **109**, 5724 (1987).
58. V. D. Parker, *J. Am. Chem. Soc.* **98**, 98 (1976).
59. S. F. Nelsen, P. A. Petillo, H. Chang, T. B. Frigo, D. A. Dougherty and M. Kaftory, *J. Org. Chem.* **56**, 613 (1991).
60. S. F. Nelsen, M. F. Teasley, A. J. Bloodworth and H. J. Eggelte, *J. Org. Chem.* **50**, 3299 (1985).
61. L. Eberson, *Adv. Phys. Org. Chem.* **18**, 79 (1982).
62. R. J. Klingler and J. K. Kochi, *J. Am. Chem. Soc.* **102**, 4790 (1980).
63. W. N. Nau, W. Adam, D. Klapstein, C. Sahin and H. Walter, *J. Org. Chem.* **62**, 5128 (1997).
64. R. T. Boerè and K. H. Moock, *J. Am. Chem. Soc.* **117**, 4755 (1995).
65. K. H. Moock, S. A. Macgregor, G. A. Heath, S. Derrick and R. T. Boerè, *J. Chem. Soc. Dalton*, 2067 (1996).
66. R. H. Holm, P. Kennepohl and E. I. Solomon, *Chem. Rev.* **96**, 2239 (1996).
67. D. E. Richardson and P. Sharpe, *Inorg. Chem.* **30**, 1412 (1991).
68. A. Szabo and N. S. Ostlund, *Modern Quantum Chemistry*, Macmillan Publishing Co., Inc., New York (1982).
69. J. A. Connor, E. M. Jones, G. K. McEwan, M.K. Lloyd and J. A. McCleverty, *J. Chem. Soc. Dalton Trans.* 1246 (1972).
70. P. M. Treichel, *Adv. Organomet. Chem.* **11**, 21 (1973).
71. A. C. Sarapu and R. F. Fenske, *Inorg. Chem.* **14**, 247 (1975).
72. E. W. Tsai, L. Throckmorton, R. McKellar, M. Baar, D. S. Marynick, K. Rajeshwar and A. L. Jr. Ternay, *J. Electroanal. Chem.* **210**, 45 (1986).
73. P. G. Gassman, M. J. Mullins, S. Richtsmeier and D. A. Dixon, *J. Am. Chem. Soc.* **101**, 5793 (1979).
74. G. J. Hoijtink, *Rec. Trav. Chim. Pays-Bas.* **77**, 555 (1958).
75. A. Zweig, W. G. Hodgson and W. H. Jura, *J. Am. Chem. Soc.* **86**, 4124 (1964).
76. A. Zweig, J. E. Lancaster, M. T. Neglia and W. H. Jura, *J. Am. Chem. Soc.* **86**, 4130 (1964).
77. B. E. Bursten, M. R. Green, V. Katovic and D. Jr. Lightner, *Inorg. Chem.* **25**, 831 (1986).
78. P. M. Treichel, H. J. Mueh and B. E. Bursten, *Isr. J. Chem.* **15**, 253 (1977).
79. L. J. Lyons, S. L. Pitz and D. C. Boyd, *Inorg. Chem.* **34**, 316 (1995).
80. A. B. P. Lever, unpublished observations.
81. Y. F. Yang, F. Arias, L. Echegoyen, L. P. F. Chibante, S. Flanagan, A. Robertson and L. J. Wilson, *J. Am. Chem. Soc.* **117**, 7801 (1995).
82. F. Barigelletti, A. Juris, V. Balzani and A. von Zelewsky, *Inorg. Chem.* **26**, 4115 (1987).
83. C. H. Martin and M. Zerner, Ch. 10, in *Inorganic Electronic Structure and Spectroscopy, Vol. I, Methodology*, E. I. Solomon and A. B. P. Lever, Eds., John Wiley & Sons, Inc.: New York (1999).
84. R. G. Parr, *Quantum Theory of Molecular Electronic Structure*, Benjamin, 1963.
85. A. J. L. Pombeiro, *Inorg. Chim. Acta* **103**, 95 (1985).

86. B. E. Bursten, *J. Am. Chem. Soc.* **104**, 1299 (1982).
87. F. L. Wimmer, M. R. Snow and A. M. Bond, *Inorg. Chem.* **13**, 1617 (1974).
88. C. E. Schaeffer and C. K. Jorgensen, *Mol. Phys.* **9**, 401 (1965).
89. A. B. P. Lever and E. I. Solomon, Ch. 1, in *Inorganic Electronic Structure and Spectroscopy, Vol. I, Methodology*, E. I. Solomon and A. B. P. Lever, Eds., John Wiley & Sons, Inc.: New York (1999).
90. D. Datta, *J. Chem. Soc. Dalton Trans.* 1907 (1986).
91. D. Datta, *Proc. Ind. Acad. Sci.* **108**, 101 (1996).
92. D. Datta, *Proc. Ind. Acad. Sci.* **98**, 269 (1987).
93. G. Butler, J. Chatt, G. J. Leigh and C. J. Pickett, *J. Chem. Soc. Dalton Trans.* 113 (1979).
94. J. Chatt, G. J. Leigh, H. Neukomm and C. J. Pickett, *J. Chem. Soc. Dalton Trans.* 121 (1980).
95. J. Chatt, C. T. Kan, G. J. Leigh, C. J. Pickett and D. R. Stanley, *J. Chem. Soc. Dalton Trans.* 2032 (1980).
96. J. Chatt, *Coord. Chem. Rev.* **43**, 337 (1982).
97. A. B. P. Lever, *Inorg. Chem.* **30**, 1980 (1991).
98. M. F. N. N. Carvalho and A. J. L. Pombeiro, *J. Chem. Soc. Dalton Trans.* 1209 (1989).
99. D. Datta, *J. Chem. Soc. Dalton Trans.* 1907 (1986).
100. (a) Molecular Electrochemistry of Inorganic, Bioinorganic and Organometallic Compounds. M. T. A. Ribeiro, A. J. L. Pombeiro, G. Facchin, M. Mozzon and R. A. Michelin. NATO ASI Series. Kluwer Academic Publishers, Dordrecht. 1992. p. 57; (b) G. Facchin, M. Mozzon, R. A. Michelin, M. T. A. Ribeiro and A. J. L. Pombeiro. *J. Chem. Soc. Dalton Trans.* 2827 (1992).
101. A. J. L. Pombeiro, C. J. Pickett and R. L. Richards, *J. Organomet. Chem.* **224**, 285 (1982).
102. M. F. N. N. Carvalho, A. J. L. Pombeiro, A. Hills, D. L. Hughes and R. L. Richards, *J. Organomet. Chem.* **469**, 179 (1993).
103. Molecular Electrochemistry of Inorganic, Bioinorganic and Organometallic Compounds. Y. Wang, A. J. L. Pombeiro, L. Kaden and M. Wahren. NATO ASI Series. Kluwer Academic, Dordrecht (1993) p. 63.
104. S. S. P. R. Almeida and A. J. L. Pombeiro, *Organomet.* **16**, 4469 (1997).
105. M. F. N. N. Carvalho and A. J. L. Pombeiro, *J. Organomet. Chem.* **410**, 347 (1991).
106. A. J. L. Pombeiro, *Port. Electrochim. Acta* **5**, 41 (1985).
107. A. J. L. Pombeiro, *Nouv. J. Chem.* **21**, 649 (1997).
108. L. Kaden, A. J. L. Pombeiro, Y. Wang and U. Abram, *Inorg. Chim. Acta* **230**, 189 (1995).
109. M. F. N. N. Carvalho, M. T. Duarte, A. M. Galvao and A. J. L. Pombeiro, *J. Organometallic Chem.* **511**, 163 (1996).
110. A. J. L. Pombeiro, *An. Quim.* **89**, 428 (1993).
111. Some recent references citing the $E_L(L)$ parameter: R. Alsfasser and R. Van Eldik, *Inorg. Chem.* **35**, 628 (1996); J. Barrera, A. K. Burrell and J. C. Bryan, *Inorg. Chem.* **35**, 335 (1996); M. J. Clarke and M. Stubbs, Metal ions in biological systems **32**, 727 (1996); F. Hartl and A. Vlcek, *Inorg. Chem.* **35**, 1257 (1996); P. I. Kvam, M. V. Puzyk, V. S. Cotlyr, J. Songstad and K. P. Balashev, *Acta Chem. Scand.* **50**, 6 (1996); I. Vernik and D. V. Stynes, *Inorg. Chem.* **35**, 1093 (1996); E. Z. Jandrasics and F. R. Keene, *J. Chem. Soc. Dalton Trans.* 153 (1997); J. F. Endicott, M. A. Watzky, X. Song and T. Buranda, *Coord. Chem. Rev.* **159**, 295 (1997).

112. H. Masui and A. B. P. Lever, *Inorg. Chem.* **32**, 2199 (1993).
113. S. Lu, V. V. Strelets, M. F. Ryan, W. J. Pietro and A. B. P. Lever, *Inorg. Chem.* **35**, 1013 (1996).
114. S. X. Lu, Ph. D. Thesis. York University (1999).
115. S. X. Lu and A. B. P. Lever, to be published (1999).
116. A. J. Blake, R. O. Gould, K. J. Taylor and L. J. Yellowlees, private communication.
117. J. P. Chang, E. Y. Fung and J. C. Curtis, *Inorg. Chem.* **25**, 4233 (1986).
118. W. Linert, V. Gutmann, B. Pouresmaeil and R. F. Jameson, *Electrochim. Acta* **33**, 975 (1988).
119. G. Gritzner, K. Danksagmuller and V. Gutmann, *J. Electroanal. Chem.* **72**, 177 (1976).
120. U. Mayer, A. Kotocova, V. Gutmann and W. Gerger, *J. Electroanal. Chem.* **100**, 875 (1979).
121. K. M. Kadish, K. Das, D. Schaeper, C. Merrill, B. R. Welch and L. J. Wilson, *Inorg. Chem.* **19**, 2816 (1980).
122. (a) G. A. Heath and D. G. Humphrey, *J. Chem. Soc. Chem. Commun.* 1668 (1991). (b) In fact, a closer look at the data in Reference 122(a) shows that it may be incorrect to assume that $[OsBr_5(CO)]^{2-}$ is anomalous. Rather, there appear to be two independent regression lines, one for the carbonyl species and one for the non-carbonyl species. These are not quite collinear. In this dual line approach, the mono-carbonyl species is well behaved with respect to both its redox potentials. The explanation of why there are two lines requires further exploration.
123. M. R. Mason, C. M. Duff, L. L. Miller, R. A. Jacobsen and J. G. Verkade, *Inorg. Chem.* **31**, 2746 (1992).
124. R. A. Metcalfe and A. B. P. Lever, *Inorg. Chem.* **36**, 4762 (1997).
125. D. Venegas-Yazigi, F. Smeriglio, E. S. Dodsworth, M. Ebadi and A. B. P. Lever, paper in preparation.
126. E. S. Dodsworth, A. A. Vlcek and A. B. P. Lever, *Inorg. Chem.* **33**, 1045 (1994).
127. S. R. Johnson, T. D. Westmoreland, J. V. Caspar, K. R. Barqawi and T. J. Meyer, *Inorg. Chem.* **27**, 3195 (1988).
128. H. B. Ross, M. Boldajii, D. P. Rillema, C. B. Blanton and R. P. White, *Inorg. Chem.* **28**, 1013 (1989).
129. P.-I. Kvam, M. V. Puzyk, V. S. Cotlyr, K. P. Balashev and J. Songstad, *Acta Chem. Scand.* **49**, 645 (1995).
130. A. B. P. Lever, work in progress (1998).
131. J. T. Hupp and T. J. Meyer, *Inorg. Chem.* **26**, 2332 (1987).
132. D. P. Rillema, G. Allen, T. J. Meyer and D. Conrad, *Inorg. Chem.* **22**, 1617 (1983).
133. S. I. Gorelsky, E. S. Dodsworth, A. A. Vlcek and A. B. P. Lever, *Coord. Chem. Rev.* **174**, 469 (1998).
134. Y. Kawanishi, N. Kitamura and S. Tazuke, *Inorg. Chem.* **28**, 2968 (1989).
135. W. Kaim, H. E. A. Kramer, C. Vogler and J. Rieker, *J. Organomet. Chem.* **367**, 107 (1989).
136. V. Skarda, M. J. Cook, A. P. Lewis, G. S. G. McAuliffe, A. J. Thomson and D. J. Robbins, *J. Chem. Soc. Perkin*, 1309 (1984).
137. J. R. Bradbury and F. A. Schultz, *Inorg Chem.* **25**, 4416 (1986).

138. L. Della Ciana, W. J. Dressick, D. Sandrini, M. Maestri and M. Ciano, *Inorg. Chem.* **29**, 2792 (1990).

139. Vos, R. Hage and J. Reedijk, *Inorg. Chim. Acta* **245**, 237 (1996).

140. T. M. Gallotti, I. Koba, H. E. Toma and N. Y. Iha, *IV Simp. Bras. Eletroquim. Eletroanal. Anais* 455 (1984).

141. C. R. Johnson and R. E. Shepherd, *Inorg. Chem.* **22**, 2439 (1983).

142. J. L. Brisset and M. Biquard, *Inorg. Chim. Acta* **53**, L125 (1981).

143. M. E. N. P. R. A. Silva, A. J. L. Pombeiro, J. J. R. Frausto da Silva, R. Hermann, N. Deus and R. E. Bozak, *J. Organomet. Chem.* **421**, 75 (1991).

144. L. Slep, L. M. Baraldo and J. Olabe, *Inorg. Chem.* **35**, 6327 (1996).

145. K. N. Soe, A. Ichimura and T. Kitagawa, *Inorg. Chim. Acta* **207**, 21 (1993).

146. J. L. Brisset and V. Ilimbi, *Can. J. Chem.* **58**, 1250 (1980).

147. K. J. Moore, L. Lee, G. A. Mabbott and J. D. Petersen, *Inorg. Chem.* **22**, 1108 (1983).

148. P. C. Ford, De F. P. Rudd, R. Gaunder and H. Taube, *J. Am. Chem. Soc.* **90**, 1187 (1968).

149. A. Nurhadi, E. Graf and M. Gross, *Electrochim. Acta* **40**, 1939 (1995).

150. R. F. Handy and R. L. Lintvedt, *Inorg. Chem.* **13**, 89 (1974).

151. R. N. Mukherjee, A. O. Rajan and A. Chakravorty, *Inorg. Chem.* **21**, 785 (1982).

152. R. N. Mukherjee and A. Chakravorty, *J. Chem. Soc. Dalton Trans.* 955 (1983).

153. A. K. Jain, R. N. Goyal and D. D. Agarwal, *J. Electroanal. Chem.* **93**, 145 (1978).

154. A. Giraudeau, H. J. Callott, J. Jordan, I. Ezhar and M. Gross, *J. Am. Chem. Soc.* **101**, 3857 (1979).

155. S. Goswami, R. Mukherjee and A. Chakravorty, *Inorg. Chem.* **22**, 2825 (1983).

156. D. P. Rillema, J. K. Nagle, L. F. Jr. Barringer and T. J. Meyer, *J. Am. Chem. Soc.* **103**, 56 (1981).

157. G. S. Patterson and R. H. Holm, *Inorg. Chem.* **11**, 2285 (1972).

158. Y. H. Tse, H. Lam, J. J. Zhang, W. J. Pietro and A. B. P. Lever, *J. Porph. Phthaloc. Chem.* **1**, 3 (1997).

159. M. F. C. G. da Silva, J. J. R. F. da Silva and A. J. L. Pombeiro, *J. Chem. Soc. Dalton Trans.* 3299 (1994).

160. C. Tsiamis, S. Karageorgiou, M. Lalia-Kantouri and G. Manoussakis, *Gazz. Chim. Ital.* **117**, 317 (1987).

161. R. N. Mukherjee, A. O. Rajan and A. Chakravorty, *Inorg. Chem.* **21**, 785 (1982).

162. J. A. Connor, C. Overton and N. El Murr, *J. Organomet. Chem.* **277**, 277 (1984).

163. S. D. Cummings and R. Eisenberg, *J. Am. Chem. Soc.* **118**, 1949 (1996).

164. A. Juris, V. Balzani, F. Barigelletti, S. Campagna, P. Belser and A. von Zelewsky, *Coord. Chem. Rev.* **84**, 85 (1988).

165. P. Ghosh and A. Chakravorty, *Inorg. Chem.* **23**, 2242 (1984).

166. A. B. P. Lever, *Inorg. Chim. Acta* **203**, 171 (1992).

167. M. Handa, A. Suzuki, S. Shoji, K. Kasuga and K. Sogabe, *Inorg. Chim. Acta* **230**, 41 (1995).

168. J. K. Hino, L. della Ciana, W. J. Dressick and B. P. Sullivan, *Inorg. Chem.* **31**, 1072 (1992).

169. L. Sacksteder, A. P. Zipp, E. A. Brown, J. Streich, J. N. Demas and B. A. DeGraff, *Inorg. Chem.* **29**, 4335 (1990).

170. T. Saji and S. Aoyagui, *J. Electroanal. Chem.* **58**, 401 (1975).
171. G. G. Ferenczy, C. A. Reynolds and W. G. Richards, *J. Comput. Chem.* **11**, 159 (1990).
172. S. S. Fielder, M. C. Osborne, A. B. P. Lever and W. J. Pietro, *J. Am. Chem. Soc.* **117**, 6990 (1995).
173. W. J. Hehre, L. D. Burke, A. J. Shusterman and W. J. Pietro, *Experiments in Computational Chemistry*, Wavefunction Press, Irvine, CA (1993).
174. C. Creutz, M. D. Newton and N. Sutin, *J. Photochem. Photobiol. A* **82**, 47 (1994).
175. A. B. P. Lever, *Inorganic Electronic Spectroscopy*, 2nd Edition. Elsevier Science Publishers, Amsterdam (1984).
176. B. N. Figgis, *Introduction to Ligand Fields*, Interscience, New York (1996); C. J. Ballhausen, "Introduction to Ligand Field Theory," McGraw Hill, New York (1962).
177. A. B. P. Lever, *Inorganic Electronic Spectroscopy*, 1st Edition. Elsevier Science, Amsterdam (1968).
178. H. L. Schläfer and G. Gliemann, *Basic Principles of Ligand Field Theory*, Wiley Interscience, New York (1969).
179. A. B. P. Lever, *J. Chem. Educ.* **51**, 612 (1974).
180. C. K. Jørgensen, *Absorption Spectra and Chemical Bonding in Complexes*, Pergamon Press, London (1962).
181. F. A. Cotton, *Chemical Applications of Group Theory*, 3rd Edition. Wiley Interscience, New York (1990).
182. S. B. Piepho and P. N. Schatz, *Group Theory in Spectroscopy*, Wiley Interscience, New York (1983).
183. R. H. Magnuson and H. Taube, *J. Am. Chem. Soc.* **97**, 5129 (1975).
184. Y. Shin, B. S. Brunsschwig, C. Creutz, M. D. Newton and N. Sutin, *J. Phys. Chem.* **100**, 1104 (1996).
185. Y.-G. Shin, B. S. Brunsschwig, C. Creutz and N. Sutin, *J. Phys. Chem.* **100**, 8157 (1996).
186. J. Endicott, Ch. 5, in *Inorganic Electronic Structure and Spectroscopy, Vol. II, Applications and Case Studies*, E. I. Solomon and A. B. P. Lever, Eds., John Wiley & Sons, Inc.: New York (1999).
187. X. Soug and J. F. Endicott, *Chem. Phys. Lett.* **204**, 400 (1993).
188. S. Sakami and M. J. Weaver, *J. Electroanal. Chem.* **122**, 171 (1981).
189. A. B. P. Lever, S. Licoccia, P. C. Minor, B. S. Ramaswamy and S. R. Pickens, *J. Am. Chem. Soc.* **103**, 6800 (1981).
190. D. F. Shriver, *J. Am. Chem. Soc.* **88**, 1672 (1966).
191. D. V. Stynes, personal communication.
192. A. Juris, S. Campagna, V. Balzani, G. Gremaud and A. von Zelewsky, *Inorg. Chem.* **27**, 3652 (1988).
193. T. Matsubara and P. C. Ford, *Inorg. Chem.* **15**, 1107 (1976).
194. A. A. Vlcek, *Electrochim. Acta* **13**, 1063 (1968).
195. A. A. Vlcek, *Disc. Faraday Soc.* 164 (1958).
196. A. A Vlcek, *Revue de Chimie Minerale* 299 (1968).
197. C. M. Duff and G. A. Heath, *Inorg. Chem.* **30**, 2528 (1991).
198. A. Juris, S. Campagna, I. Bidd, J. M. Lehn and R. Ziessel, *Inorg. Chem.* **27**, 4007 (1988).
199. B. P. Sullivan, J. V. Caspar, S. R. Johnson and T. J. Meyer, *Organomet.* **3**, 1241 (1984).

200. R. J. Crutchley, K. McCaw, F. L. Lee and E. J. Gabe, *Inorg. Chem.* **29**, 2576 (1990).

201. R. A. Cipriano, W. Levason, D. Pletcher, N. A. Powell and M. Webster, *J. Chem. Soc. Dalton Trans.* 1901 (1987).

202. K. J. LaChance-Galang, P. E. Doan, M. J. Clarke, U. Rao, A. Yamano and B. M. Hoffman, *J. Am. Chem. Soc.* **117**, 3529 (1995).

203. Y. Ohsawa, K. W. Hanck and M. K. DeArmond, *J. Electroanal. Chem.* **175**, 229 (1984).

204. A. A. Vlcek, E. S. Dodsworth, W. J. Pietro and A. B. P. Lever, *Inorg. Chem.* **34**, 1906 (1995).

205. R. Billing and D. E. Khoshtariya, *Inorg. Chem.* **33**, 4038 (1994).

206. R. Billing, *Coord. Chem. Rev.* **159**, 257 (1997).

207. A. Haim, *Comm. Inorg. Chem.* **4**, 113 (1985).

208. L. A. Lucia, K. Abboud and K. S. Schanze, *Inorg. Chem.* **36**, 6224 (1997).

209. A. Juris, S. Compagna, V. Balzani, G. Gremaud and A. von Zelewsky, *Inorg. Chem.* **27**, 3652 (1988).

210. H. Abruna, *J. Electrochem. Soc.* **132**, 842 (1985).

211. M. Furue, M. Naiki, Y. Kanematsu, T. Kushida and M. Kamachi, *Coord. Chem. Rev.* **111**, 221 (1991).

212. R. H. Morris, *Inorg. Chem.* **31**, 1471 (1992).

# 5 The Photophysics and Photochemistry of Coordination Compounds

JOHN F. ENDICOTT
Department of Chemistry
Wayne State University
Detroit, MI 48202, USA
E-mail: jfe@fourcroy.chem.wayne.edu



*Inorganic Electronic Structure and Spectroscopy, Volume II: Applications and Case Studies.*
Edited by E. I. Solomon and A. B. P. Lever.
ISBN 0-471-32682-8. 

## 1 INTRODUCTION

In this chapter we will define photophysics as the composite of all processes involved in the time evolution of a molecular system following its interaction with electromagnetic radiation. Photochemistry will be considered to be the net result of one or more of these processes. The focus will be primarily on electronic excited states, and the examples will be drawn largely from studies of transition metal complexes. Thus this chapter will deal with the properties of the electronic excited states of coordination complexes, and it will address some of the issues related to how these properties evolve after the initial excitation event. There are many such issues and a considerable literature dealing with most of them. Space constraints have resulted in a representative, rather than a comprehensive, treatment of these issues or their experimental investigation. Similarly, large areas of inorganic photochemistry have been omitted, with the most notable omissions in areas of organometallic photochemistry. The general texts and reviews cited below may be consulted for broader treatments. This chapter will sketch the basic physical principles governing events following light absorption by molecules (for more detail, see other chapters in this series), the characterization and characteristics of electronic excited states of transition metal complexes (also treated in additional detail in other chapters) and survey some observed patterns of behavior.

## 2 SOME GENERAL PROPERTIES AND FEATURES OF ELECTRONIC EXCITED STATES

The attraction of photochemical studies is based on the premise that chemical reactivity is a function of electronic configuration (i.e., that species with different populations of electronic orbitals have different chemical properties) and that light may be used to generate unique electronic configurations. In practice, the rules governing molecular photophysical behavior greatly restrict the accessible excited electronic configurations and the range of their reactivities. Before entering into a discussion of transition metal systems, it is necessary to establish some general guidelines based on basic physical principles. The characterization of an electronically excited molecule involve detailed knowledge of its: (a) electronic distribution; (b) molecular geometry; and (c) pathways for relaxation and reaction. The direct absorption of light generates an excited electronic configuration that has the ground state nuclear coordinates (or geometry) but is not a well defined, metastable state of the system. The pathways of relaxation of these initial, Franck–Condon (FC) excited states can determine which metastable states are formed, the yields of products, etc. As a consequence the complete description of a photo-excited molecule should involve an elaboration of the above list of necessary information:

a′. A configurational description of the state(s) populated by light absorption (the FC state).
b′. How the relaxation of the FC state is partitioned between competing pathways which lead to configurationally different excited states, the ground state or product species.
a″. The proper configurational description of the relevant electronic excited states.
b″. The efficiency of vibrational relaxation within the potential energy surface (PES) of these electronic excited states and the changes of nuclear geometry which accompany them.
c″. The relaxation or reaction pathways of the vibrationally equilibrated lower energy excited states.

Each issue in this list can be very complex. We will start with the simplest, most general limits. These probably do not apply in exact detail to any real system. Nevertheless, an understanding of even the most perversely complicated photosystem must begin with a comparison of its behavior to that expected for some idealized limit. This discussion will be restricted to condensed phase systems.

## 2.1 An Idealized General Limit: Some Definitions and the Most Basic General Principles

Figure 1 is a useful focus for discussion of the most basic aspects of electronically excited molecules.[1–3] The absorption of radiation with energy ($h\nu$) generally results in formation of an electronically and vibrationally excited "state" of the system (rotational excitation will not be explicitly discussed). This initial "state" (more correctly, it should be regarded as a superposition of electronic excited states since it does not correspond to a stationary state of the system) is called the Franck–Condon (FC) excited state because the process of light absorption is much faster than the rate at which nuclei can move. As a consequence, the FC excited state has a molecular structure which is nearly identical to that of the ground state. In the simplest, two state limit, vibrational relaxation (or nuclear reorganization) of the FC state is

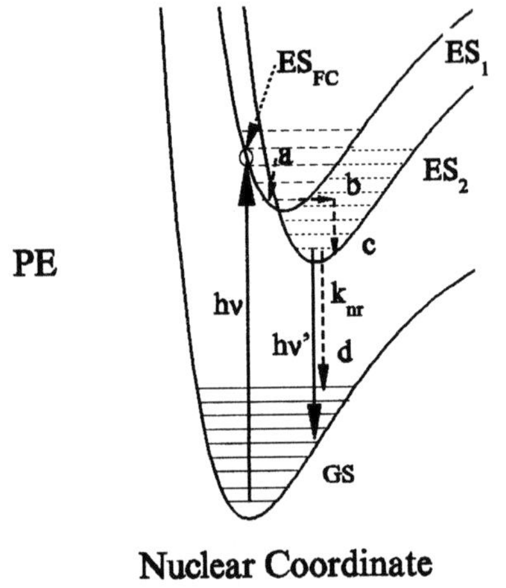


**Figure 1.** Qualitative potential energy surfaces illustrating the Franck–Condon excited state, $ES_{FC}$, in a system with two distinct electronic excited states, $ES_1$ and $ES_2$. Vibrational relaxation processes are indicated by **a** and **c**, internal conversion (or intersystem crossing) by **b**, non-radiative relaxation of the VEqES to the ground state by **d** ($k_{nr}$) and emission by $h\nu'$.

fast compared to relaxation processes which regenerate the ground state (or form other electronic excited states in a multistate system). However, this is an issue of the kinetic competition between the rates for vibrational relaxation within the initial electronic excited state (ES) compared to the rates for "prompt" radiative or non-radiative relaxation of the FC state to regenerate the (vibrationally excited) ground state (GS) or relaxation of the FC state to other electronic excited states. Each of the competing rates is proportional to a transition probability which depends on a combination of selection rules and the efficiency of energy transfer between the degrees of freedom of the system (including the solvent). It is particularly important to realize that all vibrational relaxation processes require transfers of energy (as heat) to the condensed phase matrix, and that the coupling of the excited system to its environment is one major factor in determining which of the competitive relaxation processes will be dominant. If vibrational relaxation of FC is faster than "prompt" regeneration of the ground state, then one can treat the nuclear and electronic relaxation processes separately, and thus discuss the ES in terms of a potential energy surface (PES) which is based on the configurational electronic wave function for this state. The simplest photophysical limit consists of assigning a distinct PES to each relevant electronic excited state. The system within each PES is treated as a distinct chemical entity with properties which can be discussed in terms of its configurational electronic wavefunction. This limiting approach is only useful when vibrational relaxation within a PES manifold (processes **a** and **c** in Figure 1) is more rapid than any competing process which generates a different electronic configuration (e.g., **b** in Figure 1). Vibrational relaxation times tend to vary from subpicosecond to 10's of ps,[4] so this limiting approach is most useful for excited states with lifetimes longer than about $10^2$ ps.

Discussions of electronic excited states are often formulated in terms of the simplest of electronic configurational models (such as one electron models). Even when such configurational electronic wavefunctions are reasonable initial approximations to the state wavefunctions, they do not generally provide exact descriptions of the electronic state and the resulting PES's should be treated as diabatic surfaces. In such cases a more reasonable PES can often be constructed by symmetry allowed, perturbational mixing with other electronic states of the system. This usually becomes a very important issue when there is some set of nuclear coordinates for which two PES's have the same potential energy. At such a point the surfaces may mix and split into upper and lower components. The mixing coefficient can be a factor in the surface crossing probability.

In light atom systems, the spin multiplicities of FC and GS will be the same (e.g., both may be singlets) and in many instances there will exist some lower energy electronic state with the same electronic orbital configuration, but a different spin multiplicity (e.g., a triplet state), as could be the case for $ES_2$ in Figure 1. The crossing, **b**, between two such states is isoenergetic, and it is often assumed to occur without exchange of energy with the solvent (if more rigorously viewed as a vibration-to-vibration transfer process, $\sim \pm k_BT$ energy can be exchanged with the solvent during the crossing). When a crossing such as b for $ES_1 \rightarrow ES_2$, is between states of the same spin multiplicity it is called internal conversion; however, if the crossing is between states which differ in spin multiplicity it is called intersystem crossing. The relaxation of the lowest energy electronic excited state may involve a variety of processes, and these may be distinguished as either radiative ($h\nu'$) or non-radiative ($k_{nr}$). The non-radiative relaxation of the lowest energy excited state must involve a combination of crossings between the excited and ground state PES's and ground state vibrational relaxation processes and may include photoproduct as well as substrate PES's.

In summary, the simplest model for the behavior of an electronically excited molecule in the condensed phase typically involves a sequence of distinct steps: (a) initial generation of a Franck–Condon excited state with an electronic configuration different from that of the ground state; (b) rapid vibrational relaxation of this initial state to the lowest vibrational level of the initial electronic configuration; (c) internal conversion and/or intersystem crossing from the initial electronic excited state to a lower energy excited electronic state; (d) rapid vibrational relaxation of all the lower energy excited states to their potential energy minima; and (e) emission, relaxation and/or reaction of this lowest energy excited state. A real molecule will have many excited electronic configurations. Internal conversion processes in light atom systems are usually significantly more rapid than intersystem crossing processes, so that a highly excited molecule tends to relax first to its lowest energy excited electronic configuration of the initial spin multiplicity before it intersystem crosses to lower energy states of different spin multiplicity.[1–4] Spin-orbit coupling is likely to reduce the significance of the electronic spin restriction in transition metal complexes, and intersystem crossings sometimes compete with vibrational relaxation in transition metal systems.[5,6] Furthermore, since electronic spin is not a well defined quantum number in many of the heavier metal complexes, significant absorption bands originating from nominally "spin forbidden" transitions can often be identified.[7] In such cases, it is possible to directly excite into electronic states which are nominally spin forbidden, and it may not be useful to distinguish between internal conversion and intersystem crossing.

## 2.2 Some Qualitative Features of Excited State Potential Energy Surfaces

While PES's such as those in Figure 1 are conceptually very useful in discussions of photochemistry and photophysics, many of the important details of the excited state PES's of most transition metal complexes are not known. Some features can be constructed from absorption and emission energies, band shapes and/or fine structure (see chapter by Hitchman) when these are available (see chapters by Lever and Solomon and by Güdel and Brunold). However, absorption and emission often relate to different excited state PES's, absorption bands of several transitions are often convoluted and the ground and excited states may dissociate to different products (with respect to either or both their electronic configurations and their chemical composition). If the dissociations of the ground and excited state lead to different products, then the two dimensional excited state surface may appear nearly harmonic along the ground state dissociation coordinate as illustrated in Figure 2a. In general, one expects excited state dissociation energies to be smaller, all else being equal, than those of the ground state. If the excited state dissociation energy is larger than $4k_BT$, then the state can often be treated as a "bound state" with a characteristic set of reasonably well defined properties. If the excited state dissociation energy is smaller than $k_BT$, then the excited state is "dissociative." Direct excitation into a dissociative state

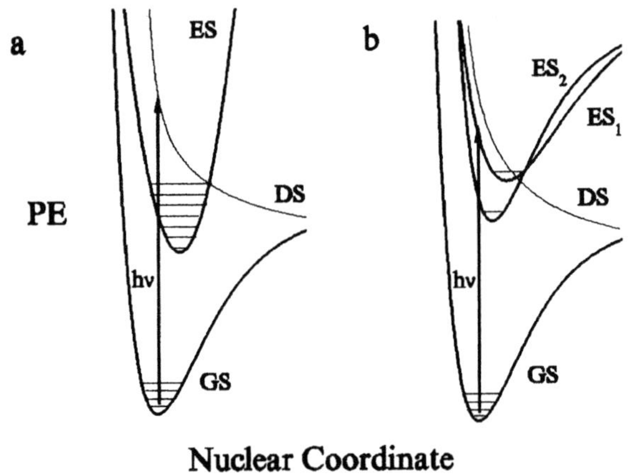


**Figure 2.** Qualitative potential energy surfaces illustrating some possible complexities of excited state relaxation processes and some possible effects of mixing with a dissociative PES: **a**, an excited state which is nearly harmonic along the GS dissociation coordinate and which crosses a dissociative excited state PES; **b**, two bound excited states and a dissociative excited state. These two figures illustrate a route for the net formation of dissociation products with and without direct excitation into the dissociative PES.

need not lead to net photochemistry if this initial electronic configuration mixes with a bound state in the region of their crossing, as illustrated in Figure 2a. Conversely, net photochemistry can result from excitation into a bound state as illustrated in Figure 2b. Finally, even if the photosystem does fragment along a dissociative nuclear coordinate, recombination in the condensed phase is possible if the kinetic energy of the separating particles is dissipated quickly enough through collisions with the solvent (the cage effect). Fast flash photolysis studies of the photodissociation of $I_2$ have nicely illustrated many of the features mentioned above (e.g., crossings from a dissociative to a bound excited state, relatively "slow" vibrational relaxation, recombination reactions to form ground and excited state species competitively).[8] The relative significance of the various possible decay pathways of an electronically excited molecule depend on factors such as the selection rules governing the mixing of electronic states and the efficiency of the transfer of vibrational energy from the molecular system to the solvent. The proposition that the rate of vibrational relaxation of electronically excited molecules is not always faster than internal conversions or intersystem crossings in the condensed phase has also been illustrated by the work on $I_2$.[8]

The simplest limit for interpretations of photophysical and photochemical behavior can be summarized in terms of one of the following pathways:

a. Photoexcitation populates a bound electronic excited state which rapidly relaxes to the lowest energy vibrational state within the electronic manifold (or PES). This vibrationally equilibrated excited state (VEqES) then internally converts and vibrationally relaxes until the lowest energy electronic excited state with the FC state spin multiplicity is reached. This lower energy VEqES will then decay through some combination of: (i) radiative relaxation (fluorescence) to the ground state; (ii) non-radiative relaxation to the ground state; (iii) the formation of chemical products; and (iv) intersystem crossing to states of different spin multiplicity followed by relaxation to the lowest energy VEqES. Such a low energy VEqES can decay by some combination of the steps (i)–(iii) listed above (radiative relaxation in this instance is called phosphorescence).
b. Photoexcitation populates a dissociative electronic excited state which proceeds smoothly to the dissociated products.

## 2.3 Characterization of Electronic Excited States

Obviously, electronic excited states are difficult to characterize. One would like to know both the molecular geometry and the distribution of electron density in each of the various relevant excited states (see the beginning of this section). Both issues are usually approached in terms of their differences from the corresponding properties of the ground state, and based on the interpretation of pertinent spectroscopic data. For example, the FC excited state has the ground state molecular geometry, but its exact electronic distribution is usually very difficult to determine. In the simplest limit, in which one is dealing with a single excited state, the electronic absorption band energy, intensity and band shape can provide much of the pertinent information (see chapter by Lever and Solomon). Thus, the molar absorptivity, $\varepsilon_{\omega}$, for light of energy $h\nu$ is given approximately by equation 1 (assuming that the electronic transition dipole moment, $\mu_{\text{ge}}$, is the same for all vibrational states),[9,10]

$$\varepsilon_{\omega} = \frac{2\pi|\mu_{\text{ge}}|^2\omega N_{\text{A}}}{(3\times 10^{10}\ln 10)nhc}\sum_{\text{w}}\frac{\Gamma\langle\chi_{\text{w}}|\chi_{\text{g}}\rangle^2}{(\omega-\omega_0-\omega_{\text{w}})^2+\Gamma^2} \tag{1}$$

where $n$ is the index of refraction, $c$ is the speed of light, $N_{\text{A}}$ is Avagadro's number, $\chi_{\text{w}}$ are excited state vibrational wavefunctions, $\chi_{\text{g}}$ is the ground state wavefunction, $\langle\chi_{\text{w}}|\chi_{\text{g}}\rangle$ is a vibrational overlap integral, $h\nu_0 = \Delta E^{00}_{\text{ge}}$, $h\nu_{\text{w}}$ is the correlated excited state vibrational quantum and $\Gamma$ is the lifetime limited line width. In the simplest limit of a single contributing vibrational mode and an approximately gaussian-shaped absorption band, the absorptivity can be represented by a simple function such as in equation 2,[11–13]

$$\varepsilon_{\nu} \cong \varepsilon_{\nu(\max)}\{\exp[-(\lambda - E_{\nu})^2/a]\} \tag{2}$$

$E_{\nu} = h\nu - \Delta E^{00}_{\text{ge}}$, $\Delta E^{00}_{\text{ge}}$ is the energy difference between the lowest energy ground and excited vibrational states, $\lambda$ is a nuclear "reorganizational" parameter and a is a band width parameter. From Equation 2 it is easily shown that the absorption maximum is given by Equation 3 and this is illustrated in Figure 3a.

$$h\nu \cong \Delta E^{00}_{\text{ge}} + \lambda \tag{3}$$

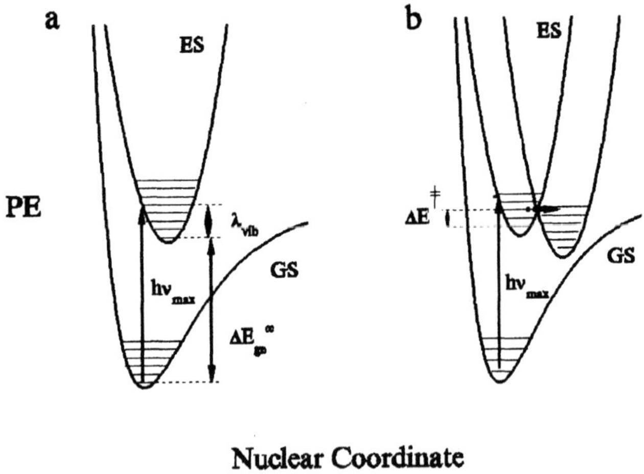


**Figure 3.** Potential energy surfaces illustrating the definition of parameters used in the text to describe excited state relaxation dynamics: **a**, $\lambda_{\text{vib}}$ and $\Delta E^{00}_{\text{ge}}$; **b**, activation energy, $\Delta E^{+}$.

In the classical limit, when $h\nu < k_B T$ (or when energy exchange between the internal, molecular vibrational modes and the solvent modes is very rapid), the full width of the absorption band for $\varepsilon_{\nu(\max)}/2$, $\Delta\nu_{1/2}$, is given by Equation 4.[12,13] The classical limit is not generally appropriate for molecular

$$\Delta\nu_{1/2} = 2[a \ln 2]^{1/2} = 2[4k_B T\lambda \ln 2]^{1/2} \tag{4}$$

systems and $a \cong 2h\nu\lambda_\nu \coth(h\nu/2k_B T)$.[11–14] The single mode limit is also rarely appropriate in large molecules, and the absorption bands very often correspond to the superposition of vibrational progressions in several modes. Consequently the intensity functions, as in Equation 2, must usually be summed over all the contributing vibrational modes. The parameters $\lambda_\nu$ are proportional to the squares of the magnitudes of the displacements of nuclei in the excited state, relative to their positions in the ground state, which contribute to the vibrational mode $\nu$ [i.e., for $\Delta Q_\nu$ the change in equilibrium nuclear positions for the normal vibrational mode $\nu$, $\Delta Q_\nu \propto (\lambda_\nu)^{1/2}$]. Thus, it is in principle possible to use the individual vibrational progressions which contribute to an absorption band intensity to construct the geometry of a molecule in an excited state in terms of the ground state geometry modified by a sum over displacements in the ground state vibrational modes. In practice, the necessary spectroscopic resolution is rarely achieved for the conditions used in photochemical studies (e.g., for ambient fluid solutions). However the spectroscopic information can almost always be used to set useful limits on the excited state structure.

If the electronic excited state luminesces, then the VEqES can be reasonably well defined (see chapter by Güdel and Brunold), and arguments similar to those summarized above for the absorption can be used to obtain $\Delta E_{ge}^{00}$, $h\nu$ and $\lambda_\nu$; an important difference is that for emission $h\nu \cong \Delta E_{ge}^{00} - \lambda$ (note that here, and in Equation 3, $\lambda$ is understood to be a sum over the relevant $\lambda_\nu$).

In principle, the molecular structure of the electronic excited state may also be inferred from resonance Raman spectroscopy (see chapter by Spiro and Czernuscowitz), which can yield some information about the nuclear distortion modes ($h\nu$) and the amplitudes of the corresponding nuclear displacements ($[\lambda_\nu]^{1/2}$).[9,10] Thus, resonance Raman spectroscopy has the potential of experimentally probing the structure of the excited state PES: (a) the vibrational lines observed can be related to the excited state distortions projected onto the ground state normal vibrational modes; and (b) the relative intensities of vibrational lines can be related to the distortion of the VEqES along the respective coordinates.[9,10] The resolution of the respective parameters is generally better by resonance Raman spectroscopy than by the absorption or emission spectroscopic methods mentioned above, and this method can often be used under the nearly the same conditions as those used in photochemical studies. Time resolved resonance Raman methods can be used to directly probe the excited state VEqES and to obtain information about excited state force constants (e.g., the shape of the PES).

These experimental approaches to excited state characterization have sometimes been complemented by theoretical calculations and almost always by more qualitative, semi-empirical arguments. This is important since the simple two electronic state models, which are the basis for most of the arguments sketched above, are rarely rigorously applicable to transition metal compounds.

Transition metal compounds typically have several electronic states separated by relatively small energy differences. The excited electronic states of photochemical interest often differ in their spin multiplicity from that of the ground state; however, some such transition metal excited state lifetimes can be very short. Absorption and resonance Raman spectroscopy most directly probe the spin allowed electronic transitions. When several electronic states are very close in energy, simple models of their electronic structures are usually inadequate (due to configurational mixing), and the spectroscopic analyses necessarily become much more complicated than sketched above.

The chemical characterization of excited states is largely based on product yields, reactions with chemical scavengers, the monitoring of photogenerated transients and spectroscopic comparisons to the ground state. The range of possible reactions includes the exchange of coordinated ligands for solvent species, changes of coordination number, homolytic bond breaking, isomerizations and rearrangements and oxidation-reduction reactions. Since the absorption of light usually results in moving an electron from one orbital to another, excited states are generally better oxidizing agents *and* better reducing agents than are their ground states. The excited state reduction potentials (see chapter by Lever and Dodsworth) are usually obtained by adding or subtracting, respectively, the VEqES/GS energy difference ($\Delta E_{ge}^{00}$) to or from the ground state reduction potential. This approach neglects an entropy contribution, which is usually argued to be small.

## 2.4 Classification of Transition Metal Excited States

For convenience of the following discussion we will divide transition metal excited states into five categories: (a) ligand field excited states in which the angular distribution of electron density is altered (with respect to the ground state) in predominately metal centered *d*-orbitals; (b) charge transfer excited states in which electron density has been moved (radially) towards (or away from) a transition metal center from (or to) a ligand moiety; (c) organometallic complex excited states in which it is difficult to distinguish clearly between metal and ligand-centered orbitals; (d) metal–metal bonded excited states in which electron density has been changed in the unique electronic states which result from the covalent interactions of two or more metal atoms; and (e) ligand-centered excited states in which the distribution of electrons has been altered in the ligand centered orbitals. Most of the discussion will be focused on the first two of these categories.

Charge transfer excited states involve donor and acceptor centers and are subdivided into categories depending on the nature of the donor and acceptor. The categories commonly distinguished are: (i) metal-to-ligand charge transfer (MLCT); (ii) ligand to metal charge transfer (LMCT); (iii) charge-transfer-to-solvent (CTTS); (iv) transition-metal-to-transition metal charge transfer (MMCT or sometimes "electron transfer"); and (v) ligand centered excited states.

## 3 EXCITED STATE RELAXATION PROCESSES

In this section, we will sketch some characteristic observations and suggest some general physical principles governing the non-radiative relaxation behavior of the excited states of transition metal complexes. We will attempt to use these postulated general principles as the basis for systematic discussions of transition metal photophysics and photochemistry. Detailed aspects of this material have been covered in other chapters of this series. The basic sequence of events is the absorption of light to generate the FC excited state, the relaxation of this state ultimately to generate the lowest energy VEqES, and then the relaxation of that state. The relaxation processes can encompass a variety of physical and chemical events at each stage. Most of the available information relates to the FC state and the lowest energy VEqES. The arguments in this chapter assume that certain parameters (discussed further below) dominate the relaxation behavior of the VEqES: (a) the rate of vibrational energy transfer, $k_{IVT}$, from high frequency ($h\nu_H$) to low frequency ($h\nu_L$) molecular or solvent modes (this includes the rate of vibrational equilibration, heat transfer, etc.); (b) the maximum frequency, $\nu_{eg}^{o}$, for crossing from VEqES into the ground state PES; (c) the magnitude of the displacement of excited state and ground state PES's ($\propto \lambda_{vib}^{1/2}$); (d) the ratio of $\lambda_{vib}$ to the difference in energy of the vibrational levels of the excited and ground electronic states ($\Delta E_{ge}^{00} = E_e^0 - E_g^0$; and (e) the coupling ($H_{ge}$) between the excited and ground state PES's. The exact significance of some of these parameters in determining the rates of non-radiative relaxation (rate constant, $k_{nr}$) is not always well established. In particular, the material in this section, and, to a significant extent, throughout this chapter, is based on the assumption that $k_{IVT}$ plays a major role in determining patterns of photophysics (and possibly of photochemistry) in coordination complexes. This assumption has not been critically evaluated, but it has proven useful in unifying and organizing much of this discussion.

Most transition metal complexes have several diabatic states which are close in energy so that the adiabatic states become complicated mixtures of the initially assumed, diabatic electronic configurations. The simplest descriptions of $k_{nr}$ assume that this mixing is intrinsically very small and that it is induced by the motion of nuclei in vibrational modes which contribute to $\lambda_{vib}$. We will start with a description of the simplest limit for a photoexcited system, then we will proceed with the discussion of observations on excited state behavior and their interpretation.

### 3.1 The Two State, Single Relaxation Channel Limit

This limit is based on diabatic PES's ($H_{ge}$ very small) and concepts basic to most excited state relaxation problems. It is convenient to consider the dynamic behavior of an electronically excited system for three different sets of conditions:

a. When $\Delta E_{ge}^{00} < \lambda_{vib}$ and/or $k_{IVT}$ is large compared to the maximum rate of crossing, $\nu_{eg}^{0} = \kappa_{el}\nu_{nu}$ (where $\kappa_{el}$ is the electronic transmission coefficient and is a function of $H_{ge}^2$ and $\nu_{nu}$ is the effective frequency of nuclear motion at the crossing point);[15] this assumption is equivalent to the equilibrium assumption of transition state theory). For this limit, the non-radiative relaxation rate constant is given by Equation 5.[15–17]

$$k_{nr}(T) = \nu^{o}_{eg} \exp\left[-\left(\lambda_{vib} - \Delta E^{00}_{ge}\right)^2 / 4\lambda_{vib} k_B T\right] \tag{5}$$

b. When $\Delta E^{00}_{ge} \geqslant \lambda_{vib}$, and when the vibrations which contribute to the displacement of the PES's (i.e., those vibrations which contribute to $\lambda_{vib}$) span a range of energies such that $h\nu_M \gg 2k_B T$ for some vibrational modes and $h\nu_L < k_B T$ for others, plus when $k_{IVT} < \nu^{o}_{ge}$, then one has Equation 6, which is a quantum mechanical variation on Equation 5, is usually applicable, where $\langle \chi^{o}_{e} | \chi_{gv}(M) \rangle$ is a vibrational overlap integral and this term can be represented by Equation 7 (below) with a small temperature dependence of $\gamma$ (Equation 8).[18–22]

$$k_{nr}(T) \cong \left[8\pi^2 H^2_{ge} / h^2 \nu_M\right] \sum_{V} \langle \chi^{o}_{e} | \chi_{gv}(M) \rangle^2$$
$$\times \exp\left\{-\left[E^{00}_{ge} - v h\nu_M - \lambda_L\right]^2 / 4\lambda_L k_B T\right\} \tag{6}$$

Relaxation processes which fall into this category depend on the condition that $k_{IVT} < \nu^{o}_{eg}$ for the vibrational modes with the largest quanta. In effect this limit treats these "acceptor" vibrational states as independent PES's, each with an intrinsic, classical activation barrier (as in Equation 5) for crossing from the VEqES to a high frequency vibrational level of the GS for which the barrier is based on the contributions of the low frequency modes to $\lambda_{vib}$.

c. If $\Delta E^{00}_{ge} \gg \lambda_{vib}$, and if $h\nu \gg k_B T$ for all the vibrational modes which contribute to $\lambda_{vib}$ (the "low temperature limit" illustrated in Figure 3a), and if $k_{IVT} < \nu^{o}_{eg}$ for all these modes, then the non-radiative relaxation rate should be temperature independent, and when a single vibrational mode contributes Equations 7 and 8

$$k^{0}_{nr} = H^2_{ge}\left[8\pi^3 / (h^3 \nu_M \Delta E^{00}_{ge})\right]^{1/2} \exp\left(-\gamma \Delta E^{00}_{ge} / h\nu_M\right) \tag{7}$$

$$\gamma = \ln\left[\Delta E^{00}_{ge} / (\lambda_M)\right] - 1 \tag{8}$$

should be used.[16,23] This is the limiting low temperature form of $k_{nr}$, and it assumes that the contributing nuclear vibrational mode has energy $h\nu_M$ and that $\lambda_{vib} = \lambda_M$.

In simple cases, the contributions of specific high frequency and low frequency vibrational modes are correlated with the contributions of these vibrational modes to absorption, emission or resonance Raman spectra. These spectroscopic methods can help identify about the vibrational modes which contribute to $\lambda_{vib}$ and their contribution to excited state distortion. However, the high frequency vibrational modes ($h\nu_M$ or $h\nu_H$) which contribute most significantly to Equations 7 and 8 are most likely those for which a $k_{IVT}$ is smallest (usually the modes with the largest quanta since these should be the slowest to relax), while the modes which are weighted most strongly in the spectroscopic measurements are the ones which contribute most to $\lambda_{vib}$.

The coupling matrix elements which are represented in Equations 5–7, both in principle and in practice, span a range of interactions. In order to address the major issues in a simple way it is necessary to assume that the state wavefunctions can be factored into electronic, $[\psi_i^{el}(r;\,Q)]$ and nuclear (or vibrational) $[\chi_j^{nu}(Q)]$ wavefunctions for $r$ and $Q$ the electronic and nuclear spatial coordinates, respectively (the standard Born–Oppenheimer approximation): $\Psi_k(r;\,Q) = \psi_k^{el}(r;\,Q)\chi_k^{nu}(Q)$. The Hamiltonian operator is a sum of kinetic and potential energy operators, $\mathcal{H} = \mathcal{T} + \mathcal{V}$, and two limiting approaches to $H_{ge}$ may be defined:

a. When the intrinsic electronic coupling between the excited and ground state wavefunctions is negligible (i.e., $\langle\psi_g^{el}|\mathcal{V}|\psi_e^{el}\rangle \cong 0$), then the nuclear motions can introduce a perturbation which results in coupling: $\langle\psi_g^{el}|\mathcal{T}|\psi_e^{el}\rangle \neq 0$. A Taylor's series expansion of PE around the equilibrium nuclear coordinates, $Q_0$, gives rise to a purely electronic matrix element, $J \propto \langle\psi_g^{el}|\partial(\text{PE})/\partial Q|\psi_e^{el}\rangle/|E_e - E_g^0|$, and $H_{ge}$ can be written as in Equation 9,[23,24] where $K_{uv}$ is the spin-orbit coupling matrix element.

$$H_{ge} \cong T_{ge} = \sum_{\gamma\neq m,n}\sum_{k}\left[K_{m\gamma}J_{\gamma n}^{k0} + J_{m\gamma}^{k0}K_{\gamma n}\right]\langle\chi_m|\partial/\partial Q_k|\chi_n\rangle \tag{9}$$

b. When there is significant electronic coupling so that $H_{ge} \cong \langle\psi_g^{el}|\mathcal{V}|\psi_e^{el}\rangle = V_{ge} \gg 0$, then this term may dominate the coupling and $V_{ge}$ may be much larger than $T_{ge}$. This limit gives rise to exchange, superexchange, etc. coupling.[25,26]

The first of these approaches is commonly used in connection with non-radiative relaxation, especially in the low temperature limit described by Equation 7. The second is commonly used with electron transfer reactions and Equations 5 and 6. This distinction is not always intrinsic. It can depend on which is the larger perturbation, and in some cases it will depend on the choice of excited and ground state wavefunctions (i.e., whether the problem is formulated in terms of diabatic or adiabatic wavefunctions). Furthermore, the contributions of $T_{ge}$ and $V_{ge}$ would not be neatly separable if, for example, the vibrational mode implicated in the mixing of $\psi_g^{el}$ and $\psi_e^{el}$ in (a) was also the principle distortion mode in a third, higher energy electronic state which contributed to $V_{ge}$ in (b). This last possibility is sometimes referred to as "vibronic coupling".[27–30]

### 3.2 FC State Relaxation in Transition Metal Complexes: Issues

Much of the photophysical and photochemical behavior which is observed depends on the evolution of the electronic distribution and changes in molecular geometry as the Franck–Condon excited state relaxes. These processes occur very quickly, usually in subpicosecond time regimes, and they are consequently difficult to investigate.

However, theoretical modeling is becoming increasingly powerful, as is discussed in other chapters of this series, and short pulsed laser systems have been increasingly used to explore the issues related to early events in excited state evolution. The details of theoretical modeling of the Franck–Condon excited state will not be treated here. However, several general questions can be raised in regard to relaxation of the FC excited state:

a. To what degree is it appropriate to treat the FC state with formalisms based on PES's?
b. Is relaxation always dominated by vibrational relaxation within each successive electronic PES, or do internal conversion and intersystem crossing compete kinetically with such vibrational relaxation (in contrast to the limit described in Section 3.1)?
c. What principles might govern the competition between internal conversion/intersystem crossing and vibrational relaxation within a PES?
d. How rapid is vibrational relaxation within an electronic PES?
e. How rapidly does the excess energy flow from the vibronically excited molecule to the condensed phase matrix?

### 3.3 FC State Relaxation: Observations on Transition Metal Complexes

So far, the experimental studies have not been altogether representative or systematically described. However, the direct probing of FC state relaxation most often requires subpicosecond flash photolysis techniques and these techniques have not become commonplace. The studies currently available do not always involve such direct probing of the FC state, and they can be divided into classes (with examples of each class to be discussed below): (i) simple molecules ($I_2$); (ii) complex molecules which relax along a dissociative coordinate [$Cr(CO)_6$]; (iii) molecules in which the absorption of light promotes a radial redistribution of electron density, or charge transfer transitions [$Ru(bpy)_3^{2+}$]; (iv) molecules in which the absorption of light promotes an angular redistribution of electron density, or ligand field (d–d) transitions [$Cr(NH_3)_6^{3+}$]; and (v) molecules in which the absorption of light promotes electron transfer between a well defined donor and acceptor [e.g., $LM^{III}(CN^-)Ru^{II}L' + h\nu \rightarrow LM^{II}(CN^-)Ru^{III}L'$; this is a special case of class iii, above]. The observations on FC relaxation in each of these classes will be summarized briefly.

The photophysical behavior of $I_2$ in solution has long played a paradigmatic role in thinking about condensed phase photophysics and photochemistry.[2,3,31–33] Picosecond flash photolysis studies have found that the initial FC excited state of $I_2$ can lie along a bound PES (labeled the B state by Mulliken[34]), but this state relaxes and/or crosses to a lower energy dissociative state ($^1\Pi_u$) very rapidly (<1 ps) and predissociation to I atoms occurs in less than 1 ps.[8] The recombination of I atoms competitively produces an intermediate excited state and vibrationally excited ground state molecules. Relaxation of the vibrationally excited ground state $I_2$ molecules takes about 100 ps.[8] In this system the crossing between states (B $\rightarrow$ $^1\Pi_u$)

is much faster than vibrational relaxation within the ground state PES. This is an example of what appears to be a common feature of FC state relaxation in heavy atom systems: Internal conversion and intersystem crossing are often faster than vibrational relaxation.

Another much studied species is $Cr(CO)_6$.[35] Irradiation of this complex efficiently produces "$Cr(CO)_5$" and CO in less than 250 fs in the condensed phase.[36,37] Solvent substitution into the coordinately unsaturated $Cr(CO)_5$ fragment takes 15–20 ps and appears to be limited by the rate of dissipation of the excess vibrational energy, in the $Cr(CO)_5$ fragment as it exits the FC state, to the solvent.[37,38] This again demonstrates that heat flow between the excited molecule and the solvent can be inefficient.

The $Ru(bpy)_3^{2+}$ complex is an example of a photosystem with a bound, well characterized VEqES.[35,39–41] This metal-to-ligand charge transfer (MLCT) state has nominal triplet spin multiplicity, although spin orbit coupling results in the splitting of this state into components,[42–44] and the spin forbidden relaxation to the singlet ground state is relatively slow (~600 ns in water under ambient conditions).[35,39–41] Excitation into the $^1$MLCT excited state has been found to produce a FC excited state which relaxes to the vibrationally excited $^3$MLCT state with $k_{nr} \sim 10^{13}$ s$^{-1}$.[6] Vibrational relaxation of the MLCT excited states seems to be significantly slower than intersystem crossing.[6]

The photophysical and photochemical behavior of the ligand field excited states of Cr(III) complexes have been studied extensively.[35,39–41,45–48] Ligand field excitation of the $^4A_{2g}$ ground state ($O_h$ symmetry notation is used for purposes of illustration and simplicity) generally results in an FC state (e.g., $^4T_{2g}$) in which an electron has been promoted from a non-bonding to an antibonding orbital. Vibrational equilibration of such a FC state within the $^4T_{2g}$ PES would result in considerable distortion of the molecular geometry, and a substantial displacement of the ES PES minimum, with respect to that of the ground state along the distortion coordinate. The reported attempts to investigate FC relaxation in Cr(III) complexes have utilized ps excitation and the evidence so far has indicated that the lowest energy $^2E_g$ excited state is formed in much less than 10 ps after excitation.[5,46] No direct observation of intersystem crossing or vibrational relaxation has been reported. Indirect evidence, usually based on the quenching of photochemical products and of $^2E_g$ luminescence, suggests that intersystem crossing ($^4T_{2g} \rightarrow {}^2E_g$ in complexes with $O_h$ symmetry) occurs more rapidly than vibrational equilibration of the higher energy ligand field excited states (e.g., the $^4T_{2g}$ state) in Cr(III) complexes in ambient solutions.[5,46–48]

Complexes in which transition metal, electron transfer donors (D) are covalently linked to transition metal acceptors (A) have been selected as the final class of compounds to be discussed in this section because some information has become available about several types of complexes. We will focus mostly on the D/A complexes in which the electron transfer partners are linked by $CN^-$. The greatest amount of information is available for complexes of the type $(NC)_5Ru^{II}(CN^-)Ru^{III}(NH_3)_5^-$ [other complexes in this class substitute Fe(II) or Os(II) for Ru(II) and/or Os(III) for Ru(III)].[49–52] The Ru(III)/Ru(II) FC excited state has been found to relax to a vibrationally excited ground state species in about $10^2$ fs,[49–52] while vibrational relaxation of the $CN^-$ stretch appears to occur over about 10 ps[49–52] and solvent relaxation times appear to range from faster to much slower than these relaxation times. Similarly, rapid ($<\sim 10$ ps) recovery of the ground state has been reported in closely related $LRu^{II}$ $(CN^-)Ru^{III}(NH_3)_5^{n+}$ complexes in which L represents a combination of polypyridyl ligands (bpy, $CN^-$ and/or tpy).[53,54] The back electron transfer of the VEqES would be nearly barrierless for the $Ru(CN)_5^{3-}$ donor and well within the inverted region ($|E_{eg}^{00}| \gg \lambda_{vib}$) for the $LRu^{2+}$ donors with polypyridyl ligands.[49–55] Related complexes in which a $Ru(NH_3)_5$ donor is $CN^-$ bridged to a Co(III) or a Cr(III) acceptor do have low energy CT excited states which differ from the ground state in spin multiplicity, and well defined, photogenerated electron transfer transients have been observed in such systems to have lifetimes in the range of $\sim 5 \times 10^{-10}$ to $\sim 10^{-8}$ s.[53,55–57] These transients apparently correspond to reasonably well defined VEqES's for the electron transfer excited states of complexes with ground state Co(III) and with Cr(III) centers. It appears that these latter complexes preferentially relax to their intermediate electronic excited states rather than to the vibrationally excited ground state despite the nominal spin forbiddeness of the net crossing from their FC states. This is probably most surprising for the complex with the Cr(III) acceptor, since the DACT absorbance of the ground state implies that the electronic coupling matrix element is large and very comparable to that of the $Ru^{II}$ $(CN^-)Ru^{III}$ analogs ($H_{ge} \sim 3 \times 10^3$ cm$^{-1}$).[56,57] This could imply that intersystem crossing occurs in the Cr(III)-acceptor complexes with $k_{isc} > 10^{13}$ s$^{-1}$, based on the rate of back electron transfer observed[49–52] in the $Ru^{II}$ $(CN^-)Ru^{III}$ complexes. Alternatively, one could infer that the coupling matrix element is so large in these complexes that the PES formalism is inappropriate and that the transient "vibrational" states should be treated as quasi-stationary vibronic states with mixed electronic character.

In contrast to the points raised by the work summarized in the previous paragraph, we note that when the electron transfer donor and acceptor are weakly coupled ($H_{ge} < 10^2$ cm$^{-1}$), as in the $(bpy)_2$ $Ru^{II}(bb)Co^{III}(bpy)_2^{5+}$ complex [bb = 1,2-*bis*(2,2′-bipyridyl-4′-yl)ethane], the simple limiting model described in Section 2 above does seem applicable: Excitation into the $^1$MLCT band of the $Ru(bpy)_2^{2+}$ moiety produces the expected $^3$MLCT excited state which is quenched by electron transfer to Co(III) ($k_q \sim 5 \times 10^9$ s$^{-1}$ under ambient conditions).[58]

### 3.4 Some Principles Which May Govern FC Excited State Relaxation

Obviously, a great deal more experimental information is necessary before one can be comfortable with any hypothesis about FC excited state relaxation processes in transition metal complexes. The ideas outlined here are presented in the spirit of a "working hypotheses" which may help to put the issues in some kind of perspective, and which may help systematize further experimental studies. One can expect much better defined patterns of FC state behavior and more rigorous statements of the rules governing them to emerge in the near future.

First of all, one can, somewhat tentatively, identify some patterns for FC state relaxation in these transition metal complexes: (i) the fast relaxation processes seem to favor population of an intermediate ES over the ground state when such kinetic competition is possible; (ii) vibrational relaxation appears to be slower than internal conversion/intersystem crossing when the coupling matrix elements (or $H_{ge}$ in Equation 7) are large; and (iii) changes of spin multiplicity do not appear to exert a retarding effect of sufficient consequence to make intersystem crossing slower than vibrational relaxation in such cases. The first of these points appears to be a straightforward manifestation of the "energy gap law" in Equation 7, and this leads to the first hypothesis: The relative rates of population of intermediate excited states (through internal conversion or intersystem crossing) and the ground state from the FC state are governed by some form of Equation 7. Since Equation 7 is usually only considered applicable to the VEqES at low temperatures, this proposal is not altogether obvious, and it will be elaborated to a limited extent below.

The second hypothesized principle governing FC state relaxation behavior is that vibrational relaxation is a highly constrained process. The constraints are: (i) energy conservation, since any vibrational energy lost within a PES must appear in either some other degrees of freedom of the system; (ii) the most probable vibrational relaxation processes involve intervibrational transfer between the dominant high frequency vibration, $h\nu_v$, which is populated near the FC state and another molecular or solvent vibration, $h\nu'_v$, which is sufficiently similar in energy that $|h\nu'_v - h\nu_v| \leqslant 200\ \text{cm}^{-1}$ (where it is assumed that the low frequency modes of the molecule and the solvent have quanta $h\nu_L \leqslant 200\ \text{cm}^{-1}$);[59] and (iii) vibrational relaxation processes which convert $h\nu_v$ into a larger number of low energy vibrational quanta are far less probable.[59] Exceptions to this pattern should occur only for molecular vibrational modes which are very strongly coupled to solvent modes. In general, molecular high frequency vibrational modes (such as $\nu_{CN}$) relax the most slowly because there are no other molecular vibrational modes within 200 cm$^{-1}$ to act as acceptor modes. Conversely, low frequency molecular modes (such as metal–ligand skeletal modes) should relax relatively rapidly since the relaxation criteria are more readily satisfied. Of course, if a great deal of vibrational energy must be deposited in the solvent, then complete relaxation to the VEqES will depend on the rate of heat flow from neighboring to bulk solvent molecules, as apparently has been observed for the $Cr(CO)_5$ fragment.[37]

The use of an expression such as Equation 7 to rationalize patterns of FC state relaxation warrants some additional comments. First of all, one expects events to occur so rapidly following FC state generation that the rates of the various relaxation processes may be much faster than the rate at which heat flows into (or, as noted above, out of) the relaxing molecule; hence, a nearly temperature independent rate constant is expected. Secondly, the probability of surface crossing will depend on vibrational overlap, and the exponential form of Equation 7 (with $\Delta E^{00}_{ge}$ understood to be the energy gap of the relevant excited states) should provide a reasonable estimate. However, this presumes that the coupling matrix element is sufficiently small that the PES's of the electronic states involved in the crossing can be meaningfully distinguished. When $H_{ij}$ is very large and/or the states are close in energy, this approach is not likely to be useful. In such instances, it may be more useful to consider the crossing to be from a vibronic state, with a wavefunction of the form given by

Equation 10

$$\Psi_{ij} \cong \psi_{ab}^{\mathrm{el}}(\chi_{a,i} + \alpha_{ab}\chi_{b,j}), \tag{10}$$

where $\psi_{ab}^{\mathrm{el}}$ is an adiabatic electronic wavefunction for the strongly mixed electronic states, $a$ and $b$, the $\chi_{k,l}$ are the respective vibrational wavefunctions and $\alpha_{ab}$ is a mixing coefficient (see also Section 3.2). Then "intersystem crossing" from such a vibronic state to an electronic state of different spin multiplicity could be considered in the usual way since the coupling matrix element should be small. The relaxation among high frequency vibronic states, $\Psi_{ab}$, which depends on the rate of energy transfer to low frequency vibrational modes (usually of the solvent), would be considered to compete with intersystem crossing in a manner which can be treated with standard kinetic approaches (see Section 3). The actual partitioning of the FC excited state relaxation between crossing to lower energy excited states and relaxation within the manifold of vibronic states is likely to depend on several factors such as: (i) vibrational overlap; (ii) the rate of transfer of energy from vibrational excited states within the molecule to the solvent; (iii) the selection rules for the vibronic matrix elements;[60] and (iv) the relative slopes of the PES's in the region of their crossing (since such a contribution is expected for $T_{if}$; Equation 11).[23]

$$T_{if} \propto \langle \psi_f^{\mathrm{el}} | \partial(\mathrm{PE})/\partial Q | \psi_i^{\mathrm{el}} \rangle / |E_i - E_f| \tag{11}$$

Current studies in several laboratories are dealing with the issues discussed here, and one can expect some resolution of the problems in the future.

### 3.5 Relaxation of the VEqES

There are very few transition metal complexes in which more than one VEqES has been identified. The basic condition for a well defined VEqES is that the rate of internal conversion and/or intersystem crossing from the PES for such a state must be significantly slower than vibrational relaxation within it. In the relatively rare instance that more than one VEqES is observed in transition metal systems, it is usually a consequence of large differences in the displacement coordinates. Particularly clear examples of systems with more than one VEqES have been found in complexes of the type $Ru(bpy)_2(MeArbpy)^{2+}$ where MeArbpy is a 5-methyl-bipyridine ligand with a pendant aromatic moiety in the 5' position. The local $(^3MLCT)Ru(bpy)_2^{2+}$ and triplet excited state of the pendant aromatic seem to be only weakly coupled, and when Ar = pyrene, the energies of the triplet states are comparable and two independent emissions are observed.[61]

Relaxation mechanisms are intrinsically very different in the nested surface ($\Delta E_{\mathrm{eg}}^{00} > \lambda_{\mathrm{vib}}$) and in the surface crossing ($\Delta E_{\mathrm{eg}}^{00} \ll \lambda_{\mathrm{vib}}$) limits. In the former, radiative and nonradiative relaxation processes may compete while in the latter only nonradiative processes are possible. The nonradiative relaxation of a VEqES in the nested surface limit can occur by means of some combination of two distinguishable classes of mechanisms (or relaxation channels, since each usually involves different nuclear coordinates): (i) relaxation to a vibrational excited state of the ground state by means of a tunneling mechanism as discussed in Section 3.1; or (ii) thermally activated relaxation. Thermally activated relaxation can involve any of several possible relaxation channels: (a) thermal population of vibrational modes of the VEqES which have better overlap with the ground state vibrations; (b) thermally promoted crossing to a neighboring electronic excited state PES which relaxes more rapidly to the ground state; or (c) thermally promoted distortion of the excited state to form a coordination environment in which the electronic configurations are combinations of the original ground and excited states and in which relaxation is relatively efficient. The first two of these are examples based on concepts mentioned above, while the last arises especially in ligand field excited states because the energy differences of the orbital configurations for these electronic states arise largely from differences in the angular distributions of electrons, and the configurational energy differences are not very large, i.e., if the L–M–L angles of the complex are changed, then values of $\Delta E_{ge}^{00}$, $\lambda_{vib}$ and the degeneracies of the electronic states also change.

Photoinduced, unimolecular chemical reactions can be regarded as special cases of nonradiative relaxation processes. In most molecules, $\Delta E_{eg}^{00}$ exceeds the energy required to break single ground state M–L bonds, to induce isomerization, etc. The rules governing electronic to vibrational energy transfer for nested excited states, sketched above, favor population of vibrations with large quanta and which contribute to $\lambda_{vib}$. *This implies a very small probability for direct population of the very low frequency vibrations which are necessarily associated with a chemical reaction coordinate.* This expectation is probably consistent with such observations as the definite patterns of photoinduced ligand substitution (which are often stereospecific and which produce distributions of products which differ from those obtained from thermal reactions of the ground state) and with the intervention of chemical reactions of solvent cage-trapped photoreaction products in many systems. On the other hand, photochemistry often accounts for less than 15% of the total of all nonradiative relaxation processes of the VEqES in 6-coordinate complexes, and there does not always need to be a correlation between photochemistry and the dominant relaxation channel.

In the following sections, we discuss examples of the relaxation of ligand field and charge transfer VEqES's in both limiting categories.

## 4 EXAMPLES OF TRANSITION METAL EXCITED STATE RELAXATION IN THE "NESTED SURFACE" LIMIT

First, we will consider relaxation processes which have been observed when the lowest energy excited state is a ligand field excited state. Then we will discuss some examples from the study of charge transfer excited states.

### 4.1 Relaxation of Ligand Field Excited States

VEqES relaxation processes have been experimentally well defined for a wide range of Cr(III) complexes ($3d^3$ electronic configurations) and for several Rh(III) complexes ($4d^6$ configurations). The qualitative differences in behavior, from the present point of view, arise from the fact that the $(^2E)$Cr(III) excited state electronic configuration is the same as that of the $(^4A_{2g})$ ground state, while the lowest energy $(^3T_{1g})$Rh(III) excited state differs in orbital configuration from the $(^1A_{1g})$ ground state by the promotion of an electron from a non-bonding to an anti-bonding orbital. As a consequence $\lambda_{vib} \sim 0$ for VEqES relaxation of Cr(III) complexes while $\lambda_{vib}$ and the correlated nuclear displacements are very large for the VEqES relaxation in Rh(III) complexes. However, for both series of complexes $\Delta E_{ge}^{00} > \lambda_{vib}$: for $(^2E_g)Cr(NH_3)_6^{3+}$, $\lambda_{vib} \sim 0$ and $\Delta E_{ge}^{00} \cong 15 \times 10^3$ cm$^{-1}$;[62] for $(^3T_{2g})Rh(NH_3)_6^{3+}$, $\lambda_{vib} \sim 3.7 \times 10^3$ cm$^{-1}$ and $\Delta E_{eg}^{00} \cong 20.7 \times 10^3$ cm$^{-1}$.[63] The contrast in VEqES lifetimes of the two hexaammines reflects their structural differences in a complex way. At 77 K, $(^2E_g)Cr(NH_3)_6^{3+}$ has about 4 times the lifetime of $(^3T_{2g})Rh(NH_3)_6^{3+}$ (75 $\mu$s compared to 19 $\mu$s)[46,48,64] while at room temperature the ratio is about 200/1.[46,48,64] To some degree this can be viewed as the consequence of the smaller width of the effective tunneling barrier for the Rh than for the Cr complex (i.e., the Franck–Condon vibrational overlap will be larger the closer that $\Delta E_{eg}^{00}/\lambda_{vib}$ is to 1; in terms of Equation 7, the larger the value of $\gamma$, the smaller the value of $k_{nr}$). These interpretations are qualitatively sound, but there appear to be some problems of detail in fitting the observations to theoretical expressions such as that in Equation 7.

The first order rate constant for $(^2E)$Cr(III) $\rightarrow$ $(^4A_2)$Cr(III) relaxation, $k_{eg} = \tau_e^{-1}$ (for $\tau_e$ the mean lifetime for the $^2E$ decay), can be represented as in Equation 12, where

$$k_{eg} = k_r + k_{nr}^0 + k_{nr}(T) \tag{12}$$

$k_r$ is the radiative rate constant, $k_{nr}^0$ is the limiting, temperature independent nonradiative rate constant (usually resolved at low temperature) and $k_{nr}(T)$ represents the thermally activated $^2E$ decay. In most systems, the emission quantum yield is small, so $k_r \ll k_{nr}^0$ and it will be neglected in most of this discussion. Theoretical arguments suggest that $k_{nr}^0$ should be described by Equation 7.[16] Strong qualitative support for this is provided by the dramatic increase in the limiting low temperature $^2E$ lifetime which has been found to accompany perdeuteration of am(m)ine ligands, e.g., $k_{nr}^0$ decreases about 70 fold upon perdeuteration of $Cr(NH_3)_6^{3+}$.[45–48] This kind of observation establishes that the high frequency N–H vibrations are usually the dominant deactivating modes, consistent with expectation since $\Delta E_{eg}^{00}/h\nu$ is larger for the ND than for the NH stretch, e.g., the observations on $Cr(NH_3)_6^{3+}$ would be consistent with Equation 7 if all the N–H bonds could contribute to deactivation and if the N–H contribution to $\lambda_{vib}$ is on the order of $10^2$ cm$^{-1}$. The fits to Equation 7 have been examined more critically for a series of simple hexa-am(m)ine and cyano-am(m)ine complexes.[65,66] For these complexes, it has been shown that $k_{nr}^0$ is roughly proportional to the number of coordinated N–H moieties, $N_{NH}$, for selected hexa-am(m)ine complexes (there are systematic deviations and possibly even a slight positive curvature for the complexes selected for the correlation). This correlation is overly simple since $N_{NH}$ should enter the expression for $k_{nr}$ through $\gamma$, defined as in Equation 8 (i.e., by replacing $\lambda_M$ in Equation 8 by $d_M\,\lambda_M$ where $d_M$ is the number of contributing vibrational modes), but it is a clear illustration of the dependence on high frequency acceptor modes. An empirical function which is more closely based on Equation 7, such as in Equation 13,[65] can better correlate the lifetimes of these

$$\left(h\nu/\Delta E^{00}_{eg}\right)\ln\left[\left(k^0_{NH}-k^0_{ND}\right)/\left(k^0_{ND}-k_r\right)\right]\cong 0.5\ln N_{NH}-0.6 \qquad (13)$$

complexes and it has been suggested that the failure of very simple correlations with the number of high frequency modes implies that: (a) there are some unique, relatively heavily weighted contributions of the low frequency modes even in the temperature independent limit; and (b) there is some variation in $H_{ge}$ through the series of Cr(III) complexes.[65,66]

As the temperature is increased, $k_{nr}(T)$ becomes increasingly important in $(^2E)$Cr(III) relaxation. Once more it seems unlikely that a single nonradiative relaxation channel can account for all the observed behavior. In many complexes, the $(^4T_2)$Cr(III) excited state is close enough to the energy of the $(^2E)$Cr(III) excited state that thermal population of the upper state becomes a possibility. Since the $^4T_2$ state is highly distorted, it is often assumed that a principle mechanism for $^2E$ relaxation and reaction involves "back intersystem crossing" from $^2E$ to populate the $^4T_2$ state. The arguments above can be used to examine the propensity of the $^4T_2$ state to relax to the ground state. Wilson and Solomon[62] have identified the principle distortion modes and the amplitudes of the displacements in the corresponding normal coordinates from the vibronic progressions resolved from the $^4A_{2g}\rightarrow{}^4T_{2g}$ absorption of $Cr(NH_3)_6^{3+}$ in the crystalline solid at 8 K. Their observations indicate that the $^4T_{2g}$ excited state of this complex distorts along $a_{1g}$ and $e_g$ vibrational coordinates (compression of 0.02 Å and expansion of 0.12 Å per CrN bond, respectively).[62] The progressions were well resolved which indicates that this is a bound state.[62] From this work, one obtains $\Delta E^{00}_{eg}=20\times10^3$ cm$^{-1}$ and $\lambda_{vib}\cong 2.2\times10^3$ cm$^{-1}$, so that $\Delta E^{00}_{eg}/\lambda_{vib}\cong 9$. As a consequence, the spectroscopy of this complex indicates that the $^4T_{2g}$ excited state is a highly nested, bound state, and it is not obvious from these considerations how this excited state could dramatically facilitate $^2E$ relaxation. Further to this point, it is useful to compare the nonradiative relaxation behavior of $Cr(NH_3)_6^{3+}$ to that of $Rh(NH_3)_6^{3+}$. An analysis of the vibronic fine structure of the $(^3T_{1g})Rh(NH_3)_6^{3+}\rightarrow(^1A_{1g})Rh(NH_3)_6^{3+}$ emission indicates that $\Delta E^{00}_{eg}\cong 20.7\times10^3$ cm$^{-1}$ and $\lambda_{vib}\cong 3.7\times10^3$ cm$^{-1}$,[63] so that $\Delta E^{00}_{eg}/\lambda_{vib}\cong 5.6$. In fluid solution and ambient temperatures $k_{nr}\cong 5\times10^7$ s$^{-1}$ for the $(^3T_{1g})Rh(NH_3)_6^{3+}$ excited state.[64] If we neglect the different contributions from $H_{ge}$, then the Franck–Condon contributions to the high temperature form of Equation 7 implies that the $(^4T_{2g})Cr(NH_3)_5^{3+}\rightarrow(^4A_{2g})Cr(NH_3)_6^{3+}$ relaxation process should be orders of magnitude less efficient than the $(^3T_{1g})Rh(NH_3)_6^{3+}\rightarrow(^1A_{1g})Rh(NH_3)_6^{3+}$ relaxation: The basic physical model requires that the more nested system relax more slowly.

However, for the distortion coordinate $h\nu_d$ is only ~400–500 $cm^{-1}$, comparable to (2–3)$k_BT$ (for ambient conditions), but the classical relaxation rate, Equation 5, in this regime is prohibitively small ($k_{nr} < k_r$) and even the thermally activated relaxation of $(^3T_{1g})Rh(NH_3)_6^{3+}$ must require the intervention of some pathway which does not correlate with the ground state spectroscopy in the way predicted by Equations 5–7. Relaxation rates in these complexes do not correlate with the efficiency of product formation,[48] but thermal activation of a distortion leading to an intermediate of different coordination geometry could function as the dominant relaxation channel. Even with spin-orbit coupling, the change in spin multiplicity probably leads to a smaller value of $H_{ge}$ for Rh than for Cr, but this would not be expected to overwhelm the Franck–Condon factors which indicate that the $(^3T_{1g})Rh(NH_3)_6^{3+}$ and $(^4T_{2g})Cr(NH_3)_5^{3+}$ excited states are bound states. A relevant observation, mentioned in Section 3.3, is that the $(^4T_{2g})$Cr(III) → $(^2E)$Cr(III) relaxation processes are very efficient and that non-radiative relaxation of the upper state directly to the ground state is not kinetically competitive.[46–48] On the other hand, there does appear to be a significant amount of $Cr(NH_3)_5OH_2^{3+}$ product formed in competition with the ultra fast $(^4T_{2g})Cr(NH_3)_6^{3+} \rightarrow (^2E)Cr(NH_3)_6^{3+}$ interstytem crossing. This part of the photogenerated product formation apparently derives from the vibrationally relaxing FC state and might be used to infer that $k_{nr} \sim 10^{12}$ $s^{-1}$ for the $^4T_{2g}$ state. (Note that this is ~$10^3$ times larger than $k_{nr}$ for the more distorted Rh(III) analog discussed above.) This inference is based on product formation not the recovery of the ground state and it may not be relevant to the VEqES. There can be no doubt that the $^4T_2$ states of many complexes are thermally populated under ambient conditions, or that the distorted $^4T_2$ excited states should be more reactive, both in the sense of product formation and the recovery of the ground state, than the much less distorted $^2E$ excited states of Cr(III) complexes. However, the activation energies inferred for $k_{nr}(T)$ rarely exceed estimates of $E^0(^4T_2) - E^0(^2E)$, which is not consistent with a bound state, and $k_{nr}(T)$ can often be largely quenched by placing the Cr(III) complex in a solid matrix,[48,66] which is consistent with the $^4T_2$ excited state being a bound state. The reconciliation of these contradictory inferences has been elusive.

An approach to reconciling the spectroscopic and excited state relaxation time observations on Cr(III) complexes would be to postulate that a dominant relaxation pathway might involve a large amplitude angular distortion of the six coordinate complex to form a coordinatively unstable intermediate in which the electronic states are mixtures of those of the octahedral precursor. An intermediate which would meet most of these requirements would be one in which the coordination geometry was trigonal prismatic (TPR). The lowest energy electronic state of a TPR Cr(III) complex would be very unstable with respect to the antiprismatic (APR) ground state, but it need not be unstable with respect to the $^2$E(APR) excited state. This possibility has been explored by synthesizing some aliphatic amine ligands which have stereochemical biases towards a trigonal prismatic geometry so that the ligand would be stressed (i.e., the intraligand steric energy would be large) when coordinated antiprismatically. Such internal ligand strain energy would tend to reduce the activation barrier for a trigonal distortion; this is qualitatively illustrated in Fig-

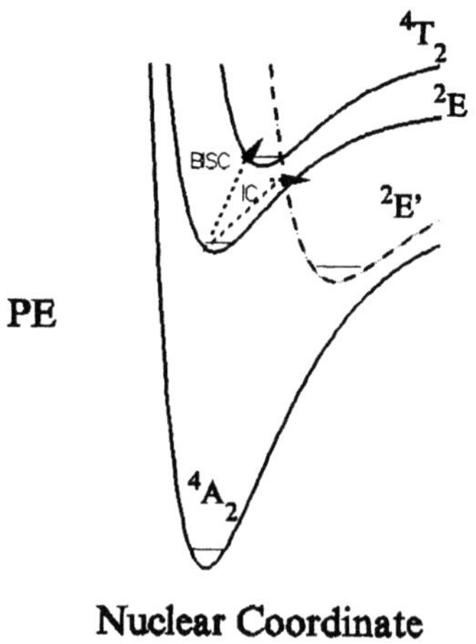


**Figure 4.** Qualitative PES's illustrating some possible pathways for ($^2E$)Cr(III) relaxation: back intersystem crossing (BISC) to a higher energy $^4T_2$ state; activated crossing (IC) to a trigonally distorted intermediate.

ure 4.[66–68] This strategy can be most simply visualized in a comparison of $Cr(en)_3^{3+}$ and $Cr(sen)^{3+}$, since the sen complex [sen = 4,4′,4″-ethylidyne-tris(3-azabutan-1-amine] differs from $Cr(en)_3^{3+}$ only in a neopentyl cap on one trigonal face of the complex. The angles around the central carbon atom of the neopentyl cap of the sen complex are larger than normal and this can readily be ascribed to the trigonal strain of the APR coordinated ligand. On the other hand, the coordination environments of Cr(III) in these complexes are nearly identical, and their absorption and emission spectra only differ slightly but in such a way that $E^0(^4T_2) - E^0(^2E)$ appears to be larger for the sen complex. Furthermore, $k_{nr}^0$ is about 30% smaller for the sen complex; however, $k_{nr}(25°C)$ is more than $10^4$ times larger for $Cr(sen)^{3+}$ than for $Cr(en)_3^{3+}$ ($\geqslant 10^{10}$ $s^{-1}$ and $8 \times 10^5$ $s^{-1}$, respectively). With more complicated ligand systems, such as (TPA)[9]ane$N_3$ (see Figure 5), it has been possible to introduce steric stress to inhibit the trigonal twist and this results in $k_{nr}(25°C) = 1.8 \times 10^3$ $s^{-1}$ (in aqueous solution) which is only about 50% larger than the value of $k_{nr}^0$ for this complex. The values of $k_{nr}(25°C)$ for several hexa-am(m)ine complexes, including those mentioned here, correlate reasonably well with the differences in internal ligand steric energies for the APR and distorted TPR geometries of the Cr(III) complexes.[66,68]

More limited studies have demonstrated that stereochemical constraints are also important in determining the magnitude of $k_{nr}(T)$ in VEqES Rh(III) complexes. While most complexes of Rh(III) have ambient solution values of $k_{nr}(T)$ in the range of $10^{10}$–$10^{12}$ $s^{-1}$, the *trans*-Rh([14]ane$N_4$)$(CN)_2^+$ complex ([14]ane$N_4$ = 1,4,8,11-tetraazacyclotetradecane) has a value of $k_{nr}(T = 25°C)$ $10^5$ $s^{-1}$.[69] The relevant distortions of the Rh(III) excited states are usually assumed to be tetragonal (see Section 6). Few other stereochemical perturbations of ($^3T_1$)Rh(III) excited state behavior have been reported.

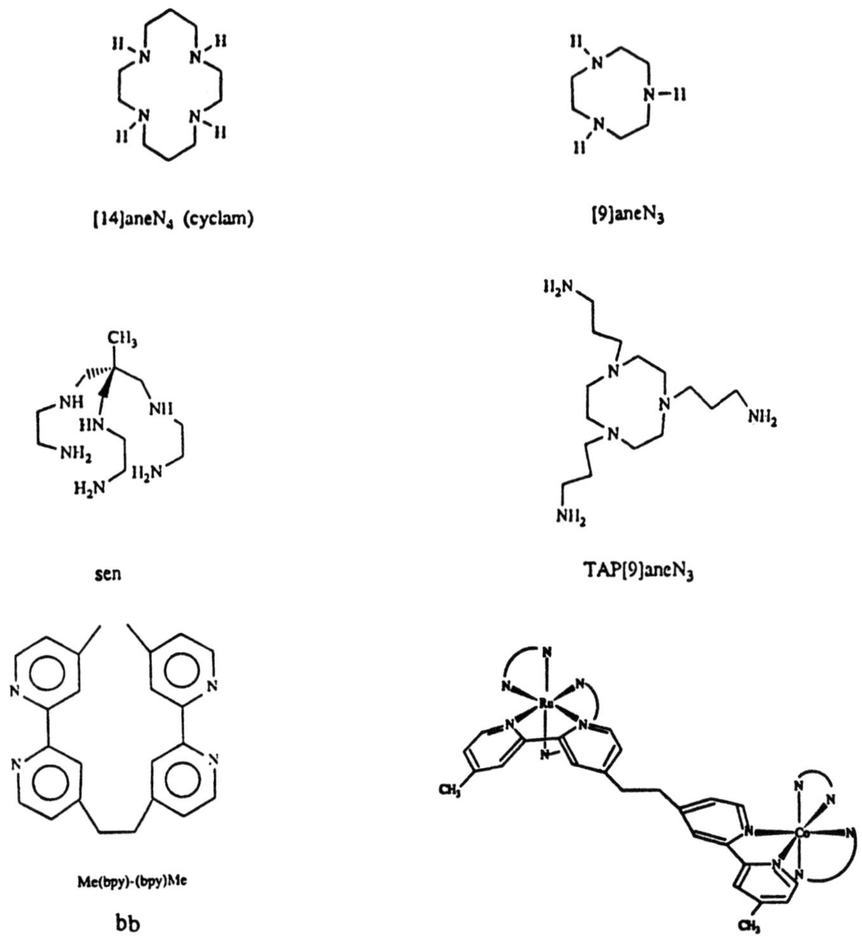


**Figure 5.** Skeletal structures of some ligands mentioned in the text.

The work cited here illustrates that the thermally activated relaxation of the ligand field excited states of coordination complexes can be very complicated. The complexities arise because the energy differences between several geometrical isomers and electronic configurations are often small compared to the amount of energy deposited in the initial excited state. Spectroscopic observations are very important in trying to characterize ligand field excited states, but their value is limited by the fact that absorption spectra inevitably sample excited state geometries near to that of the ground state; the $(^2E)$Cr(III) emission also largely corresponds to the ground state geometry. It is important to keep in mind that the ground state geometry of these complexes is stabilized by the ground state electronic configuration. When photonic excitation generates a different electronic configuration, much of that geometrical stabilization is lost and a new coordination geometry may even be more stable. If the geometrical changes are sufficiently large in amplitude, then the overall relaxation process could fall into the regime that $|\Delta E_{ge}^{00}| < \lambda_{vib}$, and in limiting cases, the species generated could be considered to be metastable intermediates. In any case, the simple two state limit is probably not adequate to describe relaxation of the ligand field excited states discussed here.

## 4.2 Charge Transfer Excited State Relaxation

The best known charge transfer VEqES is that of $Ru(bpy)_3^{2+}$.[35,39–45,70–77] Of the vast number of studies of this and closely related complexes this section will only focus on the structure and non-radiative relaxation behavior of the $Ru(bpy)_3^{2+}$ VEqES. The lowest energy absorption band is a Ru(II) $\rightarrow$ bpy MLCT transition which generates the $^1$MLCT–FC state discussed above. The $^3$MLCT–VEqES is generated on the subpicosecond time scale with approximately 100% efficiency. We will refer to the VEqES state as a "$^3$MLCT" state, despite the importance of spin-orbit coupling in this system. The electronic origin of this band is split by spin-orbit coupling into three components which are close in energy ($\sim$80 $cm^{-1}$), and an equilibrium distribution of these components is present under ambient conditions. The observed relaxation rate is a composite of the contributions of these three electronic components. The additional issue in the electronic description of the VEqES is whether the photoexcited electron is localized on one bipyridine ligand or delocalized over all three.[35,42–45,70–79] The issue of delocalization can be reduced to one of determining the magnitudes of the electronic matrix element for inter-ring coupling, $H_{LL'}$, and the effective value of $\lambda_{vib}$ for the vertical electronic excitation of an electron from one ring to another. A series of experiments first involving time resolved resonance Raman spectroscopy,[79] and more recently involving techniques such as laser line narrowing, Stark splittings of the origin, etc., with crystalline, mixed ligand and partially deuterated complexes at very low temperatures[44,76] have indicated that $H_{LL'} < 0.5$ $cm^{-1}$, so that $H_{LL} \ll \lambda_{vib}$, and that the electron density is largely localized on a single bipyridine ligand. The rate constant for electron "hopping" between bipyridine rings [more correctly, the rate constant for changing from one $Ru^{III}$ ($bpy^-$) configuration to another] was estimated to be $\sim 10^8$ $s^{-1}$.[76] The absorption and emission bands in solution under ambient conditions are broad with some poorly resolved structure, and the emission has been fitted to three components of a vibronic progression[21,22] (there are also three band origins for this emission, due to spin-orbit coupling, but these energy differences are an order of magnitude smaller).[42,44,76–78] The fitting has been used to extract Franck–Condon (or vibrational) overlap parameters, and, together with variations of Equation 6, it has been used to reasonably well reproduce the overall $^3$MLCT decay rate constant.[22] This argument treats the thermally activated, non-radiative relaxation of ($^3$MLCT)$Ru(bpy)_3^{2+}$ in the two state limit described in Section 3.1, Equation 6, with skeletal vibrations of the bipyridine ring inferred to act as acceptor modes.

A different kind of relaxation channel has been postulated for some MLCT excited states; namely, the thermal population of a higher energy metal centered ligand field excited state which then relaxes very efficiently.[80] This kind of mechanism has been examined in a series of $(bipy)_2(CN)Ru^{II}[CNRh^{III}(NH_3)_4X]^{n+}$ complexes.[81] The ligand X in these systems can be used to tune the $(^3T_1)$Rh(III) excited state to energies above and below the $(^3MLCT)Ru(bpy)_2^{2+}$ energy. The excited state relaxation times in these systems correlate with the trend expected for $\Delta E_{eg}^{00}$: X = $CN^-$, no $^3$MLCT quenching; X = $NH_3$, thermal equilibration with $^3$MLCT; X = Cl or Br, nearly 100% quenching.[81] The mean lifetime for $(^3MLCT)Ru(bpy)_2^{2+}/(^3T_1)Rh(NH_3)_5^{3+}$ (X = $NH_3$) energy transfer was estimated to be about 0.1 ns.[81] At least in this kind of model system, the ligand field excited states can act as an effective relaxation channel, but they efficiently quench the MLCT excited state only if their energy is lower than that of the MLCT excited state and these ligand field excited states do behave as bound states. The thermally activated population of a ligand field excited state has also been postulated as the principle pathway for the photosubstitutional behavior of $Ru(bpy)_3^{2+}$.[82] The subsequent description of the nonradiative relaxation of this excited state in terms of a two state model[21,22] suggests that the ligand field relaxation channel plays a minor role in these complexes.

## 5 RELAXATION OF GREATLY DISTORTED VEqES'S OF TRANSITION METAL COMPLEXES

There are not a great many well defined examples of photophysical relaxation in this regime, largely owing to the restriction that $\Delta E_{eg}^{00} < \lambda_{vib}$. This usually requires a relatively low energy excited state, and, consequently, direct detection is often difficult. There is some evidence of such behavior for several systems. The ratio of $\Delta E_{eg}^{00}/\lambda_{vib}$ can become relatively small when the excited state involves large bond length or bond angle changes. In transition metal systems this is most likely to occur when the population of metal centered antibonding orbitals (e.g., the $e_g$ orbitals in octahedral complexes) is altered by light absorption either to promote an electron from a nonbonding to an antibonding orbital or to increase the metal-centered electron density. The largest values of $\lambda_{vib}$ are expected in systems in which there are changes in coordination geometry and/or coordination number. When the values of $\lambda_{vib}$ are much larger than $\Delta E_{eg}^{00}$, the excited electronic configuration is more conventionally considered to be a transient chemical intermediate, and the relaxation channel merges with or is indistinguishable from a photochemical reaction channel. Examples of long lived, electronically excited systems with $\Delta E_{eg}^{00} < \lambda_{vib}$, thus span a range of behavior from systems which are best discussed in terms of slow excited state relaxation, arising at least in part because of a small classical activation barrier for return to the ground state, to systems in which the barrier is large enough that the nominal VEqES species could better be classified as a photochemical product. We will distinguish limits within this category, somewhat arbitrarily, in terms of the correlated structural changes.

## 5.1 Systems in Which There Is No Change of Coordination Number

Many of the excited states in this category differ from their ground states by large changes in the structures of their coordination spheres and in changes in spin multiplicity. This includes some complexes in which relatively long lived excited states are commonly classified as "spin crossover complexes",[83,84] and in which very long lived excited states can sometimes be generated.[84] These complexes are typically of the type $(^1A_1)Fe^{II}(L_6)$ in which the ligands $L_6$ are polypyridyl, pyridyl-amines, or other ligands which tend to stabilize the low spin $d^6$–Fe(II) ground state. Irradiation of a ligand field absorption band in such complexes is followed by rapid intersystem crossing, ultimately to populate a low energy quintet (or high spin) excited state. The low spin, $^1A_1$, and high spin, $^5T_2$, configurations of Fe(II) are characterized by large differences in bond lengths, usually on the order of 16–22 pm.[84] This leads to large values of $\lambda_{vib}$, probably on the order of $(3–4) \times 10^3$ cm$^{-1}$.[84] Values of $\Delta E^{00}_{eg}$ are not generally well known, but probably fall in the range of $(0.5–2.5) \times 10^3$ cm$^{-1}$, so that $\Delta E^{00}_{eg}/\lambda_{vib} \ll 1$. The nonradiative relaxation of the $^5T_2$ excited state is strongly temperature dependent and usually follows the pattern of Equation 12, but with no detected emission and no significant contributions from $k_r$. The values of $k^0_{nr}$ appear to decrease with decreasing values of $\Delta E^{00}_{eg}$. This result is consistent with expectation for this limit, and different from expectation based on Equation 7 (which should be applicable only if $\Delta E^{00}_{ge}/\lambda_{vib} > 1$). The $k_{nr}(T)$ dependence has been treated in terms of thermal population of vibrational excited states of the $^5T_2$ PES, which have better overlap with vibrational wavefunctions of the $^1A_1$ ground state than do the $v = 0$ vibrational wavefunctions of the VEqES.[84] However, McCusker, Rheingold and Hendrickson have demonstrated that a thermally activated, ligand induced trigonal distortion can very effectively relax the $^5T_2$ excited state.[85] That very long lived excited state species can be generated in this limit can be regarded to be a consequence of a combination of vibronic and spin trapping, with most of the barrier to relaxation coming from the nuclear reorganization which accompanies the change of spin multiplicity.

Cobalt(III) am(m)ine complexes may generally be systems whose photodynamic behavior places them near to this limit. Unfortunately, there is little direct information on excited state behavior and most of the pertinent information comes from the photochemical consequences of irradiation of ligand field absorption bands. The photochemical contrasts between Co(III) am(m)ine complexes and their Rh(III) analogs are very striking: Product yields are typically in the range of $10^{-3}$–$10^{-5}$ and very strongly wavelength dependent for Co, but for the comparable Rh(III) complexes the product yields are typically 0.1–0.6 and largely wavelength independent (see also Section 6). For most of the complexes of both metals the excited state lifetimes are very short in ambient solutions, although those of Rh(III) tend to be in the short ns regime while the complexes of Co(III) tend to be significantly shorter lived. Overall, the contrasts in photochemical (and where available, photophysical) behavior strongly suggest FC excited state relaxation of the Co(III) complexes by means of fast internal conversion and intersystem crossing, while the behavior of the Rh(III) complexes is of the type that one might expect for a well defined VEqES. Such a contrast would be in accord with expectation based on vibronic analyses of the absorption[86] and emission[62] bands of the respective hexa-ammines: $\Delta E_{\mathrm{ge}}^{00}$ is estimated to be about $10 \times 10^3$ cm$^{-1}$ larger for the $^3T_{2g}$ VEqES of Rh(III) than for Co(III). The much smaller energy gap should lead to much faster relaxation of $Co(NH_3)_6^{3+}$. Figure 6 is approximately based on parameters from these analyses (ligand dissociation energies are based on the activation energies observed in aqueous solutions[48]). Clearly the energy differences between the ground and excited state PES's of Co(III) are much smaller than those of Rh(III). The small ground state-excited state energy gap, as noted above, should lead to very fast internal conversion and/or intersystem crossing for the Co(III) complexes. The hypothesis that this contrast arises from differences in the ES–GS energy gaps is supported by the observations of well defined VEqES behavior in $Co(CN)_6^{3-}$ and $Co(CN)_5X^{n-}$ complexes (X typically a halide or pseudohalide).[39,41,48] It also appears that the minimum of the ($^5T_2$)Co(III) PES is lower in energy than the energy required for ligand dissociation in the ground state. Therefore, it is very unlikely that generation of the ($^5T_2$)Co(III) VEqES of the am(m)ine complexes is a major pathway for such ligand dissociation. It is more likely that the photoproducts would be produced in kinetic competition with FC excited state relaxation to the $^5T_2$ excited state.

The spectroscopic, photophysical and photochemical observations on the ligand field excited state behavior of d$^6$ complexes indicate that these systems span the full range of behavior expected for mixing a single excited state PES with a ground state PES: (i) $\Delta E_{\mathrm{eq}}^{00} \gg \lambda_{\mathrm{vib}}$ for Rh(III), so the photophysics and photochemistry are characteristic of the VEqES; (ii) $\Delta E_{\mathrm{ge}}^{00} \ll \lambda_{\mathrm{vib}}$ for Fe(II), so that the VEqES is trapped, and the excited state lifetimes can be of the magnitudes more commonly associated with a reaction intermediate; and (iii) $\Delta E_{\mathrm{ge}}^{00} \geqslant \lambda_{\mathrm{vib}}$ for Co(III) am(m)ine complexes, and this results in very short lived excited states, either because $\Delta E_{\mathrm{ge}}^{00}$ is

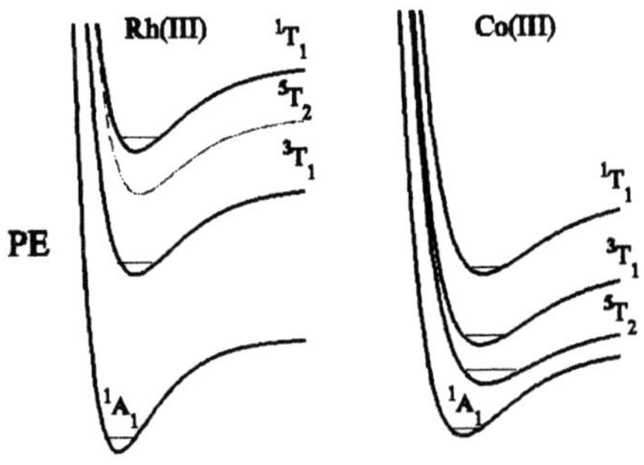


**Figure 6.** Qualitative contrasts in the lowest energy ligand field excited states of $Rh(NH_3)_6^{3+}$ (left) and $Co(NH_3)_6^{3+}$ (right). PES minima and displacements based on References 62 and 86, respectively. The dissociation limits of the ground state Morse curves have been set equal to the activation energy for substitution in aqueous solution.

so small or because of the combination of a relatively small gap and a relatively large displacement of the PES's.

### 5.2 Systems in Which There Is a Change of Coordination Number

Some of the clearest examples of excited state stabilization by change of coordination number occur among the complexes of Ni(II). Many high spin (4-coordinate tetrahedral, 5-coordinate square pyramidal or 6-coordinate octahedral) Ni(II) complexes are well known and well characterized. Most of these high spin compounds are blue or purple in color. There are also a great many well characterized low spin (4-coordinate, planar and usually yellow) complexes. In these complexes the two spin states (and different coordination geometries) are usually not much different in energy. McGarvey and co-workers[87–90] have capitalized on these properties and used photoexcitation to alter the coordination geometry of the Ni(II) complexes, by means of the assocciation or the dissociation of axially coordinated solvent species (S), thereby "trapping" the electronic excited state. If we take $Ni([14]aneN_4)^{2+}$ as an example of a low spin complex, then the lowest energy electronic excited state is a triplet state. In fluid solution the VEqES (triplet) state species is either 5-coordinate square pyramidal, $Ni^{II}([14]aneN_4)S$ or 6-coordinate, $Ni^{II}([14]aneN_4)S_2$. Such higher coordinate complexes are metastable, and they eventually relax to the four-coordinate, low spin Ni(II) ground state in 0.01 to 10 $\mu$s. Thus, the photoinduced change of spin multiplicity results in a change of coordination number which, in turn, stabilizes the excited state spin configuration. This sequence of steps is illustrated in Figure 7.

## 6 PHOTOCHEMICAL REACTION CHANNELS

In the perspective that is developed in this chapter, photophysics and photochemistry tend to merge since photochemical reactions accompany excited state relaxation along certain channels. In a small percentage of systems investigated the photoreaction product yields ($\Phi = N_p/N_{h\nu}$, where $N_p$ is the moles of product formed and $N_{h\nu}$ is the number of einsteins of photons absorbed) are high enough that the corresponding reaction channels must constitute the predominant ES relaxation modes; however, in most complexes product yields are small ($\Phi \leqslant 0.1$) and the channels leading to reaction may be a minor relaxation pathway. Photophysical considerations, such as summarized above, lead to some general inferences which may be useful in organizing approaches to the extensive photochemical observations. A photochemical reaction requires that there be a dissociative component to the relaxation channel ("dissociative" is used for simplicity here to represent any excited state PES which does not have a PE minimum nested with the ground state PES, or whose minimum is nested with the product PES and occurs at an energy which is less than that of the intersection of the initial ground state with the product PES; this includes isomerization, bond breaking, etc.; see Figure 8); i.e., the relaxation must occur at least in part along a trajectory above the ground state dissociation limit. This might occur:

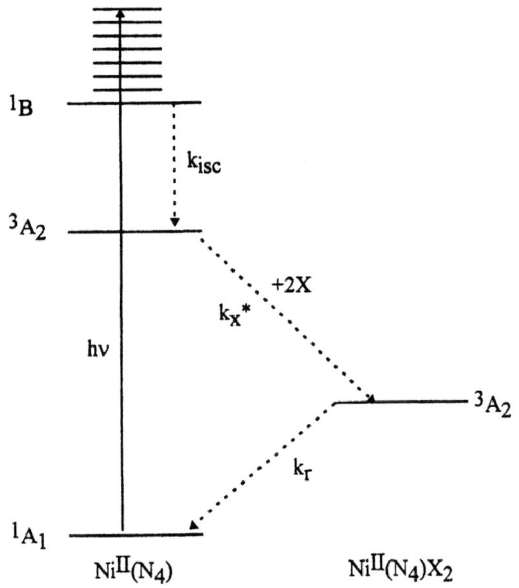


**Figure 7.** Qualitative state correlation diagram describing photoinduced substitutions of low spin Ni(II) complexes.

(i) along a dissociative PES; (ii) by a unique excited state reaction; or (iii) by non radiative relaxation of a VEqES directly into the dissociation coordinate of a ground state reaction channel (i.e., by population of the low frequency vibrational modes associated with the reaction channel above the dissociation limit). In view of the arguments presented in Section 3.3, the last of these possibilities seems relatively unlikely since the excess energy is usually large ($>10^3$ $cm^{-1}$) and the relaxation process would usually favor the distribution of this energy in a few high frequency modes.[59] Thus, a relaxing excited state should be largely forbidden from directly depositing

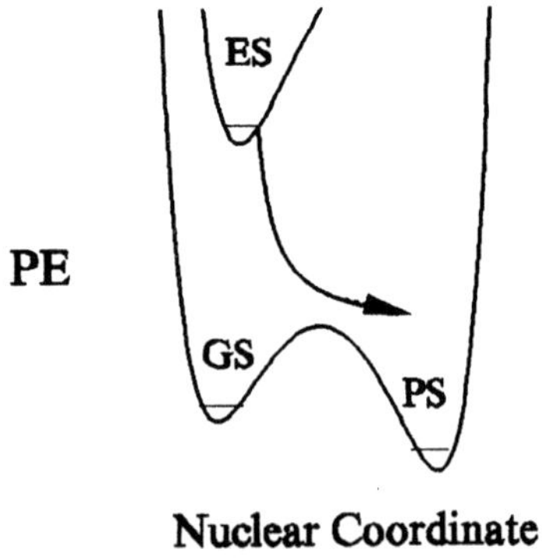


**Figure 8.** Schematic representation of the direct relaxation of an excited state (ES) into a ground state dissociation channel leading to products (PS).

sufficient energy in a ground state reaction coordinate to cause the reaction to occur. This does not preclude the possibility of a more indirect path for the entry onto a dissociative PES. For example, it should be noted that local high frequency oscillators (such as N–H, C≡N, etc.) may be transiently populated with energies in excess of the classical dissociation energy for other bonds in the molecule (or excited state), and their relaxation might lead to fragmentation of the molecule or excited state.

## 6.1 Ligand Field Photochemistry

The irradiation of ligand field absorption bands often results in ligand substitution and/or complex isomerization.[21,30] The most striking features of ligand field photochemistry are to be found in the contrasting photoinduced stereochemical changes found for the Cr(III)[45–48,91,92] and Rh(III)[48,93–96] complexes. The Cr(III) complexes tend to undergo *trans*- to *cis*-isomerizations, while the reverse tendency is exhibited by many Rh(III) complexes. Since thermally induced ground state substitution is stereoretentive in these complexes, this is a striking observation. Additionally the distribution of ligands which are replaced by solvent generally differs from that found for ground state substitution processes.

Relatively early in the course of photochemical studies of Cr(III), Adamson[33,91] suggested that the photoinduced behavior of these complexes could be summarized by a simple set of "rules" which can be rephrased somewhat as: (a) the axis containing the most weakly bonded ligands is labilized; (b) the most strongly bonded ligand on this axis is labilized; (c) the final product will contain the entering solvent species (as a ligand) in the plane orthogonal to the original axis of labilization (i.e., with reference to the metal center and the remaining ligand along that axis). This does summarize the photochemical behavior of a large number of Cr(III) complexes.[41,47,48,92] This pattern of behavior has prompted the development of theoretical models based on molecular orbital[97] or angular overlap approaches (AOM).[98] These models have attempted to define the predominant distortion coordinate of the FC excited state based on fractional orbital electronic populations or on angular overlap parameters, where both kinds of parameters are based on observed ground state absorption spectra. In either approach, the excited state coordinate with the largest tendency to distort [i.e., the coordinate with the steepest slope, $\partial(\mathrm{PE})/\partial Q_i$] is identified as the predominant dissociative reaction coordinate. The AOM has had reasonable success in rationalizing and predicting the predominant products. However, this model does not predict product ratios well,[48] and other observations do not always support a dissociative model.[48,92]

The angular overlap model approach can in principle be viewed as a means of estimating the relative Jahn–Teller linear vibronic parameters,[29,30,99] and in this sense it is consistent with observed quartet state distortions. Thus, there is evidence for more than one pathway for product formation in several Cr(III) complexes; e.g., in $Cr(NH_3)_6^{3+}$ about 1/3 of the photoaquation is formed in a pathway which cannot be quenched while 3/4 of the photoaquation is quenched when the $(^2E)Cr(NH_3)_6^{3+}$ excited state is quenched. This would be consistent with "prompt" reaction of the $(^4T_{2g})Cr(NH_3)_6^{3+}$ excited state in competition with intersystem crossing to the $(^2E_g)Cr(NH_3)_6^{3+}$ excited state, followed by $^2E_g \rightarrow {}^4T_{2g}$ back intersystem crossing, and ultimately reaction from the quartet state. This model does require that all Cr(III) photochemistry be attributed to reactions of the lowest energy quartet excited state. However, not all the available photochemical and photophysical information is consistent with this model.[48] The dominant non-radiative relaxation pathways of $(^2E)$Cr(III) do not generally appear to involve back intersystem crossing,[48,65–67] and they need not always result in chemical transformations.

The photochemistry of Rh(III) complexes has also been extensively investigated,[35,39,41,48,93–96] and theoretical models have been proposed to account for the observations.[98] The relatively definitive mechanistic work has involved isotopically labeled ligands, careful determinations of *cis/trans* ratios, volumes of activation, etc.[93–96] The pattern which has emerged is one of a largely dissociative photoreaction pathway, and the (apical/basal; see Figure 9) isomerization of and substitution into the square pyramidal 5-coordinate intermediate. The ratios of products are believed to depend on the kinetics of various competitive reactions. The formulation of the mechanism in terms of an equilibrium between electronic excited states of the 5-coordinate intermediates seems a little unlikely since one would normally expect that the distortion accompanying the isomerization is likely to mix the electronic states of the complex and induce relaxation.[100] The solvent dependence of the relative product yields and even of which product predominates[101] are not easily compatible with the MO or AOM models. To the degree that vibrationally equilibrated, 5-coordinate chemical intermediates are involved, MO or AOM models based on their 6-coordinate precursors are not likely to be accurate representations of the photoinduced reaction process. It is not yet clear whether the implicated intermediates are sufficiently long lived to vibrationally equilibrate, or whether these intermediates are vibrationally excited, as is the case of the $Cr(CO)_5$ intermediate discussed in Section 3.3. If the intermediates are not vibrationally equilibrated, then aspects of the model in Figure 9 should be re-examined.

## 6.2 Charge Transfer Photochemistry: Ligand-to-Metal-Charge-Transfer (LMCT) Photochemistry

This chapter can only touch on limited aspects of charge transfer photochemistry. Three have been selected. The reader is referred to a fine monograph on the topic for a broader and more detailed survey.[70] This section will focus on the photoreactions induced be excitation into LMCT excited states of transition metal complexes.

Most of the work on LMCT excited states has been centered on the photochemical consequences of irradiating the strongly allowed LMCT absorption bands of cobalt(III) complexes.[33,35,39,41,70,102–105] These excited states are considerably higher in energy than many of their ligand field excited states, so the chemistry is all upper state chemistry, and the photophysical observations discussed above imply that it must occur in ultrafast time regimes. Nevertheless, the quantum yields are often reasonably large.

In the simplest of cases, the irradiation of an LMCT absorption band results in the formation of a one electron oxidized ligand (a free radical) and a reduced

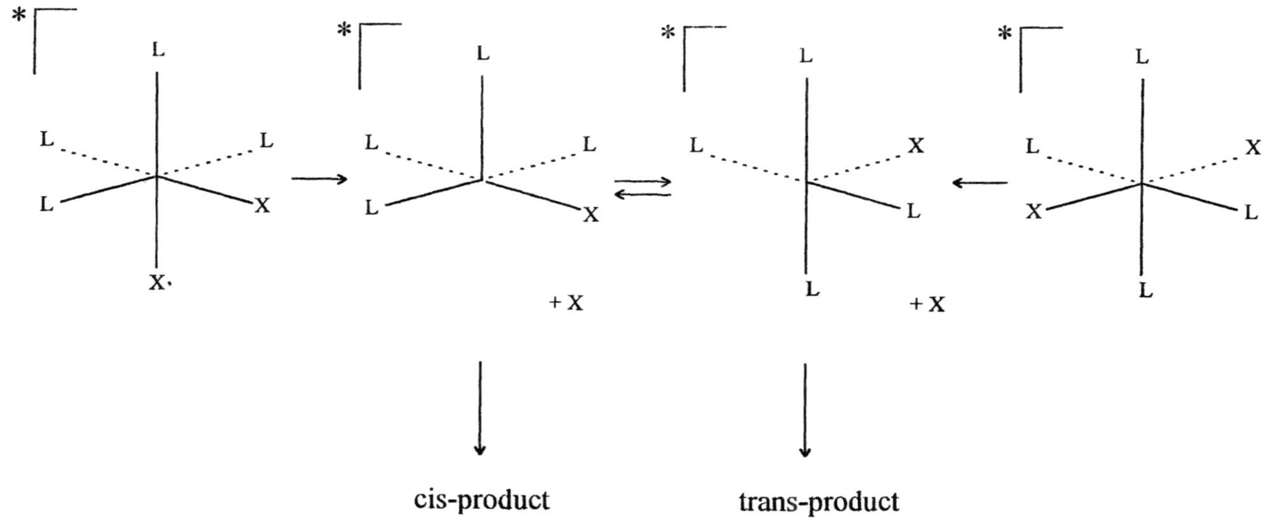


**Figure 9.** Mechanistic scheme proposed for the excited state reactions and relaxation of Rh(III) complexes, based on Reference 93. Asterisks (*) indicate electronic excited states.

metal complex. This type of behavior has been most thoroughly examined for acido-am(m)inecobalt(III) complexes,[33,102–105] and it is illustrated in Equation 14.

$$(Am)_5Co^{III}(X^-) + h\nu \xrightarrow{S,H^+} Co(S)_6^{2+} + 5AmH^+ + X\bullet \quad (14)$$

The "photoredox" processes are characterized by quantum yields, $\Phi_R$, which increase from zero for low energy excitation either: (a) to a limiting, wavelength independent value, $\Phi_R^{lim} < 1.0$; or (b) monotonically to $\Phi_R = 1.0$. In a few cases more than one limiting value of $\Phi_R$ has been observed; $\Phi_R$ can also exhibit more complex behavior. Before addressing specific systems, it is useful to consider some of the general features of $\Phi(\nu)$. The $Co(NH_3)_5Br^{2+}$ complex can be taken as reasonably characteristic. For this complex, $\Phi(\nu) \rightarrow \Phi_R^{lim} < 1$ as the excitation energy is increased.[103–106] The limiting value of $\Phi_R^{lim}$ is solvent dependent,[103–106] and while this solvent dependence is not always simple, there is a tendency of $\Phi_R^{lim}$ to decrease as solvent viscosity increases. This suggests that Equation 14 should be modified to include the initial, or solvent cage trapped "radical pair" fragmentation products, $\{Co(NH_3)_5^{2+}, X\bullet\}$, as in Equations 15 and 16.[101–105] The constant

$$Co(NH_3)_5X^{2+} + h\nu \rightarrow \left\{Co(NH_3)_5^{2+}, X\bullet\right\} \quad (15)$$

$$\left\{Co(NH_3)_5^{2+}, X\bullet\right\} \left\} \begin{array}{l} \rightarrow \text{Solvolysis products} \\ \\ \rightarrow Co(NH_3)_5X^{2+} \end{array} \right. \quad (16)$$

values of $\Phi_R^{lim} < 1$ over extended wavelength ranges suggest that the excess excitation energy does not appear to a significant extent in the Co–Br bond breaking coordinate. The properties and reactions of the $\{Co(NH_3)_5^{2+}, X\bullet\}$ species can be very important in determining the overall distribution of photoproducts. Unfortunately, very few experiments report the direct observation of such species. Numerical values of $\Phi_{lim}$ are expected to depend on the efficiency of homolytic Co–X bond cleavage and on the efficiency of radical pair recombination. In some complexes more than one wavelength region exists over which there is little change of $\Phi_R$, implying that different wavelength regions give different values of $\Phi_{lim}$.

Some relevant studies have reported on the behavior of cage trapped radical pairs formed following the picosecond and femtosecond flash photolysis of methyl- and adenosylcobalamin (vitamin $B_{12}$).[107,185,186] Although the alkyl-cobalt bonds are more covalent than ionic, these can be regarded as Co(III) complexes, and net photolysis results in alkyl radicals and Co(II)-corrin, complexes, analogous to Equation 14,[108–110] and $\Phi_R^{lim}$ behaves in a manner analogous to that described above for $Co(NH_3)_5^{2+}$. Excitation of the adenosyl–cobalamin complex appears to result in efficient and fast ($\leqslant$2 ps) Co–C homolysis to form a cage trapped radical pair species.[107,185] The radical pair lifetimes are reported to be about 250 ps with recombination accounting for about 3/4 of the radical pair decay.[107,185] Similar excitations of methylcobalamin are reported to generate a metastable chemical intermediate (lifetime ~1 ns), no radical pair recombination, and ~27% yield of Co(II)–corrin and methyl radical.[186] It appears that these radical pair species were formed without much excess excitation energy in the bond breaking coordinate, and the lifetime of the cage trapped species was long enough that the species must have been vibrationally equilibrated by the time that Equations 14 and 15 became important. Other information about the behavior of solvent cage trapped radical pair species is much less direct.

The experimental determinations of $\Phi_R$ are relatively straightforward for photoredox processes in most Co(III) complexes since most am(m)ine-cobalt(II) complexes equilibrate with acidic solutions very rapidly and most solvato-Co(II) species are not easily or rapidly oxidized by ligand radicals or their successor species (exceptions to this are found among cobalt complexes with macrocyclic ligands such as the cobalamin complexes mentioned above). Analogous photoredox processes for most other $M^{n+}$ metals are complicated by the redox lability of their one electron reduced, $M^{(n-1)+}$, complexes, and the dark reactions following photoredox processes most often regenerate the metal in its original oxidation state. Such dark reactions can be even more complicated in aerated solutions and when $M^{(n-1)+}$ is unstable with respect to disproportionation.

The increase of $\Phi_R$ from 0 to $\Phi_R^{\lim}$ can be extrapolated approximately linearly to define a threshold, or minimum excitation energy, $E_R^{th}$, required for the observation of photoredox behavior.[103,105,111] Values of $E_R^{th}$ can be approximately related to the bond dissociation energy, BDE, of the ground state.[111] The relationship of $E_R^{th}$ to the LMCT absorption spectra is different for different metals. Consider the seemingly closely related complexes $Co(NH_3)_5Br^{2+}$, $Rh(NH_3)_5I^{2+}$, $Cr(NH_3)_5Br^{2+}$ and $Ru(NH_3)_5I^{2+}$. If one defines a "threshold" energy for absorption $E_{abs}^{th}$, for the lowest energy LMCT absorption band (e.g., by taking $E_{abs}^{th}$ to be the energy where the absorptivity has ~5% of its value at the band maximum) then: (a) for the Co(III) complex, $E_R^{th} \sim E_{abs}^{th}$;[109] (b) for the Rh(III) complex, $E_R^{th} > E_{abs}^{th}$;[112] (c) for the Cr(III) complex, $E_R^{th} < E_{abs}^{th}$;[113] and (d) for the Ru(III) complex, $E_R^{th} \gg E_{abs}^{th}$ (no photoredox behavior was observed for any irradiations of the lowest energy LMCT absorption band).[114] In the simplest limits one would expect: (a) $(E_{abs}^{th} - E_R^{th}) > 0$ for a bound PES; and (b) $(E_{abs}^{th} - E_R^{th}) \leqslant 0$ for a dissociative excited state PES. The contrasts in behavior of the $M(NH_3)_5X^{2+}$ complexes can be plausibly related to expectation for $\lambda_{vib}$, $E_{eg}^{00}$ and BDE for the different excited state electronic configurations. The spectroscopic LMCT excited states of Co(III) and Rh(III), and their configurationally related triplet LMCT states, would correlate with (i.e., the dissociation limit of the excited state PES would be) low spin ($S = 1/2$) $M(NH_3)_5^{2+}$ products of reaction. The lowest energy LMCT excited states correspond to the promotion of an electron from a non-bonding ($p\pi$) orbital of $X^-$ to an antibonding ($d\sigma$) orbital of M. This suggests that in the excited state the ($^2E$)M(II)–(X•) "bond" would have about one-half the covalent character ($1/2\sigma_{cov}$); but none of the ionic character) of the ground state M(III)–($X^-$) bond,[111] so that these would be weakly bound excited states. However, the quartet (high spin) electronic configuration is thermodynamically more stable than the doublet for a $Co(NH_3)_5^{2+}$ product species while the high spin configuration is much less stable for $Rh(NH_3)_5^{2+}$. The ($^4T_2$)$Co(NH_3)_5^{2+}$ species correlates only with "non-spectroscopic" triplet or quintet LMCT excited states.[115]

At least one of these states is expected to be dissociative, and rapid crossing from the LMCT($^2E$) to the LMCT($^4T_2$) manifolds of cobalt systems would account for the contrast noted between $E_R^{th}$ and $E_{abs}^{th}$ for Co(III) and Rh(III) (see Figure 10a and b). The lowest energy $Cr(NH_3)_5^{2+}$ fragment is also expected to have a high spin ($^5E$) configuration, while the ($^3T_2$) configuration would have higher energy. In this case the LMCT($^3T_2$) state which correlates with the low spin product species should occur at lower energy than the LMCT' ($^5E/Cr^{2+}$) state because the $d\pi^4$ (or $t_2^4$) configuration of Cr(II) involves only non-bonding orbitals, and, consequently $\lambda_{vib}$ should be smaller than in the LMCT' ($^5E/Cr^{2+}$) state in which the electronic configuration of Cr(II) is $d\pi^3 d\sigma^*$ (or $t_2^3 e^*$). Based on this simple configurational argument, one expects the covalent component of the Cr(II)–(X•) bond to be about the same as that of the Cr(III)–($X^-$) bond ($\sigma_{cov}$'); hence, the LMCT($^3T_2/Cr^{2+}$) state of $Cr(NH_3)_5Br^{2+}$ should be more strongly bound (larger excited state BDE) than the LMCT($^2E/Co^{2+}$)

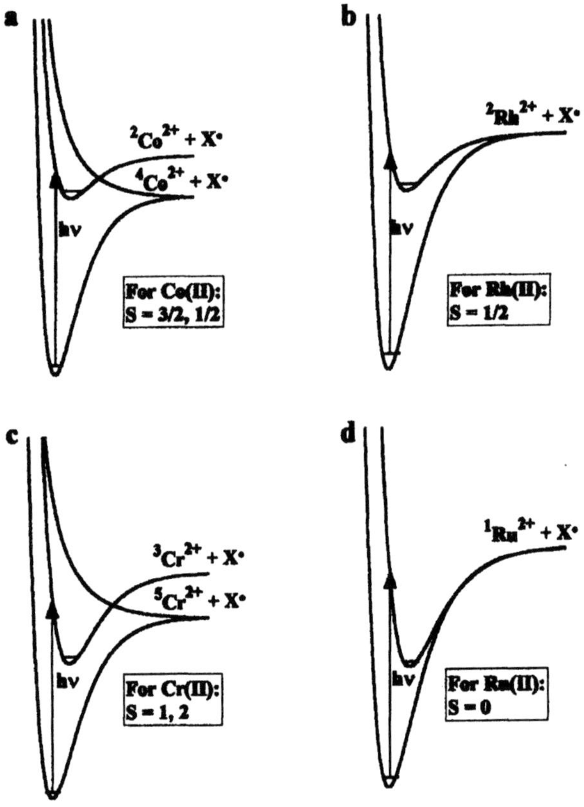


**Figure 10.** Qualitative two-dimensional PES's illustrating the relationship between the LMCT excited state bond energy and the homolytic dissociation products. Well depths and dissociation limits are chosen to be consistent with the relationships between the threshold energies for LMCT absorption and for the formation of photoredox products from irradiations of the respective $M^{II}(NH_3)_5(X^-)$ complexes.

state of $Co(NH_3)_5Br^{2+}$, consistent with the observations on $E^{th}_{abs}$ and $E^{th}_{r}$ (Figure 10a and c). One expects a more covalent $^1$Ru(II)–($^2$X•) than $^3$Cr(II)–($^2$X•) bond, but the $Ru(NH_3)_5I^{2+}$ complex is the only one of those compared here for which the low energy LMCT excited states necessarily have the same spin multiplicity as the ground state. This feature may result in ultrafast configurational relaxation in the LMCT excited state of Ru(III) complexes [compare the ultrafast configurational relaxation of $(NC)_5RuCNRu(NH_3)_5^-$ electron transfer excited states[49–52] mentioned in Section 3.3], so that the rate of bond-breaking (a relatively low frequency vibrational motion) cannot compete with the rate of regeneration of the ground state.

The work outlined above indicates that in most LMCT induced photohomolysis, the excess excitation energy does not appear in the translational energy of the M(II) and X• fragments. A preferential deposition of the excess energy in higher frequency modes would be consistent with photophysical studies described above. On the other hand, when X• is a molecular radical, such as $NO_2$, internal vibrations of the departing radical might contain some of this excess energy. There is no available information on the details of the energy flow in this type of system. There is a very crude spectroscopic correlation among Co(III) complexes, which suggests that this problem may be related to the electronic complexity of these systems: The largest values of $\Phi^{lim}_{R}$ tend to occur in complexes for which $E^{th}_{R} \leqslant h\nu_{max}$ for the lowest energy ligand field absorption band [the $^1T_{1g}$ state of octahedral complexes is split into components in the acido-am(m)ine complexes; the crude correlation is with any of these components]. This suggests that some higher energy ligand field state may provide either an efficient relaxation pathway for the LMCT excited state or an efficient recombination pathway for the radical pair species.

In some complexes LMCT excitation can result in products which may not appear to be clearly correlated with the assigned absorption spectra. For example the LMCT excitation of $Co(HEDTA)X^-$ complexes (HEDTA = singly protonated ehtylenediaminetetraacetic acid) leads to carboxalate oxidation and values of $E^{th}_{R}$ which are independent of $X^-$.[116,117] Similar LMCT irradiations of $Fe(HEDTA)OH_2^-$ generate a photochemical transient species which was correlated with carboxalate oxidation and it was suggested that this species was a metastable $RCH_2$–Fe complex formed after elimination of $CO_2$ from the photo-oxidized carboxalate of HEDTA.[117] In a relatively recent report Kunkley and Vogler have found that LMCT irradiations of $[Fe^{III}(EDTA)]_2O^{4-}$ leads to $O_2$ and $Fe^{II}(EDTA)^{2-}$.[118] These authors propose that the primary photoreaction is the generation of $Fe^{II}(EDTA)^{2-}$ and $Fe^{IV}(EDTA)O^{2-}$. An alternative mechanism would be the generation of a photo-oxidized carboxalate which reacts with the neighboring $\mu$-oxo ligand in competition with the elimination of $CO_2$. The Fe catalyzed disproportionation of the resulting aceto-peroxide species could then give the products.

Other LMCT excitations can give rise to oxidized ligand species which fragment into very stable molecules and alkyl radical fragments in the solvent cage, and such processes can result in some unusual chemical species. For example, LMCT irradiations of oxalato-cobalt(III) complexes have been shown to result in the formation $CO_2$ and $Co^{III}$-$CO_2H$ [Co(III)-C bonded] formato complexes.[119,120] In another example of this aspect of LMCT photochemistry, Reed et al.[121] have shown that irradiations of $Rh(NH_3)_5N_3^{2+}$ lead to the formation of $N_2$ and a $Rh(NH_3)_5NH^{2+}$ (nitrene) product.

## 6.3 Charge Transfer Photochemistry: Metal-to-Ligand-Charge-Transfer (MLCT) Photochemistry

Metal-to-ligand-charge-transfer photochemistry has been most extensively investigated for $Ru(bpy)_3^{2+}$.[35,39–41,70–75] The VEqES lies about 2 eV above the ground state so that the excited state is a strong oxidant ($E_{1/2} \cong 0.84$ vs. NHE) and reductant ($E_{1/2} \cong -0.86$ vs. NHE);[35,39–41,69–74] however, the estimated redox properties of the excited state differ from those of the ground state by nearly the amount of the measured enthalpy difference,[122] and this may not be the same as the free energy difference. This excited state can act as a triplet sensitizer, an electron transfer donor or acceptor, and there is a small ($\sim 10^{-2}$–$10^{-4}$) quantum yield for substitutional product formation.[82] In fact, photosubstitution is the only known way to synthesize *trans*-$Ru^{II}(bpy)_2X_2$ complexes.[82] Most of the interest in the Ru(II)-polypyridyl complexes, as well as that in their Os(II) and Re(II) analogs, has been in their potential as high energy sensitizers of chemical reactions.[35,39–41,70]

A remarkable variety of photochemical behavior is exhibited by copper(I) complexes.[50,123–129] This is a relatively labile metal which can form well defined complexes. The simplest of these are the halocuprates, and the simplest solution species are $CuX_2^-$ and $CuX_3^{2-}$. Ultraviolet excitation of these complexes produces solvated electrons (formally a charge-transfer-to-solvent process) in competition with a $CuX_3^{2-}$ excited state emission.[70,126] The lowest energy absorption of the $Cu(py)_4^+$ complex is probably MLCT in character, while the emission properties of this complex are more characteristic of a metal centered (d,s) excited state.[125] In addition to the monocuprates, a number of $Cu_nL_n$ complexes have been examined.[123,124] An example is the $Cu_4I_4(py)_4$ complex which has a Cu–Cu bonded metal core, exhibits two luminescences, each with a different ambient lifetime.[123,124,127,128] In these complexes, the lowest energy excited state appears to be a combination of metal centered (d,s) and LMCT electronic configurations.[129]

## 6.4 Charge Transfer Photochemistry: Photoinduced Electron Transfer (PET)

Photoinduced electron transfer processes have attracted a great deal of attention.[35,39–41,70–74,130–134] Much of the interest derives from possible applications of simple molecular systems to energy conversion devices,[134] or from the potential for using well designed systems to investigate fundamental chemical processes.[130–133] The donor-to-acceptor electron transfer in these systems can be the result of either: (a) sensitized transfer in which one center absorbs light to generate an electronic excited state which mediates the transfer of an electron between donor and acceptor (D/A); or (b) direct promotion of donor-to-acceptor electron transfer by the absorption of light.

The simplest limit of sensitized electron transfer would involve the absorption of light to generate an excited state followed by the bimolecular quenching of that state.

Such processes have been extensively investigated.[35,39–41,70–74] A typical reaction would be that of $(^3MLCT)Ru(bpy)_3^{2+}$ with $Co(bpy)_3^{2+}$ (Equations 17–20)[135] in which electron transfer occurs in competition with non-radiative relaxation of the $^3$MLCT excited state.

$$Ru(bpy)_3^{2+} + h\nu \rightarrow (^3MLCT)Ru(bpy)_3^{2+} \tag{17}$$

$$(^3MLCT)Ru(bpy)_3^{2+} \rightarrow Ru(bpy)_3^{2+} + \text{heat} \tag{18}$$

$$(^3MLCT)Ru(bpy)_3^{2+} + Co(bpy)_3^{3+} \xrightarrow{k_q} Ru(bpy)_3^{3+} + Co(bpy)_3^{2+} \tag{19}$$

$$Ru(bpy)_3^{3+} + Co(bpy)_3^{2+} \xrightarrow{k_{bet}} Ru(bpy)_3^{2+} + Co(bpy)_3^{3+} \tag{20}$$

For this system, the excited state electron transfer rate constant is $k_{et} \sim 2 \times 10^9\ M^{-1}\ sec^{-1}$ and the back electron transfer rate constant (Equation 20) is $k_{bet} = 2 \times 10^8\ M^{-1}\ sec^{-1}$.[134] The back electron transfer rate is more than an order of magnitude smaller than expected based on the known electron transfer self-exchange rate constants and the equilibrium constant.[134] The $Ru(bpy)_2^{2+}$ and $Co(bpy)_2^{3+}$ can be covalently linked by a bb bridge (see Figure 5 and Section 3.3) to form a molecular D/A system, $Ru(bpy)_3^{2+}$ linked by an ethane bridge to $Co(bpy)_3^{3+}$, in which PET is analogous to that in the bimolecular reactions. In this unimolecular system, intramolecular $^3$MLCT quenching is observed at low temperatures (at 77 K $k_q = 2 \times 10^6\ s^{-1}$) and is estimated to be about $10^{10}\ s^{-1}$ at 300 K.[58] This implies that this is a simple system in which D/A coupling is weak and vibrational equilibration is fast compared to electron transfer. The back electron transfer rates are identical within the estimated uncertainties for the bimolecular (corrected for reactant association) and unimolecular systems.[58] This implies that for both systems $\kappa_{el} \sim 10^{-2}$, and this conforms to expectation based on Equation 5 (Section 3.1) and a simple superexchange mechanism for electronic coupling.[58] While the overall pattern of reactivity in covalently linked D/A complexes with small $H_{DA}$ conforms to expectation based on conventional electron transfer logic,[11–13,15] it is possible to examine such complexes for possible contributions of orientational effects or specific linker properties to $H_{DA}$. Such studies are more appropriate to the electron transfer than the photochemical context (see chapter by Richardson).

A variation on this are D/A electron transfer reactions which are induced by the excitation of a chromophore other than D or A[136,137] as is schematically illustrated in Equations 21 and 22 (L and L′ are the ligands which bind the components into the supramolecular array).

$$D{-}(L){-}C{-}(L'){-}A + h\nu \rightarrow D{-}(L){-}C^*{-}(L'){-}A \tag{21}$$

$$D{-}(L){-}C^*{-}(L')A \rightarrow D^+{-}(L){-}C{-}(L'){-}A^- \tag{22}$$

Such "triads" have been reported with organic donor and acceptor moieties and a porphyrin based sensitizer.[137,138] For a hypothetical transition metal system one might use a polypyridyl-Ru(II) moiety as sensitizer. In such a complex, one would expect the $(^3LMCT)Ru(PP)^{2+}$ excited state to function as an oxidant towards D and a reductant towards A, and both reactions might be induced to follow light absorption by $Ru(PP)^{2+}$ to generate a $(D^+)$-L-$Ru^{II}$(PP)-L′-$(A^-)$ electron transfer intermediate, X. It might even be possible to design such a system so that the absorption of a photon by X would stimulate the back electron transfer process to regenerate the starting material, Equations 23 and 24. Such a molecule might be able to function as a light activated molecular switch.

$$D^+{-}(L){-}C{-}(L'){-}A^- + h\nu \rightarrow D^+{-}(L){-}C^*{-}(L'){-}A^- \quad (23)$$

$$D^+{-}(L){-}C^*{-}(L'){-}A^- \rightarrow D{-}(L){-}C{-}(L'){-}A \quad (24)$$

An alternative strategy for photoinduced charge separation is to directly irradiate a DACT absorption band. This should, in principle, result in the desired electronic configurations, but issues such as the rapid crossing of high energy to lower energy states (see Section 3), and rapid non-radiative relaxation (or back electron transfer) of the DACT excited state can result in very short lived species. When substantial reorganizational energies can be coupled with high spin/low spin configurational changes at the D or A center [e.g., at Co(II) or Cr(II) centers], it appears that back electron transfer can be relatively slow even when the ground state DACT absorption is large[55–57] (also see Section 3).

The sensitized excitation approach has been used to photoinject electrons into a $TiO_2$ semiconductor particle (in this application, the $TiO_2$ can be considered to be analogous to the acceptor of a D/A pair) from adsorbed Ru based sensitizers such as $Ru(C_2bpy)_2(NCS)_2$ ($C_2bpy$ = a dicarboxalato-bipyridine ligand which covalently links to the $TiO_2$ surface) in the design of efficient solar photocells.[139,140] Photoexcitation of the adsorbate results in the ultrafast injection of electrons into $TiO_2$[140] and the system is apparently not complicated by significant back electron transfer.

### 6.5 Energy Transfer Processes

Bimolecular reactions in which electronic excitation energy is transferred from one molecule to another, reactions of the generic type indicated in Equation 25, have been extensively studied.

$$D^* + A \rightarrow D + A^* \quad (25)$$

When D and A are part of the same molecule, then the intramolecular energy transfer process can be considered to be a special case of internal conversion (or sometimes of intersystem crossing). The intermolecular processes have been the traditional basis for definitions and theories of energy transfer. The feature which distinguishes energy transfer from most of the photoinduced processes discussed in this chapter is that it is a multielectronic process which involves changes of electronic configuration (but not electronic population) in at least two chromophoric centers. Two types of mechanisms for D/A coupling have been applied to energy transfer processes: (a) dipole-dipole (or Förster[141]) coupling; and (b) exchange (or Dexter) coupling.[1,2,4,25] The dipole-dipole coupling mechanism is dominant when the electronic transitions at the D and A chromophoric centers are dipole allowed (the mechanism involves coupling of the transition dipoles). The exchange mechanism arises from the indistinguishability of electrons,[25] and it is most important when the D*/D and A/A* transitions are dipole forbidden (or when their transition dipoles are small; see also the discussion of exchange and superexchange coupling in Section 3.1). Analogies have been drawn between energy transfer, especially for systems in which exchange coupling is operative, and one or two electron (or electron/hole) transfer.[2,142–146] To the extent that all non-radiative relaxation rate processes must depend in formally similar ways on Franck–Condon factors (i.e., their dependencies on $\lambda_{vib}$ and $\Delta E_{ge}^{00}$ can be described using similar algebraic formalisms), a point of view emphasized in this chapter, such an approach may be helpful. However, the electronic matrix elements, $H_{if}$, for energy transfer cannot be assumed to be simply related to those for electron transfer. Thus, for an exchange allowed triplet sensitization, $({}^3D^*, {}^1A)_i \rightarrow ({}^1D, {}^3A^*)_f$, the spin wavefunction is symmetric with respect to the interchange of electrons so that the orbital wave function of the transition state for the energy transfer must be antisymmetric in order for the electronic state wavefunction to be antisymmetric (Pauli principle). The electron transfer transition state is usually formulated in terms of a symmetric orbital function. In addition, that part of the reorganizational energy for electron transfer reactions which arises from the shift of charge between atomic centers will be absent in the energy transfer processes. These contrasts suggest that the electron transfer analogy should be treated with caution. Nevertheless, strong parallels have been found between the reactivity patterns for bimolecular (energy transfer) quenching of $({}^2E)$Cr(III) excited states by Co(III) complexes and the apparent contributions of electronic coupling to the outer-sphere electron transfer reaction patterns of Co(III) complexes.[142,143,147] These parallels can be treated in terms of very similar, but not identical types of superexchange contributions to the two types of reactions.

Wang and Schanze[148] have shown that, when covalently linked to a $[(Cp)Ru^{II}(arene)]^{2+}$ quencher, a series of *fac*-(PP)$Re^{I}(CO)_3$ triplet donors is quenched with rates which decrease approximately linearly with increasing $|\Delta E_{if}^{00}|$ and that the quenching rate constants are empirically correlated with the bimolecular quenching of a series of organic donors by $Ru(Cp)_2$,[149] i.e., for reactions of both series, $\ln k_q \cong 0.4E_D^0 +$ (a constant). These authors[148] did not successfully to fit the two sets of observations to a single function based on an electron transfer model[145,146] in which the Franck–Condon parameters are inferred from spectral fits of luminescence spectra. It should be noted that this electron transfer-based approach has been used for plausible correlations for a range of polypyridyl complex MLCT excited state donors and acceptors.[145,146,150]

Extremely weak D/A coupling can be achieved by encapsulating the donor (or acceptor) in such a way that intimate molecular contact is stereochemically inhibited. Some recent energy transfer studies with $|\Delta E_{if}^{00}| > \lambda_{vib}$ have found[151–153] that the decrease of $k_q$ with increasing $|\Delta E_{if}^{00}|$ is even more abrupt than one might expect from Equation 5. This is certainly in contrast to electron transfer systems[18–22] in which there appears to be a contribution from high frequency vibrational modes (as in Equation 6), and these observations do not conform to expectation based on the energy transfer models mentioned in the previous paragraph. Even if the behavior observed in these "carcerand" complexes were interpreted in terms of Equation 5, such an approach would imply that vibrational equilibration is much faster in these energy transfer systems than has been found for many electron transfer systems.[19,20,49–52] There is no obvious physical basis for such an inference. Since the solvent contribution to $\lambda_{vib}$ should be smaller for energy transfer than for electron transfer, Equation 7 may be more appropriate for the energy transfer reactions than Equation 6 but the correlations of $\ln k_q$ with $\Delta E_{ge}^{00}$ may be complicated by significant variations in $h\nu_M$ and $\gamma$ from one quencher to another. It appears that many issues of energy transfer remain to be resolved.

Dioxygen has been found to be an effective quencher of many transition metal excited states.[154–159] The quenching of ($^2E$)Cr(III) excited states should be particularly simple since the principle distortion modes are in $O_2$, yet the variations in the efficiency of ($^2E$)Cr(III) quenching by $O_2$ has been found to correlate with the ligands coordinated to Cr(III).[153–158] For example, $k_q \sim 2 \times 10^9$ $M^{-1}$ $s^{-1}$ for $O_2$ quenching of ($^2E$)Cr([9]ane$N_3$)$(NCS)_3$,[158,159] but $k_q < 10^6$ $M^{-1}$ $s^{-1}$ for $O_2$ quenching of ($^2E$)Cr([14]ane$N_4$)$(CN)_2^+$. Such observations may be consistent with the suggestion that LMCT excited states contribute, through superexchange contributions, to $H_{if}$ for energy transfer reactions.[142,143,147]

## 7 MULTIPHOTONIC PROCESSES

The photophysical processes and photochemical reactions discussed in the previous sections were presumed to follow the absorption of a single photon. When the excitation light intensity is sufficiently large, or when a long lived and strongly absorbing initial excited state is formed by the absorption of one photon, then processes characteristic of the absorption or scattering of more than one photon may occur. Such processes can interfere with photochemical transient studies when intense laser sources are used for excitation, but such processes can also be used to advantage since they can provide routes for population of non-spectroscopic states or for unusual photophysical processes. For purposes of discussion in this section, these multiphoton processes will be divided into two categories: (a) sequential (which depend on the formation of a metastable intermediate excited state); and (b) simultaneous (in which no real intermediate excited state is generated). While such processes can be a little esoteric, it should be recalled that the natural chlorophyll-based photosystems function by means of a sequential two photon process. There are also many technologically important devices, real or postulated, which are based on multiphotonic processes.

The requirements for sequential two photon processes are that the intermediate excited state have a significant lifetime relative to the experimental excitation parameters (such as pulse width or the delay between pulses), and that the absorption of this initial excited state at the excitation wavelength be appreciable. Some systematic studies have been reported for Cr(III) complexes.[160,161] The ($^2E$)Cr(III) excited states often have reasonably long lifetimes (e.g., greater than 1 ns) and they are about 1.7 V more oxidizing than the $^4A_2$ ground state, so that the LMCT absorption bands occur at $\sim 14 \times 10^3$ cm$^{-1}$ lower energy in the VEqES than in the ground state. Assuming the stationary state approximation, for an excitation laser pulse width $\Delta t_p$, greater than the excited state lifetime, $\tau_I$, for an upper state reaction efficiency, $\eta_R$, and for a small excited state absorbance (so that Beer's law can be linearized to give $N_{h\nu} \propto I_0 \varepsilon_I \delta$), the net quantum yield for the upper state reactions is given approximately by Equation 26 in which $\Phi_0$ is the

$$\Phi_R \sim \frac{\eta_R \tau_I I_0 \varepsilon_I \delta}{1 + \tau_I I_0 \varepsilon_I \delta} + \Phi_0 \tag{26}$$

singlephoton quantum yield. The qualitative features of Equation 26 have been demonstrated using $Cr(NH_3)_2(NCS)_4^-$,[160,161] for which $\tau_I$ can be systematically varied by changes in the solvent and for which CT excitation leads to $Cr^{2+}$ [which can be scavenged by $Co(NH_3)_5\ F^{2+}$] and ligand field excitation leads only to aquation of coordinated $NCS^-$. This example serves to illustrate that sequential two photon processes are only observed for: (a) sufficiently long intermediate state lifetimes ($\tau_I$); (b) sufficiently intense intermediate state absorbance ($\varepsilon_I \delta$); (c) sufficiently intense irradiation ($I_0$); and (d) significant upper state photochemistry ($\eta_R$). As $\varepsilon_I$ and/or $\tau_I$ become larger, the intensity requirement becomes correspondingly smaller. Even if $\eta_R = 0$, intermediate excited state absorbance can complicate very sensitive measurements of excited state properties, e.g., as in photoacoustic determinations of the heat released during FC state relaxation to the VEqES.[122]

It is in principle possible to find systems in which sequential two photon absorption leads to the emission of light of higher energy than that of the light absorbed ($h\nu_0$). Such a process, called upconversion,[162–166] should be possible in a bimetallic complex, as well as in the solid state (see chapter by Güdel and Brunold), provided the electronic coupling between Cr(III) centers (e.g., superexchange coupling induced by an upper LMCT state) is stronger in the excited state than in the ground state.[143] The upconversion process can be qualitatively viewed as the reverse of simultaneous pair absorption (s.p.a.),[167–170] at least in the sense that the same selection rules govern absorption and emission. For example, one of the more important intensity mechanisms proposed for s.p.a.'s involves superexchange (or CT) coupling of chromophore orbitals mediated by a bridging ligand.[164,165] This suggests that among the more likely systems in which to find such upconversion effects would be bimetallic chromium(III) complexes which are weakly coupled in their ground states, but in which the excited state superexchange interactions could lead to strong coupling. The light emission generated by such superexchange processes will in general be less than twice the excitation frequency ($\leqslant 2h\nu_0$), due to the rapid relaxation of higher energy excited states to the $^2E$ excited state in chromium(III) complexes.

The probability for simultaneous two photon absorption increases as the light incident intensity ($I_0$) increases. Since the selection rules are different for two photon and one photon absorption,[171,172] it might be possible to use this approach to gain photochemical access to excited states for which single photon excitation is forbidden. In practice internal conversion and intersystem crossing between excited states occur very rapidly in transition metal complexes, so that new photochemistry is likely for simultaneous two photon excitations only if the upper state processes are very rapid.

Multiphoton physical and chemical processes can be induced by interactions with transition metal complexes without the direct absorption of radiation. Two such processes will be mentioned here: (a) second harmonic generation; and (b) electron transfer induced by the optical electrical field.

The electric field component of light interacts with matter, even if there is no direct absorption of photons (see Chapter by Krausz). This interaction can be described as an induced, oscillatory polarization of molecular electronic charge.[173] To the extent that the induced polarization is a nonlinear function of the applied field, in this instance an optical electrical field, light scattered from the interaction will be a mixture of the incident, $h\nu_0$, and integral multiples, $nh\nu_0$, of the incident frequency.[173] The design and construction of efficient second harmonic ($n = 2$) generators has become a major technological concern.[173–177] It has been demonstrated that molecular donor/acceptor complexes are among the most efficient second harmonic generators,[174–178] and that transition metal complexes may be among the most important classes of compounds.[175–177] The coefficient for the second harmonic generation (the hyperpolarizability) contains a contribution which is analogous to the superexchange contribution in electron transfer systems[174] (discussed above and in the chapter by Richardson), which can be written as in Equation 27[174] where $h\nu_{DA}$ is the energy of the DACT transition, $f_{DA}$ its oscillator strength and $\mu_{DA}$ the change in dipole which accompanies the DACT transition.

$$\beta_{CT} = \frac{3e^2}{2m} \frac{h\nu_{DA} f_{DA} \Delta\mu_{DA}}{\left[(h\nu_{DA})^2 - (2h\nu)^2\right]\left[(h\nu_{DA})^2 - (h\nu)^2\right]} \tag{27}$$

In addition, it appears that the link between donor and acceptor is also important in determining large hyperpolarizabilities (and consistent with the superexchange coupling analogy).[174] An example of such a molecule with an unusually large hyperpolarizability is $(NC)_5Ru^{II}(CN)Ru^{III}(NH_3)_5^-$,[177] a molecule whose photophysical behavior has been discussed in Section 3.

An alternative to the optical electric field-induced oscillation of electronic charge model, discussed in the preceding paragraph, is to describe the interaction of the molecule with the optical electric field in terms of an oscillatory variation in the difference in energy between the ground and DACT excited state, $\Delta E_{DA}^{00}$, i.e., as an a.c. or optical Stark effect.[179–184] When the optical electric field gradient is sufficiently large, the electron density may be "trapped" at the acceptor. This may be most likely when the initial electron transfer state relaxes more quickly to a distorted and weakly coupled product state than it returns to the ground state (i.e., when $k_{isc} > k_{bet}$). Under such circumstances an intense radiation field might induce electron transfer without the direct absorption of light.[179–181] In the simple classical limit of such processes, when $\kappa_{el}\nu_{nu} < \nu_0$, Equation 5 functions in a manner similar to a delta function, and the yield of electron transfer products would be directly proportional to $\kappa_{el}\nu_{nu}$.

This section has very briefly summarized some very current and emerging topics in the photophysics and photochemistry of transition metal complexes. By the nature of the material, much of this is currently very speculative.

## 8 OVERVIEW

This chapter has surveyed aspects of the photophysics and photochemistry of coordination complexes. Experiments based on contemporary laser techniques as well as theoretical developments have led to what appear to be some general principles governing excited state relaxation behavior in transition metal complexes: (a) excess photonic energy tends to be deposited in high frequency vibrational modes and not in the low frequency modes associated with bond breaking (either heterolytic or homolytic bond breaking) coordinates; (b) internal conversion and intersystem crossing from one excited state to another are very fast and often efficient processes; and (c) when electronic coupling between different diabatic states is large the description of the excited states in terms of electronic PES's is often not useful and a vibronic description with mixed electronic character seems more useful. These are currently useful as unifying concepts and working hypotheses for most of the systems studied. They will no doubt be modified by future work. In particular, the study of the relaxation trajectories of FC excited states has barely begun, and the identification of the specific vibrational modes populated by the various relaxation events is so far mostly based on inference not direct observation. Similarly, there have been few direct investigations of the immediate, solvent trapped products of photoreactions. Such studies will inevitably require modification of mechanistic arguments presented in this chapter.

The range and variety of chemical and physical processes induced by the interaction of coordination complexes with electromagnetic radiation is awesome. This chapter has examined only some aspects of the observations and possibilities. We can expect even more amazing things of future studies.

## ACKNOWLEDGMENTS

Partial support of the research at Wayne State University by the Division of Chemical Sciences, Office of Basic Energy Sciences, Office of Energy Research, U.S. Department of Energy is gratefully acknowledged. The author also thanks Mr. P. McNamara for a careful reading of the manuscript and for many helpful comments.

## REFERENCES

1. Birks, J.B. *The Photophysics of Aromatic Molecules*; Wiley: New York; 1970.
2. Turro, N.J. *Modern Molecular Photochemistry*; Benjamin-Cummings: Menlo Park, CA; 1978.

3. Calvert, J.G.; Pitts, J.N. *Photochemistry*; Wiley: New York; 1967.
4. Yardley, J.T. *Introduction to Molecular Energy Transfer*; Academic: New York; 1980.
5. Serpone, N.; Jamieson, M.A. Coord. Chem. Rev. **1989**, *87*.
6. Damrauer, N.H.; Cerullo, G.; Yeh, A.; Boussie, T.R.; Shank, C.V.; McCusker, J.K. *Science* **1997**, *275*, 54.
7. Lever, A.B.P. *Inorganic Electronic Spectroscopy*; 2nd ed.; Elsevier: New York; 1984.
8. Harris, A.C.; Brown, S.K.; Harris, C.B. *Annu. Rev. Phys. Chem.* **1988**, *39*, 341.
9. Myers, A.B. *Laser Techniques in Chemistry*; Myers, A.B.; Rizzo, T.R.; Eds.; *Techniques of Chemistry Series Vol. XXIII*; Wiley: New York; 1995; Chapter X; p. 325.
10. Myers, A.B.; Mathies, R.A. In *Biological Applications of Raman Spectroscopy*; Spiro, T.G.; Ed.; Wiley: New York; 1982; Vol. 2; p. 1.
11. Marcus, R.A. *J. Phys. Chem.* **1989**, *93*, 3078.
12. Hush, N.S. *Prog. Inorg. Chem.* **1967**, *8*, 391.
13. Hush, N.S. *Electrochim. Acta* **1968** *13*, 1005.
14. Solomon, E.I. *Comments on Inorg. Chem.* **1984**, *3*, 227.
15. Newton, M.D.; Sutin, N. *Annu. Rev. Phys. Chem.* **1984**, *35*, 437.
16. Englman, R.; Jortner, J. *Mol. Phys.* **1970**, *18*, 145.
17. Marcus, R.A. *Annu. Rev. Phys. Chem.* **1964**, *15*, 155.
18. Kestner, N.; Logan, J.; Jortner, J. *J. Phys. Chem.*, **1974**, *64*, 2148.
19. Closs, G.L.; Calcaterra, L.T.; Green, N.J.; Denfield, K.W.; Miller, J.R. *J. Phys. Chem.* **1986**, *90*, 3722.
20. Closs, G.L.; Miller, J.R. *Science* **1988**, *240*, 440.
21. Kober, E.M.; Casper, J.V.; Lumpkin, R.S.; Meyer, T.J. *J. Phys. Chem.* **1986**, *90*, 3722.
22. Graff, D.; Claude, J.P.; Meyer, T.J. In *Electron Transfer Reactions*; Isied, S.S., Ed.; A.C.S. Advances in Chemistry Series No. 253; American Chemical Society: Washington, D.C.; 1997; p. 183.
23. Bixon, M.; Jortner, J. *J. Chem. Phys.* **1968**, *48*, 715.
24. Lin, S.H. *J. Chem. Phys.* **1966**, *44*, 3759.
25. Dexter, D.L. *J. Chem. Phys.* **1953**, *21*, 836.
26. Newton, M.D. *Chem. Rev.* **1991**, *91*, 767.
27. Ferretti, A.; Lanie, A.; Ondrechen, M.J.; Villani, G. *J. Phys. Chem.* **1995**, *99*, 10484.
28. Watzky, M.A.; Endicott, J.F.; Song, X.; Lei, Y.; Macatangay, A. *Inorg. Chem.* **1996**, *35*, 3463.
29. Ballhausen, C.J. In *Vibronic Processes in Inorganic Chemistry*; Flint, C.D., Ed.; Kluwer Academic Publishers: Dordrecht, 1989; p. 53.
30. Bersuker, I.B. *The Jahn–Teller Effect and Vibronic Interactions in Modern Chemistry*; Plenum: New York; 1984.
31. Noyes, R.M. *Progress in Reaction Kinetics*; Vol. 1; Porter, G.; Ed.; Pergamon: New York; 1961; p. 129.
32. Lorand, J.P. *Progress in Inorg. Chem.* **1972**, *17*, 207.
33. Fleischauer, P.D.; Adamson, A.W.; Sartori, G. *Progress in Inorg. Chem.* **1972**, *17*, 1.
34. Mulliken, R.S. *J. Phys. Chem.* **1971**, *55*, 288.
35. Roundhill, D.M. *Photochemistry and Photophysics of Metal Complexes*; Plenum: New York; 1984.

36. Joly, A.G.; Nelson, K.A. *J. Phys. Chem.* **1989**, *93*, 2976.
37. Lee, M.; Harris, C.B. *J. Am. Chem. Soc.* **1989**, *111*, 8963.
38. Sprague, J.R.; Arrivo, S.M.; Spears, K.G. *J. Phys. Chem.* **1991**, *95*, 10528.
39. Ferraudi, G.J. *Elements of Inorganic Photochemistry*; Wiley: New York; 1988.
40. *Inorganic and Organometallic Photochemistry*, Wrighton, M.S., Ed.; A.C.S. Advances in Chemistry Series No. 168; American Chemical Society: Washington, D.C.; 1978.
41. Sykora, J.; Sima, J. *Coord. Chem.Rev.* **1990**, *107*, 1.
42. Crosby, G. A. *Acc. Chem. Res.* **1975**, *8*, 231.
43. Yersin, H.; Braun, D.; Hensler, G.; Gallhuber, E. In *Vibronic Processes in Inorganic Chemistry*; Flint, C.D.; Ed; Kluwer: Dordrecht; 1989.
44. Riesen, H; Krausz, E. *Comments Inorg. Chem.* **1995**, *18*, 27.
45. Balzani. V.; Carassitti, V. *Photochemistry of Coordination Compounds*; Academic: New York; 1970.
46. Forster, L.S. *Chem.Rev.* **1990**, *90*, 331.
47. Kirk, A.D. *Coord. Chem. Rev.* **1981**, *39*, 225.
48. Endicott, J.F.; Ramasami, T.; Tamilarasan, R.; Lessard, R.B.; Ryu, C.K.; Brubaker, G.B. *Coord. Chem. Rev.* **1987**, *77*, 1.
49. Doorn, S.K.; Stoutland, D. O.; Dyer, R.B.; Woodruff, W.H. *J. Am. Chem. Soc.* **1992**, *114*, 3133.
50. Doorn, S.K.; Dyer, R.B.; Stoutland, D. O.; Woodruff, W.H. *J. Am. Chem. Soc.* **1993** *115*, 6398.
51. Tominaggi, K.; Kliner, D.A.V.; Johnson, A.E.; Levinger, N.E.; Barbara, P.F. *J. Phys. Chem.* **1993**, *97*, 1228.
52. Reid, P.J.; Silva, C.; Barbara, P. F.; Karki, L.; Hupp, J.T. *J. Phys. Chem.* **1995**, *99*, 2609.
53. Buranda, T.; Lei, Y.; Endicott, J.F. *J. Am. Chem. Soc.* **1992**, *114*, 6916.
54. Ponce, A.; Bachrach, M.; Farmer, P.J.; Winkler, J.R. *Inorg. Chim. Acta* **1996**, *114*, 6916.
55. Endicott, J.F.; Watzky, M.A.; Song, X.; Buranda, T. *J. Photochem. Photobiol. A: Chem.* **1994**, *82*, 181.
56. Endicott, J.F.; Watzky, M.A.; Song, X.; Buranda, T.; *Coord. Chem. Rev.*, in press.
57. Endicott, J.F.; Watzky, M.A.; Macatangay, A.V.; Mazzetto, S. E.; Song, X.; Buranda, T. *Electron and Ion Transfer in Condensed Media*; Kornyshev, A.; Ed.; World Scientific: Singapore; in press.
58. Song X.; van Wallendal, S.; Perkovic, M.W.; Jackman, D.C.; Endicott, J.F.; Rillema, D.P. *J. Phys. Chem.* **1993**, *97*, 3225.
59. Kenkre, V.M.; Tokmakoff, A.; Fayer, M.D. *J. Chem. Phys.* **1994**, *101*, 10618.
60. Langford, C.H.; Hollebone, B. R.; Serpone, N.A. *Coord. Chem. Rev.* **1981**.
61. Wilson, G.J.; Launikonis, A.; Sasse, W. dl.F.; Mau, A.W.-H. *J. Phys. Chem.* **1997**, *101*, 4860.
62. Wilson, R. B.; Solomon, E.I. *Inorg. Chem.* **1978**, *17*, 1729.
63. Hakamoto, K.; Urushiyama, A.; Kupka, H. *J. Phys. Chem.* **1981**, *85*, 1983.
64. Sexton, D. A.; Skibsted, L.M.; Magde, D.; Ford, P.C. *Inorg. Chem.* **1984**, *23*, 4533.
65. Ryu, C.K.; Lessard, R.B.; Lynch, D.; Endicott, J.F *J. Phys. Chem.* **1989**, *93*, 1752.

66. Endicott, J.F.; Perkovic, M.W.; Heeg, M.J.; Ryu, C.K.; Thompson, D. In *Electron Transfer*; Isied, S.S.; Ed.; A.C.S. Advances in Chemistry Series No. 253; American Chemical Society: Washington, D.C.; 1997; p. 199.
67. Perkovic, M.W.; Endicott, J.F. *J. Phys. Chem.* **1990**, *94*, 1217.
68. Perkovic, M.W.; Heeg, J.J.; Endicott, J.F. *Inorg. Chem.* **1991**, *30*, 3140.
69. Kane-Maguire, N.A.P.; Miller, P.K.; Trzupek, L.S. *Inorg. Chim. Acta* **1983**, *76*, L179.
70. Horvarth, O.; Stevenson, K.L. *Charge Transfer Photochemistry of Coordination Complexes*; V.C.H.: New York; 1993.
71. Juris A.; Barigelletti, F.; Compagna, S.; Balzani, V.; Belser, P.; von Zelewsky, A. *Coord. Chem Rev.* **1988**, *84*, 85.
72. Kalyanasundaram, K. *Coord. Chem. Rev.* **1982**, *46*, 159.
73. Kalyanasundaram, K. *Photochemistry of Polypyridine and Porphyrin Complexes*; Academic Press: New York; 1992.
74. Meyer, T.J. *Progr. Inorg. Chem.* **1984**, *30*, 389.
75. Balzani, V.; Juris, A.; Venturi, M.; Campagna, S.; Serroni, S. *Chem. Rev.* **1996**, *96*, 759.
76. Riesen, H.; Wallace, L.; Krausz, E. *Mol. Phys.* **1996**, *87*, 1299.
77. DeArmond, M.K.; Carlin, C.M. *Coord. Chem. Rev.* **1981**, *36*, 325.
78. Braun, D.; Huber, P.; Wudy, J.; Schmidt, J.; Yersin H. *J. Phys. Chem.* **1984**, *98*, 8044.
79. Dallinger, R.F.; Woodruff, W.H. *Am. Chem. Soc.* **1979**, *101*, 4391.
80. Ford, P.C. In *Inorg. and Organometallic Photochemistry*; Wrighton, M.S., Ed.; A.C.S. Advances in Chemistry Series No. 168.; American Chemical Society: Washington, D. C.; 1978; p. 73.
81. Lei, Y.; Buranda, T.; Endicott, J.F. *J. Am. Chem. Soc.* **1990**, *112*, 8820.
82. Durham, B.; Caspar, J.V.; Nagle, J.K.; Meyer, T.J. *J. Am. Chem. Soc.* **1982**, *104*, 4803.
83. Beattie, J.K. *Adv. Inorg. Chem.* **1988**, *32*, 1.
84. Gutlich, P.; Hauser, A.; Spieling, H. *Angew. Chem. Int. Ed. Engl.* **1994**, *33*, 2024.
85. McCusker, J.K.; Rheingold, A.L.; Hendrickson, D.N. *Inorg. Chem.* **1996**, *35*, 2100.
86. Wilson, R.B.; Solomon, E.I. *J. Am. Chem. Soc.* **1980**, *102*, 4085.
87. Hirohara, H.; Ivice, K. J.; McGarvey, J.J.; Wilson, J. *J. Am. Chem. Soc.* **1974**, *96*, 4453.
88. McGarvey, J.J.; Wilson, J. *J. Am. Chem. Soc.* **1975**, *97*, 2531.
89. Campbell, L.; McGarvey, J.J. *J. Am. Chem. Soc.* **1977**, *99*, 5809.
90. Campbell, L.; McGarvey, J.J.; Samman, N.G. *Inorg. Chem.* **1978**, *17*, 3378.
91. Adamson, A.W. *J. Phys. Chem.* **1967**, *71*, 798.
92. Monsted, L.; Monsted, O. *Coord. Chem. Rev.* **1989**, *94*, 109.
93. Skibsted, L.H. *Coord. Chem. Rev.* **1989**, *94*, 151.
94. Ford, P.C. *Rev. Chem. Intermediates* **1979**, *2*, 267.
95. Ford, P.C. *Coord. Chem. Rev.* **1982**, *44*, 61.
96. Ford, P.C.; Wink, D.; Dibinedetto, J. *Progr. Inorg. Chem.* **1983**, *30*, 213.
97. Zink, J.I. *Inorg. Chem.* **1975**, *14*, 4467.
98. Vanquickenbourne, L.G.; Ceulemans, A. *Coord. Chem. Rev.* **1980**, *48*, 157.
99. Bersuker, I.B. *Electronic Structure and Properties of Transition Metal Compounds*; Wiley: New York; 1996.
100. Purcell, K.F.; Clark, S.F.; Petersen, J.D. *Inorg.Chem.* **1980**, *19*, 2183.

101. Bergkamp, M.A.; Branon, J.; Magde, D.; Watts, R.J.; Ford, P.C. *J. Am. Chem. Soc.* **1979**, *101*, 4549.
102. Adamson A.W. *Discuss. Faraday Soc.* **1960**, *29*, 163.
103. Endicott, J.F.; Ferraudi, G.J.; Barber, J.R. *J. Phys. Chem.* **1975**, *79*, 630.
104. Endicott, J.F. In *Concepts of Inorganic Photochemistry*; Adamson, A.W.; Fleischauer, P.D.; Eds.; Wiley: New York; 1975; Chapter 3; p. 81.
105. Endicott, J.F. *Survey of Progress in Chemistry* **1976**, *7*, 1976.
106. Weit, S.K.; Grutsch, P.A.; Kutal, C. *Inorg. Chem.* **1991**, *30*, 2819.
107. Endicott, J.F.; Netzel, T.L. *J. Am. Chem. Soc.* **1979**, *101*, 4000.
108. Taylor, R.T.; Smucker, L.; Hanna, M.L.; Gill, J. *Arch. Biochem. Biophys.* **1973**, *156*, 521.
109. Chen, E.; Chance, M.R. *Biochem.* **1993**, *32*, 1480.
110. Garr, C.D.; Finke, R.G. *J. Am. Chem. Soc.* **1992**, *114*, 10440.
111. Endicott, J.F. *Inorg. Chem.* **1977**, *16*, 494.
112. Kelley, T.L.; Endicott, J.F. *J. Am. Chem. Soc.* **1972**, *94*, 1797.
113. Sriram, R.; Endicott, J.F. *J. Chem. Soc. Chem. Commun.* **1976**, 683.
114. Wells, W. L.; Endicott, J.F. *J. Phys. Chem.* **1971**, *75*, 3075.
115. Endicott, J.F.; Ferraudi, G.J. *Inorg. Chem.* **1975**, *14*, 3133.
116. Natarajan, P.; Endicott, J.F. *J. Am. Chem. Soc.* **1973**, *95*, 2470.
117. Natarajan, P.; Endicott, J.F. *J. Phys. Chem.* **1973**, *77*, 2049.
118. Kunkely, H.; Vogler, A. *J. Chem. Soc. Chem. Commun.* **1994**, 2671.
119. Vaudo, A.F.; Kantrowitz, E.R.; Hoffman, M.Z.; Papaconstantinou, E.; Endicott, J.F. *J. Am. Chem. Soc.* **1972**, *294*, 6655.
120. Kantrowitz, E.R.; Hoffman, M.Z.; Schilling, K.M. *J. Phys. Chem.* **1972**, *76*, 24933.
121. Reed, J.L.; Wang, F.; Basolo, F. *J. Am. Chem. Soc.* **1972**, *94*, 7173.
122. Song, X.; Endicott, J.F. *Chem. Phys. Lett.* **1993**, *204*, 400.
123. Ford, P.C.; Vogler, A. *Acct. Chem. Res.* **1993**, *26*, 220.
124. Ford, P.C.; *Coord. Chem. Rev.* **1994**, *132*, 129.
125. Simon, J.A.; Palke, W.E.; Ford, P.C. *Inorg. Chem.* **1996**, *35*, 6413.
126. Stevensen, K.L.; Knorr, D.W.; Hovarth, A. *Inorg. Chem.* **1996**, *35*, 835.
127. Lai, D.S.; Zink, J.I. *Inorg. Chem.* **1993**, *32*, 2594.
128. Rath, N.P.; Holt, E.M.; Tanimura, K. *Inorg. Chem.* **1985**, *24*, 3934.
129. Vitale, M.; Ryu, C.K.; Palke, W.E.; Ford, P.C. *Inorg. Chem.* **1994**, *33*, 561.
130. *Photoinduced Electron Transfer*; Fox, M.A.; Channon, M., Eds.; Elsevier: Amsterdam,1988; Parts A-D.
131. *J. Photochem. Photobiol. A: Chem*; Arnout, L.G., Ed.; **1994**, *82*, 1–236.
132. Kavarnos, G.J. *Fundamentals of Photoinduced Electron Transfer*; VCH Publishers: New York; 1993.
133. Sauvage, J.-P.; Collin. J.-P.; Chambron, J.-C.; Giillerez, S.; Coudret, C.; Balzani, V.; Barigelletti, F.; De Cola, L.; Flamifni, L. *Chem. Rev.* **1994**, *94*, 993.
134. *Molecular Level Artificial Photosynthetic Materials*; Meyer, G.J., Ed. (Karlin, K.D., Series Ed.); *Progr. Inorg. Chem.* **1997**, *44*.
135. Berkoff, R.; Krist, K.; Gafney, H.D. *Inorg. Chem.* **1980**, *19*, 1.

136. Jones, W.E.; Baxter, S.M., Jr.; Mecklenburg, S.L.; Erickson, B.W.; Peek, B.M.; Meyer, T.J. In *Supramolecular Chemistry*; Balzani, V.; De Cola, L., Eds.; Kluwer: Dordrecht, The Netherlands; 1992; p. 249.
137. Huang, S.C.; Macpherson, A.N.; Lin, S.; Liddell, P.A.; Seely, G.R.; Moore, A.L.; Moore, T.A.; Gust, D. *J. Am. Chem. Soc.* **1995**, *117*, 1657.
138. Wiederrecht, G.P.; Niemczyk, M.P.; Svec, W.A.; Wasielewski, M.R. *J. Am. Chem. Soc.* **1996**, *118*, 81.
139. Gerfin, T.; Grätzel, M.; Walder, L. *Molecular Level Artificial Photosynthetic Materials*; Meyer, G.J., Ed. (Karlin, K.D., Series Ed.); *Progr. Inorg. Chem.* **1997**, *44*, 345.
140. Nazeerudin, M.K.; Kay, A.; Rodicio, I.; Humphrey-Baker, R.; Müller, E.; Liska, P.; Vlachopoulos, N.; Grätael, M. *J. Am. Chem. Soc.* **1993**, *115*, 6382.
141. Förster, T. *Naturwissenschaften.* **1946**, *33*, 166.
142. Endicott, J.F.; Ramasami, T.; Gaswick, D.L.; Tamilarasan, R.; Heeg, M.J.; Brubaker, G.R.; Pyke, S.C. *J. Am. Chem. Soc.* **1983**, *105*, 5301.
143. Endicott, J.F. *Acc. Chem. Res.* **1988**, *21*, 82.
144. Closs, G.L.; Johnson, M.D.; Miller, J.R.; Piotrowiak, P. *J. Am. Chem. Soc.* **1989**, *111*, 3751.
145. Murtaza, Z.; Zipp, A.P.; Worl, L.A.; Graff, D.K.; Jones, W.E., Jr.; Bates, W.D.; Meyer, T.J. *J. Am. Chem. Soc.* **1991**, *113*, 5113.
146. Murtaza, Z.; Graff, D.K.; Zipp, A.P.; Worl, L.A.; Jones, W.E., Jr.; Bates, W.D.; Meyer, T.J. *J. Phys. Chem.* **1994**, *98*, 10504.
147. Endicott, J.F.; Tamilarasan, R.; Brubaker, G.R. *J. Am. Chem. Soc.* **1986** *108*, 5193.
148. Wang, Y.; Schanz, K.S. *Inorg. Chem.* **1994**, *33*, 1354.
149. Chapple, A.P.; Vikesland, J.P.; Wilkinson, F. *Chem. Phys. Lett.* **1977**, *50*, 81.
150. Liang, Y.Y.; Baba, A.I.; Kim, W.Y.; Atherton, S.J.; Schmehl, R.H. *J. Phys. Chem.* **1996**, *100*, 18408.
151. Farran, A.; Deshayes, K.D.; Matthews, C.; Balanessa, I. *J. Am. Chem. Soc.* **1995**, *117*, 9614.
152. Farran, A.; Deshayes, K.D. *J. Phus. Chem. Soc.* **1996**, *100*, 3305.
153. Parola, A.J.; Dina, F.; Fereira, E.; Maestri, M.; Balzani, V. *J. Am. Chem. Soc.* **1996**, *118*, 11610.
154. Pfiel, A. *J. Am. Chem. Soc.* **1971**, *93*, 5395.
155. Demas, J.N.; Harris, E.W.; McBride, R.P. *J. Am. Chem. Soc.* **1977**, *99*, 3547.
156. Buell, S.L.; Demas, J.N. *J. Phys. Chem.* **1983**, 4675.
157. Brunschwig, B.; Sutin, N. *J. Am. Soc.* **1978**, *100*, 7568.
158. Kirk, A.D.; Namasivayam, C. *Inorg. Chem.* **1988**, *27*, 1095.
159. Kirk, A.D.; Namasivayam, C.; Riske, W; Ristic-Petrovic, D. *Inorg. Chem.* **1989**, *28*, 972.
160. Sriram, R.; Endicott, J.F.; Pyke S.C. *J. Am. Chem. Soc.* **1977**, *94*, 4824.
161. Sriram, R.; Endicott, J.F.; Cunningham, K.M. *J. Phys. Chem.* **1981**, *85*, 38.
162. Powell, R.C.; Bloss, G. *Struct. Bonding.* **1980**, *42*, 43.
163. Güdel, H.U. *Comments Inorg. Chem.* **1984**, *3*, 189.
164. Gondaira, K.-I.; Tanabe, Y. *J. Phys. Soc. Japan.* **1976**, *21*, 1527.
165. Naito, M. *J. Phys. Soc. Japan.* **1973**, *34*, 1491.
166. Vasanyi, F.; Dieke, G.H. *Phys. Rev. Letters.* **1961**, *7*, 42.

167. Stokowski, S.E.; Sel, D.D. *Phys. Rev.* **1971**, *B3*, 208.

168. Ferguson, J.; Guggenheim, H.J.; Tanabe, Y. *J. Phys. Soc. Japan.* **1966**, *21*, 692.

169. Dubicki, L. *Austr, J. Chem.* **1972**, *25*, 1141.

170. Ferguson, J.; Güdel, H.U.; Krausz, E.G. *Mol. Phys.* **1975**, *30*, 1139.

171. McClain, W.M. *Acc. Chem. Res.* **1974**, *7*, 129.

172. Campochiaro, C.; McClure, D.S.; Rabinowitz, D; Dougal, S. In *Vibronic Processes in Inorg. Chem.*; Flint, C.D., Ed.; Kluwer: Dordrecht; 1989.

173. Butcher, P.N.; Cotter, D. *The Elements of Nonlinear Optics*; Cambridge Univ. Press: Cambridge; 1990.

174. Kanis, D.R.; Ratner, M.A.; Marks, T.J. *Chem. Rev.* **1994**, *94*, 195.

175. Long, N.J. *Angew. Chem. Int. Ed. Engl.* **1995**, *34*, 21.

176. Coe, B.J.; Chamberlain, M.C.; Essex-Lapreseti, J.P.; Gaines, S.; Jeffery, J.C.; Houbrechts, S.; Persoons, A. *Inorg. Chem.* **1997**, *36*, 3284.

177. Laidlaw, W.M.; Denning, R.G.; Verbiest, T.; Chauchard, E.; Persoons, A. *Nature* **1993**, *363*, 58.

178. Albert, I.D.L.; Marks, T.J.; Ratner, M.A. *J. Am. Chem. Soc.* **1997**, *119*, 6575.

179. Dakhnovskii, Yu.; Evans, D.G.; Kim, H.J.; Coalson, R.D. *J. Phys. Chem.* **1995**, *103*, 5461.

180. Evans, D.G.; Coalson, R.D.; Kim, H.J.; Dakhnovskii, Yu. *Phys. Rev. Lett.* **1995**, *75*, 3649.

181. Evans, D.G.; Coalson R.D.; Dakhnovskii, Yu. *J. Chem. Phys.* **1996**, *104*, 2287.

182. Cukier, R.I.; Morillo, M. *Chem. Phys.* **1994**, *193*, 375.

183. Ben-Tai, N.; Moiseyev, N.; Kosloff, R. *J. Chem. Phys.* **1993**, *98*, 9610.

184. Story, J.G.; Duncan, D.I.; Gallagher, T.F. *Phys. Rev. Lett.* **1993**, *70*, 3012.

185. Walker, L.A.; Shiang, J.J.; Anderson, N.A.; Pullen, S.H.; Sension, R.J. *J. Am. Chem. Soc.* **1998**, *120*, 7268.

186. Walker, L.A.; Jarrett, J.T.; Anderson, N.A.; Pullen, S.H.; Matthews, R.G.; Sension, R.J. *J. Am. Chem. Soc.* **1998**, *120*, 3597.

# 6 Multiple Metal–Metal Bonds

VINCENT M. MISKOWSKI AND MICHAEL D. HOPKINS
Department of Chemistry
University of Pittsburgh
Pittsburgh, PA 15260, USA

JAY R. WINKLER AND HARRY B. GRAY
Beckman Institute
California Institute of Technology
Pasadena, CA 91125, USA
E-mail: hgcm@cco.caltech.edu



*Inorganic Electronic Structure and Spectroscopy, Volume II: Applications and Case Studies.*
Edited by E. I. Solomon and A. B. P. Lever.
ISBN 0-471-32682-8. 

## ABSTRACT

The electronic structures and spectra of binuclear metal complexes with $[\sigma^2\pi^4]$ $(\delta, \delta^*, \pi^*)^n$ ($n = 0$ to 7) electronic configurations are examined. One-electron $\delta - \delta^*$ orbital splittings in quadruply bonded complexes are in the range 5,000–10,000 cm$^{-1}$; the weak $\delta$-interaction leads to a $^1(\delta\delta^*)$ excited state with substantial ionic character. A simple crystal-field model is presented for estimating the energies of the states of a given $(\pi^*, \delta^*)^n$ configuration. We demonstrate that this model can account for the spin-forbidden transition energies of the quartet $(\pi^*)^2(\delta^*)^1$ and triplet $(\pi^*)^1(\delta^*)^1$ ground states, and also for the zero-field splittings of the triplet $(\pi^*)^2(\delta^*)^2$, $(\pi^*)^1(\delta^*)^1$, and $(\pi^*)^2$ ground states.

## 1 INTRODUCTION

In 1964, Cotton proposed that a quadruple metal–metal bond is present in octachlorodirhenate(III), $Re_2Cl_8^{2-}$.[1] With metal–metal molecular orbital energies increasing in the order $\sigma < \pi < \delta < \delta^* < \pi^* < \sigma^*$, the ground state of a $d^4$–$d^4$ complex would be $[\sigma^2\pi^4\delta^2]$ (Figure 1). This electronic structure nicely explains the short (2.24 Å) Re–Re distance and the eclipsed ($D_{4h}$) geometry of $Re_2Cl_8^{2-}$.[2]

The field has expanded enormously,[2] and it has become clear that many binuclear compounds can be prepared that have more than, or fewer than, the ideal metal–metal bonding total of eight $d$ electrons. As a first guess at the ground states of these molecules, we simply add or subtract electrons to or from the MO diagram; this *aufbau* approach gives the correct ground states in many cases.

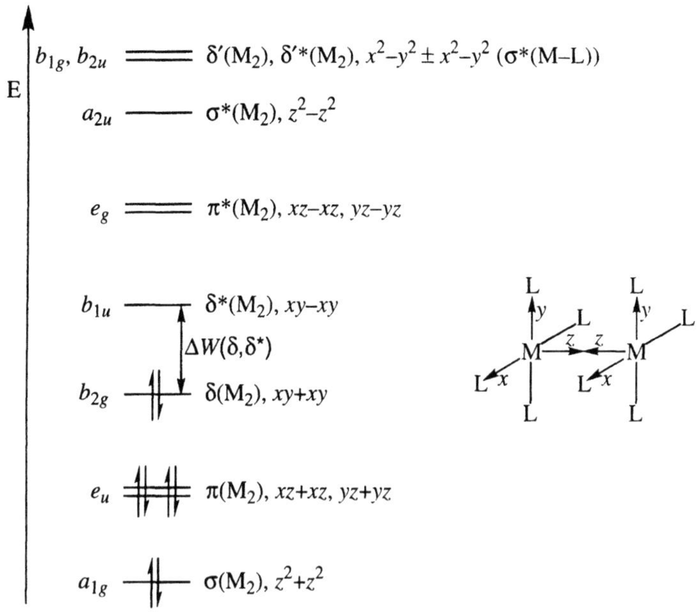


**Figure 1.** Molecular-orbital diagram of the $d$-orbital metal–metal interactions in a quadruply bonded $M_2L_8$ complex of $D_{4h}$ symmetry. Metal–ligand interactions perturb this simple picture according to the symmetries of ligand orbitals: $\sigma$(L), $a_{1g} + a_{2u} + b_{1g} + b_{2u} + e_g + e_u$; $\pi_\perp$(L) [i.e., $p$(L) orbitals directed $\perp$ to the M-M axis], $a_{1u} + a_{2g} + b_{1u} + b_{2g} + e_g + e_u$; $\pi_\parallel$(L), $a_{1g} + a_{2u} + b_{1g} + b_{2u} + e_g + e_u$. The energetically important interactions among these are: $\sigma$(L)$b_{1g}$ and $b_{2u}$ with $\delta'$, $\delta'^*$; $\pi_\perp$(L)$b_{1u}$ and $b_{2g}$ with $\delta$, $\delta^*$; $\pi_\parallel$(L)$e_g$ and $e_u$ with $\pi$, $\pi^*$. The interactions between $\pi_\perp$(L) and $\delta$ $M_2$ orbitals, in particular, may strongly perturb these levels.

## 2 THE $[\sigma^2\pi^4\delta^2]$ CONFIGURATION

A central question has concerned the strength of the $\delta$-bond. The earliest attempts along these lines were adversely affected by the primitive theoretical methods that were available at the time,[2] and inflated estimates of $\delta$-bond strength, as well as incorrect spectroscopic assignments, resulted.

Work on three different fronts ushered in a new era in this field from 1973 to 1975. On the theoretical side, calculations employing the then-novel SCF-X$\alpha$-SW method were published by Norman and Kolari,[3–5] as well as by Mortola et al.[6,7] The postulated molecular orbital ordering (Figure 1) was confirmed, although the splitting of the one-electron $\delta$- and $\delta^*$-levels was much smaller than had previously been envisioned. Meanwhile, the lowest-energy visible absorption in the electronic spectrum of $Re_2Cl_8^{2-}$ ($\nu_{max} = 14{,}700\ cm^{-1}$, $\varepsilon \approx 2{,}500$) and related quadruply bonded compounds was assigned to the $^1(\delta \rightarrow \delta^*)$ transition on the basis of its molecular $z$-polarization and pronounced vibronic structure in the metal–metal stretching mode $\nu_1$.[8] While the energy and intensity of this transition seemed to indicate a moderate $\delta$ interaction, the small decrease in $\nu_1$ in the $^1(\delta\delta^*)$ excited state relative to the ground state suggested that this bond was quite weak. Finally, the method of resonance Raman spectroscopy was applied by Clark and Franks.[9,10] Excitation into the recently assigned $^1(\delta \rightarrow \delta^*)$ band of $Mo_2Cl_8^{4-}$ resulted in very strong and selective enhancement of $\nu_1$, which was manifested as long overtone progressions in this mode. This result strongly indicated a metal–metal-localized transition, consistent with the single-crystal assignment.[8]

## 2.1 $\delta\delta^*$ Electronic States

We assume that the $\delta$-symmetry orbitals have pure metal character, thus, mixing with ligands is ignored. The $\delta$- and $\delta^*$-orbitals are then derived from symmetric and antisymmetric linear combinations of face-to-face $d_{xy}$-orbitals on atoms A and B:

$$\phi_1 = \frac{1}{\sqrt{2(1+S)}}\left[d_{xy}(\mathrm{A}) + d_{xy}(\mathrm{B})\right] \equiv \delta \tag{1}$$

$$\phi_2 = \frac{1}{\sqrt{2(1-S)}}\left[d_{xy}(\mathrm{A}) - d_{xy}(\mathrm{B})\right] \equiv \delta^* \tag{2}$$

where $S$ is the orbital overlap.

Four states, consistent with the Pauli principle, can be constructed from the $\delta$- and $\delta^*$-levels:

$$\Phi_1 = |\phi_1\phi_1| \equiv {}^1(\delta^2) \tag{3}$$

$$\Phi_2 = |\phi_2\phi_2| \equiv {}^1(\delta^{*2}) \tag{4}$$

$$\Phi_T = 2^{-1/2}\left[|\phi_1\phi_2| - |\phi_2\phi_1|\right] \equiv {}^3(\delta\delta^*) \tag{5}$$

$$\Phi_V = 2^{-1/2}\left[|\phi_1\phi_2| + |\phi_2\phi_1|\right] \equiv {}^1(\delta\delta^*) \tag{6}$$

We now write out the state energies, taking care to include the off-diagonal interaction between $\Phi_1$ and $\Phi_2$:

$$E_V = W_1 + W_2 + J_{12} + K_{12} \tag{7}$$

$$E_T = W_1 + W_2 + J_{12} - K_{12} \tag{8}$$

$$E(\Phi_1, \Phi_2) : \begin{vmatrix} 2W_1 + J_{11} - \lambda & K_{12} \\ K_{12} & 2W_2 + J_{22} - \lambda \end{vmatrix} = 0 \tag{9}$$

The $W_i$ are core integrals consisting of one-electron terms, $J_{ii}$ and $J_{ij}$ are one- and two-center coulomb repulsion integrals, and $K_{12}$ is the exchange integral.

Invoking the zero-differential-overlap approximation[11–13] results in $J_{11} = J_{22} = J_{12}$. If we additionally adopt the average energy of the one-electron $\delta$- and $\delta^*$-levels, $1/2(W_1 + W_2)$, as a reference point, we eliminate the $J$'s from the expressions and arrive at:

$$E_V = K \tag{10}$$

$$E_T = -K \tag{11}$$

$$E(\Phi_1, \Phi_2) : \begin{vmatrix} -\Delta W - \lambda & K \\ K & \Delta W - \lambda \end{vmatrix} = 0 \tag{12}$$

where we have dropped the subscript on $K = K_{12}$ and set $\Delta W = W_2 - W_1$. We finally note that $K$ is given by:

$$K = 1/2\big[(aa|aa) - (aa|bb)\big] \tag{13}$$

where $(aa|aa)$ is the energy required to transfer a $\delta$-electron from metal center A to metal center B at infinite A–B distance ($R$). The $(aa|bb)$ term is an electrostatic correction[11–14] to finite $R$, which is approximately given by $e^2/R$ in the point-charge limit.

In Figure 2 we show the internuclear-distance dependence of the $\delta$, $\delta^*$-manifold of states according to this simple theory. The strong dependence of $\Delta W$ on $R(M_2)$ (decreasing to larger $R$) and the weaker but much longer range $1/R$ dependence of $K$ (increasing to larger $R$) have both been qualitatively indicated. It is well established that $K$ is large for bonding–antibonding transitions in strongly bonded dimers; using the above analysis, it is estimated to be on the order of 30,000 $cm^{-1}$ for ethylene.[12] The situation for weakly bonding cases is of interest here. Even in the absence of bonding between A and B ($\Delta W = 0$), there will be a singlet "bonding–antibonding" (A-to-B charge-transfer) transition of energy $2K$; for $Cu_2(O_2CCH_3)_4(OH_2)_2$, the archetypal weakly metal–metal-interacting dimer, $K$ has been estimated (neglecting ligand interactions) to be 50,000 $cm^{-1}$.[14] * The ionic and covalent limits indicated in Figure 2 are most easily visualized by contrasting the ionic contribution of $K$ to the $^1(\delta\delta^*)$ state with the $\delta^2$ and $^3(\delta\delta^*)$ states:

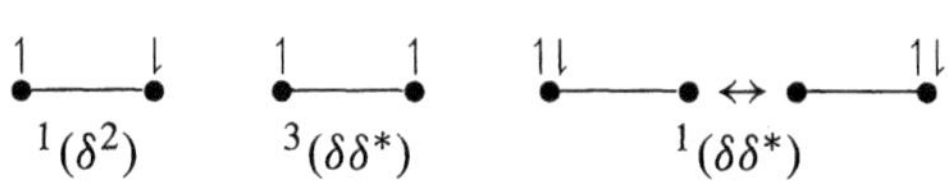


In order to attempt to fit this simple theory to experiment, we need at least two transition energies. Fortunately, there are many examples of metal–metal bonded complexes with $\delta^1$ or $\delta^2\delta^{*1}$ ground states. At the same theoretical level as our preceding analysis, the $^2(\delta \rightarrow \delta^*)$ transition of such molecules appears at exactly $\Delta W$, since the contributions of the two-electron terms to the transition energy cancel each other. Numerous structural studies have shown that there is very little effect on metal–metal bond length as a result of such one-electron oxidation or reduction;[2] indeed, even two-electron oxidation or reduction does not produce large changes.[15,16] Since $\Delta W$ can accordingly be expected to be little affected, we have performed theoretical

* $K$ should be strongly reduced by ligand covalency and is expected to be substantially larger for first-row transition metals than for second- and third-row metals.

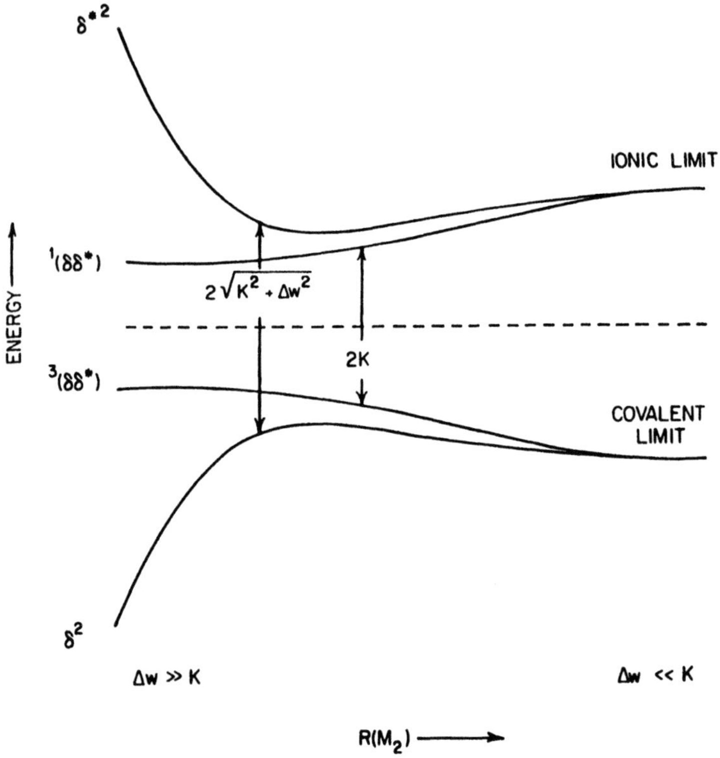


**Figure 2.** Dependence of the energies of the $\delta, \delta^*$-manifold of electronic states upon metal–metal distance.

fits of the $^1(\delta \rightarrow \delta^*)$ [$^1(\delta^2)$ ground state] energies using values of $\Delta W$ derived from the $^2(\delta \rightarrow \delta^*)$ transition energies of their redox congeners.

Table 1 shows our results for selected experimental data. Also included are some comparisons with results from SCF-X$\alpha$ calculations. It has been observed in the past that calculations of the $\delta \rightarrow \delta^*$ orbital energy differences for doublet ground-state molecules are in excellent agreement with observed $\delta \rightarrow \delta^*$ transition energies, and this can be traced to the good accounting of valence one-electron terms, but poor accounting of correlation energy, that this theory achieves. We note that such calculations for $\delta^2$ ground-state molecules also give good agreement with the $^2(\delta \rightarrow \delta^*)$ transition energies of one-electron-oxidized or one-electron-reduced species, in agreement with one of our assumptions.

A direct test of our assumption that $\Delta W$ for quadruply bonded dimers can be taken from the $^2(\delta \rightarrow \delta^*)$ transition energies of their one-electron oxidized or reduced congeners is provided by recent spectroscopic data for compounds containing the $Mo_2X_4(PMe_3)_4^+$ [$^2(\delta)$ ground state] and $Mo_2X_4(PMe_3)_4^-$ [$^2(\delta^2\delta^*)$ ground state] ions (X = Cl, Br, I).[17] In this approximation, the $^2(\delta \rightarrow \delta^*)$ energy should be the same (for a given halide) for the cation and anion radicals. The vibroni-

**Table 1 $\delta \rightarrow \delta^*$ Electronic Transition Parameters for Selected Quadruply Bonded Complexes [a]**

| | Observed | | | Calculated | |
|---|---|---|---|---|---|
| Compound | $^1(\delta \rightarrow \delta^*)$ | $^2(\delta \rightarrow \delta^*) \equiv \Delta W$ | $\Delta W$[b] (X$\alpha$) | $K$ | $^3(\delta \rightarrow \delta^*)$ |
| $Mo_2(O_2CPr^n)_4^{0/+}$ | 22,700[c] | 13,300[c] | 12,200[d] | 7,450 | 7,790 |
| $Mo_2(SO_4)_4^{4-/3-}$ | 19,420[e] | 7,140[e] | — | 8,400 | 2,630 |
| $Mo_2Cl_4(PMe_3)_4^{0/-}$ | 17,090[f] | 7,970[o] | 8,070[g,h] 8,390[h,i] | 6,690 | 3,710 |
| $Mo_2Cl_4(PMe_3)_4^{0/+}$ | 17,090[f] | 6,350[o] | 8,070[g,h] 8,390[h,i] | 7,370 | 2,350 |
| $Tc_2Cl_8^{2-/3-}$ | 14,750[j] | 6,500[k] | 5,820[l] | 5,940 | 2,860 |
| $Re_2(O_2CBu^t)_4Cl_2^{0/-}$ | 20,200[m] | 14,490[n] | — | 4,900 | 10,390 |
| $W_2Cl_4(PMe_3)_4^{0/+}$ | 15,150[f] | 7,350[o] | 7,340[g,p] | 5,790 | 3,570 |
| $Re_2Cl_4(PMe_2Ph)_4^{2+/+}$ | 13,890[q] | 7,350[q] | 5,650[r] | 5,000 | 3,890 |
| $Re_2Cl_6(PEt_3)_2^{0/-}$ | 13,950[s] | 6,560[n] | — | 5,430 | 3,090 |

[a] All energies ($cm^{-1}$) refer to vertical transitions, thus to band maxima. Calculated values are based on the observed $\delta \rightarrow \delta^*$ transition energies. [b] Calculated values of $\Delta W$ are for the $[\sigma^2\pi^4\delta^2]$ electronic configuration, except as noted. [c] Reference 124. [d] References 4 and 31. Calculated for $Mo_2(O_2CH)_4$. [e] References 37,125,126. [f] Reference 23. [g] Reference 24. [h] Calculated for $Mo_2Cl_4(PH_3)_4$. [i] Reference 25. [j] Reference 127. [k] Reference 112. [l] Calculated for the $[\sigma^2\pi^4\delta^2\delta^{*1}]$ electronic configuration, Reference 128. [m] References 129,130. [n] Reference 131. [o] Reference 17. [p] Calculated for $W_2Cl_4(PH_3)_4$. [q] Reference 16. [r] Calculated for the $Re_2Cl_4(PH_3)_4^+$ ion, Reference 108. [s] V. M. Miskowski, unpublished results.

cally structured $^2(\delta \rightarrow \delta^*)$ absorption spectra of these ions at 15 K exhibit electronic origins that are somewhat shifted to lower energy for cations versus anions; for the $Mo_2Cl_4(PMe_3)_4^{+/-}$ ions, for example, the electronic origins are observed at 7,970 $cm^{-1}$ for the anion and 6,350 $cm^{-1}$ for the cation (Table 1). The small difference between cation and anion $^1(\delta \rightarrow \delta^*)$ energies may be a reflection that the $\delta - \delta^*$ orbital splitting is smaller for the former, due to the contracton of the $d$-orbitals with increasing positive charge, in which case the best estimate of $\Delta W$ for the neutral molecule is an average of the cation and anion values. It is also probable that mixing of the $\delta \rightarrow \delta^*$ transition with LMCT transitions[18] is much larger for the cations than for the anions, leading to red shifts of the $^2(\delta \rightarrow \delta^*)$ bands of the cations.

The derived parameters in Table 1 are values of $K$ and predictions of $^3(\delta \rightarrow \delta^*)$ transition energies. As to the former, we observe that, despite a very large range of $^1(\delta \rightarrow \delta^*)$ transition energies, the calculated $K$'s are all roughly similar: ~6,000–8,000 $cm^{-1}$ for second transition series examples, and 5,000–6,000 $cm^{-1}$ for the third row. Inasmuch as these values are of comparable magnitude to the corresponding $\Delta W$s, configuration interaction (CI) is clearly of importance in determining the ground-state energy. Nonetheless, $^1(\delta^2)$ is still a fairly good representation of the ground state. For example, the values of $K$ and $\Delta W$ for $Re_2Cl_4(PMe_2Ph)_4^{2+}$ can be used to calculate the CI-mixed ground-state wavefunction as $0.956\Phi_1 - 0.295\Phi_2$, where $\Phi_1$ and $\Phi_2$ are the pure $^1(\delta^2)$ and $^1(\delta^{*2})$ state wavefunctions defined earlier (Equations 3 and 4). The occupation numbers (twice the squares of the coefficients) of the $\delta$- and $\delta^*$-orbitals are thus 1.83 and 0.17.

The 3,890 $cm^{-1}$ $^3(\delta\delta^*)$ energy calculated for $Re_2Cl_4(PMe_2Ph)_4^{2+}$ is in good agreement with an *ab initio* estimate by Hay of 3,200 $cm^{-1}$ for $Re_2Cl_8^{2-}$;[19] Hay's estimate of the magnitude of CI mixing is also similar to ours. There is also experimental evidence that bears out the large singlet-triplet $\delta\delta^*$ splitting inferred from these calculations.[20–22] As our present analysis shows, this is *not* a consequence of CI but of the large magnitude of $K$, as illustrated in Figure 3. While CI leaves the singlet-triplet $\delta\delta^*$ splitting unaffected, the effects on excitation energies are dramatic. If it were not for CI, the triplet state would be close to being the ground state. But CI will make a singlet the ground state, even in the case of almost no $\delta$ interaction, such as for copper acetate, despite small terms neglected in our analysis that would favor a triplet ground state.[14]

A second and completely different experimental approach to these questions can be taken. When diphosphines (e.g., $R_2PCH_2CH_2PR_2$) bridge dimers of the form $\beta$-$M_2X_4(P\text{-}P)_2$, where X is a halide, ligand-backbone strain enforces noneclipsed geometries. The torsional angle ($\chi$) is equal to 0° for the eclipsed and 45° for the

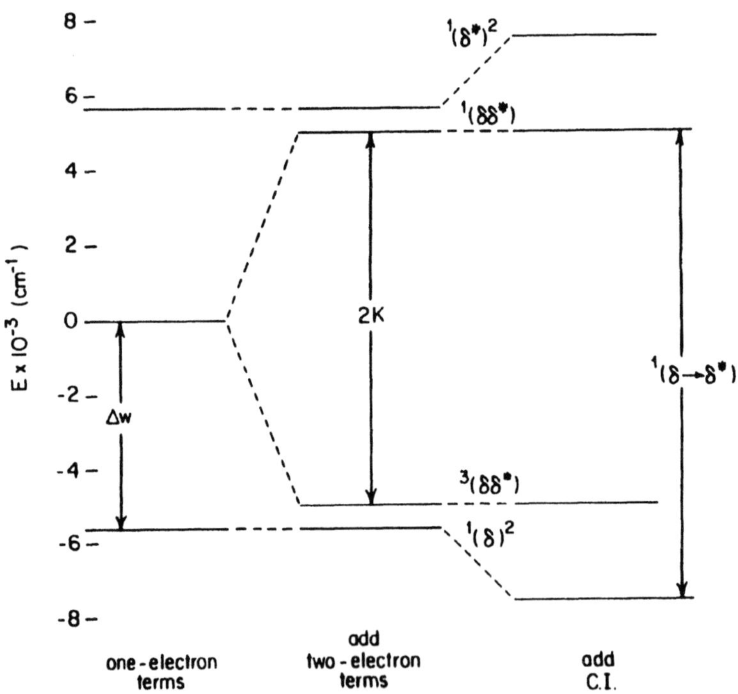


**Figure 3.** Relative energies of the $\delta$, $\delta^*$-manifold of electronic states with and without configuration interaction. Parameters used are those for $Re_2Cl_4(PMe_2Ph)_4^{n+}$ (Table 1).

staggered ($\Delta W = 0$) geometry. A theoretical prediction of the behavior of the $\delta$, $\delta^*$ states is very similar to that shown in Figure 2,[22] but $K$ is independent of rotation angle, according to the approximations defined earlier. The observed singlet "$\delta \rightarrow \delta^*$" transition energy of the nearly staggered ($\chi = 40°$) compound $\beta$-$Mo_2Cl_4(dmpe)_2$ [dmpe = 1,2-bis(dimethylphosphino)ethane] is 12,450 cm$^{-1}$, which should be very near $2K$; we thus estimate that $K$ = 6,200 cm$^{-1}$. Moreover, the rigorously eclipsed but otherwise very similar compound $Mo_2Cl_4(PMe_3)_4$ displays $^1(\delta \rightarrow \delta^*)$ at 17,090 cm$^{-1}$.[23] In combination with the above estimate of $K$, this value yields $\Delta W(\chi = 0°)$ = 8,950 cm$^{-1}$. The agreement of these derived parameters, both with calculated one-electron $\delta - \delta^*$ splittings of the model complex $Mo_2Cl_4(PH_3)_4$ obtained by SCF-X$\alpha$-SW (8,070 cm$^{-1}$)[24] and LCAO-HFS transition-state methods (8,390 cm$^{-1}$),[25] is very good.

Nocera and coworkers have assigned a band at 455 nm in the two-photon fluorescence excitation spectrum of $Mo_2Cl_4(PMe_3)_4$ to the $\Phi_1 \rightarrow \Phi_2$ $^1(\delta^2 \rightarrow \delta^{*2})$ transition.[26,27] This observation of a second transition in the $\delta\delta^*$ manifold permits a direct experimental evaluation of $\Delta W$ (9,133 cm$^{-1}$) and $K$ (6,111 cm$^{-1}$). These experimentally determined parameters are within 10% of the values estimated previously on the basis of the $^1(\delta \rightarrow \delta^*)$ excitation energy and SCF-X$\alpha$-SW calculations. The results are consistent with a weak $\delta$-interaction in the ground state, and a zwitterionic $^1(\delta\delta^*)$ excited state.[26,27] The detailed mechanism for the zwitterionic distortion (sudden polarization) involves mixing between the closely spaced $^1(\delta\delta^*)$ and $^1(\delta^{*2})$ excited states along a polar coordinate ($A_{2u}$ in $D_{4h}$ symmetry, $B_2$ in $D_{2d}$ symmetry), which makes both states totally symmetric.[28] A detailed analysis of the vibronic structure of the $^1(\delta \rightarrow \delta^*)$ absorption and $^1(\delta \leftarrow \delta^*)$ emission bands of $Mo_2X_4(PMe_3)_4$ (X = Cl, Br, I) compounds has yielded compelling support for modest double-well character (barriers of $10^2$ cm$^{-1}$) in the $^1(\delta\delta^*)$ potential energy surface.[28] Because nontotally symmetric vibrations have significant Franck–Condon factors in $^1(\delta \rightarrow \delta^*)$ and $^1(\delta \leftarrow \delta^*)$ transitions, vibronic structure is much more complex than for the $^2(\delta \rightarrow \delta^*)$ transitions of compounds with $^2(\delta)$ or $^2(\delta^2\delta^*)$ ground states, for which the $\delta \rightarrow \delta^*$ excited states are not subject to sudden polarization by the above mechanism.

The expected correlations between $^1(\delta \rightarrow \delta^*)$ transition energies and intensities and the strength of the $\delta$-bond have been elusive.[23,29] A key example is presented by $Mo_2(O_2CCH_3)_4$ and related molybdenum carboxylates, which have among the shortest metal–metal distances of all second and third transition series quadruply bonded dimers.[2] Although the $^1(\delta \rightarrow \delta^*)$ transition of this compound appears at relatively high energy (23,000 cm$^{-1}$), which seems intuitively consistent with its short bond length, it is also weak ($\varepsilon = 150$) — so weak, in fact, that the vibronically induced, $x, y$-polarized intensity is of comparable magnitude to the orbitally allowed, $z$-polarized intensity.[30] Our analysis shows that Mo(II) carboxylates, and also Re(III) carboxylates, are exceptional in having $\Delta W$ about twice as large as the other complexes. This has been previously predicted by SCF-X$\alpha$-SW calculations.[4,31] Of key importance is that these calculations indicate that the effect is due primarily to the interaction of $\delta$ and $\delta^*$ with carboxylate orbitals, rather than to an exceptionally large metal–metal $\delta$-interaction.

Many attempts have been made in the past to correlate $^1(\delta \rightarrow \delta^*)$ energies with the strength of the metal–metal interaction. We conclude that there is a correlation of the one-electron $\delta - \delta^*$ splitting with the $^1(\delta \rightarrow \delta^*)$ energy, but that it is not a simple one because two-electron terms are important. Certainly, small shifts in $^1(\delta \rightarrow \delta^*)$ cannot be casually attributed solely to variation in metal–metal interaction, and comparisons between compounds of metals in different transition series are particularly treacherous, since two-electron integrals such as $K$ typically decrease down a column.[32].

## 2.2 Other Electronic Excited States

Under $D_{4h}$ symmetry, the only other dipole-allowed metal–metal electronic transitions are $\sigma \rightarrow \sigma^*$ and $\pi \rightarrow \pi^*$, both of which are expected to display molecular $z$-polarization. An early study assigned the $^1(\pi \rightarrow \pi^*)$ transition of $Re_2Cl_8^{2-}$ at 39,200 $cm^{-1}$.[7] However, a subsequent *ab initio* calculation by Hay[19] predicted the dipole-allowed $^1(\pi \rightarrow \pi^*)$ transition to lie in the vacuum UV. A comprehensive study of the intense UV absorptions ($\nu$ = 25,000–50,000 $cm^{-1}$) of $M_2X_4(PMe_3)_4$ (M = Mo, X = Cl, Br, I; M = W, X = Cl) complexes failed to reveal any candidates for $^1(\pi \rightarrow \pi^*)$:[33] all of the absorptions in this region could be plausibly assigned to LMCT transitions based on their characteristic energy shifts as a function of ligand and metal (I < Br < Cl, Mo < W). A discussion of these LMCT transitions $[P(\sigma) \rightarrow \delta^*, X(\pi, \sigma) \rightarrow \delta^*]$ lies outside of the scope of this review, but we note that mixing of LMCT transitions with $^1(\delta \rightarrow \delta^*)$ has been proposed to play a key role in determining the $^1(\delta \rightarrow \delta^*)$ energy and intensity.[18] It appears likely that the intense UV absorptions of $Re_2X_8^{2-}$ ions (X = Cl, Br, I) are also LMCT in origin, and that $^1(\pi \rightarrow \pi^*)$ lies in the vacuum UV region for quadruply bonded metal dimers.

There is a plethora of forbidden (under $D_{4h}$ symmetry) electronic transitions, such as $\pi \rightarrow \delta^*$ and $\delta \rightarrow \pi^*$, that may occur at relatively low energies. In the case of $D_{2d}$-symmetry $Mo_2X_4(PMe_3)_4$ compounds, the $^1(\pi \rightarrow \delta^*)$ transition becomes dipole-allowed (with $x, y$-polarization); it has been definitively assigned to lie in the region $\nu_{max}$ = 20,000–23,000 $cm^{-1}$ ($\varepsilon_{max}$ = 200–500 $M^{-1}$ $cm^{-1}$) on the basis of polarization measurements and the insensitivity of the transition energy to halide.[34] The low extinction coefficients of these bands reflect the Laporte-forbidden nature of the transition in the parent $D_{4h}$ symmetry.

The $Re_2Cl_8^{2-}$ ion displays several weak ($\varepsilon \leqslant 100$ $M^{-1}$ $cm^{-1}$) transitions in the 16,500–25,000 $cm^{-1}$ range,[35,36] but assignments for these are not firm because the presence of disorder in the lattice of the $[N(Bu^n)_4]^+$ salt that was studied complicates the interpretation of single-crystal polarization data, and because vibronic selection rules apply to the fully dipole-forbidden metal–metal transitions of interest. We note, as an additional complication, that the two-photon $^1(\delta^2 \rightarrow \delta^{*2})$ transition[26] might appear weakly in this region in the one-photon spectra of this and other quadruply bonded complexes. Indeed, weak but sharply structured features in one-photon spectra of $Mo_2X_4(PMe_3)_4$[34] and $Re_2Cl_8^{2-}$[35] with origins near, respectively, 20,000 $cm^{-1}$ and 17,000 $cm^{-1}$ are candidates for such an assignment.

## 3 THE $[\sigma^2](\pi, \delta)^n$ ($n = 4, 5$) CONFIGURATIONS

For $n = 5$, all examples, to our knowledge, conform to the $^2(\pi^4\delta^1)$ configuration mentioned in Section 2.1 (Table 1). For example, the $Mo_2(SO_4)_4^{3-}$ ion displays the $^2(\delta \rightarrow \delta^*)$ transition at 7,140 $cm^{-1}$.[37] Additionally, the $^2(\pi \rightarrow \delta)$ and $^2(\pi \rightarrow \delta^*)$ transitions of this ion have been assigned at 16,810 and 23,980 $cm^{-1}$, respectively;[38] the good agreement between the difference in energy of these two transitions and the directly measured $^2(\delta \rightarrow \delta^*)$ energy is noteworthy.

While many compounds are known with $n = 4$, most do not possess the tetragonal geometry upon which the conventional MO description of metal–metal multiple bonding is based (Figure 1). A prominent exception is $Mo_2(HPO_4)_4^{2-}$, which has been shown to possess the $^1(\pi^4)$ ground state.[38] Based on a polarized single-crystal electronic absorption study, the $^1(\pi \rightarrow \delta)$ and $^1(\pi \rightarrow \delta^*)$ transitions were assigned at 18,450 and 23,800 $cm^{-1}$, respectively, suggesting a value of $\Delta W(\delta, \delta^*)$ very similar to that of $Mo_2(SO_4)_4^{3-}$. Additionally, an intense band is observed at 42,200 $cm^{-1}$,[38,39] and assignment to $^1(\pi \rightarrow \pi^*)$ has been suggested.

## 4 THE $[\sigma^2\pi^4\delta^2](\pi^*, \delta^*)^n$ ($n = 1$–$5$) CONFIGURATIONS

Analyses of the electronic structures of $(\pi^*, \delta^*)^n$ ($n = 1$–$5$) binuclear compounds are not as straightforward as those described in the preceding sections. The mixed-valence $Ru_2$(II,III) tetrakis($\mu$-carboxylato) compounds, for example, have spin-quartet ground states[40] rather than the spin-doublet $(\pi^*)^1(\delta^*)^2$ ground state that the simple MO diagram predicts (Figure 1). A conclusive explanation arose from theoretical calculations by Norman et al.;[41,42] the $\pi^*$ and $\delta^*$ levels are nearly degenerate for the ruthenium carboxylates, and a "high-spin" $^4B_{2u}(\pi^*)^2(\delta^*)^1$ ground state results, as would be expected from elementary considerations if the one-electron $\pi^*/\delta^*$ energy splitting is smaller than the spin-pairing energy. In order to understand this situation we need to consider deviations from pure metal–metal orbitals caused by $\pi$-symmetry metal–ligand interactions. The carboxylate HOMO is an internally nonbonding orbital of $a_2$ symmetry (local $C_{2v}$ symmetry) consisting essentially of oxygen $\pi$-symmetry lone pairs oriented perpendicular to the $CO_2$ plane. The $D_{4h}$ molecular orbitals ($e_g + a_{1u} + b_{1u}$) derived from these $\pi$-donor ligand orbitals include one ($b_{1u}$) with the right symmetry and spatial disposition to interact strongly with $\delta^*$ (Figure 1), destabilizing it relative to $\pi^*$, and this results in the calculated near-degeneracy. For quadruply bonded $Mo_2(\mu\text{-}O_2CR)_4$, stronger metal–metal interactions result in a "normal" (in a metal–metal sense) energetic ordering, $\pi^* > \delta^*$,[2] whereas for $Rh_2(\mu\text{-}O_2CR)_4$, which has all metal–metal orbitals filled except for $\sigma^*$, the energy of $\delta^*$ is similar to that of $\pi^*$, four weak metal–metal $\delta$- and $\pi$-symmetry interactions being overpowered by the metal–ligand interaction.[18] It has been concluded that the $\delta/\delta^*$ splitting of $(\pi^*)^4(\delta^*)^2$ $Rh_2(\mu\text{-}O_2CR)_4$ is *predominantly* due to metal–ligand interactions,[18] the metal–metal $\delta$-symmetry bonding interaction being insignificant for a metal–metal distance of ~2.4 Å.

Tetrakis($\mu$-carboxylato) binuclear complexes are easily synthesized for many metal ions,[2] and structural and spectroscopic data are now available for every possible value of $n$ of the metal–metal antibonding orbital occupancy $(\pi^*, \delta^*)^n$, $n = 0{-}6$. An enormous amount of data is also available for complexes of related three-atom bridging ligands. Replacement of one or both of the donor oxygen atoms of carboxylate, for example, gives related bridging ligands such as diphenylformamidinate [dpf$^-$, PhNC(H)NPh$^-$] or diphenyltriazenate [dptz$^-$, PhNNNPh$^-$] that are much stronger $\pi$-donors to $\delta^*$, so that the energy of $\delta^*$ may as much as 1 eV *above* that of $\pi^*$.[43,44]

For halide or pseudohalide ligands ($M_2X_8$), or mixed halide/phosphine ligand sets, ligand $\pi$-bonding effects on the $\pi^*$ and $\delta^*$ orbitals appear to balance out so that the energetic ordering $W(\pi^*) > W(\delta^*)$ of the pure metal–metal MO model is maintained (Figure 1).[2] However, to the best of our knowledge, no complex of this type with a $(\pi^*, \delta^*)^n$ configuration greater than $n = 2$ has ever been isolated and fully characterized. Planar or pseudo-planar nonbridging ligands such as porphyrins or Schiff bases may *favor* $W(\pi^*) > W(\delta^*)$, because their filled $\pi$-symmetry orbitals are oriented favorably for interaction with $\pi^*$ rather than $\delta^*$. Note also that hard ligands such as $SO_4^{2-}$ and $HPO_4^{2-}$ stabilize high oxidation states in binuclear complexes.[2,39,45]

Combining all of the ligand types, many compounds with a $(\pi^*, \delta^*)^n$ electron count have been prepared, and several ground states, sensitive to ligand type, have been inferred,[2] including every conceivable occupancy of the $\pi^*$ and $\delta^*$ orbitals. The questions we address now are as follows: Given a widely variable $\pi^*/\delta^*$ energy splitting, what are the possible ground states for the various $n$ of $(\pi^*, \delta^*)^n$; how do these relate to $W(\pi^*) - W(\delta^*)$, and, what are the excitation energies for the other states?

## 4.1 Relationship of the $(\pi^*, \delta^*)^n$ Problem to the Mononuclear $(t_{2g})^n$ Problem

Let us consider electrostatic one- and two-electron energy terms that will arise in the $(\pi^*, \delta^*)^n$ problem. Assuming tetragonal symmetry, we have only a single one-electron parameter, which we define in Equation (14):

$$\Delta = W(\pi^*) - W(\delta^*) \tag{14}$$

Note that in this context we are ignoring the absolute energies of these levels, hence their destabilization relative to their bonding counterparts $\pi$ and $\delta$.

Consider, for example, the $\left|\begin{matrix} + & + & + \\ \pi_x^* & \pi_y^* & \delta^* \end{matrix}\right|$ state [which is actually the $m_s = 3/2$ component of the $^4B_{2u}$ ground state of $(\pi^*)^2(\delta^*)^1$]. We can write the two-electron energy terms by inspection:

$$\begin{aligned} &J(\pi_x^*, \pi_y^*) + J(\pi_x^*, \delta^*) + J(\pi_y^*, \delta^*) \\ &\quad - K(\pi_x^*, \pi_y^*) - K(\pi_x^*, \delta^*) - K(\pi_y^*, \delta^*) \end{aligned} \tag{15}$$

This is exactly what we would get for the single-center "crystal-field" wavefunction $\begin{vmatrix} + & + & + \\ xz & yz & xy \end{vmatrix}$ with the replacements $\pi_x^* \leftrightarrow xz$, $\pi_y^* \leftrightarrow yz$, and $\delta^* \leftrightarrow xy$, and this is true for any wavefunction that we can write within $(\pi^*, \delta^*)^n/(t_{2g})^n$. The two problems are isomorphic. We should emphasize that if we tried simultaneously to consider partially occupied *bonding* levels $\pi$ and $\delta$, we would introduce energy terms that have no analogue in the $(t_{2g})^n$ problem, namely two-electron integrals (e.g., Equation 13) representing (in the weak-interaction limit) transfer of an electron from one metal center to the other (ionic dissociation).

We next examine the relationship of the two-electron integrals in the two problems. We initially assume pure $d$-orbital $M_2$ molecular orbitals:

$$\pi_x^* = \frac{1}{\sqrt{2(1 - S_\pi)}}[xz_A - xz_B] \tag{16}$$

$$\delta^* = \frac{1}{\sqrt{2(1 - S_\delta)}}[xy_A - xy_B] \tag{17}$$

The subscripts $A$ and $B$ refer to the two metal centers. Consider a typical two-electron integral:

$$\begin{aligned} J(\pi_x^*, \pi_y^*) &= \iint \pi_x^*(1)\pi_x^*(1)\left(\frac{1}{r_{12}}\right)\pi_y^*(2)\pi_y^*(2)d\tau_1 d\tau_2 \\ &= \frac{1}{4(1 - S_\pi)^2}\big[2(xz_A xz_A|yz_A yz_A) + 2(xz_A xz_A|yz_B yz_B) \\ &\quad + 4(xz_A xz_B|yz_A yz_B) - 8(xz_A xz_A|yz_A yz_B)\big] \end{aligned} \tag{18}$$

If we introduce the zero-differential overlap (ZDO) approximation,† this cumbersome integral reduces to:

$$J(\pi_x^*, \pi_y^*) = \frac{1}{2}\big[(xz_A xz_A|yz_A yz_A) + (xz_A xz_A|yz_B yz_B)\big] \tag{19}$$

This is seen to be the average of a single-center integral that appears in crystal-field theory, $(xz_A xz_A|yz_A yz_A) = J(xz, yz)$, and a two-center integral. Each of the six (in $D_{4h}$ symmetry) different two-electron integrals that appears in the $(\pi^*, \delta^*)^n$ problem reduces in the same way.

We now need to consider the magnitude of the two-center integral. If we let $R_{AB} \to 0$, the two-center integral goes to the single-center value,[13,46] which is the

---

† Reasons why the ZDO approximation works very well even when overlap integrals are fairly large have been presented in detail within the context of a thorough analysis of simple molecular orbital calculations on unsaturated organic molecules.[13] The key point is that the molecular orbitals in the semiempirical analysis can be supposed to be Löwdin orbitals,[46] and molecular integrals evaluated using these orbitals satisfy the ZDO approximation nearly exactly. To be consistent with this approach, overlap integrals *must* be taken equal to zero in semiempirical calculations; overlap effects manifest themselves in the scaling of other integrals in this type of approach.

united-atom limit. As $R_{AB} \to \infty$, the two-center integral goes smoothly to zero, as $(R_{AB})^{-1}$.[13,46] The conclusion from Equation 19, that we get one-half of the mononuclear value in this limit should be viewed with scepticism, because the MO wavefunctions based on Equations 16 and 17 do not dissociate correctly as $R_{AB} \to \infty$.[47] However, the important point is that we can conclude that *the binuclear integral is smaller than the mononuclear integral.* How much less is to be decided, in a semi-empirical approach, by comparison to experiments.

A realistic approach to the $(\pi^*, \delta^*)^n$ problem would be to assign "orbital reduction factors"[48] (analogous to those often employed in mononuclear problems to handle metal–ligand covalency effects) to each independent integral in order to proportionately relate them to empirical mononuclear $(t_{2g})^n$ values. Unfortunately, the number of parameters introduced (i.e., four) would make application to the limited experimental data hopeless, even in very favorable cases. We have taken the approach of assuming that all of the relevant binuclear integrals are reduced from the single-center values in exact proportion; hence, we get just one two-electron empirical parameter for the $(\pi^*, \delta^*)^n$ problem. Our approximations are entirely consistent with the various NDO calculation methods,[49] indeed, at a rather high level of such calculations.

The final approximation above may well be inadequate in detail, since the metal–metal $\pi$-overlaps are likely to be substantially larger than the $\delta$-overlaps,[2] roughly proportionately affecting two-center integrals involving $\pi^*$ and $\delta^*$ levels, and deviations from our predictions will result (see Appendix A). Another problem is that we have assumed pure $d$-orbital wavefunctions. Since the $\pi^*$ and $\delta^*$ levels may well have very different amounts of ligand mixing,[18,41,42,44] the relative values of integrals would be substantially modified. This is, however, no different from the problem in $d-d$ spectroscopy of describing $(t_{2g})^n(e_g)^m$ states.[48,50]

Collectively, the theoretical deficiencies are not very different from those facing mononuclear crystal-field theory. That crystal-field theory works rather well seems to result from the fact that it is group-theoretically consistent, and therefore gives good parametric fits (even though the values of the parameters sometimes prove to be inconsistent with the assumptions of the theory).[48] Our present effort is presented in hopes of achieving some modest success in a new application.

## 4.2 Application of the Crystal-Field Model to the $(\pi^*, \delta^*)^n$ Problem

In Tables 2 and 3, we summarize our crystal-field predictions for $(\pi^*, \delta^*)^n$ states. The parameter $\Delta$ is given by Equation 14 while the parameter $K$ is $K(\pi_x^*, \delta^*) \approx K(\pi_x^*, \pi_y^*)$. The mononuclear crystal-field integral $K = K(xz, xy) = K(xz, yz)$ is equal to $3\boldsymbol{B} + \boldsymbol{C}$ in terms of conventional Racah parameters.[50] In Appendix C, we summarize magnetic properties for the various possible ground states. The zero-field splitting $D$ obtained from magnetic measurements is an important observable that we will be attempting to fit.

The tabulated results can be obtained for the most part directly from the mononuclear $O_h(t_{2g})^n$ crystal-field results that are given by Griffith[50] (and many other textbooks). The $O_h$ electrostatic energies are subjected to perturbation by a strong

**Table 2 Electrostatic State Energies for $(\pi^*, \delta^*)^n$ Configurations**

$(\pi^*, \delta^*)^1$

$^2B_{1u}[(\delta^*)^1]: \ -\Delta$

$^2E_g[(\pi^*)^1]: \ 0$

$(\pi^*, \delta^*)^2$

$^3A_{2g}[(\pi^*)^2]: \quad \Delta$

$^1B_{1g}, {}^1B_{2g}[(\pi^*)^2]: \ \Delta + 2K$

$^3E_u[(\pi^*)^1(\delta^*)^1]: \quad 0$

$^1E_u[(\pi^*)^1(\delta^*)^1]: \quad 2K$

$$^1A_{1g}: \begin{matrix}(\pi^*)^2 \\ (\delta^*)^2\end{matrix} \begin{vmatrix} \Delta + 4K - W & \sqrt{2}K \\ \sqrt{2}K & -\Delta + 3K - W \end{vmatrix} = 0$$

$(\pi^*, \delta^*)^3$

$^4B_{2u}[(\pi^*)^2(\delta^*)^1]: \quad 0$

$^2A_{1u}, {}^2A_{2u}, {}^2B_{2u}[(\pi^*)^2(\delta^*)^1]: \ 3K$

$^2B_{1u}[(\pi^*)^2(\delta^*)^1]: \quad 5K$

$$^2E_g: \begin{matrix}(\pi^*)^1(\delta^*)^2 \\ (\pi^*)^3\end{matrix} \begin{vmatrix} -\Delta + 4K - W & K \\ K & \Delta + 4K - W \end{vmatrix} = 0$$

$(\pi^*, \delta^*)^4$

$^3A_{2g}[(\pi^*)^2(\delta^*)^2]: \quad -\Delta$

$^1B_{1g}, {}^1B_{2g}[(\pi^*)^2(\delta^*)^2]: \ -\Delta + 2K$

$^3E_u[(\pi^*)^3(\delta^*)^1]: \quad 0$

$^1E_u[(\pi^*)^3(\delta^*)^1]: \quad 2K$

$$^1A_{1g}: \begin{matrix}(\pi^*)^2(\delta^*)^2 \\ (\pi^*)^4\end{matrix} \begin{vmatrix} -\Delta + 4K - W & \sqrt{2}K \\ \sqrt{2}K & \Delta + 3K - W \end{vmatrix} = 0$$

$(\pi^*, \delta^*)^5$

$^2B_{1u}[(\pi^*)^4(\delta^*)^1]: \ \Delta$

$^2E_g[(\pi^*)^3(\delta^*)^2]: \ 0$

tetragonal field, and one-to-one correlations of the resultant $D_{4h}$ energies [neglecting all configuration interaction with states outside the $(t_{2g})^n$ configuration] to the $(\pi^*, \delta^*)^n$ state energies follows from our approximations. Similarly, the method of Stone[51] leads to the spin-orbit matrix elements being identical for the two problems except that some non-zero off-diagonal terms of the mononuclear problem vanish for $(\pi^*, \delta^*)^n$ because the correlating states may have different symmetries (gerade/ungerade) with respect to inversion in the binuclear problem.

There are two important points that must be kept in mind. First, if there is more than one state of a given symmetry and total spin for a configuration [i.e., the two $^1A_{1g}$ states for $(\pi^*, \delta^*)^2$ and $(\pi^*, \delta^*)^4$, and the two $^2E_g$ states for $(\pi^*, \delta^*)^3$], their energies must be reevaluated because they mix. Since the tetragonal perturbation in the $(\pi^*, \delta^*)^n$ problem is intrinsically "strong" in that the $\pi^*$ and $\delta^*$ levels do not mix even for $\Delta = 0$, by symmetry, strong tetragonal-field wavefunctions must be used, and the off-diagonal elements evaluated.

**Table 3 Spin-Orbit Energies for $(\pi^*, \delta^*)^n$ Configurations[a]**

$(\pi^*, \delta^*)^1$

$$E_{1/2g}(^2E_g): \ -1/2\zeta$$
$$E_{3/2g}(^2E_g): \ +1/2\zeta$$
$$E_{3/2u}(^2B_{1u}): \ 0$$

$(\pi^*, \delta^*)^2$

$$E_g(^3A_{2g}): \ 0$$
$$B_{1g}(^1B_{1g}): \ 0$$
$$B_{2g}(^1B_{2g}): \ 0$$
$$A_{1u}(^3E_u): \ +1/2\zeta$$
$$A_{2u}(^3E_u): \ +1/2\zeta$$
$$B_{1u}(^3E_u): \ -1/2\zeta$$
$$B_{2u}(^3E_u): \ -1/2\zeta$$

$$E_u: \begin{matrix} ^3E_u \\ ^1E_u \end{matrix} \begin{vmatrix} 0 & \frac{\zeta}{2} \\ \frac{\zeta}{2} & 0 \end{vmatrix}$$

$$A_{1g}: \begin{matrix} ^3A_{2g} \\ ^1A_{1g}[(\pi^*)^2] \\ ^1A_{1g}[(\delta^*)^2] \end{matrix} \begin{vmatrix} 0 & -\zeta & 0 \\ -\zeta & 0 & 0 \\ 0 & 0 & 0 \end{vmatrix}$$

$(\pi^*, \delta^*)^3$

$$E_{1/2u}(^4B_{2u}, {}^2A_{1u}, {}^2A_{2u}): \ 0$$
$$E_{3/2u}(^4B_{2u}, {}^2B_{1u}, {}^2B_{2u}): \ 0$$

$$E_{1/2g}: \begin{matrix} ^2E_g[(\pi^*)^1(\delta^*)^2] \\ ^2E_g[(\pi^*)^3] \end{matrix} \begin{vmatrix} -\frac{\zeta}{2} & \frac{\zeta}{\sqrt{2}} \\ \frac{\zeta}{\sqrt{2}} & \frac{\zeta}{2} \end{vmatrix}$$

$$E_{3/2g}: \begin{matrix} ^2E_g[(\pi^*)^1(\delta^*)^2] \\ ^2E_g[(\pi^*)^3] \end{matrix} \begin{vmatrix} \frac{\zeta}{2} & \frac{\zeta}{\sqrt{2}} \\ \frac{\zeta}{\sqrt{2}} & -\frac{\zeta}{2} \end{vmatrix}$$

$(\pi^*, \delta^*)^4$

Same as for $(\pi^*, \delta^*)^2$, but with the sign of $\zeta$ changed.

$(\pi^*, \delta^*)^5$

Same as for $(\pi^*, \delta^*)^1$, but with the sign of $\zeta$ changed.

[a] These are combined with the energies from Table 2 for the calculation of total energies.

Second, state symmetries need to be evaluated carefully. The symmetry of the $\delta^*$ orbital is not, of course, the same as that of a single-center $xy$ orbital. Because of this, while our energies converge to the diagonal energies of the $(t_{2g})^n$ states as $\Delta \rightarrow 0$, the symmetries of the states do *not* correlate to the $O_h$ limit in the same way as they do for a mononuclear tetragonally distorted $(t_{2g})^n$ problem. For example, the $^4B_{1g}$

and $(^2E_g + {}^2B_{2g})$ states of tetragonally distorted $(t_{2g})^n$ that correlate to $^4A_{2g}$ and $^2T_{2g}\,O_h$ states, respectively, correlate to $^4B_{2u}$ and $(^2E_g + {}^2B_{1u})$ states of $(\pi^*, \delta^*)^3$.

Similar procedures can be employed for other configurations (Appendix B).

What values can we expect for the parameters $\Delta$ and $K$? A wide range of values of $\Delta$ has been predicted by theoretical calculations on molecules having $(\pi^*, \delta^*)^n$ configurations, ranging from $+13{,}000\ \text{cm}^{-1}$, from a calculation on eclipsed $Os_2Cl_8^{2-}$,[52] to $-8{,}390\ \text{cm}^{-1}$, from a calculation on $Rh_2[\mu\text{-}HC(NH)_2]_4$.[43] For $K$, we could make estimates based on the *nd* Racah parameters $\boldsymbol{B}$ and $\boldsymbol{C}$ for mononuclear compounds, or better, perhaps, from the energies of the $^2E_g$, $^2T_{1g}$ spin-flip transitions of $(t_{2g})^3$ compounds, which theoretically occur at $3(3\boldsymbol{B} + \boldsymbol{C}) = 3K$. Results for first transition series complexes[53] indicate $K \approx 4{,}500$–$5{,}000\ \text{cm}^{-1}$, whereas values of $K$ for second and third transition series complexes are 2,000–3,000 $\text{cm}^{-1}$.[53–55] These are *upper* bounds for the $(\pi^*, \delta^*)^n$ value of $K$ (*vide supra*).

We now consider individual configurations.

## 4.3 $(\pi^*, \delta^*)^3$ Configurations

Most binuclear compounds with this electron count (a total of 11 $d$ electrons) are those of $Ru_2$(II,III), and nearly all of them have $(\pi^*)^2(\delta^*)^1$ ground states. Table 4 summarizes available spectroscopic and magnetic data.

**4.3.1 $(\pi^*)^2(\delta^*)^1$** The best-studied compounds having this ground state are the $Ru_2$(II,III) carboxylates. The $Ru_2(O_2CR)_4^+$ unit has $Ru_2$ distances in the narrow range of 2.248–2.290 Å,[2] and has been characterized both in discrete binuclear complexes with one or two anionic or neutral axial ligands, and in linear or zig-zag chain structures in which the cationic binuclear units are symmetrically bridged by axial ligands.

Room-temperature magnetic moments (Table 4) are typically near 4 B.M., which by itself is sufficient to establish the $^4B_{2u}$ ground state. All reported EPR spectra that we are aware of show effective $g$ values of $g_\perp \approx 4$ and $g_\parallel \approx 2$ (Table 4). This is diagnostic (Appendix C) of a zero-field splitting $D$ such that the $E_{1/2u}$ Kramers doublet derived from the $^4B_{2u}$ state lies at lowest energy. Our crystal-field model does not predict any zero-field splitting for the $^4B_{2u}$ state; as proposed by Norman et al.,[42] the fairly large values of $D$ that are experimentally required likely have their origin in high-order spin-orbit terms[51] that we are neglecting. The very low-energy doublet excited states that we do predict are a likely source of high-order off-diagonal spin-orbit terms. There is nothing in our current theoretical understanding to prevent $D$ from having the opposite sign, but a $g_\parallel \approx 6$, $g_\perp \approx 0$ EPR spectrum ($E_{3/2u}$ ground state, Appendix C) has never been reported.

The theoretical temperature dependence of the magnetism for $1/2g_\perp = g_\parallel = g_e$ is shown in Figure 4. The most thorough magnetic study available is that of Telser and Drago[56] on $Ru_2(O_2CPr^n)_4Cl$. The $g_\perp$ is 4.40, resulting in magnetic moments at high temperatures [$\mu_{\text{eff}}(294\ \text{K}) = 4.05$ B.M.] and low temperatures [$\mu_{\text{eff}}(5\ \text{K}) = 3.21$ B.M.] a little higher than those of the $g_e$ predictions, which are $\sqrt{15} \approx 3.87$ and 3 B.M., respectively. The data for other compounds (Table 4) are generally very similar. There are a few cases of slightly anomalous magnetic values for related

**Table 4 Experimental Data for $(\pi^*, \delta^*)^3$ Ground-State Molecules**

| Compound | $d(M_2)/Å$ | $\mu_{eff}$/B.M. (T/K) | $g_\perp$ | $g_\parallel$ | VIS-NIR Solvent: $\lambda$/nm ($\varepsilon$/$M^{-1}$ $cm^{-1}$) |
|---|---|---|---|---|---|
| $Ru_2(O_2CPr^n)_4Cl$ | 2.281(4)[a] | 4.05 (299), 3.21 (5)[b] | 4.40[b] | 1.95[b] | $CH_3CN$: 1,080 (36), 610 (120 sh), 458 (1,170)[c] |
| $Ru_2(O_2CCH_3)_4Cl$ | 2.287(2)[d] | 4.10 (293), 3.55 (98.6)[e] | | | $H_2O$: 1,110 (60), 425 (699)[c,f] |
| $Ru_2(O_2CH)_4Br$ | 2.2899(7)[g] | 4.12 (293)[e] | | | c |
| $Ru_2(O_2CPh)_4Cl$ | | 4.19 (292), 3.58 (103.8)[e] | | | DMF: 580 (sh), 451 (2,200)[h] |
| $Ru_2(O_2CCMe_2Ph)_4Cl$ | 2.2893(9)[i] | 4.4 (300), 3.58 (103.8)[i] | | | $CH_2Cl_2$: 485 (1,637) |
| $[Ru_2(O_2CCF_3)_4](O_2CCF_3)$ | 2.278(1)[j] | 4.1 (RT)[k] | | | MeOH: 438[j] |
| $[Ru_2(O_2CEt)_4(phz)](BF_4)$ | 2.2756(6)[l] | 4.3 (300), 1.72 (2.7)[l] | | | |
| $[RuCl(NCCH_3)_4(PPh_3)]$ $[Ru_2(O_2C(p\text{-}C_6H_4(OMe))_4Cl_2]$ | 2.299(1)[m] | 3.99 (RT)[m] | 4.30[m] | | $CHCl_3$: 1,160 (45), 470 (940)[m] |
| $K_3[Ru_2(OCO)_4]\cdot 4H_2O$ | 2.2509(5)[n] | 4.1 (RT)[k] | 4.2[k] | 2.1[k] | $H_2O$: 413 (873)[k] |
| $Ru_2(CF_3C(O)NH)_4Cl$ | | 4.06 (297)[o] | 4.119[o] | 2.06[o] | $CH_3CN$: 601 (270), 499 (sh), 454 (720)[o] |
| $Ru_2(acam)_4Cl$ | | 4.34 (296)[p] | 4.23[p] | 1.98[p] | DMSO: ~440[p] |
| $Ru_2(bzam)_4Cl$ | 2.293(2)[q] | | | | DMSO: 445 (9,800)[r] |
| $Ru_2(p\text{-}ClC_6H_4C(O)NH)_4Cl$ | 2.296(1)[s] | | | | DMSO: 558 (340), 425 (1,000)[s] |
| $Ru_2(hp)_4Cl(Hhp)$ | 2.2869(1)[t] | 4.61 (297)[t] | | | $CH_3CN$: 480 (4,690), 368 (2,500 sh)[t] |
| $Ru_2(6\text{-}Cl\text{-}hp)_4Cl\cdot CH_2Cl_2$ | 2.281(1)[u] | | | | $CH_2Cl_2$: 680 (980 br), 536 (5,740), 440 (sh), 380 (sh)[u] |

**Table 4 (Continued)**

| Compound | $d(M_2)$/Å | $\mu_{eff}$/B.M. (T/K) | $g_\perp$ | $g_\parallel$ | VIS-NIR Solvent: λ/nm ($\varepsilon$/$M^{-1}$ $cm^{-1}$) |
|---|---|---|---|---|---|
| $Ru_2(6\text{-F-hp})_4Cl\cdot CH_2Cl_2$ | 2.284(1)[v] | | | | $CH_2Cl_2$: 620 (sh), 522 (4,720), 420 (sh)[v] |
| $[(Ru_2(6\text{-Cl-hp})_4)_2(pyz)](BF_4)_2 4CH_2Cl_2$ | 2.2671(7)[w] | 3.87 (295), 2.91 (6)[w] | | | |
| $[Ru_2(6\text{-Cl-hp})_4(py)](BF_4)\cdot$hexane$\cdot$py | 2.2697(8)[w] | 3.99 (295), 2.32 (6)[w] | | | |
| $Ru_2(O_2CCH_3)(6\text{-Cl-hp})_3Cl\cdot CH_2Cl_2$ | 2.282(4)[x] | | | | $CH_2Cl_2$: 535 (3,990), 445 (sh), 380 (sh)[x] |
| $Ru_2(O_2CCH_3)_2(6\text{-Me-hp})_2Cl\cdot 0.5(CH_2Cl_2)$ | 2.278(2)[y] | | | | $CH_2Cl_2$: 550 (3,370), 460 (sh), 378 (1,540)[y] |
| $K_2(H_5O_2)[Ru_2(SO_4)_4(OH_2)_2]$ | 2.303(1)[z] | 4.13 (292)[z] | 4.580[z] | 1.937[z] | 10% $H_2SO_4$ (aq): 435 (675)[z] |
| $K_4[Ru_2(HPO_4)_3(PO_4)(OH_2)_2]$ | 2.3052(4)[aa] | | | | $H_2O$: 440[aa] |
| $Ru_2(ap)_4Cl$ | 2.275(3)[t] | 3.94 (298)[t] | | | $CH_2Cl_2$: 764 (6,910), 415 (5,770)[t] |
| $Ru_2(2\text{-}(CH_3N)\text{-py})_4Cl$ | | 3.93 (293)[bb] | | | refl: 748, 482[bb] |
| $Ru_2(2\text{-}(PhCH_2N)\text{-py})_4Cl$ | | 4.36 (293)[bb] | | | refl: 736, 480[bb] |
| $Ru_2(O_2CCH_3)_2(ap)_2Cl\cdot Hap\cdot CH_2Cl_2$ | 2.308(1)[x] | | | | $CH_2Cl_2$: 695 (6,560), 420 (2,160)[x] |
| $Ru_2(ap)_4(C_2H)\cdot 2(CH_2Cl_2)$ | 2.319(2)[cc] | 3.92 (308)[cc] | | | $CH_2Cl_2$: 735 (1,050), 480 (1,630)[cc] |
| $Ru_2(dpf)_4Cl$ | 2.339(1)[nn] | 3.89 (RT)[nn] | | | $CH_2Cl_2$: 665, 570, 465[nn] |
| $Ru_2(dpf)_4(C_2Ph)$ | 2.369(1), 2.431(1)[nn] | 3.87 (RT)[nn] | | | $CH_2Cl_2$: 1,150, 760, 530, 380 (sh)[nn] |

**Table 4** ***(Continued)***

| Compound | $d(M_2)$/Å | $\mu_{eff}$/B.M. (T/K) | $g_\perp$ | $g_\parallel$ | VIS-NIR Solvent: $\lambda$/nm ($\varepsilon$/$M^{-1}$ $cm^{-1}$) |
|---|---|---|---|---|---|
| $[Ru_2(OEP)_2](BF_4)$ | 2.29(2)[dd] | paramagnetic[ee] | | | $CH_2Cl_2$: 664 (7,400), 572 (10,500)[ee] |
| $[Ru_2(tmtaa)_2](BF_4)$ | 2.267(3)[ff] | 1.55 (RT)[ff] | | | |
| $Ru_2(dptz)_4$ $(OPF_4)$ | 2.385(2)[gg] | | | | $CH_2Cl_2$: 710[gg] |
| $[Ru_2(dttz)_4(NCCH_3)](BF_4)$ | 2.373(2)[hh] | 1.88 (298)[hh] | | | $CH_2Cl_2$: 710 (15,300)[hh] |
| (4,0)-$Ru_2(F_5ap)_4(C_2Ph)^-$ | | | $g_1 = 2.156$; $g_2 = 2.132$[oo] | $g_3 = 1.882$; ($A = 4.71 \times 10^{-3}$ $cm^{-1}$)[oo] | $CH_2Cl_2$: 847, 534[pp] |
| $Os_2(6\text{-F-hp})_4Cl$ | 2.341(1)[ii] | 3.70 (RT)[ii] | | | $CH_2Cl_2$: 413 (4,290)[ii] |
| $Os_2(6\text{-Cl-hp})_4Cl$ | 2.348(1)[jj] | 2.90 (RT)[jj] | | | |
| $Os_2(hp)_4Cl_2^-$ | | 2.68 (308)[kk] | 4.28, 4.40[kk] | ~ 1.9[kk] | |
| $Os_2(O_2CEt)_4Cl_2^-$ | | 2.47 (275)[kk] | see Ref. 72 | | $CH_2Cl_2$: 500 (sh), 380 (1,900)[kk] |
| $Os_2(O_2CPr^n)_4Cl_2^-$ | | 2.73 (298)[kk] | | | $CH_2Cl_2$: 850 (⩽100) 620 (⩽100 sh), 510 (sh), 385 (2,600)[kk] |
| $Os_2Cl_4(O_2CCH_3)((2\text{-}Ph_2P)py)_2$ ·2(acetone)·$2(CH_2Cl_2)$ | 2.388(1), 2.395(1)[ll] | | | | |
| $[Os_2(OEP)_2](SbCl_6)$ | 2.31[mm] | paramagnetic[ee] | | | $CH_2Cl_2$: 596 (9,330)[ee] |

[a] Reference 132. [b] Reference 56. [c] References 61–63; low-temperature solid-state absorption spectra are also presented in these references. [d] Reference 133. [e] Reference 134. [f] Reference 135. [g] Reference 136. [h] Reference 137. [i] Reference 60. [j] Reference 138. [k] Reference 78. [l] Reference 59. [m] Reference 90. [n] Reference 139. [o] Reference 140. [p] Reference 112. [q] Reference 141. [r] Reference 113. [s] Reference 142. [t] Reference 57. [u] Reference 143. [v] Reference 144. [w] Reference 58. [x] Reference 145. [y] Reference 146. [z] References 69 and 71. [aa] Reference 70. [bb] Reference 147. [cc] Reference 113. [dd] Reference 148. [ee] Reference 149. [ff] Reference 75. [gg] Reference 150. [hh] Reference 79. [ii] Reference 74. [jj] Reference 73. [kk] Reference 72. [ll] Reference 82. [mm] Reference 77. [nn] Reference 83. [oo] Reference 84. [pp] Reference 151.

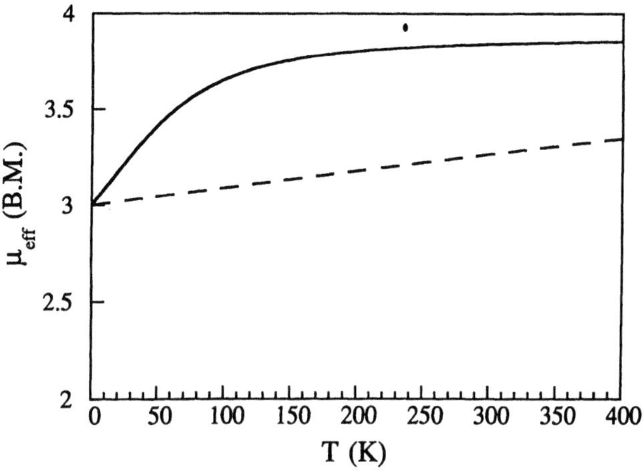


**Figure 4.** Magnetism temperature dependence predicted for the ${}^4B_{2u}(\pi^*)^2(\delta^*)^1$ state with $g_{iso} = 2$ and $D = 154$ (——) or 1,500 $cm^{-1}$ (- - - - - -).

compounds, the rather high room-temperature moment of 4.61 B.M. reported[57] for $Ru_2(hp)_4Cl(Hhp)$ being a prominent example, but none of these are so unusual as to arouse any doubt about the assignment of its ground state.

$Ru_2(O_2CPr^n)_4Cl$ has a zig-zag infinite-chain structure with bent Ru–Cl–Ru bridges, and it was therefore somewhat surprising that no evidence was found for interdimer magnetic coupling.[56] The data were very convincingly fit to isolated spin 3/2 dimers, yielding a value for $D = W(E_{3/2u}) - W(E_{1/2u}) = 154\ cm^{-1}$. [Note that our definition of $D$ (Appendix C) is equal to twice that of Reference 56.]

Cotton and coworkers[58–60] have reported several other magnetic studies that were extended to sufficiently low temperatures to address the question of interdimer magnetic coupling in chain or oligomeric structures. In one of these,[58] no interdimer coupling was seen, consistent with the results for $Ru_2(O_2CPr^n)_4Cl$. However, in a study of a $\mu$-phenazine chain-structured compound,[59] strong deviation from isolated dimer behavior was observed [$\mu_{eff}(2.7\ K) = 1.72$ B.M.], which suggests interdimer antiferromagnetic coupling. In another study, similar low-temperature deviations from isolated dimer behavior were observed for the compound $Ru_2(O_2CCMePh_2)_4Cl$, which has *linear* chloride-bridged chains, and a fit was made to the one-dimensional Ising model, $J = -10.9\ cm^{-1}$.[60] Unfortunately, these studies[58–60] did not treat $D$ as a free-fitted parameter, so it remains uncertain how typical the value obtained in Reference 56 really is.

Figure 5 shows the predictions of our crystal-field model for $(\pi^*, \delta^*)^3$ states with no spin-orbit coupling ($\zeta = 0$), and energy units of $K$ (note that $K$ is intrinsically positive). The $\Delta = 0$ limit corresponds to "octahedral" states with energies identical with those of $(t_{2g})^n$ at this level of calculation, and these have been indicated in the figure. There is a strongly avoided crossing at $\Delta = 0$ between the two ${}^2E_g$ states. The region in which ${}^4B_{2u}$ is the ground state is strikingly large, extending between $\Delta = \pm 3.85K$, which explains why the high-spin ${}^4B_{2u}$ ground state is ubiquitous.

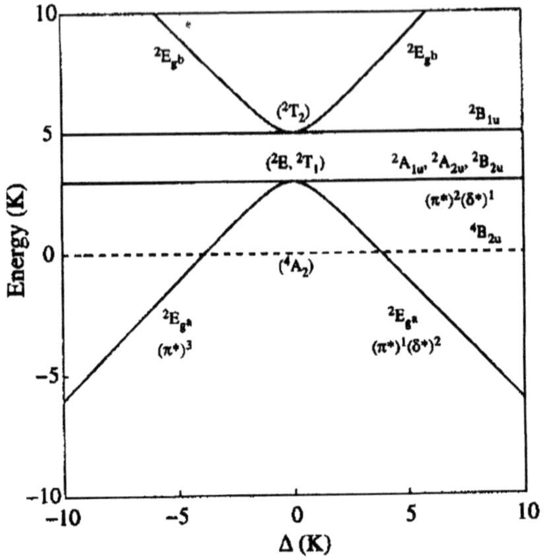


**Figure 5.** Crystal-field predictions of the state energies for the $(\pi^*, \delta^*)^3$ configuration as a function of $\Delta = W(\pi^*) - W(\delta^*)$. Energies are in units of $K = K(\pi_x^*, \delta^*) = K(\pi_x^*, \pi_y^*)$. The spin-orbit coupling parameter $\zeta$ has been set equal to 0.

There is a total of 6 doublet excited states of $(\pi^*, \delta^*)^3$, four of which are vertical "spin-flip" excitations that are all both spin- and dipole-forbidden. The remaining two excited states are $^2E_g$ states derived from the $(\pi^*)^3$ and $(\pi^*)^1(\delta^*)^2$ configurations, and transitions to these states are dipole-allowed with $x, y$-polarization. It is noteworthy, however, that these two configurations are very strongly mixed by the avoided crossing over the range of Δ values for which $^4B_{2u}$ is the ground state, so that the vertical transitions to $^2E_g$ states are not to single-configuration states.

In a thorough low-temperature electronic absorption study of $Ru_2(O_2CR)_4Cl$ and related compounds,[61–63] we located two very weak near-infrared absorption bands (in addition to more intense bands to be discussed below) and assigned them to quartet-doublet transitions. One of these, observed for all of the compounds investigated, has an extinction coefficient of 2–4 $M^{-1}$ $cm^{-1}$, is predominantly $x, y$-polarized, and has an electronic origin in the range 6,854 [$Ru_2(O_2CPr^n)_4Cl$][61] to 7,037 $cm^{-1}$ [$Ru_2(O_2CH)_4Cl$ in KBr matrix].[63] It is vibronically structured, showing a long progression in $\nu(Ru_2)$ with an excited-state frequency slightly higher than the ground-state value, which led us to propose the assignment $\pi^* \rightarrow \delta^*$, $^4B_{2u} \rightarrow {}^2E_g$. In view of the drastic mixing of the $^2E_g$ states that is indicated by our present analysis, we no longer feel that the one-electron assignment is meaningful, but the evidence that this is a $^4B_{2u} \rightarrow {}^2E_g$ transition remains firm.

A second quartet-doublet transition was only established for $Ru_2(O_2CPr^n)_4Cl$, and was observed at 5,048 $cm^{-1}$. It is very weak ($\varepsilon \sim 1$) and *vertical,* showing little trace of vibronic side-band structure, and was therefore assigned to a spin-flip transition. Because of the nonlinear chains in the crystal structure of this compound,

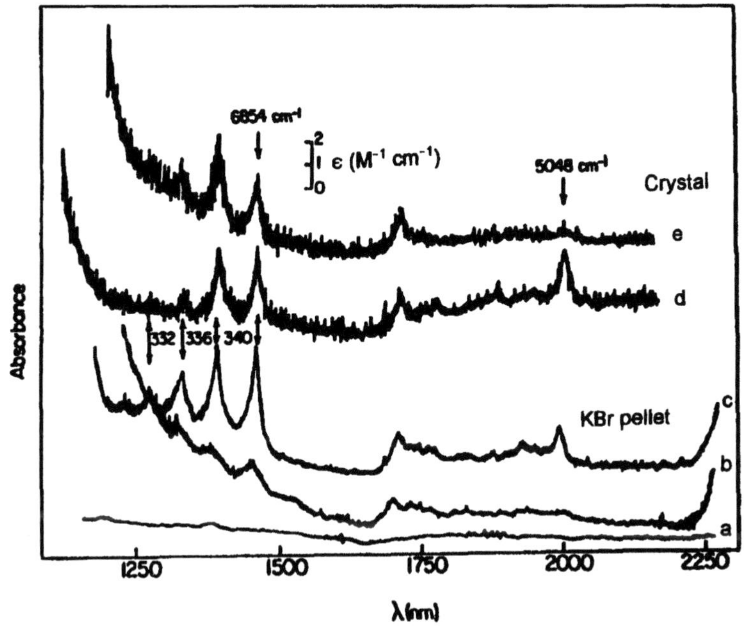


**Figure 6.** Absorption profile of the $I\bar{4}2d$ polymorph of $Ru_2(O_2CPr^n)_4Cl$: (a) KBr pellet absorption baseline; spectra of 53.1 mg $Ru_2(O_2CPr^n)_4Cl$/200 mg KBr, at 300 K (b) and 18 K (c); single-crystal spectra at 20 K polarized $\parallel$ (d) and $\perp$ (e) to the crystallographic $c$-axis. Adapted from Reference 61.

the site symmetry is effectively low and presumably induces intensity into this very forbidden electronic transition; polarized single-crystal measurements supported this hypothesis.[61] The spin-forbidden bands are illustrated in Figure 6.

We first ask whether the observed ${}^4B_{2u} \rightarrow {}^2E_g$ transition is to the higher or lower energy ${}^2E_g$ state of Figure 5. If it is to the lower energy state ${}^2E_g^a$, then the transition energy must be $\leqslant 3K$. This implies $K \geqslant$ 2,300 cm$^{-1}$ (the minimum occurring for $\Delta = 0$), which is close to the mononuclear value of $K$. Since we expect $K$ to be substantially smaller than for the mononuclear value (*vide supra*), this seems unlikely. Moreover, this assignment would leave us with no assignment for the 5,048 cm$^{-1}$ spin-flip, according to Figure 5.

The alternative assignment of the ~7,000 cm$^{-1}$ band to ${}^4B_{2u} \rightarrow {}^2E_g^b$ requires $5K \leqslant$ 7,000 cm$^{-1}$ $\leqslant 8.0K$ (hence, 875 cm$^{-1}$ $\leqslant K \leqslant$ 1,400 cm$^{-1}$), which seems more reasonable. Furthermore, the 5,048 cm$^{-1}$ band can be plausibly assigned to the ${}^4B_{2u} \rightarrow {}^2B_{1u}$ transition, which is predicted to lie slightly below ${}^4B_{2u} \rightarrow {}^2E_g^b$. This assignment implies $K$ = 1,010 cm$^{-1}$. Together, these assignments indicate $K \approx$ 1,000 cm$^{-1}$, and the range of $\Delta$ for which ${}^4B_{2u}$ is the ground state is therefore $|\Delta| <$ 3,850 cm$^{-1}$.

With this estimate for $K$ in hand we attempt to improve the theory by including spin-orbit coupling, and Figure 7 shows the calculated state energies for

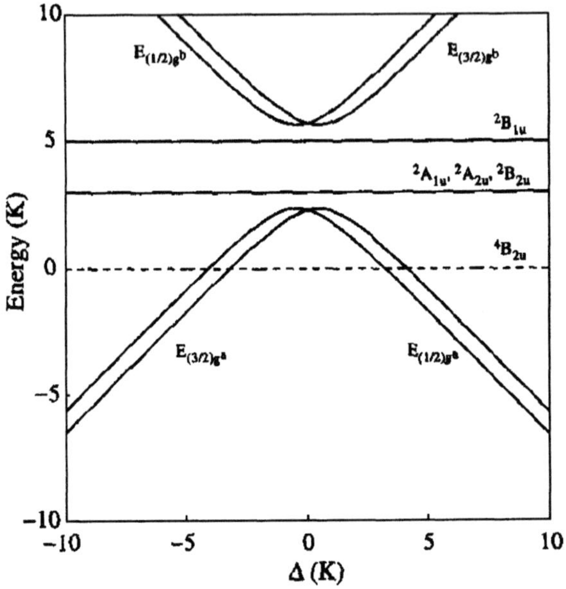


**Figure 7.** Crystal-field predictions of the state energies for the $(\pi^*, \delta^*)^3$ configuration as a function of $\Delta$: $K = 1{,}000\ \text{cm}^{-1}$ and $\zeta = 900\ \text{cm}^{-1}$.

$\zeta = 900\ \text{cm}^{-1}$, as is appropriate for Ru(II)/Ru(III). While this and following state energy diagrams continue to give energies in units of $K$, note that the form of the diagram depends on the ratio $\zeta/K$ ($900\ \text{cm}^{-1}/1{,}000\ \text{cm}^{-1}$ in this example).

Spin-orbit coupling induces a zero-field splitting (ZFS) of the $^2E_g$ states, $D = \zeta$, far from the avoided crossing, but does not, for this magnitude of $\zeta$, greatly reduce the energy region in which $^4B_{2u}$ is the ground state, $|\Delta| < 3.20K$. Fitting the two observed quartet-doublet transitions to Figure 7, we estimate $|\Delta| = 2.95K \approx 2{,}950\ \text{cm}^{-1}$.

Since the ZFS is a symmetric function of $\Delta$ (Figures 5 and 7), the algebraic sign of this parameter cannot be determined solely on the basis of the fit. However, we can make a clear choice for the sign of $\Delta$ based on the fact that if it were negative, then a small increase in the $\pi$-donor strength of the bridging ligand should make $^2E_g(\pi^*)^3$ the ground state, by destabilizing $\delta^*$, whereas experimentally (Table 4) it is found that even the $Ru_2$(II,III) complexes of the rather strong $\pi$-donor $\mu$-[2-(RN)py]$^-$ ligands have $^4B_{2u}$ ground states; only the ultra-strong[44] $\pi$-donor ligands $\mu$-$(RN_3R)^-$ induce $^2E_g[(\pi^*)^3]$ ground states (*vide infra*). Therefore, $\Delta$ must be *positive*, $+3{,}000\ \text{cm}^{-1}$, so that significant increases of the bridging-ligand $\pi$-donor strength, while decreasing $\Delta$, will still leave $^4B_{2u}$ as the ground state.

Our estimate of $\Delta$ may be compared with calculated values of $-2{,}130\ \text{cm}^{-1}$ for $Ru_2(O_2CH)_4^+$ and $-230\ \text{cm}^{-1}$ for $Ru_2(O_2CH)_4Cl_2^-$.[42] The agreement is not very good. We will show in our discussion of $Ru_2$(II,II) carboxylates that there are other reasons to believe that the SCF-X$\alpha$ calculational method underestimates $\Delta$.

Another comparison of interest is to the photoelectron spectroscopic study of $Ru_2(O_2CCF_3)_4$.[64] The ground state is $(\pi^*)^2(\delta^*)^2$ and the assigned $\pi^*$ ionization is observed 0.51 eV (4,100 $cm^{-1}$) above the $\delta^*$ ionization (which is the lowest-energy ionization); the energy difference should be the $\delta^* \rightarrow \pi^*$ excitation energy of $(\pi^*)^2(\delta^*)^1$ $Ru_2(O_2CCF_3)_4^+$.[64] This agrees acceptably well with our observed $^2E_g^b$ excitation energy, considering that the bridging carboxylato ligands and, especially, the axial ligands are not equivalent. Moreover, our picture of the low-lying states (Figure 7) suggests that the photoelectron spectrum may be somewhat more complicated than the current assignments would have it; we expect gross violations of Koopman's Theorem because of the substantially different two-electron contributions to the various states.

Given our estimates for $\Delta$ and $K$, we can further estimate that the ZFS of $^2E_g^b$ should be 780 $cm^{-1}$. A second electronic origin is not clearly defined for most of the compounds studied,[61–63] but the band of $Ru_2(O_2CH)_4Cl$ in a KBr matrix[63] shows two well-defined origins separated by 533 $cm^{-1}$, which may be the ZFS. The reduction of the ZFS $D$ relative to the calculated value possibly indicates that $\Delta$ is a little closer to 0 for the formate complex than our estimate of 2,950 $cm^{-1}$ for the butyrate complex.

The remaining quartet-to-doublet transitions within the $(\pi^*)^2(\delta^*)^1$ configuration are all calculated to lie at 3,000 $cm^{-1}$ or less, hence to be inaccessible to electronic absorption spectroscopy because of intense vibrational absorption in this region. We have previously suggested[61] that a resonance Raman feature at ~1,600 $cm^{-1}$ reported by others[65] might be an electronic Raman transition to one of the doublet excited states. Assignment to one of the other spin flips is not possible with the present model, but is easily allowed if we relax the restriction $K(\pi_x^*, \pi_y^*) = K(\pi_x^*, \delta^*)$ (Appendix A). Alternatively, the Raman feature could be a transition to a component of $^2E_g^a$, or the electronic Raman assignment, which has never been experimentally tested, might be wrong.

The VIS-NIR electronic spectra of the $Ru_2(O_2CR)_4^+$ compounds show two main absorption bands, at ~1100 ($\varepsilon \sim 50$) and at ~450 nm ($\varepsilon \sim 1000$ $M^{-1}cm^{-1}$).[61–63] The former of these has been shown by low-temperature studies to be the $^4(\delta \rightarrow \delta^*)$ transition, $^4B_{2u} \rightarrow {}^4B_{1g}$. It is molecular $z$-polarized and it shows very well-resolved vibronic structure in $\nu(Ru_2)$, with a frequency reduced ~8% from that of the ground state and a Franck–Condon factor corresponding to a ~0.05 Å excited-state increase in $d(Ru_2)$.[61] It occurs at nearly exactly the theoretically predicted energy.[42]

The major visible absorption band is also $z$-polarized but is unstructured at low temperature. A theoretical calculation[42] assigns it to a $\pi(Ru–O,Ru_2) \rightarrow \pi^*(Ru_2)$ transition of mixed LMCT and metal–metal $\pi \rightarrow \pi^*$ character. Metal–metal character has been established for this band by resonance Raman studies that show pronounced resonance enhancement of $\nu(Ru_2)$,[61,65,66] but the amount of LMCT character remains questionable.

The spin-forbidden quartet-doublet transitions are too weak ($\varepsilon < 5$ $M^{-1}cm^{-1}$) to be observable in solution spectra because of solubility limitations.[62]

For the complexes of Table 4 (other than the carboxylates), with one exception to be noted later, the near-infrared spectra have not been determined. [At least, they have not been investigated with attention paid to the possibility of weak $^4(\delta \rightarrow \delta^*)$ transitions with intensities of the order of only ~50 $M^{-1}cm^{-1}$.] Thus, available data are limited to visible spectra, and the prominent features in all of the spectra are attributable to $\pi$(Ru–X,$Ru_2$) → $\pi^*$($Ru_2$) transitions, X = O or N.

The spectra of $Ru_2$(II,III) $\mu$-(6-R-2-hp)$^-$ and $\mu$-(OC(R)NH)$^-$ complexes each display a band with a maximum (425–455 nm) very similar to that of the $\pi$(Ru–O, $Ru_2$) → $\pi^*$($Ru_2$) band of the $\mu$-carboxylato complexes. The strength of this band in the $\mu$-hydroxypyridinato complexes is enhanced by intensity stealing ($\varepsilon$ = 3,000–6,000 $M^{-1}cm^{-1}$); the oxygen donor of these ligands is phenolic, directly conjugated with an aromatic ring, so that mixing with aromatic $\pi \rightarrow \pi^*$ transitions can occur. Related but smaller intensification appears to occur when an aromatic substituent is attached at a position adjacent to the ligating atom, as in the $\mu$-benzoato complex (Table 4). Chloride LMCT transitions have been located in the UV region by their characteristic shifts upon substitution by other halides for the $\mu$-carboxylato complexes,[62] and most likely do not contribute to the visible absorption of any of these complexes. In some of these spectra, relatively weak features are reported near 600 nm; the $\mu$-carboxylato complexes show similar features, and possible metal–metal assignments [$\delta \rightarrow \pi^*$; $\delta^*, \pi^* \rightarrow \sigma^*$($M_2$ and/or M-L)] have been discussed.[62]

The $Ru_2$(II,III) $\mu$-(RNpy)$^-$ complexes have visible spectra that are somewhat different in that the lowest intense band is located at 700–800 nm ($\varepsilon$ = 6,500 $M^{-1}cm^{-1}$) and a second similarly intense visible absorption is found near 450 nm. There is an interesting comparison available to the $Rh_2$(II,III) complex $Rh_2(ap)_4^+$, having a $(\pi^*)^4(\delta^*)^1$ ground state, which displays extremely low energy, intense $\pi$(N) → $\delta^*$($Rh_2$) LMCT bands at 1,540 nm ($\varepsilon$ = 2,700 $M^{-1}cm^{-1}$) and 890 nm ($\varepsilon$ = 3,300 $M^{-1}cm^{-1}$).[67,68] This type of ligand evidently places $\pi$(N) levels at exceptionally high energies, and the 700–800 nm band of the $Ru_2$(II,III) complexes is therefore reasonably assigned as $\pi$(Ru–N,$Ru_2$) → [$\pi^*$($Ru_2$) or $\delta^*$($Ru_2$)].

The $Ru_2(SO_4)_4(OH_2)_2^{3-}$ ion is reported to show a visible absorption maximum at 435 nm ($\varepsilon$ = 675 $M^{-1}cm^{-1}$)[69] and assignment to $\pi$(Ru–O,$Ru_2$) → $\pi^*$($Ru_2$) is once again reasonable. The spectrum of the analogous $\mu$-phosphato complex is nearly identical.[70] The relatively low intensity of the visible band may signal relatively small Ru–O character in the $\pi$(Ru–O,$Ru_2$) level, hence, low LMCT character in the electronic transition. It was also reported[71] that the $^4(\delta \rightarrow \delta^*)$ transition of this complex is *not* located in the NIR near that of $Ru_2(O_2CR)_4^+$ at 1,100 nm,[61–63] and the $^4(\delta \rightarrow \delta^*)$ transition was attributed to a weak absorption at ~2,100 nm. This implies that the $\delta/\delta^*$ one-electron energy splitting $\Delta W$ (roughly equal to the transition energy)[18] is a factor of two smaller for the $\mu$-$SO_4^{2-}$ complex than for the $\mu$-carboxylato complexes, which is reasonable if $SO_4^{2-}$ is a weak *specific* $\pi$-donor to $\delta^*$; recall that a significant part of the $\delta/\delta^*$ splitting of $\mu$-carboxylato complexes is due to ligand $\pi$-donation.[18,41,42] The opposite effect, greatly increased $\delta/\delta^*$ splittings, is evident in the theoretical calculations on $\mu$-($HN_3H$)$^-$ and $\mu$-[HNC(H)NH]$^-$ complexes.[43,44]

There are a few $Os_2$(II,III) compounds known.[2] The possible range of $\Delta$ values allowing a $^4B_{2u}$ ground state is greatly reduced for third-row transition metal ions because the ZFS of the $^2E_g$ states is greatly increased. For $\zeta$ = 2,800 $cm^{-1}$ [appropriate for Os(III)] and $K = 1{,}000\ \text{cm}^{-1}$ we calculate (neglecting the ZFS of $^4B_{2u}$, Figure 7) that the $^4B_{2u}$ state is the ground state for $|\Delta| < 1{,}270\ \text{cm}^{-1}$, making the requirement of a small $|\Delta|$ much more stringent than for second transition series metal ions. There are nonetheless several clear examples of such ground states.

Magnetic properties are the key to identifying this ground state. $Os_2(hp)_4Cl_2^-$ and $Os_2$(6-Cl-hp)$_4$Cl have both been reported[72,73] to display an EPR $g_\perp \approx 4$ signal, which is completely diagnostic of the $E_{1/2u}(^4B_{2u})$ ground state. The former of these is also reported to show a weak signal at $g_\parallel \approx 1.9$. Assuming that the zero-field splitting of $^4B_{2u}$ should scale with $\zeta^2$, the 154 cm$^{-1}$ $D$ determined for $Ru_2(O_2CPr^n)_4Cl$[56] predicts $D \approx 1{,}500\ \text{cm}^{-1}$ for the Os analogue, in which case the magnetic behavior should be similar to the dashed line in Figure 4: The room-temperature magnetic moment is predicted to be not much above the low-temperature limiting value of 3 B.M. (for $g_{iso} = g_e$), increasing gently with temperature.

The reported room-temperature moment of 2.90 B.M. for $Os_2$(6-Cl-hp)$_4$Cl agrees well with this prediction. That of $Os_2(hp)_4Cl_2^-$, 2.68 B.M., is low but the reported temperature dependence is reasonable; the experimental measurements on this compound (as well as on carboxylate analogues) were reported to be particularly difficult to perform because of air sensitivity,[72] and some sort of systematic error may be involved. The reported room-temperature moment of $Os_2$(6-F-hp)$_4$Cl,[74] 3.70 B.M., surprisingly indicates a $^4B_{2u}$ ground state with $D <$ kT at room temperature. Since we cannot actually calculate $D$ (*vide supra*), this is certainly not impossible; a more detailed study of this compound seems called for, since it may shed light on the origin of $D$ of the $^4B_{2u}$ state in general. In any case, the $^4B_{2u}$ ground state is again established.

The Os–Os distances have been determined to be 2.341(1) and 2.348(1) Å for the 6-F-hp[74] and 6-Cl-hp[73] complexes. These are ~0.06 Å longer than those of their Ru congeners, which is smaller than the ~0.1 Å bond length increases that are found between second and third transition series quadruply bonded congeners,[2] but is probably reasonable for bond order 2.5.

The compounds $Os_2(O_2CR)_4Cl_2^-$, R = Et and Pr$^n$,[72] exhibit temperature-dependent magnetic behavior very similar to $Os_2(hp)_4Cl_2^-$. However, the weak EPR signals that were recorded (reported in footnote 29 of Reference 72) do not obviously correspond to *any* candidate ground state (Appendix C). We think that the carboxylato complexes *probably* have $^4B_{2u}(\pi^*)^2(\delta^*)^1$ ground states. The magnetic results do not fit either of the $^2E_g$ states (*vide infra*). There is a possibility of a thermal equilibrium between a $^2E_g$ ground state and a very low-lying $^4B_{2u}$ excited state, which would account for magnetic moments less than 3, but considerably greater than 2 B.M. and increasing with temperature. While the available data allow such an interpretation, they do not demand it.

The visible absorption spectrum of $Os_2$(6-F-hp)$_4$Cl has been reported;[74] the prominent feature at 413 nm ($\varepsilon = 4{,}290\ \text{M}^{-1}\text{cm}^{-1}$) is likely analogous to the previously discussed $\pi(\text{M-O,M}_2) \rightarrow \pi^*(\text{M}_2)$ transition of the Ru complexes. The $Os_2(O_2CR)_4Cl_2^-$ complexes have a possibly analogous band at ~500 nm.[72] A weak ($\varepsilon \leqslant 100\ \text{M}^{-1}\text{cm}^{-1}$) band is reported at 850 nm for $Os_2(O_2CPr^n)_4Cl_2^-$.[73] This might be the $^4(\delta \rightarrow \delta^*)$ transition, in which case the $\delta/\delta^*$ splitting ($\Delta W$) is ~2,500 cm$^{-1}$

larger for $Os_2$(II,III) than for $Ru_2$(II,III), but there are other possible assignments for a weak band in this wavelength region.[61–63] NIR absorption spectra have not been reported for any of the other compounds, unfortunately. They would be of great interest, both because the $^4(\delta \rightarrow \delta^*)$ transition, with its characteristic $\nu(M_2)$ vibronic structure, presumably lies in the NIR, and also because the large $\zeta$ of Os should intensify the quartet-doublet transitions by about an order of magnitude relative to the Ru compounds, as judged by comparisons to mononuclear compounds.[53] Because these transitions would still be weak ($\varepsilon$ = 10–50), studies of the compounds in the solid state might be necessary to resolve the absorptions.

**4.3.2 $(\pi^*)^3$ *and* $(\pi^*)^1(\delta^*)^2$** Figure 5 shows that $^2E_g$ $Ru_2$(II,III) ground states can exist for $|\Delta| > 3{,}850\ \text{cm}^{-1}$, and there are some clear experimental examples. Figure 8 shows that the predicted temperature-dependent magnetic behaviors of the two possible $^2E_g$ ground states are very different. We emphasize that this has nothing to do with the metal–metal bonding; rather, it is a matter of the sign of $D$, depending upon whether the $\pi^*$ level is more than or less than half-filled, resulting in either the $E_{1/2g}$ or $E_{3/2g}$ spin-orbit levels of the $^2E_g$ state (which undergoes a first-order, hence easily calculable, spin-orbit splitting) being of lowest energy.

The $(\pi^*)^1(\delta^*)^2$ ground state that should occur for large, positive $\Delta$ features a magnetic moment less than the $S = 1/2$ spin-only value and a metal–metal distance shorter than that for $(\pi^*)^2(\delta^*)^1$ owing to depopulation of the more strongly metal–metal antibonding level. The $[Ru_2(\text{tmtaa})]_2^+$ ion, $\mu_{\text{eff}}$ = 1.55 B.M. (room temperature),[75] fits these criteria, if we allow that unfavorable steric effects probably tend to increase the metal–metal distance, as is further discussed in Section 4.4. The $[M_2(\text{OEP})_2]^+$ (M = Ru, Os) ions are also indicated to have doublet ground states, and Raman data support a $(\pi^*)^1(\delta^*)^2$ ground state.[76,77] Unfortunately, there are no

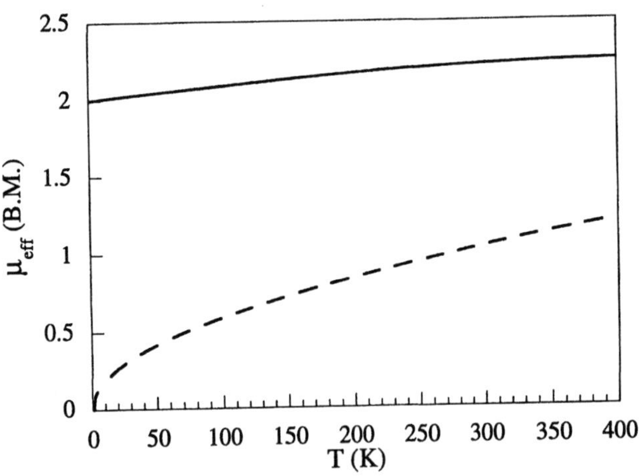


**Figure 8.** Magnetism temperature dependence predicted for the $^2E_g(\pi^*)^3$ (——) and $^2E_g(\pi^*)^1(\delta^*)^2$ (- - - - - -) states, both with $g_{\text{iso}} = 2$ and $D = 800\ \text{cm}^{-1}$.

detailed magnetic studies available, and intense ligand-localized electronic absorption makes assignment of metal–metal electronic transitions difficult if not impossible. A temperature-dependent magnetic study of a compound of this type would clearly be of considerable interest. We note that the macrocyclic ligands known to promote this type of ground state lack $\pi$-orbitals properly oriented to interact with $\delta^*$, which should favor a large, positive value of $\Delta$.

Before we leave the subject of the ${}^2E_g(\pi^*)^1(\delta^*)^2$ ground state, we should mention that it is a little surprising that $Ru_2(O_2CCF_3)_4^+$ *does not* have this ground state. Since we have found that $Ru_2(O_2CR)_4^+$ complexes have large, positive $\Delta$ values, the fact that $O_2CCF_3^-$ is a considerably poorer $\pi$-donor than $O_2CCH_3^-$ could reasonably result in sufficient stabilization of $\delta^*$ relative to $\pi^*$ to yield a $(\pi^*)^1(\delta^*)^2$ ground state; instead a $(\pi^*)^2(\delta^*)^1$ ground state is found experimentally (Table 4). However, $Ru_2(O_2CCF_3)_4^+$ is an extraordinarily strong oxidant [$E_{1/2}(Ru_2^{5+}/Ru_2^{4+}) = +1.17$ V vs. SCE[78]] and otherwise very chemically reactive; this implicates a very energetic $\delta^*$ hole. Thus, this compound is likely very close to the edge of $(\pi^*)^2(\delta^*)^1$ stability.

The two $Ru_2$(II,III) compounds with bridging diaryltriazenato ($RN_3R^-$) ligands in Table 4 show $Ru_2$ distances nearly 0.1 Å longer than those of other $Ru_2$(II,III) complexes, so they can be confidently assigned ${}^2E_g(\pi^*)^3$ ground states. The reported room-temperature magnetic moment of $[Ru_2(dttz)_4(NCCH_3)](BF_4)$, 1.88 B.M.,[79] is significantly greater than the $S = 1/2$ spin-only value, as expected (Figure 8). The compounds are EPR silent at 77 K, which is understandable in terms of relaxation effects. Kawamura et al.[80] successfully measured EPR spectra of $(\pi^*)^3(\delta^*)^2$ ground state $Rh_2(O_2CR)_4^+$ complexes at liquid helium temperature, and determined effective $g$ values close to the theoretical values (Appendix C, $k \approx 1$) of $g_\parallel = 4$, $g_\perp = 0$. EPR measurements at 4.2 K would be of interest for the $Ru_2(\mu\text{-}RN_3R)_4^+$ complexes.

The key bonding factor for the $\mu$-triazenato ligands [and also for $\mu$-formamidinato ligands $RNC(H)NR^-$] is that they are extremely strong $\pi$-electron donors. Theoretical calculations on $M_2(\mu\text{-L})_4$, M = Ru(II) and Rh(II), show $\Delta \approx -1$ eV for L = $HN_3H^-$ or $HNC(H)NH^-$,[43,44] which is clearly sufficient to enforce the $(\pi^*)^3$ ground state (Figure 7). For $\Delta \approx -8{,}000$ cm$^{-1}$, we predict, with $K \approx 1{,}000$ cm$^{-1}$ and $\zeta \approx 900$ cm$^{-1}$, that the ZFS $D$ of the ${}^2E_g(\pi^*)^3$ state should be 882 cm$^{-1}$, $W(E_{1/2g}) > W(E_{3/2g})$. The spin-forbidden $E_{3/2g}({}^2E_g) \rightarrow {}^4B_{2u}(\pi^* \rightarrow \delta^*)$ excitation is predicted at 4,600 cm$^{-1}$, while the spin-allowed $\pi^* \rightarrow \delta^*$ excitations are predicted at 7,600 (${}^2A_{1u}$, ${}^2A_{2u}$, and ${}^2B_{2u}$) and 9,600 cm$^{-1}$ (${}^2B_{1u}$).

Experimentally, the lowest-energy band reported is at 756 nm ($\varepsilon$ = 15,300 M$^{-1}$cm$^{-1}$), and it has been tentatively assigned to $\pi^* \rightarrow \delta^*$.[79] In view of the high intensity of the band, we believe that it is better assigned to an LMCT transition of the type $\pi(N) \rightarrow \delta^*$. Such transitions should be very intense because of the large $\pi$-bonding interaction that is responsible for the large negative value of $\Delta$. For comparison, $Rh_2(dpf)_4^+$ exhibits an intense LMCT band at 890 nm,[81] and the preferred assignment is similar to $\pi(N) \rightarrow \delta^*$.[67]

The metal–metal $\pi^* \rightarrow \delta^*$ transitions are likely to be much weaker, although the extensive mixing of the $\delta^*$ orbital with ligand orbitals that is indicated by the theoretical calculations[43,44] may make them somewhat more intense than the typical

$\varepsilon \sim 10^2\ M^{-1}cm^{-1}$ observed for related transitions such as $\pi \rightarrow \delta$.[38,39,45] Moreover, according to our predictions, they should lie in the NIR region, which has not been carefully investigated for a compound of this type.

It is possible that $^2E_g(\pi^*)^3$ might be the ground state, or possibly a thermally accessible excited state, of the $Os_2(O_2CR)_4Cl_2^-$ complexes[72] discussed earlier. It is also possible that the compound $Os_2Cl_4(O_2CCH_3)(2\text{-}(Ph_2P)py)_2$ has this ground state, in view of its rather long $Os_2$ distance, 2.388 and 2.395 Å for two different solvates,[82] but no magnetic studies have been reported. It is not at all obvious why the ligands of this complex would promote a large, negative $\Delta$; indeed, the chloride and phosphine ligands might be expected to favor a large, positive $\Delta$, hence a $^2E_g(\pi^*)^1(\delta^*)^2$ ground state, so the long $Os_2$ distance may have some other explanation involving the steric constraints of the 2-($Ph_2P$)py ligand.

Bear et al. recently reported ions of the type $Ru_2(\mu\text{-}L)_4X_2^-$ (L = dpf, $F_5$ap; X = CN, C≡CPh).[83,84] These are reported to exhibit rhombic $S = 1/2$ EPR spectra, which excludes a $(\pi^*)^2(\delta^*)^1$ ground state. Unfortunately, assignment to either $^2E_g$ ground state seems to be precluded at first glance by the range of rhombic $g$ values $g_1$–$g_3$ (*ca.* 1.9–2.15). A possible explanation invokes the well-known fact that a strong rhombic distortion will quench angular momentum in an orbitally degenerate doublet ground state;[80] the neutral parent molecules $Ru_2(\mu\text{-}L)_4X_2$, in fact, show very large rhombic distortions by X-ray crystallography.[83,84] Structural data are unavailable for the anions, however, precluding both confirmation of this mechanism and any further discussion of the nature of the ground state.

### 4.4 $(\pi^*, \delta^*)^4$ Configurations

Except for $Os_2$(porphyrin)$_2$ and a few $Rh_2$(III,III) complexes that have only been observed in solution, all the characterized compounds with this $d$-electron count are $Ru_2$(II,II) complexes. Table 5 summarizes available data. In Figure 9 we show the predictions of our crystal-field model for $K = 1{,}000\ cm^{-1}$ and $\zeta = 900\ cm^{-1}$, appropriate for Ru(II). There is a double avoided crossing of $A_{1g}$ states near $\Delta = 0$. Two of these states are of singlet parentage, $A_{1g}(\pi^*)^4$ and $A_{1g}(\pi^*)^2(\delta^*)^2$, while the third is a spin-orbit component of a triplet state, $A_{1g}(^3A_{2g})$, $(\pi^*)^2(\delta^*)^2$.‡ There are three possible ground states: $^3A_{2g}(\pi^*)^2(\delta^*)^2$ is the ground state for $\Delta > 230\ cm^{-1}$ ($\Delta > 0$ for $\zeta \rightarrow 0$), $^1A_{1g}(\pi^*)^4$ is the ground state for $\Delta < -3{,}180\ cm^{-1}$, and, in the intermediate range of $-3{,}180 < \Delta < 230\ cm^{-1}$, the ground state is $A_{1u}$, $A_{2u}(^3E_u)$, $(\pi^*)^3(\delta^*)^1$.

**4.4.1 $(\pi^*)^2(\delta^*)^2$** This is the most commonly observed ground state by far for $(\pi^*, \delta^*)^4$. It is observed both for the porphyrin-type ligands for which $\Delta$ is expected

‡ Throughout, we use the conventions employed by Herzberg (Herzberg, G. *Molecular Spectra and Molecular Structure*; Van Nostrand: New York; 1966) for double-group (spin-orbit) states. For integral spin, the representations are the same as the usual ones, but the orbital symmetry of a given state is multiplied by the spin symmetry (e.g., in $D_{4h}$, $S = 0$ is $A_{1g}$, $S = 1$ is $A_{2g} + E_g$) to yield the symmetries of the spin-orbit states.

**Table 5 Experimental Data for $(\pi^*, \delta^*)^4$ Ground-State Molecules**

| Compound | $d(M_2)$/Å | $\mu_{eff}$/B.M.[a] (T/K) | VIS-NIR Solvent: λ/nm ($\varepsilon$/$M^{-1}$ $cm^{-1}$) |
|---|---|---|---|
| $Ru_2(O_2CCH_3)_4$ | | 3.05 (310), 0.60 (6)[b] | |
| $Ru_2(O_2CCH_3)_4(thf)_2$ | 2.261(3)[c] | 3.0 (295)[c] | thf: 438 (580)[c] |
| $Ru_2(O_2CH)_4(thf)_2$ | | 3.0 (295)[c] | thf: 448 (620)[c] |
| $Ru_2(O_2CEt)_4(acetone)_2$ | 2.260(3)[c] | 3.0 (295)[c] | thf: 441 (750)[c] |
| $Ru_2(O_2CCF_3)_4(thf)_2$ | 2.276(3)[d] | 3.0 (RT)[d] | thf: 456 (950)[d] |
| $Ru_2(O_2CPh)_4(thf)_2$ | | 3.0 (295)[c] | thf: 440 (sh)[c] |
| $Ru_2(O_2CPh)_4(OH_2)_{1.2}(EtOH)_{0.8}$ | | 2.90 (300), 0.80 (6)[b] | |
| $Ru_2(L\text{-}O_2CCH(OH)Ph)_4(OH_2)_2$ | 2.266(1)[b] | | thf: 443[b] CD: 534, 443, 390, 343[b] |
| $Ru_2(palmitate)_4$ | | 3.22 (320)[e] | |
| $[RuCl(NCCH_3)_3(PPh_3)_2]_2$ $[Ru_2(O_2C(p\text{-}C_6H_4Me))_4Cl_2]$ | 2.291(2)[f] | 3.07 (RT)[f] | $CHCl_3$: 497 (1,400)[f] |
| $Ru_2(HNC(CF_3)O)_4$ | | | $CH_2Cl_2$: 546 (780)[g] $CH_3CN$: 628 (1,040)[g] |
| $Ru_2(6\text{-}Me\text{-}hp)_4$ | 2.235(1)[h] | 2.23 (294), 0.58 (5)[h] | |
| $Ru_2(6\text{-}Cl\text{-}hp)_4 \cdot CH_2Cl_2$ | 2.247(1)[h] | 2.77 (300), 0.94 (5)[h] | |
| $Ru_2(6\text{-}Br\text{-}hp)_4 \cdot 1.5(CH_2Cl_2)$ | 2.259(1)[h] | 2.58 (305), 0.77 (5)[h] | |
| $Ru_2(6\text{-}F\text{-}hp)_4(thf)$ | 2.274(1)[i] | 2.29 (295), 0.61 (5)[i] | no VIS absorption maxima[i] |
| $Ru_2(tmtaa)_2$ | 2.379(1)[j] | 2.88 (RT)[j] | |
| $Ru_2(OEP)_2 \cdot 2(pentane)$ | 2.408(1)[k] | 2.8 (RT)[k] | $C_6H_6$: 724(5,250), 650 (4,570)[k] |
| $Os_2(OEP)_2$ | 2.39[l] | paramagnetic[m] | $C_6H_6$: 858(5,370), 606 (6,170)[m] |
| $Ru_2(Pc)_2$ | 2.40[n] | 2.54 (RT)[n] | |
| $Ru_2(dttz)_4(NCCH_3)$ | 2.407(1)[o] | diamagnetic[o] | |
| $Ru_2(dttz)_4 \cdot 3(toluene)$ | 2.417(2)[p] | diamagnetic[p] | $CH_2Cl_2$: 810(3,120), 540 (13,580)[p] |
| $Ru_2(dptz)_4$ | 2.3994(6)[o] | diamagnetic[o] | |

**Table 5** ***(Continued)***

| Compound | $d(M_2)/Å$ | $\mu_{eff}$/B.M.[a] (T/K) | VIS-NIR Solvent: λ/nm ($\varepsilon/M^{-1}\ cm^{-1}$) |
|---|---|---|---|
| $Ru_2(dtf)_4 \cdot (C_6H_6)$ | 2.474(1)[q] | diamagnetic[q] | $C_6H_6$: 905 (3,005), 540 (4,900 sh), 492 (8,780), 434 (6,250), 396 (7,150)[q] |
| $[Ru_2(dmpm)_2(ap)_2Cl](BPh_4)$ | 2.340(2)[r] | 2.6 (308)[r] | $CH_3CN$: 395(2,600)[r] |

[a] Low-temperature values have significant contributions from paramagnetic impurities; see Reference 86. [b] Reference 86. [c] Reference 91. [d] Reference 78. [e] Reference 152. [f] Reference 90. [g] Reference 140. [h] Reference 153. [i] Reference 88. [j] Reference 75. [k] Reference 123. [l] Reference 77. [m] Reference 149. [n] Reference 93. [o] Reference 79. [p] Reference 94. [q] Reference 95. [r] Reference 97.

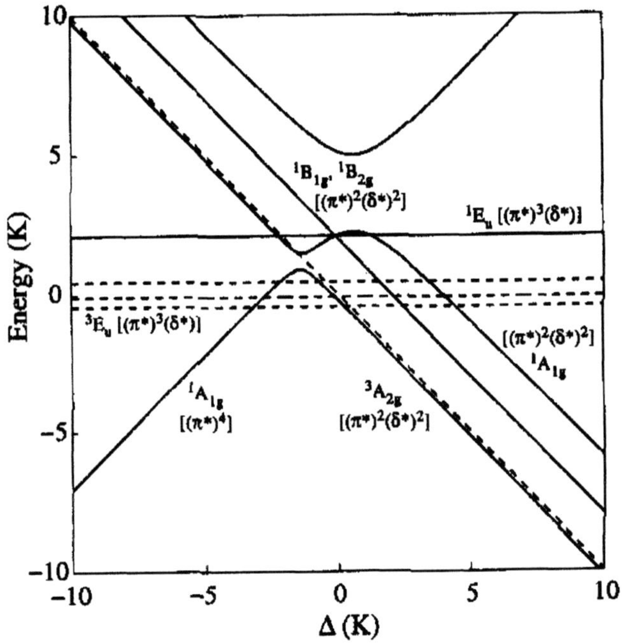


**Figure 9.** Crystal-field predictions of the state energies for the $(\pi^*, \delta^*)^4$ configuration as a function of Δ: $K = 1{,}000\ \text{cm}^{-1}$ and $\zeta = 900\ \text{cm}^{-1}$. Solid and dashed lines are singlet and triplet states, respectively.

to be quite large and positive, and for carboxylato and related bridging ligands, except for the most potent $\pi$-donating ones.

The key to identifying this ground state is again magnetic behavior. The $^3A_{2g}$ state undergoes a substantial ZFS because of the off-diagonal spin-orbit matrix element that connects the $A_{1g}(^3A_{2g})$ and $A_{1g}(^1A_{1g})(\pi^*)^2(\delta^*)^2$ states (Table 3). The resulting calculated ZFS of $^3A_{2g}$ is $D = W(E_g) - W(A_{1g}) \approx 200\ \text{cm}^{-1}$ for the parameters

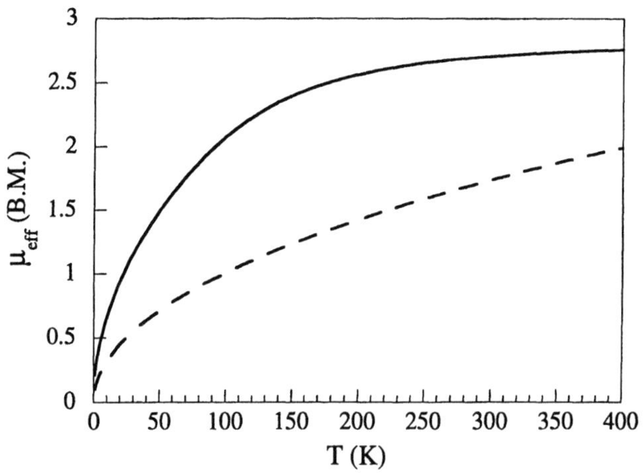


**Figure 10.** Magnetism temperature dependence predicted for the ${}^3A_{2g}(\pi^*)^2$ and ${}^3A_{2g}(\pi^*)^2(\delta^*)^2$ states with $g_{iso} = 2$ and $D = 250$ (———) or 1,100 (- - - - - -) $cm^{-1}$.

of Figure 9 [197 $cm^{-1}$ at $\Delta = +10{,}000$ $cm^{-1}$; 219 $cm^{-1}$ at the crossing with $A_{1u}$, $A_{2u}(^3E_u)$]. Since there are no spin-orbit couplings to the $E_g(^3A_{2g})$ state among the $(\pi^*, \delta^*)^4$ states, the sign of $D$ is established, and the magnitude of $D$ is insensitive to $\Delta$ except near $\Delta = 0$, where ${}^3A_{2g}$ is not the ground state.

Figure 10 shows the calculated magnetic behavior (Appendix C) for $g_{iso} = 2$ and $D = 250$ $cm^{-1}$, as is indicated by the experimental data to be roughly correct for second transition series metal complexes. The high-temperature-limiting magnetic moment is the triplet spin-only value, $\sqrt{8} \approx 2.83$ B.M., while the low-temperature-limiting value is 0 B.M.! This result occurs because the $A_{1g}(^3A_{2g})$ state is the "nonmagnetic" $m_S = 0$ component of the triplet state; however, the predicted magnetic susceptibility at low temperatures is actually substantial, but *temperature-independent* because it arises from "high-frequency" terms[85] due to magnetic-field induced coupling between the $A_{1g}(^3A_{2g})$ and $E_g(^3A_{2g})$ triplet components. The $A_{1g}(^3A_{2g})$ state is, of course, EPR-silent.

From the calculation, we can generally expect room-temperature magnetic moments for $Ru_2$(II,II) complexes with this ground state to be near the triplet spin-only value, and many examples are known (Table 5). Several detailed temperature-dependent studies have been performed, and the predicted temperature dependence has been confirmed in each case. Derived values of $D$ and $g_{iso}$ are summarized in Table 6. It can be seen that our earlier estimate of $D$ is reasonably close to experiment. The experimental values range as high as 300 $cm^{-1}$. The perturbation theory estimate for $D$ is given by:[§]

[§] The expression for $D$ given in Reference 86 is incorrect. The correct expression can be obtained by comparison to that established for the isomorphous problem of the $(\pi^{*2})$ ground state of the $O_2$ molecule[87] or, as has been done in this work, by proper correlation to the octahedral expressions of Griffith.[50]

**Table 6 Parameters $D$ and $g_{iso}$ Derived from Fits of the Temperature Dependence of the Magnetic Susceptibility of $^3A_{2g}(\pi^*)^2(\delta^*)^2$ Ground-State Molecules**

| Compound | $D$ (cm$^{-1}$) | $g_{iso}$ | Reference |
|---|---|---|---|
| $Ru_2(O_2CCH_3)_4$ | 244(10) | 2.08(5) | 86 |
| $Ru_2(O_2CPh)_4(OH_2)_{1.2}(EtOH)_{0.8}$ | 215(25) | 2.1(1) | 86 |
| $Ru_2(O_2CPr^n)_4$ | 278 | 2.07 | 152 |
| $Ru_2$(laurate)$_4$ | 287 | 2.46 | 152 |
| $Ru_2$(palmitate)$_4$ | 299 | 2.32 | 152 |
| $Ru_2$(6-Me-hp)$_4$ | 208(5) | 1.55(2) | 153 |
| $Ru_2$(6-Cl-hp)$_4$ | 248(4) | 1.77(2) | 153 |
| $Ru_2$(6-Br-hp)$_4$ | 232(5) | 1.86(2) | 153 |
| $Ru_2$(6-F-hp)$_4$ | 263(4) | 1.64(1) | 88 |
| $Ru_2$(Pc)$_4$ | 235 | 1.81(1) | 93 |

$$D \approx \frac{\zeta^2}{W(^1A_{1g}) - W(^3A_{1g})} \approx \frac{\zeta^2}{4K} \tag{20}$$

From this equation, we estimate that $D \approx 300$ cm$^{-1}$ could be achieved for slightly larger $\zeta$ ($\sim$1,100 cm$^{-1}$) or smaller $K$ ($\sim$680 cm$^{-1}$), or some combination thereof. Alternatively, higher-order spin-orbit terms, or spin-orbit terms involving states outside of the $(\pi^*, \delta^*)^4$ configuration, small terms that we are neglecting, could be responsible for the larger values of $D$.

$D$ tells us nothing about $\Delta$ except that it is positive. Theoretical calculations have had mixed success. Norman's[42] SCF-X$\alpha$ calculation on $Ru_2(O_2CH)_4$ predicted a $(\pi^*)^3(\delta^*)^1$ ground state, owing to a very small negative value of $\Delta$, but the $(\pi^*)^2(\delta^*)^2$ state was calculated to lie only 700 cm$^{-1}$ to higher energy. SCF-X$\alpha$ calculations by Cotton et al.[88] on the $D_{2d}$ and $C_{4v}$ isomers of $Ru_2(HNC(H)O)_4$ gave $\Delta = -1{,}100$ and $-1{,}630$ cm$^{-1}$, respectively, and predicted a $(\pi^*)^4$ ground state [although it is not clear if these workers explicitly considered the $(\pi^*)^3(\delta^*)^1$ configuration].

However, an RHF calculation on $Ru_2(O_2CH)_4$ correctly predicted the $(\pi^*)^2(\delta^*)^2$ ground state, and placed $^3A_{2g}$ 5,300 cm$^{-1}$ below $^3E_u(\pi^*)^3(\delta^*)^1$.[64] This is a theoretical estimate for the $\delta^* \rightarrow \pi^*$ excitation energy (neglecting spin-orbit coupling), and also for $\Delta$ (Table 2). Interestingly, semiempirical methods (extended Hückel and Fenske–Hall) give consistent results for $Ru_2(O_2CH)_4$, $\Delta \approx$ 4,000–5,000 cm$^{-1}$.[89]

We crudely estimate excitation energies from Figure 9 by assuming that $\Delta$ has the same value that we arrived at earlier for $Ru_2(O_2CPr^n)_4Cl$, +3,000 cm$^{-1}$. Since the $(\pi^*)^2(\delta^*)^1$ $Ru_2$(II,III) and $(\pi^*)^2(\delta^*)^2$ $Ru_2$(II,II) compounds have very similar Ru–Ru bond distances (Tables 4 and 5), owing to the very weak $\delta$-symmetry metal–metal bonding interaction, this is a reasonable approximation. We then predict the spin-forbidden $\delta^* \rightarrow \pi^*$ excitation, $A_{1g}(^3A_{2g}) \rightarrow {}^1E_u$, at 5,300 cm$^{-1}$, which is in an accessible spectral region. All eight remaining one-electron excitations within the $(\pi^*, \delta^*)^4$ configuration, including the spin-allowed $\delta^* \rightarrow \pi^*$ excitations, are predicted to be clustered in the 2,200–4,000 $cm^{-1}$ region, which is not a spectroscopically accessible region because strong vibrational absorption obscures it.

Experimentally, one report[90] explicitly shows that the absorption spectrum of $Ru_2(O_2CC_6H_4\text{-}p\text{-}Me)_4Cl_2^{2-}$ lacks the $^4(\delta \rightarrow \delta^*)$ transition of the $Ru_2$(II,III) complexes at ~1,100 nm, as it must for a $(\pi^*)^2(\delta^*)^2$ ground state. No near-infrared absorption is reported, nor is any expected. The $A_{1g}(^3A_{2g}) \rightarrow {^1E_u}$ transition should be extremely weak ($\varepsilon \approx 2$ $M^{-1}cm^{-1}$), like the analogous spin-forbidden excitations of the $Ru_2$(II,III) complexes. It is possible that a low-temperature solid-state absorption measurement would reveal it.

The visible absorption spectra of the $Ru_2$(II,II) carboxylato complexes are very similar to those of the $Ru_2$(II,III) carboxylato complexes, showing a moderately intense ($\varepsilon \approx 600$ $M^{-1}cm^{-1}$) band at ~450 nm. It has been suggested that the assignments should be similar.[91] If so, then the $\pi(Ru\text{–}O,Ru_2) \rightarrow \pi^*(Ru_2)$ assignment discussed earlier for the $Ru_2$(II,III) complexes should be revised to $\pi(Ru_2) \rightarrow \pi^*(Ru_2)$, since an LMCT transition should be strongly blue-shifted for the $Ru_2$(II,II) complexes. Because of the well-known basis-set sensitivity of Mulliken population analysis, the composition of the $\pi$ orbital that was calculated by Norman,[42] indicating dominant $\pi(Ru\text{–}O)$ contributions, could be incorrect. Alternatively, the transitions of the $Ru_2$(II,II) and $Ru_2$(II,III) complexes might be unrelated. A possible assignment of the $Ru_2$(II,II) band along these lines is to MLCT from $\delta^*$ or $\pi^*$ to a carboxylato $\pi^*$ level. The slight red-shift of the band for $Ru_2(O_2CCF_3)_4(thf)_2$ does suggest some MLCT character in the excitation (certainly *not* LMCT character). However, the assigned $\delta \rightarrow \pi^*$ (carboxylato) MLCT transition of $W_2(O_2CR)_4$ is *strongly* red-shifted for R = Ph (18,100 $cm^{-1}$, $\varepsilon$ = 10,000 $M^{-1}cm^{-1}$) versus R = Et (27,000 $cm^{-1}$, $\varepsilon$ = 15,000 $M^{-1}cm^{-1}$),[92] and no such shift is seen for the visible bands of the corresponding $Ru_2$(II,II) compounds (Table 5). On balance, the $\pi(Ru_2) \rightarrow \pi^*(Ru_2)$ assignment, with a small amount of LMCT or MLCT mixing [that is, carboxylato $\pi$ mixing with $\pi(Ru_2)$ or carboxylato $\pi^*$ mixing with $\pi^*(Ru_2)$], seems to be the preferable one for the major visible absorption band of both $Ru_2$ oxidation states. We emphasize that metal–ligand mixing in the $\pi$ level may be more significant for higher oxidation states of the binuclear complexes such as $Ru_2$(III,III) and $Os_2$(III,III).

The lowest-energy metal–metal excitation outside of the $(\pi^*, \delta^*)^4$ configuration should be $\delta \rightarrow \pi^*$, $A_{1g}(^3A_{2g}) \rightarrow {^3E_g}$. With $\Delta$ = 3,000 $cm^{-1}$ and taking the $\delta/\delta^*$ splittting to be ~9,000 $cm^{-1}$ (*vide supra*), we predict this excitation at ~12,000 $cm^{-1}$ (830 nm). No such band has been reported, but this dipole-forbidden excitation is likely to be weak ($\varepsilon \leqslant 100$ $M^{-1}cm^{-1}$) and might be obscured by the tail of the main visible absorption. The CD spectrum reported[86] for the optically active derivative $Ru_2(L\text{-}O_2CCH(OH)Ph)_4(OH_2)_2$ shows several features that are not resolved in the absorption spectrum, but all of these are probably of too high energy to be the $\delta \rightarrow \pi^*$ transition.

The visible band of $Ru_2(HNC(CF_3)O)_4(thf)_2$ (Table 5) is similar to that of the carboxylato complexes, and assignment to $\pi(Ru_2) \rightarrow \pi^*(Ru_2)$ seems reasonable. Curiously, brown $Ru_2(6\text{-Fhp})_4$ is reported[88] to lack any distinct visible absorption bands. Very intense UV absorption (287 nm, $\varepsilon$ = 25,520 $M^{-1}cm^{-1}$) (MLCT?) apparently tails sufficiently into the visible to obscure the relatively weak metal–metal transitions.

The ground states of $Ru_2$(II,II) porphyrin and tmtaa complexes are interpreted as $A_{1g}(^3A_{2g})$ states according to magnetic data, although the magnetic temperature dependence has been determined only for $Ru_2(Pc)_2$.[93] The Ru-Ru distances of these compounds are over 0.1 Å longer than those of the $\mu$-carboxylato complexes with this ground state, a finding that is attributable to steric effects involving ligand–ligand repulsion. This in turn suggests that the $^2E_g[(\pi^*)^1(\delta^*)^2]$ ground states of the $Ru_2$(II,III) complexes of these ligands (*vide supra*) likewise have $Ru_2$ bond lengths that are considerably elongated by steric factors.

**4.4.2 $(\pi^*)^4$** This ground state is unequivocally identified in several compounds by the combination of experimentally determined diamagnetism and a long M-M distance ($\geqslant$ 2.4 Å) similar to those exhibited by $(\pi^*)^4(\delta^*)^2$ ground-state singly-bonded complexes such as $Rh_2(O_2CR)_4^2$. It is observed for the $Ru_2$(II,II) complexes of the very strongly electron-donating amidinato and triazenato bridging ligands, and model theoretical calculations predict $\Delta \approx -8{,}000\ cm^{-1}$.[43,44]

Figure 9 indicates that $\Delta \approx -8{,}000\ cm^{-1}$ is more than sufficient to establish the $^1A_{1g}(\pi^*)^4$ ground state. The spin-allowed $^1A_{1g} \rightarrow {}^1E_u$ $\pi^* \rightarrow \delta^*$ excitation is predicted at 7,200 $cm^{-1}$ while spin-forbidden $\pi^* \rightarrow \delta^*$ excitations are predicted at 4,700, 5,000, and 5,600 $cm^{-1}$. The latter excitations may just barely be observable in a low-temperature solid-state measurement, depending upon how intense they are relative to vibrational overtone absorption.

The $(\pi^*)^4$ ground-state molecules have rich visible absorption spectra (Table 5), but the intensities of the bands appear to be too high for assignments to intrinsically weak metal–metal transitions such as $\pi^* \rightarrow \delta^*$. Cotton et al.[79,94,95] have nonetheless assigned the characteristic band at 800–900 nm ($\varepsilon \sim 3{,}000\ M^{-1}cm^{-1}$) to $\pi^* \rightarrow \delta^*$.¶ The calculations[43,44] indicate such large mixing of the $\delta^*$ orbital with $\pi$(N) (~30%) for these compounds that this assignment cannot be excluded; the $\pi^* \rightarrow \delta^*$ transition, as well as transitions such as $\delta \rightarrow \delta^*$, acquire a great deal of charge-transfer character, and pure "metal–metal" assignments may not be appropriate. Careful measurements of NIR spectra would help to settle this question, and in any case might locate the spin-forbidden excitations.

Several $Rh_2$(III,III) complexes have been prepared electrochemically in solution from $Rh_2$(II,II) or $Rh_2$(II,III) precursors.[68,81,96] They all involve strongly $\pi$-donating bridging ligands, and have very intense absorption extending into the NIR [e.g., 1,650 nm ($\varepsilon = 8{,}600\ M^{-1}cm^{-1}$) for $Rh_2(ap)_4(C_2H)^+$],[68] which is undoubtedly due

¶ In Reference 95, some of the visible absorption bands of $Ru_2(dtf)_4$, including the 907 nm band, were assigned to spin-forbidden excitations on the basis of comparisons to the literature assignments of several third-transition series complexes. This is not valid. Spin-orbit induced spin-forbidden transition intensities are about an order of magnitude smaller for second-transition series complexes.[53] Assignment of the 907 nm absorption of $Ru_2(dtf)_4$ to the singlet-singlet $\pi^* \rightarrow \delta^*$ excitation is possible, however, if mixing with LMCT transitions is very substantial. $Os_2$ complexes may show relatively intense singlet-triplet parentage transitions, as discussed elsewhere in this chapter.

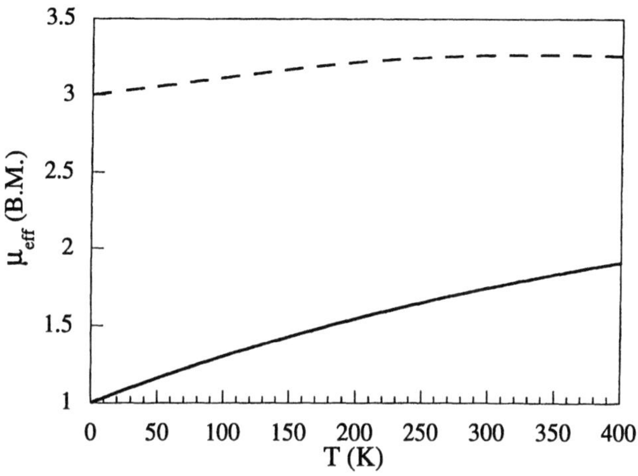


**Figure 11.** Magnetism temperature dependence predicted for the $^3E_u(\pi^*)^1(\delta^*)^1$ (——) and $^3E_u(\pi^*)^3(\delta^*)^1$ (- - - - - -) states, both with $g_{iso} = 2$ and $D = 800$ cm$^{-1}$.

to LMCT transitions. The ground states are likely to be $(\pi^*)^4$, but there is no direct evidence available.

**4.4.3 $(\pi^*)^3(\delta^*)^1$** While this ground state has been much discussed,[41,42,64,89] no established cases are known. Figure 9 shows that the $A_{1u}$, $A_{2u}(^3E_u)$, $(\pi^*)^3(\delta^*)^1$ ground state should be accessible for bridging ligands that have intermediate $\pi$-donor strength, 2-anilinopyridinate (ap$^-$) being a plausible possibility. We show the theoretical magnetic behavior in Figure 11. It is evident that this state *cannot* be distinguished from $^3A_{2g}(\pi^*)^2(\delta^*)^2$ on the basis of the room-temperature magnetic moment, although the temperature behavior obviously provides an easy distinction, the $(\pi^*)^3(\delta^*)^1$ state being predicted to have $\mu_{eff}$ nearly temperature-independent.

The other distinction of this ground state is that the metal–metal distance should be roughly intermediate between those found for $(\pi^*)^2(\delta^*)^2$ and $(\pi^*)^4$ ground states, hence ~2.35 Å. Just one candidate emerges from the structurally characterized compounds, $[Ru_2(dmpm)_2(ap)_2Cl](BPh_4)$, which has a Ru-Ru distance of 2.340(2) Å.[97] The ligands of this complex are somewhat different from those of other $Ru_2$(II,II) complexes, so the metal–metal bond distance correlation must be viewed cautiously in this case. A temperature-dependent magnetic study of this compound would be of interest.

## 4.5 $(\pi^*, \delta^*)^2$ Configurations

Figure 12 shows the crystal-field predictions for $(\pi^*, \delta^*)^2$ with $\zeta = 2{,}800$ cm$^{-1}$, appropriate for third transition series metal ions, most of the known examples being $Os_2$(III,III) complexes. The $(\pi^*, \delta^*)^n$ energy expressions (Tables 2 and 3) are identical for $n = 2$ and $n = 4$, except for a change in sign of both $\Delta$ and $\zeta$. The only effect of the change in sign of $\zeta$ is to invert the spin-orbit splitting of the state. We

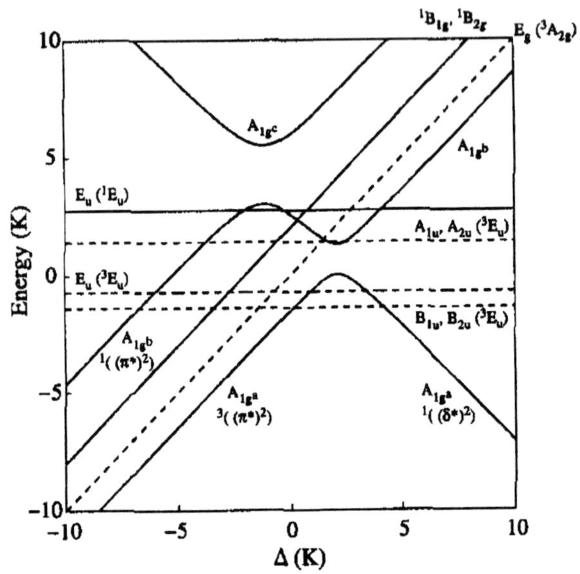


**Figure 12.** Crystal-field predictions of the state energies for the $(\pi^*, \delta^*)^2$ configuration as a function of Δ: $K = 1{,}000$ cm$^{-1}$ and $\zeta = 2{,}800$ cm$^{-1}$. Solid and dashed lines are singlet and triplet states, respectively.

can therefore use, with care, Figure 9 for $n = 2$ and $\zeta = 900$ cm$^{-1}$, or Figure 12 for $n = 4$ and $\zeta = 2{,}800$ cm$^{-1}$.

From Figure 12, $^1A_{1g}(\delta^*)^2$ is the ground state for $\Delta > 4{,}150$ cm$^{-1}$, while $A_{1g}(^3A_{2g})(\pi^*)^2$ is the ground state for $\Delta < +145$ cm$^{-1}$. The range over which $B_{1u}$, $B_{2u}(^3E_u)(\pi^*)^1(\delta^*)^1$ is the ground state [the ZFS of $^3E_u(\pi^*)^1(\delta^*)^1$ is inverted from that of $^3E_u(\pi^*)^3(\delta^*)^1$] is a little larger than the range for which it is the ground state for $\zeta = 900$ cm$^{-1}$. The ZFS of $^3A_{2g}$ is much larger than for $\zeta = 900$ cm$^{-1}$, $D = 1{,}460$ cm$^{-1}$ for $\Delta = -10{,}000$ cm$^{-1}$, and 1,545 cm$^{-1}$ at the crossing with $B_{1u}$, $B_{2u}(^3E_u)$.

The predicted magnetic temperature dependences for the two possible paramagnetic ground states are shown as the dashed line in Figure 10 [$^3A_{2g}(\pi^*)^2$] and the solid line in Figure 11 [$^3E_u(\pi^*)^1(\delta^*)^1$]. Both states yield low magnetic moments at room temperature; while their limiting low-temperature behavior is different, they may nonetheless be hard to distinguish even with full temperature-dependent magnetic data sets because, for such low magnetic moments, traces of impurities can obscure the low-temperature behavior. Since compounds having the nominally diamagnetic $^1A_{1g}(\delta^*)^2$ ground state may also have paramagnetic impurities, room-temperature magnetic moments of the order of 1–1.5 B.M. are not conclusive for any of the three possible ground states, although determination of the full temperature dependence may allow a distinction to be made for pure samples.

Experimental data are summarized in Table 7.

**Table 7 Experimental Data for $(\pi^*, \delta^*)^2$ Ground-State Molecules**

| Compound | $d(M_2)$/Å | $\mu_{eff}$/B.M. (T/K) | VIS-NIR Solvent: λ/nm ($\varepsilon/M^{-1}\ cm^{-1}$) |
|---|---|---|---|
| $Os_2(dtf)_4Cl_2$ | 2.467(1)[a] | 1.4 (300), 0.35 (5)[a] | $C_6H_6$: 496, 419[a] |
| $Os_2(HNC(Ph)O)_4Cl_2$ | 2.367(2)[b] | 1.76 (298)[b] | $CH_2Cl_2$: 750 (1,050), 635 (1,665), 410 (sh), 345 (sh)[b] |
| $Os_2(HNC(Ph)O)_4Br_2$ | 2.384(2)[b] | | $CH_2Cl_2$: 762 (850), 643 (1,290), 427 (5,850), 350 (7,400)[b] |
| $Os_2(ap)_3Cl_3$ | 2.392(1)[c] | 1.46 (308)[c] | $CH_2Cl_2$: 868 (1,880), 615 (4,920)[c] |
| $Os_2(hp)_4Cl_2{\cdot}(Et_2O)$ | 2.344(2)[i], | 2.04 (300), 1.73 (195);[e] | $CH_2Cl_2$: 488 (5,700), 362 (sh), 343 (6,700)[e] |
| $Os_2(hp)_4Cl_2{\cdot}2(CH_3CN)$ | 2.357(1)[d] | 1.65 (262), 0.53 (6)[f] | $CH_2Cl_2$: 1100 (50 sh), 848 (450), 715 (520), 494 (7,540), 354 (5,450)[g] |
| $Re_2(dtf)_4$ | 2.344(2)[h] | | $C_6H_6$: 692 (1,300), 576 (2,280), 367 (16,400)[h] |
| $Ru_2(hpp)_4Cl_2$ | 2.321(1)[ii] | 2.78 (RT)[ii] | $CH_2Cl_2$: 770, 610, 430[ii] |
| $[Ru_2(ap)_4Cl](FeCl_4){\cdot}2.5(CH_2Cl_2)$ | 2.301(1)[jj] | | $CH_2Cl_2$: 975, 505, 430[jj] |
| $[Ru_2(ap)_4F](PF_6)$ | | 2.75 (RT, solid); 2.90 (RT, $CH_2Cl_2$)[jj] | $CH_2Cl_2$: 940, 490, 395[jj] |
| $[Ru_2(ap)_4(OH_2)]^{2+}$ | | 2.81 (RT, $CH_2Cl_2$)[jj] | $CH_2Cl_2$: 940, 490, 395[jj] |
| $Os_2(6\text{-Cl-hp})_2Cl_4(py)$ | 2.293(1)[i] | 1.65 (RT)[i] | $CH_2Cl_2$: 555 (2,820), 330 (9,090)[i] |
| $Os_2(6\text{-Cl-hp})_2Cl_4(OH_2)$ | 2.322(1)[i] | | acetone: 549 (3,440), 390 (sh), 330 (10,200)[i] |
| $Os_2$(2-phenylbenzoate)$_4Cl_2$ | 2.3173(6)[f] | 1.93 (304), 0.41 (1.5)[f] | $CH_2Cl_2$: 887 (53), 412 (8,600)[f] |
| $Os_2(O_2CCH_3)_4Cl_2$ | 2.314(1)[j] | 1.65 (288), 1.34 (209);[e] 1.95 (297)[k] | solid (refl): 370[e] |
| $Os_2(O_2CEt)_4Cl_2$ | 2.316(2)[j] | 1.60 (300), 1.44 (188)[e] | $CH_2Cl_2$: 392 (10,800), 364 (sh)[e] |
| $Os_2(O_2CPr^n)_4Cl_2$ | 2.301(1)[e] | 1.63 (300), 1.02 (188)[e] | $CH_2Cl_2$: 392 (10,400), 360 (sh)[e] |
| $Os_2(O_2CBu^t)_4Cl_2$ | | 1.99 (300), 0.69 (6)[k] | $CH_2Cl_2$: 1,120 (15), 850 (70), 630 (90 sh), 500 (130 sh), 394 (11,600), 360 (5,000 sh)[k] |
| $Os_2(O_2CBu^t)_4Br_2$ | | | $CH_2Cl_2$: 1,250 (17), 850 (110), 455 (13,100), 385 (3,000 sh)[k] |

**Table 7 (*Continued*)**

| Compound | $d(M_2)$/Å | $\mu_{eff}$/B.M. (T/K) | VIS-NIR Solvent: λ/nm ($\varepsilon$/$M^{-1}$ $cm^{-1}$) |
|---|---|---|---|
| $Os_2(O_2CCH_3)_2(Ph_2PC_6H_4)_2Cl_2$ | 2.271(1)[m] | 0.29 (295)[m] | |
| $Os_2(O_2CEt)_2(Ph_2PC_6H_4)_2Cl_2$ | 2.272(1)[m] | 0.07 (295)[m] | |
| $Os_2(Ph_2PC_6H_4)_2Cl_4$ | 2.231(1)[n] | | |
| $[N(Bu^n)_4]_2[Os_2Cl_8]$ | 2.182(1)[o] | 0.49 (295);[p] 0.77 (295)[q] | $CH_2Cl_2$: 729 (180), 466 (130)[r,s] |
| $[N(Bu^n)_4]_2[Os_2Br_8]$ | 2.196(1)[o] | 1.10 (295)[q] | $CH_2Cl_2$: 950 (sh), 695 (258), 500 (sh), 430 (sh), 340 (13,000)[t,s] |
| $Re_2Cl_4(PMe_3)_4$ | 2.247(1)[u] | diamagnetic[v] | $CH_2Cl_2$: 732 (sh), 635 (sh), 613 (210), 519 (100), 466 (105), 401 (180)[v] |
| $Re_2Cl_4(PPr^n_3)_4$ | 2.252(2)[u] | diamagnetic[w] | pentane: 760 (sh), 645 (138), 542 (sh), 498 (100), 415 (85), 361 (sh);[w] KBr pellet (5 K): 815 (sh), 725, 647, 598 (sh), 531, 485, 411, 355 (sh)[x] |
| $Re_2Br_4(PMe_3)_4$ | | diamagnetic[v] | $CH_2Cl_2$: 785 (sh), 654 (210), 625 (205), 485 (90), 415 (150)[v] |
| $Re_2Br_4(PPr^n_3)_4$ | 2.253(4)[u] | | |
| $Tc_2Cl_4(PEt_3)_4$ | 2.133(3)[y] | diamagnetic[y] | $C_6H_6$: 730 (30), 648 (30), 528 (sh), 496 (56)[y] |
| $Os_2(OEP)_2^{2+}$ | 2.27[z] | diamagnetic[aa] | $CH_2Cl_2$: 537 (sh)[aa] |
| $Ru_2(OEP)_2^{2+}$ | 2.22[bb] | diamagnetic[aa] | $CH_2Cl_2$: 560 (10,500)[aa] |
| $Ru_2(tmtaa)_2^{2+}$ | | diamagnetic[cc] | |
| $Ru_2(HNC(CH_3)O)_4Cl_2$ | | | DMSO/0.1 M LiCl: 740, 480 (sh), 450, 395[dd] |
| $Ru_2(F_5ap)_4(C_2Ph)_2$ | 2.451(1)[ll] | diamagnetic[ll] | $CH_2Cl_2$: >1,000, 803, 720 (sh), 647, 582 (sh), 441[ll] |
| $Ru_2((p\text{-}ClPh)NCHN(p\text{-}ClPh))_4(C_2Ph)_2$ | 2.5554(12)[mm] | diamagnetic[mm] | $CH_2Cl_2$: 690 (sh), 526 (17,310), 504 (sh), 424 (sh)[mm] |
| $Ru_2(dpf)_4(C_2Ph)_2$ | 2.556(1)[kk] | diamagnetic[kk] | $CH_2Cl_2$: 990 (1,890), 693 (3,150), 535 (17,900), 510 (18,200), 410 (6,090 sh)[kk] |
| $Ru_2(dpf)_4(CN)_2$ | 2.539(1)[kk] | diamagnetic[kk] | |
| $Cs_2[Ru_2(SO_4)_4(OH_2)_2]$ | 2.343(1)[ee] | | |

**Table 7** ***(Continued)***

| Compound | $d(M_2)/Å$ | $\mu_{eff}$/B.M. (T/K) | VIS-NIR Solvent: $\lambda$/nm ($\varepsilon$/$M^{-1}$ $cm^{-1}$) |
|---|---|---|---|
| $K_2[Ru_2(SO_4)_4(OH_2)_2]$ | | 4.58 (292);[ff] 3.90 (10)[gg] | 10% $H_2SO_4$ (aq): 500 (1,000 sh), 355 (4,200)[ff] |
| $K_n[Ru_2(phosphate)_4]$[hh] | 2.320(3)[gg] | | $H_2O$: 490[gg] |

[a] Reference 98. [b] Reference 101. [c] Reference 154. [d] Reference 155. [e] Reference 156. [f] Reference 99. [g] Miskowski, V. M., unpublished work. See Figure 13. [h] Reference 100. [i] Reference 157. [j] Reference 158. [k] Reference 111. [l] Low-temperature solid-state absorption spectra are presented in Reference 111. [m] Reference 159. [n] Reference 160. [o] Reference 52. [p] Evans method, $CH_2Cl_2$ solution, Reference 104. [q] Solid state, Reference 105. [r] Reference 104. [s] Low-temperature data for solid films are presented in Reference 110. [t] Reference 105. [u] Reference 161. [v] Reference 162. [w] Reference 163. [x] Reference 108. [y] Reference 109. [z] Estimated from Raman data, Reference 77. [aa] Reference 149. [bb] Reference 148. [cc] Reference 75. [dd] Reference 114. [ee] Reference 71. [ff] Reference 69. [gg] Reference 70. [hh] Disordered crystal structure (Reference 111); protonation states of the phosphate ligands are uncertain. [ii] Reference 102. [jj] Reference 103. [kk] Reference 83. [ll] Reference 151. [mm] Reference 164.

**4.5.1 $(\pi^*)^2$** The metal–metal distance is expected to provide an important clue as to the correct ground state. If we neglect the weak $\delta$-symmetry metal–metal interaction, the $(\pi^*)^2$ and $(\delta^*)^2$ ground states have, respectively, double and triple metal–metal bonds, for which we may expect third transition series metal ions to show metal–metal distances of about 2.35 and 2.2 Å.[2] The collected experimental data in Table 7 actually show a rather continuous distribution of $d(Os_2)$ ranging from 2.467 to 2.182 Å.

The longest bond distance is found for the $Os_2$(III,III) compound bridged by four of the very strongly electron-donating $dtf^-$ ligands, and this must surely be a $(\pi^*)^2$ ground state. The exceptionally long bond distance is attributable to strong $\pi$-donation by this ligand into the metal–metal antibonding orbitals of the electron-deficient binuclear complex. The $(\pi^*)^2$ ground state has been confirmed by a temperature-dependent magnetic study,[98] a fit to the theory being achieved for $g_{eff} = 1.802$ and $D = 1{,}350$ $cm^{-1}$. Magnetic data for $Os_2(hp)_4Cl_2$ have also been fit to this ground state, $g_{eff} = 1.922$ and $D = 1{,}196$ $cm^{-1}$.[99] These values of $D$ are in reasonable agreement with our prediction of $D \approx 1{,}500$ $cm^{-1}$ (Figure 12). The slightly smaller experimental values very likely reflect large covalency, hence reduced $\zeta$,[48,50,53] for the strongly $\pi$-donating bridging ligands. For $K \approx 1{,}000$ $cm^{-1}$, values of $\zeta \approx 2{,}600$ $cm^{-1}$ will fit the experimental ZFS.

The $Os_2$(III,III) entries in Table 7 with bridging $HNC(Ph)O^-$ or 2-$(PhN)py^-$ ligands have metal–metal distances intermediate between those of the two examples above, so likely also have $(\pi^*)^2$ ground states. The metal–metal distance for the compound $Re_2(dtf)_4$ falls at the lower end of this range.[100] The electron-rich $Re_2$(II,II) unit should be less affected by bridging-ligand $\pi$-donation, so a relatively short metal–metal distance is reasonable. However, no magnetic measurements have been reported. SCF-X$\alpha$ theoretical calculations on $Os_2(HNC(H)NH)_4Cl_2$ and $Re_2(HNC(H)NH)_4$ are consistent with $(\pi^*)^2$ ground states, the former giving $\Delta = -9{,}100$ $cm^{-1}$,[98] and the latter giving $\Delta = -2{,}020$ $cm^{-1}$.[100]

For the $Re_2$(II,II) compound with $\Delta \approx -2{,}000\ cm^{-1}$, Figure 12 predicts the spin-allowed (triplet-triplet) $\pi^* \rightarrow \delta^*$ excitations to fall at $< 5{,}000\ cm^{-1}$, hence to be experimentally inaccessible by absorption spectroscopy. Experimentally accessible spin-forbidden excitations are predicted at 6,210 $cm^{-1}$ [$A_{1g}(^3A_{2g}) \rightarrow E_u(^1E_u)$, $\pi^* \rightarrow \delta^*$] and 6,250 $cm^{-1}$ [the $A_{1g}(^3A_{2g}) \rightarrow A_{1g}(^1A_{2g})\pi^*$ spin-flip]. The reported visible absorption bands (Table 7) have been assigned to metal–metal transitions, including $\pi^* \rightarrow \delta^*$ at 692 nm ($\varepsilon = 1{,}300\ M^{-1}cm^{-1}$); this assignment is inconsistent with our analysis. The spin-allowed $\delta \rightarrow \delta^*$ transition should occur in the visible region because the calculated $\delta/\delta^*$ splitting is 14,800 $cm^{-1}$;[100] assuming the $\delta/\delta^*$ exchange integral has a magnitude similar to that evaluated for third transition series quadruply bonded binuclear complexes, ~5,000 $cm^{-1}$ (Table 1),[18] the spin-allowed $\delta \rightarrow \delta^*$ transition should occur ~1,000 $cm^{-1}$ to higher energy of the one-electron energy difference. The $\delta \rightarrow \delta^*$ band might also have moderately high intensity because of extensive metal–ligand mixing (25% calculated for $\delta^*$).[100] The $\pi \rightarrow \pi^*$ transition should, as for other $(\pi^*, \delta^*)^n$ configurations discussed earlier, also occur in the visible region. The 692 and 576 nm bands are possibly due to these two spin- and dipole-allowed metal–metal transitions. The very high intensity of the band observed at 367 nm indicates that it must be a charge-transfer transition.

For the $Os_2$(III,III) $(\pi^*)^2$ ground-state complexes, we crudely estimate $\Delta \approx -9{,}100\ cm^{-1}$ from the $Os_2(HNC(H)NH)_4Cl_2$ calculation[98] and thereby predict the spin-allowed (triplet-triplet) $\pi^* \rightarrow \delta^*$ excitations at 9,160 ($B_{1u}$, $B_{2u}$), 9,840 ($E_u$) and 11,960 $cm^{-1}$ ($A_{1u}$, $A_{2u}$). The [$A_{1g}(^3A_{2g}) \rightarrow E_u(^3E_u)$] and [$A_{1g}(^3A_{2g}) \rightarrow A_{2u}(^3E_u)$] transitions are dipole-allowed and should be separated by ~2,000 $cm^{-1}$ as a result of a combination of first- and second-order spin-orbit coupling effects in the $^{1,3}E_u$ states. The calculated splitting is given by:

$$W\left[A_{1u}, A_{2u}(^3E_u)\right] - W\left[E_u(^3E_u)\right] = \frac{\zeta}{2} - K + \sqrt{K^2 + \frac{\zeta^2}{4}} \tag{21}$$

The compounds $Os_2(HNC(Ph)O)_4X_2$ (X = Cl, Br) are reported[101] to display moderately intense ($\varepsilon \approx 10^3\ M^{-1}cm^{-1}$) bands at about 650 and 750 nm, which are plausible candidates for the allowed $\pi^* \rightarrow \delta^*$ transitions. These bands are fairly narrow, half-widths 1,500 $cm^{-1}$, consistent with the modest excited-state metal–metal contraction expected for the $\pi^* \rightarrow \delta^*$ excited states. These assignments imply $\Delta \approx -14{,}000\ cm^{-1}$. The compound $Os_2(hp)_4Cl_2$ has two similar absorption bands at 715 and 848 nm (Figure 13) for which we also suggest $\pi^* \rightarrow \delta^*$ assignments, the implied value of $\Delta$ for this compound being $-12{,}500\ cm^{-1}$. Relatively intense bands at 400–500 nm are likely due to $\pi$[Os-(N,O),$Os_2$] $\rightarrow \pi^*(Os_2)$ transitions. The spin-forbidden $A_{1g}(^3A_{2g}) \rightarrow A_{1g}(^1A_{1g})$ spin-flip excitation is predicted to lie at 6,800 $cm^{-1}$ (1,470 nm). This transition should be extremely sharp, but may be very weak because it is highly forbidden. It has not been observed experimentally.

The reported visible absorption bands (Table 7) of $Os_2(dtf)_4Cl_2$ are probably analogous to those discussed above for $Re_2(dtf)_4$, and we assign them to charge-transfer and/or metal–metal $\pi \rightarrow \pi^*$ transitions. It is likely that $\pi^* \rightarrow \delta^*$ transitions of this and related $Os_2$(III,III) complexes occur as relatively weak low-energy bands

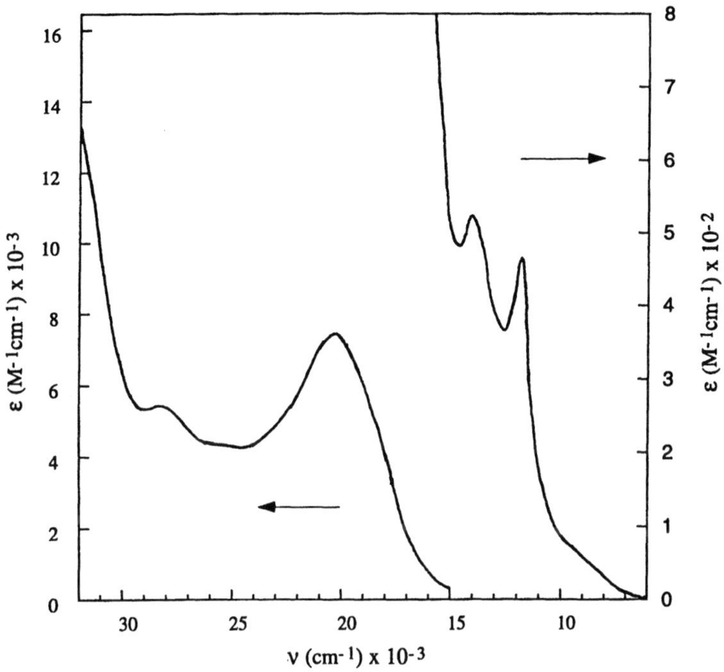


**Figure 13.** The electronic absorption spectrum of $Os_2(hp)_4Cl_2$ in $CH_2Cl_2$ solution. An additional maximum (not shown) at 279 nm ($\varepsilon$ = 19,500 $M^{-1}cm^{-1}$) is probably due to a ligand-localized transition.

that have not yet been identified. The reported spectrum of $Os_2(ap)_3Cl_3$ (Table 7) is similar to those of the other $(\pi^*)^2$ ground-state molecules, but because of the low symmetry of this compound, we will not attempt assignments.

Very recently, the first examples of second transition series compounds that possess this ground state have been reported, each of them containing the $Ru_2^{6+}$ core.[102,103] Room-temperature magnetic moments are in the range 2.75–2.90 B.M., indicating a triplet state, and no EPR signals are observed. Two crystallographic determinations are available: the $Ru_2(ap)_4Cl^+$ cation[103] and the compound $Ru_2(hpp)_4Cl_2$[102] display $Ru_2$ distances of 2.321(1) and 2.301(1) Å, respectively. These are at the long end of the range found for $(\pi^*)^2(\delta^*)^2$ $Ru_2^{4+}$ complexes (*vide supra*) and are much longer than expected for a $(\pi^*)^1(\delta^*)^1$ configuration, so $(\pi^*)^2$ ground states are indicated; a temperature-dependent magnetic study would confirm this. Electronic spectra have not been reported in sufficient detail to allow assignments to be attempted.

**4.5.2 $(\delta^*)^2$** There are many examples of diamagnetic binuclear compounds with this ground state,[2] and only a few key examples are included in Table 7. Os-Os or Re-Re distances near 2.2 Å provide a definitive assignment of this ground state. Room-temperature magnetic susceptibility measurements on $Os_2X_8^{2-}$ (X = Cl, Br)[104,105] salts raise a cautionary flag as to the interpretation of magnetic data in that magnetic moments as high as 1.1 B.M. are reported (Table 7). Most measured moments are considerably smaller than this extremum value, and there is little doubt that the $Os_2X_8^{2-}$ complexes have the $^1A_{1g}(\delta^*)^2$ ground state. Outside of the persistent problem of paramagnetic impurities, these complexes are likely to possess significant temperature-independent paramagnetic susceptibilities ($\chi_{TIP}$)[50,85] owing to magnetic-field-induced coupling of the $^1A_{1g}(\delta^*)^2$ ground state to $^1E_g$ excited states such as $\delta \rightarrow \pi^*$ (contributing to $\chi_\perp$) and to $^1A_{2g}$ excited states such as $\delta^* \rightarrow (x^2 - y^2)$ (contributing to $\chi_\parallel$); there is a strict analogy to the significant $\chi_{TIP}$ of the $^1A_{1g}$ ground state of mononuclear low-spin $d^6$ complexes.[18] Such $\chi_{TIP}$ terms are actually likely to be of significance for most $(\pi^*, \delta^*)^n$ ground states, because excited states of the appropriate symmetries to couple magnetically to the ground state typically lie at fairly low energies regardless of the ground-state configuration. However, $\chi_{TIP}$ terms have been largely neglected in previous work. An experimental determination of $\chi_{TIP}$ for a $(\delta^*)^2$ ground-state molecule via a temperature-dependent magnetic study would accordingly be valuable. The terms that show up in $\chi_{TIP}$ are the same that lead to diamagnetic anisotropy that can be determined by NMR methods,[106,107] and are therefore of considerable interest.

$\Delta$ has been theoretically estimated to be $+15{,}980\ \text{cm}^{-1}$ for $D_{2d}$ $Re_2Cl_4(PH_3)_4$,[108] and $+12{,}985\ \text{cm}^{-1}$ for $D_{4h}$ $Os_2Cl_8^{2-}$.[52] The $\delta^* \rightarrow \pi^*$ singlet-singlet transition, $^1A_{1g}(\delta^*)^2 \rightarrow {}^1E_u\ (\pi^*)^1(\delta^*)^1$, is estimated from our model to lie at $(\Delta - 2K + [K^2 + (\zeta/2)^2]^{1/2})$, including the second-order spin-orbit coupling for the $^1E_g$ state (Table 3) but neglecting off-diagonal effects on the $^1A_{1g}$ state (Table 2), which are negligible for such large values of $\Delta$; this quantity is nearly equal to $\Delta$ for our estimate of $K \approx 1{,}000\ \text{cm}^{-1}$ and the values of $\zeta$ characteristic of third transition series metal ions. The $^1(\delta^* \rightarrow \pi^*)$ transition has been assigned to a band of $Re_2Cl_4(PPr^n_3)_4$ at 647 nm (15,460 cm$^{-1}$), and a singlet-triplet $\delta^* \rightarrow \pi^*$ assignment has been suggested for a shoulder at 815 nm (12,270 cm$^{-1}$) that was resolved in a 5 K solid-state spectrum.[108] The singlet-triplet splitting of 3,190 cm$^{-1}$ may be compared to a $E_u(^1E_u)/E_u(^3E_u)$ splitting of 3,440 cm$^{-1}$ (independent of $\Delta$) calculated for $\zeta = 2{,}800\ \text{cm}^{-1}$ (Figure 12). There is a complication in that another (orbitally forbidden) transition, $\delta^* \rightarrow \sigma^*$, is expected from the calculation[108] to lie near the $\delta^* \rightarrow \pi^*$ transitions, and the singlet-singlet $\delta^* \rightarrow \sigma^*$ transition has been assigned to a feature of $Re_2Cl_4(PPr^n_3)_4$ at 725 nm in the low-temperature spectrum. The $Tc_2$(II,II) compound $Tc_2Cl_4(PEt_3)_4$ displays bands at 770 nm ($\varepsilon = 30\ \text{M}^{-1}\text{cm}^{-1}$) and 648 nm ($\varepsilon = 30\ \text{M}^{-1}\text{cm}^{-1}$)[109] that are probably analogous to the 725 and 647 nm transitions of the $Re_2$(II,II) compounds. Singlet-triplet transitions should be much weaker for the Tc compound, so the observed bands are presumably spin-allowed, which supports the assignments for the Re compounds.

The bands of $Os_2X_8^{2-}$ (X = Cl, Br, I)[52,104,105,110] at ~700 nm ($\sim -14{,}000\ \text{cm}^{-1}$) have been assigned to $\delta/\delta^* \rightarrow \pi^*$, where it is assumed that the $\delta$ and $\delta^*$ levels coalesce to a degenerate pair of levels in a staggered ($D_{4d}$) structure, the observed solid-state geometry for the $[N(Bu^n)_4]^+$ salts.[52,110] In support of this assignment, thin-film low-temperature spectra of the $[N(Bu^n)_4]^+$ salts revealed vibronic structure for this band with $\nu(Os_2)$ reduced about 30% from the ground-state value.[110] The frequency reduction, together with the long observed progressions (large Huang–Rhys factors), are consistent with large excited-state metal–metal elongations, as expected for the $\delta/\delta^* \rightarrow \pi^*$ excitations. It is quite interesting that $Os_2Cl_8^{2-}$ can be crystallized in an eclipsed conformation with the bis(triphenylphosphine)iminium cation,[104,105] and this (solid) salt is reported to be different in color (pink) from the green crystalline salts that contain the staggered conformation. Unfortunately, solid-state electronic absorption data have not been reported for a salt containing an eclipsed $Os_2X_8^{2-}$ anion.

The visible spectra of the $Re_2$(II,II) and $Os_2$(III,III) $(\delta^*)^2$ ground-state molecules are rich with additional weak bands (Table 7); specific assignments to various metal–metal transitions terminating in $\pi^*$, $\sigma^*(M_2)$, and $\sigma^*$(ML) levels have been suggested for $Re_2Cl_4(PPr_3^n)_4$.[108] An assignment of the weak band of $Os_2Cl_8^{2-}$ at 465 nm ($\varepsilon$ = 130 $M^{-1}cm^{-1}$) to $\pi$(Os–Cl, $Os_2$) $\rightarrow$ $\pi^*(Os_2)$ was suggested,[52] based solely upon the agreement of the observed energy with the calculated $\pi/\pi^*$ splitting. This assignment is very unlikely to be correct because such a transition is expected to be considerably more intense, especially if the calculated ligand character in the $\pi$ orbital is real (because of the resultant LMCT character of the transition); even for pure metal–metal $\pi$ and $\pi^*$ orbitals this transition should be intense because of the short metal–metal bond length, hence relatively large metal–metal $\pi$-overlap. The $\pi \rightarrow \pi^*$ transition probably lies at $\lambda < 350$ nm, where intense absorption that is (mostly) attributable to LMCT transitions is observed for $Os_2Cl_8^{2-}$, red-shifting for the bromide and iodide complexes.[110] Because the metal–metal and LMCT transitions are likely to mix,[33] detailed assignments of the intense bands are not readily made, particularly because no polarization information is available.

Several $Os_2$(III,III) compounds have been reported having bridging ligands derived from orthometallated triphenylphosphine, $Ph_2P(C_6H_4)^-$ (Table 7). They display Os–Os distances that are a little longer than the metal–metal distances of the compounds discussed above, but still quite short, <2.28 Å, and two of them have been shown to be essentially diamagnetic, so they very likely possess $(\delta^*)^2$ ground states.

$Os_2$(III,III) and $Ru_2$(III,III) complexes with porphyrin-type ligands have been reported (Table 7) and are, as expected, diamagnetic, consistent with $(\delta^*)^2$ ground states. The generalization that $\Delta$ is large and positive for porphyrin ligands does not fail for this electron count.

**4.5.3 $(\pi^*)^1(\delta^*)^1$** Whereas there was little evidence for the "intermediate $\Delta$" ground state ${}^3E_u(\pi^*)^3(\delta^*)^1$ of the $(\pi^*, \delta^*)^4$ configuration, there is considerable evidence for the ${}^3E_u(\pi^*)^1(\delta^*)^1$ ground state of $(\pi^*, \delta^*)^2$. The established compounds contain $Os_2$(III,III) bridged by four carboxylato ligands. They display Os–Os bond distances near 2.3 Å (Table 7). Theory predicts that this ground state should have a magnetic moment near 2 B.M. at room temperature (Figure 11), like that for the ${}^3A_{2g}(\pi^*)^2$ ground state with large $\zeta$, but decreasing to ~1 B.M. in the low-temperature limit rather than to zero; this results from partial cancellation of the spin and orbital angular momenta in the lowest-energy, $B_{1u}$, $B_{2u}$ spin-orbit levels of the ${}^3E_u$ state. This type of magnetic behavior was observed for the compound

$Os_2$(2-phenylbenzoate)$_4Cl_2$ by Cotton and coworkers[99] but was not recognized as such because correct theoretical equations for the $^3E_u$ state were not available. It was, however, correctly concluded that the observations did not fit a $(\pi^*)^2$ ground state (nor any other available model). Measurements on $Os_2(O_2CBu^t)_4Cl_2$ have been fit to the theoretical dependence for the $^3E_u$ state (Appendix C), yielding values of $D = 331(9)$ cm$^{-1}$ and $g_{iso} = 1.91(1)$.[111] ‖ The reported data for the 2-phenylbenzoate complex[99] also fit the theory reasonably well; we obtain $D = 600$ cm$^{-1}$ and $g_{iso} = 2.02$.[111]

$$D = W[E_u(^3E_u)] - W[B_{1u}, B_{2u}(^3E_u)] = \frac{\zeta}{2} + K + \sqrt{K^2 + \frac{\zeta^2}{4}} \qquad (22)$$

The interpretation of $D$ requires that both first- and second-order spin-orbit coupling effects be considered. Our model predicts that $D$ is given by Equation 22. The predicted $D$ is 680 cm$^{-1}$ for the parameters of Figure 12, considerably reduced from the first-order spin-orbit splitting of $\zeta/2 = 1{,}400$ cm$^{-1}$. The effective values of either $K$ or $\zeta$ (or both) are evidently slightly smaller for the $\mu$-carboxylato $Os_2$(III,III) complexes than the values used in Figure 12. We note that the $A_{1u}, A_{2u}(^3E_u)$ states are located to higher energy of the $B_{1u}, B_{2u}(^3E_u)$ levels by $\zeta$ (see Appendix C) and make little or no contribution to the predicted magnetic temperature dependence because they lie at such high energy.

Detailed electronic spectral studies, including low-temperature single-crystal polarized measurements on $Os_2(O_2CBu^t)_4Cl_2$, Figure 14, have been performed on the $\mu$-carboxylato $Os_2$(III,III) complexes $Os_2(O_2CR)_4Cl_2$, R = $CH_3$, $Pr^n$, and $Bu^t$.[111] The $\delta \rightarrow \delta^*$ transitions were identified by their characteristic $\nu(Os_2)$ vibronic structure and molecular $z$-polarization; the electronic origin is found in the range 11,579 (R = $Bu^t$) to 11,792 cm$^{-1}$ (R = $CH_3$). The observation of a $\delta \rightarrow \delta^*$ transition eliminates the $(\delta^*)^2$ ground state as a possibility. The transition energy is nearly as low as that observed for the $(\pi^*)^2(\delta^*)^1$ ground-state $Ru_2(O_2CR)_4Cl$ molecules (*vide supra*), and *lower* than that observed for the $(\delta^*)^1$ ground state $Tc_2(hp)_4Cl$ (12,194 cm$^{-1}$).[112,113] This is consistent with the assignment of the ground state as $(\pi^*)^1(\delta^*)^1$, since the observed $\delta \rightarrow \delta^*$ transition energies fall in the expected order based on metal–metal bond order and bond distance, and also because the $\delta \rightarrow \delta^*$ transition for a $(\pi^*)^2$ ground state is expected to be shifted to higher energy of the one-electron $\delta/\delta^*$ energy difference by a substantial two-electron term.[18]

A second NIR band is observed in the range 7,700–8,600 cm$^{-1}$ (Figure 14). It is also $z$-polarized, but distinct from the $\delta \rightarrow \delta^*$ transition in that it shows a

‖ For $Os_2(O_2CBu^t)_4Cl_2$ and $Os_2$(2-phenylbenzoate)$_4Cl_2$, magnetic data do not fit the theory for the $(\pi^*)^1(\delta^*)^1$ ground state below ~30 K and ~10 K, respectively. Below these temperatures, the magnetic moments are observed to decrease abruptly, possibly as a result of energetic $B_{1u}, B_{2u}(E_u)$ splittings involving rhombic distortion, which would result in quenching of the magnetic moment of this pseudo-doublet state. However, intermolecular antiferromagnetic coupling is an alternative explanation for the anomalous very-low-temperature magnetic behavior that cannot currently be experimentally distinguished.

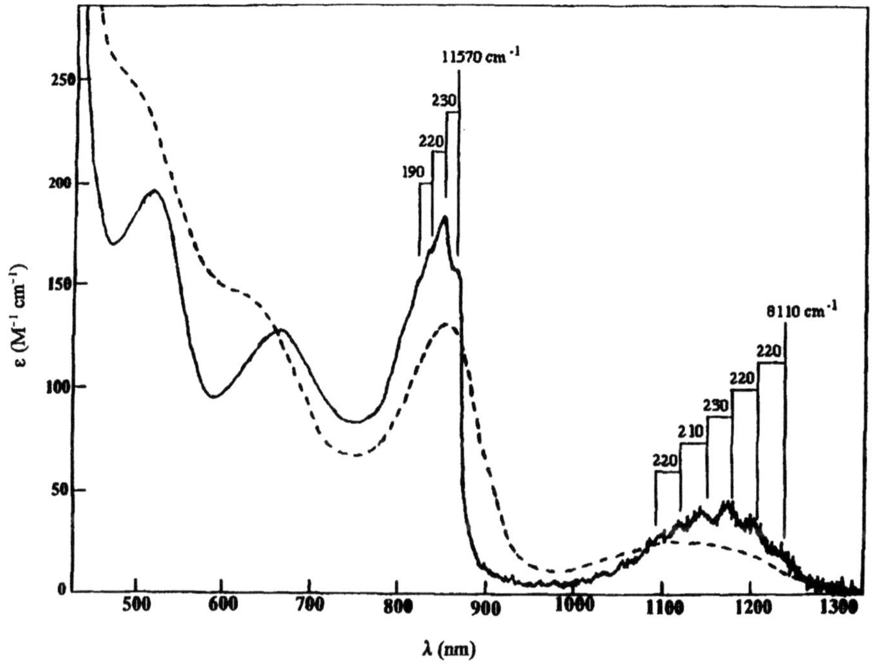


**Figure 14.** Absorption spectrum of a single crystal of $Os_2(O_2CBu^t)_4Cl_2$ for the polarization $\|$(Os–Os) at 295 K (- - - - - -) and 20 K (———). Adapted from Reference 111.

longer vibronic progression in $\nu(Os_2)$, indicating a larger excited-state distortion. Over the range of $\Delta$ values that allows the $(\pi^*)^1(\delta^*)^1$ ground state, Figure 12 predicts that the three highest energy electronic transitions within the $(\pi^*, \delta^*)^2$ manifold are $B_{1u}, B_{2u}(^3E_u) \rightarrow {}^1A^c_{1g}(\pi^* \rightarrow \delta^*)$, $\rightarrow {}^1B_{1g}, {}^1B_{2g}(\delta^* \rightarrow \pi^*)$, and $\rightarrow E_g(^3A_{2g})(\delta^* \rightarrow \pi^*)$, with energy ranges of, respectively, 7,500 to 11,200, 3,600 to 7,600, and 1,500 to 5,600 cm$^{-1}$. Note that the one-electron designations of these transitions are crude because of considerable mixing among the various states for small values of $\Delta$. The remaining excitations are all calculated to lie at energies too low to be experimentally observable.

The best fit is to the $B_{1u}, B_{2u}(^3E_u) \rightarrow {}^1B_{1g}, {}^1B_{2g}$ excitations, since these are dipole-allowed with molecular $z$-polarization. This assignment requires that $\Delta$ must be near the upper limit of the possible range for the $(\pi^*)^1(\delta^*)^1$ ground state, ~4,000–5,000 cm$^{-1}$. It is encouraging that this is close to the estimate of ~3,000 cm$^{-1}$ that we arrived at earlier for $\mu$-carboxylato $Ru_2$(II,III) complexes. The assignment is to several pseudo-degenerate spin-forbidden transitions, and the modest observed intensity reflects the large spin-orbit coupling parameter for Os(III). The intensity is, as expected, about an order of magnitude higher than that of an otherwise similar spin-forbidden transition of the $Ru_2$(II,III) complexes (Figure 6).

The transition to $^1A^c_{1g}$ is both spin- and dipole-forbidden, and is probably obscured by the other transitions. The spin- and dipole $(x, y)$-allowed transition to $E_g(^3A_{2g})$ likely lies at too low energy to be observed because of interfering vibrational absorption.

Several other bands are observed to higher energy of the $\delta \rightarrow \delta^*$ transition. An intense ($\varepsilon = 11{,}000\ M^{-1}cm^{-1}$) $z$-polarized feature at ~395 nm has been assigned to an axial-chloride LMCT transition, $\pi(Cl^-) \rightarrow \pi^*(Os_2)$, because it is strongly red-shifted for the bromide analogue (Table 7).[111] Several weaker bands in the 500–700 nm range (Figure 14) have not been assigned.

The room-temperature (low-resolution) electronic spectra in the NIR region reported for $(\pi^*)^2$ ground-state $Os_2$(II, III) complexes (Figure 13, for example) are rather similar to those observed for the $(\pi^*)^1(\delta^*)^1$ ground-state cases (except for somewhat higher intensities, which could conceivably be a function of metal–ligand mixing). The assignments we have given rely heavily upon the magnetic assignment of the ground state, and low-temperature electronic spectroscopic studies of one or more of the $(\pi^*)^2$ ground-state molecules should be done to test our model.

Several $Ru_2$(II,III) molecules with strongly $\pi$-donating bridging ligands exhibit reversible one-electron electrochemical oxidations at accessible (<1 V vs. SCE) potentials,[114–117] and spectroelectrochemical data have been reported for the $\mu$-acam$^-$ complex.[114] The $Ru_2$(III,III) complexes may be expected to have either $(\pi^*)^2$ or $(\pi^*)^1(\delta^*)^1$ ground states, but neither X-ray structural nor magnetic data are available.

**4.5.4** ***Other Ground States for Formally $(\pi^*, \delta^*)^2$ Molecules*** There is precedent for ground states that deviate from those of $(\pi^*, \delta^*)^n$ in that a $\sigma$(axial ligand) level may lie at sufficiently high energy to be the HOMO for $(\pi^*, \delta^*)^6$ molecules such as $Rh_2(O_2CR)_4(PR_3)_2$, so that one-electron oxidation yields radicals for which the ground state is $(\pi^*)^4(\delta^*)^2(a_{1g}, \sigma_{ax})^1$ rather than $(\pi^*, \delta^*)^5$.[67,118–120] Ground states of this unusual type have not been identified for other values of $n$ in $(\pi^*, \delta^*)^n$, but are plausible for highly reducing axial ligands such as phosphines or iodide in combination with a high-oxidation-state metal ion such as Ru(III).

Bear and co-workers[83,84] have characterized an unusual class of $Ru_2$(III,III) compounds of the type $Ru_2(\mu\text{-}L)_4X_2$ (L = dpf, $F_5$ap; X = CN, C≡CPh). These compounds are diamagnetic and feature exceptionally long $Ru_2$ distances (2.441–2.558 Å). The two possible ground states that have been discussed in the literature[83,84,103] are $\pi^4\delta^2\pi^{*4}$ and $\sigma^2\pi^4\pi^{*4}$, both of which deviate from a simple separation of metal–metal bonding and antibonding orbitals. However, a correct description of the bonding in these molecules may be beyond the limits of the simple metal–metal localized picture, as indicated by unusual structural distortions that include features such as $C_{ax}$–Ru–Ru angles of 160°; delocalization involving the axial ligands may be significant. Ren and co-workers[164] have specifically proposed, based on Fenske–Hall calculations, that electronic mixing of $\pi^*(Ru_2)$ and $\sigma^*$(Ru-C) levels occurs upon distortion. Because of this large $\sigma/\pi$ mixing, there is a very large splitting of the $\pi^*$ level so that the usual metal–metal MO diagram does not apply.

An enigmatic problem presently exists for the ground states of $Ru_2$(III,III) complexes bridged by sulfate or phosphate ligands.[69–71] These compounds have been found to display Ru–Ru distances indicative of metal–metal double bonds, and room-temperature magnetic moments suggestive of spin-*quintet* ground states (Table 7). The ground states have been assigned to $(\delta)^1(\pi^*)^2(\delta^*)^1$,[69–71] which is surprising because simple theoretical considerations indicate that a $(\delta)^1(\delta^*)^1$ configuration cannot be a ground state so long as the $\delta/\delta^*$ two-electron exchange integral is nonzero.[18] The presence of unpaired spins in the $\pi^*$ level does not change the theoretical prediction; a ${}^3A_{2g}(\delta^*)^2(\pi^*)^2$ or $(\delta)^2(\pi^*)^2$ state must lie to lower energy of the ${}^5A_{1u}(\delta)^1(\pi^*)^2(\delta^*)^1$ state, the $(\delta^*)^2(\pi^*)^2$ state being the ground state if the one-electron $\delta/\delta^*$ splitting $\Delta W$ is negative.

There is no obvious way to explain the magnetic observations for these latter complexes within the context of $(\pi^*, \delta^*)^2$ states; the logical ground state consistent with the X-ray structural data is $(\pi^*)^2$, but this configuration cannot produce magnetic susceptibilities as high as are reported.[69–71] The visible absorption band of the complexes at ~500 nm (Table 7) is likely due to metal–metal $\pi \rightarrow \pi^*$, regardless of the ground-state formulation. The apparent absence of low-energy absorption bands in the reported spectra argues against a $(\pi^*)^1(\delta^*)^1$ ground state, which must feature low-energy metal–metal transitions. Preparation of aryl phosphate derivatives, which might be soluble in nonaqueous solvents [as proved fruitful for $Mo_2$(III,III) complexes],[45] could possibly shed some light on this puzzling matter.

### 4.6 $(\pi^*, \delta^*)^1$ and $(\pi^*, \delta^*)^5$ Configurations

For both of these configurations, the ground states are either ${}^2E_g$ ($\pi^*$ electron or hole) or ${}^2B_{1u}$ ($\delta^*$ electron or hole), and the energy difference between the two states is just $\Delta$.

The only known ground state for $n = 1$ is $(\delta^*)^1$, and $\Delta$ is generally in excess of 10,000 $cm^{-1}$ according to calculations,[2] regardless of ligand type. Many examples of this ground state are known; the results are well summarized in Reference 2, and some data are set out in Table 1. For the $(\delta^*)^1$ ground-state compounds $Tc_2Cl_8^{3-}$ and $Tc(hp)_4Cl$, for example, $\delta^* \rightarrow \pi^*$ transitions have been assigned at, respectively, 15,700 ($\varepsilon = 172$ $M^{-1}cm^{-1}$)[112] and 17,000 $cm^{-1}$.[113] These assignments are plausible by comparison to the results for $(\delta^*)^2$ ground-state molecules. The most characteristic electronic transition of this ground state is, however, the low-energy (NIR) and vibronically structured [in $\nu(M_2)$] ${}^2(\delta \rightarrow \delta^*)$ transition that was discussed in Sections 2 and 3.[2,112,113]

The $(\pi^*, \delta^*)^5$ configuration is more complicated. Just as for other $(\pi^*, \delta^*)^n$ configurations, strongly $\pi$-donating bridging ligands make $\Delta$ large and negative, enforcing a $(\pi^*)^4(\delta^*)^1$ ground state. However, $\mu$-carboxylato and closely related ligands make $\Delta$ quite small, ~$10^3$ $cm^{-1}$, and it can be either positive or negative.[42,67] The ${}^2E_g(\pi^*)^3(\delta^*)^2$ state has properties similar to those previously discussed for the ${}^2E_g(\pi^*)^3$ state of the $(\pi^*, \delta^*)^3$ configuration.

## 5 CONCLUDING REMARKS

Much is known about multiple metal–metal bonds. Yet, there is much left to do. Although we can make rather detailed predictions about the manifolds of states deriving from each $(\pi^*, \delta^*)^n$ configuration, and give good predictions of ground-state ZFS parameters, available NIR electronic absorption data that could further test the predictions are lacking in many cases. Our crystal-field model could be improved, as suggested by Appendix A, but there is little point in doing so in the absence of experimental results that would justify a higher level theory. One area that clearly needs attention is the experimental identification of weak NIR absorptions.

## ACKNOWLEDGMENT

Our work on the electronic structures of complexes with metal–metal bonds has been supported by the National Science Foundation.

## APPENDIX A. SEMIEMPIRICAL ANALYSIS WITH INDEPENDENT TWO-ELECTRON INTEGRALS

Consider the $(\pi^*)^2(\delta^*)^1$ states. If we write out wavefunctions for these states and evaluate energies, we find that all one-electron and coulomb integrals contribute identically to each state. [It is important to note that, by rotational symmetry, $J(\pi_x^*, \pi_x^*) = J(\pi_x^*, \pi_y^*) + K(\pi_x^*, \pi_y^*)$]. Subtracting the energy of the $^4B_{2u}$ state from each of these we get:

$$\begin{aligned}
W(^4B_{2u}) &= 0 \\
W(^2A_{2u}) &= K(\pi_x^*, \delta^*) + 2K(\pi_x^*, \pi_y^*) \\
W(^2B_{2u}) &= 3K(\pi_x^*, \delta^*) \\
W(^2A_{1u}) &= 3K(\pi_x^*, \delta^*) \\
W(^2B_{1u}) &= 3K(\pi_x^*, \delta^*) + 2K(\pi_x^*, \pi_y^*)
\end{aligned} \tag{A-1}$$

If $K(\pi_x^*, \delta^*) = K(\pi_x^*, \pi_y^*) = K$, as is true by symmetry for the $O_h$ integrals $K(xz, xy) = K(xz, yz)$, then we get the results of Table 2. We can alternatively let $K = K(xz, xy) = K(xz, yz)$ and use different "orbital reduction factors" for the two exchange integrals as in

$$\begin{aligned}
K(\pi_x^*, \delta^*) &= kK \\
K(\pi_x^*, \pi_y^*) &= k'K
\end{aligned} \tag{A-2}$$

State energies in units of $kK$ for this case as a function of the ratio $k'/k$ are shown in Figure 15. One of the accidental state degeneracies is removed for $(k'/k) \neq 1$. It

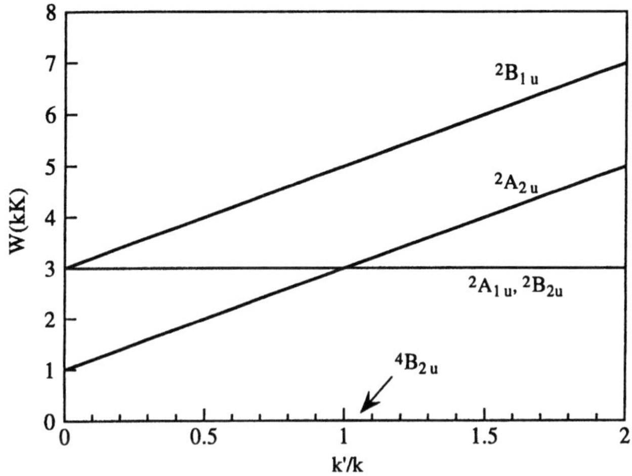


**Figure 15.** Energies of $(\pi^*)^2(\delta^*)^1$ states as a function of the ratio of orbital reduction factors $k'/k$. The energy of the $^4B_{2u}$ state is set equal to zero, and spin-orbit coupling is neglected. See Appendix A for details.

can be seen, however, that a very large difference in the values of the two integrals is required to cause significant deviation from the $k = k'$ pattern of states.

Consideration of states involving other orbital occupancies would introduce additional integrals [e.g., $J(\delta^*, \delta^*)$ for $^2E_g(\pi^*)^1(\delta^*)^2$], hence additional reduction factors, but the problem is still not completely hopeless if sufficient experimental information on state energies is available.

## APPENDIX B. APPLICATION OF THE CRYSTAL–FIELD MODEL TO OTHER METAL–METAL CONFIGURATIONS

The integral $K(xz, x^2 - y^2)$ is equal to $K(xz, xy)$ in the mononuclear crystal-field theory. Therefore the energies of $(\pi^*)^n(\delta')^1$ configurations ($\delta'$ is a $d_{x^2-y^2}$ derived molecular orbital) can be treated the same as $(\pi^*)^n(d)^1$ configurations, but with a new one-electron parameter, $\Delta W(\delta', \pi^*)$, which is additionally different for the gerade and ungerade $\delta'(D_{4h})$ levels. Since ligand covalency is much larger for the $x^2 - y^2$-derived levels than for the $xy$-derived levels, we would actually expect the effective $K(xz, x^2 - y^2)$ to be reduced from the value for $K(xz, xy)$, which results in a situation similar to the $z^2$ case discussed below. Occupancy of more than one of the $\delta'$ levels introduces charge-transfer integrals[18] that we are not considering here.

The configurations $(\pi^*)^n(\sigma^*)^1$ (where $\sigma^*$ is derived from $z^2$) introduce an exchange integral, $K(\pi^*, \sigma^*)$, proportional to the mononuclear integral $K(xz, z^2) = \boldsymbol{B} + \boldsymbol{C}$;[49] this is different in magnitude from $3\boldsymbol{B} + \boldsymbol{C}$ for the exchange integrals of $(t_{2g})^n$, about 70 % as large for mononuclear systems.[53] The states of the $(\pi^*)^2(\sigma^*)^1$ configuration are illustrated in Figure 16, in a way that emphasizes the coupling of

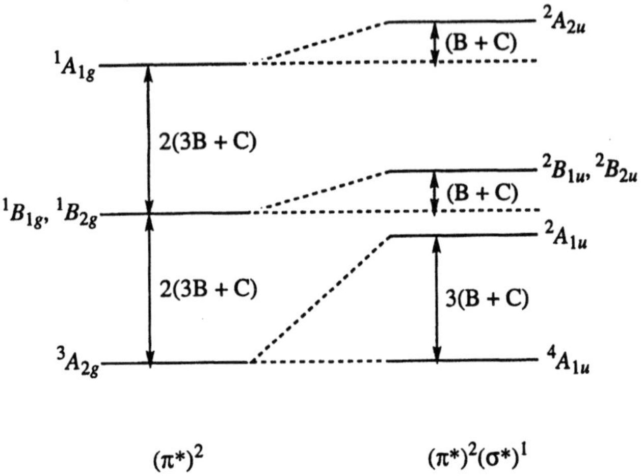


**Figure 16.** Energies of $(\pi^*)^2(\sigma^*)^1$ states showing their derivation from the energies of $(\pi^*)^2$ states by addition of the $\sigma^*$ coupling. Spin-orbit coupling is neglected. See Appendix B for details.

the $\sigma^*$ electron to the triplet $(\pi^*)^2$ configuration. The similarity to the $(\pi^*)^2(\delta^*)^1$ state pattern (Figure 15) is clear.

## APPENDIX C. MAGNETIC PROPERTIES

The EPR "effective" $g$ values ($h\nu/\beta\mathbf{H}$) for the various states of interest in $D_{4h}$ symmetry are summarized in Table 8. The intrinsic gyromagnetic ratio has been taken to be the free-electron value, $g_e \approx 2$. The factor $\zeta$ is the spin-orbit coupling constant for a given metal ion. The orbital reduction factor $k$ arises from the result of the angular momentum operator $\boldsymbol{l}_z$, and is equal to unity for orbitals comprised of pure $d$ orbitals.[48] The diamagnetic $^1A_{1g}$ states of the $(\delta^*)^2$ and $(\pi^*)^4$ configurations have, of course, $g_\parallel = g_\perp = 0$. The $E_u(^1E_u)$ state arising from the $(\pi^*)^1(\delta^*)^1$ and $(\pi^*)^3(\delta^*)^1$ configurations has the same $g$ values as the $E_u(^3E_u)$ state of the same configuration.

The top sign in the rightmost column of Table 8 applies to $(\pi^*)^3$, $(\pi^*)^3(\delta^*)^2$, and $(\pi^*)^3(\delta^*)^1$ states, and the bottom sign to $(\pi^*)^1$, $(\pi^*)^1(\delta^*)^2$, and $(\pi^*)^1(\delta^*)^1$ states. The zero-field splittings derived from these energies do not include second-order spin-orbit and rhombic contributions. The latter may be significant for orbitally degenerate states ($^2E_g$ or $^3E_u$).[80] Among the various states, $E_u(^3E_u)$ can never be an electronic ground state, and $E_g(^3A_{2g})$ is never expected to be a ground state, although it could be in principle if there were some other low-energy $E_g$-symmetry state to which it was strongly coupled. Second-order spin-orbit coupling effects will generally cause the gyromagnetic ratio to deviate from $g_e$, to different degrees for $g_\parallel$ and $g_\perp$, with proportionate effects upon the EPR $g$ values. The pseudo-doublets of the $^3E_u$ state, $(B_{1u}, B_{2u})$ and $(A_{1u}, A_{2u})$, can be ground states, but since second-

**Table 8 EPR Effective $g$ Values for $(\pi^*, \delta^*)^n$ States**

| State (Orbital Configuration) | Spin-Orbit State | Effective $g_{\parallel}$ | Effective $g_{\perp}$ | First-Order Spin-Orbit Energy |
|---|---|---|---|---|
| $^2B_{1u}[(\delta^*)^1; (\pi^*)^4(\delta^*)^1]$ | $E_{3/2u}$ | 2 | 2 | 0 |
| $^4B_{2u}[(\pi^*)^2(\delta^*)^1]$ | $E_{1/2u}$ | 2 | 4 | 0 |
| | $E_{3/2u}$ | 6 | 0 | 0 |
| $^2E_g[(\pi^*)^3; (\pi^*)^1(\delta^*)^2; (\pi^*)^3(\delta^*)^2; (\pi^*)^1]$ | $E_{1/2g}$ | $2(1-k)$ | 0 | $\pm k\zeta/2$ |
| | $E_{3/2g}$ | $2(1+k)$ | 0 | $\mp k\zeta/2$ |
| $^3A_{2g}[(\pi^*)^2; (\pi^*)^2(\delta^*)^2]$ | $A_{1g}$ | 0 | 0 | 0 |
| | $E_g$ | 4 | 0 | 0 |
| $^3E_u[(\pi^*)^1(\delta^*)^1; (\pi^*)^3(\delta^*)^1]$ | $B_{1u}, B_{2u}$ | $2(2-k)$ | 0 | $\pm k\zeta/2$ |
| | $E_u$ | $2k$ | 0 | 0 |
| | $A_{1u}, A_{2u}$ | $2(2+k)$ | 0 | $\mp k\zeta/2$ |

order spin-orbit and/or low-symmetry effects can split them, they are very unlikely to show EPR $g$ values similar to the idealized values that are given. (The magnitude of the terms splitting the pseudo-doublets need only be comparable to the magnitude of $\beta\mathbf{H}$ to seriously perturb the EPR $g$ values.)

Equations for the temperature dependence of the magnetic susceptibility of these states are given below. In each equation, the gyromagnetic ratio has been taken to be 2, the parameter $x$ is $D/k_BT$, where $k_B$ is the Boltzmann constant, and $D$ is the zero-field splitting, consistently defined as the zero-field energy difference between spin-orbit states. For the $^4B_{2u}$ state, a key reference[56] defines $D$ as one-half of our value, and, in general, care should be taken in comparisons to the literature. The effective magnetic moment $\mu_{\text{eff}}$ is related to $\chi_M$ by $(\mu_{\text{eff}})^2 = \chi_M(3k_BT/N\beta^2)$, where $N$ is Avogadro's number and $\beta$ is the Bohr magneton.

$$
\begin{aligned}
^2B_{1u}: \chi_M &= N\beta^2/k_BT \\
^4B_{2u}: \chi_M &= \left(N\beta^2/3k_BT\right)\left[\left(9+9\exp(-x)\right.\right. \\
&\quad \left.\left.+ (12/x)\left(1-\exp(-x)\right)\right)/\left(1+\exp(-x)\right)\right] \\
^2E_g: \chi_M &= \left(N\beta^2/3k_BT\right)\left[4\left(1+(1/x)\left(1-\exp(-x)\right)\right)/\left(1+\exp(-x)\right)\right] \qquad \text{(C-1)} \\
^3A_{2g}: \chi_M &= \left(N\beta^2/3k_BT\right)\left[8\left(\exp(-x)\right.\right. \\
&\quad \left.\left.+ (2/x)\left(1-\exp(-x)\right)\right)/\left(1+2\exp(-x)\right)\right] \\
^3E_u: \chi_M &= \left(N\beta^2/3k_BT\right)\left[\left(1+\exp(-x)+9\exp(-2x)\right.\right. \\
&\quad \left.\left.+ (8/x)\left(1-\exp(-2x)\right)\right)/\left(1+\exp(-x)+\exp(-2x)\right)\right]
\end{aligned}
$$

The equation for the $^3E_u$ state neglects second-order coupling between the $E_u(^3E_u)$ and $E_u(^1E_u)$ states of the $(\pi^*)^1(\delta^*)^1$ or $(\pi^*)^3(\delta^*)^1$ configurations that generate this state. As discussed in the text, we can crudely correct for this by replacing $-2x$ by $-z$ in each of the three places in the $\chi_M$ equation in which $-2x$ appears, but retaining $x$ in the three places it appears alone; $z$ and $x$ are then separate parameters. The new parameter $z$ is expected to be $\approx k\zeta/k_BT$ and $x$ may be significantly less than the first-order value of $(1/2)k\zeta/k_BT$.

The sign of $D$ is expected to be positive according to theory, as the equations are given, for the $^3A_{2g}(\pi^*)^2$ and $(\pi^*)^2(\delta^*)^2$ states, for the $^2E_g(\pi^*)^3$ and $(\pi^*)^3(\delta^*)^2$ states, and for the $^3E_u(\pi^*)^1(\delta^*)^1$ state; it should be negative for the $^2E_g(\pi^*)^1(\delta^*)^2$ and $(\pi^*)^1$ states, and for the $^3E_u(\pi^*)^3(\delta^*)^1$ state. The equation for the $^4B_{2u}(\pi^*)^2(\delta^*)^1$ state is independent of the sign of $D$; a small dependence of the susceptibility upon the sign of $D$ results if $g$ is anisotropic.[56,121] The susceptibility of the $^2B_{1u}(\delta^*)^1$ or $(\pi^*)^4(\delta^*)^1$ states is simple Curie Law, $\chi_M = C/T$, as there is no zero-field splitting of a Kramers doublet.

Rhombic distortions ($x \neq y$) will affect orbitally degenerate states ($^2E_g$ and $^3E_u$) by quenching the orbital angular momentum, causing EPR $g$ values to approach $g_e$ and tending $\chi_M$ towards the spin-only value. Defining $\Delta_\pi = W(\pi_x^*) - W(\pi_y^*)$, we find, for the example of the $^2E_g$ state, a zero-field splitting $D$ of $(\Delta_\pi^2 + k^2\zeta^2)^{1/2}$ and EPR $g$ values for the $E_{3/2g}$ spin-orbit state of $g_\parallel = 2[1 + (k\zeta/D)]$ and $g_\perp = 2|\Delta_\pi/D|$. Magnetic susceptibility expressions for the rhombic-distorted $^2E_g$ state are presented elsewhere.[67]

Since the magnetic susceptibilities of all of the states except $^2B_{1u}$ have intrinsic temperature dependences that deviate from the Curie Law, fits of data to a Curie–Weiss equation, $\chi_M = C/(T+\theta)$, are meaningless,[122] and we have ignored several examples of such fits in the literature.

The effect on the magnetic susceptibility of deviations of the gyromagnetic ratio from 2 can be accounted for in an isotropic approximation simply by multiplying each $\chi_M$ equation by $(g_{\text{iso}})^2/4$. Some anisotropic magnetic susceptibility equations are given in the literature.[70] The $\chi_\perp$ $[\chi = (\chi_\parallel + 2\chi_\perp)/3]$ is comprised exclusively of the term in our equations that has an $(n/x)$ multiplying factor (a "high-frequency" term in the terminology of Van Vleck),[85] except for the $^2B_{1u}$ state [for which $\chi_\parallel$ and $\chi_\perp$ are equal except for possibly different $(g)^2/4$ multiplying factors], and the $^4B_{2u}$ state.[56,121] Thus, for example, anisotropic equations for the $^2E_g$ state are:

$$\begin{aligned}(\chi_\parallel)_M &= \left(N\beta^2/k_BT\right)(g_\parallel)^2\left[1/\left(1+\exp(-x)\right)\right]\\ (\chi_\perp)_M &= \left(N\beta^2/k_BT\right)(g_\perp)^2(1/2x)\left[\left(1-\exp(-x)\right)/\left(1+\exp(-x)\right)\right]\end{aligned} \tag{C-2}$$

We emphasize that the values of $g_\parallel$ and $g_\perp$ to be used in the above and literature susceptibility equations are generally expected to be close to $g_e$, and are *not* the EPR $g$ values, although they are closely related to them. For example, for the $E_{1/2u}(^4B_{2u})$ state, $g_\parallel(\text{EPR}) \approx g_\parallel$, and $g_\perp(\text{EPR}) \approx 2g_\perp$. For the $A_{1g}(^3A_{2g})$ state, EPR $g$ values are, of course, zero, but the gyromagnetic ratios to be used in magnetic susceptibility calculations are global for the triplet state and are expected to be, as usual, close to $g_e$

for $A_{1g}(^3A_{2g})$. For $Ru_2(OEP)_2$, analysis of the anisotropy of the magnetic susceptibility of the $^3A_{2g}$ state has yielded such $g$ values.** There is some inconsistency in the literature regarding reported values of $g_\perp$(EPR) for the $^4B_{2u}$ state; sometimes the reported value is the observed value divided by two, which has justification in the theory for this particular state[56] but, we feel, may lead to confusion. EPR values should be reported as the actually observed numbers, not as theoretically related values.

## APPENDIX D. ABBREVIATIONS

The following uncommon ligand abbreviations are used in this chapter. Different ligand abbreviations are used in some of the cited references.

| | |
|---|---|
| $dpf^-$ | $PhNC(H)NPh^-$ |
| $dtf^-$ | $(4\text{-}MePh)NC(H)N(4\text{-}MePh)^-$ |
| $dptz^-$ | $PhN_3Ph^-$ |
| $dttz^-$ | $(4\text{-}MePh)N_3(4\text{-}MePh)^-$ |
| $acam^-$ | $HNC(O)CH_3^-$ |
| $bzam^-$ | $HNC(O)Ph^-$ |
| $hp^-$ | 2-oxypyridinate$^-$ |
| $hpp^-$ | 1,3,4,6,7,8-hexahydro-2H-pyrimido[1,2-$\alpha$]pyrimidinate$^-$ |
| $ap^-$ | 2-(PhN)-py$^-$ |
| $F_5ap^-$ | 2,3,4,5,6-pentafluoro-2-anilinopyridinate$^-$ |
| dmpm | bis(dimethylphosphino)methane |
| tmtaa | dianion of 7,16-dihydro-6,8,15,17-tetramethyldibenzo[*b*,*i*][1,4,8,11] tetraazacyclotetradecine; sometimes named tetramethyldibenzotetraaza[14] annulene in the literature |
| $OEP^{2-}$ | 1,2,3,4,5,6,7,8-octaethylporphyrin$^{2-}$ |
| $Pc^{2-}$ | phthalocyanine$^{2-}$ |
| pyz | pyrazine |
| phz | phenazine |

## REFERENCES

1. Cotton, F. A.; Curtis, N. F.; Harris, C. B.; Johnson, B. F. G.; Lippard, S. J.; Mague, J. T.; Robinson, W. R.; Wood, J. S. *Science* **1964**, *145*, 1305.
2. Cotton, F. A.; Walton, R. A. *Multiple Bonds between Metal Atoms*; 2nd ed.; Oxford Press: Oxford, 1993.

** The ZFS of the $^3A_{2g}$ state was neglected in the treatment of the magnetic anisotropy of $Ru_2(OEP)_2$ in Reference 123. Assuming $D \approx 250\ cm^{-1}$ and a temperature of 295 K, we calculate (Reference 121 and Appendix C) $g_\parallel = 2.31$ and $g_\perp = 1.99$ from the reported anisotropic susceptibilities, versus $g_\parallel = 1.71$ and $g_\perp = 2.10$ calculated neglecting $D$.[123] Since the experimental susceptibilities were actually derived from measurements over a range of temperatures, a more detailed treatment of this sort of data is required before the $g$ values can be considered to be reliable.

3. Norman, J. G., Jr.; Kolari, H. J. *J. Chem. Soc., Chem. Commun.* **1974**, 303.
4. Norman, J. G., Jr.; Kolari, H. J. *J. Chem. Soc., Chem. Commun.* **1975**, 649.
5. Norman, J. G., Jr.; Kolari, H. J. *J. Am. Chem. Soc.* **1975**, *97*, 33.
6. Mortola, A. P.; Moskowitz, J. W.; Rösch, N. *Int. J. Quant. Chem.* **1974**, *8*, 161.
7. Mortola, A. P.; Moskowitz, J. W.; Rösch, N.; Cowman, C. D.; Gray, H. B. *Chem. Phys. Lett.* **1975**, *32*, 283.
8. Cowman, C. D.; Gray, H. B. *J. Am. Chem. Soc.* **1973**, *95*, 8177.
9. Clark, R. J. H.; Franks, M. L. *J. Chem. Soc., Chem. Commun.* **1974**, 316.
10. Clark, R. J. H.; Franks, M. L. *J. Am. Chem. Soc.* **1975**, *97*, 2691.
11. Coulson, C. A.; Fischer, I. *Phil. Mag.* **1949**, *40*, 386.
12. Parr, R. G. *The Quantum Theory of Molecular Electronic Structure*; W. A. Benjamin: New York, 1964.
13. Ohno, K. *Adv. Quantum Chem.* **1967**, *3*, 239.
14. Hansen, A. E.; Ballhausen, C. J. *Trans. Faraday Soc.* **1965**, *61*, 631.
15. Bino, A.; Cotton, F. A. *Inorg. Chem.* **1979**, *18*, 3562.
16. Cotton, F. A.; Dunbar, K. R.; Falvello, L. R.; Tomas, M.; Walton, R. A. *J. Am. Chem. Soc.* **1983**, *105*, 4950.
17. John, K. D.; Miskowski, V. M.; Hopkins, M. D., unpublished results.
18. Hopkins, M. D.; Gray, H. B.; Miskowski, V. M. *Polyhedron* **1987**, *6*, 705.
19. Hay, P. J. *J. Am. Chem. Soc.* **1982**, *104*, 7007.
20. Zietlow, T. C.; Hopkins, M. D.; Gray, H. B. *J. Solid State Chem.* **1985**, *57*, 112.
21. Miskowski, V. M.; Goldbeck, R. A.; Kliger, D. S.; Gray, H. B. *Inorg. Chem.* **1979**, *18*, 86.
22. Hopkins, M. D.; Zietlow, T. C.; Miskowski, V. M.; Gray, H. B. *J. Am. Chem. Soc.* **1985**, *107*, 510.
23. Hopkins, M. D.; Schaefer, W. P.; Bronikowski, M. J.; Woodruff, W. H.; Miskowski, V. M.; Dallinger, R. F.; Gray, H. B. *J. Am. Chem. Soc.* **1987**, *109*, 408.
24. Cotton, F. A.; Hubbard, J. L.; Lichtenberger, D. L.; Shim, I. *J. Am. Chem. Soc.* **1982**, *104*, 679.
25. Ziegler, T. *J. Am. Chem. Soc.* **1984**, *106*, 5901.
26. Engebretson, D. S.; Zaleski, J. M.; Leroi, G. E.; Nocera, D. G. *Science* **1994**, *265*, 759.
27. Nocera, D. G. *Acc. Chem. Res.* **1995**, *28*, 209.
28. Hopkins, M. D.; Miskowski, V. M.; Gray, H. B. unpublished results.
29. Manning, M. C.; Trogler, W. C. *J. Am. Chem. Soc.* **1983**, *105*, 5311.
30. Martin, D. S.; Newman, R. A.; Fanwick, P. E. *Inorg. Chem.* **1979**, *18*, 2511.
31. Norman, J. G., Jr.; Kolari, H. J.; Gray, H. B.; Trogler, W. C. *Inorg. Chem.* **1977**, *16*, 987.
32. Jørgensen, C. K. *Modern Aspects of Ligand Field Theory*; North-Holland: Amsterdam, 1971, Ch. 23.
33. Miskowski, V. M.; Gray, H. B.; Hopkins, M. D. *Inorg. Chem.* **1992**, *31*, 2085.
34. Hopkins, M. D.; Miskowski, V. M.; Gray, H. B. *J. Am. Chem. Soc.* **1988**, *110*, 1787.
35. Trogler, W. C.; Cowman, C. D.; Gray, H. B.; Cotton, F. A. *J. Am. Chem. Soc.* **1977**, *99*, 2993.

36. Bursten, B. E.; Cotton, F. A.; Fanwick, P. E.; Stanley, G. G. *J. Am. Chem. Soc.* **1983**, *105*, 3082.
37. Erwin, D. K.; Geoffroy, G. L.; Gray, H. B.; Hammond, G. S.; Solomon, E. I.; Trogler, W. C.; Zagars, A. A. *J. Am. Chem. Soc.* **1977**, *99*, 3620.
38. Hopkins, M. D.; Miskowski, V. M.; Gray, H. B. *J. Am. Chem. Soc.* **1986**, *108*, 959.
39. Chang, I.-J.; Nocera, D. G. *J. Am. Chem. Soc.* **1987**, *109*, 4901.
40. Stephenson, T. A.; Wilkinson, G. *J. Inorg. Nucl. Chem.* **1966**, *28*, 2285.
41. Norman, J. G., Jr.; Kolari, H. J. *J. Am. Chem. Soc.* **1978**, *100*, 791.
42. Norman, J. G., Jr.; Renzoni, G. E.; Case, D. A. *J. Am. Chem. Soc.* **1979**, *101*, 5256.
43. Rizzi, G. A.; Casarin, M.; Tondello, E.; Piraino, P.; Granozzi, G. *Inorg. Chem.* **1987**, *26*, 3406.
44. Cotton, F. A.; Feng, X. *Inorg. Chem.* **1989**, *28*, 1180.
45. Chang, I.-J.; Nocera, D. G. *Inorg. Chem.* **1989**, *28*, 4309.
46. Löwdin, P.-O. *Svensk Kem. Tidskr.* **1955**, *67*, 365.
47. DeKock, R. L.; Gray, H. B. *Chemical Structure and Bonding*; Benjamin/Cummings: Menlo Park, California, 1980, Ch. 4.
48. Ballhausen, C. J. *Molecular Electronic Structures of Transition Metal Complexes*; McGraw-Hill: New York, 1979.
49. Pople, J. A.; Beveridge, D. L. *Approximate Molecular Orbital Theory*; McGraw-Hill: New York, 1970, Ch. 3.
50. Griffith, J. S. *The Theory of Transition-Metal Ions*; Cambridge University Press: Cambridge, 1961.
51. Stone, A. J. *Proc. Roy. Soc. London, Ser. A* **1964**, *271*, 424.
52. Agaskar, P. A.; Cotton, F. A.; Dunbar, K. R.; Falvello, L. R.; Tetrick, S. M.; Walton, R. A. *J. Am. Chem. Soc.* **1986**, *108*, 4850.
53. Lever, A. B. P. *Inorganic Electronic Spectroscopy*; 2nd ed.; Elsevier: Amsterdam, 1984.
54. Stiefel, E. I. *Progr. Inorg. Chem.* **1977**, *22*, 1.
55. Allen, G. C.; Warren, K. D. *Structure and Bonding* **1974**, *19*, 105.
56. Telser, J.; Drago, R. S. *Inorg. Chem.* **1984**, *23*, 3114.
57. Chakravarty, A. R.; Cotton, F. A.; Tocher, D. A. *Inorg. Chem.* **1985**, *24*, 172.
58. Cotton, F. A.; Kim, Y.; Ren, T. *Inorg. Chem.* **1992**, *31*, 2608.
59. Cotton, F. A.; Kim, Y.; Ren, T. *Inorg. Chem.* **1992**, *31*, 2723.
60. Cotton, F. A.; Kim, Y.; Ren, T. *Polyhedron* **1993**, *12*, 607.
61. Miskowski, V. M.; Loehr, T. M.; Gray, H. B. *Inorg. Chem.* **1987**, *26*, 1098.
62. Miskowski, V. M.; Gray, H. B. *Inorg. Chem.* **1988**, *27*, 2501.
63. Miskowski, V. M.; Loehr, T. M.; Gray, H. B. *Inorg. Chem.* **1988**, *27*, 4708.
64. Clark, D. L.; Green, J. C.; Redfern, C. M.; Quelch, G. E.; Hiller, I. H.; Guest, M. F. *Chem. Phys. Lett.* **1989**, *154*, 326.
65. Clark, R. J. H.; Franks, M. L. *J. Chem. Soc., Dalton Trans.* **1976**, 1825.
66. Clark, R. J. H.; Ferris, L. T. H. *Inorg. Chem.* **1981**, *20*, 2759.
67. Miskowski, V. M.; Gray, H. B., unpublished results.
68. Yao, C.-L.; Park, K. H.; Khokhar, A. R.; Jun, M.-J.; Bear, J. L. *Inorg. Chem.* **1990**, *29*, 4033.

69. Zhilyaev, A. N.; Fomina, T. A.; Kuz'menko, I. V.; Rotov, A. V.; Baranovskii, I. B. *Russ. J. Inorg. Chem.* **1989**, *34*, 532.
70. Cotton, F. A.; Datta, T. L.; Labella, L.; Shang, M. *Inorg. Chim. Acta* **1993**, *203*, 55.
71. Kuz'menko, I. V.; Zhilyaev, A. N.; Fomina, T. A.; Porai-Koshits, M. A.; Baranovskii, I. B. *Russ. J. Inorg. Chem.* **1989**, *34*, 1457.
72. Tetrick, S. M.; Coombe, V. T.; Heath, G. A.; Stephenson, T. A.; Walton, R. A. *Inorg. Chem.* **1984**, *23*, 4567.
73. Cotton, F. A.; Dunbar, K. R.; Matusz, M. *Inorg. Chem.* **1986**, *25*, 1585.
74. Cotton, F. A.; Matusz, M. *Polyhedron* **1987**, *6*, 1439.
75. Warren, L. F.; Goedken, V. L. *J. Chem. Soc., Chem. Commun.* **1978**, 909.
76. Tait, C. D.; Garner, J. M.; Collman, J. P.; Sattelberger, A. P.; Woodruff, W. H. *J. Am. Chem. Soc.* **1989**, *111*, 7806.
77. Tait, C. D.; Garner, J. M.; Collman, J. P.; Sattelberger, A. P.; Woodruff, W. H. *J. Am. Chem. Soc.* **1989**, *111*, 9072.
78. Lindsay, A. J.; Wilkinson, G.; Motevalli, M.; Hursthouse, M. B. *J. Chem. Soc., Dalton Trans.* **1987**, 2723.
79. Cotton, F. A.; Falvello, L. R.; Ren, T.; Vidyasagar, K. *Inorg. Chim. Acta* **1992**, *194*, 163.
80. Kawamura, T.; Katayama, H.; Nishikawa, H.; Yamabe, T. *J. Am. Chem. Soc.* **1989**, *111*, 8156.
81. Bear, J. L.; Yao, C.-L.; Lifsey, R. S.; Korp, J. D.; Kadish, K. M. *Inorg. Chem.* **1991**, *30*, 336.
82. Cotton, F. A.; Matusz, M. *Inorg. Chim. Acta* **1988**, *143*, 45.
83. Bear, J. L.; Han, B.; Huang, S.; Kadish, K. M. *Inorg. Chem.* **1996**, *35*, 3012.
84. Bear, J. L.; Li, Y.; Han, B.; Van Caemelbecke, E.; Kadish, K. M. *Inorg. Chem.* **1997**, *36*, 5449.
85. van Vleck, J. H. *The Theory of Electric and Magnetic Susceptibilities*; Oxford University Press: London, 1932, Ch. 10.
86. Cotton, F. A.; Miskowski, V. M.; Zhong, B. *J. Am. Chem. Soc.* **1989**, *111*, 6177.
87. Kayama, K.; Baird, J. C. *J. Chem. Phys.* **1965**, *43*, 1082.
88. Cotton, F. A.; Ren, T.; Eglin, J. L. *Inorg. Chem.* **1991**, *30*, 2552.
89. Clark, D. L.; Green, J. C.; Redfern, C. M. *J. Chem. Soc., Dalton Trans.* **1989**, 1037.
90. Das, B. K.; Chakravarty, A. R. *Inorg. Chem.* **1992**, *31*, 1395.
91. Lindsay, A. J.; Wilkinson, G.; Motevalli, M.; Hursthouse, M. B. *J. Chem. Soc., Dalton Trans.* **1985**, 2321.
92. Chisholm, M. H.; Clark, D. L.; Huffman, J. C.; van der Sluys, W. G.; Kober, E. M.; Lichtenberger, D. L.; Bursten, B. E. *J. Am. Chem. Soc.* **1987**, *109*, 6796.
93. Capobianchi, A.; Paoletti, A. M.; Pennesi, G.; Rossi, G.; Caminiti, R.; Ercolani, C. *Inorg. Chem.* **1994**, *33*, 4635.
94. Cotton, F. A.; Matusz, M. *J. Am. Chem. Soc.* **1988**, *110*, 5761.
95. Cotton, F. A.; Ren, T. *Inorg. Chem.* **1991**, *30*, 3675.
96. Bear, J. L.; Yao, C.-L.; Liu, L.-M.; Capdeveille, F. J.; Korp, J. D.; Albright, T. A.; Kang, S.-K.; Kadish, K. M. *Inorg. Chem.* **1989**, *28*, 1254.
97. Chakravarty, A. R.; Cotton, F. A.; Falvello, L. R. *Inorg. Chem.* **1986**, *25*, 214.
98. Cotton, F. A.; Ren, T.; Eglin, J. L. *Inorg. Chem.* **1991**, *30*, 2559.

99. Cotton, F. A.; Ren, T.; Wagner, M. J. *Inorg. Chem.* **1993**, *32*, 965.
100. Cotton, F. A.; Ren, T. *J. Am. Chem. Soc.* **1992**, *114*, 2495.
101. Chakravarty, A. R.; Cotton, F. A.; Tocher, D. A. *Inorg. Chem.* **1985**, *24*, 1334.
102. Bear, J. L.; Li, Y.; Han, B.; Kadish, K. M. *Inorg. Chem.* **1996**, *35*, 1395.
103. Cotton, F. A.; Yokochi, A. *Inorg. Chem.* **1997**, *36*, 567.
104. Fanwick, P. E.; King, M. K.; Tetrick, S. M.; Walton, R. A. *J. Am. Chem. Soc.* **1985**, *107*, 5009.
105. Fanwick, P. E.; Tetrick, S. M.; Walton, R. A. *Inorg. Chem.* **1986**, *25*, 4546.
106. San Filippo, J., Jr. *Inorg. Chem.* **1972**, *11*, 3140.
107. Cotton, F. A.; Kitagawa, S. *Polyhedron* **1988**, *7*, 1673.
108. Bursten, B. E.; Cotton, F. A.; Fanwick, P. E.; Stanley, G. G.; Walton, R. A. *J. Am. Chem. Soc.* **1983**, *105*, 2606.
109. Burns, C. J.; Burrell, A. K.; Cotton, F. A.; Haefner, S. C.; Sattelberger, A. P. *Inorg. Chem.* **1994**, *33*, 2257.
110. Preetz, W.; Hollman, P.; Thiele, G.; Hillebrecht, H. *Z. Naturforsch.* **1990**, *45b*, 1416.
111. Miskowski, V. M.; Gray, H. B. *Top. Cur. Chem.* **1997**, *191*, 41.
112. Cotton, F. A.; Fanwick, P. E.; Gage, L. D.; Kalbacher, B.; Martin, D. S. *J. Am. Chem. Soc.* **1977**, *99*, 5642.
113. Cotton, F. A.; Fanwick, P. E.; Gage, L. D. *J. Am. Chem. Soc.* **1980**, *102*, 1570.
114. Chavan, M. Y.; Feldmann, F. N.; Lin, X. Q.; Bear, J. L.; Kadish, K. L. *Inorg. Chem.* **1989**, *23*, 2373.
115. Chakravarty, A. R.; Cotton, F. A.; Tocher, D. A. *J. Am. Chem. Soc.* **1984**, *106*, 6409.
116. Chakravarty, A. R.; Cotton, F. A. *Inorg. Chim. Acta* **1986**, *113*, 19.
117. Ryde, K.; Tocher, D. A. *Inorg. Chim. Acta* **1986**, *118*, L49.
118. Kawamura, T.; Fukamachi, K.; Sowa, T.; Shigeru, H.; Yoneawa, T. *J. Am. Chem. Soc.* **1981**, *103*, 364.
119. Sowa, T.; Kawamura, T.; Shida, T.; Yonezawa, T. *Inorg. Chem.* **1983**, *22*, 56.
120. Bursten, B. E.; Cotton, F. A. *Inorg. Chem.* **1981**, *20*, 3042.
121. Telser, J.; Miskowski, V. M.; Drago, R. S.; Wong, N. M. *Inorg. Chem.* **1985**, *24*, 4765.
122. O'Connor, C. J. *Progr. Inorg. Chem.* **1982**, *29*, 203.
123. Collman, J. P.; Barnes, C. E.; Swepston, P. N.; Ibers, J. A. *J. Am. Chem. Soc.* **1984**, *106*, 3500.
124. Hopkins, M. D.; Ph.D. Thesis; California Institute of Technology: Pasadena, CA, 1986.
125. Cotton, F. A.; Martin, D. S.; Fanwick, P. E.; Peters, T. J.; Webb, T. R. *J. Am. Chem. Soc.* **1976**, *98*, 4681.
126. Fanwick, P. E.; Martin, D. S.; Webb, T. R.; Robbins, G. A.; Newman, R. A. *Inorg. Chem.* **1978**, *17*, 2723.
127. Cotton, F. A.; Daniels, L.; Davison, A.; Orvig, C. *Inorg. Chem.* **1981**, *20*, 3050.
128. Cotton, F. A.; Kalbacher, B. J. *Inorg. Chem.* **1977**, *16*, 2386.
129. Martin, D. S.; Huang, H.-W.; Newman, R. A. *Inorg. Chem.* **1984**, *23*, 699.
130. Collins, D. M.; Cotton, F. A.; Gage, L. D. *Inorg. Chem.* **1979**, *18*, 1712.
131. Dunbar, K. R.; Walton, R. A. *Inorg. Chem.* **1985**, *24*, 5.
132. Bennet, M. J.; Caulton, K. G.; Cotton, F. A. *Inorg. Chem.* **1969**, *8*, 1.

133. Martin, D. S.; Newman, R. A.; Vlasnik, L. M. *Inorg. Chem.* **1980**, *19*, 3404.
134. Mukaida, M.; Nomura, T.; Ishimori, T. *Bull. Chem. Soc. Jpn.* **1972**, *45*, 2143.
135. Wilson, C. R.; Taube, H. *Inorg. Chem.* **1975**, *14*, 2276.
136. Kimura, T.; Sakurai, T.; Shima, M.; Togano, T.; Mukaida, M.; Nomura, T. *Bull. Chem. Soc. Jpn.* **1982**, *55*, 3927.
137. Das, B. K.; Chakravarty, A. R. *Polyhedron* **1988**, *7*, 685.
138. Cotton, F. A.; Matusz, M.; Zhong, B. *Inorg. Chem.* **1988**, *27*, 4368.
139. Cotton, F. A.; Labella, L.; Shang, M. *Inorg. Chem.* **1992**, *31*, 2385.
140. Malinski, T.; Chang, D.; Feldmann, F. N.; Bear, J. L.; Kadish, K. M. *Inorg. Chem.* **1983**, *22*, 3225.
141. Chakravarty, A. R.; Cotton, F. A. *Polyhedron* **1985**, *4*, 1957.
142. Chakravarty, A. R.; Cotton, F. A.; Tocher, D. A. *Polyhedron* **1985**, *4*, 1097.
143. Chakravarty, A. R.; Cotton, F. A.; Tocher, D. A. *Inorg. Chem.* **1985**, *24*, 1263.
144. Chakravarty, A. R.; Cotton, F. A.; Schwotzer, W. *Polyhedron* **1986**, *5*, 1821.
145. Chakravarty, A. R.; Cotton, F. A. *Inorg. Chim. Acta* **1985**, *105*, 19.
146. Chakravarty, A. R.; Cotton, F. A.; Tocher, D. A. *Inorg. Chem.* **1985**, *24*, 2857.
147. Tocher, D. A. *Inorg. Chim. Acta* **1986**, *115*, 51.
148. Asahina, H.; Zisk, M. B.; Hedman, B.; McDevitt, J. T.; Collman, J. P.; Hodgson, K. O. *J. Chem. Soc., Chem. Commun.* **1989**, 1360.
149. Collman, J. P.; Prodolliet, J. W.; Leidner, C. R. *J. Am. Chem. Soc.* **1986**, *108*, 2916.
150. Cotton, F. A.; Matusz, M. *Chimia* **1989**, *43*, 167.
151. Li, Y.; Han, B.; Kadish, K. M.; Bear, J. L. *Inorg. Chem.* **1993**, *32*, 4175.
152. Maldivi, P.; Giroud-Godquin, A.-M.; Marchon, J.-C.; Guillon, D.; Skoulios, A. *Chem. Phys. Lett.* **1989**, *157*, 552.
153. Cotton, F. A.; Ren, T.; Eglin, J. L. *J. Am. Chem. Soc.* **1990**, *112*, 3439.
154. Chakravarty, A. R.; Cotton, F. A.; Tocher, D. A. *Inorg. Chem.* **1984**, *23*, 4693.
155. Cotton, F. A.; Thompson, J. L. *J. Am. Chem. Soc.* **1980**, *102*, 6437.
156. Behling, T.; Wilkinson, G.; Stephenson, T. A.; Tocher, D. A.; Walkinshaw, M. D. *J. Chem. Soc., Dalton Trans.* **1983**, 2109.
157. Cotton, F. A.; Dunbar, K. R.; Matusz, M. *Inorg. Chem.* **1986**, *25*, 1589.
158. Cotton, F. A.; Chakravarty, A. R.; Tocher, D. A.; Stephenson, T. A. *Inorg. Chim. Acta* **1984**, *87*, 115.
159. Chakravarty, A. R.; Cotton, F. A.; Tocher, D. A. *Inorg. Chem.* **1984**, *23*, 4697.
160. Cotton, F. A.; Dunbar, K. R. *J. Am. Chem. Soc.* **1987**, *109*, 2199.
161. Cotton, F. A.; Jennings, J. G.; Price, A. C.; Vidyasagar, K. *Inorg. Chem.* **1990**, *29*, 4138.
162. Root, D. R.; Blevins, C. H.; Lichtenberger, D. L.; Sattelberger, A. P.; Walton, R. A. *J. Am. Chem. Soc.* **1986**, *108*, 953.
163. Ebner, J. R.; Walton, R. A. *Inorg. Chem.* **1975**, *14*, 1987.
164. Lin, C.; Ren, T.; Valente, E. J.; Zubkowski, J. D. *J. Chem. Soc., Dalton Trans.* **1998**, 571.

# 7 Transition Metal Nitrosyls

BARRY L. WESTCOTT* AND JOHN H. ENEMARK
Department of Chemistry
The University of Arizona
Tucson, AZ 85721, USA
E-mail: westcottb@ccsu.edu
jenemark@u.arizona.edu



* Present address: Department of Chemistry, Central Connecticut State University, New Britain, CT 06050, USA.

*Inorganic Electronic Structure and Spectroscopy, Volume II: Applications and Case Studies.*
Edited by E. I. Solomon and A. B. P. Lever.
ISBN 0-471-32682-8. 

## ABBREVIATIONS

| | |
|---|---|
| Cp | cyclopentadienyl |
| Cp* | pentamethylcyclopentadienyl |
| das | *o*-phenylenebis(dimethylarsine) |
| DSC | differential scanning calorimetry |
| EDTA | ethylenediaminetetraacetic acid |
| EPR | electron paramagnetic resonance |
| EXAFS | extended X-ray absorption fine structure |
| FAD | flavin adenine dinucleotide |
| FMN | flavin mononucleotide |
| HOMO | highest occupied molecular orbital |
| IR | infrared |
| LUMO | lowest unoccupied molecular orbital |
| MCD | magnetic circular dichroism |
| NACysMe | L-*N*-acetylcysteinate methyl ester |
| NADPH | nicotinamide adenine dinucleotide phosphate |
| NLO | non-linear optical |
| NOS | nitric oxide synthase |
| OEP | octaethylporphyrin |
| PES | photoelectron spectroscopy |
| RR | resonance Raman |
| SBL | soybean lipoxygenase |
| SCF | self consistent field |
| SNP | sodium nitroprusside |
| SOMO | singly occupied molecular orbital |
| tacn | $N,N',N''$-trimethyl-1,4,7-triazacyclononane |
| tbp | trigonal bipyramidal |
| Tp* | hydrotris(3,5-dimethylpyrazolyl)borate |
| TPP | tetraphenylporphyrin |
| TS | transition state |
| TTP | tetratolylporphyrin |
| XAS | X-ray absorption edge spectroscopy |
| XPS | X-ray photoelectron spectroscopy |

## 1 INTRODUCTION

Nitric oxide has long been of interest to inorganic chemists because of its ability to coordinate to a metal center to afford either a linear or bent MNO geometry.[1,2]

General interest in the chemistry of nitric oxide and its complexes has increased dramatically in recent years with reports of NO activity in biological systems.[3] In 1992, NO was named the *Science* Molecule of the Year,[4] and the chemical and biological aspects of nitric oxide continue to be an active area of research. The biological importance of NO coupled with its reactivity toward metalloproteins has reawakened interest in the structure, bonding, and reactivity of transition metal nitrosyl complexes. A relatively recent comprehensive review of metal nitrosyls has been provided by Richter-Addo and Legzdins.[5] Here we give a brief background of some recent highlights of NO chemistry, review the important features of bonding in metal nitrosyl complexes, and survey several topics and specific examples at the forefront of current research on metal nitrosyls and their electronic structures.

## 1.1 Nitric Oxide in Biology

Current evidence indicates that nitric oxide is an obligatory intermediate in the four-step denitrification pathway of certain bacteria.[6,7] In this process of the inorganic nitrogen cycle, nitrate ion is reduced to dinitrogen gas via the sequence:[7,8]

$$NO_3^- \longrightarrow NO_2^- \longrightarrow NO \longrightarrow N_2O \longrightarrow N_2$$

Studies have shown that the enzyme, nitric oxide reductase, responsible for the third step acts in a similar manner and has a similar structure to the heme centers of cytochrome oxidase, the enzyme involved in reduction of $O_2$ to water in aerobic respiration.[7,8]

In mammals, NO is necessary for such diverse functions as inhibiting tumor growth,[3a] acting as a neurotransmitter,[3e,f,k] and translating sexual excitement into potency.[3c] Since NO is not normally available in the atmosphere for direct inhalation, and in fact, would be toxic if it were, the body must have some way to manufacture NO at the site where it is needed for these vital operations, as well as a mechanism to regulate its production. The nitric oxide synthase (NOS) enzymes are responsible for producing the quantity of NO needed by the body[3h,i,j] and are among the most tightly regulated enzymes in biology.[9] Brain (bNOS),[10] macrophage (macNOS) and endothial (eNOS), have been cloned. These complex proteins possess a reductase domain that is similar to cytochrome p450 reductase, and this domain contains conserved sequences for binding NADPH, FAD, and FMN. In addition, there is a heme/tetrahydrobiopterin binding domain where arginine binds and NO is produced and a calmodulin binding domain.[9]

The production, transport, and utilization of NO in biological systems involve metal centers. Several review articles are available on the role of NO in biology.[3k,6,11–13] This review of transition metal nitrosyls is intended to provide fundamental background material for better understanding of the biological processes.

## 1.2 Bent and Linear Nitrosyls

As X-ray crystallography became a more widely used technique in the 1960's, metal nitrosyl complexes attracted considerable interest because two distinctly different

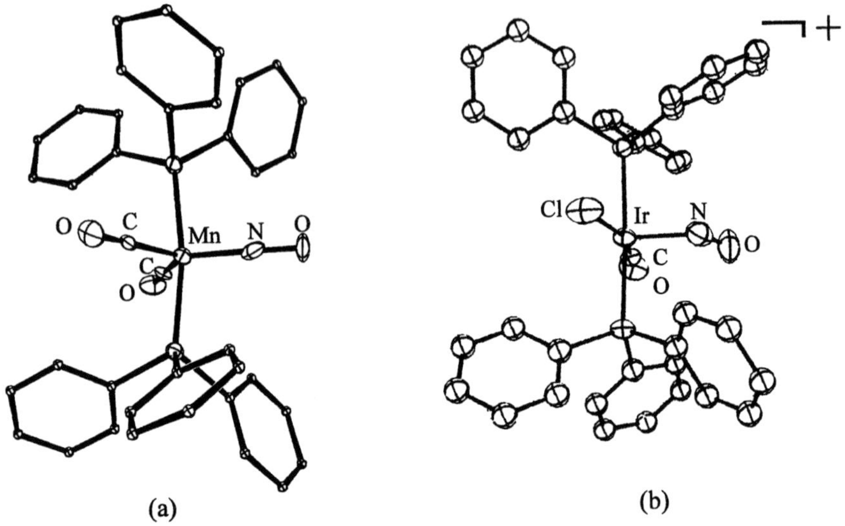


**Figure 1.** (a) Structure of $Mn(NO)(CO)_2(PPh_3)_2$. (b) Structure of $[Ir(NO)(CO)(Cl)(PPh_3)_2]^+$. Reproduced with permission from References 15,16.

limiting geometries were observed for the MNO unit.[14] Figure 1 shows structures of a manganese complex and an iridium complex, both of which are five-coordinate. However the manganese complex (1a) exhibits a linear MNO group and assumes a trigonal bipyramidal geometry,[15] whereas the iridium complex (1b) has a strongly bent MNO group and approximate square pyramidal geometry.[16] These distinctly different MNO angles for formally isoelectronic molecules (*vide infra*) necessitated the development of a formalism to describe the metal–ligand interaction in these systems.

As early as 1934, Sidgwick realized that there were two different types of metal nitrosyls, which he classified as derivatives of either $NO^+$ or $NO^-$.[17] Compounds with a bent MNO moiety were classified as derivatives of $NO^-$, and complexes with a linear MNO unit were classified as derivatives of $NO^+$. This formal oxidation state description is still used to a large extent today.[18] As a pair of illustrative examples, we take the two complexes shown in Figure 1. The Mn complex would now formally be $Mn^{-I}(d^8)(NO^+)$, while the Ir could be called $Ir^{III}(d^6)(NO^-)$. This assignment can be easily made since the geometries of both complexes are known based on the crystallographic data available. Without structural data, however, this assignment becomes more difficult to make.

Early studies of the electronic spectra of six-coordinate, pentacyanonitrosyl complexes led to a molecular orbital description in which a very strong M–NO bond dominated the overall ligand field, and the $\pi$-antibonding orbitals were *lower* in energy than the metal $\sigma$-antibonding orbitals.[19] Computational studies on a similar series of six-coordinate mononitrosyl complexes were in agreement with this conclusion[20] (Section 4). While this molecular orbital description seemed to accurately describe the bonding in complexes with linear MNO units, the bent complexes were still not well understood, and the bending of NO in five-coordinate species was attributed to a second order Jahn–Teller effect.[21]

### 1.3 The $\{MNO\}^n$ Formalism

In this description, the nitrosyl ligand is not treated as an independent unit which has either +1 or −1 formal charge. Instead, the entire metal-nitrosyl moiety is treated as a singular unit. Since early molecular orbital calculations showed significant mixing between the metal $d$ and $\pi^*$(NO) orbitals,[19,20] it makes intuitive sense not to separate the metal and NO in the bonding description. To this end Enemark and Feltham developed the $\{MNO\}^n$ formalism for describing mononitrosyl complexes,[1] where $n$ is the total additive number of electrons contained in the metal d and $\pi^*$(NO) orbitals. Under this formalism both complexes in Figure 1 are $\{MNO\}^8$ derivatives. If it is valid to regard the MNO unit as a covalent triatomic species, then comparison to triatomic molecules of main group elements should provide insight into the differing geometries of the $\{MNO\}^8$ complexes of Figure 1.

Walsh considered the structure and bonding of triatomic species in detail, but he was concerned only with interactions between $s$ and $p$-type orbitals in the valence shell.[22] The extension of Walsh's concepts to the $\{MNO\}^n$ moiety has been presented in detail elsewhere,[1] so it is only briefly summarized here.

Before attempting to describe the $\{MNO\}^n$ unit, we briefly review the Walsh diagram for $XY_2$ molecules as applied to the series O–N–$O^m$, where m is +1, 0, −1. For the series $NO_2^+$, $NO_2$, and $NO_2^-$ the bond angles are 180°, 134°, and 115°, respectively. Figure 2, a modified version of Walsh's diagram, explains the geometries of these species. The $NO_2^+$ ion has 16 valence electrons in the $2s$ and $2p$ orbitals, and the molecular orbitals through $1\pi_g$ will be filled. The $1\pi_g$ orbital is non-bonding and becomes destabilized by decreasing the O–N–O bond angle; therefore, $NO_2^+$ is linear.

For $NO_2$ and $NO_2^-$, one and two electrons, respectively, are added to the $2\pi_u$ molecular orbital of $NO_2^+$. The $2\pi_u$ is antibonding with respect to all three atoms, and Figure 2 shows that a deviation from linearity causes a splitting of the $2\pi_u$ and stabilization of the occupied $6a_1$ orbital, a non-bonding orbital localized on the central atom. Thus, the bent geometries of $NO_2$ and $NO_2^-$ are a direct consequence of the addition of one and two electrons, respectively, to the totally antibonding $2\pi_u$ orbital of the linear triatomic system.

Replacement of one oxygen atom by a d-block metal gives rise to a slightly modified molecular orbital description. The new molecular orbital description for the $\{MNO\}^n$ unit in $C_{\infty v}$ symmetry is shown in Figure 3. The $4\sigma$ is analogous to the $5\sigma_g$ in $NO_2^+$, the $2\pi$ corresponds to the $1\pi_g$, and the $\delta$ is non-bonding with respect to the NO moiety and has no analogous orbital in $NO_2^+$. Replacement of one terminal oxygen atom in $NO_2^m$ with a $d$-block metal leads to a triatomic unit which now requires 18 electrons to fill all bonding and non-bonding orbitals of the MNO triatomic species (through $\delta$). Additional electrons will occupy the totally antibonding

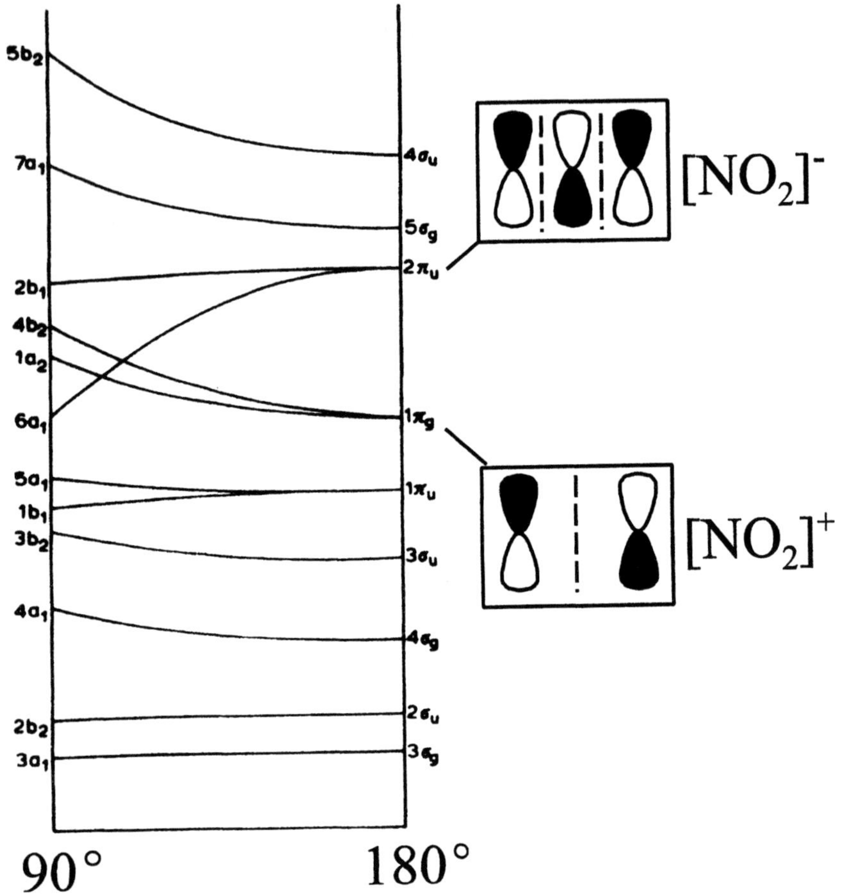


**Figure 2.** Walsh Diagram for the $NO_2^m$ moiety showing the nodal properties of the HOMO for $NO_2^+$ and $NO_2^-$.

$3\pi$ orbital. If Walsh's arguments are also applicable to the $\{MNO\}^n$ unit, then partial occupation of the $3\pi$ orbital should lead to bending of the $\{MNO\}^n$ group.

### 1.4 $\{MNO\}^8$ as an Example

Five-coordinate metal nitrosyl complexes of the type $L_4M(NO)$ which possess an $\{MNO\}^8$ unit tend toward two limiting geometries: trigonal bipyramidal with linear MNO units, and square pyramidal with bent MNO units.[2] Insight into the bonding differences in these two systems can be obtained by considering the correlation diagram shown in Figure 4. The possible orderings of the orbitals in a tetragonal complex with a linear MNO unit are shown in Figure 4b–c. For a tetragonal $\{MNO\}^8$ complex adopting the electronic structure of Figure 4b, the HOMO is $4a_1$, which is primarily $d_z^2$ and $\sigma$-antibonding. Distortion of the molecule to trigonal bipyramidal

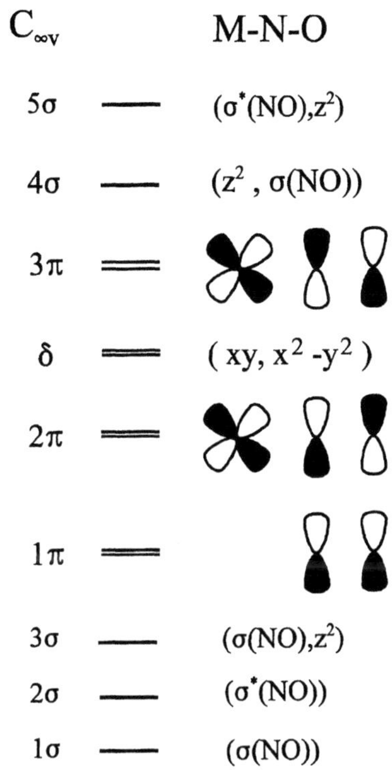


**Figure 3.** Molecular orbital diagram for the $\{MNO\}^n$ unit in $C_{\infty v}$ symmetry, showing the nodal properties of the $\pi$ orbitals.

geometry (Figure 4a) will make $4a_1$ less antibonding and also allow direct delocalization of the electron density from the ligand $\sigma$ orbitals in the equatorial plane into the $\pi^*$ orbitals of the NO group. On the other hand, for a tetragonal $\{MNO\}^8$ complex adopting the electronic structure of Figure 4c the HOMO is $3e$, and two electrons will occupy the totally antibonding $\pi$-type orbital analogous to the $3\pi$ of Figure 3. This should lead to bending of the MNO unit and localization of a pair of electrons on the nitrogen atom as shown in Figure 4d. Thus, most five-coordinate $\{MNO\}^8$ complexes can be described by one of the two limiting structures: a trigonal bipyramidal complex with a linear MNO group (Figure 4a), or a square pyramidal complex with a strongly bent MNO group (Figure 4d).[2] Extended Huckel calculations on five-coordinate metal nitrosyls are consistent with the qualitative bonding picture of Figure 4.[23]

Figure 4 implies that at the crossing point between b and c, the energies of the $4a_1$ and $3e$ orbitals will be accidentally degenerate. The properties of a tetragonal system with a $(4a_1, 3e)^2$ electron configuration at the crossing point have been discussed in detail elsewhere.[1] A simple ligand field approach shows that this two-electron, three-orbital system can lead to a five-fold degeneracy of the singlet states at the crossing point. Distorted or intermediate molecular structure(s) could result from vibronic

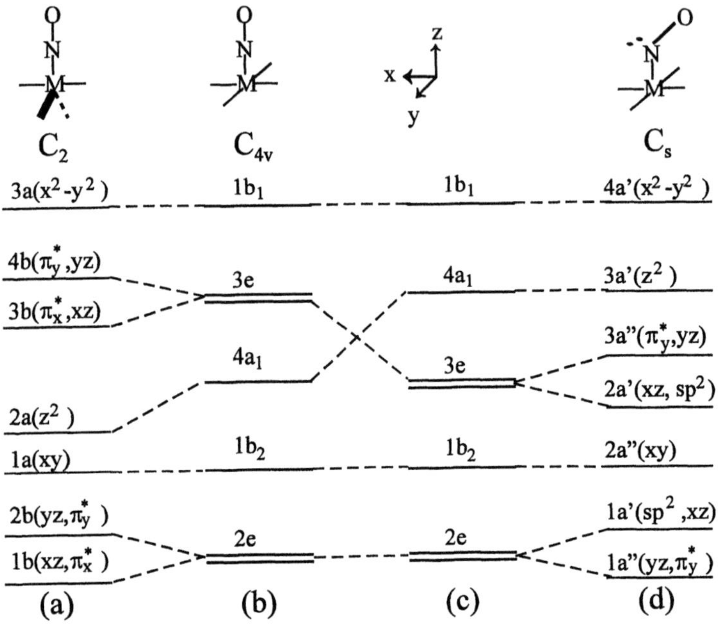


**Figure 4.** Correlation diagram for five coordinate $\{MNO\}^8$ complexes: (a) shows the expected electronic structure for trigonal bipyramidal complexes with a linear MNO; (b, c) show how the relative energies of the $4a_1$ and $3e$ orbitals in $C_{4v}$ give rise to the two limiting structures; (d) represents the expected electronic structure for square pyramidal complexes with a bent MNO.

coupling among these approximately degenerate states and the vibrational modes of the compound. An unusual group of complexes that may fall into this category are the $CoCl_2(NO)(PR_3)_2$ molecules which exhibit two NO stretching modes in both solution and the solid state.[24] The authors concluded that there must exist two discrete molecular structures—one with a bent MNO unit ($C_s$), the other with a linear unit ($C_2$)—that are in dynamic equilibrium in solution. This suggestion was supported by $^{15}N$ substitution, and the temperature dependence of the IR intensities in solution which showed that as one NO band grew, the other diminished by an equivalent amount such that the summation of the two absorbances was constant at all temperatures. This process was reversible with the higher frequency form, attributed to a linear CoNO group, being favored at low temperatures.

One unusual aspect of the IR studies of the solid was that the intensity of the peaks also seemed to be dependent on pressure; that is, the band attributed to the bent form seemed to increase in intensity as more pressure (qualitatively) was used to grind the sample. The X-ray photoelectron spectra of the solid samples showed two different N 1$s$ ionizations, but their relative intensities suggested that the bent form (lower binding energy) predominated, exactly the opposite of the conclusion derived from the IR studies. The X-ray crystal structure of $CoCl_2(NO)[P(CH_3)(C_6H_5)]_2$ showed distorted tbp coordination geometry with a single well resolved NO ligand in the equatorial plane with Co–N–O = 165°. There was no evidence for a square pyramidal form with a strongly bent Co–N–O group, and the disorder at the Cl positions was less than 0.5%.

One qualitative conclusion that is consistent with all of the solid state data is that the form of the complex at the surface differs from that in the bulk material. XPS is a surface technique whereas IR and single crystal X-ray structure determination are bulk techniques. This explanation also accounts for the apparent sensitivity of the observed IR spectrum to pressure, because grinding would increase the surface area, and hence the relative amount of the low frequency complex that is assumed to have a bent CoNO geometry. The unusual properties of the $CoCl_2(NO)(PR_3)_2$ compounds warrant reinvestigation by $^{15}N$ NMR (*vide infra*).

Although there are relatively few examples of metal nitrosyl complexes that appear to exist close to the "crossover" point in the ground state, the reactions of metal nitrosyl complexes may lead to crossover electronic structures somewhere along the reaction coordinate. Specific examples are discussed in the next section (2.3, 2.4).

## 2 STRUCTURE AND REACTIVITY OF MNO COMPLEXES

Having reviewed the relationship between structure and bonding for the two limiting MNO geometries, we next address the factors that determine which geometry will be exhibited by a particular complex, and how the reactivity of a complex depends upon its molecular and electronic structure.

### 2.1 Reactions of Coordinated Nitrosyls

Both electrophilic and nucleophilic reactions of the nitrosyl unit have been observed.[5] The linear $\{FeNO\}^6$ unit of the nitroprusside anion exhibits electrophilic character and reacts with hydroxide ion to give the pentacyanonitro complex according to Equation 1:[25]

$$\left[Fe(CN)_5(NO)\right]^{2-} + 2OH^- \longrightarrow \left[Fe(CN)_5(NO_2)\right]^{4-} + H_2O \qquad (1)$$

Other nucleophiles like $SH^-$ and ketones also react with the nitroprusside anion,[25] and with other six-coordinate complexes containing linear $\{MNO\}^6$ groups, such as $[Fe(NO)X(das)_2]^{2+}$ (X = halide, $CN^-$) and $[RuCl(NO)(das)_2]^{2+}$.[26]

While many linear units react as electrophiles, the bent MNO moiety shows nucleophilic character.[5] Complexes with bent MNO units can react with molecular oxygen or excess NO to give complexes with a coordinated nitro group.[27] Also, some of these complexes react with protons in the following manner:

$$\left[Fe(CN)_5(NO)\right]^{3-} + H^+ \longrightarrow \left[Fe(CN)_5(NOH)\right]^{2-} \qquad (2)$$

An analogous reaction is also observed for $[Co(NO)(das)_2Br]^+$.[28]

Since there are two limiting geometries for the MNO unit, then a reaction of particular interest is interconversion of one form into the other. For instance, the five-coordinate complex $[Co(NO)(das)_2]^{2+}$ possesses a nearly linear $\{CoNO\}^8$ unit (179°) in which the geometry about the metal center is trigonal bipyramidal; however, upon reaction with a ligand X ($X = NCS^-$, $Cl^-$), the resulting six-coordinate complex $[Co(NO)X(das)_2]^+$ exhibits a strongly bent (134°) $\{CoNO\}^8$ unit.[29] This bending can be explained by Figure 4. In the linear case (Figure 4a), the HOMO should be the $4a_1$ orbital ($d_z^2$). The addition of the sixth ligand will produce a pseudo tetragonal complex with approximate $C_{4v}$ symmetry, and can sufficiently raise the energy of the $4a_1$ so that the totally antibonding, doubly occupied $3e$ orbital will become the HOMO. This electronic structure will lead to a strongly bent $\{CoNO\}^8$ group in order to lower the energy of the system as shown in Figure 4d, and as discussed previously for the MNO triatomic unit. This transformation of the $\{CoNO\}^8$ group upon addition of a ligand can also be formally described as an intramolecular redox reaction in which the linear $Co^Id^8$–$NO^+$ center is converted to the bent $Co^{III}d^6$–$NO^-$ center.[29]

Nitrosyl complexes are often compared to carbonyl complexes because both ligands have the ability to act as $\pi$-acceptor molecules.[18] If the NO ligand is taken to be a neutral donor, then it may be considered a 3 electron donor (*vide supra*), while the carbonyl is only a 2 electron donor. Thus, because NO has one more electron than CO, complexes of NO have a central metal whose atomic number is lower by one than the analogous carbonyl complex, as evidenced by the series $Ni(CO)_4$, $Co(CO)_3(NO)$, $Fe(CO)_2(NO)_2$, and $Cr(NO)_4$.[18] Comparison of the ligand substitution reactions of some formally isoelectronic complexes of CO and NO provides insight into the effect that the NO ligand has on the overall chemical reactivity of complexes. The mechanisms of Reactions 3 and 4 illustrate the effect of coordinated NO upon CO substitution in six-coordinate systems.

$$Cr(CO)_6 + L \longrightarrow Cr(CO)_5L + CO \qquad (3)$$

Reaction 3 proceeds via a dissociative mechanism; that is, a CO ligand dissociates before the attacking group, L, binds to the metal center. However, in the case of the isoelectronic $\{VNO\}^6$ complex, kinetic studies suggest that the analogous Reaction (4) occurs via an associative pathway.[30]

$$V(CO)_5(NO) + L \longrightarrow V(CO)_4(NO)L + CO \qquad (4)$$

This trend also holds for four coordinate complexes. The complex $Ni(CO)_4$ undergoes a dissociative substitution reaction with a ligand L,[31] whereas $Co(CO)_3(NO)$ undergoes an associative substitution reaction.[32] If we invoke 18-electron rule arguments, then the intermediates of the associative mechanism would be unfavorable 20-electron species if the NO group remained linear. However, if a pair of electrons is allowed to shift to the NO, then the resulting intermediate (or transition state) will still have an 18-electron metal center, but with a bent MNO unit. This follows the arguments developed above for the conversion of a five-coordinate complex with a linear $\{CoNO\}^8$ group to a six-coordinate complex with a bent $\{CoNO\}^8$ moiety, i.e., Figure 4. Subsequent loss of CO from the associative pathway intermediate allows the pair of electrons to shift back to the metal center, regenerating a linear MNO moiety. Reaction 5 shows the overall process:

$$\underset{18e}{(CO)_nM\text{–}N\text{–}O} \xrightarrow{+L} (CO)_n\underset{L}{M}\text{–}\ddot{N}\diagdown_{O} \longrightarrow (CO)_{(n-1)}\underset{L\ \ 18e}{M}\text{–}N\text{–}O + CO \quad (5)$$

## 2.2 Experimental Evidence for NO Bending

Two lines of experimental evidence support the bending of NO during a chemical reaction. The isolation and characterization of five-coordinate linear $\{CoNO\}^8$ and six-coordinate bent $\{CoNO\}^8$ have been discussed above. Additional evidence comes from the reduction of formally 18-electron linear metal nitrosyl complexes. The reversible one-electron reduction of $CpM(CO)_2(NO)$ complexes ($M = Cr$, Mo) produces paramagnetic $\{MNO\}^7$ species that can be studied by EPR.[33] In theory, changing the geometry of the complex should change the observed $g$ values in the EPR experiment because the g tensor values are dependent on the nature of the MO's by Equation 6:

$$g_i = g_e - \frac{\lambda \sum |\langle 0|\mathbf{L_i}|k\rangle|^2}{\Delta} \quad (6)$$

where $g_i$ is the tensor element in question; $g_e$ is the free electron $g$ value (2.0023); $\lambda$ is the spin orbit coupling of the metal; $\mathbf{L_i}$ is the orbital angular momentum operator; and $\Delta$ is the positive energy difference between the ground and excited states.[34–36] If the ground and excited states of the reduced species contain substantial NO character, as predicted by Figure 4c and Huckel calculations, then the $g$ values should be sensitive to the M–N–O angle. The highly anisotropic $A(^{14}N)$ hyperfine splittings ($A_\parallel = 24.5 \times 10^{-4}\ cm^{-1}$, $A_\perp = 2.0 \times 10^{-4}\ cm^{-1}$) for the reduced species indicate that about 40% of the spin density of the SOMO resides on the N atom, primarily in a $2p$ orbital.

The geometry of the reduced species (bent or linear) can be determined by analysis of Equations 7–9 for the $g$-tensors for these types of complexes, where $R$ and $N$ are weighting factors for the wavefunctions.

$$g_x = g_e - (\lambda/\Delta)N_1^2N_2^2(R_1 - R_2)^2 \quad (7)$$

$$g_y = g_e \quad (8)$$

$$g_z = g_e - (\lambda/\Delta)N_1^2N_2^2(2 - R_1R_2)^2 \quad (9)$$

If the MNO fragment were linear then the only contribution to the SOMO from the nitrogen would be from an electron in an sp hybrid orbital directed along the metal–NO axis. In any other case, (Figure 4d) the unpaired electron will be in a molecular orbital that is M–NO $\sigma^*$ and mostly $2p$ in character.[23] The dependence of the $g$ values upon the MNO angle was simulated by determining new $N$ and $R$ values for each geometry and then evaluating the $g$ tensor equations (7–9). The simulation with an MNO angle of about 130° best reproduced the experimental $g$ values. Thus, these studies show that one-electron reduction of an 18-electron metal nitrosyl center leads to the bending of the MNO unit and localization of electron density on the nitrogen atom. These one-electron reduction experiments further support the proposed bending of the NO moiety in the intermediate step in associative ligand substitutions (Equation 5). The electronic and molecular structures of these transient organometallic $\{MNO\}^7$ species also serve as a preview for the discussion of the properties of stable $\{FeNO\}^7$ systems (Sections 4.2 and 5.2).

## 2.3 Computational Evidence for NO Bending

Recent computational studies corroborate the proposed bending of the NO group in ligand substitution mechanisms of 18-electron metal nitrosyl complexes. Song and Hall reported a computational study of the ligand substitution reaction of phosphine ligands (L) on the $\{WNO\}^6$ complex *trans*-$W(CO)_4(NO)Cl$.[37] Using $PMe_3$ as the nucleophile, they monitored the bond lengths and angles in the complex to determine, first, if the reaction proceeds via a seven-coordinate intermediate, and second, if the MNO group assumes a bent geometry in this intermediate.

The potential energy surface for this reaction is shown in Figure 5. As the reaction proceeds from right to left, two distinct saddle points are apparent in the potential energy surface. These saddle points correspond to two distinct transition states in which seven-coordinate species exist, one in which the $PMe_3$ is initially bonding to the metal center, and one in which the CO group is leaving. In both of these transition states, the MNO moiety has a slightly bent configuration (156.2° and 152.8°, respectively). The local minimum between the saddle points corresponds to the intermediate complex. In this intermediate complex the MNO angle is 135.7°, which is consistent with values observed for bent nitrosyl complexes. These calculations suggest that a pair of electrons that were metal $d$ and $\pi^*$(NO) in character in the starting complex are nearly entirely N $2p$ in character in the intermediate complex. This theoretical investigation supports the electrochemical and EPR evidence discussed in Section 2.2. The optimized geometries along the associative pathway are shown in Figure 6, and the appropriate bond distances and angles are labeled.

The authors went on to explore the ligand replacement in an isoelectronic carbonyl complex. In $Re(CO)_5Cl$, no bending of the CO group which corresponds to the NO group in $W(CO)_4(NO)Cl$ is observed, demonstrating the unique role that the NO ligand plays in associative ligand substitution reactions of coordinatively saturated complexes.

An aside to this argument is that even though the general trend is for reaction of metal–NO complexes to undergo an associative pathway upon nucleophilic attack, this is not always the case. With very weak nucleophiles such as $PH_3$ or $PPh_3$ these reactions have been observed to proceed via a dissociative reaction pathway.[38] This

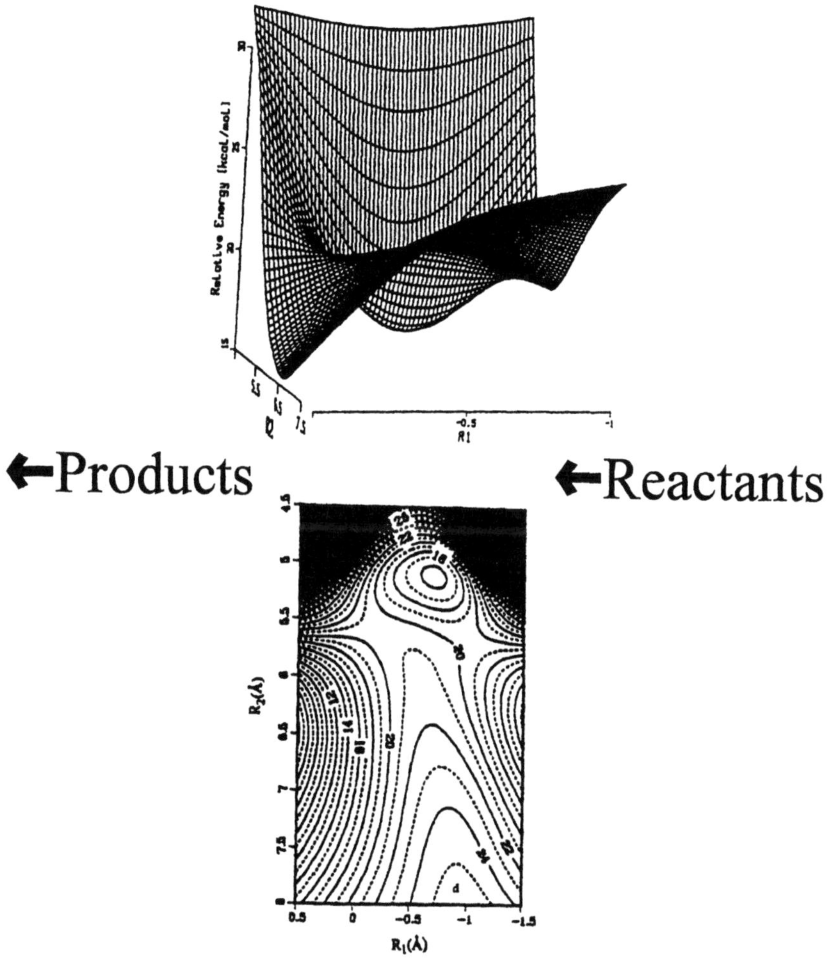


**Figure 5.** Potential energy surface and contour plot for the substitution of CO by $PMe_3$ as a function of the $R_1$ and $R_2$ distances. Reproduced with permission from Reference 37.

is borne out by the computational analysis for $PH_3$ substitution. However, it is especially interesting that these calculations indicate that the dissociative path for $PH_3$ substitution also involves bending of the MNO unit to an angle of 136.5°.[37]

## 2.4 Stereo-Electronic Properties of Linear $\{MNO\}^6$ and $\{MNO\}^4$ Complexes

Sections 1.2 and 1.4 showed that bending of the MNO group can occur in the ground state structures of molecules whenever the orbital that is totally $\pi$-antibonding ($3e$ of Figure 4) is partially occupied. Such bending is also implicated for the intermediate (transition state) along the reaction coordinate of bimolecular substitution reactions of metal nitrosyl compounds (Figures 5 and 6). In this section we consider the stereo-

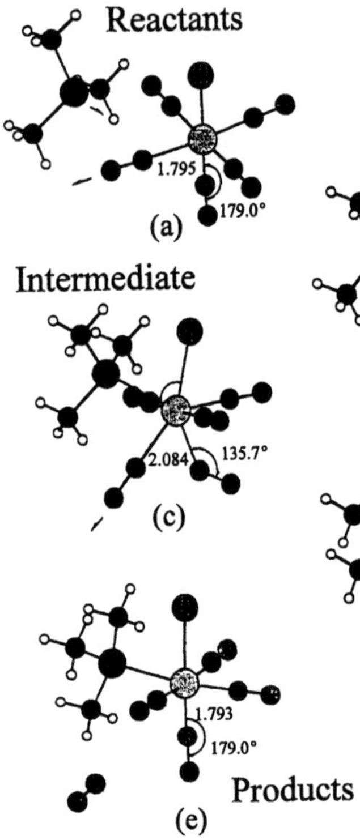


**Figure 6.** Optimized geometries along the associative path for the reaction in Figure 5 showing the reactants, early transition state, intermediate, late transition state, and products. The M–N distance and the M–N–O angle are given for each species. Reproduced with permission from Reference 37.

electronic consequences of MNO groups which are linear in the ground state and which appear to remain linear along the reaction coordinate.

From Figures 3 and 4, it is clear that $\{MNO\}^6$ complexes will be linear so long as the HOMO, the metal orbital that is analogous to the $d_{xy}$ orbital of Figure 4 and the $\delta$ orbitals of Figure 3, is lower in energy than the totally $\pi$-antibonding $3e$ orbitals of Figure 4b–c. It has long been known that chiral CpMn(NO)($PAr_3$)(COR) complexes, which contain the $\{MnNO\}^6$ group, undergo substitution of phosphine ligands by a dissociative mechanism with retention of configuration at the metal center.[39] Molecular orbital calculations on the related CpRu(NO)($alkyl)_2$ complexes show that the HOMO in $\{MNO\}^6$ complexes of the type CpM(NO)($ligand)_2$ is an energtically isolated lone pair that is localized on the metal and that cannot interact with the $\pi^*$ orbitals of the nitrosyl group[40] (Figure 7). Gladysz has shown that reaction of CpRe(NO)($CH_3$)($PPh_3$) with a non-coordinating acid generates the $[CpRe(NO)(PPh_3)]^+$ cation, a coordinatively unsaturated $\{MNO\}^6$ species (Figure 8) which binds enantioselectively to aldehydes.[41] Subsequent reduction of the complex generates chiral alkoxides in high yield. Recently Schenk and cowork-

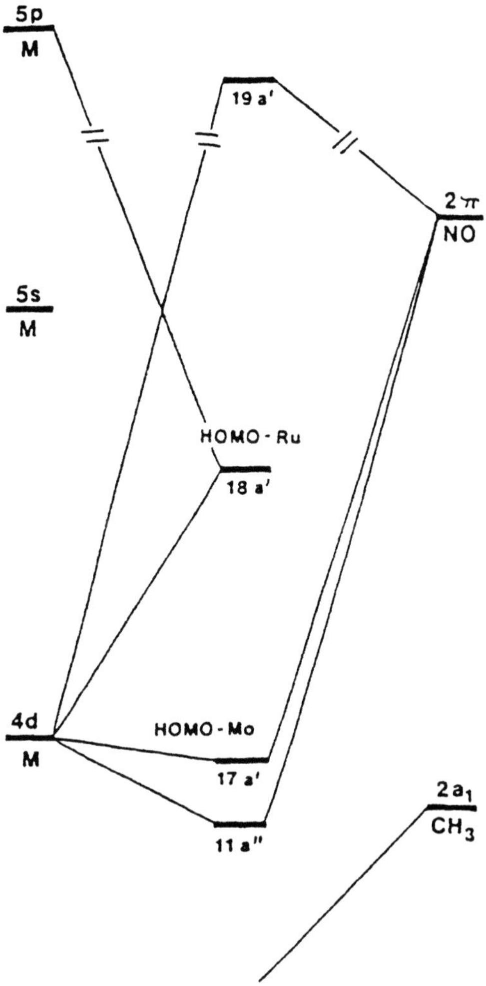


**Figure 7.** Molecular orbital diagram for the upper valence orbitals of $CpM(NO)Me_2$ (M = Mo, Ru) showing the energetically isolated frontier orbital (18a′) which is the HOMO when M = Ru and the LUMO when M = Mo. The shape of this orbital is shown in Figure 8. Reproduced with permission from Reference 40.

ers have prepared chiral $CpRe(NO)(SR)(PPh_3)$ complexes,[42] and subsequent treatment of these thiolate complexes with a non-coordinating acid yields isolable chiral $\eta^2$-thioaldehyde complexes in high enantiomeric excess. Major factors in the high stereoselectivity of these reactions are the localized $\pi$-donor orbital on the metal that is generated by the covalent $\{MNO\}^6$ group (Figure 8) and the steric properties of the Cp ring and the phosphine ligand.[41,42]

The $\{MNO\}^4$ group should be linear under all conditions. The four electrons will exactly fill the orbitals which are involved in $d\pi{-}p\pi$ bonding between the metal $d$

and the $\pi^*$(NO) orbitals (Figure 7), and thus the totally antibonding orbitals (3$e$ of Figure 4) will always be empty. Six-coordinate,[43] seven-coordinate, and binuclear $\{MNO\}^4$ complexes are known.[44] The six-coordinate $\{MNO\}^4$ complexes of the type CpM(NO)(alkyl)$_2$ with M = Mo, W are formally 16-electron species, but they are surprisingly stable. The NO ligand is the only $\pi$-acceptor ligand, and for these complexes the LUMO is the same non-bonding $d$ orbital that was the HOMO in the $\{MNO\}^6$ complexes discussed above (Figures 7 and 8). The access of nucleophiles to this non-bonding energetically isolated and spatially localized LUMO[45] can be controlled by the steric properties of the alkyl ligands.[40]

It has long been known that CpMo(NO)(SR)$_2$ compounds are monomeric in solution,[46] and these complexes do not readily react with nucleophiles, even though they are formally 16-electron species.[47] This stablilty is a direct consequence of the conformations of the two thiolate ligands which are oriented so that the filled $p\pi$ orbitals on the sulfur atoms can $\pi$-donate into the empty $d_{xy}$ orbital of the metal, as shown in Figure 9a. Thus, the electronic structure of the $\{MoNO\}^4$ group (Figure 7) controls the rotational conformation of the single-sided $\pi$-donor ligand.[47] The geometric constraints of chelating dithiolates preclude $\pi$-donation to the empty $d$ or-

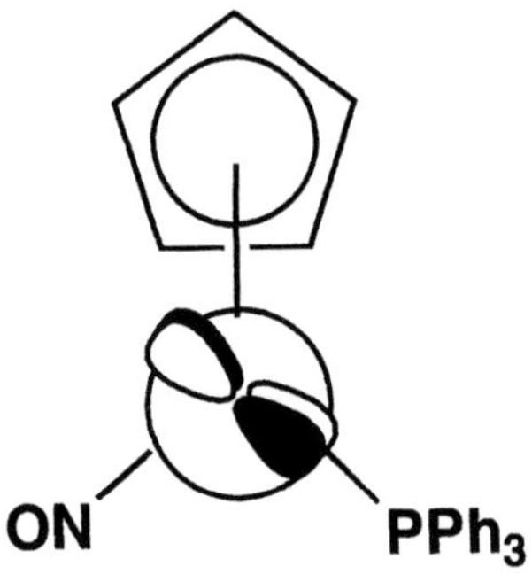


**Figure 8.** Stereo-electronic properties of the key frontier molecular orbital (18a′ of Figure 7) of $\{MNO\}^6$ and $\{MNO\}^4$ complexes.

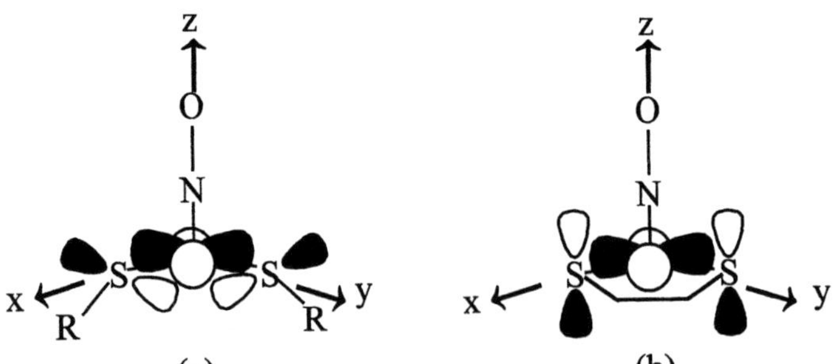


(a) (b)

**Figure 9.** Interaction of the filled S $p$–$\pi$ orbitals with the empty metal $d_{xy}$ orbital for: (a) two monodentate thiolates; (b) a chelating dithiolate.

bital (Figure 9b), and such compounds readily react with nucleophiles to give seven-coordinate $\{MNO\}^4$ complexes.[48]

Six-coordinate $\{MNO\}^4$ complexes can also be stablized by coordination of bulky tridentate ligands such as hydrotris(3,5-dimethylpyrazolyl)borate (Tp*) to give complexes of the general formula Tp*M(NO)(X,Y) where X and Y are monoanions and M = Mo, W.[46] The steric bulk of Tp* inhibits coordination by a seventh ligand and thereby stabilizes the 16-electron complexes. The Mo complexes can be reduced by one electron to give paramagnetic $\{MoNO\}^5$ centers,[49] and EPR spectroscopy is consistent with the unpaired electron being primarily localized in a non-bonding metal $d$ orbital (analogous to $d_{xy}$ of Figure 4),[52] as would be expected from the preceding discussion of CpM(NO)(alkyl)$_2$ compounds.[40] McCleverty and Jones and their coworkers have prepared a variety of homo- and heterobimetallic complexes in which the redox active Tp*M(NO)(X,Y) center is linked to a second redox active metal center through polyene bridges. These bimetallic systems with a reducible $\{MoNO\}^4$ center are of general interest due to their possible utility as non-linear optical (NLO) materials.[50,51] Of particular current interest is [{Tp*Mo(NO)Cl}$_2$(4,4′-$NC_5H_4C_5H_4N$)] which shows strong coupling between the two equivalent metal centers and splitting of the two one-electron couples by 765 mV.[52]

Although the $\{MNO\}^4$ group is linear and does not change geometry during chemical reactions, the electronic structure of the $\{MNO\}^4$ group has important consequences for chemical reactivity. The strong covalent bonding of the $\{MNO\}^4$ group leads to an energetically isolated LUMO which is a non-bonding metal $d$ orbital (Figure 7). This orbital can function as a site for electrochemical reduction, coordination of a nucleophile or coordination by a $\pi$-donor ligand. The stereochemical properties of this acceptor orbital (Figure 8) enable the $\{MNO\}^4$ group to control the orientation of a $\pi$-donor ligand relative to the other ligands in the coordination sphere (Figure 9).

The Tp*Mo(NO)(X,Y) compounds discussed above are isostructural to the Tp*MoO(X,Y) complexes that have been extensively studied as models for the oxomolybdenum(V) centers of enzymes.[53] The volatility of both types of compounds allows the bonding in their respective $[MoX]^{3+}$ units (X = NO, O) to be experimentally compared by gas phase photoelectron spectroscopy (PES).[54] Figure 10 shows the HeI PES spectra of the isostructural complexes Tp*Mo(X)(OEt)$_2$. From these spectra it is clear that the oxo complex is nearly 1 eV more easily ionized than the NO complex. This result seems counterintuitive; typical oxidation state assignments for the two centers are $[Mo^{II}(NO)^+]^{3+}$ and $[Mo^{V}O^{2-}]^{3+}$, yet the formally $Mo^V$ center is much more easily oxidized. The experimental PES data show that $NO^+$ is indeed a very effective $\pi$-acceptor that removes substantial electron density from the metal center onto the NO ligand. Likewise, the $O^{2-}$ ligand is a strong $\pi$-donor that transfers substantial electron density to the metal center. Another clue to the relationship between these two systems is that one-electron reduction of the diamagnetic Tp*Mo(NO)(X,Y) complexes gives paramagnetic $\{MNO\}^5$ species whose EPR spectra are similar to those of the neutral paramagnetic Tp*MoO(X,Y) complexes which have one unpaired electron in Mo $d_{xy}$ orbital.[53] Thus, the neutral Tp*Mo(NO)(X,Y) complexes actually contain one less valence electron than their

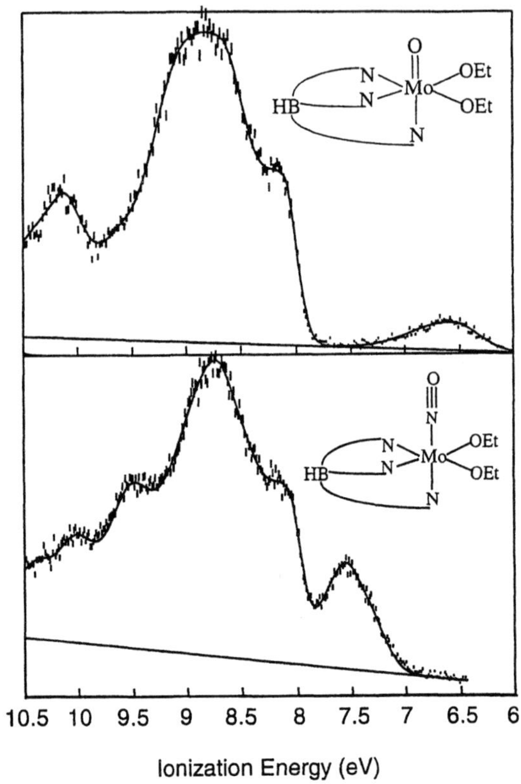


**Figure 10.** Gas phase HeI PES spectra of Tp*Mo(X)$(OEt)_2$ (X=O, NO).

isostructural oxomolybdenum counterparts. If the electron counting formalism developed for metal nitrosyls is applied to the Tp*MoO(X,Y) complexes they become $\{MoO\}^5$ species by virtue of having one metal *d* electron and four $p\pi$ electrons from the $O^{2-}$ group. The PES data of Figure 10 show that it is much easier to remove a non-bonding *d* electron from a $\{MoO\}^5$ group than it is to remove a strongly $\pi$-bonding electron from a $\{MoNO\}^4$ group of an analogous complex.

Bursten and Cayton explored the differences between the interaction of NO and oxo groups with transition metal centers via computational techniques.[55] Descending from idealized $O_h$ symmetry, the orbitals important to this interaction come from the $t_{2g}$ set. The interaction of the $\pi^*$(NO) orbitals with the metal *d* orbitals stabilizes the $d_{xz}$ and $d_{yz}$ orbitals, while the $d_{xy}$ remains non-bonding. However, in the case of an oxo donor, $d_{xy}$ is still a non-bonding orbital, but $d_{xz}$ and $d_{yz}$ now become strongly *antibonding* due to interaction with the filled oxo $\pi$ orbital (Figure 11). For the Tp*Mo(NO)(X,Y) complexes described above, the HOMO is the filled $d_{xz,yz}$ orbitals. The PES data of Figure 10 show that these orbitals are substantially stabilized relative to the non-bonding $d_{xy}$ orbital, the HOMO in the Tp*Mo(O)(X,Y) compounds.

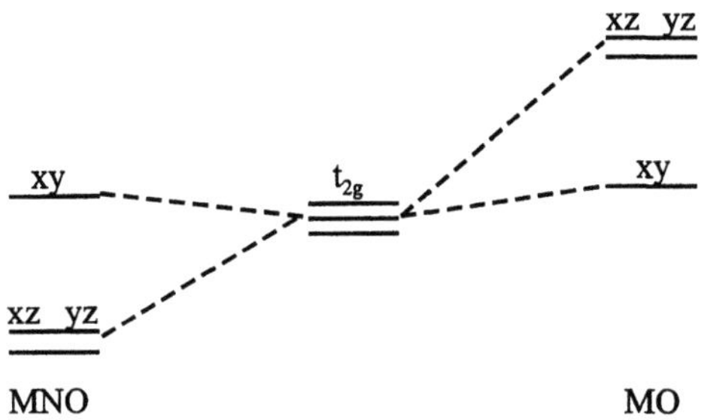


**Figure 11.** Splitting diagram for the metal $t_{2g}$ orbital for a strong $\pi$-acceptor (NO) and a strong $\pi$-donor (oxo).

# 3 SPECTROSCOPIC AND STRUCTURAL TECHNIQUES

A variety of methods have been used to experimentally investigate the molecular and electronic structures of metal nitrosyl complexes. X-ray crystallography is the most definitive method for determining the geometry of an MNO group when suitable crystals are available. Recently it has been shown that multiple scattering analysis of EXAFS data can be used to determine the M–N–O angle of non-crystalline samples.[56] This technique is especially sensitive for M–N–O angles between 150° and 180° (Section 4.2) Infrared, Raman, and nitrogen NMR spectroscopies have also been employed to probe the nature of MNO groups. Gas phase PES spectra provide an experimental method for the determination of the relative amount of NO and metal character in valence ionizations, which is important in understanding the $\pi$-acceptor ability of the NO ligand. The discussion of these techniques focuses upon their use in determining the nature of the M–NO interaction. This section also provides the background for the specific examples discussed in later sections.

## 3.1 X-Ray Crystallography

This technique has been discussed indirectly in the first few sections and is the best way to determine the geometry of a MNO moiety, provided that a suitable single crystal can be obtained. However, even for crystalline materials, the geometry of the MNO moiety cannot always be unambiguously determined by X-ray crystallography due to disorder in the crystal,[57] or a mixture of species within the lattice.[24] Such complications require that other techniques also be used to provide additional information about the geometry of the MNO group in complexes.

## 3.2 Infrared Spectroscopy

The conversion of the neutral nitric oxide molecule to the nitrosonium ion ($NO^+$) involves removal of an electron from a $\pi^*$(NO) orbital. This process will increase the NO bond order, thus $\nu_{NO}$ is expected to increase.[5] Indeed, $\nu_{NO}$ for free NO is 1,876 $cm^{-1}$, whereas that for $NO^+$ it is 2,200 $cm^{-1}$.[58] In metal complexes the

$NO^+$ formalism is frequently used for electron counting purposes; however, back-bonding from the metal center into the $\pi^*$(NO) orbital has a profound effect on the observed value of $\nu_{NO}$.[18] In general, linear complexes have $\nu_{NO}$ values which fall in the range of 1,600–2,000 $cm^{-1}$ while bent complexes range from 1,400–1,650 $cm^{-1}$.[5] One expects to also observe a $\nu_{MN}$ and $\delta_{MNO}$, but these are weaker and may be obscured by other vibrations of the molecule;[59] the $\nu_{NO}$ band is usually very strong and easy to identify.

The correlations between $\nu_{NO}$ and MNO angle are helpful, but by no means absolute. For instance, $Ir(NO)(PPh_3)_3$ exhibits $\nu_{NO}$ at 1,600 $cm^{-1}$, but X-ray studies show an MNO angle of 180°.[60] On the other hand, $[Ir(CO)I(NO)(PPh_3)]^+$ has $\nu_{NO}$ of 1,720 $cm^{-1}$ and an MNO angle of 124°.[61] These two examples show that while for many cases IR spectroscopy can give information about the MNO angle, additional techniques should be used before making a conclusion.

### 3.3 Resonance Raman Spectroscopy

Raman spectroscopy has seen limited use in the identification of metal nitrosyl complexes. However, resonance Raman (RR) spectroscopy has proven particularly useful for probing the binding of NO to iron porphyrins and hemoproteins.[62–65] RR of $Fe^{II/III}$–NO porphyrin systems, excited in the Soret band, has enabled $\nu_{MN}$ to be detected for several complexes.[66]

Lipscomb et al., explored how the base *trans* to the NO group affected $\nu_{Fe-NO}$.[67] In isostructural $O_2$, $CN^-$, or CO complexes, the nature of the Fe–L interaction varied greatly as a function of the proximal base. However, the Fe–NO complexes seemed insensitive to either electronic or steric changes in the proximal base, as X-ray crystallographic studies confirm.[67] While complexes of $O_2$, $CN^-$, and CO have bond length differences of 0.07–0.15 Å,[68] the Fe–N distance in the iron-nitrosyl porphyrins changes by only 0.02 Å from 4-methylpiperidine to *N*-methylimidazole, suggesting that the NO group accommodates the change in electron density. Interestingly, the $\nu_{Fe-NO}$ is about 30 $cm^{-1}$ higher in the native proteins than in the model complexes.[62]

For $\{FeNO\}^7$ complexes with $S = 3/2$ electronic states the RR enhancement of the $\delta$(FeNO) vibration has been especially helpful for interpreting the electronic absorption and MCD spectra[69] (Section 4.2).

### 3.4 NMR spectroscopy

Nitrogen NMR is another spectroscopic technique that has been used to obtain information about the geometry of MNO groups. The large quadrupole moment for $^{14}N$ ($I = 1$) leads to broad lines in the NMR,[70,71] limiting the applicability of NMR studies of naturally occurring $^{14}N$; therefore, the majority of research in this field has been with $^{15}N$ ($I = 1/2$) enriched samples. Mason described both the general theory of $^{15}N$ NMR, and its application to nitric oxide chemistry.[72]

In the first extended studies of nitrosyl complexes by $^{15}N$ NMR, Bell et al., compared the $^{15}N$ chemical shifts of bent MNO moieties to those of related ligands with strongly bent nitrogen functionalities, such as diazenido complexes.[73] In such species, the low-lying $n_N$–$\pi^*$ electronic circulations in the magnetic field lead to a greater deshielding in the bent geometry, and dramatic fingerprint differences between linear and bent environments at the nitrogen atom. The same held true for $\{MNO\}^n$ complexes. The signals from the $^{15}N$-labeled complexes with a linear MNO group occurred in the $-50$ ppm $\leqslant \delta \leqslant$ 50 ppm range, whereas bent nitrosyls were observed at $\delta \geqslant$ 350 ppm with respect to neat liquid nitromethane. Values as high as 740 ppm were observed for certain cobalt complexes with bent NO groups.[73]

The advantage of $^{15}N$ NMR over vibrational spectroscopy, i.e., chemical shifts versus $\nu_{NO}$ frequencies, for determining the MNO geometry is apparent from the following examples. For [CpCr(NO)(CO)$_2$], a linear complex, $\nu_{NO} = 1{,}680\ cm^{-1}$,[74] but for *trans*-[RhCl(CO)(NO)(PPr$^i_3$)$_2$]$^+$, a bent complex, $\nu_{NO}$ occurs at $1{,}712\ cm^{-1}$.[73] Thus, the IR data alone do not enable unambiguous structural assignments to be made for this pair of complexes. The $^{15}N$ NMR spectra show shifts of 49.0 ppm for the chromium complex and 368.2 ppm for the rhodium, clearly indicating that the Cr compound has a linear configuration whereas the Rh complex is bent. The power of $^{15}N$ NMR for assigning MNO geometries in solution is partially negated by the expense of preparing the $^{15}N$ labeled compounds required for optimum sensitivity.

### 3.5 Photoelectron Spectroscopy

Many examples of PES and XPS spectra of MNO complexes are available in the literature.[75] The common thread amongst the spectra is the implication of significant mixing between the metal $d$ and $\pi^*$(NO) orbitals. Section 6 presents a detailed analysis of one particular example (*vide infra*).

## 4 ELECTRONIC STATES OF $\{FeNO\}^6$ AND $\{FeNO\}^7$ COMPLEXES

Nitroprusside is the prototypical example of a metal nitrosyl complex, and it has been studied extensively.[19] The geometry of the complex is well known from X-ray crystallographic studies, and theoretical studies, assuming idealized $C_{4v}$ geometry,[20] give the MO diagram for the anion shown in Figure 12. The HOMO is the $2b_2$ orbital which is primarily $d_{xy}$ in character, and the LUMO is the $9e$ orbital which is primarily $\pi^*$(NO) in character. From our earlier arguments (Section 1.3), this $\{FeNO\}^6$ complex will have a linear NO group in the ground state because the totally antibonding $9e$ orbital is much higher in energy than the $2b_2$.

Nitroprusside has become of particular recent interest because it exhibits two distinct metastable states upon irradiation at low temperatures.[76] The discovery of the states was the result of years of studies on the photo-excitation of nitroprusside, and is briefly discussed here. One impetus for further understanding of these metastable states is the possible utility of nitroprusside as an information storage system due to the color change exhibited by the crystal.[77]

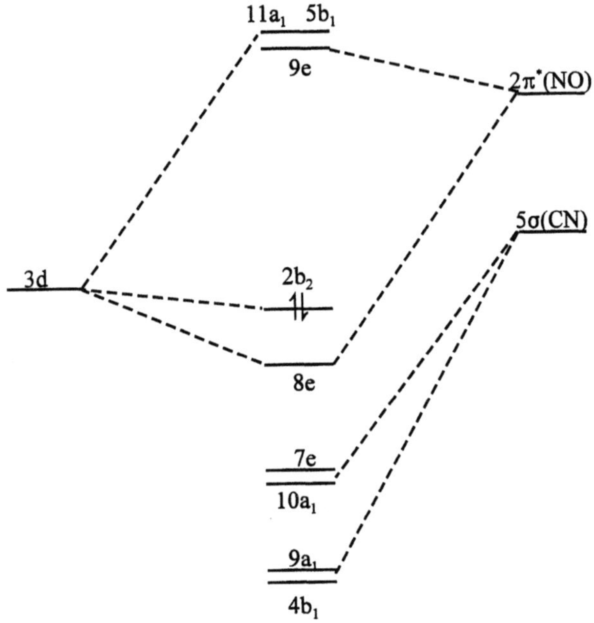


**Figure 12.** Molecular orbital diagram for $[Fe(NO)(CN)_5]^{2-}$.

## 4.1 Metastable States of Nitroprusside

In 1977, a study was undertaken to investigate the effect of polarized light on transparent Mössbauer substances.[78] The experimental results for sodium nitroprusside (SNP) did not agree with theory. It seemed that SNP was exhibiting a new molecular state that appeared to have an infinite lifetime upon photo-irradiation; however, this new state was thermally unstable. The Mössbauer spectra (Figure 13) showed that this new state had a substantially different quadrupole splitting than the ground state at 100 K (2.736 mm/s and 1.717 mm/s, respectively), suggesting a change in the electron density at the metal center. Furthermore, the absence of any line broadening showed that the metastable state was a physically well-defined state. The authors concluded from analysis of the data that a direct transition from the ground state to the excited state was impossible, and that an intermediate state must lie between the ground state and the new state.

In 1985, Yang and Zink explored the excited state by Raman spectroscopy and found that $\nu_{NO}$ for the new state was lower than that for the ground state, suggesting the new state had a bent Fe–N–O unit.[79] In two papers in 1986 and 1988, Güida et al. explored the infrared spectra of the metastable state.[80] They reported seeing four lines in the infrared which were assigned to the metastable state; however, upon heating the sample, these bands did not disappear at the same temperature. Instead, there appeared to be two distinct sets of bands. The first set disappeared at ~90 K and the second at ~140 K. This was the first spectroscopic evidence of *two* metastable states for SNP, although the authors did not identify them as such at that time.

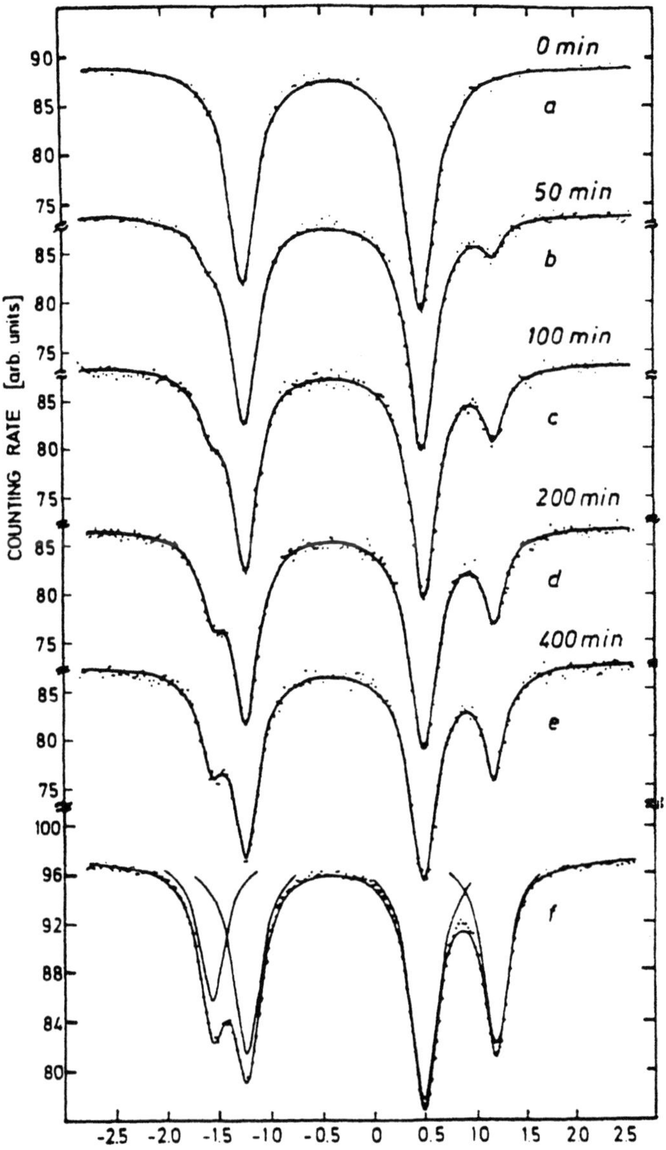


**Figure 13.** Time dependent Mössbauer spectra of SNP at 100 K: (a) normal groundstate spectrum; (b–f) growth of the additional quadruple spectrum with increasing laser irradation ($\lambda = 454$ nm). Reproduced with permission from Reference 78.

In 1989, Zöllner et al. performed differential scanning calorimetry (DSC) experiments on SNP and concluded that there were indeed two distinct, observable metastable states for the nitroprusside anion.[76] They studied the dependence of the decay temperatures of these two states as a function of counterion and water molecules per formula unit, and concluded that these variables had little effect on the temperatures at which the two metastable states appeared and decayed. Further, they revisited the assignment of the of the Raman and IR spectra previously reported, and explained the observed spectra in terms of two states instead of one. Güida et al. also reexamined their IR data from the viewpoint of determining the molecular geometry of the states.[81] They used isotopically labeled species and concluded that the Fe–N–O unit does not bend "appreciably" in either metastable state.

Since "appreciable" is difficult to quantify, the true structural nature of the metastable states remained a mystery without conclusive structural evidence for either state. To this end, several neutron diffraction studies were reported on the complex and its metastable states.[82,83] These studies supported the conclusions from the spectroscopic studies, and showed an elongation of the Fe–N bond in the metastable states. However, the reported result was an average of the two states because both states were present in the crystal.

Several preliminary X-ray studies were performed which were primarily concerned with the structure of the first metastable state ($MS_I$).[84,85] The conclusions from these studies were that the Fe–N bond did indeed lengthen, and that there was negligible bending of the FeNO group for this state, as compared to the ground state. The explanation is that there is a depopulation of the $2b_2$ orbital by one electron, and thus a population of the $9e$ by one electron. This leads to an overall $E$ state, and the degeneracy is lifted by a Jahn–Teller distortion; however, this distortion is seen in the C–Fe–N angle, and not the Fe–N–O angle in $MS_I$. This still did not describe the structure of the second metastable state ($MS_{II}$).

A recent study further explores the nature of $MS_I$ and addresses the structure of $MS_{II}$.[77] The results of this study are quite surprising. First, they do confirm the initial assumption from the Mössbauer studies that there must be an intermediate state between the ground and excited states. $MS_{II}$ appears to be intermediate between the geometries of the ground state and $MS_I$, and the linearity of $MS_I$ is confirmed, but that is not the whole story.

A very elegant experiment was carried out to determine the structure of $MS_{II}$ and further understand $MS_I$.[77] The reason $MS_I$ is further explored is due to the unsettling result of an excited state which appeared to have little geometric variance from the ground state, yet IR and Raman data show a decrease in the NO stretch of 105 $cm^{-1}$. This would suggest either a change in the NO bond length, or a bending of the Fe–N–O moiety, neither of which was previously observed. The results of this most recent study give both the structure of $MS_{II}$ and a reasonable explanation for the shift in NO stretch for $MS_I$.

In the groundstate SNP has $C_{4v}$ symmetry and a linear Fe–N–O unit (Figure 14a). Spectroscopic studies have suggested that the symmetry of the system is not destroyed in either metastable state; therefore, the X-ray data should confirm these results. For $MS_{II}$, the authors report a side bound, $\eta^2$(NO) unit (Figure 14b). This can be disordered in the unit cell (as shown in the figure) such that the overall symmetry averages to $C_{4v}$. Side on ($\eta^2$) geometry is well known for other small molecules,[86] and $\eta^2$-dioxygen complexes are well known.[87] Therefore, this proposed structure for the intermediate state seems reasonable.

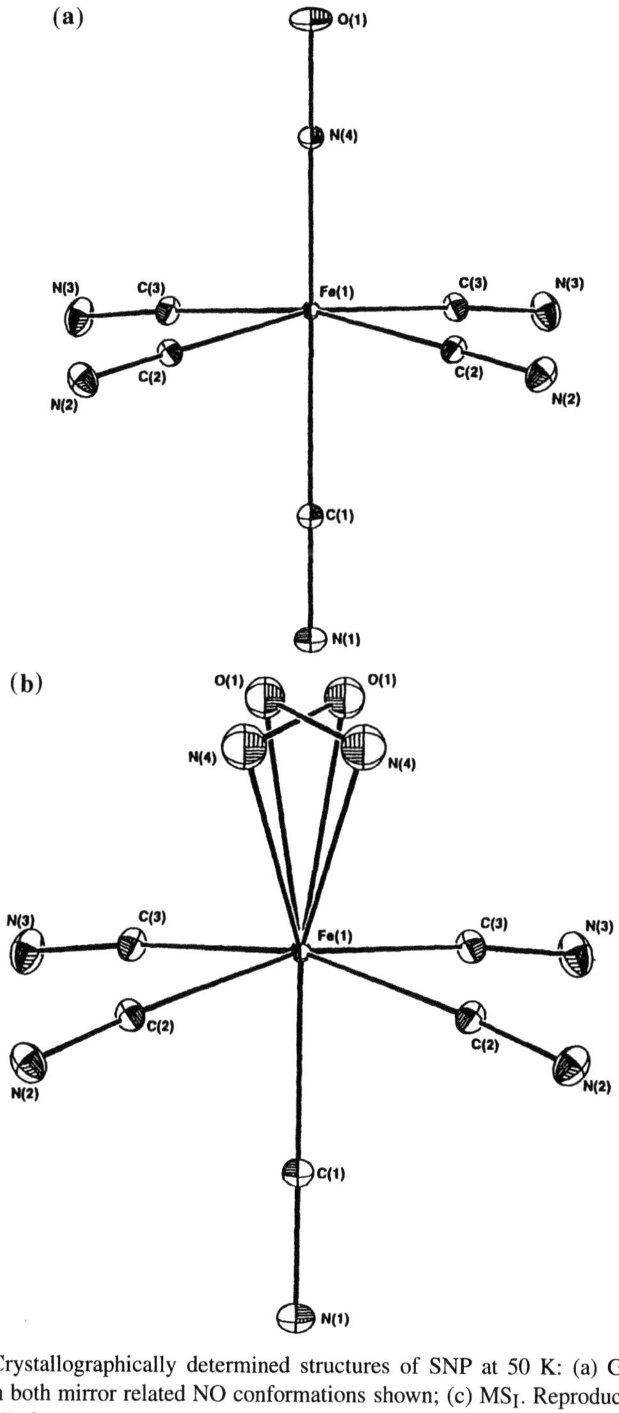


**Figure 14.** Crystallographically determined structures of SNP at 50 K: (a) Ground state; (b) $MS_{II}$ with both mirror related NO conformations shown; (c) $MS_{I}$. Reproduced with permission from Reference 77.

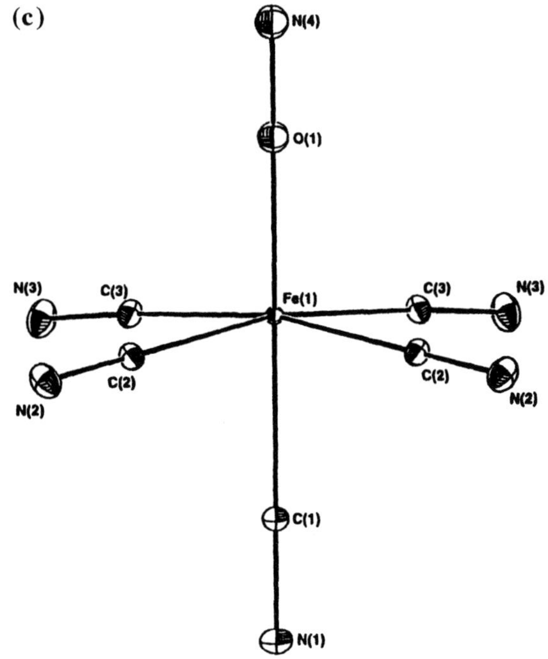


**Figure 14.** *(Continued.)*

As for $MS_I$, the data showed a linear Fe–N–O unit with some lengthening in the Fe–N bond, but little to no effect on the N–O bond length. This result is puzzling, unless we consider it in terms of the intermediate state. If $MS_{II}$ is an intermediate, then $MS_I$ may be an O-bound state. This would explain the linearity of the Fe–O–N unit, and the lengthening of the Fe–N (Fe–O) bond, without a lengthening of the N–O bond. Furthermore, the NO stretch would be expected to shift when going from an N-bound to an O-bound complex. Also, isonitrosyl complexes are not unknown, with several reports of O-bound nitrosyls.[5] Analysis of the data as the isonitrosyl analog yielded a featureless difference map and acceptable thermal and structural parameters. An ORTEP of this complex is shown in Figure 14c, and Figure 15 shows the results of an Extended Hückel calculations to model the relationship between the ground state and the two metastable states.

The recent results for SNP illustrate that the excited state properties of metal nitrosyl complexes can be even more complicated than the ground state structures discussed in Section 1. Complementary spectroscopic and structural techniques are essential for understanding the novel metastable states of SNP.

## 4.2 Magnetic States of $\{FeNO\}^7$ Complexes

A series of five- and six-coordinate complexes which contain a $\{FeNO\}^7$ unit have been synthesized, structurally characterized, and their magnetic characteristics have

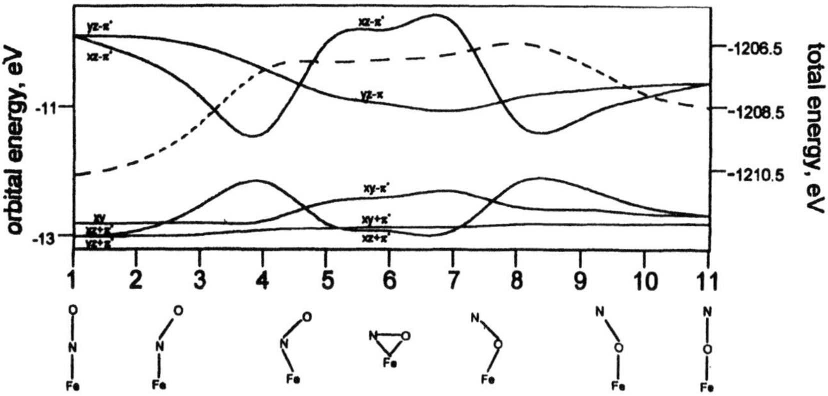


**Figure 15.** Walsh diagram of the end-to-end interconversion of nitrosyl to isonitrosyl in nitroprusside. The dotted line gives the total filled orbital energy; solid lines give individual orbital energies. Steps 1, 5, and 11 correspond to the groundstate, $MS_{II}$, and $MS_I$, respectively. Reproduced with permission from Reference 77.

been explored. A number of the five-coordinate complexes have an $S = 3/2$ state (from EPR) at room temperature and exhibit a spin crossover at low temperature to $S = 1/2$.[88–91] The six-coordinate complexes generally have low spin iron centers, but an exception to this has been reported. The complex (tacn)Fe(NO)$(N_3)_2$ (tacn = $N,N',N''$-trimethyl-1,4,7-triazacyclononane) exhibits an $S = 3/2$ ground state with no spin crossover.[92] The aqueous FeEDTA-NO complex also shows an $S = 3/2$ ground state at all $T$.[93] The behavior of this $S = 3/2$ state mimics that of the NO bound form of several non-heme enzymes, specifically soybean lipoxygenase (SBL).[94]

SBL is a representative example of a class of non-heme enzymes which bind and activate dioxygen. The NO forms of these species have been studied as reversible analogues of the enzyme-$O_2$ intermediates which may be formed during the catalytic cycle.[95] In order for the NO derivatives to be used as probes for the real dioxygen intermediates, their electronic structure must be clearly understood. Table 1 lists six different possible bonding descriptions and corresponding magnetic couplings that could give rise to an $S = 3/2$ ground state for an {FeNO}$^7$ center.

**Table 1 Possible Bonding Description for $S = 3/2$**

| | Description | Coupling |
|---|---|---|
| 1 | $[Fe^{+}d^7(S = 3/2)\text{-}NO^{+}(S = 0)]$ | N/A |
| 2 | $[Fe^{2+}d^6(S = 2)\text{-}NO^{0}(S = 1/2)]$ | Antiferromagnetic |
| 3 | $[Fe^{2+}d^6(S = 1)\text{-}NO^{0}(S = 1/2)]$ | Ferromagnetic |
| 4 | $[Fe^{3+}d^5(S = 1/2)\text{-}NO^{-}(S = 1)]$ | Ferromagnetic |
| 5 | $[Fe^{3+}d^5(S = 3/2)\text{-}NO^{-}(S = 0)]$ | N/A |
| 6 | $[Fe^{3+}d^5(S = 5/2)\text{-}NO^{-}(S = 1)]$ | Antiferromagnetic |

X-ray structure determination has shown that the $\{FeNO\}^7$ group of (tacn)-Fe(NO)$(N_3)_2$ is slightly bent with an FeNO angle of 155.5°.[92] For the FeEDTA-NO complex, the FeNO angle was found to be 156° from GNXAS multiple scattering analysis of the EXAFS data.[56] The current description of the bonding in the $\{FeNO\}^7$ centers in these $S = 3/2$ complexes has been obtained from a comprehensive study involving MCD, RR, EPR, XAS, magnetic susceptibility measurements, and SCF-X$\alpha$ theoretical calculations.[69,93]

The XAS Fe K-edge spectrum FeEDTA-NO (Figure 16) is most consistent with a ferric complex. In addition, the $1s \rightarrow 3d$ pre-edge feature (inset of Figure 16) appears as a single feature near 7113 eV that is characteristic of other high spin ferric complexes, and distinctly different from the split pre-edge peak found in the ferrous species, $[Fe^{II}(OH_2)EDTA]^{2-}$. The greater intensity of the pre-edge peak in FeEDTA-NO is consistent with a strong axial distortion of the iron site due to the short Fe–N(O) bond of ~1.7 Å.

EPR spectroscopy and temperature dependent magnetic susceptibility studies of FeEDTA-NO and (tacn)Fe(NO)$(N_3)_2$ show that both molecules are pure $S = 3/2$ spin state systems over the temperature range 2–260 K, with no spin crossover and no population of higher total spin states. Both complexes exhibit axial EPR spectra with $g_{\parallel} = 2.0$, $g_{\perp} \approx 4$, and E/D rations of less than 0.02. The overall $S = 3/2$ spin state for FeEDTA-NO and (tacn)Fe(NO)$(N_3)_2$ and the indication from XAS that the iron center is best described as high spin ferric, result in general bonding description 6 of Table 1 ($[Fe^{3+}d^5(S = 5/2)\text{-}NO^-(S = 1)]$) being favored, i.e., high spin ferric antiferromagnetically coupled to $NO^-(S = 1)$. The lower energy of the Fe edge in FeEDTA-NO compared to $[Fe^{III}(OH_2)EDTA]^-$ (Figure 16) indicates that the $NO^-$ ligand reduces the effective nuclear charge on the iron atom.

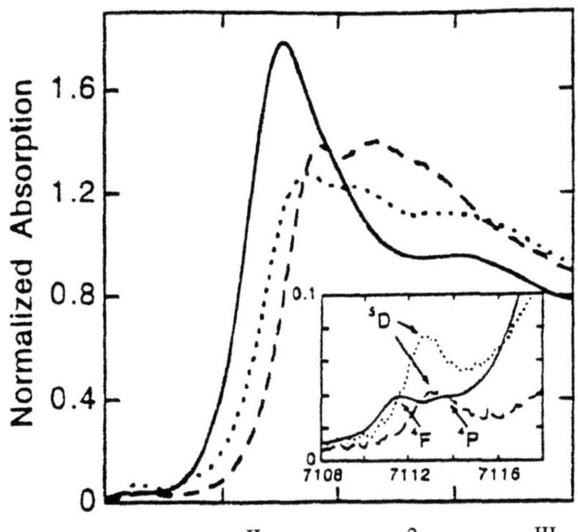


**Figure 16.** XAS Fe K-edge spectra of $[Fe^{II}(OH_2)EDTA]^{2-}$ (—), $[Fe^{III}(OH_2)EDTA]^-$ (- - -), and FeEDTA-NO (• • •). The inset shows the $1s \rightarrow 3d$ pre-edge region. Reproduced with permission from Reference 69.

Additional information about the binding of the NO ligand in FeEDTA-NO and (tacn)Fe(NO)($N_3$)$_2$ was obtained by resonance Raman (RR) spectroscopy of the $^{14}N^{16}O$, $^{15}N^{16}O$, and $^{14}N^{18}O$ complexes. Three vibrations were sensitive to isotope substitution and were assigned to $\upsilon$(NO), $\delta$(FeNO), and $\upsilon$(FeN). Normal coordinate analysis using a simple three-body model[59] resulted in experimental force constants that were between those of neutral NO and $NO^-$, consistent with the XAS results that the NO ligand is partially reduced. The values of $\upsilon$(NO) and $\upsilon$(FeN) for (tacn)Fe(NO)($N_3$)$_2$ are larger that those for FeEDTA-NO, indicating that more electron density has been donated from the NO ligand to the Fe center in (tacn)Fe(NO)($N_3$)$_2$, thereby strengthening both the FeN and NO bonds (*vida infra*).

The bonding in these {FeNO}$^7$ systems with $S = 3/2$ has been investigated by SCF-X$\alpha$-SW calculations[69] assuming both linear and bent (156°) FeNO units with the remaining ligands modeled by $NH_3$ and OH ligands. The general ordering of the energy levels for the spin restricted calculations parallels previous molecular orbital calculations on {FeNO}$^6$ complexes[19,20] with the NO $\pi^*$ orbitals similar in energy to the $\sigma^*$ orbitals on the metal (Figure 12). For a linear {FeNO}$^7$ group, the highest occupied level is predicted to be primarily NO $\pi^*$, and analogous to the totally antibonding $3\pi$ orbital of Figure 3. Bending of the FeNO unit splits the degeneracy of the NO $\pi^*$ level and significantly stabilizes the occupied component. This stabilization of the {FeNO}$^7$ group by bending is consistent with the observed FeNO geometry and with the qualitative bonding diagram of Figure 4c. The $S = 3/2$ spin state is reproduced by the spin restricted calculations, but the formal description is an intermediate spin $Fe^{2+}$ $d^6$ ($S = 1$) center ferromagnetically coupled to $NO^0$ ($S = 1/2$) (description 3 of Table 1), rather than the [$Fe^{3+}d^5(S = 5/2)$-$NO^-(S = 1)$] antiferromagnetically coupled electronic structure (description 6 of Table 1) favored by the XAS results.

The bonding in these $S = 3/2$ complexes was also investigated by spin unrestricted calculations, which better reproduce exchange effects in open shell compounds by allowing individual spatial orbitals for "spin up" and "spin down" electrons.[96] Figure 17 shows the spin unrestricted energy level diagrams for the linear and bent species of the calculated model. The substantial stabilization of the spin up Fe $3d$ orbitals relative to their spin down counterparts (~6 eV) commonly occurs in high spin ferric compounds.[96,97] The NO $\pi^*$ orbitals also split, and the spin down orbitals are stabilized by spin pairing with the spin up $d_{xz}$ and $d_{yz}$ orbitals of the Fe through direct overlap. Thus, the spin unrestricted calculations give $d^5$ $Fe^{3+}$ $S = 5/2$ antiferromagnetically coupled to $NO^-$ $S = 1$ (description 6, Table 1) for the electronic structure of the $S = 3/2$ state, in agreement with the XAS results. Further experimental support for this description is provided by the EPR[98] and Mössbauer spectra.[99] The spin pairing of the two electrons on the $NO^-(S = 1)$ ligand with the two Fe $d_{xz,yz}$ electrons accounts for the absence of hyperfine splittings by N or O, and the $(d_{xy})^1$, $(d_{x^2-y^2})^1$, $(d_{z^2})^1$ electrons would produce anisotropic spin distribution in the $xy$ plane.

The spin unrestricted calculations for linear and bent structures of Figure 17 have nearly identical energy level orderings, in contrast to the spin restricted calculations

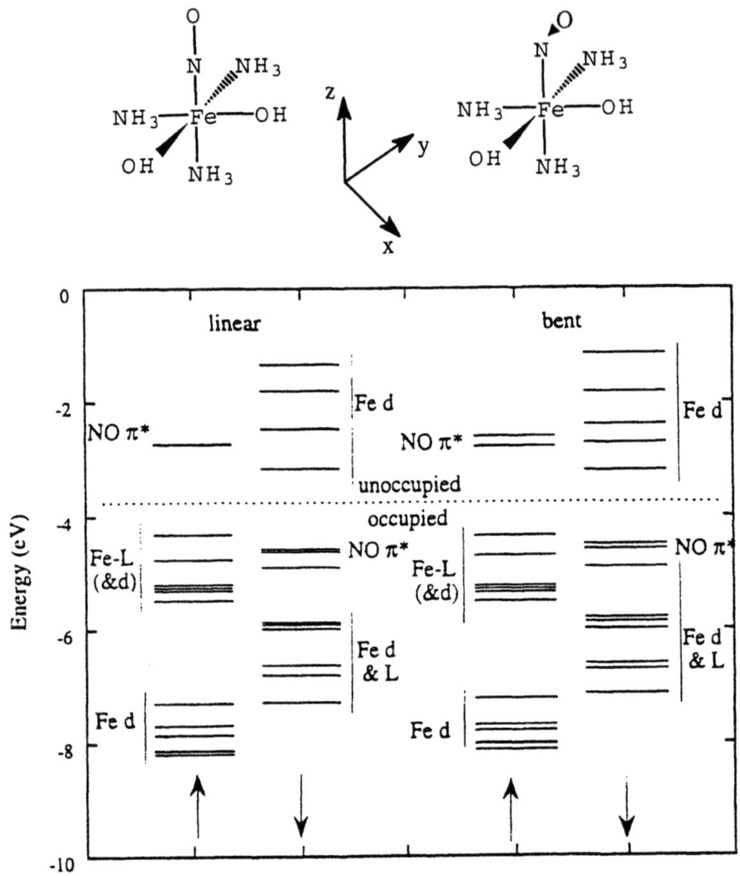


**Figure 17.** Energy level diagrams for spin unrestricted calculations on model six-coordinate complexes with both linear (left) and bent (right) FeNO groups. Reproduced with permission from Reference 69.

which showed significant energy stabilization for the bent form. Figure 18 shows the contour plots for the orbitals that are predominantly NO $\pi^*$ for the spin unrestricted calculation for the linear structure. The two singly occupied spin down orbitals are *π-bonding* with respect to Fe and N and *π-antibonding* with respect to NO. The Fe character of these orbitals changes little upon bending the FeNO unit, but bending increases the $\sigma$ bonding by mixing in additional Fe 4*s* and 4*p* character (Figure 19). The Fe–N bonding character for these singly occupied NO $\pi^*$ orbitals contrasts with the simple qualitative spin restricted descriptions of Figures 3 and 4c, which predict that the HOMO (e.g., $3\pi$ of Figure 3) should be totally $\pi$-antibonding with respect to Fe, N and O. In the spin unrestricted calculation the totally antibonding levels are unoccupied (Figures 18 and 19).

The predominant bonding interactions involving the NO orbitals found from the spin unrestricted calculations are shown in Figure 20. Of particular importance are

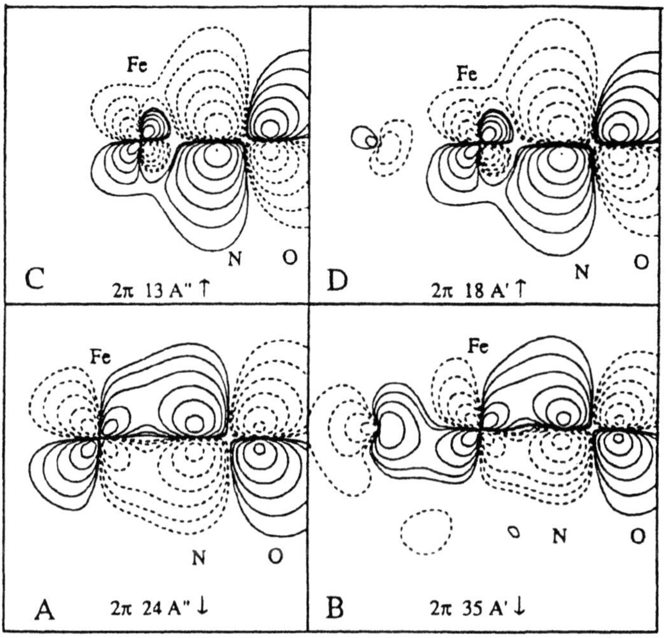


**Figure 18.** Contour plots of the predominatly NO $\pi^*$ orbitals of the linear calculation of Figure 17. Spin down orbitals A and B are each occupied by one electron; spin up orbitals C and D are empty. Reproduced with permission from Reference 69.

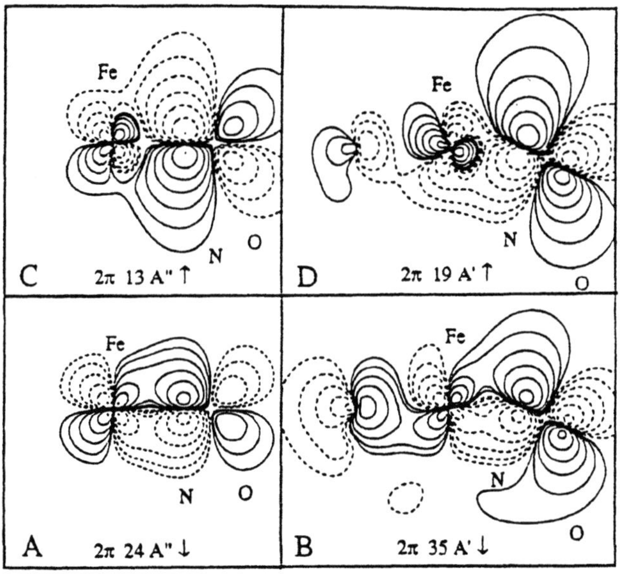


**Figure 19.** Contour plots of the predominantly NO $\pi^*$ orbitals of the bent calculation of Figure 17. Spin down orbitals A and B are each occupied by one electron; spin up orbitals C and D are empty. Reproduced with permission from Reference 69.

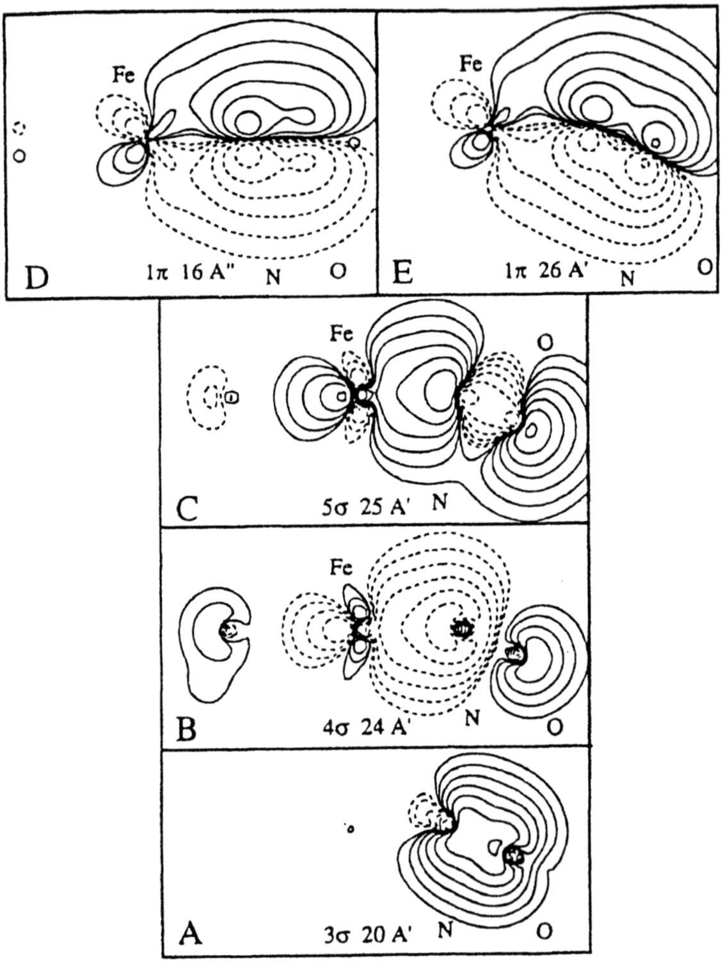


**Figure 20.** Contour plots of the predominantly bonding NO orbitals for the bent calculation of Figure 17. A is primarily NO $\pi$-bonding; B and C are strongly Fe–N $\sigma$-bonding (see text); D and E are FeNO $\pi$-bonding. Reproduced with permission from Reference 69.

orbitals B and C which are Fe–N $\sigma$ bonding and at least 20% Fe in character and which change little upon FeNO bending. Orbital B primarily involves $\sigma$ donation to the Fe $4s$, $4p$, $3d$ orbitals from the antibonding NO $2s$ orbitals; orbital C involves $\sigma$ donation from the bonding combination of $p_z$ orbitals for the NO. Although B and C of Figure 19 both contribute to Fe–N $\sigma$ bonding, they have opposite effects on the NO bond strength. Interaction B transfers charge from an antibonding NO orbital and should increase the NO bond order; interaction C transfers charge from a bonding NO orbital and should have the reverse effect. The observed force constants for the FeNO groups of FeEDTA-NO and (tacn)Fe(NO)$(N_3)_2$ are larger than those for $NO^-$, implying that interaction B is more important than C. These covalent $\sigma$ donation interactions from the $NO^-$ group are also reflected in the 2 eV lower Fe K-edge energy of FeEDTA-NO relative to $[Fe^{III}(OH_2)EDTA]^-$ (Figure 16). The major contribution of $\sigma$ bonding to $\{FeNO\}^7$ complexes with $S = 3/2$ states contrasts with the results found for linear $\{NiNO\}^{10}$ compounds (Section 6).

The electronic absorption and MCD spectra of FeEDTA-NO and (tacn)Fe(NO)-$(N_3)_2$ are also consistent with the electronic structure derived from the spin unrestricted calculations. The lowest energy spectral feature of (tacn)Fe(NO)$(N_3)_2$ (9,230 $cm^{-1}$) is assigned to a ligand field transition of the $Fe^{3+}$ center whose intensity is enhanced by coupling with the NO group. The bands between 13,450 and 18,700 $cm^{-1}$ are assigned to NO $\pi^*$ to Fe charge transfer bands, consistent with the enhancement of the Fe–N–O bend in the RR spectrum upon excitation in these bands.

The combined spectroscopic and theoretical studies on FeEDTA-NO and (tacn)-Fe(NO)$(N_3)_2$, as well as related studies on SBL-NO,[93] provide a self consistent picture of the electronic structures of the $S = 3/2$ state that is best described as high spin $Fe^{3+}$ antiferromagnetically coupled to $NO^-$ ($S = 1$). This system is the most thoroughly characterized $\{MNO\}^n$ group, and illustrates the value of spin unrestricted calculations for interpreting the properties of paramagnetic nitrosyl compounds for which multiple formal electronic structures can be written.

## 5 NITROSYL DERIVATIVES OF HEMES AND METALLOPORPHYRINS

Heme proteins play a key role in the biological synthesis and transport of nitric oxide. As noted earlier (Section 1.1), nitric oxide synthases (NOS) are heme-containing enzymes that produce NO from arginine, the other product being citruline.[3h,i,j,k] Nitrophorins are heme proteins that reversibly bind NO and which are found in the salivary glands of certain bloodsucking insects, including *Rhodnius prolixus*.[100] The importance of heme proteins in the biochemistry of nitric oxide has reawakened interest in nitrosyl derivatives of metalloporphyrins.[139]

### 5.1 Reversible Binding of NO

The salivary glands of the bloodsucking insect *Rhodnius prolixus* have a deep red color due to an abundance of heme proteins.[100] The Soret band of the isolated proteins occurs at 422 nm under exposure to NO and shifts to 406 nm when flushed with argon. Under an atmosphere of NO the protein is EPR silent, whereas argon-flushed samples exhibit an EPR spectrum characteristic of a high-spin Fe(III) heme center. A detailed study of the reversible binding of NO to NP1, including the pH dependence of the dissociation constants, has recently appeared.[137] These data are most consistent with an Fe(III) heme protein that can reversibly bind NO as a function of pH. At the low pH of the saliva glands the NO is strongly bound, however, at the higher pH of the blood and tissues of the victim, the NO is released. The known vasodilator and anti-coagulant properties of the released NO help insure the insect of a good blood meal from the victim.

Nitric oxide binds much less strongly to Fe(III) hemes than to Fe(II) hemes and the off-rates of NO from Fe(III) hemes are much larger than those from Fe(II) hemes.[3a,101,102] Thus, the pH dependent reversible transport of NO by nitrophorins can be reasonably explained. Nonetheless, reversible binding of nitric oxide by a transition metal center is a relatively rare and poorly studied phenomenon.

Very recently the three-dimensional structure of nitrophorin 1 (NP1) from *Rhodnius prolixus* has been determined to 2.0 Å resolution by X-ray crystallography.[103] The protein is nearly all beta-sheet and evolutionarily different from the globins. NP1 is structurally similar to lipocalins, proteins involved in binding lipophilic molecules.[104] The heme is six-coordinate with a histidine in the proximal site; in the absence of NO the spacious pocket on the distal side can be occupied by several different ligands ($NH_3$, $CN^-$, histamine).

The binding of NO to a nitrophorin involves bringing together a high-spin Fe(III) center ($S = 5/2$) and the NO molecule ($S = 1/2$) to give a diamagnetic ($S = 0$) system which contains the $\{FeNO\}^6$ unit. Table 2 lists 9 different possible descriptions for an $\{FeNO\}^6$ unit that are the analogues of those presented in Table 1 for $\{FeNO\}^7$ systems. For the $\{FeNO\}^6$ unit of the nitrophorins the end description must be either 1 or 2 because they are the only diamagnetic possibilities ($S = 0$). However, it is unclear if the reaction coordinate progresses from the $S = 5/2$ (unligated form) to the $S = 0$ (NO bound form) in a single step, or if it proceeds through a sequence of the spin state descriptions shown in Table 2. Moreover, does the reaction coordinate for reversible NO binding to form the $\{FeNO\}^6$ group of nitrophorins involve the side-on or O-bonded geometries found for the metastable states of SNP (Figure 14)? The nitrophorins and their NO binding properties present a unique opportunity for extensive studies of the fundamental properties of metal nitrosyl centers by a wide range of spectroscopic methods.

Few $\{MNO\}^6$ complexes of porphyrins have been structurally characterized by X-ray crystallography. The five- and six-coordinate $\{MNO\}^6$ complexes that result

**Table 2 Possible Bonding Descriptions for $\{FeNO\}^6$**

| | Description | Total Spin ($S$) | Coupling |
|---|---|---|---|
| 1 | $[Fe^{2+}d^6(S = 0)\text{-}NO^+(S = 0)]$ | 0 | N/A |
| 2 | $[Fe^{3+}d^5(S = 1/2)\text{-}NO^0(S = 1/2)]$ | 0 | Antiferromagnetic |
| 3 | $[Fe^{2+}d^6(S = 1)\text{-}NO^+(S = 0)]$ | 1 | N/A |
| 4 | $[Fe^{3+}d^5(S = 1/2)\text{-}NO^0(S = 1/2)]$ | 1 | Ferromagnetic |
| 5 | $[Fe^{3+}d^5(S = 3/2)\text{-}NO^0(S = 1/2)]$ | 1 | Antiferromagnetic |
| 6 | $[Fe^{2+}d^6(S = 2)\text{-}NO^+(S = 0)]$ | 2 | N/A |
| 7 | $[Fe^{3+}d^5(S = 3/2)\text{-}NO^0(S = 1/2)]$ | 2 | Ferromagnetic |
| 8 | $[Fe^{3+}d^5(S = 5/2)\text{-}NO^0(S = 1/2)]$ | 2 | Antiferromagnetic |
| 9 | $[Fe^{3+}d^5(S = 5/2)\text{-}NO^0(S = 1/2)]$ | 3 | Ferromagnetic |

from reaction of Mn(II) and Fe(III) porphyrinates with NO all exhibit essentially linear MNO groups.[105,106] The chemical studies of Richter-Addo and coworkers have resulted in several $\{RuNO\}^6$ porphyrinate complexes.[107] Their investigations of the reactions of *S*-nitroso compounds with ruthenium(II) porphyrins as models for the interaction between heme proteins and *S*-nitroso compounds in biology led to the first structural characterization of a nitrosyl thiolate derivative of a metalloporphyrin, Ru(OEP)(NO)(*S*-NACysMe);[108] the Ru–N–O angle is 175°. They have also prepared the first structurally characterized organometallic nitrosyl porphyrin, Ru(TTP)(NO)($p$-$C_6H_4F$).[109] The Ru–N bond is inclined 9.8° from the normal to the plane of the four porphyrin nitrogen atoms, and the Ru–N–O angle is 152°. The authors suggest that these distortions of the $\{RuNO\}^6$ group from linearity may be due to crystal packing forces. However, electronic effects may also play a role. The presense of a *single* axial aromatic ligand lowers the overall symmetry to $C_s$ and splits the degeneracy of metal $d_{xz}$, $d_{yz}$ orbitals because only one of these orbitals can participate in $d\pi - p\pi$ bonding with the $\pi$-orbitals of the aryl group.[110] Such splitting of the $d_{xz}, d_{yz}$ orbitals could in turn affect the Ru–NO $\pi$-bonding interaction by lifting the degeneracy of the $2e$ level of Figure 4. This perturbation of the electronic structure of the $\{RuNO\}^6$ group could make it easier to bend the group in a specific direction under the influence of crystal packing forces.

### 5.2 Properties of $\{FeNO\}^7$ Porphyrins

Nitric oxide has long been used as paramagnetic probe of heme proteins involved in oxygen transport and metabolism.[111] This original interest in nitrosyl derivatives of heme proteins as models for metal–oxygen binding is now being superseded by the direct relationship of nitrosyl hemes to the transport and metabolism of nitric oxide itself.[112] Reaction of NO with Fe(II) hemes or reductive nitrosylation of Fe(III) hemes by NO generate $\{FeNO\}^7$ centers with one unpaired electron ($S = 1/2$) that can be studied by EPR spectroscopy. A comprehensive survey of this extensive EPR literature is beyond the scope of this review, but one feature of common interest has been the $^{14}N$ hyperfine splitting due to the nitrogen atom of the nitrosyl group and the implications of this splitting for the coordination environment about the iron atom. It has long been know from studies of discrete $\{FeNO\}^7$ complexes of protoporphyrin IX[113] and TPP[114] that the experimental $g$-values and the $^{14}N$ hyperfine splitting of the nitrosyl group are both dependent upon the axial ligand that is trans to the NO group.

Several crystal structures of five- and six-coordinate $\{FeNO\}^7$ complexes of porphyrins have been determined.[115,116] The Fe–N–O angles range from 138–149°; the substantial bending of the FeNO group indicates that the electronic structures of these complexes are best described by Figure 4d. The detailed geometry of the FeNO group in such complexes is often partially obscured by crystallographic disorder of the bent NO group. However, very recently Ellison and Scheidt have reported two well-ordered polymorphs of Fe(OEP)(NO) with essentially identical Fe–N–O angles of 143 and 144°.[117] The Fe–N(O) vector in these ordered structures is not perpendicular to the mean heme plane, but is tilted off axis by 6.5–8.2°, in the same direction

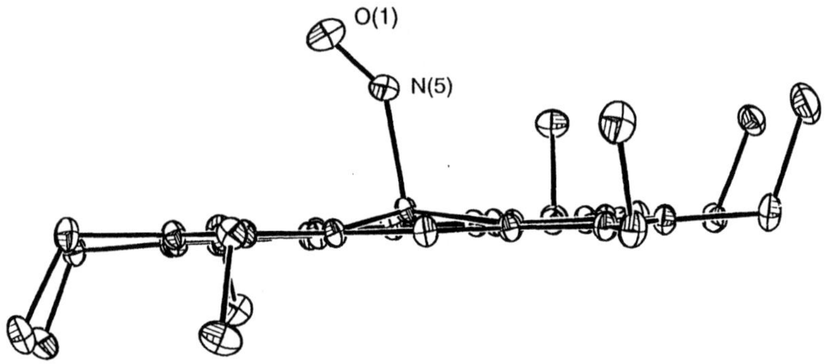


**Figure 21.** An edge-on view of one form of Fe(OEP)(NO) showing the bending of the FeNO group and angle of the Fe–N bond relative to the mean porphyrin plane. Reproduced with permission from Reference 117.

as the O atom (Figure 21). The two Fe–N(porphyrin) distances closest to the tilted NO group are slightly shorter (0.02 Å) than the other pair of Fe–N(porphyrin) distances. This difference is present in both polymorphs and is about 5 to 10 times the estimated errors on the Fe–N bond lengths. The EPR spectra of Fe(OEP)(NO) in frozen solution resemble those originally reported by Wayland and Olson[114] for Fe(TPP)(NO) and shown in Figure 22, and therefore the structure provides a firm basis for interpreting the EPR spectra such complexes.

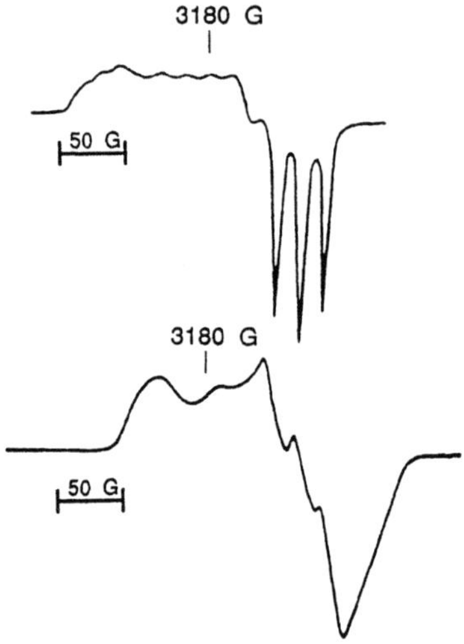


**Figure 22.** X-band EPR spectra in toluene glass (120° K); the units for the $A(^{14}N)$ hyperfine splittings are $10^{-4}$ $cm^{-1}$: (top) five-coordinate Fe(TPP)(NO) [$g_1 = 2.102$, $A_1(^{14}N) = 12.4$; $g_2 = 2.064$, $A_2(^{14}N) = 16.6$; $g_3 = 2.010$, $A_3(^{14}N) = 16.2$]; (bottom) six-coordinate Fe(TPP)(NO)(pip) [$g_1 = 2.08$; $g_2 = 2.04$; $g_3 = 2.003$, $A_3(^{14}N) = 20.3$]. Reproduced with permission from Reference 114.

The rhombic nature of the frozen solution EPR spectrum of Fe(TPP)(NO) is consistent with the $C_s$ symmetry of the molecule. All five-coordinate $\{FeNO\}^7$ porphyrin complexes show $^{14}N$ ligand hyperfine splitting by the coordinated NO group, with the sharp three-line hyperfine splitting of $g_3$ being especially distinctive (Figure 22). The $^{14}N$ hyperfine splitting from the nitrosyl group implies that the unpaired electron resides in an orbital that is primarily $d_{z^2}$ and N$\sigma$ (i.e., $2a'$ of Figure 4d). Addition of a sixth ligand trans to the NO group will raise the energy $d_{z^2}$ and decrease its contribution to the SOMO, thereby changing both the $g$-values and the $^{14}N$ hyperfine splitting pattern (Figure 22), as discussed in Section 2.2. If the additional ligand is a nitrogen donor then the $^{14}N$ ligand hyperfine pattern will be complicated by overlap of the unequal ligand hyperfine splittings from the two kinds of axial nitrogen donors.[113,114] The simultaneous presence of both five- and six-coordinate nitrosyl species can further complicate the EPR spectra.[113,118]

The sensitivity of the EPR parameters of $\{FeNO\}^7$ porphyrin centers to the nature of the axial ligand (*vida supra*)[113,114] is also consistent with the qualitative molecular orbital diagram of Figure 4d because the HOMO will be a mixture of $d_{z^2}$ and a nitrogen $\sigma$-orbital whose exact composition will depend upon the donor ability of the axial ligand trans to the NO group. However, $\{FeNO\}^7$ complexes of porphryins and hemes all appear to have strongly bent FeNO units and $S = 1/2$ ground states. This behavior contrasts with the wide range of FeNO geometries and electronic structures found for $\{FeNO\}^7$ complexes of other ligand systems,[1,2] such as those described in Section 4.2.

As mentioned in Section 5.1, nitric oxide is known to bind very strongly to Fe(II) in porphyrin systems. In an attempt to determine what factors may change the affinity of Fe(II) for NO, Bohle and Hung undertook a kinetic study of a series of Fe(II) porphyrin systems.[115] Using a variety of substituents in the meso and pyrrole positions, and using either pyridine or $N$-methylimidazole as the displacing ligand, the authors found that the pseudo-first order rate constant for NO displacement varied from $4.17 \times 10^{-5}$ s$^{-1}$ to 37.1 s$^{-1}$ depending on the nature of the substituent groups. The reasons for the extreme variations in the rate constants are not completely understood, but all of these systems have a bent FeNO unit and a severely distorted porphyrin (saddle shaped). Ellison and Scheidt have speculated that distortions may play an important role in dissociation rates.[117] Rate constants that differ by six orders of magnitudes further demonstrate that there is still much to learn about the interaction of NO with metalloporphyrins.

The interconversion of $\{FeNO\}^7$ and $\{FeNO\}^6$ derivatives of porphyrins has also been observed. The reaction of Fe(TPP)(NO) with excess NO produces Fe(TPP)(NO)($NO_2$) and results in an increase of $\nu$(NO) from 1,675 to 1,884 cm$^{-1}$,[119] consistent with converting a bent $\{FeNO\}^7$ group into a linear $\{FeNO\}^6$ group. *In situ* IR and UV/vis studies of a series of Fe(porphyrin)(NO) compounds by spectroelectrochemistry provided evidence for Fe(porphyrin)(NO), [Fe(porphyrin)-(NO)]$^+$, Fe(porphyrin)(NO)X (X = $Cl^-$, $Br^-$, $I^-$), and [Fe(porphyrin)(solvent)]$^+$ species with $\nu$(NO) frequencies ranging from 1,667 to 1,925 cm$^{-1}$.[120]

## 6 ELECTRONIC STRUCTURE OF CpNiNO

The electronic structure of CpNiNO has been extensively studied because of the simplicity of the molecule, its high symmetry, and ease of synthesis.[121] The complex is a liquid at room temperature, and its structure in the gas phase has been determined by microwave spectroscopy,[122] a useful technique for obtaining the structures of volatile metal nitrosyls in the absence of perturbing crystal lattice forces.[123] The $\{NiNO\}^{10}$ moiety is linear and the molecular symmetry is $C_{5v}$.

Figure 23 shows a qualitative molecular orbital diagram for the linear $\{NiNO\}^{10}$ fragment with $C_{\infty v}$ symmetry. The numbering of the orbitals in Figure 23 parallels that in Figure 3, but Figure 23 has the $4\sigma\,[d_{z^2}, \sigma(\text{NO})]$ orbital lower in energy than the totally antibonding $3\pi\,[d_{xz}, d_{yz}, \pi^*(\text{NO})]$ orbital. This will make the $4\sigma$ the HOMO of the fragment, in accord with the linear $\{NiNO\}^{10}$ group in CpNiNO. The energy of $4\sigma$ (i.e., $d_{z^2}$) relative to the formally non-bonding $\delta(d_{xy}, d_{x^2-y^2})$ orbital is one of the primary questions of interest in the electronic structure of CpNiNO (*vide infra*). Therefore, in Figure 23 the energy of the $4\sigma$ has been represented by a box which spans the energy gap between the $\delta$ and $3\pi$ levels. This box has been carried over to Figure 24, which shows the qualitative energy level diagram for CpNiNO that results from interaction of the linear $\{NiNO\}^{10}$ group with the filled $\pi$ orbitals of $Cp^-$. The HOMO is most likely $e_1$, but the energy of the $a_1(d_{z^2})$ relative to $e_1$ and the non-bonding $e_2(d_{xy}, d_{x^2-y^2})$ orbital remains a fundamental question. Moreover, the electron density of the Cp ring lies primarily in the node of $d_{z^2}$ so there is little

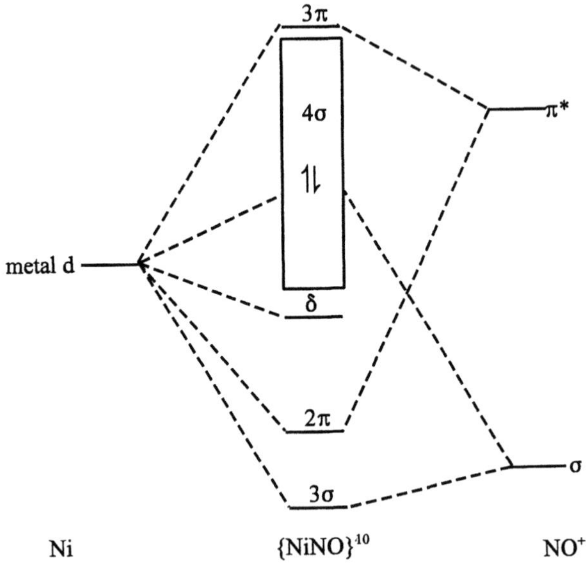


**Figure 23.** Molecular orbital diagram for the $\{NiNO\}^{10}$ unit in $C_{\infty v}$ symmetry. The $4\sigma$ level has been shown as a box to represent the possible range of $\sigma$-donor interaction of the NO ligand.

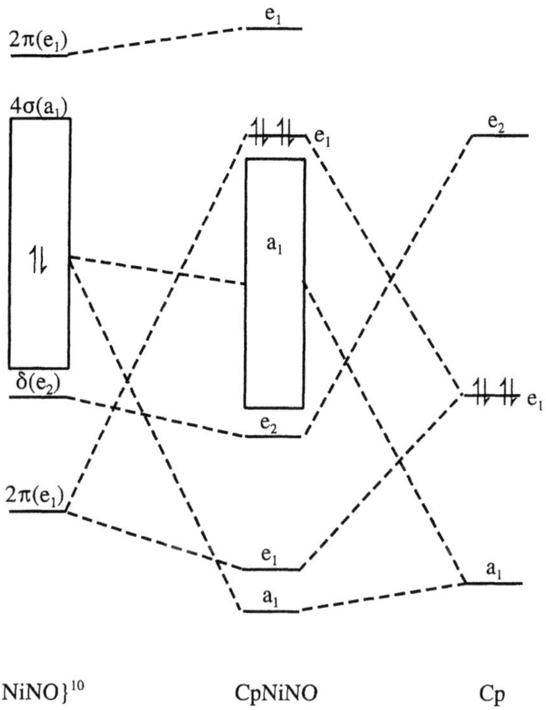


**Figure 24.** Molecular orbital diagram for CpNiNO. The $a_1$ level is shown as a box to represent the uncertainty of the energy of this filled orbital (see text).

overlap between $d_{z^2}$ and the appropriate filled Cp orbital.[124] Thus, the primary factor perturbing the energy of the $a_1(d_{z^2})$ orbital of CpNiNO in Figure 24 is expected to be $\sigma$ donation from the NO ligand. Consequently, the energy of the $a_1$ should provide an excellent opportunity to assess the $\sigma$-donor ability of a linear NO ligand.

## 6.1 Photoelectron Spectra

Gas phase photoelectron spectroscopy (PES) is a powerful method for experimentally probing the electronic structures of organometallic molecules.[125] The first PES study of CpNiNO was reported in 1974 by Evans et al.[126] using HeI radiation. More recently three separate groups have reinvestigated the PES of CpNiNO using several excitation sources.[127–129] The PES of CpNiNO exhibits three major features in the 8–11 eV region (Figure 25). Evans et al. used the results of *ab initio* molecular orbital calculations to assign the lowest energy feature (peak A) as arising from both the $e_1$ and $a_1$ ionizations, with the $e_2$ ionization assigned to peak B.[126] This original assignment of the spectrum suggests that NO is a strong $\sigma$ donor because the energy of the $a_1(d_{z^2})$ orbital is much higher than that of the essentially non-bonding $e_2$.

The three recent reinvestigations[127–129] of the PES of CpNiNO have taken advantage of the relative differences in photoionization cross sections of different types

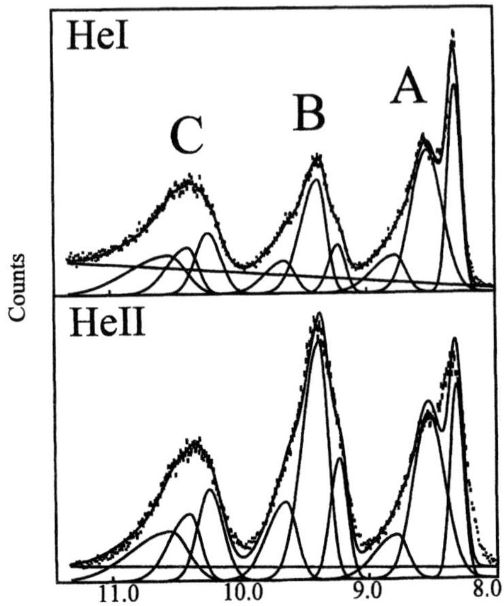


**Figure 25.** Gas phase HeI and HeII PES spectra of CpNiNO showing the three principal spectral features (A, B, C). Reproduced with permission from Reference 129.

of atoms as a function of the energy of the ionizing radiation.[125] Table 3 shows the cross-sections for the atoms of CpNiNO for HeI (21.2 eV) and HeII (40.8 eV); cross-section values for other energies are also available.[130] From Table 3, it is clear that upon changing from HeI to HeII radiation the relative intensity of peaks due to ionizations from orbitals that are primarily Ni in character should increase with respect to ionizations that have substantial ligand character. The spectra of Figure 25 show that bands A and B increase in intensity relative to C and therefore bands A and B must be largely metal in character. Peak B, which shows the greatest increase in intensity, can be assigned to ionization from the $e_2(d_{xz}, d_{yz})$ orbital which is localized on the Ni atom (Figure 24). The shoulder on peak A also appears to grow in intensity in the HeII spectrum, and Bancroft and coworkers assigned this feature to

**Table 3 Photoionization Cross-Sections for Selected Elements (Mbarns)[a]**

| | HeI (21.2 eV) | HeII (40.8 eV) |
|---|---|---|
| Ni | 3.984 | 8.357 |
| N | 9.688 | 4.351 |
| C | 6.128 | 1.875 |
| O | 10.67 | 6.816 |

[a]Data from Reference 128.

ionization from $a_1(d_{z^2})$, as originally proposed by Evans et al.[126] However, Green and coworkers claim that the apparent increase in the intensity of the shoulder of peak A with HeII radiation is an artifact of the decreased resolution at higher ionization energy.[128] They have assigned the shoulder on peak A to an NO stretching vibration on the $e_1$ ionization. They further propose that the ionizations from the $a_1$ and $e_2$ orbitals are essentially degenerate and that both are located under peak B. This assignment would mean that a linear NO ligand is a very weak $\sigma$ donor.

Pollard and Lichtenberger have carried out HeI and HeII studies of CpNiNO under experimental conditions that enabled both spectra to be obtained with the same resolution.[129] These experiments showed that the relative intensities of the main feature of peak A (Figure 25) and its high energy shoulder were nearly independent of the photoionization energy, consistent with $a_1$ and $e_2$ being nearly degenerate, as proposed by Green and coworkers. In addition, the PES spectrum of Cp*NiNO has also been investigated.[129,131] Comparative PES studies of analogous Cp and Cp* complexes can be very helpful for elucidating the electronic structures of metal complexes.[132] The Cp* ligand is more electron donating than Cp, and thus metal ionizations that also contain substantial Cp character should shift to lower energy upon Cp* substitution. If band A contains the ionization from $a_1(d_{z^2})$ and $e_1(d_{xz}, d_{yz}, \pi\text{Cp})$ then substition of Cp* should destablize the $e_1$ (HOMO) but leave $a_1(d_{z^2})$ essentially unchanged, and an increased splitting between the main feature and the shoulder of band A would be observed. On the other hand, if band A is a vibrationally split ionization of $e_1(d_{xz}, d_{yz}, \pi\text{Cp})$ then substitution by Cp* will move the entire envelope of band A to lower energy, but leave the splitting essentially unchanged. Band B is not expected to change with Cp* substituion because it is primarily metal based in both assignments. A decrease in the overall ionization energy of the envelope of band A upon Cp* substitution was observed, but little change was seen in the other ionizations.[129,131] The PES results for Cp*NiNO are most consistent with the suggestion that the $a_1$ and $e_2$ orbitals are nearly degenerate, thereby implying that a linear NO ligand is a very weak $\sigma$ donor in this $\{NiNO\}^{10}$ system. The weak $\sigma$ donor ability of a linear NO ligand in this system contrasts with the major importance of $\sigma$ bonding in bent $\{FeNO\}^7$ systems (Sections 4.2 and 5.2).

## 6.2 Metastable State

All of the models for the bonding in CpNiNO agree that the LUMO is the $e_1$ molecular orbital that is totally $\pi$-antibonding with respect to Ni and NO and the analogue of the 3$e$ orbital of Figure 4. Thus, the LUMO of CpNiNO is electronically very similar to the LUMO of the nitroprusside ion, $[Fe(NO)(CN)_5]^{2-}$, an $\{FeNO\}^6$ center discussed in Section 4.1 Like the nitroprusside anion, CpNiNO exhibits a photo induced excited state, and one metastable state has been observed.[133] Electronic and infrared spectra were presented as evidence for this new state, and the authors proposed that a shift in electron density to the nitrosyl ligand from the metal center produced either a bent NiNO unit or an isonitrosyl complex. Recently the structure of the metastable state of Cp*NiNO has been determined at low temperature and shown to be an $\eta^2$-NO complex, similar to the metastable state II of SNP.[134,135]

## 7 SUMMARY

The $\{MNO\}^n$ bonding formalism, developed over 20 years ago to relate ground state geometry and electronic structure, is still a useful and simple framework for discussing the properties of most metal nitrosyls. This formalism assumes that the electronic structures of transition metal nitrosyl compounds are dominated by the $\pi$-interactions between the metal $d$ orbitals and the $\pi^*$(NO) orbitals, and that bending of the MNO group occurs when the orbital that is totally $\pi$-antibonding becomes occupied. For paramagnetic nitrosyl complexes, such as the $\{FeNO\}^7$ complexes with an $S = 3/2$ electronic state, spin unrestricted density functional calculations have provided insight concerning their electronic structures and emphasized the importance of metal–N $\sigma$ bonding in the overall stability of these complexes. In striking contrast, recent PES studies of CpNi(NO) and Cp*Ni(NO) indicate that a linear NO group is a weak $\sigma$-donor, and that Ni–NO $\pi$ interactions dominate the bonding in these $\{NiNO\}^{10}$ species.

The ground state properties of transition metal nitrosyl complexes seem fairly well understood; however, understanding of the excited states of $\{MNO\}^n$ complexes and their ramifications for chemistry, materials science, and biological activity is still in its earliest stages. Experimental and theoretical studies of ligand substitution reactions of metal nitrosyl complexes with linear nitrosyl groups often exhibit bimolecular kinetics that are consistent with an intramolecular redox reaction and the formation of an intermediate with a bent MNO unit. The novel MNO geometries found for the metastable excited states of SNP and CpNiNO raise questions about the detailed reaction pathway(s) for coordination and dissociation of nitric oxide. Reversible binding of NO is biologically important, but the details of the process are not yet well understood. Stable $\{FeNO\}^7$ complexes of porphyrins are well known; all are low spin ($S = 1/2$) species with bent nitrosyl groups, and their EPR properties are sensitive to coordination number. However, to date there appear to have been no detailed electronic structure calculations on these paramagnetic systems using density functional methods.

This review has highlighted how the electronic structures of transition metal nitrosyls are related to their geometry and reactivity. Several spectroscopic methods for studying the ground and excited states of $\{MNO\}^n$ complexes have been discussed as well. It is hoped that this review will provide a background for those interested in the interactions of NO with metal centers.

## ACKNOWLEDGMENTS

We would like to acknowledge Drs. P. Coppens, W. R. Montfort, G. Richter-Addo, and W. R. Scheidt for preprints of results prior to publication. We further acknowledge Drs. R. D. Feltham, N. Gruhn, D. L. Lichtenberger, J. Pollard, and F. A. Walker, and H. Phinney and K. Winfield for helpful and stimulating discussions related to this review. Finally, we thank the University of Arizona for financial support.

## Note Added in Proof

The importance of NO in biological systems has been underscored by the award of the 1998 Nobel Prize in medicine to Furchgott, Ignarro, and Murad for their discoveries concerning NO as a signaling molecule in the cardiovascular system. The structure of the heme/tetrahydrobiopterin domain of NOS has been reported.[136] The binding of NO to the ferri- and ferroheme states of nitrophorin 1 (NP1, Section 5.1) has been investigated by protein crystallography, electrochemistry, and EPR, NMR, UV/vis, and FTIR spectroscopies. NP1 readily binds NO, but the protein is not easily autoreduced by excess NO. The reduction potential of NP1 is ~300 mV more negative than that of metmyoglobin. The N–O stretching frequencies are typical of other six-coordinate hemes with a coordinated NO ligand.[137] The subtle energy relationships in $\{RuNO\}^8$ complexes (Section 1.4) have been demonstrated by studies of several five-coordinate $RuX(NO)(CO)(P^tBu_2Me)_2$ species. X = $H^-$ gives trigonal bipyramidal geometry with a linear NO group in the equatorial plane (Figure 4a); whereas X = $Cl^-$ or $BF_4^-$ produces square pyramidal geometry with a strongly bent apical NO (Figure 4d). *Ab initio* computations on model compounds show that for $RuH(CO)(NO)(PH_3)_2$ the square pyramidal/bent NO and trigonal bipyramidal/linear NO structures are essentially isoenergetic. For $[Ru(NCH)(CO)(NO)(PH_3)_2]^+$ the calculations predict a square pyramidal/bent NO isomer and a distorted trigonal bipyramidal structure with Ru–N–O = 154.2°.[138]

## REFERENCES


1. Enemark, J. H.; Feltham, R. D., *Coord. Chem. Rev.*, **1974**, *13*, 339.
2. Feltham, R. D.; Enemark, J. H., *Top. Stereochem.*, **1981**, *12*, 155.
3. (a) Traylor, T. G.; Sharma, V. S., *Biochemistry*, **1992**, *31*, 2847; (b) Moncada, S.; Palmer, R. M. J.; Higgs, E. A., *Pharmacol. Rev.*, **1991**, *43*, 109; (c) Ignarro, L. J., *Biochem. Pharmacol.*, **1991**, *41*, 485; (d) Nathan, C. S.; Hibbs, J. B., Jr., *Curr. Opin. Immunol.*, **1991**, *3*, 65; (e) Bredt, D. S.; Snyder, S. H., *Neuron.*, **1992**, *8*, 3; (f) Varma, A.; Hirsch, D. J.; Glatt, C. E.; Ronnett, G. V.; Snyder, S. H., *Science*, **1993**, *259*, 381; (g) Gorbunov, N. V.; Osipov, A. N.; Day, B. W.; Zayas-Rivera, B.; Kagan, V. E.; Elsayed, N. M., *Biochem.*, **1995**, *34*, 6689; (h) Wang, T.; Xie, Z.; Lu, B., *Nature*, **1995**, *374*, 262; (i) Jia, L.; Bonaventura, C.; Bonaventura, J.; Stamler, J. S., *Nature*, **1996**, *380*, 221; (j) Arancio, O.; Kielder, M.; Lee, C. J.; Lev-Ram, V.; Tsien, R. Y.; Kandel, E. R.; Hawkins, R. D., *Cell*, **1996**, *87*, 1025; (k) Vincent, S., Ed., *Nitric Oxide in the Nervous System*, Academic Press, New York, NY, **1995**.
4. Culotta, E.; Koshland, D. E., *Science*, **1992**, *258*, 1862.
5. Richter-Addo, G. B.; Legzdins, P., *Metal Nitrosyls*, Oxford University Press, New York, **1992**.
6. Averill, B. A., *Chem. Rev.*, **1996**, *96*, 2951.
7. Zumft, W. G., *Arch. Microbiol.*, **1993**, *160*, 253.
8. Försterman, U.; Schmidt, H. H. W.; Pollock, J. S.; Sheng, H.; Mitchell, J. A.; Warner, T. D.; Nakane, M.; Murad, F., *Biochem. Pharmacol.*, **1991**, *42*, 1849.

9. Bredt, D. S., "Molecular Characterization of Nitric Oxide Synthase," in *Nitric Oxide in the Nervous System*, Vincent, S., Ed., Academic Press, New York, 1995, p. 1.
10. Bredt, D. S.; Hwang, P. M.; Glatt, C. E.; Lowenstein, C.; Reed, R. R.; Snyder, S. H., *Nature,* **1991**, *351*, 714.
11. (a) Moncada, S.; Palmer, R. M. J.; Higgs, E., *Pharmacol. Rev.*, **1991**, *43*, 109; (b) Snyder, S. H., *Science*, **1992**, *257*, 494; (c) Stamler, J. S.; Singel, D. J.; Loscalzo, J., *Science*, **1992**, *258*, 1898; (d) Lancaster, J. R., Jr., *Amer. Scientist*, **1992**, *80*, 248; (e) Butler, A. R.; Williams, D. L. H., *Chem. Soc. Rev.*, **1993**, 233; (f) Lancaster, J. R., Jr., "Nitric Oxide in Biology," in *Encyclopedia of Inorganic Chemistry*, King, R. B., Ed., John Wiley & Sons, Ltd., Chichester, **1994**.
12. Snyder, S. H.; Bredt, D. S., *Sci. Am.*, **1992**, 68.
13. Marletta, M. A., *J. Biol. Chem.*, **1993**, *268*, 12231.
14. Frenz, B. A.; Ibers, J. A., *M. T. P. Int. Rev. Sci., Inorg. Chem., Ser. One*, **1972**, *5*, 229.
15. Enemark, J. H.; Ibers, J. A., *Inorg. Chem.*, **1967**, *6*, 1575.
16. Hodgson, D.; Ibers, J. A., *Inorg. Chem.*, **1968**, *7*, 2345.
17. Sidgwick, N. V.; Bailey, R. W., *Proc. Roy. Soc. London,* **1934**, *A144*, 521.
18. Collman, J. P.; Hegedus, L. S.; Norton, J. R.; Finke, R. G., *Principles and Applications of Organotransition Metal Chemistry*, University Science Books, Mill Valley, CA, **1987**.
19. Manorohan, P. T.; Gray, H. B., *Inorg. Chem.*, **1966**, *5*, 823.
20. Fenske, R. F.; DeKock, R. L., *Inorg. Chem.*, **1972**, *11*, 437.
21. Mingos, D. M. P., *Inorg. Chem.*, **1973**, *12*, 1209.
22. Walsh, A. D., *J. Chem. Soc.*, **1953**, 2266.
23. Hoffmann, R.; Chen, M. M. L.; Elian, M.; Rossi, A. R.; Mingos, D. M. P., *Inorg. Chem.*, **1974**, *13*, 2666.
24. Brock, C. P.; Collman, J. P.; Dolcetti, G.; Farnham, P. H.; Ibers, J. A.; Lester, J. E.; Reed, C. E., *Inorg. Chem.*, **1973**, *12*, 1304.
25. Swinehart, J. H., *Coord. Chem. Rev.*, **1967**, *2*, 385.
26. Douglas, P. G.; Feltham, R. D.; Metzger, H. G., *J. Am. Chem. Soc.*, **1971**, *93*, 84.
27. Clarkson, S. G.; Basolo, F., *Inorg. Chem.*, **1973**, *12*, 1528.
28. Enemark, J. H.; Feltham, R. D.; Riker-Nappier, J.; Bizot, K. F., *Inorg. Chem.*, **1975**, *14*, 625.
29. Enemark, J. H.; Feltham, R. D., *Proc. Nat. Acad. Sci. USA*, **1972**, *69*, 3534.
30. Shi, Q. Z.; Richmond, T. G.; Trogler, W. C.; Basolo, F., *Inorg. Chem.*, **1984**, *23*, 957.
31. Basolo, F.; Wojcicki, A., *J. Am. Chem. Soc.*, **1961**, *83*, 525.
32. Thornstein, E.; Basolo, F., *J. Am. Chem. Soc.*, **1966**, *88*, 3929.
33. Geiger, W. E.; Reiger, P. H.; Tulyathan, B.; Rausch, M. D., *J. Am. Chem. Soc.*, **1984**, *106*, 7000.
34. Atherton, N. M., *Electron Spin Resonance*, Ellis Howard, Chichester, England, **1973**.
35. Weil, J. A.; Bolton, J. R.; Wertz, J. E., *Electron Paramagnetic Resonance*, John Wiley and Sons, New York, **1994**.
36. Mabbs, F. E.; Collison, D., *Electron Paramagnetic Resonance of d Transition Metal Compunds*, Elsevier, Amsterdam, **1992**.
37. Song, J.; Hall, M. B., *J. Am. Chem. Soc.*, **1993**, *115*, 327.
38. Sulfab, Y.; Basolo, F.; Rheingold, A. L., *Organometallics*, **1989**, *8*, 2139.

39. Brunner, H., *Adv. Organometal. Chem.*, **1980**, *18*, 151.

40. Legzdins, P.; Rettig, S. J.; Sanchez, L.; Bursten, B. E.; Gatter, M. G., *J. Am. Chem. Soc.*, **1985**, *107*, 1411.

41. (a) Fernandez, J. M.; Gladysz, J. A., *Organometallics*, **1989**, *8*, 207; (b) Garner, C. M.; Mendez, N. Q.; Kowalczyk, J. J.; Fernandez, J. M.; Emerson, K.; Laresen, R. D.; Gladysz, J. A., *J. Am. Chem. Soc.*, **1990**, *112*, 5146; (c) Gladysz, J. A.; Boone, B. J., *Angew. Chem. Int. Ed. Engl.*, **1997**, *36*, 550.

42. (a) Schenck, W. A.; Burzlaff, N., *Z. Naturforsch*, **1994**, *49b*, 1633; (b) Schenk, W. A.; Frisch, J.; Durr, M.; Burzlaff, N.; Stalke, D.; Fleischer, F.; Walderman, A.; Frank, P.; Smerz, A. K., *Inorg. Chem.*, **1997**, *36*, 2372.

43. McCleverty, J. A., *Chem. Soc. Rev.*, **1983**, *12*, 331.

44. (a) James, T. A.; McCleverty, J. A., *J. Chem. Soc. A*, **1971**, 1068; (b) King, R. B., *Inorg. Chem.*, **1967**, *6*, 90; (c) King, R. B.; Kapoor, R. N., *J. Organomet. Chem.*, **1968**, *15*, 457; (d) James, T. A.; McCleverty, J. A., *J. Chem. Soc. A*, **1970**, 3308.

45. Block, T. F.; Fenske, R. F.; Casey, C. P., *J. Am. Chem. Soc.*, **1976**, *98*, 441.

46. McCleverty, J. A., *Chem. Soc. Rev.*, **1983**, *12*, 331.

47. Ashby, M. T.; Enemark, J. H., *J. Am. Chem. Soc.*, **1986**, *106*, 730.

48. Templeton, J. L.; Ward, B. C., *J. Am. Chem. Soc.*, **1980**, *102*, 6568.

49. AlObaidi, N. J.; Salam, S. S.; Beer, P. D.; Bush, C. D.; Hamor, T. A.; McQuillan, F. S.; Jones, C. J.; McCleverty, J. A., *Inorg. Chem.*, **1992**, *31*, 263.

50. Coe, B. J.; Jones, C. J.; McCleverty, J. A.; Bloor, D.; Kolinsky, P. V.; Jones, R. J., *J. Chem. Soc., Chem. Commun.*, **1989**, 1485.

51. AlObaidi, N.; Jones, C. J.; McCleverty, J. A., *Polyhedron*, **1989**, *8*, 371.

52. McWhinnie, S. L. W.; Thomas, J. A.; Hamor, T. A.; Jones, C. J.; McCleverty, J. A.; Collison, D.; Mabbs, F. E.; Harding, C. T.; Yellowlees, L. J.; Hutchings, M. G., *Inorg. Chem.*, **1996**, *35*, 760.

53. Cleland, W. E., Jr.; Barnhart, K. M.; Yamanouchi, K.; Collison, D.; Mabbs, F. E.; Ortega, R. B.; Enemark, J. H., *Inorg. Chem.*, **1987**, *26*, 1017.

54. Westcott, B. L.; Enemark, J. H., *Inorg. Chem.*, **1997**, *36*, 5404.

55. Bursten, B. E.; Cayton, R. H., *Organometallics*, **1987**, *6*, 2004.

56. Westre, T. E.; Di Cicco, A.; Filipponi, A.; Natoli, C. R.; Hedman, B.; Solomon, E. I.; Hodgson, K. O., *J. Am. Chem. Soc.*, **1994**, *116*, 6757.

57. (a) Mingos, D. M. P.; Ibers, J. A., *Inorg. Chem.*, **1971**, *10*, 1035; (b) Johnson, P. L.; Enemark, J. H.; Feltham, R. D.; Swedo, K. B., *Inorg. Chem.*, **1976**, *15*, 2989.

58. Greenwood, N. N.; Earnshaw, A., *Chemistry of the Elements*, Pergamon Press, **1984**.

59. Quinbey-Hunt, M.; Feltham, R. D., *Inorg. Chem.*, **1978**, *17*, 2515.

60. Albano, V. G.; Bellon, P.; Sansoni, M., *J. Chem. Soc. A*, **1971**, 2420.

61. Hodgson, P. J.; Ibers, J. A., *Inorg. Chem.*, **1969**, *8*, 1282.

62. Chottard, G.; Mansuy, D., *Biochem. Biophys. Res. Commun.*, **1977**, *77*, 133.

63. Tsubaki, M.; Yu, N.-T., *Biochemistry*, **1982**, *21*, 1140.

64. Benko, B.; Yu, N.-T., *Proc. Natl. Acad. Sci. USA*, **1983**, *80*, 7042.

65. Choi, I.-K.; Yanming, L.; Feng, D.-W.; Paeng, K.-J.; Ryan, M. D., *Inorg. Chem.*, **1991**, *30*, 1832.

66. Yu, A. E.; Hu, S.; Spiro, T. G.; Burstyn, J. D., *J. Am. Chem. Soc.*, **1994**, *116*, 4117.

67. Lipscomb, L. A.; Lee, B. S.; Yu, N.-T., *Inorg. Chem.*, **1993**, *32*, 281.

68. (a) Scheidt, W. R.; Haller, K. J.; Fons, M.; Mashiko, T.; Reed, C. A., *Biochemistry*, **1981**, *20*, 3653; (b) Collman, J. P.; Gagne, R. R.; Reed, C. A.; Robinson, W. T.; Rodley, G. A., *Proc. Nat. Acad. Sci., USA*, **1974**, *71*, 1326; (c) Jameson, G. B.; Robinson, W. T.; Gagne; Reed, C. A.; Collman, J. P., *Inorg. Chem.*, **1978**, *17*, 850; (d) Jameson, G. B.; Molinaro, F. S.; Ibers, J. A.; Collman, J. P.; Brauman, J. I.; Rose, E.; Suslick, K. S., *J. Am. Chem. Soc.*, **1978**, *100*, 6769.

69. Brown, C. A.; Pavlosky, M. A.; Westre, T. E.; Zhang, Y.; Hedman, B.; Hodgson, K. O.; Solomon, E. I., *J. Am. Chem. Soc.*, **1995**, *117*, 715.

70. Drago, R. S., *Physical Methods for Chemists, Second Ed.*, Saunders College Publishing, Ft. Worth, TX, **1992**.

71. Young, C. G.; Minelli, M.; Enemark, J. H.; Hussain, W.; Jones, C. J.; McCleverty, J. A., *J. Chem. Soc., Dalton Trans.*, **1987**, 619.

72. Mason, J., *Chem. Rev.*, **1981**, *81*, 205.

73. Bell, L. K.; Mason, J.; Mingos, D. M. P.; Tew, D. G., *Inorg. Chem.*, **1983**, *22*, 3497.

74. Botto, R. E.; Kolthammer, B. W. S.; Legzdins, P.; Roberts, J. D., *Inorg. Chem.*, **1979**, *18*, 2049.

75. (a) Quelch, G. E.; Hillier, I. H.; Guest, M. F., *J. Chem. Soc., Dalton Trans.*, **1990**, 3075; (b) Morris-Sherwood, B. J.; Kolthammer, B. W. S; Hall, M. B., *Inorg. Chem.*, **1981**, *20*, 2711; (c) Hubbard, J. L.; Lichtenberger, D. L., *Inorg. Chem.*, **1980**, *19*, 1388; (d) Copperthwaite, R. G.; Reimann, R. H.; Singleton, E., *Inorg. Chim. Acta*, **1978**, *28*, 107; (e) Miller, I. H.; Guest, M. F.; Higginson, B. R.; Lloyd, D. R., *Mol. Phys.*, **1974**, *27*, 215.

76. Zöllner, H.; Kraser, W.; Woike, T.; Haussühl, S., *Chem. Phys. Lett.*, **1989**, *161*, 497.

77. Carducci, M. D.; Pressprich, M. R.; Coppens, P., *J. Am. Chem. Soc.*, **1997**, *119*, 2669.

78. Hauser, U.; Oestreich, V.; Rohrweck, H. D., *Z. Phys. A*, **1977**, *280*, 17; 125.

79. Yang, Y. Y.; Zink, J. I., *J. Am. Chem. Soc.*, **1985**, *107*, 4799.

80. (a) Güida, J. A.; Piro, O. E.; Aymonino, P. J., *Solid State Comm.*, **1986**, *57*, 175; (b) Güida, J. A.; Piro, O. E.; Aymonino, *P. J., Solid State Comm.*, **1988**, *66*, 1007.

81. Güida, J. A.; Aymonino, P. J.; Piro, O. E., Castellano, E. E., *Spectrochimica Acta*, **1993**, *A49*, 535.

82. Rüdlinger, M.; Schefer, J.; Chevrier, G.; Furer, N.; Gudel, H. U.; Haussühl, S.; Heger, G.; Schweiss, P.; Vogt, T.; Woike, T.; Zöllner, H., *Z. Phys. B*, **1991**, *83*, 125.

83. Rüdlinger, M.; Schefer, J.; Vogt, T.; Woike, T.; Haussühl, S.; Zöllner, H., *Physica*, **1992**, *B180*, 293.

84. Pressprich, M. R.; White, M. A.; Coppens, P., *J. Am. Chem. Soc.*, **1993**, *115*, 6444.

85. Pressprich, M. R.; White, M. A.; Vekhter, Y.; Coppens, P., *J. Am. Chem. Soc.*, **1994**, *116*, 5233.

86. Tatsumi, K.; Titsui, M., *J. Mol. Catalysis*, **1982**, *13*, 117.

87. (a) Vaska, L., *Acc. Chem. Res.*, **1976**, *9*, 175; (b) Hoffman, R.; Chen, M.-L.; Thorn, D. L., *Inorg. Chem.*, **1977**, *16*, 503.

88. Hodges, K. D.; Wollman, R. G.; Kessel, S. L.; Hendrickson, D. N.; Van Derveer, D. G.; Barefield, E. K., *J. Am. Chem. Soc.*, **1979**, *101*, 906.

89. Wells, F. V.; McCann, S. W.; Wickman, H. H.; Kessel, S. L.; Hendrickson, D. N.; Feltham, R. D., *Inorg. Chem.*, **1982**, *21*, 2306.

90. Haller, K. J.; Johnson, P. L.; Feltham, R. D.; Enemark, J. H., *Inorg. Chim. Acta*, **1979**, *33*, 119.

91. Earnshaw, A.; King, E. A.; Larkworthy, L. F., *J. Chem. Soc. A*, **1969**, 2459.

92. Pohl, K.; Wieghardt, K.; Nuber, B.; Weiss, J., *J. Chem. Soc., Dalton Trans.*, **1987**, 187.

93. Zhang, Y.; Pavlosky, M. A.; Brown, C. A.; Westre, T. E.; Hedman, B.; Hodgson, K. O.; Solomon, E. I., *J. Am. Chem. Soc.*, **1992**, *114*, 9189.

94. Slappendel, S.; Veldink, G. A.; Wliegenthart, J. F. G.; Aasa, R.; Malmström, B. G., *Biochim. Biophys. Acta*, **1983**, *747*, 32.

95. Nelson, M. J., *J. Biol. Chem.*, **1987**, *262*, 12137.

96. Noodleman, L.; Baerands, E. J., *J. Am. Chem. Soc.*, **1984**, *106*, 2316.

97. Butcher, K. D.; Gebhard, M. S.; Solomon, E. I., *Inorg. Chem.*, **1990**, *29*, 2067.

98. Arciero, D. M.; Orville, A. M.; Lipscomb, J. D., *J. Biol. Chem.*, **1985**, *260*, 14035.

99. Orville, A. M.; Chen, V. J.; Kriauciunas, A.; Harpel, M. R.; Rox, F. G.; Münck, E.; Lipscomb, J. D, *Biochemistry*, **1992**, *31*, 4602.

100. Ribeiro, J. M. C.; Hazzard, J. M. H.; Nussenzveig, R. H.; Champagne, D. E.; Walker, F. A., *Science*, **1993**, *260*, 539.

101. (a) Antonini, E.; Brunori, M.; Wyman, J.; Noble, R. W., *J. Biol. Chem.*, **1966**, *241*, 3236; (b) Cassoly, R.; Gibson, Q. H., *J. Mol. Biology*, **1975**, *91*, 301.

102. Anderson, J. F.; Champagne, D. E.; Weichsel, A.; Ribeiro, J. M.; Ralfour, C. A.; Dress, V.; Montfort, W. R., *Biochemistry*, **1997**, *36*, 4423.

103. Weichsel, A.; Andersen, J. F.; Champagne, D. E.; Walker, F. A.; Montfort, W. R., *Nature Struct. Biol.*, **1998**, *5*, 304.

104. Flower, D. R., *Biochem. J.*, **1996**, *318*, 1.

105. Scheidt, W. R.; Hatano, K.; Rupprecht, G. A.; Piciulo, P. L., *Inorg. Chem.*, **1979**, *18*, 292.

106. Scheidt, W. R.; Lee, Y. J.; Hatano, K., *J. Am. Chem. Soc.*, **1984**, *106*, 3191.

107. Yi, G.-B.; Khan, M. A.; Richter-Addo, G. B., *Inorg. Chem.*, **1996**, *35*, 3453.

108. Yi, G.-B.; Khan, M. A.; Richter-Addo, G. B., *Chem. Commun.*, **1996**, 2045.

109. Hodge, S. J.; Wang, L.-S.; Khan, M. A.; Young, V. G.; Richter-Addo, G. B., *Chem. Commun.*, **1996**, 2283.

110. Basu, P.; Shokhirev, N. V.; Enemark, J. H.; Walker, F. A., *J. Am. Chem. Soc.*, **1995**, *117*, 9042.

111. Kon, H., *J. Biol. Chem.*, **1968**, *243*, 4350.

112. Henry, Y.; Ducrocq, C.; Drapier, J.-C.; Servent, D.; Pellat, C.; Guissani, A., *Eur. Biophys. J.*, **1991**, *20*, 1.

113. Yoshimura, T., *Inorg. Chimica Acta*, **1982**, *57*, 99.

114. Wayland, B. B.; Olson, L. W., *J. Am. Chem. Soc.*, **1974**, *96*, 6037.

115. Bohle, D. S.; Hung, C.-H., *J. Am. Chem. Soc.*, **1995**, *117*, 9584.

116. (a) Scheidt, W. R.; Brinegar, A. C.; Ferro, E. B.; Kirner, J. F., *J. Am. Chem. Soc.*, **1977**, *99*, 7315; (b) Scheidt, W. R.; Frisse, M. E., *J. Am. Chem. Soc.*, **1975**, *97*, 17; (c) Piciulo, P. L.; Scheidt, W. R., *J. Am. Chem. Soc.*, **1976**, *98*, 1913; (d) Nasri, H.; Haller, K. J.; Wang, Y.; Huynh, B. H.; Scheidt, W. R., *Inorg. Chem.*, **1992**, *31*, 3459; (e) Nasri, H.; Ellison, M. K.; Chen, S.; Huyn, B. H.; Scheidt, W. R., *J. Am. Chem. Soc.*, **1997**, *119*, 6274.

117. Ellison, M. K.; Scheidt, W. R., *J. Am. Chem. Soc.*, **1997**, *119*, 7404.
118. Szabo, A.; Perutz, M. F., *Biochemistry*, **1976**, *15*, 4427.
119. Yoshimura, T., *Inorg. Chim. Acta,* **1984**, *83*, 17.
120. Mu, X. H.; Kadish, K. M., *Inorg. Chem.*, **1988**, *27*, 4720.
121. Piper, T. S.; Cotton, F. A.; Wilkinson, G., *J. Inorg. Nucl. Chem.*, **1955**, *1*, 165.
122. (a) Kukolich, S. G.; Rund, J. V.; Pauley, D. J.; Bumgarner, R. E., *J. Am. Chem. Soc.*, **1988**, *110*, 7356; (b) Cox, A. P.; Brittain, A. H., *Trans. Faraday Soc.*, **1970**, *66*, 557; (c) Cox, A. P.; Thomas, L. E.; Sheridan, J., *Nature (London)*, **1958**, *181*, 1157.
123. Domenicano, A.; Hargittai, I., *Accurate Molecular Structures: Their Determination and Importance*, Oxford Press, Oxford, UK, **1992**.
124. Albright, T. A.; Burdett, J. K., *Problems in Molecular Orbital Theory*, Oxford University Press, New York, **1992**.
125. Green, J. C., *Acc. Chem. Res.*, **1994**, *27*, 131.
126. Evans, S.; Guest, M. F.; Hillier, I. H.; Orchard, A. F., *J. Chem. Soc., Faraday Trans. 2*, **1974**, 417.
127. Li, X.; Tse, J. S.; Bancroft, G. M.; Puddephatt, R. J.; Tan, K. H., *Inorg. Chem.*, **1996**, *35*, 2515.
128. Field, C. N.; Green, J. C.; Mayer, M.; Nasluzov, V. A.; Rösch, N.; Siggel, R. F., *Inorg. Chem.*, **1996**, *35*, 2504.
129. (a) Pollard, J. R., Ph. D. Dissertation, University of Arizona, **1996**; (b) Pollard, J. R.; Lichtenberger, D. L., *Abstract of Papers*, 213$^{th}$ National Meeting of the American Chemical Society; San Francisco, CA; American Chemical Society: Washington, D.C., **1997**; INOR 453.
130. Yeh, J. J.; Landau, I., *At. Nucl. Data Tables*, **1985**, *32*, 7.
131. Green, J. C.; Underwood, C., *J. Organometal. Chem.*, **1997**, *91*, 528.
132. Lichtenberger, D. L.; Kellogg, G. E., *Acc. Chem. Res.*, **1987**, *20*, 379.
133. Crichton, O.; Rest, A. J., *J. Chem. Soc., Dalton Trans.*, **1977**, 986.
134. Fomitchev, D. V.; Furlani, T. R.; Coppens, P., *Inorg. Chem.*, **1998**, *37*, 1519.
135. Coppens, P.; Fomitchev, D. V.; Carducci, M. D.; Culp, K., *J. Chem. Soc., Dalton Trans.*, **1998**, 865.
136. Crane, B. R.; Arvai, A. S.; Gachhui, R.; Wu, C.; Ghosh, D. K.; Getzoff, E. D.; Stuehr, D. J.; Tainer, J. A., *Science*, **1997**, *278*, 425.
137. Ding, X. D.; Weichsel, A.; Andersen, J. F.; Shokhireva, T. Kh.; Balfour, C.; Pierik, A. J.; Averill, B. A.; Montfort, W. R.; Walker, F. A., *J. Am. Chem. Soc.*, **1999**, *121*, 128.
138. Ogasawara, M.; Huang, D.; Streib, W. E.; Huffman, J. C.; Planas-Gallego, N.; Maseras, F.; Eisenstein, O.; Caulton, K. G., *J. Am. Chem. Soc.*, **1997**, *119*, 8642.
139. Scheidt, W. R.; Ellison, M. K., *Acc. Chem. Res.*, **1999**, in press.

# 8 Electronic Structure of Heme Sites

GILDA LOEW
Molecular Research Institute
845 Page Mill Road
Palo Alto, CA 94304, USA
E-mail: loew@montara.molres.org



*Inorganic Electronic Structure and Spectroscopy, Volume II: Applications and Case Studies.*
Edited by E. I. Solomon and A. B. P. Lever.
ISBN 0-471-32682-8. 

## 1 BACKGROUND AND SCOPE

Heme proteins are a class of biologically important macromolecules that have a unique, common active site the iron–protoporphyrin-IX complex shown in Figure 1. This "heme" unit is a nearly independent entity linked to the remainder of the globular protein only by one or two amino acids that serve as axial ligands to the heme iron. Despite the common active site, heme proteins have three distinct biological functions: reversible oxygen transport, reversible electron transfer, and true metabolizing proteins converting substrates to products.

The globins, typified by myoglobin and hemoglobin, serve as reversible oxygen transport proteins. In performing this biological function, as shown in Figure 2A, they alternate between two ferrous heme species with and without the $O_2$ molecule directly bound to the heme iron. In addition to molecular oxygen, a variety of small molecules and ions such as CO, NO, and $CN^-$ bind to the heme iron in the same "distal" binding pocket that binds the $O_2$. The only link of the heme unit in the

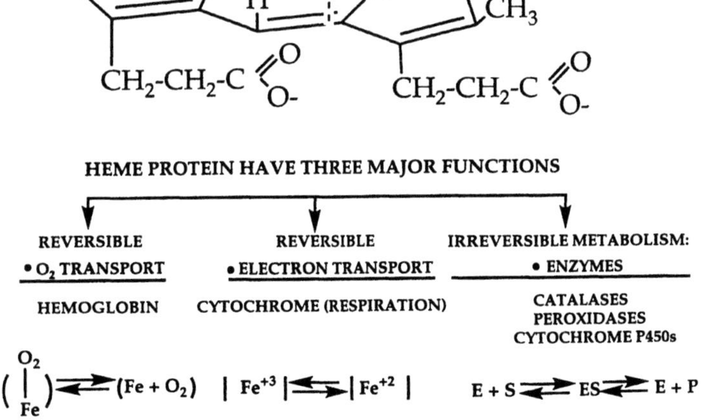


**Figure 1.** Common active site of heme proteins: a [Fe-Protoporphyrin-IX] complex; despite this commonality, heme proteins have different functions as indicated.

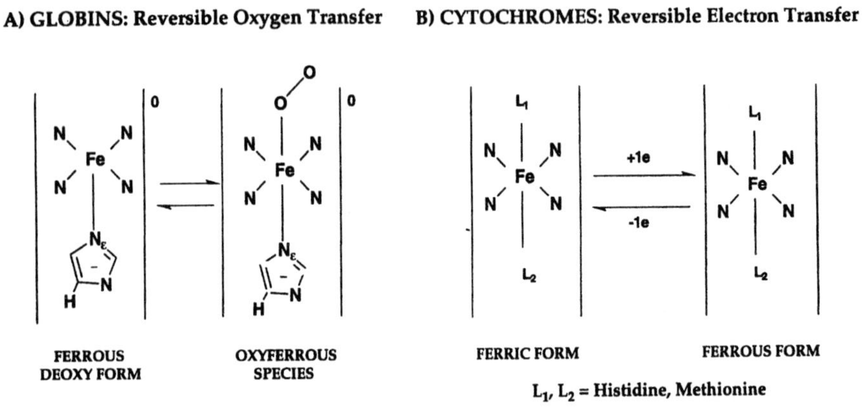


**Figure 2.** Heme species involved in biological function of a) globins and b) cytochromes.

globins to the remainder of the protein is by binding of an imidazole from a nearby proximal conserved histidine residue as an axial ligand to the heme iron.

A second function of heme proteins, typified by the cytochrome b and c families is as reversible one electron transfer agents. In this function, as shown in Figure 2B, the heme unit alternates between a six coordinated ferric and ferrous species with no change in axial ligands. Two nearby amino acids, typically histidine and methionine, in some combination serve as axial ligands to the heme iron linking it to the surrounding globular portion of the cytochromes.

The third function of heme proteins is as metabolizing enzymes, transforming diverse substrates into oxidized products. The major families of metabolizing heme proteins are catalases, peroxidases, and cytochrome P450s. The heme unit in each of these families has only one proximal axial ligand linking it to the remainder of the protein: a phenolate for catalases, an imidazolate for typical peroxidases, and a cysteinate for cytochrome P450s. Substrates bind and are transformed in or near the distal binding pocket on the opposite side of the heme unit from the proximal axial ligand. Unlike the globins and the cytochromes, the resting form of each of these metabolizing heme proteins is not the biologically active form. Instead, as shown in Figure 3, they are thought to share a common catalytically active form a so-called Compound I, ferryl, Fe=O species, two oxidation states above the ferric resting form. Thus, there are two parts to the enzymatic cycle of these metabolizing heme proteins. The first is the pathway from the inactive ferric resting form to the catalytically active form. The second is the mechanism of oxidation of substrates by the active form. There are differences in one or both of these parts of the enzymatic cycle among the three families of metabolizing heme proteins.

As shown in Figures 4A and 4B peroxidases and catalases with a different axial ligand are thought to share a very similar pathway from their resting form to Compound I. Both require peroxide to convert the inactive ferric resting form to the putative enzymatically active Compound I, Fe=O species that performs the oxidation of substrates. The main difference between the peroxidases and catalylases

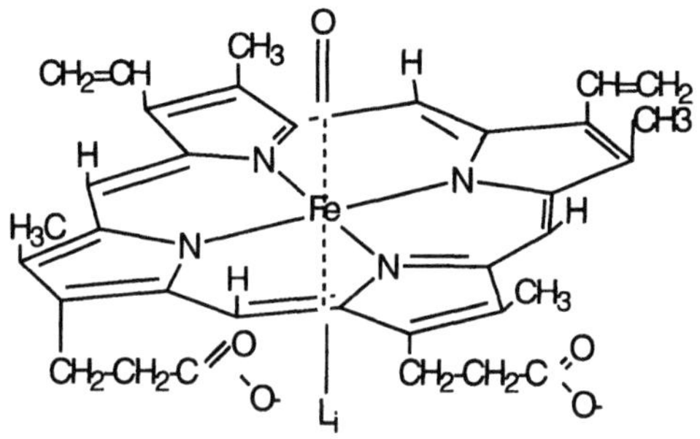


Compound I

Two oxidation equivalents above "Resting forms"

**Figure 3.** Common catalytically active species shared by metabolizing heme proteins: a ferryl, Fe=O, "Compound I," species, two oxidation states above their ferric resting forms.

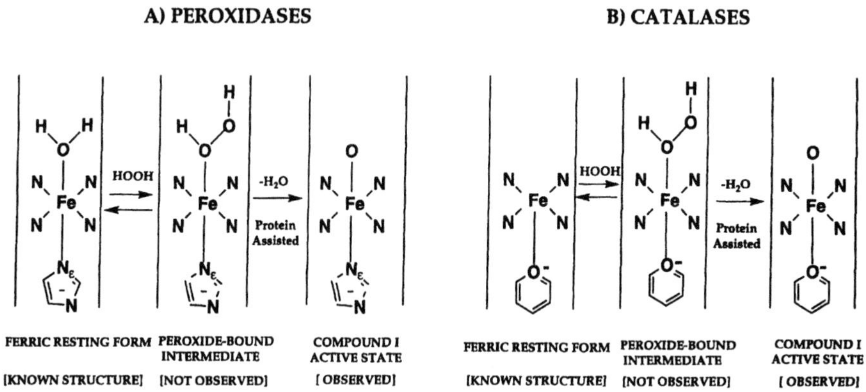


**Figure 4.** Heme species involved in biological function of: a) peroxidases and b) catalases.

is the mechanism of oxidation and substrate specificity. Peroxidases have diverse substrates, typically aromatic phenols, and perform two sequential one electron oxidations with the intermediary of a second reduced form of the oxy ferryl species called Compound II.

$$[\text{Peroxidase Compound I}] + \text{ROH} \rightarrow [\text{Peroxidase Compound II}] + \text{RO}$$
$$[\text{Peroxidase Compound II}] + \text{ROH} \rightarrow [\text{Peroxidase Resting Form}] + \text{RO}$$

Catalases have peroxide itself as a substrate and Compound I is the only identified intermediate performing simultaneous two electron oxidations of it.

$$[\text{Catalase Compound I}] + \text{HOOH} \rightarrow [\text{Catalase Resting Form}] + H_2O + O_2$$

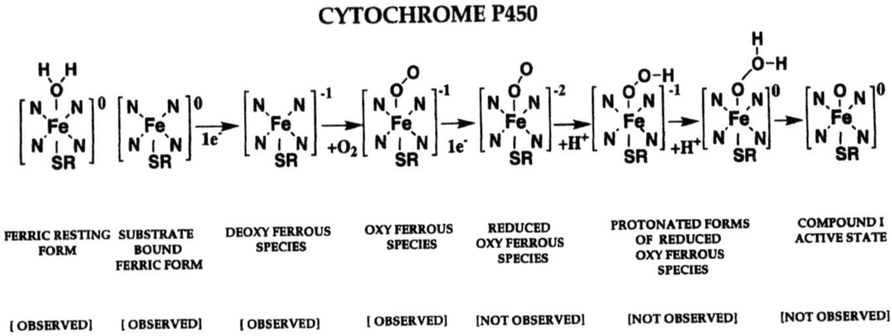


**Figure 5.** Heme species involved in biological function of the cytochrome P450s.

The ubiquitous cytochrome P450s differ from the peroxidases and catalases in both the pathway to formation of the catalytically active species from the ferric resting form and the underlying mechanism of oxidation. As shown in Figure 5, the enzymatic activation pathway of cytochrome P450s is much more complex and many features are still unresolved. Subsequent to substrate binding to the ferric resting form, one electron reduction occurs and then molecular oxygen binds to the ferrous heme iron forming a species similar to the oxygenated form of the globins. However, instead of reversible dioxygen binding, the oxyferrous cytochrome P450s undergo a second electron reduction, forming an unusual and unique iron-dioxygen species that is twice reduced from the ferric state. This unusual species is then thought to undergo proton assisted dioxygen bond cleavage, forming the same catalytically active ferryl Fe=O Compound I species as the peroxidases. However, the P450s function as monoxygenases, transferring the single iron bound oxygen atom to a variety of largely non-polar substituents, typically leading to aromatic and aliphatic hydroxylations, epoxidation, and heteroatom oxide formation.

$$[\text{P450 Compound I}] + \text{RH} \rightarrow [\text{P450 Resting Form}] + \text{ROH}$$

Neither peroxidases nor catalases transfer their ferryl oxygen to substrates. Thus, although they share a common ferryl oxygen enzymatically active species, the mechanism of oxidation of substrates by this species differs among these three families of metabolizing heme proteins. The origin of these differences is not well understood.

The putative determinants of heme proteins function can be divided into three components:

i) The first is the nature and intrinsic properties of the specific heme species themselves and transformations among them required for each function. While the biological activity is centered on the heme unit and all types of heme proteins have the identical iron-porphyrin heme complex, shown in Figure 1, they differ in the specific forms of the heme complex that are involved in their biological function. As shown schematically in Figures 2–5, the heme site of each heme protein family undergoes different transformations in oxidation state, ligands, and spin states unique to its different function. The properties of the heme species implicated in each function should be consistent with their postulated role.

ii) The second possible factor in determining properties of the heme species important for function is the nature of the axial ligand that links the heme unit to the protein. The relationship between axial ligand and function is not obvious. For example, metmyoglobin and peroxidases have a common axial ligand, an imidazole from a highly conserved histidine and yet have different function. The three metabolizing heme proteins all have different axial ligands: imidazolate, phenolate, and mercaptate respectively for the peroxidases, catalases, and cytorchrome P450s. Yet they are thought to have a common bioactive species. Compound I and the pathway to formation of Compound I is similar for the peroxidases and catalases. The extent to which axial ligands contribute to differences in function among these heme proteins is still unresolved.

iii) The third factor is the role of the protein environment itself. The protein environment, particularly the residues around the distal binding site and proximal axial ligand, appears to be a key modulator of the type of function of the heme protein. For example, there is increasing evidence that the difference in function between globins and peroxidases that share a common axial ligand could be due to such differences in protein environment as the redox potential that keep the globins in a ferrous resting state; the presence of polar residues in the distal binding pocket and the presence of a conserved aspartate near the proximal imidazole in the peroxidases that is absent in the globins.

Both experimental and computational methods have been used over many years to address the role of each of these factors in elucidating heme protein structure, spectra, and function. Essentially two types of goals have been addressed. One is the elucidation of the steps and species involved in the function of each heme protein family. The second and even more challenging goal is the identification of the origin of differences among them.

Not only biological activity, but also spectroscopic properties of the intact heme proteins are largely centered on the heme unit, again mediated by the axial ligand and protein environment. A variety of experimental methods have been used to probe the nature of the active site in both model compounds and intact heme proteins. The experimental methods used include X-ray and neutron diffraction structure determinations and diverse spectroscopic measurements including electronic, electron spin resonance, nuclear spin resonance, Mössbauer resonance, magnetic circular dichroism, infrared, and resonance Raman spectra. Specific characteristics of each of these types of spectra are sensitive indicators of the changes in ligand, oxidation, and spin states of the iron bound heme complexes. Thus, spectroscopic properties can be used to help identify and characterize both stable and transient heme species involved in the biological cycle of heme proteins as well as model compounds. However, these spectroscopic methods cannot by themselves yield a detailed molecular description of the heme active site. The techniques of computational chemistry combined with results of experimental measurements provide the most robust strategies for probing the relationship between structure, spectra, and function of the heme site in model compounds and intact proteins by providing a correspondence between an explicit model for the heme complex and observable properties.

It is not possible to describe all of the experimental and computational studies made addressing the electronic structure of heme sites. Instead, this chapter focuses on six topics and the extent to which they have been elucidated by computational methods, coupled with spectroscopic observations. While influenced by the authors work in this field over the past 30 years, we have tried to include as broad a representation as possible; apologies are offered for any omissions we might have made.

### 1.1 Properties Common to All Heme Sites

These properties include significant electron delocalization, the presence of low lying states of different spin multiplicities, the nature of the highest occupied and lowest empty orbitals, and common features of the electronic spectra of all heme complexes.

### 1.2 Heme Sites of the Globins

i) Characterization of the electronic structure, ground state electromagnetic properties, and spectra of the two biologically relevant ferrous heme species of the globins (Hb) involved in oxygen transport, the five coordinated deoxyHb species, and the six coordinated oxyHb species as well as the competing complexes formed by CO and NO. The mode of binding of these three small ligands, as well as their relative effect on observed properties will be described to elucidate the origin of the differences in their electromagnetic properties.

ii) Characterization of the electronic structure and spectra of the aquo resting form (MetMb) and CN complex of metmyoglobin, (MetMbCN), the oxidized, ferric form of myoglobin. These are typical high and low spin six coordinated ferric heme complexes. The goals of these studies were to calculate and compare the electronic structure of MetMb with similar ferric resting states of the peroxidases and catalases and to understand the origin of the extra features found in the electronic spectra of low spin ferric species absent in the low spin ferrous CO complex.

### 1.3 Heme Sites of the Peroxidases

Characterization of the electronic structure and spectra of the four heme species in the enzymatic cycle of peroxidases. These are the ferric resting state, the putative peroxide intermediate, and the biologically active Compound I and Compound II species. The goals of these calculations were: i) to understand the origin of the observed spectral properties of these species and ii) to further probe and assess the putative role of the transient peroxide species in the enzymatic cycle of these metabolizing heme proteins.

### 1.4 Heme Sites of the Cytochrome P450s

Characterization of the electronic structure and spectra of the heme species in the enzymatic cycle of the cytochrome P450s.

**1.4.1** ***The Substrate Free Ferric Resting Form and Substrate Bound Ferric Species*** The goals of these calculations were to understand the origin of the spin state change upon substrate binding and the relationship between this change and observed changes in the electronic spectra used to monitor substrate binding.

**1.4.2** ***The Oxy Ferrous and Carboxyl Ferrous P450 Heme Complexes*** The goals of these studies were: i) to calculate and compare the electronic structure of the two complexes with each other and with their counterparts in the globins; ii) to identify the origin of the unique signature spectra of the ferrous carbonyl P450 heme species that distinguished the P450s from the globins; and iii) to determine to what extent this feature persists in other P450 species.

**1.4.3** ***The Putative Transient Species Invoked in the Pathway to Compound I Formation*** These species include the reduced ferrous dioxygen species and its once and twice protonated tautomers. All of these species are too transient to have been definitively characterized experimentally. The goals of these studies were: i) to determine if the electronic properties and stability of these species were consistent with their proposed role in this least understood portion of the enzymatic cycle and ii) to help decide if transient spectra observed could be assigned to any of these species.

### 1.5 Comparisons of the Electronic Structure of Heme Sites with Varying Axial Ligands

The goal of these comparative studies was to determine the extent to which the axial ligand modulates function and the observed spectroscopic properties of heme proteins with different axial ligands.

**1.5.1** ***Comparative Studies of the Ferric Resting State of Heme Models*** These include models for metmyoglobin, peroxidase, catalase, and cytochrome P450s. The goal of these studies was to determine the exent to which changes in proximal axial ligand could modulate electronic structure and spin distributions.

**1.5.2** ***Comparative Studies of the Compound I Form of Heme Species*** These species have varying axial ligands and include heme models for metmyoglobin, peroxidase, catalase, cytochrome P450s, and model compounds with biological activity. The goals of these studies were to determine the extent to which changes in proximal axial ligand affect electronic structure and electronic spectra and the relationship, if any, to function.

### 1.6 Investigation of the Role of the Protein in Function

**1.6.1** ***The Mode of Binding of Small Ligands to the Globins*** Globins have an essentially hydrophobic binding pocket, appropriate for binding molecular oxygen. The participation of the residues in this binding site is not required to transform the protein to a functional form. The full protein simulations of the globins have therefore focused on further elucidation of the details of small ligand binding.

**1.6.2** ***The Role of the Protein Environment in Peroxidases in Formation of Compound I*** Since globins and peroxidases have the same axial ligand, the role of the protein in forming and preserving the different species required for function must be considered. As shown in Figure 4A, it is Compound I and not the ferric resting form of the peroxidases that is its biologically active species. Polar residues found in the distal binding pocket of peroxidases, not present in the globins are thought to play a key role in transforming the putative peroxide intermediate to Compound I. The full protein study reported probes the postulated role of these residues in Compound I formation.

**1.6.3** ***The Role of the Protein Environment of the Cytochrome P450s in Compound I Formation*** As shown in Figure 5, the proposed pathway to Compound I formation in the cytochrome P450s is a proton assisted dioxygen bond cleavage. The role of the protein in providing and delivering these protons has been addressed using the known X-ray structure of a specific P450 isozyme.

## 2 PROPERTIES COMMON TO ALL IRON PORPHYRIN (HEME) COMPLEXES

Heme complexes consist of three covalently linked components: the iron, the porphyrin macrocycle, and axial ligands binding directly to the iron. There is a great deal of both experimental and computational evidence that in no state of any model heme site or intact heme protein are the electromagnetic properties centered on the heme iron close to those in the free ion form. These deviations from free ion character are manifest in many observed properties. For example:

i) the presence of low lying states of different spin multiplicities. The free ion, in which these energy differences are very large, appears to be perturbed in heme complexes in such a way that spin state changes occur readily and are modulated by changes in the axial ligand and small changes in the Fe-porphyrin interactions. The near degeneracy of these spin states in turn leads to a set of characteristic properties such as large zero field splittings and the large anisotropy of $g$-values observed in electron spin resonance (ESR) spectra.
ii) The large electric field gradients observed in Mössbauer resonance spectra even for some low spin ferrous and high spin ferric porphyrins that would be zero in their ionic states. Such deviations are consistent with strong interactions between

the iron and the porphyrin ring and in many cases between the iron and axial ligands.

Many levels of calculations have been used to characterize the strong interactions occurring in heme systems. Those that treat the perturbations of the central metal ion phenomenologically include the spin-Hamiltonian formalism, and the crystal and ligand field formalisms. The studies made using these methods have elucidated the origin of a variety of observable electromagnetic properties. These include zero field splittings, spin-orbit coupling, $g$-values observed in electron spin resonance (ESR) spectra, hyperfine splitting observed in ESR and NMR spectra, quadrupole splitting observed in Mössbauer resonance spectra, and magnetic susceptibilities. However, molecular orbital methods are required to characterize explicit models of the entire heme unit because it is, in fact, a covalently bound molecular system. The results of such studies thus provide a more realistic description of electron and spin distributions of the ground and low lying electronic states in the heme complex. Observable electromagnetic properties can be calculated from these more realistic and complete descriptions of electronic structure. Thus, molecular orbital calculations of heme complexes can in principal provide the most robust link between structure, spectra, and function.

The minimum level of molecular orbital calculations useful in describing transition metal porphyrin systems must include at least all valence electrons. One of the first such methods to be applied to these systems was the semi empirical all valence electron methods called extended Huckel (EH) and iterative extended Huckel (IEHT) procedures. More recently another semi empirical all valence electron method, INDO, has been extensively used. As computational methods and capabilities increase, two types of *ab initio* methods have also been increasingly used; the more traditional Hartree Fock (HF) self consistent field (SCF) method and procedures based on density functional theory (DFT). A number of properties common to all iron-porphyrin (heme) complexes have emerged from the combined results of these different types of computational studies, together with spectroscopic measurements.

## 2.1 Removal of the Degeneracy of the *d* Orbitals of the Iron

As a consequence of spherical symmetry of the isolated iron atom, the five $3d$ atomic orbitals are all essentially degenerate. In complexes with porphyrins, this degeneracy is removed because of the deviation from spherical symmetry of the molecular system and the covalent nature of the bonds between the iron and its ligands. Figure 6 is a schematic representation of the single $d$-orbital energies of iron in molecular complexes of different symmetry. As shown in Figure 6, in complexes with octahedral (Oh) symmetry, the 5-fold degenerate $d$ orbitals are split into a triply degenerate $t_{2g}$ and doubly degenerate $e_g$ manifold. This same splitting but inverted is a consequence of tetrahedral (Td) symmetry. In complexes with tetragonal (D4h) axial symmetry, there is a further lifting of the degeneracy with only the $d_{xz}$ and $d_{yz}$ orbitals remain-

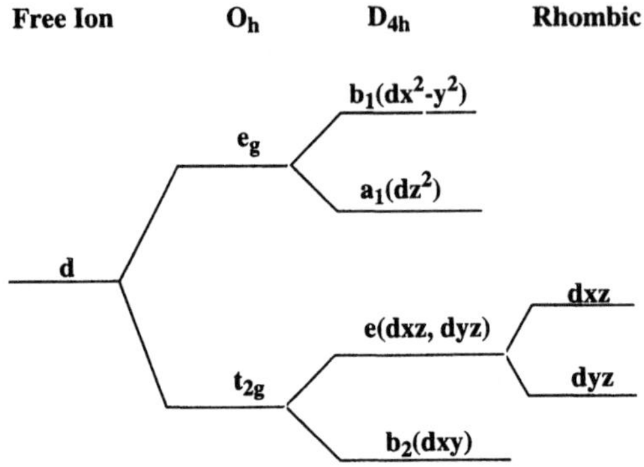


**Figure 6.** Splitting of degeneracy of the 3*d* orbitals of Fe in complexes with different symmetry.

ing degenerate. In all lower symmetries, usually generically designated as rhombic, there is a further splitting of these two orbitals.

Four coordinated metal-porphyrin complexes in which the two inner hydrogens of the free base porphyrin are replaced by iron and other transition metals, have fourfold symmetry (D4h), in which the direct ligands of the metal ion are the four pyrrole nitrogens resulting in the corresponding *d*-orbital pattern shown in Figure 6.

The formation of metal ion complexes with unsubstituted porphyrins does not in itself appear to cause significant deviations from planarity of the metal free porphyrin. *Ab initio* local density functional calculations for Mg and Pd complexes[1] and for a more extensive[2] series of 4 coordinated metal porphyrins with Mg, Ca, Cd, Fe, Ni, Zn, Ru, and Pd dications have been reported. In the optimized geometries of all of these four coordinated unsubstituted porphyrin metal complexes, the porphyrin ring remains essentially planar and the metal ion is in the plane of the porphyrin with very little distortion of the porphyrin geometry.

Deviations from D4h symmetry can be caused by substituents on the porphyrin ring, by the addition of axial ligands to the heme iron, and by the environment of the protein. The geometry of porphyrinoid containing systems has been an active area of both experimental and computational investigation. The additional of peripheral substituents to metal containing unsubstituted porphyrins can have a pronounced effect on porphyrin ring distortions. A large amount of structural information exists for synthetic metalloporphyrins with a variety of substituents.[3,4] In addition, a number of studies have been reported using molecular mechanics methods together with resonance Raman spectra and X-ray diffraction to systematically study the effect of metal ions,[5–7] substituents,[8–13] and axial ligands[9,14] on ruffling of metalloporphyrins. However, these studies are very largely concerned with synthetic porphyrins.

The substituents shown in Figure 1 corresponding to a protoporphyrin IX ring system common to almost all the heme protein families do not cause extreme deviations from planarity of the porphyrin ring. Although there is some distortions of the porphyrin ring, in specific heme proteins these deviations *per se* do not appear to play a significant role in modulating biological function. More relevant to function and observed electromagnetic properties are the nature and number of axial ligands that, in general, lower the formal D4h symmetry of the four coordinated species.

## 2.2 Extensive Electron Delocalization

There is a great deal of both experimental and computational evidence that in heme proteins and model heme systems there is significant electron delocalization among and between the three components of the heme unit, the iron, the porphyrin macrocycle, and axial ligands binding directly to the iron. Electron distributions are obtained in a straightforward manner from the results of Hartree–Fock quantum mechanical calculations employing molecular orbitals expressed as linear combinations of atomic orbitals. The results of density functional calculations also yield the total electron distribution in the molecule. From these calculations, descriptors of electronic structure can be calculated in the form of:

i) atomic charges on each atom
ii) electron occupation numbers in each atomic orbital
iii) bond orders that represent bond strength
iv) unpaired spin distributions in radical species.

Taken together they describe the effect of covalent bond formation on each separate component of the heme unit.

The porphyrin macrocycle itself has an electron density characteristic of a highly delocalized system. The extent of electron delocalization of the porphyrin macrocycle in free base unsubstituted porphyrin has been characterized using both semiempirical and *ab initio* quantum chemical methods.[15–17] The calculated electronic properties for free base porphyrin could be rationalized in terms of one monocyclic conjugated $\pi$-system isoelectronic with 18-annulene connected to two essentially isolated double bonds between the two $C_\beta$ carbons of the pyrrole rings that do not have the protonated nitrogen and two equally isolated amino nitrogens of the other two rings (Figure 7). Although such an electronic structure had been proposed,[18–20] the *semiempirical* $AM_1$/UHF (Unrestricted Hartree–Fock) calculations[15] provided the first theoretical evidence in support of it.

In metal containing porphyrins with equivalent pyrrole rings, porphyrin $\pi$ orbitals are completely delocalized. In contrast to the metal free system, there are no longer any isolated porphyrin double bonds or pyrrole nitrogens. This more complete charge delocalization in the porphyrin macrocycle is clearly manifest in the four coordinated iron-porphyrin complexes using both semi-empirical and *ab initio* molecular orbital methods. In one such study,[2] calculations were performed using the local density functional (LDF) method for neutral Ni, Pd, Ca, Mg, F, Zn, Cd, and Ru porphyrins. Optimized geometries had D4h symmetry with no significant porphyrin ring distortion or out of planarity of the metals. Electron distributions showed completely equivalent charge densities on the same type of atoms of the porphyrin ring, i.e., the four pyrrole nitrogens, eight $\alpha$ carbons, four $\beta$ carbons, and four meso carbons.

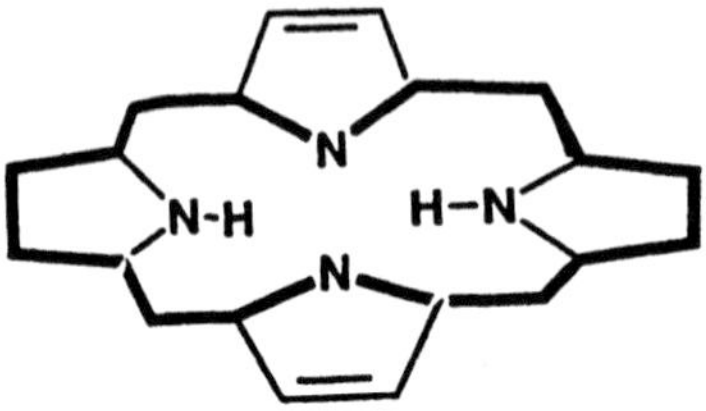


**Figure 7.** Extent of delocalization of the $\pi$ electron network in metal free porphine. The delocalized annulene $\pi$ electron network is shown in bold. Two regions of this macrocycle do not share in this electron delocalization. These are shown in plain type. They are: a) the $C\alpha$–$C\beta$ double bonds in the two deprotonated pyrrole rings and b) the protonated pyrrole nitrogens.

Not only is there electron delocalization within the porphyrin macrocycle in metal porphyrin complexes, but also significant charge transfer is found between the metal and the porphyrin due to covalent interactions. Using both semi-empirical and *ab initio* molecular orbital methods, the results demonstrate that extensive charge transfer from the porphyrin to the metal occurs with a reduction of both the formal [+2] charge on the metal ion and [−2] charge on the porphyrin. For example, from the LDF calculation cited above, the extent of electron transfer from the dianionic porphyrin to the dicationic ligand was found to be in the order Ni > Fe > Pd > Mg > Ru > Zn > Cd > Ca and ranged from 1.77 to 0.7 e. Thus, in all cases there is significant charge transfer from the porphyrin to the metal via covalent bond formation. Although actual values calculated depend on the method and definition used, this trend is found independent of method and definition of net atomic charges. This delocalization is manifest in the nature of the molecular orbitals as well as in the net atomic charges calculated from them. In such molecular systems, the *d* orbitals shown in Figure 6 are no longer localized on the iron, but are transformed to molecular orbitals with varying degrees of participation of the porphyrin ring atomic orbitals. In addition, the molecular orbitals that are primarily centered on the porphyrin atoms have varying amounts of iron centered atomic orbitals.

The ability of the porphyrin macrocycle to participate in efficient redistribution of net charge and unpaired spin can play a key role in formation and stabilization of the various heme species needed for function. Changes in the formal oxidation and spin state of the heme iron occur during many heme protein functions. For example, although the enzymatically active Compound I Fe=O species is formally two oxidation states above the ferric form because of extensive electron redistribution, the iron is more like an Fe(IV) species with the charge distribution of the porphyrin ring that of a $\pi$ cation radical.

## 2.3 Low Energy States of Different Spin Multiplicities

One of the most characteristic properties of transition metal containing porphyrin systems is the presence of several low energy states with different spin multiplicities. For example, ferrous porphyrins have quintet, triplet, and singlet state species and ferric porphyrins have sextet, quartet, and doublet state species in close energetic proximity. This behavior is in marked contrast to inorganic ferrous and ferric containing compounds with very well defined ground states and large energy separations between states of different spin multiplicities. The presence of low lying states of differing spin multiplicity are manifest in a number of signature spectroscopic properties of heme sites.

**2.3.1** ***Significant Spin-Orbit (L.S.) Coupling*** This interaction occurs among these low lying states of different spin multiplicities. This spin orbit coupling is the underlying effect responsible for: a) large values of zero field splittings; b) anisotropy in $g$ values in electron spin resonance; and c) deviations from spin only values of magnetic moments.

**2.3.2** ***Large "Zero" Field Splittings*** Zero field splitting is the lifting of degeneracy of states that are part of the same total spin manifold but have different values of the $z$ component of the total spin, even in the absence of a magnetic field. For example, there is only one possible sextet state of ferric heme complexes, a $^6A_1$ state. It corresponds to a unique Fe $3d$ configuration with one electron is each of the five $d$ orbitals of the iron. As shown schematically in Figure 8a, in ferric heme complexes, this sextet state is split into three separate doubly degenerate states, so-called Kramer's doublets. The zero field splittings are the energy difference between the multiplets of the sextet state. In high spin ferric heme complexes, as shown in Table 1, these splittings are very large (10 and 50 $cm^{-1}$) compared to typical values

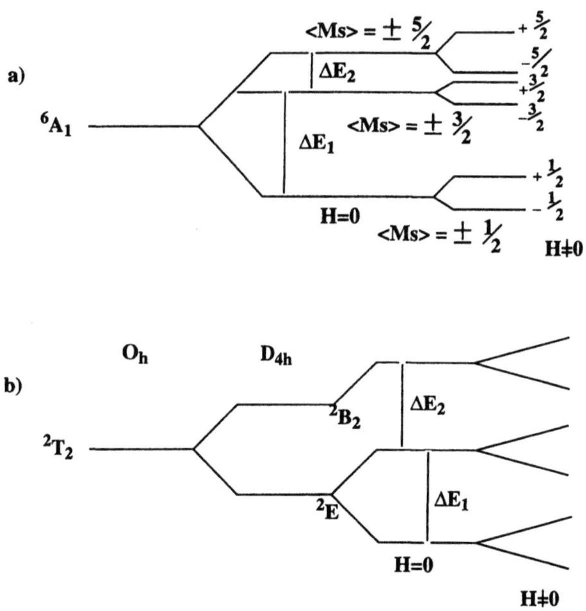


**Figure 8.** Schematic representation of zero field splitting in sextet and doublet states of ferric heme complexes.

**Table 1 Zero Field Splitting, g Values, and Effective Magnetic Moments in 8 High Spin Ferric Porphyrin Compounds**

| Ligand | $OCH_3^-$ | $OC_6O_5^-$ | $F^-$ | $CH_3COO^-$ | $N_3^-$ | $Cl^-$ | $Br^-$ | $I^-$ |
|---|---|---|---|---|---|---|---|---|
| $\Delta E^{a}_{1\,exp}$ | 9.3 | 10.9 | 11.1 | 13.8 | 14.8 | 17.9 | 23.6 | ~33 |
| $\Delta E_{1calc}$ | 9.3 | 10.9 | 11.1 | 13.8 | 14.8 | 17.9 | 23.6 | 33 |
| $\Delta E^{b}_{2calc}$ | 19.0 | 22.5 | 23.0 | 28.9 | 30.2 | 37.4 | 49.8 | 78 |
| $g\|$ | 2.00 | 2.00 | 2.00 | 2.00 | 2.00 | 2.00 | 2.00 | 2.00 |
| $g\perp$ | 5.99 | 5.98 | 4.98 | 5.97 | 5.97 | 5.95 | 5.92 | 5.85 |
| $u_{eff}(T)$ | | | | | | | | |
| $T = 293°$ K | 4.91 | 5.90 | 5.90 | 5.90 | 5.89 | 5.88 | 5.86 | 5.80 |
| $T = 77°$ K | 5.90 | 5.89 | 5.89 | 5.88 | 5.87 | 5.86 | 5.81 | 5.70 |
| $T = 4°$ K | 5.29 | 5.20 | 5.18 | 5.05 | 5.01 | 4.90 | 4.74 | 4.57 |

[a] $\Delta E_1 = E\,(^6A_1 \pm 3/2) - E\,(^6A_1 \pm 1/2)$.
[b] $\Delta E_2 = E\,(^6A_1 \pm 5/2) - E\,(^6A_1 \pm 3/2)$.

of 0.2 $cm^{-1}$ in inorganic iron compounds such as $Fe_2O_3$ and $FeCl_4^-$. The origin of this large zero field splitting in ferric heme complexes was shown in early calculations based on a crystal field model of heme complexes[21] to be the large spin orbit coupling between components of this sextet state and low lying quartet states. In the crystal field model used, the metal ion is assumed to be surrounded by a regular array of ligands, which create an electric field of a given symmetry.[22] As also shown in Table 1, changing only the axial ligand in a series of five-coordinated ferric porphyrin complexes can alter the magnitude of the zero-field splitting. Calculations of $\Delta E_1$ with the crystal field model, in which the strength of the interactions with the axial ligand were allowed to vary, could completely account for the observed values of $\Delta E_1$ for this series of compounds and made predictions of the corresponding values of $\Delta E_2$. These variations in zero field splittings are an indication of the effect of the axial ligands on the relative energies of the low lying states of different multiplicities.

Unlike the sextet state, there are many possible Fe 3*d* configurations that can lead to a doublet state. There is a great deal of evidence to suggest that the state that becomes the low spin ground state in ferric heme complexes arises from a $^2T_2$ ($t_2^5$) manifold of states. The doublet states can interact with excited sextet and quartet states by spin orbit coupling. We have in the past[23] calculated the *g* values to be expected in such a case, and they do not correspond to the observed values. Therefore interactions with excited quartet and sextet states can be ignored in considering the behavior of the low-lying doublet states originating from the six fold degenerate $^2T_2$ manifold. As shown in Figure 8b, these six degenerate states interact among themselves under the influence of spin orbit-coupling and deviations from octahedral symmetry. They also split into three separate doubly degenerate states. Thus, this low spin state with orbital degeneracy also undergo zero field splitting. Since the magnitude of the zero field splittings $\Delta E_1$ and $\Delta E_2$ do not depend on interactions with excited states of different spin multiplicity, the values calculated for a series of low spin ferric heme complexes using the crystal field model[24] are in the range of 400 and 1,600 $cm^{-1}$, respectively both much greater than for high spin ground states.

**2.3.3** ***Large Anisotropy in g Values Observed Electron Spin Resonance Spectra*** As shown in Figure 8a, the three doublets in high spin ferric heme complexes already split in zero magnetic fields by the effect of spin orbit coupling are further split in the presence of a magnetic field. Because of the spin orbit coupling, the magnetic field energy differences between these states lead to characteristic anisotropic $g$ values observed in electron spin resonance spectra of high spin ferric heme species, deviating significantly from the isotropic value of 2 observed in organic free radicals. For high spin systems, these $g$ values are typically 6 ($g_{xx}$, $g_{yy}$) and 2 ($g_{zz}$) for systems with axial symmetry as illustrated, for example, in Table 1 for the same series of 5 coordinated high spin ferric heme complexes for which zero field splittings were calculated. The calculated $g$ values[21] are in excellent agreement with experiment. The two symmetry equivalent $g$ values of 6 can be further split for heme systems with significant deviations from axial symmetry.

For low spin ferric heme systems g values are closer to two, with two values greater and one value less than two for the three major components of the $g$ tensor. Typical experimental $g$-values for five unique types of ferric heme proteins that exist in essentially low spin state are given in Table 2. Using crystal field theory, early studies in our laboratory provided a robust connection between the nature of the ligands binding to the iron and the observed anisotropy in $g$ values due to spin orbit coupling among the components of the low-lying doublet states originating from the six fold degenerate $^2T_2$ manifold of states in low spin ferric heme complexes.[25]

**2.3.4** ***Deviations From the Spin Only Value of Effective Magnetic Moments*** These are also caused by spin orbit mixing of nearly degenerate states of different multiplicities. Variations in axial ligand in the same heme complex can modulate the relative energy of these states and hence the extent of spin orbit coupling. This variation in spin orbit coupling in turn is manifest by deviations from the spin only value of magnetic moment measured at the same temperature in the same ferric heme system with different axial ligands. As shown in Table 3, the value of magnetic moments varied dramatically with the change in a single axial ligand. Binding of ligands such as $F^-$, $H_2O$, formate ($HCOO^-$), $SCN^-$, $NO_2$, azide ($N_3^-$), and $SH^-$ to hemoglobin causes a continuous decrease in magnetic moment from a high spin ($S = 5/2$) to a low spin ($S = 1/2$) value.[26,27] Using a crystal field formalism for the ferric heme complex, early studies in our laboratory[28] could consistently account for these continuously varying values of magnetic moment by the extent of spin-orbit mixing. This

**Table 2 Experimental $g$-Values for Five Unique Types of Low Spin Ferric Heme Proteins**

| Type | Example | Number of Compounds | $g_z$ | $g_y$ | $g_x$ |
|---|---|---|---|---|---|
| *c* | Cytochrome C | 25 | 3.15 | 2.25 | 1.25 |
| *b* | Cytochrome B; salycylate Hb | 50 | 2.95 | 2.26 | 1.47 |
| *b* | Histidine Hb | 100 | 2.80 | 2.26 | 1.67 |
| *o* | (OH) Hb, Mb | 150 | 2.55 | 2.17 | 1.85 |
| *p* | P450; pyridine Hb | 100 | 2.41 | 2.26 | 1.93 |

**Table 3 Effective Magnetic Moments of Ferric Ion in Hemoglobin Derivatives**

| | Ligand | $\mu_{eff}$ (exp.) |
|---|---|---|
| 1 | $F^-$ | 5.76 |
| 2 | $H_2O$ | 5.65 |
| 3 | $HCOO^-$ | 5.44 |
| 4 | $OCN^-$ | 5.40 |
| 5 | $SCN^-$ | 5.06 |
| 6 | $OH^-$ | 4.66 |
| 7 | $NO_2$ | 4.13 |
| 8 | $SeCN^-$ | 3.88 |
| 9 | imidazole | 2.87 |
| 10 | $CN^-$ | 2.50 |
| 11 | azide$^-$ | 2.35 |
| 12 | $SH^-$ | 2.26 |

spin orbit coupling results in quantum-mechanically spin mixed states. The different extent of mixing of states with quartet and doublet character with the sextet state by spin-orbit coupling, rather than just thermal mixing of pure spin states resulted in the continuous variation observed for the derivatives shown in Table 3. However, low lying states of different spin multiplicity can also undergo further thermal mixing as indicated in observed temperature dependence of effective magnetic moments ($u_{eff}$) shown for the high spin compounds in Table 1.

**2.3.5** ***Facile Changes in Ground State Spin of Heme Complexes*** Because of the close proximity of states of multiplicity, the relative spin state energy ordering is very sensitive to the nature of the metal ion and the number and nature of axial ligands. While five coordinated heme complexes, in general, have high spin ground states, six coordinated species can vary in their spin state ordering depending on the nature of the axial ligands. One of the most striking biological manifestation of this modulation is the binding of molecular oxygen to the 5-coordinated ferrous heme site of globins. The addition of $O_2$ and other small ligands such as CO transforms a high spin, quintet species to a singlet species. Another significant effect on function is the transformation of the low spin (doublet) ferric resting form of many cytochrome P450s to a five coordinated high spin (sextet) system upon substrate binding. The change in spin state facilitates the next step in the enzymatic cycle, one-electron reduction.

While various spectroscopic techniques are helpful in identifying the spin multiplicity of the ground state, it is not always possible to clearly determine this property from experiment alone. It is even more difficult to obtain the rank order of energy of states of different multiplicities for a given heme species by either experimental or computational methods. However, a number of computational studies have been made over the years with just such a goal. These studies involved use of *ab initio*

Hartree–Fock and density functional theory as well as semiempirical methods, particularly the INDO/ROHF method.[29,30]

The relative spin state ordering in a 4-coordinated unsubstituted iron porphyrin [Fe(II)–P] complex has been studied by *ab initio* methods.[2,31–36] In all of these studies a pair of nearly degenerate triplet states were found a $^3A_{2g}$ state or a $^3E_g$ state, with one or the other predicted to be the ground state. Mössbauer resonance spectroscopy indicates that the [Fe(II)–P] complex should be a quintet state,[37–39] while proton NMR spectra suggest that [Fe(II)–P] has a triplet ground state[40,41] as found in these calculations.

In another *ab initio* study,[32] the Restricted Hartree–Fock (RHF) method was used to characterize the neutral complexes of the first row transition metal ions Cr(III), Mn(III), Fe(III), Co(II), Co(III), Ni(II), Cu(II), and Zn(II) with unsubstituted porphyrins [M-P]. The results of the RHF calculations indicate that while the ground state of [Zn-P] was a singlet and of the [Cu(II)-P] a doublet, all other metal porphyrins possessed a ground state electronic distribution with a high spin multiplicity that equals the spin multiplicity of the bare metal ion. Comparisons with observed electromagnetic properties were not systematically discussed, so that is was difficult to determine the accuracy of these predictions. Accurate determination of the spin state ordering by any *ab initio* method remains a very challenging current problem.

The difficulties of obtaining the correct spin state ordering by *ab initio* methods led to exploration of an alternative procedure, the semiempirical quantum chemical INDO method with the capability of performing restricted Hartree–Fock (RHF) calculations for open shell radical species. The ability of this INDO/ROHF method to correctly predict the ground state spin was systematically tested by calculations of the relative energies, electron and spin distributions, and Mössbauer quadrupole splittings for the low lying spin states of a series of eighteen ferric porphryin complexes with varying axial ligands for which crystal geometries were known. Among these were ferric heme complexes known to have high (sextet), intermediate (quartet), and low (doublet) spin ground states states that could be inferred from the observed electromagnetic properties.[42] In these calculations, the crystal structure geometry of the heme complex was used for all the spin states with the porphyrin substituents replaced by H atoms. The calculated ground spin state, the calculated energies of the other spin states relative to it, and the calculated values of quadrupole splitting in Mössbauer resonance spectra were quantitatively consistent with the observed effective magnetic moments, ESR spectra, and the Mössbauer quadrupole splitting spectra for the 18 compounds studied.

For all the high spin compounds studied, the doublet states were predicted to be thermally inaccessible being 14–35 kcal/mol higher in energy than the sextet state. The calculated energy separation between the quartet and sextet states were significantly higher in the five coordinated compounds (about 11 kcal/mol) decreasing to less than 5 kcal/mol in the six coordinated heme complexes. For diaquo and diethanol ferric heme complexes, near degenerate quartet and sextet states were calculated. The intermediate spin compounds were calculated to have a quartet ground state consistent with experimental observations with an energy separation of 2–7 kcal/mol to the sextet and about 20 kcal/mol to the doublet states. The calculated properties for the known doublet complexes were also consistent with observation.

The consistency between calculated and observed properties obtained for all complexes studied provide strong support for the reliability of the INDO/RHF method to obtain relative energy ordering of states of different multiplicity for heme systems with known geometries and assuming the same geometry for all spin states of the same complex. In addition, comparisons of the results obtained for the 18 complexes studied allowed insights into the effect of the geometry and axial ligand(s) on the relative energy and observable properties of the three spin states of ferric heme complexes.

## 2.4 Characteristic Values of Quadrupole Splittings in Heme Sites

These sites are of different oxidation state and different spin multiplicity. The magnitude and anisotropy of the quadrupole splittings ($\Delta E_Q$) in Mössbauer resonance spectra is a direct measure of the deviation from spherical symmetry of the electron distribution around the iron. Low spin ferrous compounds, in which the three *d*-orbitals of the $t_{2g}$ manifold, shown in Figure 6, are doubly occupied as well as high spin ferric species in which each *d* orbital is half occupied are associated with the most spherically symmetric electron distributions around the iron. They should therefore have little or no quadrupole splitting. By contrast, low spin ferric heme complexes in which two of the $t_{2g}$ orbitals are doubly occupied and one is singly occupied and high spin ferrous heme complexes in which one *d* orbital is double occupied and four are singly occupied should have large deviations from spherical symmetry and hence larger values of quadrupole splittings. Thus, experimental values of the quadrupole splittings ($\Delta E_Q$) observed in Mössbauer resonance spectra are often used to deduce the oxidation and spin state in ferrous and ferric heme sites. However, this idealized behavior is based on the isolated metal ions and does not take into account the possible effects of electron delocalization when the Fe is covalently bound to the heme unit and axial ligands. These effects can cause significant perturbation of the electron distribution around the ferric and ferrous ion compared to the free ion. Thus, inferences drawn about oxidation and spin state from measured values of ($\Delta E_Q$) can be misleading.

A particularly striking example of the effect of different axial ligands on the charge distribution around the iron is the binding of $O_2$ and CO to the ferrous heme unit of the globins. Both of these complexes are formally low spin ferrous heme species and should have little or no quadrupole splitting. However, the extent of perturbation of the electron distribution around the iron by these two of them is very different.[43,44] As shown in Table 4, when $O_2$ binds, there is significant charge delocalization resulting in a *d* orbital population that is essentially ferric, with five electrons in the three *d* orbitals that comprise the manifold of $t_{2g}$ orbitals that are degenerate in octahedral symmetry. This charge delocalization is a result of covalent interaction between the Fe and the $O_2$ ligand and hence is not accompanied by change in either local or total unpaired spin density. The quadrupole splitting calcu-

**Table 4 Comparison of Properties of Ferrous Porphine Model of Carbonyl Hb and Oxy Hb Complexes**

| | HbCO | $HbO_2$ |
|---|---|---|
| $q$(Fe)[a] | +1.47 | +1.41 |
| (L) | +0.03 | −0.27 |
| (O) | −0.54 | −0.48 |
| ($N_{pyr}$) (Porph) | −0.55 | −0.25 |
| ($N_{imid}$) | −0.41 | −0.41 |
| | Orbital Population[a] | |
| $d_{z^2}$ | 0.44 | 0.57 |
| $d_{x^2-y^2}$ | 0.42 | 0.55 |
| $d_{xy}$ | 1.99 | 1.98 |
| $d_{xz}$ | 1.79 | 1.96 |
| $d_{yz}$ | 1.80 | 1.01 |
| $\Delta E_Q$(calcd) | 0.56 mm/sec | −2.16 mm/sec |
| $\Delta E_Q$ (obsd) | 0.30 mm/sec | −2.10 mm/sec |
| | Bond Overlap Density[a] | |
| $\rho(\text{Fe–L})_{calc}$ | 0.66 | 0.48 |
| $P1/2_{obs}$ (Torr)[b] | $8.9 \times 10^{-3}$ | 38 |
| $\rho$ $(\text{L–O})_{calc}$ | 1.08 | 0.56 |
| $\nu(\text{L–O})_{obs}$ $(cm^{-1})$[c] | 1970 | 1160 |

[a] Net atomic charges, orbital population and bond densities calculated from Mulliken population analyses.
[b] Pressure for 50% maximinum binding of ligand.
[c] Observed L–O stretching frequency.

lated from this charge distribution also given in Table 4 resulted in a large negative value in excellent agreement with experiment.

Prior to this calculation, the interpretation of this large quadrupole splitting was made in a spin Hamiltonian formalism assuming that the Fe and the $O_2$ were separate entities. In this simplified model, transfer of an electron to the oxygen must be coupled to creation of unpaired spin density on the iron and on the oxygen. Since there is no net magnetic moment in this system, these unpaired spins were further assumed to be antiferromagnetically coupled. However, heme complexes are molecular species. As demonstrated by this molecular orbital calculation of an explicit model [Fe(II)–P–Im–$O_2$] of the oxyferrous globin ($HbO_2$) heme complex, electron transfer can occur by delocalization without causing unpaired spin on either partner. These results thus explain how an anomolously large value of quadrupole splitting can be associated with a formal low spin diamagnetic oxyferrous Hb system.

The calculated charge distribution found for the ferrous heme carbonyl [Fe(II)–P–Im–CO] complex, could also explain the observed value of this property in HbCO and its large difference from the $HbO_2$ complex. As shown in Table 4, in contrast to the oxy complex, CO binding to the ferrous heme species preserves the low spin ferrous spherical symmetry with nearly two electrons in each of the $d_{xy}$, $d_{xz}$, and $d_{yz}$ orbitals. The quadrupole splitting calculated from this charge distribution also given in Table 4 results in the very small value of quadrupole splitting observed.

## 2.5 Nature of the Highest Occupied and Lowest Unoccupied Orbitals

The nature of the highest energy occupied orbitals in porphyrin containing complexes has been an active area of both computational and a variety of spectroscopic investigations. Among the spectroscopic techniques used to probe this property are UV-visible photoelectron and electronic,[45,46] NMR,[47–49] ESR,[50–52] MCD,[53] EXAFS,[54,55] ENDOR (electron nuclear double resonance),[56–58] Mössbauer,[51,59,60] and resonance Raman[61–64] spectra. Both semi-empirical and *ab initio* quantum chemical methods have also been used to characterize the nature of the highest occupied molecular orbitals for a number of specific porphyrin systems.

The optimized structures and ionization potentials (IPs) of free base unsubstituted trans-porphyrin have been calculated using both Hartree–Fock and localized density functional (LDF) methods.[17] The absolute values of LDF IPs of porphyrins are generally in excellent agreement with gas-phase photoelectric spectra with discrepancies between calculation and experiment being only about 0.0–0.3 ev.[65,66] From the results of both the HF and LDF calculated IPs, the two lowest IPs correspond to $a_{1u}$ type ($^2A_u$) and $a_{2u}$ type($^2B_{1u}$) cationic final states and are well separated from other IPs.

The effect of adding peripheral substituents to the porphine macrocycle such as the widely studied meso tetraphenyl porphyrin (TPP)[67,68] and *B*-octa alkyl porphyrin (OEP)[69] on the two lowest energy $\pi$ cationic states in metal free systems has been investigated by both experimental gas-phase photoelectron spectroscopy (UPS) and computational methods.[70] The results clearly indicate that the two lowest IPs correspond to electron abstraction from the $a_{1u}$ and $a_{2u}$ like orbitals with the $a_{2u}$-like $\pi$ cation radical the lowest energy species.

The effect of metal ion complexation on the ionization potential of free base unsubstituted porphyrins has also been studied by both experimental and computational methods. The first question addressed was whether the two lowest IP remained the $a_{2u}$-like and $a_{1u}$-like porphyrin $\pi$ orbitals upon metal ion complexation, as they are in the metal free species. Computational studies have been reported using *ab initio* local density functional theory for Mg and Pd porphyrins[1] and for a more extensive series of 4 coordinated metal porphyrins with Mg, Ca, Cd, Fe, Ni, Zn, Ru, and Pd dications.[2,33] In another *ab initio* study,[32] the Restricted Hartree–Fock (RHF) method was used to characterize the first row transition metal complexes with unsubstituted porphyrins Cr(III), Mn(III), Fe(III), Co(II), Co(III), Ni(II), Cu(II), and Zn(II).

The interesting characteristic that emerges from these studies is that, in general, in porphyrin containing systems, including free base unsubstituted and substituted porphyrins and metal containing porphyrins, two porphyrin $\pi$ orbitals an $a_{1u}$-like orbital and $a_{2u}$-like orbital are consistently found to be the two highest energy filled orbitals (HOMO and HOMO-1) with a significant separation from the others. Typical electron density distributions in these orbitals for metal containing porphyrin complexes are shown in Figure 9. Electron density in the $a_{1u}$ type orbital is mainly in the $C_\alpha$

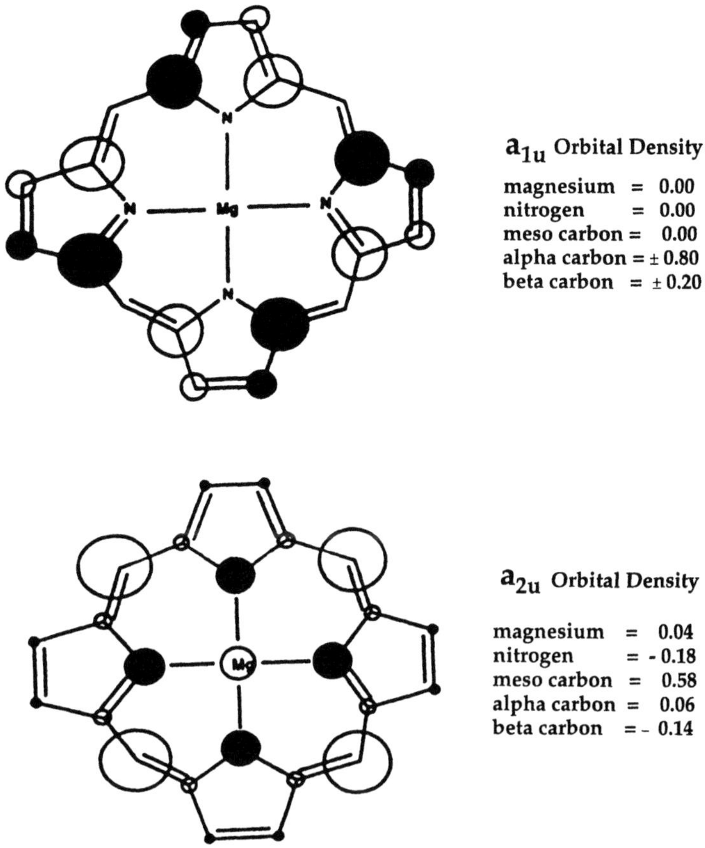


**Figure 9.** Typical electron distribution in $a_{1u}$ and $a_{2u}$ porphyrin $\pi$ orbitals in metal porphyrin complexes. The size of the circles are proportional to the square of the orbital coefficients of each atom. The signs for the phases were retained from the sign of the atomic orbital coefficients in the molecular orbital. Filled circles correspond to a (+) sign and open circles to a (−) sign.

and $C_\beta$ atoms of the pyrrole rings while electron density in the $a_{2u}$-like orbitals is mainly on the pyrrole nitrogens and meso carbons. In transition metal complexes, the occupied molecular orbitals with essentially *d* orbital metal character have energies that are more negative than the double occupied $a_{1u}$ and $a_{2u}$ porphyrin $\pi$ orbitals. It requires less energy to remove an electron from either of these orbitals than from one of the *d* orbitals. As for the highest occupied orbitals, the lowest lying unoccupied orbitals are also similar for metal free and metal containing porphyrins.[1] They are two nearly degenerate eg($\pi^*$) type porphyrin $\pi^*$ orbitals.

A typical example of the energy ordering of molecular orbitals is given in Table 5 for a model ferrous carbonyl Hb [Fe(II)–P–Im–CO] complex from a semiempirical quantum chemical INDO calculation.[71] As seen in Table 5, the two highest energy occupied orbitals are almost pure porphyrin $\pi$ $a_{1u}$ and $a_{2u}$ orbitals with the molecular

**Table 5 Filled and Virtual Orbitals of Model Carbonyl-Heme Complexes**

| | Carbonyl-Heme |
|---|---|
| 90. | $(59\%\ d_{xy}, 31\%\ b_{2u}\sigma)^*$ |
| 89. | $\{(94\%\ e_g, 3\%\ \text{CO}\pi)^*\}$ |
| 88. | $\{(88\ \%\ e_g, 5\%\ d\sigma, 3\%\ \text{CO}\pi)^*\}$ |
| 87. | $\{43\%\ d_{x^2}, 14\%\ \text{Imid}\sigma, 12\%\ a_{1u}\sigma, 12\%\ e_g, 7\%\ \text{CO}\sigma\}^*$ |
| 86. | $\{65\%\ \text{CO}\pi, 21\%\ \text{Imid}\pi, 7\%\ d\pi)^*$ |
| 85. | $(55\%\ \text{CO}\pi, 37\%\ \text{porph}\pi)^*, 5\%\ d\pi)^*$ |
| 84. | $(67\%\ \text{porph}\pi, 30\%\ \text{CO}\pi)^*$ |
| 83. | $\{(78\%\ \text{Imid}\pi, 18\%\ \text{CO}\pi)^*\}$ |
| 82. | $100\%\ b_{2u}^*$ |
| 81. | $\{98\%\ \text{Imid}\pi\}$ |
| 80. | $(98\%\ e_g, 2\%\ d\pi)^*$ |
| 79. | $(98\%\ e_g, 2\%\ d\pi)^*$ |
| 78. | $100\%\ a_{1u}$ |
| 77. | $98\%\ a_{2u}$ |
| 76. | $(62\%\ d\pi, 31\%\ e_g, 4\%\ \text{CO}\pi)$ |
| 75. | $(62\%\ d\pi, 31\%\ e_g, 4\%\ \text{CO}\pi)$ |
| 74. | $96\%\ d_{x^2-y^2}$ |
| 73. | $(70\%\ \text{Imid}\pi, 29\%\ \text{porph}\pi)$ |
| 72. | $(93\%\ a_{2u}, 3\%\ \text{CO}\sigma)$ |
| 71. | $(77\%\ b_{1u}, 22\%\ \text{Imid}\pi)$ |

*Note*: Unless noted otherwise, all orbital designations refer to porphyrin $\pi$ orbitals in $D_{4h}$ symmetry. Those orbitals enclosed in braces are not significant components of the respective spectra.

orbitals predominantly *d* atomic orbital in nature of lower energy. Also illustrated by this table is the characteristic result that the two lowest lying unoccupied orbitals are the eg($\pi^*$) type porphyrin $\pi^*$ orbitals.

The persistence of the $a_{1u}$ $\pi$ and $a_{2u}$ $\pi$ orbitals as the highest occupied orbitals and eg($\pi^*$) orbitals as the lowest empty ones in diverse metal porphyrins can account for the facile formation of porphyrin $\pi$ cation radicals that play an important role in the biological and catalytic function of porphyrins. For example, oxidative metabolism involving heme proteins and model heme sites proceeds via formation of the highly oxidized porphyrin-iron-oxo (Compound I) radical cation species. In addition, $\pi$ cation radical formation is critical to the photooxidation of chlorophyll, in which Mg-porphyrins play a central role.

## 2.6 Nature and Modulation of Porphyrin $\pi$ Cation Radicals

The fact that the $a_{1u}$ and $a_{2u}$ orbitals remain the highest occupied nearly degenerate orbitals in metal containing porphyrins has led to both experimental and computational investigations of the extent to which changes in metal ions, peripheral ring substituents, and axial ligands favor formation of $a_{1u}$ or $a_{2u}$ cation radical species.

One striking result of metal complexation is that while unsubstituted metal free porphyrin favors an $a_{2u}$-like cation radical, complexation with Mg-P,[72] Zn-P,[73] and Fe-P[2,33] lead to an $a_{1u}$ like ground state. These results are consistent with previous *ab initio* calculations of unsubstituted four coordinated Fe(II) porphyrins using both LDF[2] and SCF methods.[33] Both report the $a_{1u}$ $\pi$ orbital as HOMO and the $a_{2u}$-$\pi$ orbital as HOMO-1 reversing the order found in the metal free species.

Another significant effect of metal ion complexation is found on tetra meso and $\beta$-octa-substituted porphyrins. While metal free $\beta$-octa porphyrins consistently favor an $a_{2u}$-like cation species, Mg–OEP and Zn–OEP[52,74] have been found to have a $^2A_{1u}$ ground state. By contrast, the $^2A_{2u}$ ground state found for metal free tetra meso substituted porphyrins is retained for Mg–TPP, Mg–Tn-propylP, Zn–TPP, and Zn–TMeP as deduced mainly from EPR studies,[52,74] and for Zn–TFP[73] and Ni–T(n-Pr)P[75] as deduced from semiempirical quantum chemical calculations and spectroscopic studies. In one of these studies,[75] mixing of these two highest occupied orbital was found in the crowded Ni–T(t-Bu)P analog in response to a lowering of symmetry by extreme porphyrin distortions.

Although Mg–OEP+ is known to have a $^2A_{1u}$ cation radical ground state, and Mg–TPP+ is known to have $^2A_{2u}$ cationic ground state, no systematic experimental studies had been reported for the effect of varying substituents in these positions in the Mg complex. Such a systematic study was performed[76] using the semiempirical quantum chemical (INDO/RHF) method[29,30] for four coordinated Mg porphyrins with $NH_2$, $OH^-$, phenyl, $CH_3$, $Cl^-$, and $CN^-$ substituents at the four meso positions and with these same substituents with the exception of phenyl, in the 8-$\beta$ positions. To assess the reliability of this method for such predictions, four systems were included that have been experimentally characterized. These are the unsubstituted Mg-porphyrin (Mg–P) and octaethyl Mg-porphyrin (Mg–OEP) both known to have a $^2A_{1u}$ cationic ground state and Mg-tetraphenyl (Mg–TPP) and Mg–T(n-propyl) P both experimentally known to have a $^2A_{2u}$ ground state. For all four species, the calculated results were consistent with experiment. For the remaining systems, all of the $\beta$-substituted Mg-porphyrins were predicted to preserve the $^2A_{1u}$ ground state of the unsubstituted Mg–P found in these and subsequent *ab initio* LDF calculations.[1] The substituents had a modest effect [(+/−)3 kcal/mol] on the relative energies of the $^2A_{1u}$ and $^2A_{2u}$ cationic states compared to that in the unsubstituted Mg-porphyrin. In contrast, the effect of the tetra meso substituents was more pronounced. Electron withdrawing meso substituent such as $Cl^-$ and $CN^-$ preserved the $^2A_{1u}$ ground state of the unsubstituted Mg–P. However, electron donating or neutral meso substituents $NH_2$, phenyl, OH, and methyl substituent led a $^2A_{2u}$ ground state with decreasing energy differences (15–2 kcal/mol) to the higher $^2A_{1u}$ cationic state. These predictions can be further tested by use of experimental techniques such as NMR, ESR, and ENDOR spectra that monitor the extent of unpaired spin on the substituents and the porphyrin ring atoms.

The combined results of all these experimental and computational studies clearly indicate that a wide variety of $\beta$-substituents in metal containing porphyrins favor an $a_{1u}$-like $\pi$ cation radical species. Since protoporphyrin IX, the most commonly occuring porphyrin group found in heme proteins, (Figure 1) have only $\beta$ substituents, intact heme proteins would be expected to favor $^2A_{1u}$ $\pi$ cation radicals. However, the substitution of Fe for Mg; changes in oxidation state and addition of axial ligands can be further modulators of these states possibly leading lead to different ground states and properties. These possibilities were addressed in models for the biologically active Compound I species of the metabolizing heme proteins discussed in subsequent sections.

## 2.7 Common Features in the Electronic Spectra of Porphyrins

The persistence of the $a_{1u}$ $\pi$ and $a_{2u}$ $\pi$ orbitals as the highest occupied orbitals and eg($\pi^*$) orbitals as the lowest empty ones in diverse metal porphyrins can also account for the common features observed in the electronic spectra of heme sites both in intact proteins and model compounds. All heme complexes have two characteristic $\pi$–$\pi^*$ absorption regions, the so called $Q$ bands, weak visible transitions ($Q_o$, $Q_v$) at 550–570 nm and the intense UV Soret transitions ($B$) at 408–450 nm. The higher UV bands ($L$ and $N$) appearing in a third region are broad and diffuse. A typical spectra, that for the ferrous carboxy complex in sperm whale myoglobin, illustrating these common features is shown in Figure 10.

The earliest and most intensive theoretical investigations of porphyrins centered on explanations of these observed features of the electronic absorption spectra common to all porphyrins. The origin of the $Q$ and $B$ bands was first explained by Gouterman using the so-called four-orbital model. This model was based on the per-

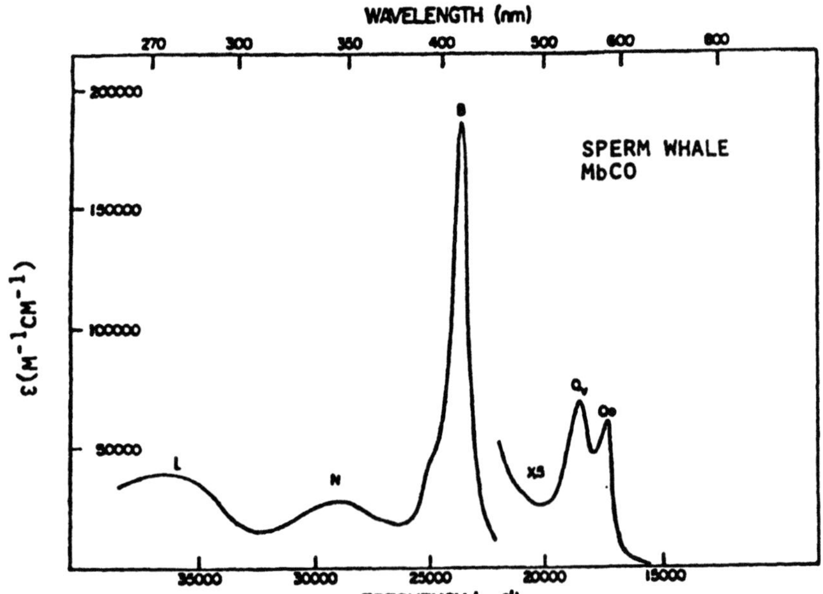


**Figure 10.** Typical electronic absorption spectra of a heme protein exemplified by the carbonyl complex of sperm whale myoglobin. The $Q$, $B$, $N$, and $L$ bands are common features in spectra of both model and intact protein heme sites.

sistence of the $a_{1u}$ $\pi$ and $a_{2u}$ $\pi$ orbitals as the highest occupied orbitals and eg($\pi^*$) orbitals as the lowest empty ones. As shown in Figure 11, using the four-orbital model,[7,78–80] without configuration interaction, excitations from the filled $a_{1u}$ $\pi$ and $a_{2u}$ $\pi$ orbitals to the unoccupied as the eg($\pi^*$) orbitals leads to the prediction of two strong, nearly degenerate bands. This result is inconsistent with the properties of the *B* and *Q* bands that have well resolved frequencies and very different intensities. However, as also shown schematically in this figure, when these four orbitals are allowed to mix by configuration interaction the resulting transitions have the observed properties, a weak *Q* band in the visible and a very strong *B* or Soret band in the near UV. In addition to explaining the origin of the *Q* and *B* band, as shown in this figure, the *N* and *L* bands could be assigned to transitions from other occupied $\pi$ orbital to these same lowest unoccupied eg($\pi^*$) orbitals.

A number of reviews have described the success of the four orbital model based primarily on simple Huckel molecular orbital models as well as $\pi$-electron only methods such as the SCF–PPP–CI method and the all valence electron CNDO/S methods to account for these regular $\pi$–$\pi^*$ transitions.[81–83] Additional early alternative approaches based on semiempirical quantum chemical methods applied to the calculation of the spectra of iron-porphyrins were the IEHT,[82,84–87] the PPP/CI,[88] and the X$\alpha$/CI[89] methods. While these methods have proven extremely useful in qualitative interpretation of electronic absorption spectra, quantitative assignment of both the typical *Q* and *B* bands and extra features often present in heme spectra and the effect of changes in spin state, ligands, and oxidation state on heme spectra require more accurate methods.

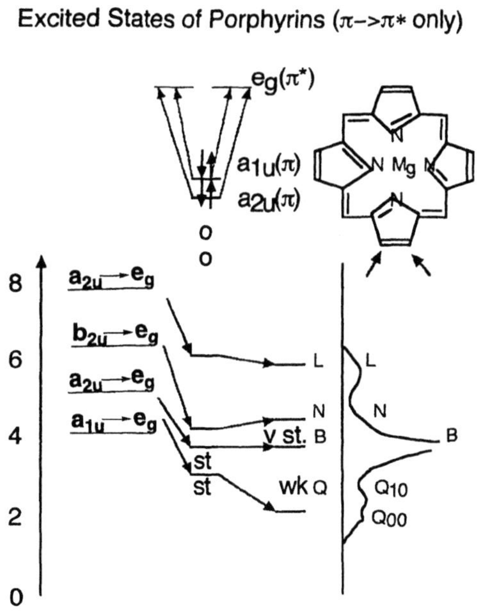


**Figure 11.** Initial explanation of the *Q* and *B* band, common features in the electronic spectra of heme complexes, based on the "four-orbital" model.

In this chapter, we shall focus primarily on the calculation of electronic absorption spectra of transition metal porphyrins and related compounds using an all valence electron INDO/RHF/CI semiempirical method developed in the laboratory of Dr. Michael Zerner.[29,30] This method allows the assignment not only of $\pi$–$\pi^*$ but of $d$–$d^*$ and all types of charge-transfer transitions that appear as extra features in many iron porphyrin compounds.

The development of an INDO/SCF/CI method, which includes both open and closed shell transition metals and extensive configuration interaction,[29,30] filled a very important gap between the IEHT and *ab initio* methods for the calculation of spectra since the use of *ab initio* methods calculations of transition metal porphyrin complexes has been limited. *Ab initio* configuration interaction calculations of the electronic spectrum of free base porphyrin[90,91] and of a closed shell singlet [Fe(II)–P–Im–CO] complex[92] clearly indicate the use of split valence type basis sets and the inclusion of correlation effects of both $\pi$ and sigma electrons are important to obtain good agreement with the observed $Q$ and $B$ band energies. However, these methods are very difficult to reliably extend to complex systems containing metal ions and unpaired spin density and for calculation of other than $Q$ or $B$ type excitations.

The INDO/SCF/CI program developed primarily in the laboratory of Dr. Michael Zerner, was first limited to closed shell species, i.e., those with no unpaired electron spins. In this form (INDO/UHF/CI), it proved extremely useful for the calculation of electronic absorption spectra of a variety of closed shell iron porphyrins. First tested for ferrocene[93] it was then successfully used for the calculation of transition frequencies, oscillator strengths, the $x$-$y$-$z$ components of transition moments, and the assignment of the electronic excitations contributing to each observed absorption for a variety of heme models for intact proteins. Among these studies have been the calculations of the spectra of oxy[94] and carbonyl[44] ferrous Hb complexes and oxy, carbonyl,[95] and carbene[96] ferrous P450 complexes. As will be discussed in Section 3, each of these spectrum present very interesting unresolved questions, which these calculations will help resolve.

The INDO method was later extended to include restricted open shell, i.e., INDO/ROHF/CI capabilities. This extremely significant advance allowed the application of this method for the first time to calculation of the spectra of radical transition metal species, a capability which remains unique. This method is now reliable enough to be used in identification and assignment of both the typical $Q$ and $B$ bands and extra features often present in heme spectra. It can also be used to accurately predict the effect of changes in spin state, axial ligands, and oxidation state on heme spectra. Thus, it can also be used to calculate the spectra of putative transient heme species to help in their identification, provided a reliable geometry can be obtained for them.

## 3 HEME MODELS FOR THE GLOBINS

### 3.1 Heme Model [Fe(II)–P–Im] for the Deoxy Form of the Globins [deoxyHb]

Several experiments have demonstrated that in the deoxy form of intact hemoglobin (Hb)[97] the Fe is about 0.6 Å from an apparently planar porphyrin ring while in a model deoxy Hb species, a [Fe(II)–TPP–2–Me–Im] complex, the Fe out of planarity is reduced to 0.42 Å.[98] In an early semiempirical quantum chemical calculation,[99] the relative energies of the quintet, triplet, and singlet state of a model [Fe(II)–P–Im] species were calculated as a function of the Fe out of planarity. As shown in Figure 12, the quintet was found to be the lowest energy state at the experimentally determined values for both the intact Hb protein and model deoxy Hb unit. Moreover, the calculated values of the quadrupole splitting $\Delta E_Q = 2.44$ mm/sec and 2.23 mm/sec for the ground state quintet at these distances are both in very good agreement with the measured low temperature value of $\Delta E_Q = 2.38$ mm/sec[100,101] and with the known direction of the principal field gradient, in the heme plane.[101] The singlet state is preferred as the Fe moves further into the porphyrin plane (i.e., <0.20 Å). A recent density functional calculations for a model Fe–P–Im[102] complex with geometry optimization of each spin state, led to a triplet ground state with the Fe only 0.15 Å out of the plane. The quintet state was 6.5 kcal/mol higher in

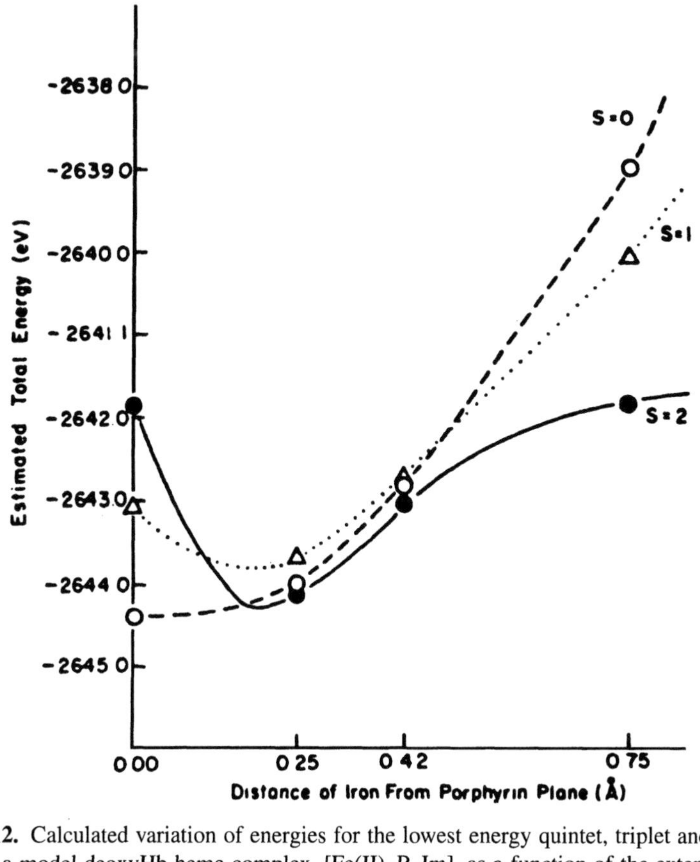


**Figure 12.** Calculated variation of energies for the lowest energy quintet, triplet and singlet states of a model deoxyHb heme complex, [Fe(II)–P–Im], as a function of the extent of displacement of the iron out of the plane of the porphyrin macrocycle. Results are from use of the IEHT method with exchange corrections.

energy with an optimized out of planar distance of the iron of 0.33 Å. These geometry optimized structures illustrate that increased out of planarity is favored by the high spin state. However, the optimized geometry of the quintet state obtained from these calculations underestimates the out of planarity observed both in model compounds and intact proteins and this method incorrectly predicts a triplet ground state that is inconsistent with known spectroscopic properties. This favoring of the lower spin state is a frequent outcome of DFT calculations, particularly with use of non-hybrid exchange-correlation functionals. However, consistent with the much earlier findings using the IEHT method, the authors suggest that the quintet state would be the ground state at greater out of plane distances of the heme iron than found in the optimized geometries. A plot of the relative spin state energies of the model deoxyheme as a function of Fe out of plane distance, similar to that made earlier using the IEHT method[99] supports this possibility. Thus, there is general agreement from both types of calculations that a quintet ground state results at experimental out of plane Fe distances in intact deoxyHb and model deoxyHb complexes.

## 3.2 5-Coordinated Heme Models for the Liganded Form of the Globins

In the same density functional study of the 5 coordinated deoxy [Fe(II)–P–Im] heme site,[102] 5-coordinated heme models of ligand bound ferrous species without the axial Im ligand, i.e., [Fe(II)–P–L] with L = $O_2$, CO, and NO were characterized. The ground state of the [Fe(II)–P–$O_2$] complex was found to be an open shell singlet and remained so when the Im ligand was added. This result is in contrast to previous *ab initio* and semiempirical studies that found the closed shell singlet to be the ground state.[103–105] However, the favored mode of $O_2$ binding to the heme iron found in these studies, a bent end on structure, is consistent with that determined in all prior computational studies. Interestingly, molecular dynamic simulations of this binding process[102] indicated rapid end on binding of the $O_2$ with no significant energy barrier associated with the reaction in the gas phase. The 5-coordinated Fe–P–CO complex was found to be a closed shell singlet with a linear mode of binding consistent with experiment for model compounds and with previous calculations using INDO/SCF semi-empirical methods.[44,106] In the optimized 5-cordinated [Fe(II)–P–NO] complex, the NO binds in a bent end on geometry with a Fe–N–O bond angle of 150° and the ground state is a doublet with single unpaired spin mainly in the $d_{z^2}$ Fe orbital. These results are very similar to those of an earlier calculation for the same system using the semi empirical INDO/UHF method.[107] In that study, the energy of the competing NO centered radical in the five-coordinated complex was also calculated and found to be only 2 kcal/mol higher in energy.

## 3.3 6-Coordinated Ferrous Heme Models for the Liganded Form of the Globins

**3.3.1** ***Model Ferrous Dioxygen Heme ($HbO_2$) Complex*** The characterization of the ferrous dioxygen heme complex in model heme complexes and intact proteins has been an active and long standing area of both experimental and theoretical investigations. Among the major unresolved characteristics that have been addressed are:

i) the geometry and extent of rotational freedom of the $O_2$ ligand
ii) the nature of the ground and low lying spin states, particularly whether the ground state is diamagnetic singlet formally corresponding to a [Fe(II)($d^6$)–$O_2$] singlet or a spin paired biradical singlet formally corresponding to a [Fe(III)($d^5$)–$O_2^-$] electronic configuration. Because of the similarities between the electronic spectra of oxyhemoglobin and alkaline metmyoglobin[108–111] and the anomolously large quadrupole splitting in the Mössbauer resonance spectra of intact and model compounds,[112] it has been suggested that the spin paired biradical species is favored
iii) the possible presence of any low-lying triplet states that could account for the room-temperature paramagnetism observed for oxyHb[113,114]
iv) the origin of the weak $xy$-polarized transitions observed at 10,200 cm$^{-1}$ and the two broad $z$-polarized bands observed in the single crystal polarized absorption spectra of oxymyoglobin[86] and oxyhemoglobin.[87]

Over the years the semiempirical iterative extended Huckel (IEHT),[99,106,115] INDO/UHF/CI[116] methods, *ab initio* CAS SCF,[117,118] and DFT methods[102] have been used to address these questions for a model six coordinated [Fe(II)–P–$O_2$–Im] system, an oxyferrous heme model for the globins (oxyHb). The combined results of all of these studies definitively indicate a bent end-on mode of binding of $O_2$ with significant rotational freedom about the Fe–$O_2$ bond in agreement with both model compound and intact protein structures rather than a bridged structure found for other transition metal five coordinated porphyrin compounds. All of the results except the recent DFT calculation find the diamagnetic closed shell singlet as the predominant ground state configuration. Even in this DFT calculation, however, this closed shell singlet is only 3 kcal/mol above the open shell singlet ground state. No attempt was made to calculate a directly observable property with this ground state. By contrast, as illustrated in Table 4, previous calculations of the quadrupole splittings $\Delta E_Q$[99,115,16] clearly demonstrated that a totally spin paired diamagnetic Fe(II) dioxygen configuration ground state species could account for the magnitude and sign of the observed quadrupole splitting in the Mössbauer spectra of oxyhemoglobin. Although a formal low spin Fe(II) configuration [$(d_{xy})^2$, $(d_{xz})^2$, $(d_{yz})^2$] would yield a spherically symmetric charge distribution around the iron nucleus and result in a vanishingly small quadrupole splitting, extensive electron delocalization including backbonding from the iron to the dioxygen ligand and forward bonding from the porphyrin and imidazole ligands to the iron results in a d orbital electron distribution around the Fe resembling a $d^5$ configuration. This electron delocalization without any unpaired spin density on the iron results in the calculated large negative value for $\Delta E_Q$ in agreement with experiment. Moreover, as shown in Table 6, the temperature dependence of the quadrupole splitting observed in the model oxygen heme and intact proteins could be accounted for by rotational averaging of the accessible Fe–$O_2$ rotamers. These calculations[116] also led to a very low lying

**Table 6 Comparison between Calculated and Observed Temperature Dependence of the Quadrupole Splitting of Ferrous Porphine Models of Oxy Hb**

| | $\Delta E_Q$(low $T$)[a] | $\Delta E_Q$(high $T$)[a] | High $T$/Low $T$ |
|---|---|---|---|
| | | Experimental | |
| Model[c] | −2.10 | 1.34 | 0.64 |
| Protein[d] | −2.24 | 1.89 | 0.84 |
| | | Calculated | |
| Model | −2.67 | 1.80 ($V_{zz}$)[b] | 0.67 |
| Protein | −2.67 | −2.40 ($V_{xx}$, $V_{yy}$)[c] | 0.90 |

[a] $\Delta E_Q$ in millimeters per second.
[b] Largest component of the field gradient is in the porphine plane.
[c] Collman, J. P.; Gagne, R. R.; Reed, C. A.; Halbert, T. R.; Lang, G.; Robinson, W. T. *J. Am. Chem. Soc.*, **97**, 1427, 1975.
[d] Lang, G.; Marshall, W. *Proc. Phys. Soc.*, **87**, *3*, 1966.

triplet in agreement with the observed room temperature paramagnetism of oxyhemoglobin.[113,114] Thus, the combined results of these calculations using semiempirical molecular orbital methods clearly demonstrate that the observable properties of ground and low lying states of the ferrous oxy heme site are obtained with a diamagnetic singlet ground state.

In addition to the characterization of the ground and low lying states of model oxyHb complexes, the electronic spectra was calculated from the diamagnetic singlet ground state of the model six coordinated [Fe(II)–P–$O_2$–Im] system, using the INDO/SCF/CI method.[94] As shown in Figure 13 and Table 7 in the calculated spectrum in addition to the weak $Q$ and intense $B$ bands common to all heme proteins, the four weak $xy$-polarized transitions observed only in oxy heme complexes have been calculated and their origin identified. They arise from [($a_{1u}$, $a_{2u}$) to (O$\pi$, $d\pi$)*] excitations involving the lowest empty, highly delocalized (O$\pi$, $d\pi$)* orbitals unique to such complexes. The two broad $z$-polarized bands observed in the single-crystal polarized absorption spectra of oxymyoglobin and oxyhemoglobin have also been calculated. Controversy existed over the assignment of these transitions and, in particular over the extent of involvement of the oxygen ligand. The calculations assign the weaker near IR-visible band at 22,000 $cm^{-1}$ mainly to [($d\sigma$, $d\pi$) to $d\sigma$*] excitations, and the more intensive u.v. band at 31,000 $cm^{-1}$ mainly to [$a_{2u}$ to $d\sigma$*] excitations. While significant participation (25%) of the highly delocalized (O$\pi$, $d\pi$)* virtual orbital is also found, these $z$-polarized transitions need not be totally unique to oxy heme complexes, in keeping with experimental observations. From Figure 13 and Table 7, it is seen that in addition to the "extra bands" the regular $Q$, $B$ and $N$ bands are well described. Both the $Q$ bands and the $B$ bands involve primarily [($a_{1u}$, , $a_{2u}$) $\rightarrow$ $e_g$*] transitions in keeping with the four orbital model. Thus, the spectrum calculated from a diamagnetic closed-shell singlet ground state of a model oxyHb complex is in excellent agreement with all of the major features of the observed circular dichroism and single crystal polarized absorption spectra of oxymyoglobin(oxyMb) and oxyhemoglobin (oxyHb). These results lend further support to the validity of such a ground state.

**Table 7 Comparison between Calculated and Observed Electronic Spectra of Ferrous Porphine Oxy Hb**

| | Frequency | | Osc. Strenght | | Polarization | | Assignment |
|---|---|---|---|---|---|---|---|
| Band | Obs. | Calc. | Obs. | Calc. | Obs. | Calc. | Calculated |
| I | 7,700 | 6,820 | ~0 | ~0 | | $y, z$ | $d\sigma, e_g \to (d\pi, \mathrm{O}\pi)^*$ (40%, 30%)<br>$d\sigma, e_g \to e_g^*$ (10%) |
| II | 8,700 | 9,700 | 0 | 0.002 | | $x, y$ | $(\mathrm{O}\pi, \mathrm{por}\pi), a_{1u} \to (d\pi, \mathrm{O}\pi)^*$ (30%, 20%)<br>$(d\pi, \mathrm{O}\pi), a_{2u} \to (d\pi, \mathrm{O}\pi)^*$ (10%, 10%) |
| III | 10,200 | 9,790 | 0.003 | 0.004 | $y$ | $x$ | { $a_{1u} \to (d\pi, \mathrm{O}\pi)^*$ (65%)<br>$(\mathrm{O}\pi, \mathrm{por}\pi) \to (d\pi, \mathrm{O}\pi)^*$ (10%) |
| IV | 12,800 | 11,590 | 0.002 | 0.001 | $x, z$ | $y(z)$ | { $a_{2u} \to (d\pi, \mathrm{O}\pi)^*$ (65%)<br>$(\mathrm{O}\pi, \mathrm{por}\pi) \to (d\pi, \mathrm{O}\pi)^*$ (10%) |
| $Q_0, Q_y$ | 17,300<br>18,500 | 16,750<br>17,010 | 0.120 | 0.042<br>0.005 | $x, y$ | { $x$<br>$y$ | $a_{1u}, a_{2u} \to e_g^*$<br>$a_{1u}, a_{2u} \to e_g^*$ |
| VI($Z$) | 22,000 | 22,050 | 0.022 | 0.002 | $z$ | $z$ | { $d\sigma \to d\sigma^*$ (50%)<br>$e_g \to d\sigma^*$ (30%) |
| | | 22,530<br>24,620 | | 0.081<br>0.124 | | $z$<br>$z$ | $(d\pi, \mathrm{O}\pi, \mathrm{por}\pi) \to d\sigma^*$ (55%)<br>$(d\pi, \mathrm{O}\pi, \mathrm{por}\pi) \to (d\pi, \mathrm{O}\pi)^*$ (25%) |
| $B$(Soret) | 24,100 | { 28,300<br>29,100 } | 1.30 | { 1.54<br>1.56 } | $x, y$ | { $x$<br>$y$ | $a_{1u}, a_{2u} \to e_g^*$<br>$b_{1u}(\mathrm{Im}\pi) \to (d\pi, \mathrm{O}\pi)^*$ } (10%)<br>$a_{1u} \to (\mathrm{Im}\pi)^*$ } |
| $N$ | 29,000 | { 32,000<br>32,400<br>32,600 } | 0.54 | { 0.37<br>0.18<br>0.10 } | | $x$<br>$x$<br>$y$ | { $(d\pi, \mathrm{O}\pi, \mathrm{por}\pi) \to e_g^*$ (40%)<br>$b_{1u} \to (d\pi, \mathrm{O}\pi)^*$ (25%)<br>$a_{1u} \to b_{2u}(\pi)^*$ (25%)<br>$d\pi \to (d\pi, \mathrm{O}\pi)^*$ (20%) |
| VII($Z$) | 31,000 | 35,300<br>36,300 | 0.15 | { 0.02<br>0.16 } | | $z$<br>$z$ | $a_{2u} \to d\sigma^*$ (50%)<br>$e_g, d\sigma \to (d\pi, \mathrm{O}\pi)^*$ (25%) |

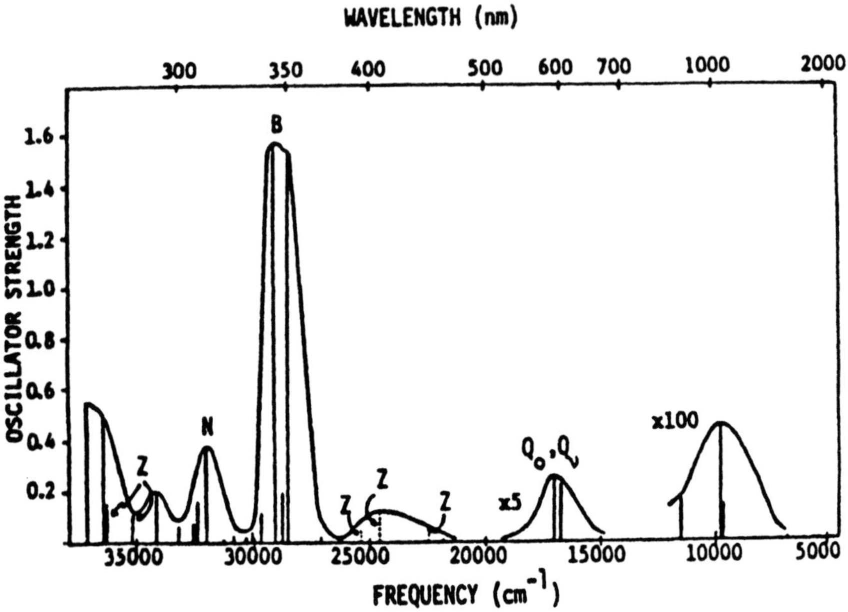


**Figure 13.** Calculated electronic absorption spectrum for a heme model of oxyHb, [Fe(II)–P–Im–$O_2$], using the INDO/SCF/CI method. The position of the vertical lines are the calculated frequencies and the height if the lines are the calculated intensity in terms of oscillator strength. Gaussian envelopes have been drawn around these stick figures for ease of comparison of calculated and experimental spectra.

In addition to the characterization of the electronic spectra of model oxyHb complexes, the INDO/SCF method has been used to identify the photoactive states involved in the photodissociation of oxyhemoproteins.[119] Excited-state energies and electron distributions were calculated as a function of iron oxygen distance. The results were interpreted in terms of a two-channel photodissociation of oxygen above and below the visible $Q$ band. The higher energy channel proceeds through excited states of a $d\pi$–$d^*_{z^2}$ nature. The lower energy channel involves excited charge transfer states. The proposed mechanism is consistent with and accounts for many of the features observed in picosecond photodissociation experiments.

The description of the ground state electronic structure and geometry of model ferrous oxyHb complex first obtained by semiempirical methods[99,115,116] have been largely confirmed by subsequent CASSCF calculations reported.[113,114] Specifically, there is a predominance of the spin paired singlet configuration in the ground state and extensive electron delocalization leading to a significant quadrupole splitting. The additional insight provided by the DFT calculations[102] is that the addition of the Im ligand greatly strengthens the Fe–$O_2$ bond from a dissociation energy of 9 kcal/mol to 15 kcal/mol while not significantly changing its geometry and rotational mobility.

**3.3.2** ***Model Ferrous Carbonyl Heme (HbCO) Complex*** There is general agreement from both semiempirical,[44,106] *ab initio* SCF,[92,120,121] and DFT[102] computational studies of a model six coordinated [Fe(II)–P–CO–Im] system for the carbonyl hemoglobin (HbCO) that when CO binds to the ferrous heme iron of the globins, a closed shell singlet ground state results as it does in the $O_2$ complex. The main controversy surrounding the ground state of the ferrous CO heme species was the geometry of the bound CO in model compounds and intact proteins and the biological significance of possible differences.

In model compounds,[122,123] the CO binds in a linear mode while in the X-ray structure of MbCO[124] and HbCO[125] the CO is kinked. The C atom is about 6° from the heme norm and the Fe–C–O bond angle is about 170°. This deviation from a linear Fe–C–O bond in Hb and Mb is generally cited as a major factor in reducing the affinity to CO for the heme iron making it less competitive with $O_2$ binding under physiological conditions.[126,127] Geometry optimization of model [Fe(II)–P–CO–Im] complexes using both SCF SAC (symmetry adapted cluster) CI methods[92,121] and DFT[102] both obtain a linear Fe–C–O geometry consistent with observed model compound structures. Moreover, the DFT study provided additional important insight into the feasibility of deviations from linearity observed on the intact proteins by demonstrating a small energy cost for bending up to 15° well within with the observed values from both X-ray and IR solution studies of the intact proteins.[128] This explanation appears more reasonable than one offered in another *ab initio* study[120] that the proximal histidine geometry leads to the kinked CO geometry. This latter calculation suffers from the inadequacy of the chemical model used for the ferrous CO heme, replacing the entire porphyrin ring with two simply amide moieties.

Whether or not the observed kink in the ferrous carbonyl complex of intact protein is the main factor in the observed reduction in CO affinity in the intact protein affinity is still unresolved although the calculated[44] and observed[129] reduction in the CO stretching frequency upon kinking tend to support this hypothesis. Another more recent computational[130] result is also consistent with the subtle modulation of CO stretching frequency by the protein environment. In this study, the effect of the electric field of the protein on the CO stretching frequency in two closely related heme proteins, horse, and yeast cytochrome was calculated and a value very close to the observed shift of 8 $cm^{-1}$ was obtained.

In model ferrous carbonyl (HbCO) and di-oxy ($HbO_2$) heme complexes, in which the linear binding mode of the CO and bent end on binding mode of the $O_2$ are well established, comparative properties calculated using the INDO/SCF method are in very good agreement with the relative behavior observed. As shown in Table 4, the calculated Fe–$O_2$ bond overlap density is less than the Fe–CO in keeping with its lower observed binding affinity. As indicated in this table, a much higher pressure of $O_2$ than CO is required for 50% maximum binding ($P_{1/2}$). The calculated C–O bond overlap density in also greater than the O–O bond density, consistent with its much larger stretching frequency. Most strikingly, the origin of the large quadrupole splitting in $HbO_2$ and the very small one in HbCO has been explained. In contrast to the effect of $O_2$ binding, a nearly spherically symmetric electron distribution, typical of low spin ferrous ion results when CO binds. As illustrated in

Table 4, from an INDO/SCF calculation, the nearly equal population of the three $t_{2g}$ orbitals, $d_{xy}$, $d_{xz}$, and $d_{yz}$, and equal electron donation to the two formally unoccupied eg orbitals, $d_{z^2}$ and $d_{x^2-y^2}$ found[44] results in a small positive calculated values of $\Delta E_Q$ (0.5 mm/sec). This value is consistent with the experimental value of 0.30–0.37 found respectively in model compounds and intact proteins.[129,131] These values are the smallest experimentally detectable values of the quadrupole splitting corresponding to the minimum observable deviation from spherical symmetry of the electron distribution around the heme iron. By contrast, in the $HbO_2$ model complex extensive electron delocalization including backbonding from the iron to the dioxygen ligand and forward bonding from the porphyrin and imidazole ligands to the iron results in a $d$ orbital electron distribution around the Fe resembling a $d^5$ configuration. This electron delocalization without any unpaired spin density on the iron results in the calculated large negative value for $\Delta E_Q$ in agreement with experiment. Despite this difference the iron in the two complexes has a very similar total charge due to compensatory increased electron transfer from the porphyrin to the iron in the oxyHb complex.

While a closed shell singlet is well established as the ground state of model and intact carbonyl Hb complexes, there has been some question about the role of triplet states in observed properties. The presence of a low lying triplet state in intact CO complexes of hemoglobin has been suggested by a possible room temperature paramagnetism reported for some species of HbCO,[132,133] but not for others.[134,135] There was also evidence that excited triplet states are involved in the photodissociation pathway of CO from Hb. To address both of these possibilities, the lowest energy triplet states in a heme model of HbCO was calculated using the INDO/UHF/CI method.[136] The results clearly show that the calculated energy difference of 11,000 $cm^{-1}$ between the lowest triplet state and the reference singlet ground state of HbCO, independent of the Fe–C–O geometry, is much too high to accommodate a thermally populated paramagnetic state in HbCO.

In order to identify the photoactive excited states involved in the photodissociation of carbonyl hemoproteins, the excited state energies from the diamagnetic reference state of carbonyl heme complexes were calculated. Compounds with both bent and linear CO geometries, corresponding the hemoproteins and model compounds, respectively, were investigated[119] using the same INDO/UHF/CI method. For both types of complexes, only excited states involving ($d\pi$ to $d^*_{z^2}$) and ($d_{x^2-y^2}$ to $d^*_{z^2}$) transitions showed a significant decrease in energy with increasing iron-ligand distance, making them the definitive candidates for the photoactive states. Of the two, the ($d\pi$ to $d^*_{z^2}$) state with significant oscillator strength and more bond weakening characteristics appears to be the main photodissociating state. The previously suggested possibility, that states resulting from ($\pi$ to $\pi^*$) transitions are directly photodissociating, is ruled out. The results also show that triplet partners of these states need not be involved in the initiation of dissociation. The calculated energies of the singlet ($d\pi$ to $d^*_{z^2}$) transitions in the bent carbonyl heme complex are in excellent agreement with circular dichrosism spectra[87] and form the basis of a proposed mechanism of photodissociation, which is consistent with the observed high and wavelength independent quantum yield of CO photodissociation from hemoproteins.[137]

Calculation of the electronic spectrum of model carbonyl ferrous imidazole porphyrin [Fe(II)–P–Im–CO] complexes was of interest because it has the classic absorption profile typical of all heme proteins: weak visible ($Q_0$, $Q_\nu$), strong Soret ($B$), and moderately higher energy ultraviolet transitions ($L$, $N$) all polarized in the $xy$ heme plane. There are no "extra" $xy$ or $z$-allowed transitions as observed in the oxy-ferrous-imidazole heme [Fe(II)–P–Im–$O_2$] complex. Thus, it was of interest to understand how a change of ligand from $O_2$ to CO could lead to such a "normal" spectra composed of the universal $Q$, $B$, $L$, and $N$ bands. The calculated electronic spectrum for a model carbonyl heme complex with a linear CO and imidazole ligand is presented in Figure 14, and the assignment of the transitions given in Table 8. The agreement between the energies and strengths of the observed and calculated bands is apparent from the results presented in Table 8.

The relatively simple nature of the optical spectrum of the carbonyl ferrous heme complex (Figure 14) as compared to that of the ferrous oxy ferrous heme complex (Figure 13) was elucidated by the assignment of each spectrum. Unlike the situation

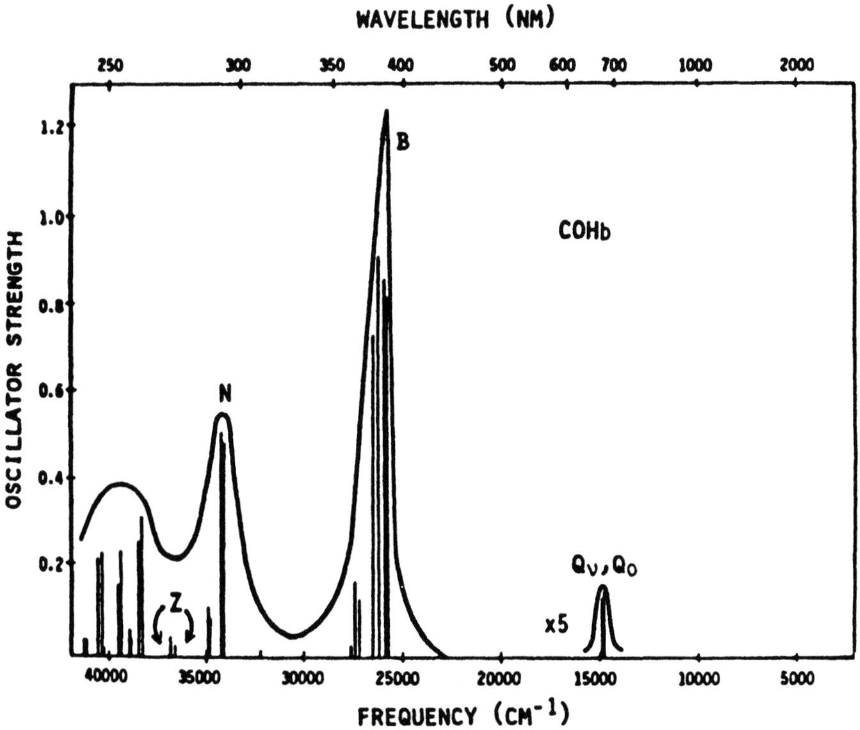


**Figure 14.** Calculated electronic absorption spectrum for a heme model of carbonyl Hb, [Fe(II)–P–Im–CO], using the INDO/SCF/CI method. The position of the vertical lines are the calculated frequencies and the height of the lines are the calculated intensities in terms of oscillator strength. Gaussian envelopes have been drawn around these stick figures for ease of comparison of calculated and experimental spectra.

**Table 8 Comparison between Calculated and Observed Electronic Spectra of Carbonylheme Complex; Assignment of the Optical Spectrum of Carbonylheme**

| | | Calculated | | | |
|---|---|---|---|---|---|
| Band | Exp. Frequency[a] | Frequency[a] | Oscill. Strength | Polarization | Major Components[b] |
| $Q_0$ | 17,500 | 14,870 | 0.030 | $xy$ | $a_{1u}, a_{2u} \rightarrow e_g^*$ (53%, 39%) |
| $Q_y$ | 18,400 | 14,920 | 0.027 | $xy$ | $a_{1u}, a_{2u} \rightarrow e_g^*$ (53%, 40%) |
| | | 25,800 | 0.819 | $xy$ | $a_{1u}, a_{2u}, d_{x^2-y^2} \rightarrow e_g^*$ (17%, 23%, 13%)<br>$(d_{yz}, d_{xz}) \rightarrow d_{xy}^*, d_{z^2}^*$ (20%, 12%) |
| | | 25,850 | 0.865 | $xy$ | $a_{1u}, a_{2u}, d_{x^2-y^2} \rightarrow e_g^*$ (14%, 23%, 13%)<br>$(d_{xz}, d_{yz}) \rightarrow d_{xy}^*, d_{z^2}^*$ (21%, 11%) |
| | | 26,350 | 0.912 | $xy$ | $d_{x^2-y^2}, a_{1u}, a_{2u} \rightarrow e_g^*$ (34%, 16%, 21%) |
| $B$ | 23,400 | 26,520 | 0.729 | $xy$ | $d_{x^2-y^2}, a_{1u}, a_{2u} \rightarrow e_g^*$ (44%, 10%, 17%) |
| | | 27,260 | 0.123 | $xy$ | $d_{x^2-y^2} \rightarrow e_g^*$ (44%)<br>$(d_{yz}, d_{xz}) \rightarrow d_{z^2}^* d_{x^2-y^2}^*$ (26%, 13%) |
| | | 27,490 | 0.167 | $xy$ | $d_{x^2-y^2} \rightarrow e_g^*$ (32%)<br>$(d_{xz}, d_{yz}) \rightarrow d_{z^2}^*, d_{x^2-y^2}^*$ (22%, 18%) |
| $N$ | 29,000 (broad) | 34,170 | 0.483 | $xy$ | $a_{2u}$, (Imid $\pi, b_{1u}) \rightarrow e_g^*$ (31%, 20%)<br>$(d_{xz}, d_{yz}) \rightarrow b_{2u}^*$ (27%) |
| | | 34,210 | 0.500 | $xy$ | $a_{2u}$, (Imid $\pi, b_{1u}) \rightarrow e_g^*$ (16%, 29%)<br>$(d_{yz}, d_{xz}) \rightarrow b_{2u}^*$ (34%) |
| | | 34,850 | 0.085 | $(x)y$ | (Imid $\pi, b_{1u}), a_{2u} \rightarrow e_g^*$ (40%, 24%) |
| | | 34,880 | 0.104 | $x(y)$ | $a_{2u}$, (Imid $\pi, b_{1u}) \rightarrow e_g^*$ (35%, 24%) |
| $L$ | 37,000 (broad) | 38,360 | 0.303 | $xy$ | $(d_{xz}, a_{1u}) \rightarrow e_g^*$ (45%) |
| | | 38,520 | 0.251 | $xy$ | $(d_{yz}, a_{1u}) \rightarrow e_g^*$ (44%) |
| | | 39,410 | 0.226 | $xy$ | $(d_{xz}, a_{1u}), (d_{xz}, a_{2u}) \rightarrow e_g^*$ (39%, 28%) |
| | | 39,540 | 0.156 | $xy$ | $(d_{yz}, a_{1u}), (d_{yz}, a_{2u}) \rightarrow e_g^*$ (46%, 25%) |

[a] in $cm^{-1}$. [b] Assignments: major excitations contributing to each transition frequency.

that prevails for ferrous oxyHb all of the calculated transitions are $xy$-polarized. This result is consistent with the singly crystal spectrum of carbonylmyoglobin.[86] The $Q_0$ and $Q_\nu$ bands arise from $[(a_{1u}, a_{2u})$ to $e_g^*]$ excitations, as in oxyferrous Hb. However, in contrast to the oxyHb complex, the Soret band is not due only to these four orbital excitations. For the ferrous carbonyl Hb complex, in addition to $[(a_{1u}, a_{2u})$ to $e_g^*]$ excitations two other major types of excitations contribute to the intensity of the Soret band. These are $[d\pi$ to $d\sigma^*]$ and $[d\sigma$ to $eg(\pi^*)]$ excitations. These results indicate that while the four orbital model was the minimum necessary to explain the $Q$ and $B$ bands, the more complete spectral calculations indicate that the excitations contributing to even the universally observed $B$ band can be much more complex and differ among closely related heme complexes.

**3.3.3** ***Model Ferrous Nitrosyl Heme (HbNO) Complexes*** Until recently, NO was not considered as important a biological ligand for heme proteins as $O_2$ or CO. Extensive spectroscopic studies reported for the ferrous-NO-heme site in both model and intact heme species were made primarily because the odd electron of this ligand allowed more detailed monitoring of its binding to the iron porphyrin unit than for the singlet state species resulting from binding of either $O_2$ or CO. Thus, electron spin resonance,[138–141] Mössbauer resonance,[142] and resonance Raman data[143] as well as IR and optical spectra have been obtained for both model compounds and intact proteins.[144,145] The recent identification of a role for NO as a neurotransmitter and its proposed role in regulation of heme proteins such as guanylate cyclase has enhanced interest in this species. In addition to spectroscopic properties, the early experimental observation that addition of a NO significantly weakens the Fe-Imidazole bond[146–149] has new significance since this same trans-repulsive effect has been proposed as a mechanism of NO regulation of guanylate cyclase.[149]

The first computational study of a 6-coordinate [Fe(II)–P–Im–NO] heme species was made using the semiempirical IEHT method.[106] In that study, the variation of calculated configuration energies, quadruple splittings, bond overlap densities, net atomic charges, and unpaired spin density distributions were calculated as a function of ligand conformations. As shown in Table 9, three geometries were considered. The second one in this table was obtained from the X-ray crystal structure of a model compound.[150] It is a bend end-on N–O structure with a Fe–N bond length of 1.75 Å, a N–O bond length of 1.12 Å, a Fe–N–O bond angle of 142°, and a Fe–N bond length 2.18 Å. The other two geometry variations investigated were a linear Fe–N–O bond, similar to that found in the Fe–C–O compound and a variation of the bent end on X-ray structure with a shorter Fe–N bond of 2.06 Å as found in the carbonyl and dioxygen ferrous heme model compounds. The results shown in this table indicated that the crystal structure geometry has the lowest energy. In this geometry, the calculated value of the quadrupole splitting of 1.48 mm/sec is in excellent agreement with the observed value of $\Delta E_Q = 1.50$ mm/sec for HbNO.[151] Also, in agreement with observed ESR spectra,[138,139] the results show that although the unpaired spin density is primarily in an Fe $d_{z^2}$ orbital, there is also unpaired spin density on both the NO and imidazole N$\varepsilon$ nitrogen atoms in about a 3 to 1 ratio, consistent with the relative magnitude of the observed nitrogen hyperfine splitting.

**Table 9 Variation of Calculated Configuration Energies, Quadrupole Splittings ($\Delta E_Q$), Bond Overlap Densities, and Net Atomic Charges for Model Nitrosyl Hb as a Function of Ligand Conformation**

| | 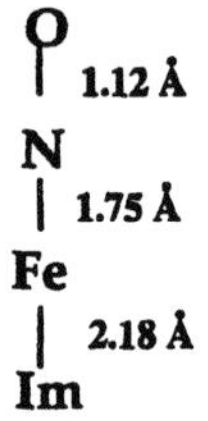 | 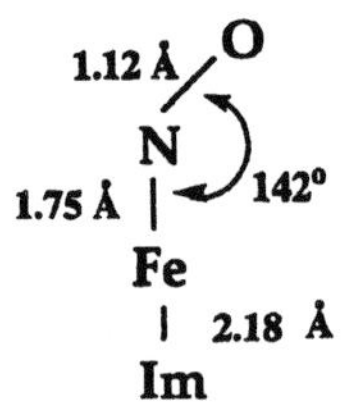 | 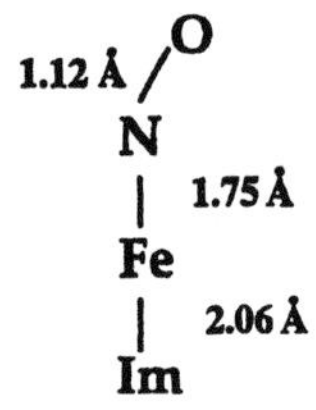 |
|---|---|---|---|
| $E_{conf}$(eV) | −2880.0 | −2880.4 | −2880.2 |
| $\Delta E_{Qcalc}$ (mm/sec) | 1.64 | 1.48 | 1.48 |
| | | Bond overlap[a] | |
| Fe–N | 0.57 | 0.55 | 0.55 |
| N–O | 1.12 | 1.17 | 1.17 |
| Fe–$N_\varepsilon$ | 0.28 | 0.27 | 0.28 |
| | | Net charges[a] | |
| Fe | 0.23 | 0.24 | 0.24 |
| N | −0.05 | −0.03 | −0.04 |
| O | −0.24 | −0.23 | −0.23 |

[a] Calculated from a Mulliken population analysis.

In another computational study,[107] the semiempirical INDO/SCF method was used to further probe the ground state electronic structure of a model [Fe(II)–P–Im–NO]–heme complex. The results demonstrate the existence of two distinct low energy electronic states corresponding to the formal assignment of the unpaired electron to either NO or iron atomic orbitals. It was shown that these two states are nearly degenerate at axial ligand distances close to the observed values in heme protein and model compounds. It was further shown that both the extent of degeneracy and the relative energy order of these two electronic states are sensitive to minor changes in the axial ligand bond distances and angles. The implication of these results provided a coherent interpretation of additional details of the observed ESR spectra of complexes of NO with various heme proteins. These spectra can be understood in terms of varying contributions from the two electronic states of the same heme complex and mediated via small perturbations of the ligand geometry by the rest of the protein.

The finding of a preferred bent end on mode of NO binding to ferrous heme complexes with a longer Fe–Ne bond than in the CO or $O_2$ ferrous heme complexes, first obtained using semiempirical quantum chemical methods, have been recently confirmed by a geometry optimization study of the same model compound [Fe(II)–P–Im–NO] using the DFT method.[102] In that study, the optimized geometry was essentially the same as the X-ray structure. Short molecular dynamic simulations performed indicated large amplitude oscillations in the Fe–N–O bond angle, but no other significant structural variations. This mode of NO binding should be sterically feasible in the ligand pocket of the intact globins. The energy required to break the Fe–imidazole bond was calculated to be only 6 kcal/mol, half the value of the Fe–imidazole bond energy calculated by the same methods for the $O_2$ and CO complex. This important finding provides support for the observed weakening of the Fe–imidazole bond in NO complexes of intact heme proteins compared to $O_2$ and CO.

### 3.4 6-Coordinated Ferric Heme Models for the Liganded Form of the Globins (MetMb)

Metmyoglobin (MetMb) is the oxidized ferric form of myoglobin. While the resting state of the ferrous form of the globins is 5-coordinate, the resting state of the ferric form, MetMb has a water as a sixth ligand in its X-ray structure with an Fe–$OH_2$ distance of 2.00 Å.[152] Using the crystal structure geometry of the heme site from the intact proteins and also a regularized planar porphyrin ring, calculations were made[153] for the 6-coordinated resting form using the INDO/RHF method. The MetMb complex [Fe(III)–P–$H_2O$–Im] was found to have a sextet ground state, but with a very low lying quartet state 2.7 kcal/mol higher in energy in the X-ray structure with a doming of the porphyrin macrocycle. The energy difference was even smaller (1.5 kcal/mol) in the regularized porphyrin model. The finding of a sextet ground state with a low lying quartet could explain the observed temperature dependence of magnetic moment in which the $u_{\text{eff}}$ values were founds to decrease from a pure high spin value with increasing temperature.[154–157] In addition, these results could account for the origin of the rather large experimental value of the quadrupole splitting of 1.33 mm/sec.[158] The calculated value of quadrupole splitting is 0.76 mm/sec for the ground state sextet and the calculated value for the low lying quartet is 2.44 mm/sec. Thus, thermal contributions from both the sextet ground state and low lying quartet state can account for the experimental value.

**3.4.1** ***Ferric Heme Complexes with Strong Axial Ligands*** While the aquo complex of the oxidized MetMb is a sextet state, the binding of stronger electron donating ligands like $CN^-$ usually leads to a low spin state. To understand the effect of this change in axial ligand on observed electromagnetic properties, the relative energies of different spin states, electronic structure, and spectra of a model [Fe(III)–P–Im–$CN^-$] heme complex (cyanoMetMb) were calculated[159] using the INDO/ROHF/CI method. In these calculations, the geometry of the heme complex was taken from the crystal structure of the appropriate protein.[44,160] The Fe–CN and Fe–Im distances used were 1.91 Å and 2.06 Å respectively; the FeCN bond angle was 180° and the Fe(III) atom was kept in the plane of a regularized porphyrin ring with an Fe–$N_{\text{pyr}}$ distance of 2.01 Å. A $^2E_g$ state with a $d$-orbital occupancy of $[d_{xy}^2, d_{xz}^{1.5}, d_{yz}^{1.5}, d_{x^2-y^2}^0, d_{z^2}^0]$ was found to be the ground state. Another doublet state, $^2B_{2g}$ with a d orbital occupancy of $[d_{xy}^1, d_{xz}^2, d_{yz}^2, d_{x^2-y^2}^0, d_{z^2}^0]$, was found to be nearly degenerate with this doublet. The ground $^2E_g$ state was, however, well separated from the lowest quartet (1.07 ev) and sextet states (1.39 ev). The calculated value of $\Delta E_Q$ of the $^2E_g$ state of 1.44 mm/sec agreed well with the experimental value of 1.46 mm/sec.[161]

In addition to characterizing the electronic structure and relative spin state energies of the ground and low lying states of different multiplicity, the INDO/ROHF/CI methods were used to elucidate the origin of extra features in the electronic spectra in addition to the $Q$ and $B$ bands. A major unresolved aspect of both high and low spin ferric heme proteins is the origin of the extra absorption bands of lower energy than the $Q$ band observed in single crystal polarized absorption spectra.[162] Such spectra have been reported for water, fluoride, formate, cyanide, and azide complexes of MetMb and MetHb.[163–166] The spectra of cyanoMet myoglobin is an example of a typical low spin ferric heme complex, exhibiting four such extra $xy$-polarized transitions in the near IR region from 5,000 to 13,000 $cm^{-1}$. In order to explain the origin of these features, the INDO/ROHF/CI method was used to calculate the spectra of a model [Fe(III)–P–Im–$CN^-$] heme complex of cyanoMetMb.[159] In these calculations, the $^2E_g$ state ground state with equal electron population of the two $d\pi$ orbitals was used as the reference state. The four extra $xy$-polarized absorptions with frequencies lower than the visible $Q$ bands were found in the calculated spectra in agreement with the single crystal polarized absorption spectra. These calculations allowed the assignment of these bands and hence the identification of their origin that could not be determined from the experimental spectra alone. These absorptions originate primarily from excitations from the highest two occupied porphyrin $\pi$ orbitals ($3a_{2u}$ and $1a_{1u}$) to the two lowest unoccupied $\pi^*$ orbitals ($4e_g$). However, mixing of porphyrin to Fe(III) charge transfer excitations ($3a_{2u}$ to $d\pi$) and ($1a_{1u}$ to $d\pi$) into these wave functions results in nonzero oscillator strengths for these extra, low energy transitions. This assignment, based on spin coupling, differs from that suggested prior to the computations,[167] in which charge transfer was thought to be the primary nature of these transitions.

## 4 HEME MODELS OF PEROXIDASE SPECIES

### 4.1 Ferric Aquo Resting State

The heme unit of the resting form of peroxidases and metmyoglobin (MetMb), two functionally distinct heme proteins are identical consisting of a ferric protoporphyrin IX complex with a water and imidazole axial ligand. Subtle differences in the protein environment appear to mediate function. Specifically, the current hypothesis is that peroxidase activity is favored by anionic character of the imidazole ligand conferred by proton transfer from the proximal histidine to a nearby conserved aspartate residue absent in myoglobin. To determine the effect of deprotonation of the imidazole ligand on the electronic structure of the resting state of peroxidases, INDO/SCF calculations were made for heme models with both the neutral imidazole [Fe(III)–P–$H_2O$–Im]$^{+1}$ and ionic imidazolate ligand, [Fe(III)–P–$H_2O$–Im]. In these calculation, both the X-ray structure of the porphyrin macrocycle as in cytochrome *c* peroxidase[168] and a regularized planar porphyrin macrocycle was used[153] with the observed Fe–$OH_2$ distance of 2.4 Å for this six coordinated resting forms. For the heme unit of the peroxidase with a neutral Im, a quartet state was found to be the lowest energy in both the X-ray geometry and with a regularized porphyrin ring but with a very low lying sextet only 1.5 kcal/mol higher. In the deprotonated [Fe–P–$OH_2$–Im] resting form, the energy ordering of these states was reversed. A sextet state was the ground state with a very low lying quartet. This sensitivity of the spin state ordering is consistent with the observed temperature dependence of magnetic susceptibility[154,156,169] of quadrupole splitting in Mössbauer resonance,[170–172] electron spin resonance,[173–176] resonance Raman spectra,[177–181] and electronic spectra[175,180,181] of horseradish peroxidase, cytochrome *c* peroxidase that indicate contributions to these properties from both a sextet and quartet ground state.

The results obtained using the INDO/SCF method for the common [Fe–P–$H_2O$–Im] heme model for MetMb and peroxidase with different Fe–$OH_2$ distances indicates the sensitivity of the spin state ordering to subtle differences in axial ligand geometry even in heme complexes with the same axial ligands. It is very difficult to experimentally determine changes in geometry for different spin states of the same heme complex. To further explore the relationship between geometry and spin state, in a very recent effort,[182] *ab initio* methods were used for the first time to calculate optimized geometries and electronic structure of the lowest lying sextet, quartet, and doublet states of this [Fe–P–Im–$H_2O$] heme model, using the heme site in CCP for an initial geometry. In this study, the single configuration (HF) and multiconfiguration complete active space (CAS) self consistent field (SCF) *ab initio* methods were used with extended basis sets. The results obtained indicate that allowing geometry optimization of each spin state, the sextet state is the lowest energy. There is also a significant decrease in Fe-water distances in the low spin form of this complex. Thus, these results suggest that the environment of the protein could modulate spin state changes by imposing geometric changes on this mobile Fe–$H_2O$ ligand interaction. Experimental determination of spin state populations by a number of spectroscopic methods in wt and mutants of the Fe(III) resting form of cytochrome *c* peroxidases (CCP) with known Fe-water distances from crystal structures provide strong support for this hypothesis.

## 4.2 Transient Putative Peroxide Intermediates of Peroxidases

As shown in Figure 4, in the enzymatic cycle of peroxidases, hydrogen peroxide is required to transform the ferric resting state to the catalytically active ferryl Fe=O, Compound I species.[183] While a peroxide complex has been proposed as a key intermediate in this reaction,[184] this intermediate species is too transient to have thus far been definitively characterized. The validity of these mechanistic inferences depends to a large extent on whether the peroxide does indeed form a stable complex with the heme iron and the nature of the preferred binding mode of this peroxide.

To further explore whether peroxide forms a stable complex with the heme site of peroxidases and the nature of its binding mode, in a recent study, *ab initio* quantum chemical methods were used[185] to determine the optimized geometry and stability of a peroxide complex with a model heme peroxidase [Fe–P–HOOH–Im]. Two stable minima were identified with a peroxide binding energy of 9.7 kcal/mol. In both of these complexes, the peroxide binds in an end-on mode with alternative oxygen atoms as the Fe ligand. These two minima were connected by a low energy ridge. No minimum corresponding to a bridge structure could be found. Thus, the quantum chemical results provide further support for the existence and nature of a stable bound peroxide complex with peroxidases.

These *ab initio* calculations were made for the sextet state of the model peroxidase–peroxide complex since this was the ground state found for the resting form with a water as a ligand. To probe the validity of this assumption, the relative energies of the sextet, quartet, and doublet states of the peroxide–peroxidase heme complex were calculated[186] using the optimized structure from the *ab initio* study. These calculations made using the INDO/RHF method[30,187] confirmed the sextet ground state for the peroxide species with the doublet and quartet about 5 kcal/mol higher in energy.

Because of the assumed importance of a peroxide complex as a key intermediate in the formation of the catalytically active form of peroxidases, attempts to trap and identify this species by rapid scan spectrophotometric studies of the reaction between peroxide, the WT[188–190] and R38L HRP[191] have been made. In these studies, spectra attributed to transient peroxide intermediates have been detected. There are, however, significant qualitative differences in these spectra in the 300–450 nm region. A hyper or split Soret (B) band is observed in the wt spectrum and a normal Soret, not very different from the resting state, found in the R38L HRP spectrum. In the absence of any additional information, it is not possible from these reported spectra alone to identify the species that give rise to them or to understand these differences.

In order to identify the origin of these spectra and their differences, the INDO/ROHF/CI method was used to calculate the electronic structure and spectra of the resting state and two plausible forms of the peroxide intermediate of a model peroxidase[186] [Fe(III)–Protoporphyrin-IX–$Im^-$–L] L $=$ [HOOH], [$OOH^-$]. Both candidate intermediates share a common protoporphyrin-IX heme unit and an imidazolate proximal ligand. However, one candidate has a neutral HOOH and the other an anionic form, $OOH^-$ as the heme Fe ligand. Formation of the anion is possible by proton transfer to a nearby histidine residue already implicated in Compound I formation.

The spectra calculated for these species were distinctly different. For the high spin peroxide species, a single intense broad Soret (B) band was found with a maximum at 360 nm similar to that calculated for the resting state. These calculated spectra were very similar to those observed for the resting state and transient intermediate spectra in the studies of peroxide with the mutant R38L HRP. These similarities between calculated and observed spectra allowed the identification of the neutral peroxide intermediate as the likely origin of the observed mutant spectra. The calculated spectra of the low spin peroxy-anion form of the transient peroxidase species exhibited the same unusual hyperporphyrin spectra, two intense peaks with maxima at 300 and 378 nm as the observed spectral peaks at 330 and 400 nm reported for the transient intermediate in the experiments of wt HRP and a soluble model of it. This similarity of this very unusual spectra feature in the calculated and observed spectra, allowed the identification of the peroxy-anionic species as the transient intermediate observed in the wt HRP experiments. Thus, by relating species to spectra and by comparison of the calculated and observed spectra, a plausible identification has been made of transient intermediate species in the pathway from the resting state to Compound I of peroxidases in both in the wild type and mutant HRP and the differences between them explained.

The trapping of the neutral peroxide species in R38L mutant and the peroxy-anion form in WT HRP implies that the slower rate of Compound I formation found in the mutant is due to interference with proton transfer from the peroxide to the imidazole of the histidine residue, and provides additional evidence that this process is important in the rapid formation of Compound I.

### 4.3 Model Compound I Peroxidase (HRP-I) Heme Species

Both semi-empirical (INDO/ROHF/CI) and *ab initio* methods as well as a variety of spectroscopic techniques have been used to characterize Compound I intermediates of horseradish peroxidase (HRP-I). One of the main questions addressed in these studies was whether this species is an $a_{2u}$ or $a_{1u}$ $\pi$ cation radicals. Evidence from a variety of experimental and theoretical studies indicates that HRP-I is a quartet ground state with two of the unpaired spins delocalized on the Fe=O group leading to an Fe (IV, $S = 1$)=O moiety. However, the question of whether the third unpaired spin is predominantly distributed in a porphyrin $\pi$ $a_{2u}$ or $a_{1u}$ orbital, i.e., whether these species are $a_{2u}$ or $a_{1u}$ $\pi$ cation radicals is still unresolved despite much effort. Over the years a variety of properties including optical absorption spectra, hyperfine splitting in NMR and ESR spectra, resonance Raman spectra, and quadrupole splitting in Mössbauer resonance spectra have been examined and used to try to identify the preferred cation radical for HRP-I and CAT-I. For many years there has been a widely accepted hypothesis that the differences in oxidation mechanism of catalases and peroxidases are due to the differences in their axial ligand that lead to different ground state Compound I species: an $a_{1u}$ $\pi$ cation radical in catalases (CAT-I) and an $a_{2u}$ $\pi$ cation radical in the Compound I species of peroxidases such as horse radish peroxidase (HRP). In principal, the $a_{1u}$ and $a_{2u}$ $\pi$ cation radicals can be distinguished by ESR and NMR spectroscopy because of the difference in unpaired electron density in the two orbitals as illustrated in Figure 9. Unfortunately, the ESR signal of the porphyrin radical cation in Compound I species is broadened by the coupling of the unpaired porphyrin $\pi$ electron with the ferryl iron ($S = 1$) resulting in a broad featureless band.[187,192] Similar difficulties were encountered in the NMR experiments, in which the chemical shift caused by the radical densities on the meso and $\beta$ carbons could not be observed because of protein interference.[193,194] Thus, despite a great deal of effort, no one experimental method has been able to definitively elucidate the nature of the preferred $\pi$ cation state in Compound I, Fe=O type heme species.

Density function theory has been used to characterize both five and six coordinated simple porphine models of Compound I. In one such study, a five coordinated simple porphine model of Compound I preserving the Fe=O moiety, but with no second axial ligand was characterized using local density functional calculations including geometry optimization.[195] A quartet state with two unpaired spins clearly localized on the ferryl moiety and the third in an $a_{2u}$ $\pi$ orbital was found. However, no attempt was made to characterize the alternative $a_{1u}$ $\pi$ cation quartet species. In addition, the five coordinated species used in the LDF calculation[195] does not take into account the possible effect of axial ligands on the Compound I heme site electronic structure.

In a very recent study using non-local density functional theory,[196] the electronic structure of both cation radical states of the same Compound I species with an unsubstituted porphine ring an added imidazole ligand [O=Fe–P–Im] was characterized as a model for Compound I of peroxidases. The $a_{2u}$ $\pi$ cation radical state ($^4A_{2u}$) was determined to be the ground state with the $^4A_{1u}$ $\pi$ cation radical state about 0.15 ev higher. As in all calculations the spins were localized in the porphyrin ring ($S = 1/2$) and the Fe=O center ($S = 1$). In another recent study,[197] using a much more primitive, approximate self consistent charge (SCC) $X_\alpha$ local density method,[198] the $a_{2u}$ was also found to be the ground $\pi$ cation radical species in an unsubstituted iron-porphine ferryl complex with chloride, imidazole, and methyl mercaptide ligands as models for Compound I of hemin, peroxidases, and cytochrome P450. Thus, all levels of *ab initio* density functional methods appear to favor an $a_{2u}$ $\pi$ cation radical quartet for a model heme Compound I peroxidase species with a simple unsubstituted porphyrin.

In contrast to these results, both *ab initio* studies using a CAS SCF (complete active space electron correlation SCF) method[199] and studies using the semiempirical INDO/RHF method[200] lead to the prediction of an $a_{1u}$ $\pi$ cationic ground state species for unsubstituted iron porphine Compound I models with a pyridine[199] imidazole or imidazoleate[200] axial ligand.

It is clear from the combined computational and experimental efforts thus far discussed that the two competing quartet $\pi$ cation radial states are very close in energy.

To further address the long-standing question of whether HRP-I favors $a_{2u}$ $\pi$ and CAT-I $a_{1u}$ $\pi$ cation radicals, additional theoretical studies of the electronic structure and spectra of models for the ferric resting state and Compound I intermediates of HRP-I and CAT-I, each in an $a_{1u}$ and $a_{2u}$ $\pi$ cation state, have been performed using the INDO/ROHF/Cl method.[201] In these studies both a simple porphine model and a fully substituted protoporphyrin-IX Compound I model were characterized. Properties such as the quadrupole splitting in Mössbauer resonance spectra and electronic spectra were calculated for each cationic species and assessed to determine which, if any, were most consistent with experiment. For both HRP-I and CAT-I, the $a_{1u}$ porphyrin $\pi$ cation state is the lowest energy state. Moreover, it is further stabilized by the anionic form of the ligand an imidazolate for the peroxidase species and a phenolate for the catalate species as well as by the addition of the porphyrin ring substituents of protoporphyrin-IX. The calculated values of quadrupole splitting observed in the Mössbauer resonance and the calculated electronic spectra for these

$a_{1u}$ $\pi$ cation radical ground states give results that were more consistent with experiment than those calculated using the alternative $a_{2u}$ $\pi$ cation radical. Thus, there is no compelling reason to invoke the higher energy $a_{2u}$ $\pi$ cation radical as the favored state in HRP-I. It can hence be concluded that the distinct function of catalases and peroxidases cannot be attributed to different ground states of their Compound I species. Other factors, such as different protein environments in the catalytic centers, may play a more direct role in determining function.

In addition to addressing the nature of the preferred $\pi$ cation radical in HRP-I, the INDO/SCF/CI methods were used to elucidate the origin of the dramatic decrease in intensity of the Soret $B$ band in the HRP-I compared to its ferric resting form. One of the most useful spectroscopic probes for determining the kinetics of Compound I formation in a variety of wild type and mutant peroxidases is to monitor the dramatic reduction in the intensity of the Soret band observed during the course of Compound I formation from the resting form. However, while frequently used to determine enzyme kinetics, the origin of the reduced intensity was unresolved. This question was addressed by calculation of the electronic structure and spectra for the sextet resting state and quartet Compound I model peroxidases with both a neutral imidazole and an anionic imidazolate ligand. The imidazolate models, but not the neutral imidazole liganded species, had calculated electronic spectra of the resting and Compound I species consistent with the experimentally observed significant reduction in the intensity of the Soret band.

The origin of the observed reduction in the Soret band intensity in the spectra of the imidazolate Compound I model of peroxidases and the reasons for its absence in the neutral imidazole form were examined and identified. The lower intensity of the imidazolate model of Compound I of peroxidases is caused by the mixing of charge transfer transitions such as porphyrin to metal and ligand to metal transitions into the predominantly porphyrin $\pi \rightarrow \pi^*$ transitions in the $B$ band. The greater the calculated percent contribution of charge transfer transition components in the $B$ band, the lower its calculated intensity. This extent mixing of charge transfer character is greatest in the imidazolate form of the peroxidase Compound I heme species and leads to the observed reduction in intensity compared to the imidazolate resting form.

The combined results of these studies strongly support a heme site in typical model Compound I peroxidase species that is an $a_{1u}$ $\pi$ cation radical species with an imidazolate axial ligand. The calculated ground state properties for this model are all consistent with observed electromagnetic properties. Moreover, the dramatic decrease in intensity of the Soret band in Compound I of peroxidases compared to the ferric state was obtained only for the imidazolate species and the origin of this decrease with identified. These results provide compelling evidence for significant proton transfer to the conserved Asp found near the proximal histidine ligand of peroxidases. This feature of the peroxidases is absent in the globins. Although they share a common imidazole ligand with peroxidases, they lack a nearby proton acceptor residue. This distinction could therefore be important for functional specificity of the two heme families.

### 4.4 Model Compound II Peroxidase Heme Species

As already indicated, Iron Oxo, Fe=O, Compound I species two oxidation states above the ferric resting form, are proposed to be the common catalytically active form of the three families of metabolizing heme proteins, peroxidases, catalases, and cytochrome P450s. However, in general, only peroxidase substrate oxidation occurs in two sequential one electron oxidations. The first is performed by Compound I and the second is performed by the reduced, Fe=O species called Compound II. As discussed above, the electronic structure of Compound I complexes of intact heme proteins as well as model heme complexes has been a subject of extensive theoretical and experimental investigations. These combined results all indicate that the Compound I species is a quartet. One of the three unpaired electron spins is centered on the porphyrin ring leading to either an $a_{1u}$ or $a_{2u}$ porphyrin $\pi$ cation radical. The other two unpaired electrons are centered on the Fe=O moiety, in molecular orbitals consisting mainly of [Fe($3d\pi$)$O_2 p\pi$] atomic orbitals and the iron is in a formal Fe(IV) state. Previous calculations of the electronic structure of the peroxidase Compound II model heme unit indicates it is a triplet state, in which the added electron reduces the porphyrin ring, eliminating its radical character, leaving the electron and spin distribution around the Fe=O moiety intact. This result[202] explains the very similar Fe nuclear properties such as quadrupole splitting, observed for Compounds I and II.[170,199]

The electronic absorption spectra of Compound II heme species exhibit interesting transitions in addition to the normal heme $Q$ and $B$ bands. Specifically, the spectra of intact peroxidase Compound II species such as CCP-II, with an imidazole or imidazolate ligand, as well as model compounds with a chloride (hemin) or a pyridine ligand show common extra absorptions in the frequency range between $Q$ and $B$ bands, i.e., between 15,000 and 25,000 $cm^{-1}$. In order to understand the origin of these extra spectral features, the electronic structure and spectra of these model Compound II heme species [Fe=O–P–L] with L = imidazole, imidazolate, $Cl^-$, and pyridine have been calculated using the INDO/RHF/CI method.[203] These calculations confirm the nature of the triplet ground state as one in which the two unpaired electrons occupy the [Fe$\pi$–O$\pi$] orbitals. About 20% of the unpaired spin density is calculated to be on the oxygen atom in agreement with the experimental estimate of about 25%. Consistent with experimental observation, the calculated spectra show that charge transfer (CT) transitions exist for all of the four Compound II complexes with calculated excitation energies for these CT transitions between the Q and B bands and with moderate oscillator strengths. The identification of the excitations responsible for these transitions not directly obtainable from experiment, allowed the elucidation of their origin. Although a description of the CT excited state for each of the model systems is complex, an electronic transition from the porphyrin $\pi$ orbitals to the Fe–O $\pi^*$ orbitals is a common component of all of them.

## 5 HEME MODELS FOR THE CYTOCHROME P450s

### 5.1 Resting Form of Cytochrome P450s

The resting form of the substrate free cytochrome P450s in isozymes with a known X-ray structure[204–209] is a ferric aquo heme complex, in common with that of peroxidases and metmyoglobin, described above, but with a mercaptide ligand from a nearby conserved cysteine instead of an imidazole ligand. In an attempt to independently verify the presence of a water as the second axial ligand, the ESR spectrum of the substrate free resting form of P450cam in 65% $^{17}O$ enriched $H_2O$ has been determined.[210] In this initial study, no hyperfine splittings of the electron spin resonance due to nuclear-electron spin interactions were observed. These negative results were in apparent conflict with the X-ray structure that indicated that P450cam has a water molecule as a sixth ligand. The question of whether negative results in such ESR experiments need always be interpreted as the absence of water was investigated by computational studies of the expected unpaired spin density on the six coordinated ferric aquo P450 resting form in the high spin and low spin states using the INDO/ROHF method.[211] The results of these studies predicted negligible unpaired spin density on the water oxygen in both of these spin states. The calculated value was greatly reduced by a factor of over 100 compared to the calculated value for MetMb, for which broadening of the ESR is signal is barely observed. Thus, these results resolved the conflict between the esr and X-ray spectra by illustrating that the absence of broadening or splitting of the ESR signal is not *a priori* evidence for the lack of a water ligand. More recently, the esr spectrum of the substrate free resting form of P450cam in 65% $^{17}O$ enriched $H_2O$ has been reexamined using a more sensitive electron spin echo spectroscopic technique that was able to better detect some residual broadening that confirmed the presence of water.[212]

However, another enigma remained, the origin of the low spin resting state so unusual for a ferric aquo heme complex. While the observed temperature dependence of magnetic susceptibility[156,168,213] of quadrupole splitting in Mössbauer resonance,[170–172] electron spin resonance,[173–176] resonance Raman spectra,[178–181,214] and electronic spectra[175,180,181] of horseradish peroxidase, cytochrome *c* peroxidase and MetMb indicate either a sextet or quartet ground state albeit with subtle variations of populations of different spin states, the preferred ground state spin of the substrate free ferric resting form of most cytochrome P450s is low spin. This preference for a low spin state in ferric aquo heme complexes is unprecedented not only in intact proteins, but in the one known model structure, a ferric diaquo-porphyrin complex.[215] This enigma is compounded by the observation of very similar Fe-water distances of about 2.2–2.4 in the ferric resting forms of the P450s and the peroxidases.

Two possible origins for this disparity, change in axial ligand to a mercaptide in P450s, or the effect of the environment of the protein have been explored for the heme model of the substrate-free ferric resting form of cytochrome P450cam. In this study the combined techniques of restricted open shell Hartree–Fock INDO/ROHF methods together with molecular dynamics were used.[216] For the heme site by itself, [Fe(III) Protoporphyrin IX–$H_2O$–$SCH_3^-$], a sextet state ground state was found albeit with a very small energy separation of 3.8 kcal/mol to the doublet state and 5.80 kcal/mol to the quartet state. Thus the presence of the two axial ligands alone, water and mercaptide, are not by themselves sufficient to explain the experimentally observed low spin state of the intact P450cam protein based on ESR and temperature dependent magnetic susceptibility measurements.[217] To investigate the role of the protein environment as a modulator, the relative spin state energies of the [Fe(III) Protophyrin IX–$H_2O$–$SCH_3^-$], heme site were again calculated in the presence of the electrostatic field of the protein. The field of the protein was included by substituting the net atomic charges for every atom in P450cam in the position occupied by that atom in the optimized X-ray structure of substrate free P450cam. For this purpose, the electrostatic potential derived partial charges found in the AMBER suite of programs[218] was used. The inclusion of the electrostatic field of the protein in the INDO/ROHF Hamiltonian resulted in a reversal of spin states with a low spin resting form of cytochrome P450cam in agreement with experiment. The high spin state was 2.35 kcal/mol higher in energy. Molecular dynamic simulations of the optimized X-ray structure using the Amber suite of programs[218] further supported the role of the protein in modulating the spin state equilibrium. The dynamic motion of the heme unit alone is not sufficient to account for the predominance of the low-spin state, while the dynamic effect of the field of the protein favors the low-spin sate. These results have revealed a heretofore unidentified role of the protein in modulating spin equilibrium, a property of the ferric heme unit that is central to the maintenance of the enzymatic function of the cytochrome P450s.

### 5.2 Heme Model for the Substrate Bound Ferric P450s

Substrate binding to the distal binding pocket of intact P450 isozymes appears to result in a 5-coordinated species with the heme iron bound only to the proximal cysteine ligand. The X-ray structures of two substrate bound P450s, P450cam,[205] and P450-eryf[204] confirm this trend. INDO/RHF calculations made for the heme unit of this species, a [Fe(III)–Protoporphyrin-IX–$SCH_3^-$] complex,[216] using the X-ray coordinates from the crystal structure of P450cam resulted in a high spin ground state with the quartet and doublet states 3 and 18 kcal/mol above the sextet state. In contrast to the resting form, the high spin state remains the ground state even n the field of the camphor bound cytochrome P450cam. These results are consistent with the high spin ground state inferred from a number of spectroscopic measurements of substrate bound P450cam including ESR and optical spectra.

When substrates bind to the ferric resting form of the cytochrome P450s, a small shift to higher frequency is observed in the maximum of the Soret B band.[217,219–222] This shift is used to monitor the binding of substrates to the cytochrome P450s. Moreover, this shift has been attributed to the spin state change from a low spin to a high spin ferric species, which generally occurs upon substrate binding to the P450s. Thus, for many years now, a shift in frequency of the maximum in the Soret band in the UV-visible spectra has been routinely used to quantify the amounts of high and low spin forms of ferriheme proteins. Moreover, since this spin state change has been shown to be functionally important in the enzymatic cycle of this family of metabolizing heme proteins, characterization of the percentage of high-spin form in this manner resulting from binding of different substrates or from mutations of the enzyme has become a commonly used indicator of enzymatic efficiency.

While monitoring a small change in the frequency of the Soret band is a very efficient way of inferring a spin-state change in heme proteins, a mechanistic link between the observed blue shift in the Soret band and a change in the spin state of the iron had yet to be established, since optical spectra are not a priori a probe of the spin states of a chromophore. Another confounding aspect is that, in all cases, Soret band shifts are monitored between two heme proteins that differ in some respect in addition to a spin state change, for example, in axial ligand, in substrate, inhibitor, or a site-specific mutation of the protein. Few of the many heme systems thus studied have had both their magnetic and optical properties characterized. Thus, it was possible that the observed shift in the B band maximum was a direct result of other differences between the two protein species being monitored, without the obligatory involvement of a spin state change.

The techniques of computational chemistry are ideally suited to address both of these concerns. Thus, the INDO/RHF/CI method has been used[223] to determine whether there is an underlying mechanism coupling the spin-state change with the observed shifts in the Soret band of the optical spectra of heme proteins. The calculations were made for the identical heme complex in both a high and low spin state without changing any other feature of the heme unit. The model used in the study was the active site of substrate free P450cam, an [Fe(III)–Protoporphyrin-IX–aquo mercaptide] complex, as found in the 2.2 Å resolution crystal structure.[224] Such an idealized comparison is possible only in a computational approach, since in a real system some change has to occur to cause the spin-state change and one can never completely decouple the two possible sources of the effect on the Soret band.

Comparison of the calculated spectra of the high and low spin states revealed that, upon change in the Fe(III) from a low to a high spin reference state, without any other change, in the heme complex, a blue shift does indeed occur in the strong Soret (B) band absorptions found in the 25,000 $cm^{-1}$ region that consist primarily of porphyrin $\pi$–$\pi^*$ transitions. These results establish that a spin-state change alone leads to the observed shift in the Soret band. They also elucidated the origin of the observed shift. It was shown to be a direct consequence of the spin-state change. In the low-spin state there is enhanced mixing of the iron $d$ and porphyrin 4eg ($\pi^*$) orbitals resulting in a lowering of the eg ($\pi^*$) states and a consequent shift to the red of the Soret band. These results clearly demonstrate an underlying physical basis for the observed correlation between ferric heme spin states and measured spectral shifts associated with ($\pi$ to $\pi^*$) heme transitions.

### 5.3 Heme Models for Carboxy and Oxy Ferrous P450s

In common with myoglobin and hemoglobin, the reduced ferrous form of the cytochrome P450s also binds $O_2$, CO, and other small ligands. In fact as shown in Figure 5, the binding of molecular oxygen to the ferrous P450 heme site is required in the enzymatic cycle of these metabolizing heme proteins. Competitive binding of CO at this stage inhibits the function of the protein, preventing it from continuing its normal enzymatic cycle. CO binding was also found to result in an unusual red shift of the Soret band to a maximum of 450 nm, a signature spectral feature giving the name to this entire family of heme proteins. In the early stages of characterization of the P450s, a heme protein was identified as a P450 if CO inhibited its function and

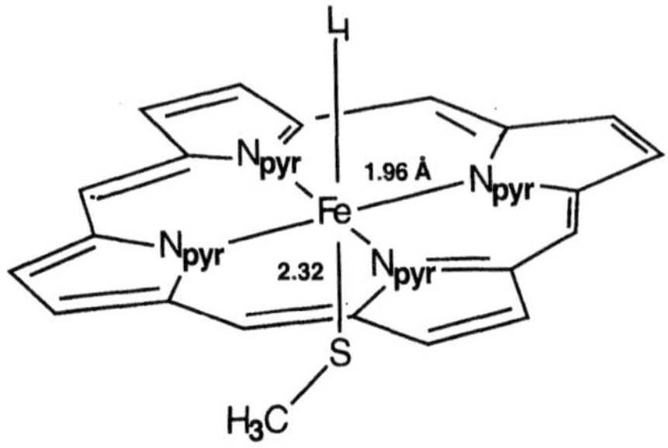


**Figure 15.** The model heme complex used for the three ferrous P450 heme site calculations with $L_1 = O_2$ and CO, $CHCF_3$.

resulted in this signature spectrum that is absent in the corresponding $O_2$ and CO complexes of globins.

The main difference in the heme site in these proteins and in model compounds is the nature of the axial ligand. In P450s, the universal axial ligand is a mercaptide group from a proximal conserved cysteine residue, while in the globins it is an imidazole from a conserved histidine residue. With this small difference, the globins function as reversible oxygen transport proteins, while the P450s proceed by subsequent transformations to function as true oxidative metabolizing heme proteins. The question thus arises as to what extent the intrinsic properties of the oxyferrous heme site of these two proteins contribute to the difference in function. The answer to that question is provided by the combined result of experimental and computational studies.

The observed ground state properties of oxy and carbonyl ferrous P450 complexes are very similar to those of their hemoglobin and myoglobin counterparts. In particular, oxyferrous P450 has a large negative value of $\Delta E_Q = -2.15$ mm/sec very similar to that of oxyHb and carbonyl ferrous P450 has a small positive value of $\Delta E_Q = 0.32$ mm/sec very similar to that of carbonyl Hb. To address the origin of these similarities and differences, computational studies were made for 6-coordinate ferrous P450 heme systems [Fe(II)–P–$SCH_3$–L] with L $= O_2$ and CO shown in Figure 15. These studies were first made using the IEHT method and subsequently the INDO/SCF/CI method. In common with the model globin counterparts with an imidazole ligand, the ground state of both of these complexes were found to be a diamagetic singlet. The charge distributions, bonding and calculated values of $\Delta E_Q$ obtained from the INDO method for the oxy and carbonyl ferrous P450 heme sites in this diamagetic singlet states are given in Table 10.

As seen in Table 10, there is good agreement between observed and calculated properties for both of these complexes and with the observed differences between them. The calculated value of $\Delta E_Q = -2.46$ mm/sec for a diamagnetic singlet Fe(II) dioxygen P450 heme complex is in excellent agreement with the experimental value. Similar to the model oxyHb heme complex, the large electric field gradient observed in the model oxyP450 complex is due to extensive electron delocalization leading to a significant net transfer of electrons to the dioxygen ligand, and a ferric-

**Table 10 Calculated Bond Overlap Densities, Net Atomic Charges, Iron Atomic Orbital Populations, and Quadrupole Splittings ($\Delta E_Q$) for Model Carbonyl and Oxy P450 from INDO SCF Method**

| $L_1 =$ | 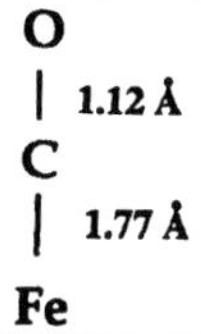 | 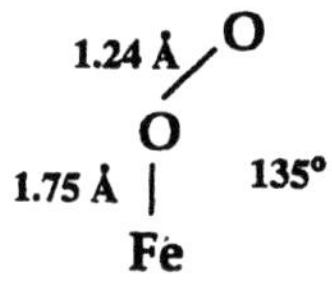 |
|---|---|---|
| | Bond overlap | |
| Fe–$L_1$ | 0.66 | 0.48 |
| Fe–S | 0.40 | 0.39 |
| $L_1$–O | 1.07 | 0.55 |
| $\nu_{L_1-O}$ obs. ($cm^{-1}$) | 1969[b] | 1159[c] |
| | Net charge | |
| Fe | 1.61 | 1.55 |
| $SCH_3^-$ | −0.85 | −0.76 |
| $L_1$ | −0.02 | −0.58 |
| O | −0.59 | −0.70 |
| | Iron AO population | |
| $3d_{z^2}$ | 0.51 | 0.70 |
| $3d_{x^2-y^2}$ | 0.41 | 0.52 |
| $3d_{xy}$ | 1.99 | 1.98 |
| $3d_{xz}$ | 1.75 | 1.95 |
| $3d_{yz}$ | 1.75 | 0.88 |
| $\Delta E_{IQ}$ (mm/sec) | | |
| Calc. | 0.45 | −2.46 |
| Obs. | 0.32[a] | −2.15[a] |

[a] Sharrock, M.; Debrunner, P. G.; Schulz, C.; Lipscomb, J.P.; Marshall, V.; Cunsalus, I. C. *Biochem. Biopys. Acta*, **1976**, *420*, 8.
[b] Collman, J. P.; Brauman, J. I.; Halbert, T. R.; Suslik, K. S. *Proc. Nat. Acad. Sci. USA*, **1976**, *73*, 3333.
[c] Maxwell, J. C.; Caughey, W. S. *Biochemistry*, **1976**, *15*, 378.

like distribution of electrons around the iron, even in a formal Fe(II)–$O_2$ diamagnetic state. The calculated small positive $\Delta E_Q = 0.45$ mm/sec for the carboxy P450 complex is also in excellent agreement with experiment. As in the carboxy Hb model, it is the result of nearly equal occupancy of the three *d* orbitals comprising the original $t_{2g}$ manifold. The calculated bond overlap density between the iron and the ligand decreases from 0.66 to 0.48 in going from CO to $O_2$ in agreement with the larger binding affinity of CO. The origin of this enhanced affinity appears to be less charge transfer from the iron to the CO than to the $O_2$. The bond density between the iron and the sulfur ligand is very similar in the CO and $O_2$ complexes. The calculated O–O bond overlap density is about half that of the C–O bond overlap indicating that bound $O_2$ forms a much weaker bond that bound CO. These results are consistent with the relative values of their observed stretching frequencies.[129,144]

Comparing the electronic structure of the heme site in oxyP450 (Table 10) with that of oxyHb (Table 4) we see that despite their very different axial ligands all of the intrinsic properties that could modulate their differences in function are essentially identical. These include the Fe–O bond strength which is no weaker in the globins than in the P450s and the extent of charge transfer to the oxygen that is the same. Thus, there is no obvious trans effect of the endogenous proximal ligands in these two families of heme proteins and differences in function cannot then be directly attributed to differences in the intrinsic properties of the oxyferrous Hb and oxyferrous P450 heme sites. Rather, differences in protein environment among the two heme families must be a major determinant of differences in their function. Among such possible determinants is the organization of the P450s that allows the donation of a second electron to this system from nearby redox partners that is absent in the globins. Another determinant are features in the protein environment that are crucial to the transient stability of this twice reduced ferrous dioxygen species of P450 and the rapid formation of the enzymatically active Compound I species from it.

While the different axial ligand in the globins and the P450 heme sites lead to very similar ground state properties of the two heme families in their oxy ferrous and carboxy ferrous forms, these differences lead to a striking difference in the observed electronic spectra. The optical spectra of the carbonyl complex of ferrous P450, the signature spectrum for this family has been the subject of considerable experimental investigation.[225–229] In contrast to the electronic spectra of carbonyl complexes of hemoglobin and myoglobin (shown in Figure 10) with a single maximum at about 420 nm, the Soret band of the carbonyl complex of the P450s is split into two components with maxima at 364 and 450 nm. Although first thought to be unique to this complex, subsequent spectroscopic studies[230] of the dioxygen complex of these enzymes, which is part of their normal enzymatic cycle, also revealed a split Soret with broadened and blue shifted Soret components at 350 and 420 nm. Model oxy[226] and carbonyl[227,228] P450 heme complexes with mercaptide or thiolate ligands appear to have very similar optical spectra to the intact proteins. Thus, this characteristic spectral feature appears to be intimately associated with this axial ligand.

In order to understand the origin of the difference in the spectroscopic behavior, particularly the split Soret of the oxy and carbonyl complexes of P450 and its absence in the globins, a detailed calculation of the electronic spectra of model oxy and carbonylP450 complexes was made[71] and compared with previously calculated spectra of model oxy and carbonyl Hb complexes. The models used for these two six coordinated P450 heme species were the same as for the calculation of the ground state properties shown in Figure 15. The corresponding models for the myoglobin or hemoglobin heme sites were identical except for the replacement of the mercaptide ligand by an imidazole ligand. The geometries used for the ligands were based on X-ray structures of model compounds. Identical regularized porphine macrocycle geometries were used for each model to eliminate any confounding effects of small differences in them. The INDO/SCF/CI method used in both sets of calculations was identical.

The Soret-type transitions calculated for model carbonyl P450 heme sites are shown in Figure 16, together with that calculated for model carbonylHb heme sites for comparison. In this as in all spectral figures, envelopes have been drawn around the calculated frequencies and oscillator strengths to aid in visualizing the spectra. As seen in Figure 16, the Soret bond is split in the calculated spectrum of the model carbonyl P450s with calculated maxima at 430 and 325 nm. By contrast, there is only one very intense band in the model carbonylHb complex with a maximum of 380 between these two split peaks. These results are clearly in agreement with the major changes observed in the experimental spectra.

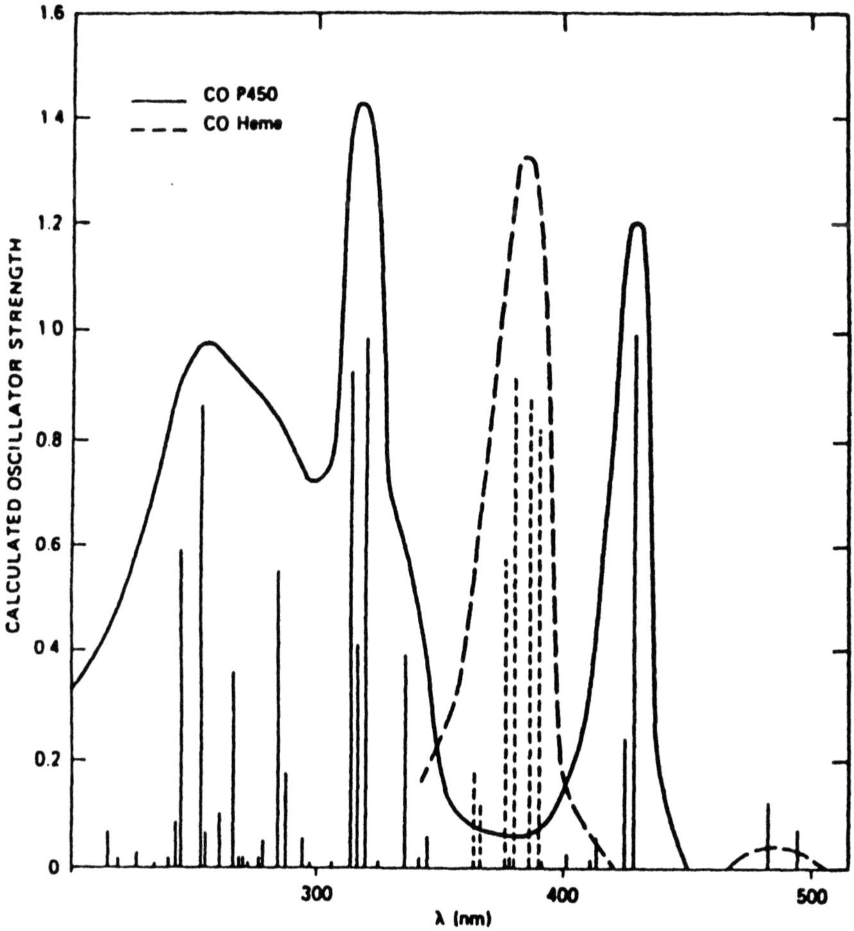


**Figure 16.** Calculated electronic absorption spectrum for a heme complex of carbonyl P450 [Fe(II)–P–$SCH_3^-$–CO] and for comparison that for a heme model of carbonylHb [Fe(II)–P–Im–CO] using the INDO/SCF/CI method. The position of the vertical lines are the calculated frequencies and the height of the lines are the calculated intensities in terms of oscillator strength. Gaussian envelopes have been drawn around these stick figures for ease of comparison of calculated and experimental spectra.

The origin of the split Soret in the model ferrous carbonyl P450 complexes can be seen by examination of the filled and virtual orbitals involved in the allowed electrons transition of the model carbonyl P450 and carbonylHb complexes shown in Table 11. The highest occupied orbital, HOMO is a pure $a_{1u}$ porphyrin $\pi$ orbital in both complexes. However, while the (HOMO-1) orbital remains a pure $a_{2u}$ porphyrin $\pi$ orbital in the HbCO complex, there is a significant mixture of sulfur S$\sigma$-type atomic orbitals orbital in the $a_{2u}$ orbital of the carbonyl P450 complex. Moreover, the next four lower-energy orbitals in the carbonyl P450 have significant contributions from the S$\pi$, S$\sigma$, or $a_{2u}$ ($\pi$) orbitals. As shown in the detailed assignment of the transitions calculated for the model P450CO complex in Table 12, this admixture of ligand S$\sigma$ orbitals with the porphyrin $a_{2u}$ $\pi$, orbitals, missing in the HbCO complex, perturbs the energy profile of the excitations contributing intensity in the Soret band region, leading to the observed split Soret. As seen in Table 12, while the $Q$ bands are assigned to pure $a_{1u}$, $a_{2u}$ to eg($\pi^*$) excitations in keeping with the four orbital model, the origin of the first (B) and second (B′) Soret bands involve multiple [$a_{1u}$, ($a_{2u}$, S$\sigma$) to eg*] excitations together with some participation by [($d\pi$, eg($\pi$)) to ($d_{xy}$, eg($\pi$))*] and [(S$\pi$, $d\pi$) to (CO$\pi$, $d\pi$*] excitations. The origin of the split Soret was hence determined from these calculations to be primarily a mixing of the unpaired sulfur S-$\sigma$

**Table 11 Filled and Virtual Orbitals Involved in Allowed Electronic Transitions of Model Carbonyl P450 and CarbonylHb Complexes[a]**

| (a) Carbonyl P450 | (b) CarbonylHb |
|---|---|
| | 90. (59% $d_{xy}$, 31% $b_{2u}\sigma$)* |
| 83. (56% $d_{xy}$, 33% $e_g\sigma$)* | 89. {(94% $e_g$, 3% CO$\pi$)*} |
| 82. (53 % $d_{z^2}$, 20% $e_g\sigma$, 11% S$\sigma$, 8% CO$\sigma$)* | 88. {(88% $e_g$, 5% $d\sigma$, 3% CO$\pi$)*} |
| 81. {100% $a_{1u}$} | 87. (43 % $d_{z^2}$, 14% Imid$\sigma$, 12% $a_{1u}\sigma$, 12% $e_g$, |
| | 7% CO$\sigma$)* |
| 80. {(96% $e_g$, 3% CO$\pi$)*} | 86. (65% CO$\pi$, 21% Imid$\pi$, 7% $d\pi$)* |
| 79. {(96% $e_g$, 3% CO$\pi$)*} | 85. (55% CO$\pi$, 37% porph$\pi$)5% $d\pi$)* |
| 78. 99% $b^*_{1u}$ | 84. (67% porph$\pi$, 30% CO$\pi$)* |
| 77. (83% CO$\pi$, 9% $d\pi$)* | 83. {(78% Imid$\pi$, 18% CO$\pi$)*} |
| 76. (83% CO$\pi$, 9% $d\pi$)* | 82. 100% $b^*_{2u}$ |
| 75. 100% $b^*_{2u}$ | 81. {98% Imid$\pi$*} |
| 74. (97% $e_g$, 3% $d\pi$)* | 80. (98% $e_g$, 2% $d\pi$)* |
| 73. (97% $e_g$, 3% $d\pi$)* | 79. (98% $e_g$, 2% $d\pi$)* |
| 72. 99% $a_{1u}$ | 78. 100% $a_{1u}$ |
| 71. (84% $a_{2u}$, 15% S$\sigma$) | 77. 98% $a_{2u}$ |
| 70. (65% S$\pi$, 25% $d\pi$, 3% CO$\pi$) | 76. (62% $d\pi$, 31% $e_g$, 4% CO$\pi$) |
| 69. (61% $d\pi$, 29% $e_g$, 5% CO$\pi$) | 75. (62% $d\pi$, 31% $e_g$, 4% CO$\pi$) |
| 68. (39% $d\pi$, 31% $e_g$, 25% S$\pi$) | 74. 96% $d_{x^2-y^2}$ |
| 67. (51% S$\sigma$, 32% $a_{2u}$, 5% $d\sigma$) | 73. (70% Imid$\pi$, 29% porph$\pi$) |
| 66. 96% $d_{x^2-y^2}$ | 72. (93% $a_{2u}$, 3% CO$\sigma$) |
| 65. 95% $b_{1u}$ | 71. (77% $b_{1u}$, 22% Imid$\pi$) |
| 64. (82% $a_{2u}$, 7% S$\sigma$) | |

[a] Unless noted otherwise, all orbital designations refer to porphyrin $\pi$–orbitals in $D_{4h}$ symmetry; those orbitals enclosed in brackets are not significant components of the respective spectra.

**Table 12 Observed and Calculated Electronic Absorption Spectra of Model Carbonyl-P450 Complexes**

| Band | Frequency ($cm^{-1}$) Obs.[a] | Calc. | Calc. Oscill. Strength | Calc. Polarization | Assignment |
|---|---|---|---|---|---|
| $Q_0$ | | 15,340 | 0.009 | $xy$ | $(a_{2u}, S\sigma), a_{1u} \rightarrow e_g^*$ (52%, 43%) |
| | 18,200 | 16,020 | 0.009 | $xy$ | $(a_{2u}, S\sigma), a_{1u} \rightarrow e_g^*$ (54%, 41%) |
| $Q_y$ | (broad) | 20,170 | 0.008 | $xy$ | $(S\pi, d\pi), (d\pi, e_g) \rightarrow e_g^*$ (58%, 27%) |
| | | 20,680 | 0.013 | $xy$ | $(d\pi, e_g) \rightarrow e_g^*$ (87%) |
| | | 23,290 | 0.605 | $(x)y$ | $a_{1u}, (a_{2u}, S\sigma), (S\sigma, a_{2u}) \rightarrow e_g^*$ (31%, 24%, 22%) |
| | 22,400 | 23,330 | 0.394 | $x(y)$ | $(S\pi, d\pi), (d\pi, e_g) \rightarrow e_g^*$ (20%, 18%) |
| | | | | | $a_{1u}, (a_{2u}, S\sigma), (S\sigma, a_{2u},) \rightarrow e_g^*$ (19%, 15%, 11%) |
| $B$ | | 23,510 | 0.241 | $x$ | $(d\pi, e_g), (S\pi, d\pi), (d\pi, e_g, S\pi) \rightarrow e_g^*$ (23%, 19%, 12%) |
| | | | | | $a_{1u}, (S\sigma, a_{2u}), (a_{2u}, S\sigma) \rightarrow e_g^*$ (13%, 12%, 9%) |
| | | 24,150 | 0.057 | $xy$ | $d_{x^2-y^2} \rightarrow e_g^*$ (52%) |
| | | | | | $(S\pi, d\pi) \rightarrow (d_{z^2}, e_g\sigma, S\sigma)^*, (d_{xy}, e_g\sigma)^*$, (10%, 5%) |
| | | | | | $(d\pi, e_g, S\pi) \rightarrow (d_{z^2}, e_g\sigma, S\sigma)^*, (d_{xy}, e_g\sigma)^*$, (10%, 6%) |
| | | 28,960 | 0.065 | $xy$ | $(S\pi, d\pi) \rightarrow b_{2u}^*$ (48%) |
| | | | | | $[(S\pi, d\pi);\ a_{1u}] \rightarrow [e_g^*]$ (18%) |
| | | | | | $(S\sigma, a_{2u}) \rightarrow e_g^*$ (11%) |
| | | 29,770 | 0.386 | $xy$ | $(d\pi, e_g) \rightarrow b_{2u}^*$ (32%) |
| | | | | | $(S\sigma, a_{2u}) \rightarrow e_g^*$ (28%) |
| | | | | | $[(d\pi, e_g);\ a_{1u}] \rightarrow [e_g^*]$ (14%) |
| $B'$ | 27,500 | 30,270 | 0.985 | $(x)y$ | $(S\sigma, a_{2u}), a_{1u}, (a_{2u}, S\sigma) \rightarrow e_g^*$ (35%, 7%, 5%) |
| | (broad) | | | | $(d\pi, e_g) \rightarrow (d_{xy}, e_g\sigma)^*$, (13%) |
| | | | | | $(S\pi, d\pi) \rightarrow b_{2u}^*$ (8%) |
| | | 31,580 | 0.413 | $xy$ | $(d\pi, e_g) \rightarrow (d_{xy}, e_g\sigma)^*, b_{2u}^*$ (40%, 9%) |
| | | | | | $(S\sigma, a_{2u},) \rightarrow e_g^*$ (11%) |
| | | 31,760 | 0.789 | $x(y)$ | $(S\sigma, a_{2u}) \rightarrow e_g^*$ (24%) |
| | | | | | $(d\pi, e_g) \rightarrow (d_{xy}, e_g\sigma)^*, b_{2u}^*$ (14%, 10%) |
| | | 31,850 | 0.141 | $xy$ | $(S\pi, d\pi), (d\pi, e_g, S\pi), (d\pi, e_g) \rightarrow (CO\pi, d\pi)^*$ (30%, 26%, 18%) |

($Spz$) orbitals with the $a_{2u}$ porphyrin $\pi$ orbital, a mixing that does not occur in the model carbonylHb complexes with an imidazole ligand.

The calculated electronic spectra of the model oxy ferrous P450 complex is shown in Figure 17. For comparison, the calculated spectra of the carbonyl P450 complex and the experimental spectra for both are given in this figure. Consistent with the experimental spectra, the calculated spectra demonstrate that the oxy P450 complex also has split Soret bands, but is blue shifted relative to those for the model carbonyl P450 complex. Another significant difference between the two calculated spectra is the appearance of $z$-polarized transitions in the oxyP450 complex. Such $z$-polarized transitions were observed in the single crystal polarized spectra of the oxyHb species. Their predicted occurrence for the oxy P450 complex could be verified by similar single crystal polarized spectroscopic studies.

Calculated spectra of another ferrous P450 complexed with a carbene ($CF_3CH$) (see References 96,231) ligand also yielded a hyper or split Soret spectra shifted to the red of the carbonyl P450 complex. In each species, the origin of the split Soret was the same type of excitation, involving a mixed [$a_{2u}$-S$\sigma$] orbital and was rela-

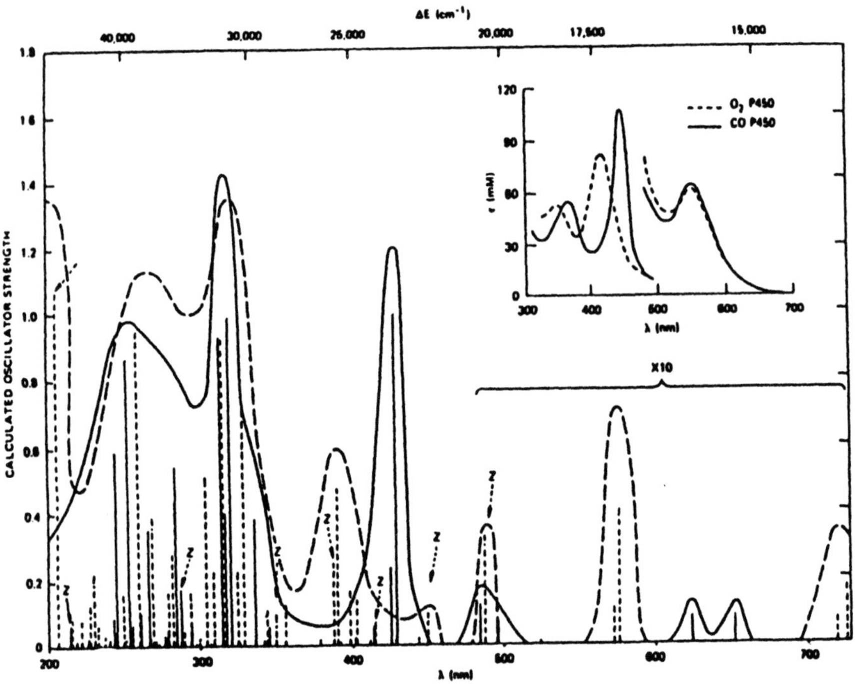


**Figure 17.** Calculated electronic absorption spectrum for a heme site of oxy and carbonyl P450 [Fe(II)–P–$SCH_3^-$–L] L = $O_2$, CO using the INDO/SCF/CI method and for comparison the experimental spectra of these species in intact P450 protein. The position of the vertical lines are the calculated frequencies and the height of the lines are the calculated intensities in terms of oscillator strength. Gaussian envelopes have been drawn around these stick figures for ease of comparison of calculated and experimental spectra.

tively independent of the nature of the second axial ligand. However, a subtle effect of the second axial ligand on the positions of the maxima in the two components of the split Soret band was identified. The finding that the presence of an intact Fe–S–cysteine bond is an essential factor in producing the signature spectra is consistent with the observation that treatments of these enzymes, such as reconstitution that make them dysfunctional also restores the single Soret peak. These findings taken together provided a robust link between structure, spectra, and function for the P450s.

One of the perturbations that diminishes function of the cytochrome P450s and restores a normal Soret peak is a decrease in the pH. In order to explore the origin of this effect and to see if a split Soret would be predicted for a ferric form of P450s, the INDO/RHF/CI method that allows calculation of the spectra of radical species was used to calculate and compare the spectra of two closely related six coordinated low spin complexes of the ferric resting form [Fe(III)–P–$L_1$–$L_2$]. In one complex, $L_1 = L_2 =$ mercaptide and in the other complex one of the mercaptide ligands is protonated, i.e., $L_1 =$ mercaptide, $L_2 =$ mercaptan.[232] The low spin ferric P450 complex with two mercaptide ligands displays a similar characteristics as the ferrous P450 complexes with oxy, carbonyl, and carbene ligands, which showed split Soret characteristics. In common with these ferrous complexes, the ferric dimercaptide complex has at its two highest occupied orbitals (HOMO) a pure porphyrin $a_{1u}$ $\pi$ and an $a_{2u}$ orbital with significant contribution from sulfur $\sigma$ type atomic orbitals ($Spz$). Consistently, the calculated spectra exhibits two intense UV-vis peaks that can be assigned as components of a split Soret, that is blue shifted from that of the [ferrous P450 CO] complex. Thus, this signature feature persists even in ferric complexes with an intact iron-mercaptide bond. Contribution of this sulfur $s$ orbital to the highest porphyrin $a_{2u}$ is much smaller for the mercaptan complex than for the mercaptide and results is a single band that can be clearly attributed to Soret-type excitations. These results provide additional support for the close link between structure, spectra, and function of the P450s by illustrating that the spectra as well as the function is modified at lower pHs, corresponding to protonation of the conserved cysteinate ligand.

## 5.4 Transient Species in the Formation of Compound I

The ferrous dioxygen P450 species is the last intermediate in the enzymatic cycle of P450 shown in Figure 5, that is stable enough to have been characterized by a variety of spectroscopic properties. The second electron reduction of the ferrous dioxygen P450 species that follows results in the formation of a reduced ferrous dioxygen species, a very transient species two oxidation states below the ferric resting form. This species rapidly forms water and Compound I, the putative catalytically active species that in turn rapidly oxidized substrates by transfer of the active oxygen atom to them. The steps in the common enzymatic cycle of all P450s leading to the formation of Compound I from the reduced ferrous dioxygen P450 species are the least understood aspects of P450 metabolism because they occur so rapidly and involve extremely transient intermediates. The proposed pathway shown in Figure 5 involving a proton assisted mechanism of dioxygen bond cleavage has only been inferred from indirect experimental evidence. This evidence includes kinetic isotope effect measurement that have definitively demonstrated the requirement of two protons to form Compound I from the reduced ferrous dioxygen species.[233] In addition, earlier studies using isotopically enriched molecular oxygen established that the oxygen that is transferred to substrates is the "proximal" one directly bound to the heme iron and that the other "distal" oxygen is involved in the concurrent formation of water.[234,235] Although consistent with these observations, it is difficult from experiment alone to directly assess the validity of this pathway. To address this question, very recent studies were made in our laboratory to determine whether the intrinsic properties of the proposed oxy intermediates are consistent with their proposed role in Compound I formation. To this end, nonlocal density functional methods were used with extended basis sets to obtain the optimized geometries, relative energies and electronic structures of each of the putative heme species in the pathway from the oxyferrous form to Compound I. The results are summarized in Figure 18. Shown in this figure are the optimized structures of each of these species with the calculated bond lengths and bond order indicated for the two crucial bonds, the Fe–O and O–O bonds. Also indicated are the relative energies of formation of each species.

The optimized geometry of the ferrous dioxygen form shown in Figure 18 is in good agreement with the experimental structure of a model compound of this species and the calculated diamagnetic singlet ground state is consistent with the reported lack of ESR spectrum.[236] Upon second electron addition, in the optimized reduced ferrous dioxygen species, the dioxygen ligand was shown to preferentially bind to the heme iron in an asymmetric "end on" geometry similar to the ferrous dioxygen species, rather than a bridged one. The ground state doublet obtained, with one unpaired electron distributed over both the distal and proximal oxygen atoms of the ligand, was consistent with a reported ESR of this transient intermediate.[237] Addition of this second electron results in a species 37 kcal/mol higher in energy that the ferrous dioxygen form. However, this reduced species is a viable intermediate form, albeit highly reactive and therefore extremely transient. Electronic spectra calculated, using the semiempirical INDO/SCF/CI procedure at the DFT optimized geometries of the ferrous dioxygen and the reduced ferrous dioxygen intermediates resulted in spectra shifts completely analogous to the experimental spectra attributed to these putative species.[238] This agreement between the calculated spectra of these species and the observed spectra provided further direct that these species were the origin of the observed spectra. The main effect of second electron reduction is elongation and weakening of the Fe–O bond with no appreciable change in the O–O bond. Thus, second electron reduction alone is not sufficient to lead to facile dioxygen bond cleavage.

Distal protonation of the reduced ferrous dioxygen form does lead to significant O–O bond weakening. As shown in Figure 18, this weakening is manifest in both an increase in O–O bond length from 1.33 to 1.46 Å and a decrease in bond order from 0.87 to 0.48. This significant O–O bond weakening upon even single protonation of the distal oxygen, and the very large calculated proton affinity of −442.1 kcal/mol provide direct support for the proposed mechanism of proton assisted dioxygen bond cleavage to formation of Compound I. However, this singly protonated species is an

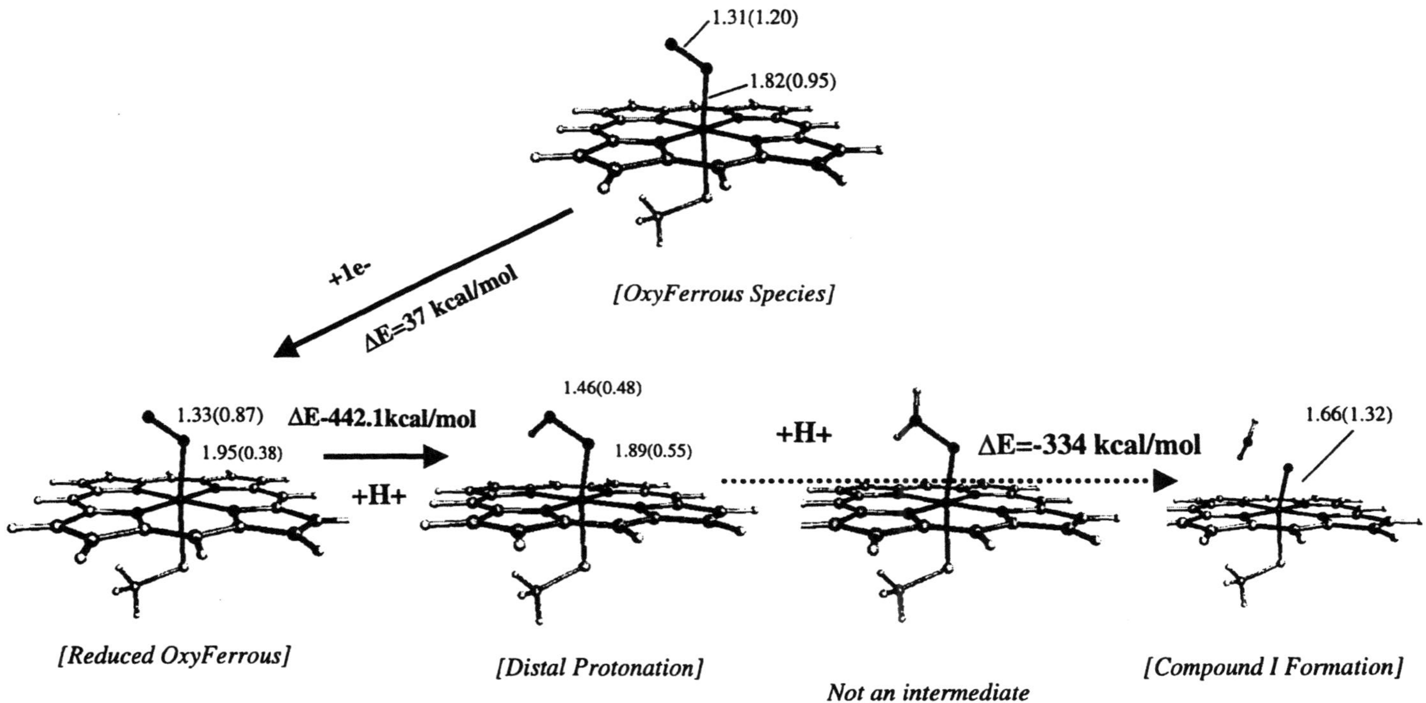


**Figure 18.** Optimized geometries, Fe–O and O–O bond orders, and relative energies of the putative transient heme species in the pathway to formation of the catalytically active Compound I. These results were obtained using *ab initio* nonlocal density functional theory methods with an extended basis set.

electronic bound state and as such does not spontaneously dissociate. Thus, neither the reduction step alone nor single protonation result in a species which spontaneously converts to the oxyferryl, Compound I species. However, addition of a second proton to the distal oxygen does not result in a stable species. No minimum was found corresponding to this twice protonated species. Rather the addition of a second proton led directly and without any apparent energy barrier to the formation of Compound I and water. As shown in Figure 18, this is also a very exothermic reaction ($\Delta E = -334$ kcal/mol) with the products, Compound I and water much more stable than the reactants, the singly protonated distal species and the second proton. These results, taken together, provide strong support for either consecutive or simultaneous double protonation of the distal oxygen atom of the reduced ferrous dioxygen P450 as the preferred pathway to Compound I formation.

## 6 ROLE OF AXIAL LIGANDS IN MODULATING ELECTROMAGNETIC PROPERTIES AND FUNCTION OF HEME SITES

### 6.1 Comparative Studies of the Resting Form of Catalases, Peroxidases and Metmyoglobin

Although these three different heme protein families share the same ferric-protoporphyrin-IX resting form, they have different axial ligands. The resting state of catalases has been shown[239,240] to be five coordinated with a tyrosine as its axial ligand thought to be deprotonated at physiological pH. Calculations using a spin restricted open shell INDO/ROHF methods were made of a model [Fe(III)–P–L] complex of the resting form with L = phenol, phenolate. The goal of these studies were to determine whether the axial ligand is best described as a phenolate or neutral phenol and also to determine the effect of porphyrin ring distortions observed in the intact protein on observed properties. Calculations were made of the relative energies of the Fe centered doublet, quartet, and sextet spin states. For the phenolate CAT model, [Fe(III)–P–Phenolate] the sextet state was calculated to be the ground state for both the X-ray structure of the macrocycle that included doming and ruffling and a regularized planar geometry. The porphyrin ring distortion resulted in a slight decrease in the energies of the quartet (2.8 vs. 2.1 kcal/mol) and doublet (6.3 vs. 5.8 kcal/mol) states. By contrast in the protonated form, [Fe(III)–P–Phenol], a quartet state was the ground state with an energy separation of 4.7 kcal/mol to the sextet and 12.2 kcal/mol to the doublet. The calculated quadrupole splitting of the sextet, $(\Delta E_Q) = 0.72$ mm/sec, but not the quartet was in excellent agreement with the experimental value of 0.84 mm/sec.[241] The agreement between experimental and measured $(\Delta E_Q)$ and the measured $\mu_{eff}$ of 5.92 BM[242] confirm the reliability of the calculated results and the electronic structure provided by the calculations. These results provide further direct evidence that in the resting form of catalase the anionic form of the axial ligand prevails and that it is in a high spin ground state.

The heme unit of the resting form of peroxidases and metmyoglobin (MetMb) two functionally distinct heme proteins are identical consisting of a ferric protoporphyrin-IX complex with a water and imidazole axial ligand. However, the current hypothesis is that differences in function can be at least partly attributed to the fact that in peroxidase there is anionic character of the imidazole ligand conferred by proton transfer from the proximal histidine to a nearby conserved aspartate residue absent in myoglobin. In the computation studies described above, the imidazolate model of the peroxidase was indeed found to be most consistent with experimental results.

The combined computational and experimental results thus provide direct evidence that the resting forms of catalase, MetMb and peroxidases differ in the nature of their proximal axial ligand, a phenolate, imidazole, and imidazolate respectively. The question then arises is whether these differences in axial ligand are reflected in significant differences in electronic structure or spin distributions that could be associated with differences in their function. Table 13 summarizes these properties for the three model ferric heme sites in the resting forms of these heme proteins.

As shown in Table 13, each of the three heme sites MetMb [Fe(III)–P–$H_2O$–Im], Peroxidase [Fe(III)–P–$H_2O$–$Im^-$], and Catalase [Fe(III)–P–Phenolate] has a sextet ground state with low lying quartet states. The net atomic charges and unpaired spin density distribution calculated for the sextet ground states of these heme sites[153] are very similar. They all demonstrate the same appreciable spin delocalization with 4.24–4.33 of the unpaired spin remaining on the Fe and 0.51–0.56 on the porphyrin macrocycle. The calculated unpaired spin density on the water ligands is very small (0.02 and 0.06 in peroxidase and MetMb, respectively). Unpaired spin on the proxi-

**Table 13 Relative Spin State Energies, Ground State Electron Structures, and Spin Distribution in Ferric Porphine Models of CCP, MetMb, and CAT**

| Property | CCP ($Im^-$) | MetMb(Im) | CAT($O\phi^-$) |
|---|---|---|---|
| Relative Spin (kcal/mol) | | State Energy | |
| $\Delta E$ (5/2 – 3/2) | 2.4 | 2.7 | 2.1 |
| $\Delta E$ (5/2 – 1/2) | 9.5 | 7.5 | 5.8 |
| | | Populations | |
| $d_{z^2}$ | 1.25[a](0.74)[b] | 1.21(0.78) | 1.26(0.69) |
| $d_{x^2-y^2}$ | 1.31(0.67) | 1.34(0.65) | 1.28(0.70) |
| $d_{xy}$ | 1.02(0.98) | 0.01(0.99) | 1.01(0.98) |
| $d_{xz}$ | 1.05(0.94) | 1.04(0.96) | 1.00(0.89) |
| $d_{yz}$ | 1.07(0.93) | 1.04(0.95) | 1.08(0.91) |
| | | Charges | |
| Fe | 1.35 | 1.37 | 1.37 |
| $L_1$[c] | 0.10 | 0.13 | – |
| $L_2$[d] | −0.63 | 0.26 | −0.55 |
| Porphyrin | −0.83 | −0.76 | −0.82 |
| | | Spin distributions | |
| Fe | 4.26 | 4.33 | 4.24 |
| $L_1$[c] | 0.02 | 0.06 | – |
| $L_2$[d] | 0.16 | 0.06 | 0.25 |
| Porphyrin | 0.56 | 0.55 | 0.51 |

[a] electron population. [b] unpaired spin population. [c] $L_1 = H_2O$. [d] $L_2$ = imidazolate for CCP, imidazole for MetMb and phenolate for CAT.

mal ligand is somewhat less on the neutral Imidazole (0.06) in the MetMb heme site than on the anionic proximal ligands being (0.16) on the imidazolate ligand in the peroxidase and (0.25) on the phenolate ligand. There is also very similar net charges on the Fe(III) of (+1.35 – +1.37) and on the porphyrin ring of −0.82–0.76 despite that presence of anionic axial ligands in two of these model systems and a neutral ligand in one of them. It is mainly the net charge on the proximal ligands themselves that reflect this difference.

The intrinsic properties of the heme site resting forms of these three heme proteins thus cannot account for their differences in function. The imidazolate form of the peroxidases and the imidazole form of metMb have very similar properties. Thus, for these proteins the amino acid environment around the heme unit, rather than the heme itself appears to be a predominant factor in controlling function. This finding lends support to the key role proposed[239] for polar residues in the ligand binding site of peroxidases, absent in MetMb, in peroxidase function. The full protein studies described below further support this hypothesis.

Except for differences in axial ligands, the heme sites of peroxidases and catalases also have very similar electronic structure and spin distribution. Thus, the differences in the ability of Compound I to perform sequential one electron oxidations in one case and simultaneous two electron oxidations in the other cannot be discerned in the resting form properties of their heme sites.

## 6.2 Comparative Studies of the Effect of Axial Ligands on Model Compound I Species

While different axial ligands do not appear to cause major differences in the resting form of different heme proteins, it is possible that the properties of the putative enzymatically active Compound I species are affected by these axial ligand differences and in such a manner as to affect function. To address this question,[243] using the INDO/ROHF method, the effect of various axial ligands on the electronic structure and relative energies of the $a_{1u}$ and $a_{2u}$ $\pi$ cation radical quartet states of Compound I species was calculated for seven model Compound I iron-porphyrin with varying axial ligands shown in Figure 19. The varying axial ligands included in thus study were models for those in intact enzymes: imidazole and imidazolate (models for peroxidases), an acetylamide ligand (model for a Gln 175 mutant of cytochrome *c* peroxidase), phenolate (model for catalase), methyl mercaptide (model for P450s), and two that have been used in biomimetics of these enzymes chloride (hemin) and thiophenylate (model P450s). In these calculations, a regularized unsubstituted porphine taken from the crystal structure of TPP[105] was used. The energies of the two $\pi$ cation radical states were calculated as the energy difference between the parent compound and the final radical species. As shown in Table 14, for all but the two P450 models the $a_{1u}$-porphyrin $\pi$ cation radical Compound I quartet species was the ground state with energy separation from the higher $a_{2u}$ $\pi$ cation radical quartet ranging from 3 to 8 kcal/mol. The results of these semiempirical calculations are consistent with the results obtained from CAS SCF studies of models for peroxidases. Although a

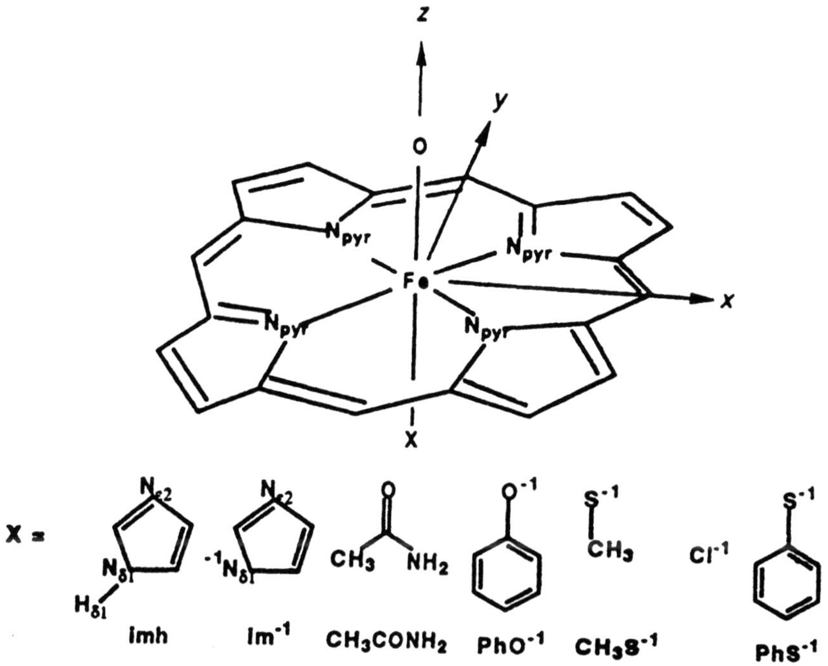


**Figure 19.** Heme model used for the comparative study of the properties of Compound I with varying axial ligands. The six axial ligands (X) included in this study are also shown.

pyridine axial ligand was used in those previous studies, it would have a very similar effect as the imidazole used here.

It is possible that the finding of a preferred $a_{2u}$ radical cation only in the P450 Compund I heme site has functional significance. For example, it is possible that the symmetry allowed overlap between the Fe=O and $^3a_{2u}$ orbitals may lower the activation energy for the Fe=O bond cleavage typical of P450-I enzymatic reactions. However, as shown in Table 15 the calculated electronic and spin properties includ-

**Table 14 Relative Energies of $3a_{2u}$ and $1a_{1u}$ Radical Cations in Model Compound I Heme Species with Varying Axial Ligand**

| Model | Ligand | L[a] | $R$(Fe–L)[b] | $\Delta E$ (kcal/mol) |
|---|---|---|---|---|
| Gln175•CCP | $CH_3CONH_2$ | O | 2.25 | 8.38 |
| HRP/CCP | imh | N | 2.02 | 7.02 |
| HRP/CCP | $im^{-1}$ | N | 2.02 | 4.64 |
| CAT | $PhO^{-1}$ | O | 1.89 | 4.50 |
| hemin | $Cl^{-1}$ | Cl | 2.20 | 2.79 |
| P450 | $CH_3S^{-1}$ | S | 2.41 | −0.51 |
| P450 | $PhS^{-1}$ | S | 2.41 | −1.26 |

[a] Ligand atom bound directly to Fe. [b] Fe–L distance in Å. [c] $\Delta E = E(3a_{2u}^{+}) - E(1a_{1u}^{+})$. A positive value represents an $a_{1u}$ ground state.

**Table 15 Properties of $3a_{2u}$ and $1a_{1u}$ Cations of Model Compound I Complexes**

| Ligand | $Cl^{-1}$ | | imh | | $im^{-1}$ | | $PhO^{-1}$ | | $CH_3S^{-1}$ | | $PhS^{-1}$ | | $CH_3CONH_2$ | |
|---|---|---|---|---|---|---|---|---|---|---|---|---|---|---|
| Rad Cat | $1a_{1u}$ | $3a_{2u}$ | $1a_{1u}$ | $3a_{2u}$ | $1a_{1u}$ | $3a_{2u}$ | $1a_{1u}$ | $3a_{2u}$ | $1a_{1u}$ | $3a_{2u}$ | $1a_{1u}$ | $3a_{2u}$ | $1a_{1u}$ | $3a_{2u}$ |
| | Net Charge | | | | | | | | | | | | | |
| Fe | 1.75 | 1.75 | 1.49 | 1.49 | 1.49 | 1.48 | 1.54 | 1.53 | 1.54 | 1.55 | 1.56 | 1.56 | 1.43 | 1.43 |
| O | −0.83 | −0.82 | −0.74 | −0.73 | −0.80 | −0.79 | −0.84 | −0.82 | −0.82 | −0.79 | −0.81 | −0.79 | −0.73 | −0.73 |
| Poryrin | 0.09 | 0.01 | 0.10 | 0.04 | 0.06 | −0.01 | 0.00 | −0.05 | −0.03 | −0.12 | 0.03 | 0.01 | 0.18 | 0.16 |
| Ligand | −1.01 | −0.94 | 0.15 | 0.28 | −0.75 | −0.68 | −0.70 | −0.66 | −0.75 | −0.64 | −0.78 | −0.76 | 0.12 | 0.14 |
| Total | 0.00 | 0.00 | 1.00 | 1.00 | 0.00 | 0.00 | 0.00 | 0.00 | 0.00 | 0.00 | 0.00 | 0.00 | 1.00 | 1.00 |
| | Spin Population (Total = 3) | | | | | | | | | | | | | |
| Fe | 1.50 | 1.52 | 1.36 | 1.36 | 1.41 | 1.40 | 1.44 | 1.43 | 1.45 | 1.47 | 1.47 | 1.47 | 1.43 | 1.45 |
| O | 0.46 | 0.44 | 0.59 | 0.60 | 0.55 | 0.55 | 0.50 | 0.51 | 0.51 | 0.50 | 0.49 | 0.49 | 0.53 | 0.51 |
| Porphyrin | 1.03 | 0.93 | 1.04 | 1.02 | 1.02 | 1.01 | 1.03 | 1.00 | 1.03 | 0.88 | 1.03 | 1.03 | 1.04 | 1.03 |
| Ligand | 0.01 | 0.11 | 0.01 | 0.02 | 0.02 | 0.04 | 0.03 | 0.06 | 0.01 | 0.15 | 0.01 | 0.01 | 0.00 | 0.01 |
| | Total Fe–O $\pi$ Bond Order | | | | | | | | | | | | | |
| Fe O | 1.956 | 1.950 | 2.095 | 2.112 | 2.054 | 2.064 | 1.996 | 2.006 | 1.965 | 2.008 | 2.001 | 1.998 | 2.136 | 2.119 |
| | Quadruple Splittings | | | | | | | | | | | | | |
| $\Delta E_Q$ (mm.sec) | 1.36 | 1.39 | 1.45 | 1.47 | 1.34 | 1.35 | 1.41 | 1.43 | 1.21 | 1.27 | 1.28 | 1.34 | 1.48 | 1.50 |
| $\eta$ | 0.00 | 0.00 | 0.00 | 0.00 | 0.00 | 0.00 | 0.00 | 0.00 | 0.00 | 0.00 | 0.00 | 0.00 | 0.00 | 0.00 |

ing the crucial Fe=O bond overlap density as well as unpaired spin and net charges on the oxygen, important determinants of the reactivity of Compound I are very similar for all model heme sites and in both cation states. The unpaired spin on the active oxygen atom is about 0.5 electrons for all model heme sites residing equally in the $2px$ and $2py$ orbitals of the oxygen. This spin distribution imparts a quasi triplet character to this oxygen atom consistent with the radical mechanism of its oxidation of substrates. Most significantly, the calculated Fe–O bond order is very similar in all these model heme sites whether they are in an $a_{1u}$ or $a_{2u}$ $\pi$ cation radical form. The net charge on the oxygen atom and the iron atom is also very similar in all forms of Compound I. Thus, there is no significant "trans" effect of these different axial ligands on the intrinsic properties of the heme site of Compound I. The similarity found in the properties of this active intermediate for all the metabolizing heme proteins cannot account for their different modes of oxidation: in the cytochrome P450s involving transfer of this active oxygen and in the peroxidases and catalases involving retention of it. Other factors, such as different protein environments in the catalytic centers, may play a more direct role in determining function.

## 7 FULL PROTEIN STUDIES

While the heme unit is the primary center of both the spectroscopic properties and function of intact heme proteins, there is evidence for a key role of the protein in modulating both of these types of properties. With the advent of X-ray structures from all of the major families of heme proteins on the one hand and the advanced in computational capabilities that allow the simulation of these proteins, it is now possible to probe the role of the protein in determining structure, spectra, and function of the heme site. Two examples of such modulation have already been described above. The anomalous low spin ground state of the resting state of the cytochrome P450s could only be explained by including the electrostatic field of the entire protein.[216] In addition, the mode of binding of CO to ferrous hemoglobins and myoglobins and its stretching frequency have also been shown to be modulated by both steric and electrostatic properties of the protein environment.[130] Two other types of full protein studies—one related to the function of the globins and the other to the function of the metabolizing heme proteins, peroxidases, and cytochrome P450s—are described below.

### 7.1 Small Ligand Binding to Globins

One of the first and most long standing uses of molecular dynamic simulations of full proteins has been to help characterize the kinetic and dynamic properties of small molecule binding to ferrous heme proteins.[244–246] These studies are closely related to the natural function of the globins, the reversible transport of oxygen and interference in this process by competition from other small ligands. Since X-ray structures of both ligand bound and ligand free forms of the globins do not reveal any channel, by which ligands can move between the heme binding site and the outside of the protein, protein motion must be involved in this path. The experimental methods used have focused on the photodissociation and rebinding of ligands in myoglobin (Mb), hemoglobin (Hb), and the computational studies on dynamic behavior. The pioneering studies on the rebinding kinetics of photodissociated CO from Mb has provided a detailed description of protein dynamics in the picosecond to millisecond time interval.[89,247,248] Additionally, computer simulations[244–246] have shown that a complete description of protein function must include the inherent flexibility of fluctuations of the protein. Continued computer simulations of binding of CO to the room temperature, fully solvated protein can help resolve features that cannot be resolved unambiguously using current experimental methods.

In one such study,[249] a reaction coordinate approach was used to characterize the diffusing of CO through leghemoglobin. Analysis of the results for three different candidate diffusion paths led to similar behavior in the close neighborhood of the ligand. The diffusion process was found to consist of two steps. In the first, the ligand jumps from the heme pocket to another cavity in the protein matrix. In the second, it hops to the protein exterior. The first barrier is dominated by a tilt of a single residue (Phe 29B9). The second barrier is more complex and includes many types of motions. In particular, global translations and rotations of helices C and G are involved.

In a very recent study,[250] classical molecular dynamic simulation was employed to study the photodissociated states of COMb for the fully solvated protein using periodic boundary conditions at 300 K. The simulated structure of the CO-heme complex shows that the ligand is localized under the C-ring of the heme. This result is consistent with the low temperature X-ray crystal structures of the photolyzed state.[251,252] The ligand is found to be parallel to the heme plane and confined to a small region displaced from the heme iron, in apparent agreement with the low temperature X-ray crystal structures and room temperature experimental studies.[128] As in a previous study,[130] Stark shifts in the ligand vibrational frequency induced by the protein electric field are calculated and compare well with experimental results. The effect on the ligand affinity of the neighboring amino acids in the distal binding pocket is assessed and both an electrostatic and steric component deduced from these results.

Other studies have determined that the barrier to ligand rebinding in the globins is a combination of electronic[253–255] and steric factors.[244,256,257] The rebinding of NO is very rapid because the barrier is greatly diminishes compared to other ligands such as $O_2$ and CO. Thus, protein relaxation to equilibrium after photodissociation can occur on the same time scale as the rebinding resulting in the observed nonexponential behavior of NO rebinding at room temperature. Experimental and computational studies have been reported[258] to further analyze the origin of this nonexponential behavior. The structural basis of non exponential geminal rebinding of NO to Mb bind is attributed to a time dependent or independent distribution of protein conformations. The proximal heme environment appears to be more important than the distal pocket region for geminate recombination. In fact, in the presence of an intact Fe–N $\varepsilon$ bond, rebinding of NO to Hb is slower. By contrast, the distal region plays the dominant role in escape of ligands from the protein.[245]

## 7.2 Formation of the Active Compound I Form of Peroxidases

A characteristic common step in the enzymatic cycle of peroxidases is the formation of the catalytically active Compound I from the inactive resting form. As shown in Figure 4, peroxide is required for this transformation and it is thought to proceed via a transient reversible peroxide-heme complex peroxide required for this transformation.[184–186,188,259] The role of the protein appears crucial in this process and is one of the least understood aspects of the enzymatic cycle of peroxidases. In an initial attempt to address the mechanism of Compound I formation, empirical energy minimizations and molecular dynamic simulations have been performed[260] using the Amber suite of programs[218] for the resting state and peroxide complex of cytochrome *c* peroxidase, the first peroxidase with a known X-ray structure.[168] The goals of this full protein study were to determine the energy optimized and dynamic modes of binding of the peroxide to the heme iron and to characterize key interactions with nearby residues and bound waters on the distal and proximal sides of the heme unit implicated in the formation of Compound I. Chief among these were His 52 and Arg 48 in the distal pocket. The results of these study indicate that solvated, full protein dynamics is required to obtain reliable results for the known resting state CCP form and hence for the uncharacterized peroxide CCP complex. In this complex, the peroxide binds to the heme iron in a dynamically averaged end-on fashion rather than a bridged structure with approximately equal probability of each oxygen serving as the ligand to the iron. Binding of the peroxide as a ligand disrupts the H-bonded network of waters in the distal binding pocket, which are present in the resting state, but there is no dramatic perturbation of the nearby amino acid residues. The dynamic behavior found provided direct support for a proposed mechanism of Compound I formation and the postulated role of the nearby His and Arg residues. Specifically, the peroxide hydrogens were found to form temporally stable H-bonds with the $N_\varepsilon$ of the distal histidine. As the O atoms of the peroxide alternate as ligands to the iron, the H atoms alternate as partners with the imidazole nitrogen of this proximal histidine. Thus, the imidazole of this His residue can function as an acid-base catalyst by abstracting a proton from the proximal oxygen and transferring it to the distal peroxide oxygen, facilitating O–O bond cleavage leading to formation of Compound I and water. The results also confirmed a role for the nearby Arg in electrostatic stabilization of this leaving group since the distal peroxide oxygen remains near the cationic Arg–$N_\varepsilon$–H during the simulation. As discussed above, subsequent *ab initio* studies of the optimized structure of a model peroxide–peroxidase heme complex[204,261] confirmed the bent end on binding mode and equal preference for either peroxide oxygen as a ligand.

## 7.3 Formation of the Active Compound I Form of Cytochrome P450s

The steps in the complex enzymatic cycle of the cytochrome P450s are shown in Figure 5. The most enigmatic part of this cycle common to all P450s is the mechanism of formation of the putative Compound I species after the second electron reduction of the substrate bound ferrous dioxygen species. This second reduction results in a very unusual heme-dioxygen heme species twice reduced from the ferric resting form. This species, like the peroxide intermediate of peroxidases, leads directly to the formation of Compound I. The mechanism of formation of Compound I in the cytochrome P450s is experimentally challenging because the reaction proceeds rapidly from the second electron reduction to Compound I and then product formation. The most plausible mechanism is the addition of two protons to the distal oxygen of the twice reduced ferrous dioxygen species leading to facile proton assisted cleavage of the O–O bond and formation of the Fe=O, Compound I species, and water. This mechanism is consistent with current experimental evidence, including isotope substitution kinetic data[262] and an observed solvent kinetic isotope effect and has been directly validated from a very recent computational study using nonlocal DFT described above and summarized in Figure 18. The questions remaining, however, are how the protons are delivered to this distal oxygen in the binding site of the intact protein and the ultimate source of these protons. In recent work,[263] molecular dynamic simulations have been performed using the recently determined X-ray structure of substrate bound P450eryF[204] to address these two aspects of the mechanism of formation of Compound I from the reduced ferrous dioxygen species. In these studies, the plausibility of the proton assisted mechanism of dioxygen bond cleavage of the twice reduced dioxygen species in the full protein was investigated as well as the possible source and mechanism of transfer of these protons. The analysis of the molecular dynamics trajectories of this species does indeed provide further evidence for this pathway and points to a source of protons. Specifically, two dynamically stable hydrogen bonds to the distal oxygen atom of the dioxygen ligand, one by the substrate and the other by a bound water, were found, consistent with the proposed proton-assisted cleavage of the O–O bond and formation of water. In addition, an extensive, dynamical stable hydrogen bond network is formed involving specific protein residues and bound waters that connects the distal oxygen to Glu 360 in P450eryF. While a distinct, highly conserved Asp(251) residue in P450cam appears to be the predominant nearby source of protons in the P450cam isozyme, this Glu is also a well conserved residue in a channel accessible to solvent that could be the ultimate source of protons in some P450s. Simulations were done for both a protonated and unprotonated Glu and led to a proposed mechanism of proton transfer by it to the distal oxygen atom. In order to validate the procedures for the simulation of this transient twice-reduced species, the same procedures were used to perform molecular dynamic simulations of two other forms of P450eryF, the ferric and ferryl substrate bound species and the results compared with experiment. The results of these comparisons lend credibility to the important mechanistic inferences from the simulations of the transient twice reduced dioxygen species. The combined results of these simulations provide further evidence for a proton assisted pathway from the twice reduced dioxygen form of P450s to the catalytically active ferryl species and for a possible source of the protons. A crucial role of the protein environment in this pathway has been clearly implicated.

## 8 CONCLUSIONS

While the active site of most protein families is composed of a specific stereochemical arrangement of a few amino acids that are an integral part of the polypeptide chain, heme proteins belong to an unusual group of proteins in which the active site is a metal-organic complex, an iron-porphyrin "heme" unit that is a nearly independent entity with totally different properties than the rest of the protein. The electronic structure of this heme unit is well suited to perform the three major functions of heme proteins and to facilitate the requisite changes that occur in this species during function. The heme unit also has unique spectroscopic properties that are sensitive indicators of the nature of the heme unit and changes in this species during function. Thus, while it is difficult to make small compounds that mimic the activity and spectroscopic properties of intact proteins in general, this is not the case for heme proteins and an enormous experimental effort has centered on the synthesis and characterization of model iron-porphyrin compounds and the extent to which they mimic the spectroscopic properties and function of intact heme proteins. In general, spectroscopic properties of the isolated heme complexes mimic those of the same heme species in the intact protein. It is much more difficult to mimic function with comparable efficiency. The spectroscopic studies have identified the unique features of the heme unit discussed here such as signature electronic spectra, anisotropic $g$ values in ESR spectra, large zero field splittings, and quadrupole splittings in Mössbauer response spectra, temperature dependent magnetic moments and deviations in their spin-only values.

Complementary computational studies focusing on the heme unit itself have elucidated the molecular origin of these observed properties. The fundamental origin of several of these properties were shown to be the presence of low lying states of different spin multiplicities that in most inorganic iron containing systems have large energy separations. For example, spin orbit coupling among these states was shown to lead directly to the large zero field splittings, anisotropic $g$ values and deviations from spin-only values of magnetic moments. Electron delocalization caused by covalent bonding of the heme iron to the porphyrin macrocycle and axial ligands were shown to modulate the perturbations of the electron distribution around the iron compared to their free ion values. The results obtained could explain the enigmatic variations of quadrupole splitting in Mössbauer response spectra with spin state and axial ligands. In addition, these calculations identified a vital role of the porphyrin macrocycle in facilitating extensive and facile electron and spin delocalization favoring the unusual heme species, such as the ferryl form, involved in function. Electronic spectra calculated from known structures identified the origin of the two common features of all heme spectra, an intense band ($B$) in the near uv and a weak band ($Q$) in the visible; as well as of the extra features such as additional $xy$ polarized and $z$ polarized bands in single crystal spectra of heme sites and of the split Soret characteristic of the cytochrome P450s state. In addition, electronic spectra calculations of putative transient species in the enzymatic cycle of the metabolizing heme proteins provided the key link between observed spectra of these transient species and their molecular origin. Most recently, *ab initio* density functional calculations have allowed characterization of transient species and assessment of long-proposed mechanisms of activation of heme proteins to their enzymatically active form, providing crucial support for these species and pathways that could not be established by experiment alone. While all levels of theory over the past 30 years have made contributions to elucidation of structure, spectra, and function of heme species, advances in computational chemical methods and increasingly efficient software and hardware have allowed more and more reliable methods and realistic models to be used. It is expected that these improvements will continue and will further expand the scope, depth, and accuracy of computational studies of the structure, spectra, and function of heme sites.

In addition to continued studies of heme species themselves, the advent of known X-ray structures of heme proteins and the complementary development of computational methods that allow the characterization of full proteins have led to the use of these computational methods to probe the role of the protein environment in heme protein function. Although function centers on the heme unit, characterization of the heme unit associated with each type of heme function is not sufficient to explain differences in function. For example, the globins involved in reversible oxygen transport and the metabolizing heme proteins peroxidases share a common prosthetic group and a common proximal axial histidine ligand. Calculations already reported indicate that it is the protein environment both around this proximal ligand and in the distal substrate binding site that modulates the differences in function of these two heme families. Results of full protein simulations have also shown that the pathway from inactive to enzymatically active species is greatly facilitated by key residues in the protein environment. The effect of mutation on function can only be addressed by such studies. A key advance still in progress is the combination of quantum mechanical and empirical energy methods that will allow simultaneous and enhanced insights into the role of both the heme site and the protein in function.

## REFERENCES


1. Lamoen, D.; Parrinello, M. *Chem Phys. Lett.*, **1996**, *248*, 309–315.
2. Matsuzawa, N.; Masafumi, A.; Dixon, D. *J. Phys. Chem.*, **1995**, *99*, 7698–7706.
3. Hoard, J. L. In *Porphyrins and Metalloporphyrins*: Smith, K. M., Ed.; Elsevier Scientific Publishing Company: Amsterdam, **1975**; Chapter 8.
4. Scheidt, W. R.; Lee, Y. J. *Struct. Bonding*, **1987**, *64*, 1.
5. Munro, O. Q.; Bradley, J.; Hancock, R. D.; Marques, H. M.; Marsicano, F.; Wade, P. W. *J. Am. Chem. Soc.*, **1992**, *114*, 7218–7230.
6. Sparks, L. D.; Medforth, C. J.; Park, M. -S.; Chamberlain, J. R.; Ondrias, M. R.; Senge, M. O.; Smith, K. M.; Shelnutt, J. A. *J. Am. Chem. Soc.*, **1993**, *115*, 581–592.
7. Anderson, K. K.; Hobbs, J. D.; Luo, L.; Stanley, K. D.; Quirke, J. M. E.; Shelnutt, J. A. *J. Am. Chem. Soc.*, **1993**, *115*, 12436–12352.
8. Hobbs, J. D.; Majumder, S. A.; Hobbs, J. D.; Sickelsmith, G. A.; Quirke, J. M. E.; Medforht, C. J.; Smith, K. M.; Shelnutt, J. A. *J. Am. Chem. Soc.*, **1994**, *116*, 3261–3270.

9. Munro, O. Q.; Marques, H. M.; Debrunner, P. G.; Mohanrao, K.; Scheidt, R. W. *J. Am. Chem. Soc.*, **1995**, *117*, 935–954.
10. Jentzen, W.; Simpson, M. C.; Hobbs, J. D.; Song, X.; Ema, T.; Nelson, N. Y.; Medforth, C. J.; Smith, K. M.; Veyrat, M.; Mazzanti, M.; Ramasseul, R.; Marchon, J.-C.; Takeuchi, T.; Gaddard, W. A., III; Shelnutt, J. A. *J. Am. Chem. Soc.*, **1995**, *117*, 11085–11097.
11. Song, X. -Z.; Jentzen, W.; Jia, S. -L.; Jaquinod, L.; Nurco, D. J.; Medforth, C. J.; Smith, K. M.; Shelnutt, J. A. *J. Am. Chem. Soc.*, **1996**, *187*, 12975–12988.
12. DiMagno, S. G.; Wertsching, A. K.; Ross, C. R. *J. Am. Chem. Soc.*, **1995**, *117*, 8279–8280.
13. LeCours, S. M.; DiMagno, S. G.; Therien, M. J. *J. Am. Chem. Soc.*, **1996**, *118*, 11854–11864.
14. Momot, K. I.; Walker, F. A. *J. Phys. Chem.*, **1997**, *101*, 2787–2795.
15. Reynolds, C. H. *J. Org. Chem.*, **1988**, *53*, 6061–6064.
16. Yamamoto, U.; Noro, T.; Ohno, K. *Int. J. Quan. Chem.*, **1992**, *42*, 1563–1575.
17. Ghosh, A. *J. Phys. Chem.*, **1997**, *101*, 3290–3297.
18. Clezy, P. S.; Fookes, J. R.; Sternhell, S. *Aust. J. Chem.*, **1978**, *31*, 639.
19. Silvers, S. J.; Tulinsky, A. *J. Am. Chem. Soc.*, **1967**, *89*, 3331.
20. Crossley, M. J.; Harding, M. M.; Sternhell, S. *J. Org. Chem.*, **1988**, *53*, 1132.
21. Harris-Loew, G. *J. Chem. Phys.*, **1968**, *48*, 2191.
22. Grifith, J. S. *The Theory of Transition Metal Ions*. Cambridge U.P.: Cambridge, **1971**.
23. Harris, G. M. *Theor. Chim. Acta*, **1968**, *10*, 155–180.
24. Harris-Loew, G. *Theor. Chim. Acta*, **1966**, *5*, 379.
25. Loew, G. H. *Biophys. J.*, **1970**, *10(2)*, 196–212.
26. Kotani, M. *Advances in Chem. Phys.*, **1964**, *7*, 159.
27. Griffith, J.; Kotani, M. In *Quantum Aspects of Polypeptides and Polynucleotides*, Weissbluth, M., Ed.; Interscience: New York, **1964**, 35–45, 67, 78.
28. Harris-Loew, G. *Theor. Chim. Acta*, **1968**, *10*, 155.
29. Ridley; Zerner, M. C. *Theor. Chim. Acta*, **1976**, *42*, 223. Bacon, A.; Zerner, M. C. *Theor. Chim. Acta*, **1979**, *53*, 21.
30. Edwards, W. D.; Zerner, M. C. *Theor. Chim. Acta*, **1987**, *72*, 347.
31. Lamoen, D.; Parrinello, M. *Chem. Phys. Lett.*, **1996**, *248*, 309–315.
32. Zwaans, R.; van Lenthe, H. H.; den Boer, D. H. W. *J. Mol. Str. (Theochem)*, **1996**, *367*, 15–24.
33. Kashiwagi, H.; Obara, S. *Int. J. Quantum Chem.*, **1981**, *20*, 843.
34. Rohmer, M. -M. *Chem. Phys. Lett.*, **1985**, *116(1)*, 44–49.
35. Obara, S.; Kashiwagi, H. *J. Chem. Phys.*, **1982**, *77*, 3155.
36. Dediew, A.; Rohmer, M. -M.; Veillard, A. *Quantum Chem.*, **1982**, *16*, 43; Rohmer, M. -M.; Dedieu, A.; Veillard, A. *Chem. Phys.*, **1983**, *77*, 449.
37. Kobayahi, H.; Maeda, Y.; Yanagawa, A. *Bull. Chem. Soc. Jpn.*, **1970**, *43*, 2342.
38. Lang, G.; Spartalian, K.; Reed, C.A.; Collman, J. P. *J. Chem. Phys.*, **1978**, *69*, 5424.
39. Hambright, P.; Bearden, A. J. In *Porphyrins and Metalloporphyrins*, Smith, K. M., Ed.; Elsevier Scientific: Amsterdam, **1975**, 539.
40. Goff, H.; La Mar, G. N.; Reed, C. A. *J. Am. Chem. Soc.*, **1977**, *99*, 3641.

41. Benecky, M. J.; Frew, J. E.; Scowen, N.; Jones, P.; Hoffman, B. M. *Biochemistry*, **1993**, *32*, 11929.

42. Axe, F.; Flowers, C.; Loew, G. H.; Waleh, A. *J. Am. Chem. Soc.*, **1989**, *111(19)*, 7333.

43. Herman, Z. S., Loew, G. H. *J. Am. Chem. Soc.*, **1980**, *102*, 1815.

44. Herman, Z. S., Loew, G. H., Rohmer, M. M. *Int. J. Quant. Chem. QBS*, **1980**, *7*, 137.

45. Dolphin, D.; Muljiani, Z.; Rousseau, K.; Borg, D.; Fajer, C.; Felton, J. *Ann. N. Y. Acad. Sci.*, **1973**, *206*, 177–198.

46. Gouterman, M. Optical Spectra and Electronic Structure of Porphyrins and Related Rings; In *The Porphyrins*; Dolphin, D., Ed.; Academic Press: New York, 1978; Vol. III.

47. Morishima, I; Takamuki, Y.; Shoro, Y. *J. Am. Chem. Soc.*, **1984**, *106*, 7666–7672.

48. Godziela, G. M.; Goff, H. M. *J. Am. Chem. Soc.*, **1986**, *108*, 2237–2243.

49. La Mar, G. N.; de Ropp, J. S.; Latos-Grazynski, L.; Balch, A. L.; Johnson, R. B.; Smith, K. M.; Parish, D. W.; Cheng, R. *J. Am. Chem. Soc.*, **1983**, *105*, 782–787.

50. Beinert, H. *Biochem. Soc. Trans.*, **1985**, *13*, 542–547. Palmer, G. *Biochem. Soc. Trans.*, **1985**, *13*, 548–560.

51. Rutter, R.; Valentime, M.; Hendrich, M. P.; Hager, L. P.; Debrunner, P. G. *Biochemistry*, **1983**, *22*, 4769–4774.

52. Fajer, J.; Davis, M. S. In *The Porphyrins*; Dolphin, D., Ed.; Academic Press: New York, 1978; Vol. III.

53. Browett, W. R.; Stillman, M. J. *Inorg. Chim. Acta*, **1981**, *49*, 69–77.

54. Kau, K.; Svastits, E. W.; Dawson, J. H.; Hodgson, K. O. *Inorg. Chem.*, **1986**, *25*, 4307–4309.

55. Chance, B.; Powers, L.; Ching, Y.; Poulos, T.; Schonbaum, G.R.; Yamazaki, I.; Paul, K. G. *Arch. Biochem. Biophys.*, **1984**, *235*, 596–611.

56. Roberts, J. E.; Hoffman, B. M.; Rutter, R.; Hager, L. P. *J. Am. Chem. Soc.*, **1981**, *103*, 7654–7656.

57. Roberts, J. E.; Hoffman, B. M.; Rutter, R.; Hager, L. P. *J. Biol. Chem.*, **1981**, *256*, 2118–2121.

58. Sandusky, P. O.; Salehi, A.; Chang, C.K.; Babcock, G. T. *J. Am. Chem. Soc.*, **1989**, *111*, 6437–6439.

59. English, D. R.; Hendrickson, D. N.; Suslick, K. S. *Inorg. Chem.*, **1983**, *22*, 368–370.

60. Monck, E. Mössbauer Spectra of Hemes; In *The Porphyrins*; Dolphin, D., Ed.; Academic Press: New York, 1978; Vol. IV.

61. Mizutani, Y.; Hashimoto, S.; Tatsuno, Y.; Kitagawa, T. *J. Am. Chem. Soc.*, **1990**, *112*, 6809–6814.

62. Oertling, W. A.; Salehi, A.; Chang, C.K.; Babcock, G. T. *J. Phys. Chem.*, **1989**, *93*, 1311–1319.

63. Czernuszewicz, R. S.; Macor, K. A.; Li, X.-Y.; Kincaid, J. R.; Spiro, T. G. *J. Am. Chem. Soc.*, **1989**, *111*, 3860–3869.

64. Bangcharoenpaurpong, O.; Champion, P. M.; Martinis, S.; Sligar, S. *J. Chem. Phys.*, **1987**, *87*, 4273.

65. Ghosh, A.; Almlof, J. *Chem. Phys. Lett.*, **1993**, *213*, 519.

66. (a) For extensive LDF SCF calculations of valence IPs of porphyrins, see: Ghosh, A. *J. Am. Chem. Soc.*, **1995**, *117*, 4691. (b) For nonlocal DFT calculations of porphyrin valence IPs, see: Ghosh, A.; Vangberg, T. *Theor. Chim. Acta* (Jan Almlof memorial issue), in press.

67. UPS of $TPPH_2$: Khandelwal, C.; Roebber, J. L. *Chem Phys. Lett.*, **1975**, *34*, 355.

68. For LDF simulations of the lowest UPS Ips of free base phthalocyanine, see: Ghosh, A.; Gassman, P. G.; Almlof, J. *J. Am. Chem. Soc.*, **1994**, *116*, 1932.

69. (a) UPS of $PEt_8H_2$: Kitagawa, S.; Morishima, I.; Yonezawa, T.; Sato, N. *Inorg. Chem.*, **1979**, *18*, 1345. (b) Dupuis, P.; Roberge, R.; Sandorfy, C. *Chem. Phys. Lett.*, **1980**, *75*, 434. This article reports the UPS of free base $PH_2$ and etioporphyrins I-III, whose lower IPs, not unexpectedly, are essentially the same as those of $PEt_8H_2$[13a].

70. Ghosh, A. *J. Am. Chem. Soc.*, **1995**, *117*, 4691–4699.

71. Loew, G. H.; Herman, Z. S.; Rohmer, M. M.; Goldblum, A.; Pudzianowski, A. In *Quantum Chemistry in Bio-Medical Sciences*, Weinstein, H.; Green, J. P., Eds.; *Ann. N.Y. Acad. Sci.*, **1980**, *367*, 192.

72. Crossley, M. J. private communication.

73. Prendergast, K.; Spiro, T. G. *J. Phys. Chem.*, **1991**, *95*, 9728–9736.

74. Dolphin, D.; Felton, R. H. *Acc. Chem. Res.*, **1974**, *7*, 26–32.

75. Lin, C.-Y.; Hu, S.; Rush, T. III; Spiro, T. G. *J. Am. Chem. Soc.*, **1996**, *118*, 9452–9453.

76. Skillman, A. G.; Collins, J. R.; Loew, G. H. *J. Am. Chem. Soc.*, **1992**, *114*, 9538.

77. Gouterman, M. *J. Mol. Spectrosc.*, **1961**, *6*, 138.

78. Falk, J. E. *Porphyrins and Metalloporphyrins*, Elsevier: Amsterdam, **1964**.

79. Smith, K. M. *Porphyrins and Metalloporphyrins*, Elsevier: Amsterdam, **1975**.

80. Weiss, C.; Kobayashi, H.; Gouterman, M. *J. Mol. Spectrosc.*, **1965**, *16*, 415; Weiss, C. Jr., *Ibid.*, **1972**, *44*, 37.

81. Chantrell, S. J.; McAuliffe, C. A.; Munn, R. N.; Pratt, A. C. *Coord. Chem. Rev.*, **1975**, *16*, 259.

82. Gouterman, M. In *The Porphyrins, Vol. III. Physical Chemistry*, Dolphin, D., Ed.; Academic: New York **1977**; Part A; Chapter 1, ii-V108.

83. Seno, Y.; Otsuka, J. *Adv. Biophys.*, **1978**, *11*, 13.

84. Serner, M.; Gouterman, M.; Kobayashi, H. *Theor. Chem. Acta*, **1966**, *6*, 363.

85. Hanson, L. K.; Eaton, W. A.; Sligar, S. G.; Gunsalus, I. C.; Gouterman, M.; Connell, C. R. *J. Am. Chem. Soc.*, **1976**, *98*, 2572.

86. Churg, A. K.; Makinen, M. W. *J. Chem. Phys.*, **1978**, *68*, 1913.

87. Eaton, W. A.; Hanson, L. K.; Stephens, P. J.; Sutherland, J. C.; Dunn, J. B. R. *J. Am. Chem. Soc.*, **1978**, *100*, 4991.

88. Petke, J. D.; Maggiora, G. M.; Shipman, L. L; Weimann, L. J. *Chem. Phys. Lett.*, **1973**, *22*, 297.

89. Austin, R. H.; Beeson, K. W.; Eisenstein, L.; Frauenfelder, H.; Gunsalus, I. C. *Biochemistry*, **1975**, *14*, 5355.

90. Yamamoto, Y.; Noro, T.; Ohno, K. *Int. J. Quantum Chem.*, **1992**, *42*, 1563–1575.

91. Merchan, M.; Orti, E.; Roos, B. *Chem. Phys. Lett.*, **1994**, *226*, 27–36.

92. Nakatsuji, H.; Tokita, Y.; Hasegawa, J.; Hada, M. *Chem. Phys. Lett.*, **1996**, *256*, 220–228.

93. Zerner, M.; Loew, G. H.; Kirchner, R. F.; Mueller-Westerhoff, U. *J. Am. Chem. Soc.*, **1980**, *102*, 589.
94. Loew, G. H.; Herman, Z.S.; Zerner, M. *Int. J. Quant. Chem.* QBS, **1980**, *7(18),* 481–492.
95. Loew, G. H.; Rohmer, M. M. *J. Am. Chem. Soc.*, **1980**, *102*, 3655.
96. Loew, G. H.; Goldblum, A. *J. Am. Chem. Soc.*, **1980**, *102*, 3657.
97. Fermi, G. *J. Mol. Biol.*, **1975**, *97*, 237–256.
98. Hoard, J. L.; Scheidt, W. R. *Proc. Natl. Acad. Sci. U.S.A.*, **1973**, *70*, 3919–3922.
99. Loew, G. H.; Kirchner, R. *Biophys. J.*, **1978**, *22*, 179–189.
100. Huynh, B. H.; Papaefthymious, G. C.; Yen, C. S.; Groves, J. L.; Wu, C. S. *J. Chem. Phys.*, **1976**, *61*, 3750–3758.
101. Trautwein, A.; Zimmerman, R.; Harris, F. E. *Theor. Chim. Acta*, **1975**, *37*, 89–104.
102. Rovira, C.; Kunc, K.; Hutter, J.; Ballone, P.; Parrinello, M. *J. Phys. Chem. A*, **1997**, *101*, 8914–1925.
103. See for instance: (a) Case, D. A.; Huynh, B. H.; Karplus, M. *J. Am. Chem. Soc.*, **1979**, *101*, 4433; Kirchner, R. F.; Loew, G. H. *J. Am. Chem. Soc.*, **1977**, *99*, 4639; Newton, J. E.; Hall, M. B. *Inorg. Chem.*, **1984**, *23*, 4627; Loew, G. H.; Kirchner, R. F. *Int. J. Quantum Chem. Quntum Biol. Symp.*, **1978**, *5*, 403; Doetschman, D. C. *Chem. Phys.*, **1980**, *48,* 307; Waleh, A.; Ho, N.; Chantranupong, L; Loew, G. H. *J. Am. Chem. Soc.*, **1989**, *111*, 2767; Edwards, W. D.; Weiner, B.; Zerner, M. C. *J. Am. Chem. Soc.*, **1986**, *108*, 2196.
104. Hoffmann, R.; Chen, M. M. -L.; Thorn, D. *Inorg. Chem.*, **1977**, *16*, 503.
105. (a) Dedieu, A.; Rohmer, M. -M.; Benard, M.; Veillard, A. *J. Am. Chem. Soc.*, **1976**, *98*, 3717, (b) Nakatsuji, H.; Hasegawa, J.; Ueda, H.; Hada, M. *Chem. Phys. Lett.*, **1989**, *161*, 85.
106. Loew, G. H.; Kirchner, R. F. *Int. J. Quant. Chem.* QBS, **1978**, *5*, 403–415.
107. Waleh, A.; Ho, N.; Chantranupong, L.; Loew, G. H. *J. Am. Chem. Soc.*, **1989**, *111*, 2767–2773.
108. Weiss, J. J. *Nature (London)*, **1964**, *202*, 83.
109. Weiss, J. J. *Nature (London)*, **1964**, *203*, 183.
110. Wittenberg, J. B.; Wittenberg, B. A.; Peisach, J.; Blumberg, W. E. *Proc. Natl. Acad. Sci. U.S.A.*, **1970**, *67*, 1846.
111. Spiro, T. G. *Biochim. Biophys. Acta*, **1975**, *416*, 169.
112. Lang, G.; Marshall, W. *Proc. Phys. Soc.*, **1966**, *87*, 3.
113. Cerdonio, M.; Congiu-Castellano, A.; Mogno, F.; Pispisa, B.; Romani, G. L.; Vitale, S. *Proc. Natl. Acad. Sci. U.S.A.*, **1977**, *74*, 398.
114. Cerdonio, M.; Congiu-Castellano, A.; Calabrese, L.; Morante, S.; Pispisa, B.; Vitale, S. *Natl. Acad. Sci. U.S.A.*, **1978**, *75*, 4916.
115. Loew, G. H.; Kirchner, R. F. *J. Am. Chem. Soc.*, **1975**, *97(25)*, 7388–7390.
116. Herman, Z. S.; Loew, G. H. *J. Am. Chem. Soc.*, **1980**, 102, 1815–1821.
117. Yamamoto, S.; Kashiwagi, H. *Chem. Phys. Lett.*, **1993**, *205(2,3)*, 306–312.
118. Yamamoto, S.; Kashiwagi, H. *Chem. Phys. Lett.*, **1989**, *161(1)*, 85–89.
119. Waleh, A.; Loew, G. H. *J. Am. Chem. Soc.*, **1982**, *104*, 2352–2356.
120. Jewsbury, P.; Yamamoto, S.; Minato, T.; Saito, M.; Kitagawa, T. *J. Phys. Chem.*, **1995**, *99*, 12677–12685.

121. Tokita, Y.; Nakatsuji, H. *J. Phys. Chem. B.*, **1997**, *101*, 3281–3289.
122. Hoard, J. L. Stereochemistry of porphyrins and metalloporphyrins, In *Porphyrins and Metalloporphyrins*, Smith, K. M., Ed.; Elsevier: New York, **1975**, 351–371.
123. Peng, S. M.; Ibers, J. A. *J. Am. Chem. Soc.*, **1976**, *98*, 8032.
124. Quillin, M. L.; Arduini, R. M.; Olson, J. S.; Phillips, G. N. *J. Mol. Biol.*, **1993**, *234*, 140.
125. Derewenda, Z.; Dodson, G.; Emsley, P.; Harris, D.; Nagai, K.; Perutz, M.; Reynaud, J. -P. *J. Mol. Biol.*, **1990**, *211*, 515.
126. Stryer, L. *Biochemistry*, 4th ed.; Freeman: New York, **1995**, 152.
127. Service, R. F. *Science*, **1995**, *269*, 921.
128. Lim, M.; Jackson, T. A.; Anfirud, P. A. *Science*, **1995**, *269*, 962.
129. Collman, J. P.; Brauman, J. I.; Halbert, T. R.; Suslick, K. S. *Proc. Nat. Acad. Sci. USA*, **1976**, *73*, 3333.
130. Laberge, M.; Vanderkooi, J. M.; Sharp, K. A. *J. Phys. Chem.*, **1996**, *100*, 10793–10801.
131. Marcolin, H. E.; Reschke, R.; Trautwein, A. *Eur. J. Biochem.*, **1979**, *96*, 119.
132. Cerdonio, M.; Morante, S.; Vitale, S.; De Young, A.; Noble, R. W. *Proc. Natl. Acad. Sci. U.S.A.*, **1980**, *77*, 1462–1465.
133. Cerdonio, M.; Morante, S.; Vitale, S. *Isr. J. Chem.*, **1981**, *21*, 76–80.
134. Cerdonio, M.; Congiu-Castellano, A.; Mogno, F.; Pispisa, B.; Romani, G. L.; Vitale, S. *Proc. Natl. Acad. Sci. U.S.A.*, **1977**, *74*, 398–400.
135. Pauling, L; Coryell, C. D. *Proc. Nat. Acad. Sci. U.S.A.*, **1936**, *22*, 210–216.
136. Waleh, A.; Loew, G. H. *J. Am. Chem. Soc.*, **1982**, *104*, 3513–3515.
137. Hoffman, B. M.; Gibson, Q. H. *Proc. Natl. Acad. Sci. U.S.A.*, **1978**, *75*, 21.
138. Henry, Y.; Banerjee, R. *J. Mol. Biol.*, **1973**, *73*, 469–482.
139. Wayland, B. B.; Olson, L. W. *J. Am. Chem. Soc.*, **1974**, *96*, 6037–6041.
140. Hori, H.; Ikeda-Saito, M.; Yonetani, T. *J. Biol. Chem.*, **1981**, *256*, 7849–7855.
141. Kobayashi, K.; Tamura, M.; Hayashi, K. *Biochim. Biopys. Acta*, **1982**, *702*, 23–29.
142. Lang, G.; Marshall, W. *J. Mol. Biol.*, **1966**, *18*, 385.
143. Scholler, D.; Wang, M.-Y. R.; Hoffman, B. M. *J. Biol. Chem.*, **1979**, *254*, 4072–4078.
144. Maxwell, J. C.; Caughey, W. S. *Biochemistry*, **1976**, *15*, 388.
145. Nagai, K.; Hori, H.; Morimoto, H.; Hayashi, A.; Taketa, F. *Biochemistry*, **1979**, *18*, 1304–1308.
146. Sahany, J. M.; Ogawa, S.; Shulman, R. G. *Proc. Natl. Acad. Sci. U.S.A.*, **1974**, *71*, 3359.
147. Perutz, M. F.; Kilmartin, J. V.; Nagai, K.; Szabo, A. *Biochemistry*, **1976**, *15*, 378.
148. Scheidt, W. R.; Piciulo, P. L. *J. Am. Chem. Soc.*, **1976**, *98*, 1913.
149. Traylor, T. G.; Vijay, S. *Biochemistry*, **1992**, *31*, 2847.
150. Piciulo, P. L.; Rupprecht, G.; Scheidt, W. R. *J. Am. Chem. Soc.*, **1974**, *96*, 5295.
151. Lang, G.; Marshall, W. *J. Mol. Biol.*, **1966**, *18*, 385.
152. Takano, T. *J. Mol. Biol.*, **1977**, *110*, 537.
153. Axe, F.U.; Waleh, A.; Chantranupong, L.; Loew, G. H. *Int. J. Quant. Chem.*, **1989**, *35*, 181–191.
154. Iizuka, T.; Kotani, M.; Yonetani, T. *J. Biol. Chem.*, **1971**, *246*, 4731.
155. Hzuka, T.; Kotani, M.; Yonetani, T. *Biochim. Biophys. Acta*, **1968**, *167*, 257.
156. Tasaki, A.; Otsuka, J.; Kotani, M. *Biochim. Biophys. Acta*, **1967**, *140*, 284.

157. Hzuka, T.; Kotani, M. *Biochim. Biophys. Acta*, **1969**, *181*, 275.

158. Maeda, Y.; Trautwein, A.; Gonser, U.; Yoshida, K.; Kikuchi-Tonii, K.; Homma, T.; Orgura, Y. *Biochim. Biophys. Acta*, **1973**, *303*, 230.

159. Du, P.; Loew, G. H. *J. Phys. Chem.*, **1991**, *95*, 6379–6383.

160. Scheidt, W. R.; Lee, Y. J.; Luangdilok, W.; Haller, K. J.; Anzai, K.; Hatano, K. *Inorg. Chem.*, **1983**, *22*, 1516.

161. Rhynard, D.; Lang, G.; Spartalian, K. *J. Chem. Phys.*, **1979**, *71*, 3715.

162. Eaton, W. A.; Hofrichter, J. *Methods in Enzymology*, Antonin, E.; Rossi-Barnardi, L.; Chiancone, E.; Eds.; Academic Press: New York, **1981**, 175, Vol. 76.

163. Makinen, M. W.; Eaton, W. A. *Nature (London)*, **1974**, *247*, 62.

164. Makinen, M. W.; Eaton, W. A. *Ann. N.Y. Acad. Sci.*, **1973**, *206*, 210.

165. Makinen, M. W.; Churg, A. K.; Shen, Y.-Y.; Hill, S. C. *International Symposium on Interaction between Iron and Proteins in Oxygen and Electron Transport*, Ho, Chien, Ed.; Elsevier/North Holland: New York, **1982**.

166. Eaton, W. A.; Hochstrasser, R. M. *J. Chem. Phys.*, **1968**, *49*, 985.

167. Cheng, J. C.; Osborne, G. A.; Stephens, P. J.; Eaton, W. A. *Nature (London)*, **1973**, *241*, 193.

168. Finzel, B. C.; Poulos, T. L.; Kraut, J. *J. Biol Chem.*, **1984**, *259*, 13027.

169. Tamura, M. *Biochim. Biophys. Acta*, **1971**, *243*, 249.

170. Moss, T. H.; Ehrenberg, A.; Bearden, A. *Biochemistry*, **1969**, *8*, 4159.

171. Lang, G.; Asakura, T.; Yonetani, T. *J. Phys. C.*, **1969**, *2*, 2246.

172. Maeda, Y.; Morita, Y.; Yoshida, K. *J. Biochem.*, **1971**, *70*, 509.

173. Tamura, M.; Hori, H. *Biochim. Biophys. Acta*, **1972**, *284*, 20.

174. Hori, H. *Biochim. Biophys. Acta*, **1971**, *251*, 227.

175. Yonetani, T.; Anni, H. *J. Biol. Chem.*, **1987**, *262*, 9547–9554.

176. Maltempo, M. M.; Ohlsson, P. L.; Paul, K. G.; Peterson, L.; Ehrenberg, A. *Biochemistry*, **1979**, *18*, 2935.

177. Evangetista-Kirkup, R.; Crisanti, M.; Poulos, T. L.; Spiro, T. G. *FEBS Lett.*, **1985**, *190*, 221.

178. Strekas, T. C.; Spiro, T. G. *Biochim. Biophys. Acta*, **1974**, *351*, 237.

179. Smulevich, G.; Wang, Y.; Manro, J. M.; Wang, J.; Fishel, L. A.; Kraut, J.; Spiro, T. G. *Biochemistry*, **1990**, *29*, 7174–7180.

180. Smulevich, G.; Manro, J.; Fishel, L. A.; English, A. M.; Kraut, J.; Spiro, T. G. *Biochemistry*, **1988**, *27*, 5477–5485.

181. Smulevich, G.; Nero, F.; Willemsen, O.; Choudhury, K; Marzocchi, M. P.; Poulos, T. L. *Biochemistry*, **1995**, *34*, 13485–13490.

182. Loew, G. H.; Dupuis, M. *J. Am. Chem. Soc.*, **1997**, *119(41)*, 9848–9851.

183. Frew, J. E.; Jones, P. *Adv. Inorg. Bioinorg. Mech.*, **1984**, *3*, 175–212.

184. Dunford, H. B. In *Peroxidases in Chemistry and Biology*; **1991**; Vol. 2. Everse, J.; Everse, K.; Grisham, M. B.; Eds.; CRC Press: Boca Raton, 1–24.

185. Dupuis, M.; Loew, G. H. *J. Am. Chem. Soc.*, **1996**, *118(43)*, 10584–10587.

186. Harris, D.; Loew, G. H. *J. Am. Chem. Soc.*, **1996**, *118*, 10588–10594.

187. Schulz, C. E.; Devaney, P. W.; Winkler, H.; Debrunner, P. G.; Doan, N.; Chiang, R.; Rutter, R.; Hager, L. P. *FEBS Lett.*, **1979,** *103*, 102–105.

188. Baek, H. K.; Van Wart, H. E. *Biochemistry*, **1989**, *28*, 5714–5719.

189. Baek, H. K.; Van Wart, H. E. *J. Am. Chem. Soc.*, **1992**, *114*, 718–725.

190. Wang, J.-S.; Baek, H. K.; Van Wart, H. E. *Biochem. Biophys. Res. Instit.*, **1991**, *179*, 1320–1324.

191. Rodrigues-Lopez, J. N.; Smith A. T.; Thorneley, R. N. F. *J. Biol. Chem.*, **1996**, *271*, 4023–4030.

192. Schulz, C. E.; Rutter, R.; Sage, J. T.; Debrunner, P. G.; Hager, L. P. *Biochemistry*, **1984**, *23*, 4743–4754.

193. Hanson, L.V.; Chang, C. K.; Davis, M. S.; Fajer, J. *J. Am. Chem. Soc.*, **1981**, *103*, 663–670.

194. La Mar, G. N.; Thanabal, V.; Johnson, R. D.; Smith, K. M.; Parish, D. W. *J. Biol. Chem.*, **1989**, *264*, 5428–5434.

195. Ghosh, A.; Almlof, J.; Que, Jr., L. *J. Phys. Chem.*, **1994**, *98*, 5576–5579.

196. Kuramochi, H.; Noodleman, L.; Case, D. A. *J. Am. Chem. Soc.*, **1997**, *119*, 11442–11451.

197. Antony, J.; Grodzicki, M.; Trautwein, A. X. *J. Phys. Chem. A*, **1997**, *101*, 2692–2701.

198. Grodzicki, M. *J. Phys. B: At. Mol. Phys.*, **1980**, *13*, 2683.

199. Yamamoto, S.; Kashiwagi, H. *Chem. Phys. Lett.*, **1988**, *145(2)*, 111–116.

200. Loew, G. H.; Axe, F. U.; Collins, J. R.; Du, P. *Inorg. Chem.*, **1991**, *30*, 2291–2294.

201. Du, P.; Loew, G. H. *Biophys. J.*, **1995**, *68*, 69–80.

202. Loew, G. H.; Herman, Z. S. *J. Am. Chem. Soc.*, **1980**, *102*, 6173–6174.

203. Du, P.; Axe, F. U.; Loew, G. H.; Canuto, S.; Zerner, M. C. *J. Am. Chem. Soc.*, **1991**, *113*, 8614–8621.

204. Cupp-Vickery, J. R.; Poulos, T. L. *Struct. Biol.*, **1995**, *2*, 144–153.

205. Poulos, T. L.; Finzel, B. C.; Howard, A. J. *J. Mol. Biol.*, **1987**, *195*, 687–700.

206. Boddupalli, S. S.; Hasemann, C. A.; Ravichandran, K. G.; Lu, J. -Y.; Coldsmith, E. J.; Deisenhofer, J.; Peterson, J. A. *Proc. Natl. Acad. Sci. U.S.A.*, **1992**, *89*, 5567–5571.

207. Hasemann, C. A.; Ravichandran, K. G.; Peterson, J. A.; Deisenhofer, J. *J. Mol. Biol.*, **1994**, *236*, 1169–1185.

208. Ravichandran, K. G.; Boddupalli, S. S.; Hasemann, C. A.; Peterson, J. A.; Deisenhofer, J. *Science*, **1993**, *261*, 731–736.

209. Li, H.; Poulos, T. L. *Acta Crystallogr. D*, **1995**, *51*, 21–32.

210. Ullrich, V.; Kroneck, P. M. H.; Beinert, H. private communication.

211. Loew, G. H.; Collins, J.; Chantranupong, L.; Waleh, A. *Int. J. Quant. Chem.* QBS, **1987**, *14*, 75–83.

212. Thomann, H.; Bernardo, M.; Goldfarb, D.; Kroneck, M. H.; Ulrich, V. *J. Am. Chem. Soc.*, **1995**, *117*, 8243–8251.

213. Iizuka, T.; Kotani, M.; Yonetani, T. *J. Biol. Chem.*, **1971**, *246*, 4731.

214. Evangetista-Kirkup, R.; Crisanti, M.; Poulos, T. L.; Spiro, T. G. *FEBS Lett.*, **1985**, *190*, 221.

215. Scheidt, W. R.; Cohen, I. A.; Kastner, M. E. *Biochemistry*, **1979**, *18*, 3546–3552.

216. Harris, D.; Loew, G. H. *J. Am. Chem. Soc.*, **1993**, *115*, 8775–8779.

217. Sligar, R. *Biochemistry*, **1976**, *15*, 5399.

218. Singh, U. C.; Weiner, P. K.; Caldwell, J. W.; Kollman, P. AMBER UCSF Version 3.0a, Department of Pharmaceutical Chemistry, University of California, San Francsco, 1986; Revision A by George Seibel, **1989**.
219. Beetlestone, J.; George, P. *Biochemistry*, **1964**, *3*, 707.
220. George, P.; Beetlestone, J.; Griffith, J. S. *Rev. Mod. Phys.*, **1964**, *36*, 441.
221. Smith, D. W.; Williams, R. J. P. *Biochemistry J.*, **1968**, *110*, 297.
222. Jefcoate, C. R. E.; Gaylor, J. L. *Biochemistry*, **1970**, *8*, 3464.
223. Harris, D.; Loew, G. H. *J. Am. Chem. Soc.*, **1993**, *115*, 5799–5802.
224. Poulos, T. L.; Finzel, B. C.; Howard, A. J. *Biochemistry*, **1991**, *25*, 5314.
225. Ullrich, V.; Ruf, H. H.; Wende, P. *Croatica Chim. Acta*, **1977**, *49(2)*, 213.
226. Chang, A. K.; Dolphin, D. *J. Am. Chem. Soc.*, **1976**, *98*, 1607.
227. Collman, J. P.; Sorrell, T. N. *J. Am. Chem. Soc.*, **1975**, *97(14)*, 4133.
228. Chang, A. K.; Dolphin, D. *Proc. Natl. Acad. Sci. U.S.A.*, **1976**, *73(10)*, 3338.
229. Stern, J. O.; Peisach, J. *J. Biol. Chem.*, **1974**, *249(23)*, 7495.
230. Peterson, J. A.; Ishimura, Y.; Criffin, B. W. *Arch. Biochem. Biophys.*, **1972**, *149*, 197.
231. Loew, G. H.; Herman, Z. S.; Rohmer, M. M.; Goldblum, A.; Pudzianowski, A. In *Quantum Chemistry in Biomedical Sciences.*, Weinstein, H.; Green, J. P., Eds.; The New York Academy of Sciences, **1981**, *367*, 192–218.
232. Waleh, A.; Collins, J. R.; Loew, G. H.; Zerner, M. C. *Int. J. Quant. Chem.*, **1986**, *29*, 1575–1589.
233. Aikens, J.; Sligar, S. *J. Am. Chem. Soc.*, **1994**, *116*, 1143.
234. Atkins, W. M.; Sligar, S. G. *J. Am. Chem. Soc.*, **1988**, *27*, 1610–1616.
235. Jones, J. P.; Rettie, A. E.; Trager, W. F. *J. Med. Chem.*, **1990**, *33*, 1242–1246.
236. Schappacher, M.; Ricard, L.; Fischer, J.; Weiss, R.; Bill, E.; Montiel-Montoya, R.; Winkler, H.; Trautwein, A. Z. *Eur. J. Biochem.*, **1987**, *168*, 419–430.
237. Davydov, R.; Kappl, R.; Huttermann, J.; Peterson, J. A. *FEBS*, **1991**, *295*, 113.
238. Benson, D. E.; Suslick, K. S.; Sligar, S. G. *Biochemistry*, **1997**, *36*, 5104–5107.
239. Poulos, T. L. *Adv. Inorg. Biochem.*, **1988**, *7*, 1.
240. Poulos, T. L.; Finzel, B. C. *Peptide and Protein Reviews*, Vol. 4, Hearn, M. T. W., Ed.; Marcel Dekker: New York, **1984**, 115.
241. Maeda, Y.; Trautwein, A.; Gonser, U.; Yoshida, K.; Kikuchi-Tonii, K.; Homma, T.; Orgura, Y. *Biochim. Biopys. Acta*, **1973**, *303*, 230.
242. Torii, K.; Ogura, Y.; Otsuka, J.; Tasaki, A. *J. Biochem.*, **1969**, *66*, 791.
243. Du, P.; Loew, G. H. *Int. J. Quant. Chem.*, **1992**, *44*, 251–261.
244. Case, D.; Karplus, M. *J. Mol. Biol.*, **1979**, *132*, 343.
245. Elber, R.; Karplus, M. *J. Am. Chem. Soc.*, **1990**, *112*, 9161.
246. Kottalam, J.; Case, D. A. *J. Am. Chem. Soc.*, **1988**, *110*, 7690.
247. Ansari, A.; Berendzen, J.; Braunstein, D.; Johnson, J. B.; Ormos, P.; Sauke, T. S.; Scholl, R.; Schulte, A. *Biophys. Chem.*, **1987**, *26*, 337.
248. Campbell, B. F.; Chance, M. R.; Friedman, J. M. *Science*, **1987**, *238*, 373.
249. Nowak W.; Czerminski, R.; Elber, R. *J. Am. Chem. Soc.*, **1991**, *114*, 5627–5637.
250. Ma, J.; Huo, S.; Straub, J. E. *J. Am. Chem. Soc.*, **1997**, *119*, 2541–2551.
251. Schlichting et al. *Nature*, **1994**, *371*, 808.

252. Teng; Srajer; Moffat. *Nat. Struct. Biol.*, **1994**, *1*, 701.

253. Chernoff, D. A.; Hochstrasser, R. M.; Steele, W. A. *Proc. Natl. Acad. Sci. U.S.A.*, **1980**, *77*, 5606.

254. Cornelius, P. A.; Hochstrasser, R. M.; Steele, W. A. *J. Mol. Biol.*, **1983**, *163*, 119.

255. Petrich, J. W.; Poyart, C.; Martin, J. L. *Biochemistry*, **1988**, *27*, 4049.

256. Szabo, A. *Proc. Natl. Acad. Sci. U.S.A.*, **1978**, *75*, 2108.

257. Frauenfelder, H.; Wolynes, P. G. *Science*, **1985**, *229*, 337.

258. Petrich, J. W.; Lambry, J. -C.; Kuczera, K.; Karplus, M.; Poyart, C.; Martin, J. -L. *Biochemistry*, **1991**, *30*, 3975–3987.

259. Jones, P.; Dunford, H. B. *J. Theor. Biol.*, **1977**, *69*, 457–470.

260. Collins, J. R.; Du, P.; Loew, G. H. *Biochemistry*, **1992**, *31*, 11166–11174.

261. Loew, G. H.; Harris, D. L.; Dupuis, M. *J. Mol. Str. (THEOCHEM)*, **1997**, *398–399*, 497–505.

262. (a) Atkins, W. M.; Sligar, S. G. *Biochemistry*, **1988**, *27*, 1610–1616; Jones, J. P.; Rettie, A. E.; Trager, W. F. *J. Med. Chem.*, **1990**, *33*, 1242–1246.

263. Harris, D.; Loew, G. H. *J. Am. Chem. Soc.*, **1996**, *118(27)*, 6377–6387.

# 9 Characterization of the Electronic Structure of Transition Metal Carbonyls and Metallocenes

NADINE E. GRUHN AND DENNIS L. LICHTENBERGER
Department of Chemistry
The University of Arizona
Tucson, AZ 85721-0041, USA
E-mail: dlichten@arizona.edu



*Inorganic Electronic Structure and Spectroscopy, Volume II: Applications and Case Studies.*
Edited by E. I. Solomon and A. B. P. Lever.
ISBN 0-471-32682-8. 

## 1 INTRODUCTION

The two most defining achievements of early organometallic chemistry were the discovery of transition metal carbonyl compounds and the synthesis of metallocenes. Over the last forty years, compounds with carbonyl and cyclopentadienyl ligands have become ubiquitous throughout organometallic chemistry. In addition to the remarkable chemical and physical properties displayed by metal complexes that contain these ligands, the fascinating bonding modes of these ligands to metals have inspired new insights into electronic structure theories and new understanding of molecular structure and reactivity.

### 1.1 Early Background

The first transition metal carbonyl complex reported was $Pt(CO)_2Cl_2$, synthesized in 1868 by Schützenberger.[1] This compound was considered a bit of an oddity, and no further attention was paid to carbonyl complexes until the serendipitous discovery of $Ni(CO)_4$ by Mond in 1890.[2] Mond[3] immediately saw the industrial potential of this new compound,[4] and a patent for the Mond process for the purification of nickel was granted to him that same year. The synthesis of several other metal carbonyls quickly followed,[5] although little was understood of the structure and bonding of this class of compounds. The geometry of this compound had been much argued, with everything from a square planar to a cyclic geometry being suggested at different times.[5] In 1935, the first electron diffraction structure of $Ni(CO)_4$ was obtained,[6] which showed the structure to be tetrahedral. The structure of $Ni(CO)_4$ also showed a very short Ni–C bond distance, which was described using a valence bond description as having some double bond character. The bonding of carbonyls with metals was not better understood until the 1950's, when Pauling,[7] Zeiss,[8] and Orgel[9] expanded the $\pi$-acid bonding theory proposed by Dewar, Chatt, and Duncanson[10,11] for transition metal olefin compounds to include metal carbonyls (*vide infra*).

The discovery of the first metallocenes occurred at almost exactly the same time as the description of the bonding of metal carbonyls. Ferrocene was first reported in late

1951 and early 1952 in papers by Kealy and Pauson,[12] and Miller,[13] however both sets of authors formulated the compound with a $\sigma$-only interaction between the iron and cyclopentadienyl ligands. The sandwich nature of ferrocene was first formulated by Wilkinson and Woodward[14] and was followed quickly by the synthesis of other metallocenes.[15]

Taken together, the electronic structure descriptions that developed for metal carbonyls and metallocenes greatly extended the concepts of bonding beyond the localized valence bond descriptions that were prevalent for organic molecules and beyond the crystal field theory descriptions that dominated discussions of the bonding and spectroscopy of transition metal coordination compounds.

While these two vast areas of organometallic chemistry have a long experimental and theoretical history, both remain active areas of research to this day. For metal carbonyls, much recent research has centered around the use of carbonyl cluster compounds[16] as models of surface adsorption of CO and other small molecules.[17] Metallocenes, especially of early transition metals, are very common as catalysts.[18,19] Extensive chemistry has also been built around the incorporation of metallocenes into conducting polymers[20] and other materials,[21] primarily because of the ease with which the electron richness at the metal center can be varied. The detailed description of the bonding of these two classes of molecules also remains an active challenge to experimental and theoretical chemists. Indeed, for inorganic chemists, the first test of any new spectroscopic or theoretical method is typically how well the approach applies to characterization of a metal carbonyl or metallocene.

## 1.2 Scope of this Chapter

Contributions to the understanding of the electronic structure of transition metal carbonyls or metallocenes are clearly too numerous to be covered completely in one chapter. Leading references are provided so that interested readers can pursue further study. Rather than provide a cursory survey of the large amount of information on the large number of molecules in these classes, this chapter focuses on a very few of the most representative molecules in these classes, and it compares and contrasts a variety of the most broad-based experimental and theoretical approaches to studying the fundamental bonding interactions in these molecules. Section 2 provides a brief survey of results from different spectroscopic methods on metal carbonyls and metallocenes. This summary of information from different methods about these classes of molecules serves as background material for the more in-depth discussions that follow in the later sections.

After the basic experimental results are introduced, the focus will turn to particular areas that have received attention in recent years. Section 3 looks at theoretical studies of the structure and bonding of ferrocene. The description of the electronic structure of metallocenes is a long and controversial story, particularly because theoretical methods have had considerable difficulty in agreeing with even the most basic experimental results for this class of compound, such as geometrical structure and ionization energy. This section will describe some of the major areas where disagreements occur, and it discusses the shortcomings of theoretical methods with this class of molecules. Section 4 looks at some aspects of the description of metal–carbonyl bonding by theoretical methods, in particular some of the problematic areas that have arisen from *ab initio* calculations. Different approaches present different views of the roles of carbonyl sigma donation and pi acceptance, and call attention to other contributions to the metal–carbonyl bond. Section 5 discusses the information that can be obtained by evaluation of the vibrational fine structure present in the electronic spectra of metal carbonyls. These results have proven to be some of the most direct experimental evidence for the applicability of the Dewar–Chatt–Duncanson bonding model.

### 1.3 Basic Picture of Metal–Carbonyl Bonding

Before discussing the spectroscopy and calculations, it is helpful to begin with a short description of the models used to discuss the bonding in these types of molecules. As mentioned, the general description that is used for the bonding of carbon monoxide to the metal is the $\pi$ acid picture of Dewar, Chatt, and Duncanson. In this model, the interaction of a single carbonyl ligand with a metal is broken into two separate components. The first of these is sigma donation from the filled $5\sigma$ orbital of carbon monoxide, which is the highest occupied orbital of CO, to an unfilled orbital on the metal. This $5\sigma$ orbital is a predominantly carbon-based lone pair orbital that is directed towards the metal. In addition, the empty $2\pi^*$ orbital of the carbonyl ligand can accept electrons from any filled metal orbital of $\pi$ symmetry with respect to the ligand. Three-dimensional orbital surface plots which show the electron distributions and nodal positions for these orbital interactions are shown below. The relative amounts to which these two bonding components contribute to the overall metal–carbonyl bond is of particular interest. Although this model is widely known and taught, questions continue to be raised as to the correctness of this description.

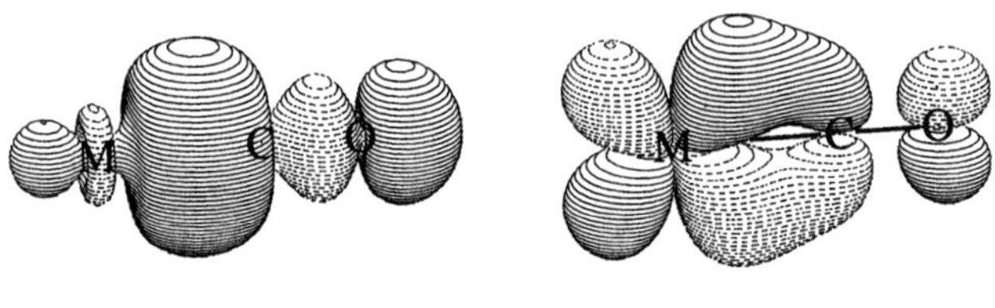


M (dσ) ← CO (5σ)
σ donor interaction

M (dπ) → CO (2π)
π acceptor interaction

It should be noted that this same bonding description is applied to many other ligands in additional to carbonyl, such as olefins (for which this model was first derived), phosphines, nitrosyl, etc. Comparison of the relative $\sigma$ and $\pi$ interactions of other ligands to carbonyl have been made both experimentally[22] and computationally.[23] Besides the terminally bonded carbonyl, carbonyls can also bridge across two or three metal atoms. These interactions can serve as models for pathways along which activation of the carbon–oxygen bond can occur in surface-catalyzed hydrogenation of carbon monoxide. Also, unsymmetrically bound "semi-bridging" carbonyl ligands that can be viewed as being intermediate between terminal and bridging modes have been observed in clusters. Descriptions of these structures and bonding continue to be in terms of overlap and electron donation and acceptance by the frontier orbitals of carbonyl.

For purposes of later discussion, a molecular orbital diagram for an 18-electron, octahedral $M(CO)_6$ complex is shown in Figure 1. These molecules have served as some of the most-studied models for metal–carbonyl bonding, and will be discussed throughout this chapter. This molecular orbital diagram is constructed using simply symmetry considerations[24] and general estimates of the relative stabilities of the metal and carbonyl orbitals based on spectroscopy.[25] On the left are shown the valence atomic orbitals of a metal at their approximate relative energies. The symmetry labels given for each are the symmetry of these particular orbitals in the $O_h$ point group. On the right are represented the molecular orbitals created by the interaction of the $5\sigma$ (sigma donor) and $2\pi^*$ (pi acceptor) of the six carbonyl ligands to form symmetry-adapted linear combinations of molecular orbitals. The symmetry-adapted carbonyl orbitals interact with the metal orbitals to form the molecular orbital picture for a $M(CO)_6$ molecule. The lowest orbitals shown are the $a_{1g}$, $t_{1u}$, and $e_g$ orbitals, which can be described as the metal–carbonyl $\sigma$-bonding orbitals. At next highest energy is the $t_{2g}$ set of orbitals, which is primarily metal-based, but by symmetry can interact with a combination of $2\pi^*$ orbitals from the carbonyls. This interaction can give some bonding character to the resulting orbitals and lower them in energy from the pure metal orbitals. These are followed by the anti-bonding and empty $e_g^*$ orbitals, the carbonyl $2\pi^*$ based $t_{1u}$, $t_{2u}$, and $t_{1g}$ orbitals, which do not have the

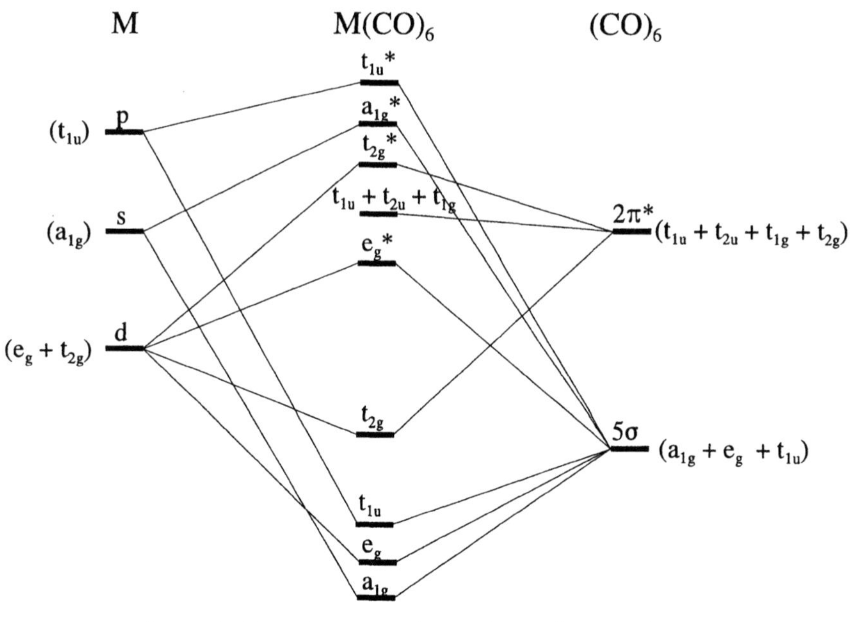


**Figure 1.** Molecular orbital diagram for octahedral $M(CO)_6$.

proper symmetry or nodal characteristics to interact with any of the metal d-based orbitals, and finally the anti-bonding $t_{2g}^*$, $a_{1g}^*$, and $t_{1u}^*$ orbitals.

This molecular orbital diagram is in many ways similar to the electronic energy level diagram that would be derived from crystal field theory for an octahedral molecule, in which the $\pi$-acceptor ability of ligands such as carbonyl is not explicitly included is a first-order description. The main difference is the relative position of the $t_{2g}$ orbitals and the delocalization of orbital characters between metal and CO. For molecules without $\pi$-acceptor ligands, the $t_{2g}$ orbital is a non-bonding orbital, while for $M(CO)_6$ this orbital may have a significant amount of bonding character and relative stabilization. The overall effect is to increase the energy splitting between the $t_{2g}$ orbital and the antibonding $e_g^*$ orbital. In crystal field theory terms, ligand $\pi$-acceptor interactions increase the apparent strength of the ligand field. Because of this interaction, carbonyl is considered to be a strong field ligand. This interaction is a major point of discussion of the experimental and computational results.

## 1.4 Basic Picture of Bonding in Metallocenes

Before discussing the basic electronic structure expected for a metallocene, it is necessary to comment on the geometry of this class of molecules. The two cyclopentadienyl rings of a metallocene may be oriented in the two extremes of either an eclipsed ($D_{5h}$) or a staggered ($D_{5d}$) conformation. The energy for rotation about the metal–cyclopentadienyl axis is very small, and ground state structures of metallocenes may show either of these conformations. There is also very little difference in electronic structure between the $D_{5h}$ and $D_{5d}$ symmetries. Much of the literature employs the $D_{5d}$ point group with reference to the solid state structure of ferrocene. We have chosen to use the $D_{5d}$ point group irreducible representations in our discussions of these molecules in order to simplify the description of electronic structure. The *gerade* and *ungerade* notation of the $D_{5d}$ point group allows for a rapid grasp of symmetry matching with metal orbitals that is not as accessible with the prime and double prime notation of the $D_{5h}$ point group.

The primary orbital interactions that form the metal–ligand bonds in metallocenes occur between the metal orbitals and the carbon-carbon $\pi$ orbitals of the cyclopentadienyl ligand. Representations of the $\pi$-based orbitals of a cyclopentadienyl ligand

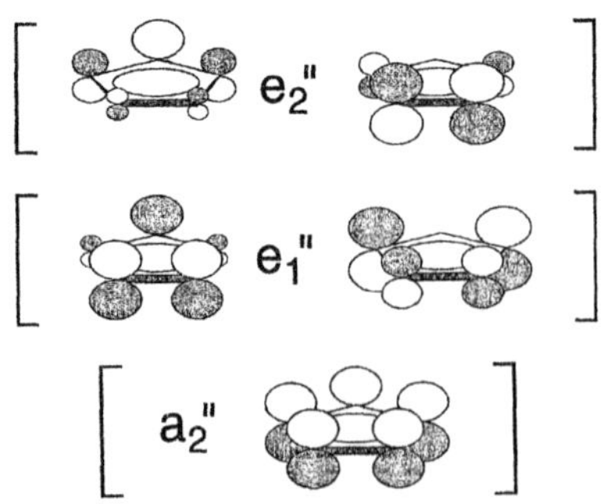

are shown below with the symmetry labels they have in $D_{5h}$ symmetry. The relative energy of these orbitals increases as the number of nodes increases as shown in Figure 2. The $a_2''$ orbital has no vertical nodes and is most stable, the $e_1''$ orbitals have one vertical node and are the highest fully occupied orbitals for the closed shell anion electron configuration, and the $e_2''$ orbitals have two vertical nodes and are net antibonding and unoccupied for the anion.

Using these orbitals, a qualitative molecular orbital diagram for a metallocene is shown in Figure 2. The $\pi$ orbitals of the two cyclopentadienyl ligands are combined pairwise to form the symmetry-adapted linear combinations shown on the right side of Figure 2, while on the left side are the metal orbitals with their respective symmetry designations.

In many ways, the molecular orbital diagram for a metallocene is very similar to that of $M(CO)_6$. The cyclopentadienyl ligands can be treated as occupying three coordination sites, and a sandwich compound can be viewed as pseudo-octahedral. For an 18-electron metallocene such as ferrocene, the six metal electrons occupy the $e_{2g}$ and $a_{1g}$ orbitals, which are descended from the $t_{2g}$ orbitals of octahedral symmetry. Both the six carbonyl ligands and the two cyclopentadienyl anions have six filled orbitals to donate to the metal center. However, the cyclopentadienyl ligands present a much weaker field ligand than carbonyls. Only the $e_{1g}$ symmetry combination of the two cyclopentadienyl rings has appreciable overlap with the metal $d$ orbitals to act as a significant donor, and the $e_{2g}$ combination of the two cyclopentadienyl rings is a relatively weak acceptor. Due to this, the highest occupied orbitals in this diagram, the $a_{1g}$ and $e_{2g}$, are primarily metal-based, and the electron occupation of these or-

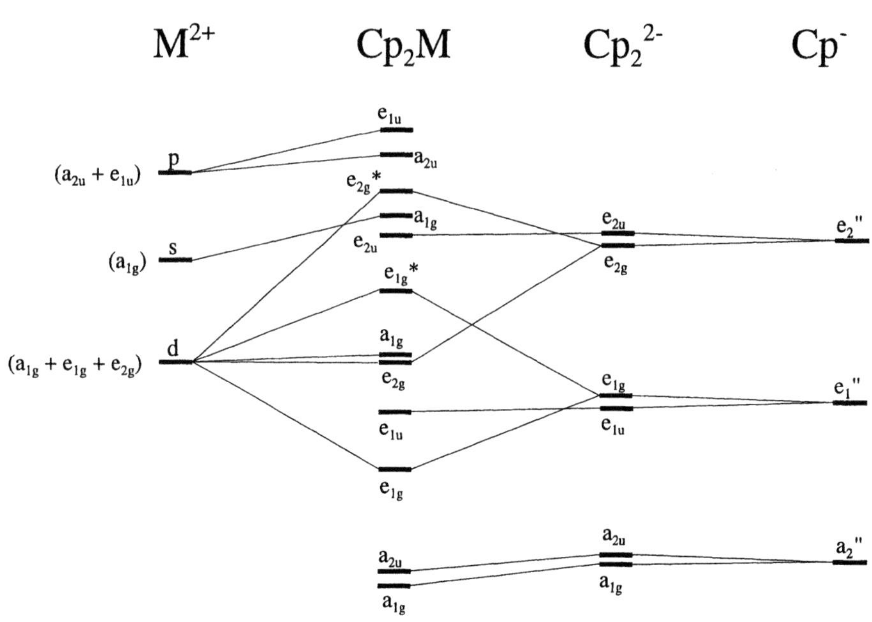


**Figure 2.** Molecular orbital diagram for a metallocene.

bitals does not drastically affect the net metal–ligand bonding. In addition, the $e_{1g}^*$ orbital is also primarily metal-based and is not as anti-bonding as for most octahedral molecules. The addition of electrons into this orbital also does not drastically affect the metal–ligand bonding. While sandwich metallocenes that have an 18 $e^-$ configuration do have the overall strongest metal–ligand bonding, metallocenes with from 15 to 20 valence electrons are also stable.

If a metallocene has fewer than 15 valence electrons, it can achieve a higher electron configuration by adding additional ligands. When additional ligands coordinate to the metal the Cp ligands "bend back," causing the symmetry to be lowered to $C_{2v}$. The effect upon the metal-based orbitals from this bending back is shown in Figure 3. All three of these orbitals are destabilized upon bending due to increased interaction with the filled Cp $\pi$ orbitals. This effect is especially noticeable for $2a_1$, which was initially nonbonding with respect to the Cp ligands because the $\pi$ orbitals of the Cp ligands were approximately in the nodal plane of the $d_{z^2}$ orbital. This orbital also rises steeply in energy due to its interaction with $1a_1$. In $C_{2v}$ symmetry, these two orbitals, which had different symmetry labels in the higher symmetry of the sandwich metallocene, are allowed to mix, and thus causing them to repel each other. This increases the additional destabilization of $2a_1$.

These three orbitals play a primary role in the coordination of additional ligands to the $Cp_2M$ fragments. The three-dimensional representations of these orbitals are

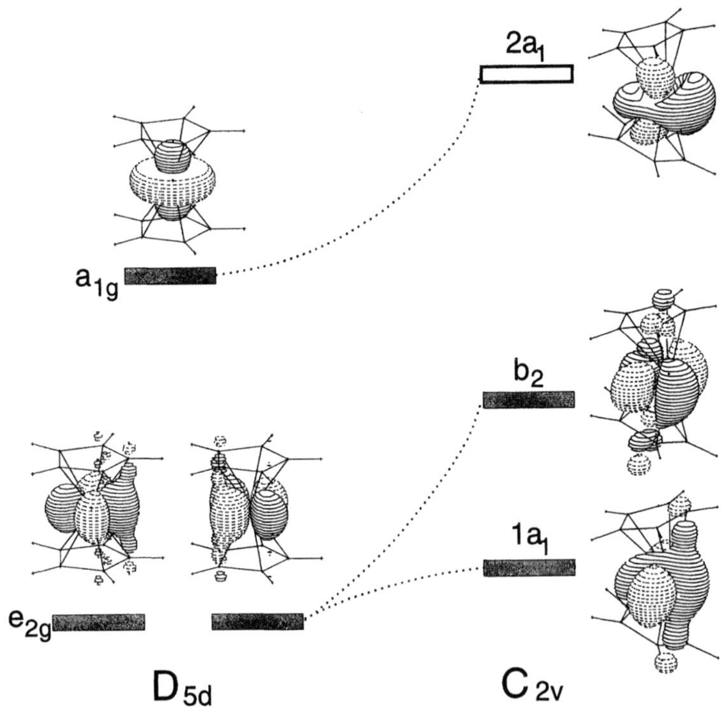


**Figure 3.** Effect of back-bending upon the metal-based orbitals of a metallocene.

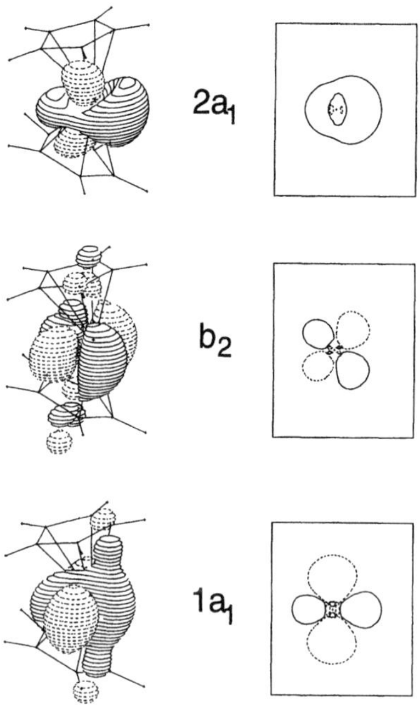


**Figure 4.** The metal-based orbitals of a bent metallocene.

reproduced on the left hand side of Figure 4, and two-dimensional plots of these three orbitals in the plane bisecting the Cp ligands are shown on the right hand side. All three of these orbitals have extensive density in this plane and are well situated for interaction with additional ligands. Bent metallocene complexes with one, two, and three additional ligands are known.

## 2 OVERVIEW OF SPECTROSCOPIC INFORMATION

This section will review some of the different spectroscopic techniques that have been used to gain bonding information about metal carbonyls and metallocenes. This information provides an overall picture of the current level of applicability of models of the electronic structure and bonding in these systems. The discussion will proceed from lower energy to higher energy spectroscopic studies of excitations, starting with vibrational spectroscopy and proceeding to different electron spectroscopic techniques.

### 2.1 Vibrational Spectroscopy

Vibrational spectroscopy is a traditionally important technique for the characterization of inorganic compounds[26] and has been particularly one of the most used tools in the study of metal carbonyls. Several review articles have appeared on this subject over the years and the subject is widely discussed in textbooks of inorganic chemistry.[27] Carbonyl complexes have strong and sharp vibrational bands associated with C–O stretches that are highly characteristic. The observed stretching frequencies of metal complexes are typically lower than the C–O stretching frequency of free CO, which is explained in terms of the synergistic bonding scheme described in the previous section. The $\sigma$-bonding component of the metal–carbonyl bond involves donation of electrons from the $5\sigma$ orbital of CO. This would tend to raise the C–O vibrational frequency because the $5\sigma$ orbital is slightly anti-bonding. The $\pi$ back donation into the C–O $\pi^*$ orbital, however, will tend to reduce the CO bond order and lower the C–O vibrational frequency. The overall result is that the C–O stretching frequency of metal carbonyl complexes is generally lower than for free CO. However, for CO complexes of high oxidation state or early transition metals back-bonding does not occur, and the C–O stretching frequencies are actually higher than those of free CO.

The C–O stretching bands in binary metal carbonyls are for the most part free from mixing with other vibrational modes of the same symmetry and are not obscured by the presence of other vibrations in this energy region. Due to these characteristics, study of the C–O stretching frequencies and intensities can provide much information about the structure and bonding of carbonyl complexes. In addition to the direct vibrational spectroscopies such as infra-red and Raman spectroscopy, carbonyl vibrations are also important to other spectroscopies, such as the electron spectroscopies discussed in Section 5.

Normal coordinate analysis has been carried out on many carbonyl complexes. Rigorous calculations on several systems including $M(CO)_6$ (M = Cr, Mo, W),[28] $Fe(CO)_5$,[29] and $Mn(CO)_5Br$[30] have been carried out by Jones. An alternative, approximate method has been developed by Cotton and Kraihanzel, which treats the C–O stretching modes only and ignores their interactions with low frequency vibrations such as M–C stretches and M–C–O wags, and also takes symmetry properties of the metal $d\pi$ orbitals into account.[31]

The CO stretching force constants obtained by the Cotton–Kraihanzel method, or a modification of this method,[32] give a quantitative evaluation of how substituents in metal carbonyls affect M–CO backbonding as a function of their own capacity to compete with the CO ligand for metal $d\pi$ electrons. This analysis can be used to make an evaluation of the relative backbonding capacity of a series of ligands. However, it must be remembered that the carbonyl stretching frequencies are determined by both $\sigma$ and $\pi$ effects. Vibrational spectroscopy does not allow observation of these two effects separately because the observed frequencies and their corresponding force constants reflect only the net result of the two counteracting components. Hall and Fenske have correlated C–O force constants with the calculated occupancies of the CO $5\sigma$ and $2\pi^*$ orbitals for a series of metal carbonyl halides.[33] This study showed the importance of considering $\sigma$-bonding effects upon the observed vibrational frequencies.

The vibrational spectra of cyclopentadienyl complexes have also been studied and reviewed extensively.[34] A normal coordinate analysis on ferrocene has also been carried out.[35] Metallocenes all share a similar type of vibrational spectra, and in partic-

**Table 1 M–Cp Vibrational Frequencies and Force Constants for Some Metallocenes**

| | Os | Fe | Ru | Cr | Co | V | Ni | Zn |
|---|---|---|---|---|---|---|---|---|
| $\nu$(M–Cp) ($cm^{-1}$) | 353 | 478 | 379 | 408 | 355 | 379 | 355 | 345 |
| $k$(M–Cp) (mdyn/Å) | 2.8 | >2.7 | >2.4 | $\gg$1.6 | ~1.5 | ~1.5 | ~1.5 | ~1.5 |

ular have a M–Cp asymmetric stretching frequency located from 300 to 500 $cm^{-1}$. From the observed M–Cp stretching frequencies for the metallocenes, force constants have been calculated that follow the trends in M–Cp bond strengths expected from the molecular orbital diagram discussed in Section 1. Some of these are shown in Table 1.[34] For example, the M–Cp stretching band for ferrocene is located at 478 $cm^{-1}$,[36,37] while that of the ferrocenium cation is located at 423 $cm^{-1}$.[38] As expected from the molecular orbital diagram, change from the 18 electron configuration due to the ionization of the $e_{2g}$ orbital electron weakens the M–Cp bond.

## 2.2 UV-Vis Spectroscopy

Electronic absorption and emission spectroscopy are much used in classical coordination inorganic chemistry and can for the most part be easily interpreted in terms of ligand field theory. However, the description of organometallic molecules such as metal carbonyls and metallocenes in terms of delocalized bonding through overlapping orbitals, such that the predominantly metal orbitals tend to contain a substantial amount of ligand character, creates problems for the simple assignment of electronic excitations as either charge transfer or metal-based (*d*-to-*d*). Also, due to the difficulties of methods for assignment of spectra that are often useful for coordination compounds, such as solvatochromism or substitution effects, study of the absorption and emission spectroscopy of organometallic compounds has lagged behind that of coordination compounds. This field finally received much attention in the late 1970's, in particular because of results showing that irradiation of organometallics could lead to catalytically and synthetically useful transformations.[39]

The one class of organometallic compounds that has been greatly studied by electronic absorption spectroscopy is metal carbonyls, particularly due to the role of metal carbonyls in photochemical processes such as the photolabilization and substitution of ligands.[40] The photolabilization of ligands such as carbonyl can be understood if the absorption spectra are evaluated based on the molecular orbital diagram for a metal hexacarbonyl shown in Figure 1. For octahedral, $d^6$ hexacarbonyls such as $[M(CO)_6]^-$ (M = V, Nb, Ta),[41] $[M(CO)_6]$ (M = Cr, Mo, W),[42] or $[M(CO)_6]^+$ (M = Mn, Re), the lowest energy absorptions have been assigned to the excited states involving $t_{2g}$ to $e_g$ one-electron excitations.[43] Despite the fact that these are *d*-to-*d* transitions, the bands are quite intense. The intensity of these bands suggested the possibility that they are charge-transfer transitions and, according to crystal field theory, the states should be reassigned.[44] However, the consistency with the absorption spectra of other metal carbonyls shows that the assignments of the symmetries of the transition are correct. The molecular orbital diagram in Figure 1 offers an explanation of the intensity of the excitation. As previously noted, the $t_{2g}$ orbitals are of $\pi$ symmetry with respect to the ligands and donate density to the CO $2\pi^*$ orbitals. Thus the $t_{2g}$ to $e_g$ excitation involves some charge transfer from the CO $2\pi^*$ component of the $t_{2g}$ orbitals to the primarily metal $e_g$ orbitals. The molecular orbital diagram also offers an explanation of the photoreactivity. The combination of the $\pi$-bonding nature of the $t_{2g}$ orbitals and the $\sigma$-antibonding nature of the excited-state $e_g$ orbitals with respect to the M–CO bond indicates that the M–CO bond strength should be significantly lower upon excitation from the $t_{2g}$ to the $e_g$. In fact, the metal hexacarbonyls are extremely photosensitive with respect to dissociative loss of CO. Compared to the ground state, the dissociative rate constant for the excited state of $d^6$ metal hexacarbonyls is 16 orders of magnitude greater.[39]

One interesting point in relation to the synergism of donation and acceptance is that for $Cr(CO)_6$, $Mo(CO)_6$, and $W(CO)_6$, the value of the ligand field splitting remains almost constant ($\sim$30,000 $cm^{-1}$ for all three). This has been understood in terms of a balancing of diminishing $\sigma$ bonding and increasing $\pi$ bonding for the heavier metal systems.[43]

In addition to the lowest energy electronic transitions, the absorption spectra of metal hexacarbonyls also show very strong absorptions resulting from transitions from the $t_{2g}$ orbitals to the predominantly CO $\pi^*$ orbitals of $t_{1u}$ symmetry. These transitions also substantially lower the M–CO bond strength, and a detailed analysis of the vibrational structure seen on these transitions for the metal hexacarbonyls will be discussed in Section 5.

A large number of cyclopentadienyl complexes have also had their photochemical properties examined. Early work in this area has been reviewed.[45] A detailed examination of the UV-visible spectrum of ferrocene crystals at 4.2 K has led to the lowest energy transition at 22,700 $cm^{-1}$ being assigned to the $d$–$d$ transition from the $a_{1g}$ to $e^*_{1g}$ orbital, and the next lowest transition at 30,800 $cm^{-1}$ has been assigned to the $d$–$d$ transition from the $e_{2g}$ to $e_{1g}$ orbital.[46,47] This experimental evidence is in agreement with the ordering of orbitals as shown in the simple molecular orbital diagram of a metallocene in Figure 2.

The absorption spectrum of the ferrocenium cation has also been reported.[46] This is of interest because it is one of the few organometallic systems for which ligand-to-metal charge-transfer transitions have been observed. Such excited states have been of relatively little importance in organometallic complexes because the unfilled metal $d$ orbitals are generally not low-lying. This is a fact that follows from the high ligand field strengths associated with organometallic complexes, but as has been earlier noted the ligand field for metallocenes is much weaker than for most organometallics. One important point is that the ligand-to-metal charge transition observed at $\sim$16,200 $cm^{-1}$ in the spectrum of ferrocenium has been assigned to a ligand $e_{1u}$ to metal $e_{2g}$ transition. This assignment indicates that when an electron is removed from ferrocene to form the ferrocenium cation, the hole caused by removal of the electron is in the $e_{2g}$ orbital. However, as seen above for the ferrocene excitation, the highest occupied molecular orbital of ferrocene is the $a_{1g}$ orbital. This indicates that ionization of ferrocene causes a change in the ordering of the metal-based orbitals. This intriguing point will be explored in detail in Section 3.

### 2.3 Photoelectron Spectroscopy

Several review articles have appeared that address the photoelectron spectroscopy of inorganic compounds, and several have focused in particular on metal carbonyls and metallocenes.[48,49] The first photoelectron spectroscopic studies of transition-metal carbonyl complexes appeared in 1969 and were of $Ni(CO)_4$ and $Fe(CO)_5$[50] and $Mn(CO)_5X$ (X = halogen, etc.).[51] The predominantly metal-based ionizations are often separated to lower ionization energy from the ligand-based ionizations and are sensitive to the electronic structure and bonding in the molecules. They reflect the basic *d* electron configurations and electronic symmetries at the metals, and they also reflect the charge distributions and overlaps with ligand orbitals. The metal carbonyl halide complexes provided classic studies of metal-heteroatom mixing in the ionizations and directly illustrated the concepts of molecular orbital theory in transition metal compounds. Core level photoelectron spectra, which are most sensitive to the charge potentials at the atoms, have also been much studied.[52] The group 6 metal hexacarbonyls were also some of the first transition metal complexes studied via variable photon energy from a synchrotron source.[53,54] A high resolution study of $Fe(CO)_5$ has been reported, which allows detailed analysis of the Jahn–Teller distortion that occurs upon photoionization.[55] For metal carbonyls, vibrational structure is also often seen on ionization bands that is associated with the carbonyl ligands. More will be said on this in Section 5.

The photoelectron spectrum of ferrocene was first reported in 1968 by Turner.[56] The results from this study and others[57,58] show that the first ionization of ferrocene corresponds to removal of an electron from the primarily metal-based $e_{2g}$ orbital. This orbital ordering is in agreement with the previously mentioned absorption spectroscopy results on ferrocenium, but differs from the ordering indicated by the absorption spectroscopy of ferrocene. The ordering of ion states for ferrocene will be discussed further in Section 3. The PES of many substituted ferrocene derivatives have also been reported.[58,59]

The PES of the ruthenium and osmium congeners of ferrocene show increased covalency as compared to ferrocene as well as the effects of spin-orbit splitting of bands.[58] Detailed high resolution studies have investigated the spin-orbit splitting observed for osmocene[60] and ruthenocene.[61,62] Comparison of the spin-orbit splitting observed for the ionizations of the metallocenes to spin-orbit splitting parameters for the metals allow a measure of the percent metal character in various orbitals. These studies indicate that the $e_{2g}$ orbitals of ruthenocene and osmocene are about 90% metal in character, while the bonding $e_{1g}$ orbitals are about 30% metal in character. Variable photon energy studies for the group 6 metallocenes have been carried out and show that ionizations from orbitals of mainly metal and mainly ligand character are clearly distinguished by their ionization cross-section behavior.[63]

The PES of many open-shell metallocenes have been studied also. Assignment of these are more difficult due to the fact that the open shell nature of the ground state of these molecules allows for multiple spin states to be produced upon photoionization. A ligand field treatment was able to give a good explanation for the energies of the ion states.[58] The spectra of the cobalt and nickel metallocenes are complicated because the ion states produced by the removal of *d* electrons overlap with those resulting from removal of ligand electrons. The spectra of these compounds were not fully assigned until variable photon studies were carried out.[64] In addition, the electronic structures of many bent metallocene complexes containing additional ligands have been studied by PES.[58,65]

Measurement of the angular distribution parameters over a range of photon energies has been performed for ferrocene and the group 6 metal hexacarbonyls.[66,67] While interesting shape resonances were observed, further experimental study on other molecules and additional theoretical work is necessary before a more complete understanding of these angle-resolved studies is possible.

## 2.4 Auger Spectroscopy

Auger spectroscopy is one of the most commonly used tools for probing the elemental composition of surfaces because the energies of Auger transitions are dependent on core level energies, which are characteristic for each atom. Auger spectroscopy is also capable of probing the valence electronic structure of molecules and solids, because the detailed line shapes of Auger transitions are dependent on the energies of the valence levels. The initial state in an Auger process is a core hole generated by photoionization or electron impact. This state decays via an electronic transition from a higher lying orbital, with release of the resultant energy of relaxation by the ejection and kinetic energy of a secondary electron, referred to as an Auger electron. If the core hole was created in a carbon or oxygen K shell, and if the two less tightly bound electrons are valence electrons, the overall process is referred to as $KV_1V_2$, where $V_1$ and $V_2$ are the valence levels.

Auger spectroscopy has some similarities to photoelectron spectroscopy. In the photoelectron process, the energy for the ejection and kinetic energy of the electron is provided by the photon, which is an essentially monoenergetic source. In the Auger process, the energy for ejection and kinetic energy of the electron comes from relaxation of a valence electron into the core hole. Because there are many different valence electron energy levels in the molecule, the source of energy for production of the Auger electron is not monoenergetic. The Auger line shape can be considered as a superposition of a series of overlapping photoelectron spectra for the molecule, where each spectrum is shifted according to the energy of relaxation from the particular valence level $V_1$. A Koopmans' approximation for the energy of a particular transition, ignoring for the moment the "repulsion" between the two valence holes in the final state, is given by the following equation, where the $\varepsilon$ values are the orbital eigenvalues of the valence

$$\text{Kinetic Energy (Auger electron)} = \varepsilon(V_1) + \varepsilon(V_2) - \varepsilon(K) \tag{1}$$

and core levels. In the low kinetic energy region of the Auger line shape there are many overlapping transitions because electrons with high binding energies (high $\varepsilon$) that are ejected by relaxation of an electron from the highest occupied valence levels (low $\varepsilon$) have kinetic energies similar to electrons with low binding energies (low $\varepsilon$) that are ejected by relaxation of an electron from more stable valence levels (high $\varepsilon$). The ionizations on the leading edge of the Auger band at high kinetic energy correspond to ejection of electrons with the lowest binding energies (lowest $\varepsilon$) by relaxation of electrons from the highest occupied valence levels (lowest $\varepsilon$) to the core level. This leading edge of the Auger band provides the most useful electronic structure and bonding information.

Additional useful information is provided as a consequence of the way Auger spectroscopy couples the core levels and the valence levels. In simple terms, the intensity of a KVV Auger transition is a measure of the valence charge available for decay into the core hole. Because a core hole is localized on a particular atom, this can give an indication of the amount of electron density in a particular valence orbital around the atom where the core hole was made.

Auger spectroscopy is typically performed on solid samples. Only a few reports exist of Auger experiments on molecular species. These have primarily focused on the bonding of simple organic molecules, but one report has been made of Auger spectroscopy of metal carbonyl and nitrosyl complexes.[68] The results of these studies show that for molecular species such as $Mo(CO)_6$ and $Fe(CO)_5$ the carbon Auger line shape can be viewed as a perturbed spectrum of carbon monoxide. For example, a comparison between the carbon and oxygen Auger spectra of $Fe(CO)_5$ and CO is shown in Figure 5. The most substantial difference between

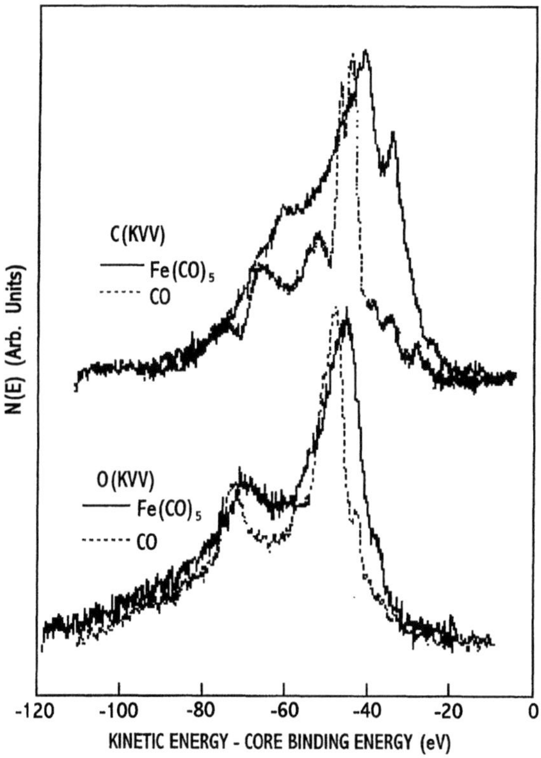


**Figure 5.** Comparison of C(KVV) and O(KVV) Auger lines for $Fe(CO)_5$ and for CO. (Reprinted with permission from Reference 68. Copyright 1982 American Chemical Society.)

the C(KVV) Auger spectra of free CO and that of the various metal carbonyls is the appearance of an additional relatively large intensity, low energy peak in the spectra of the metal carbonyls, which is located at about −35 eV in the C(KVV) spectrum of $Fe(CO)_5$ shown in Figure 5. The low energy peak corresponds to transitions where both $V_1$ and $V_2$ are the HOMO, which in the case of the metal carbonyls is the metal *d*-based orbitals that have $\pi$ symmetry with respect to the carbonyl ligands. The intensity of this transition indicates that the metal electron density is delocalized to the carbonyls and back-bonding is occurring. Auger also indicates that the $5\sigma$ orbital of CO, the carbon lone pair orbital, is still predominantly localized on the carbon. This gives an overall picture of carbonyl as a weak $\sigma$ donor and strong $\pi$ acceptor, in agreement with calculations that are discussed later.

## 2.5 Penning Ionization Electron Spectroscopy

Penning ionization electron spectroscopy (PIES) is another spectroscopy that has just begun to be used for the study of inorganic compounds.[69] In Penning ionization, electrons are ejected by collisions between target molecules and a beam of metastable atoms such as He* atoms in the ($2\ ^3S$) excited state, which have a kinetic energy of 19.82 eV. The Penning ionization process is interpreted in terms of an electron-exchange, in which an electron of the target molecule transfers to a hole of the metastable atom and the excited-state electron from the metastable atom is ejected simultaneously. The energy source for the ejection and kinetic energy of the electron is the energy of the metastable atom, which is essentially monoenergetic. The transition probability for the electron transfer depends largely on the spatial overlap of the relevant orbital of the sample molecule and the vacant orbital of the excited-state, metastable atom. In other words, the intensity of a transition will be greater if the occupied orbital of the sample molecule has electron density on the periphery of the molecule, in contrast to photoelectron spectroscopy, where the intensity of an ionization is governed by the photoionization cross-section. Comparison of the intensity of Penning ionization transitions to the intensities observed for the corresponding ionization bands in the photoelectron spectrum of the sample allows study of the spatial electron distribution of individual molecular orbitals.

The Penning ionization electron spectroscopies of several metal carbonyls have been studied.[70,71] For example, a comparison between the Penning ionization electron spectrum and He I photoelectron spectrum of $Mo(CO)_6$ is shown in Figure 6. The photoelectron bands of $Mo(CO)_6$ have been classified into three distinct groups.[72] Band 1 is attributed to ionization from the $t_{2g}$ orbitals that are primarily metal *d* in character, bands 2–8 are assigned to the CO $5\sigma$ and $1\pi$ derived orbitals, and bands 9–11 are related to the CO $4\sigma$ orbitals. This characterization is confirmed by the features of the Penning ionization electron spectrum. In the Penning spectrum, band 1 is suppressed extremely compared to the ligand bands because the $t_{2g}$ orbital is primarily localized on the metal and hence can not in-

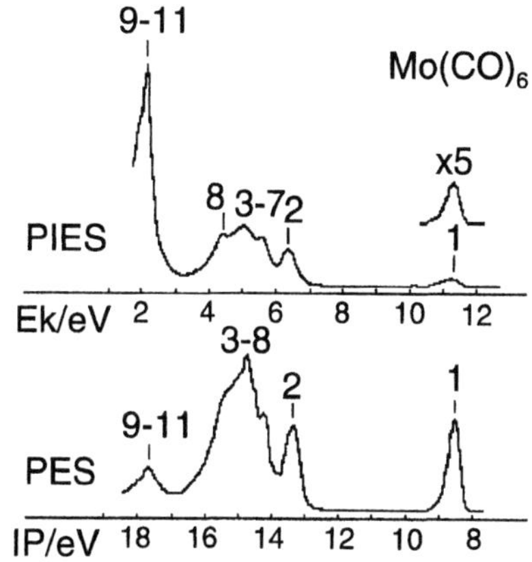


**Figure 6.** He* (2 $^3S$) Penning ionization electron spectrum and He I photoelectron spectrum of $Mo(CO)_6$. (Reprinted with permission from Reference 71. Copyright 1992 American Institute of Physics.)

teract with incident metastable atoms as effectively as the more ligand based orbitals. A similar effect has been observed in the Penning ionization electron spectrum of $Fe(CO)_5$ and CO molecules adsorbed on metal surfaces such as Ni(111),[73] Pd(111),[74] and Ru(001).[75]

In the ligand bands, the CO $4\sigma$-derived bands 9–11 are much enhanced compared to bands 2–8. The $4\sigma$ orbital of free CO is basically a non-bonding orbital and is localized largely on the oxygen atom. Therefore, the CO $4\sigma$-based orbitals are exposed outside the molecule and interact with metastable atoms preferentially. In the CO $5\sigma$ and $1\pi$ region, there should be seven ionization bands due to the three CO $5\sigma$ orbitals of $t_{1u}$, $e_g$, and $a_{1g}$ symmetries and four CO $1\pi$ orbitals of $t_{1g}$, $t_{2u}$, $t_{1u}$, and $t_{2g}$ symmetries. Except for the well-separated band 2 that has been assigned to the $t_{1u}$ orbital, it is difficult to assign the other photoelectron spectrum bands because they are not well separated. It should be noted that the highest-lying band 8 appears as a prominent peak in the PIES, although the corresponding photoelectron band is almost completely obscured. This band has been assigned to the $5\sigma$-based $a_{1g}$ orbital.

The Penning ionization electron spectroscopy of ferrocene has also been reported.[76,77] A comparison of the Penning and photoelectron spectrum of ferrocene is shown in Figure 7. Compared to the photoelectron spectrum, the intensity of the metal bands is much suppressed compared to that of the ligand bands. Such a marked suppression is not found in the case of dibenzenechromium.[77] Both of these sandwich compounds show an enhancement of the metal band due to the $a_{1g}$ orbital relative to that due to the $e_{2g}$ orbital. The difference in the comparison of the metal to ligand based transitions for ferrocene as compared to dibenzenechromium is due to the largely ionic nature of the bonding in ferrocene, which causes the metal orbitals

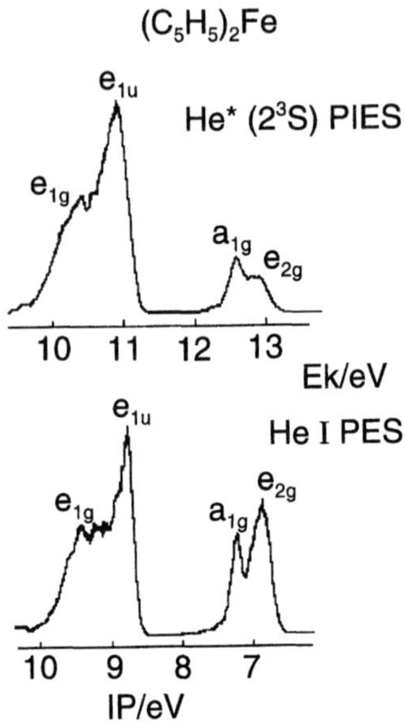


**Figure 7.** He* (2 $^3S$) Penning ionization electron spectrum and He I photoelectron spectrum of ferrocene. (Reprinted with permission from Reference 77. Copyright 1990 American Chemical Society.)

to be contracted and therefore shielded more by the ligands than the metal orbitals of the more covalent chromium compound. The fact that the $a_{1g}$ orbital is enhanced compared to the $e_{2g}$ orbital for each of these molecules indicates that the portion of the $a_{1g}$ ($d_{z^2}$) that is normal to the rings is actually considerably exposed outside the molecule.

## 2.6 Summary

This section has given some indication of the kind of information known about these classes of molecules from different spectroscopic techniques. Much information has been gained, but differences in interpretation remain.

The results of spectroscopic studies of metal carbonyls have shown that both $\sigma$ and $\pi$ bonding effects are important in an explanation of metal–carbonyl bonding. However, it is much more difficult to quantify the relative amounts of these two bonding components. One reason for this is that, other than electronic spectroscopy, the $\sigma$ and $\pi$ components can not be separated experimentally. Further evaluation of the relative importance of these two types of interactions is discussed in Sections 4 and 5.

For ferrocene, one important point is that the ionic nature of this molecule must be considered when discussing its electronic structure. Also, a not completely consistent evaluation of the ordering of the molecular orbitals of ferrocene has been pointed out and will be discussed further in the next section.

## 3 COMPARISON OF EXPERIMENTAL AND THEORETICAL RESULTS OF FERROCENE

From the experimental results that have been presented in Section 2 on ferrocene and other metallocenes, it can be seen that a reasonably consistent picture of the electronic structure, in particular the relative ordering of positive ion states, has been obtained. From photoelectron spectroscopic studies, it has been found that the two lowest ionization energies of ferrocene are due to the $^2E_{2g}$ and $^2A_{1g}$ ion states, which correspond to removal of electrons from the $e_{2g}$ and $a_{1g}$ orbitals that are for the most part localized on the iron center. The $^2E_{2g}$ is indicated to be the ground state of the positive ion. The third and fourth ionization energies are attributed to the $^2E_{1u}$ and $^2E_{1g}$ ion states, which correspond to ionization from the predominantly cyclopentadienyl-based $e_{1u}$ and $e_{1g}$ orbitals. The $^2E_{1g}$ state shows the greatest delocalization between the ligand and the metal. These assignments are consistent with all photoelectron spectroscopic studies that have been performed on ferrocene, including variable photon energy studies and evaluation of the effects of substitution of different groups on the cyclopentadienyl ligands. The electronic structure of the ferrocenium has also been investigated by EPR studies,[78] which agree with the photoelectron spectroscopy that the $^2E_{2g}$ state is the ground state configuration of the cation, and that the $^2A_{1g}$ state is an excited state of the cation.

The interpretation of the relative orders of cation states and excited electronic states in terms of the apparent order of molecular orbitals in the neutral ground state is an outgrowth of Koopmans' theorem.[79] Koopmans' theorem specifically states that the negative of the energy of an orbital that is calculated via a self-consistent-field (SCF) Hartree–Fock method is the same as the energy that would be calculated for ionization to the cation state that is defined by the removal of an electron from that orbital, provided that all orbitals are constrained to be unchanged by removal of the electron. If this concept is applied to ferrocene, the results on the positive ion states described in the preceding paragraph indicates the ordering of *orbitals* to be:

$$e_{1g}^4(\mathrm{Cp}\,\pi) < e_{1u}^4(\mathrm{Cp}\,\pi) < a_{1g}^2(d) < e_{2g}^4(d) \quad (2)$$

An extension of this concept (which further ignores changes in electron–electron repulsion and exchange energies) relates the energies of electronic excitations to the differences in energies of molecular orbitals. Absorption studies of the ferrocenium cation agree with the apparent order of the valence orbitals listed above, but absorption studies of neutral ferrocene reverse the apparent order of the $e_{2g}$ and $a_{1g}$ orbitals.

If this ordering of orbitals is compared to the prediction using the simple molecular orbital model in Figure 2, one sees that the model does a fairly good job. The energies of the $e_{2g}$ and $a_{1g}$ orbitals are switched, but they are predicted to be very close in energy. It might be thought that this is just a problem with the simple qualitative orbital analysis, and that if a more sophisticated computational method is used, an answer more in agreement with experimental results will be obtained. However, as will be seen, greater problems arise in the theoretical studies. There are large differences between the results of different methods, and even between different applications and assumptions of a given method. The more sophisticated methods are particularly limited in their ability to relate the ground state electronic structure of ferrocene to the experimental data presented above. Additionally, computational methods that perform geometry optimizations are not able to accurately reproduce structural parameters such as the metal-to-cyclopentadienyl centroid distance. Due to this, ferrocene has come to be regarded as the "enfant terrible" of quantum chemistry.[80]

The lack of agreement between experimental ionization energies and theoretical results for ferrocene is primarily due to an over-reliance on Koopmans' theorem. Experimental measurements relate positive ion states to the ground state of the neutral molecule. Koopmans' theorem estimates this relationship based only on the electronic structure of the ground state of the neutral molecule. The theorem does not take into account the changes in electronic structure of the remaining electrons due to the loss of repulsion and exchange with the electron that is removed. The energy associated with the optimization of the distribution of the remaining electrons with removal of one electron is termed the electron relaxation energy. Koopmans' *theorem* and the electron relaxation energy are precisely defined in terms of calculations of the ground state and positive ion states of molecules. We prefer to call the relationship between calculated orbital energies and *experimental* ionization energies Koopmans' *approximation*. Koopmans' approximation also neglects electron correlation and relativistic effects.

It should be remembered that the electron relaxation energy is a purely theoretical consequence of the Hartree–Fock self-consistent-field formalism and the frozen orbital approximation. There is no "right" answer for the orbital energy and ordering in the ground state except for how it relates within the construct of the particular computational method to observable energies, which are differences between total electron states. None-the-less, the language of orbitals is useful when describing these states. It also should be emphasized that experimental ionization energies, and not the orbital energies, are the important quantities which define chemical behavior through thermodynamic cycles and bonding energetics.[81] For example, suppose one wishes to know the homolytic metal–Cp bond dissociation energy of the ferrocenium ion in comparison to that of the neutral molecule. From the following thermodynamic cycle,

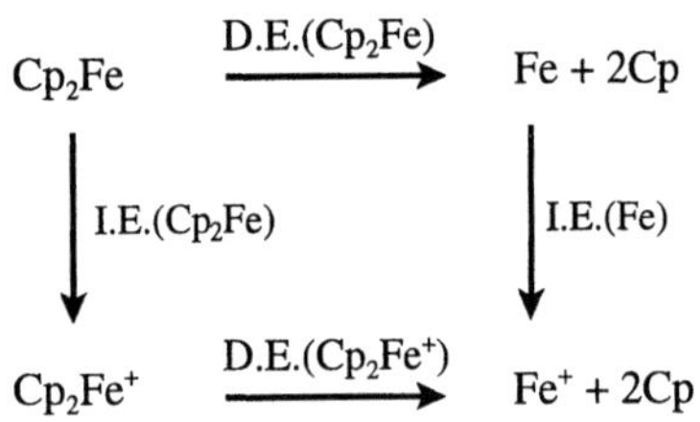


one can see that:

$$\text{D.E. }(Cp_2Fe) - \text{D.E. }(Cp_2Fe^+) = \text{I.E. }(Cp_2Fe) - \text{I.E. }(Fe) \quad (3)$$

Knowledge of the atomic and molecular ionization energies and one of the dissociation energies directly gives the other dissociation energy without recourse to considerations of orbital bonding, electron configurations, or electron relaxation energies. Knowledge of solvation energies extends gas-phase measurements of these quantities to understanding of solution chemistry. Atomic and molecular ionization energies contribute similarly to thermodynamic cycles of other processes.

### 3.1 Calculated Ionization Energies of Ferrocene

The earliest calculations reported on ferrocene were semi-empirical, and none of them were able to reproduce an ordering of orbital energies in agreement with the photoelectron spectrum. In addition, these calculations were not consistent with the results on other aspects of ferrocene. For example, these early calculations did not give similar charge distributions for ferrocene. Shustorovitch and Kyatkina obtained a positive charge of +0.69 on the iron atom,[82] while Dahl and Ballhausen obtained a negative charge of similar magnitude on the iron atom.[83]

The first *ab initio* calculation on ferrocene was reported in 1972 by Veillard.[84] An idealized, frozen geometry based on the crystal structure of ferrocene was used. The sequence of orbital energies for the ground state of the ferrocene molecule was found to be:

$$\begin{aligned} a_{1g}(d) &< e_{2g}(\mathrm{Cp}\,\sigma) < a_{2u}(\mathrm{Cp}\,\pi) < e_{2u}(\mathrm{Cp}\,\sigma) < e_{2g}(d) \\ &< e_{1g}(\mathrm{Cp}\,\pi) < e_{1u}(\mathrm{Cp}\,\pi) \end{aligned} \tag{4}$$

The calculated orbital energies are given in Table 2. The disagreement between experimental ionization energies and theoretical ionization energies through Koopmans' approximation is emphasized when it is noted that the two highest occupied orbitals in this calculation are predominantly ligand orbitals, while the two highest orbitals according to the photoelectron spectrum are predominantly metal orbitals. A series of SCF calculations were also performed with different basis sets to investigate the effect of the basis set on this sequence; except for a switch in the order of the (already close in energy) $e_{1u}$ and $e_{1g}$ levels, the ordering appeared relatively independent of the basis set. It was concluded that any larger basis set, which would lead to a wavefunction closer to the Hartree–Fock limit, would not be able to reconcile the computed orbital energies with the experimental order of energy levels.

When the results from the *ab initio* calculation on the ground state of ferrocene failed to agree with the ordering of ionizations in the photoelectron spectrum, the authors next performed a series of SCF calculations for the ferrocenium ion using various electron configurations, keeping the geometry used for the parent molecule. The total energy computed for the ground state was subtracted from the total energies calculated for the different cation states to give ionization energies to the cation states as shown in Table 2. When this was done, the computed sequence of ion states became:

$$^2E_{1g} < {}^2E_{1u} < {}^2A_{1g} < {}^2E_{2g} \tag{5}$$

This is the ordering of the ion states found in the photoelectron spectrum.

**Table 2 Comparison of Results from *Ab Initio* SCF–LCAO–MO, ΔSCF Calculations on Ferrocene**

| | | Ref. 84 | | Ref. 92 | |
|---|---|---|---|---|---|
| $Cp_2Fe$ Orbital | $[Cp_2Fe]^+$ Ion State | Energy (eV) | Computed I.P. (eV) | I.P. (eV) | Experimental I.P. (eV) |
| $e_{2g}$ | $^2E_{2g}$ | −14.42 | 8.3 | 5.69 | 6.86 |
| $a_{1g}$ | $^2A_{1g}$ | −16.57 | 10.1 | 7.46 | 7.21 |
| $e_{1u}$ | $^2E_{1u}$ | −11.67 | 11.1 | 8.79 | 8.77 |
| $e_{1g}$ | $^2E_{1g}$ | −11.86 | 11.2 | 8.85 | 9.28 |
| $a_{2u}$ | $^2A_{2u}$ | −16.02 | 15.5 | 13.03 | — |

This report made it clear that Koopmans' approximation is not valid for the ferrocene molecule. The reason for this is the different energy of electron reorganization that occurs upon ionization from each of the orbitals. A major factor in determining the magnitude of the electron relaxation energy for a given ionization is the atomic character of the molecular orbital that defines the hole state.[85] For an atom, the total electron relaxation energy is proportional to the magnitude of the electron–electron interactions in the ground state and to the total number of electrons that remain behind after ionization. Each atom experiences a different loss of electron–electron interactions and has a different number of electrons to relax. The iron atom has a higher number of valence electrons to relax than the carbon atoms, particularly when it is remembered that the $3s$ and $3p$ electrons are in the same principle shell as the $3d$ electrons, and consequently iron has a higher inherent relaxation energy than carbon when these interaction are taken into account in the calculations. For the ligand orbitals, there is little electron reorganization upon ionization and the ionization energy computed as the difference between the energies of the molecule and the ion is close to the corresponding orbital energy. For metal orbitals, there is a marked reorganization upon ionization.[86] While the predominantly metal-based orbitals were calculated to include a small amount of ligand character for the neutral molecule, they were found to be nearly pure metal in character for the ions. The computed ionization energy differed from the orbital energy by as much as 6 eV. The degree of electron reorganization also decreases with the amount of charge delocalization. Ionization from the $a_{1g}$ or $e_{2g}$ orbitals leaves a positive charge that is nearly localized on the iron atom, while the positive charge corresponding to ionization from the $e_{1u}$ or $e_{2g}$ orbitals is delocalized on the two cyclopentadienyl rings. Thus the iron-based orbitals may be calculated to be relatively stable in the ground state of the molecule, but the electron relaxation energy makes them easier to ionize.

The method used by Veillard of calculating ionization energies by calculating the difference in total energies of the ground state and cationic states of molecules is known as the ΔSCF method[87,88] and has become a standard method of estimating ionization energies for molecules that are likely to not follow Koopmans' approximation. Such molecules are usually organometallic or coordination compounds that have orbitals with much different amounts of delocalization or atomic character, as is found for ferrocene and other $\pi$-bonded complexes such as bis(allyl)nickel.[89,90] While this aspect of computational work with inorganic systems has become standard, papers still appear with high level calculations on ferrocene and other systems that try to directly correlate the ground state orbital energies and the observed ionization energies, and even question the well established assignment of the photoelectron spectrum.[91]

A similar study reporting ΔSCF calculations for ferrocene using a much larger basis set of contracted Gaussian type orbitals was reported by Almlöf.[92] The ionization energies calculated are in very good agreement with the experimental ionization energies. This study served as further evidence that the disagreement with Koopmans' approximation is not an artifact of the basis set used. This study found that because of the electron reorganization, the charges on Fe for the $^2E_{2g}$ and $^2A_{1g}$ states of ferricinium differ by less than 0.1 eV from that for ferrocene. This is in agreement with the small differences in Mössbauer isomer shifts for ferrocene and ferrocenium.[93]

A method that is similar to a ΔSCF calculation is the transition state calculation performed using the SCF X$\alpha$ method.[94] This method allows treatment of ground-state properties, ionization energies, and electronic absorption energies. To calculate ionization energies, one electron energies for transition states are calculated by removing half an electron from the orbital from which the ionization occurs, while to calculate transition energies half an electron is shifted from the occupied to the unoccupied orbital. Calculations of this type performed on ferrocene have done a fair job of agreeing with experimental data.[95,96]

Calculations that are capable of separating the reorganization energy from a ΔSCF calculation into the components that can be ascribed to relaxation and correlation effects have been reported.[97] Such calculations have been performed by Böhm using the semiempirical INDO[98] Hamiltonian employing a Green's function formalism.[99] For ferrocene, large relaxation energies are observed for the strongly metal-localized orbitals in the cationic hole state, while correlation terms are only of minor significance. The partitioning scheme into relaxation and correlation effects allows the formulation of some general rules concerning the breakdown of Koopmans' approximation in transition metal compounds. In particular, in the framework of the semiempirical INDO Hamiltonian it has been shown that the loss of ground state pair correlation (pair removal) does not lead to pronounced corrections to the calculated ionization energies in the outer valence region. However, in the case of ionization processes out of strongly localized metal orbitals, deviations are encountered that are primarily due to electron relaxation. It can therefore be expected that purely relaxational approaches (ΔSCF, transition operator method) will be most successful in this case.

## 3.2 Geometry Optimization

As can be seen from the previous section, several problems are encountered when trying to calculate electronic structure information that is consistent with spectroscopic results on ferrocene. While computational chemists are still evaluating this information with new methods, much more attention has been given in recent years to the calculation and optimization of the ground state geometry of ferrocene. The geometry (along with the total electron density distribution) is a ground state parameter that can be measured without the substantial perturbation of electronic structure found in spectroscopic techniques.

With the development of faster hardware and software for performing large *ab initio* calculations, it was found that Hartree–Fock calculations of molecular equilibrium structures were able to do a very good job of predicting the geometries of organic molecules.[100] The early results of such calculations on ferrocene did no better job of predicting geometries than they did of matching spectroscopic results. Large scale *ab initio* LCAO-SCF calculations using basis sets of better than triple zeta quality calculated an energy-optimized metal-ring distance 0.24 Å longer than the experimental value of 1.65 Å, an error of 15%.[101,102] It is also interesting to note that two Hartree–Fock calculations on ferrocene using very small basis sets were able to calculate metal-ring distances very close to experimental measure.[103,104] This agreement, however, should not be viewed as a success, but rather an accidental cancellation of errors with the insufficient basis sets used.

Calculations were reported using a basis set of double-zeta quality for a series of metallocenes,[105] and large errors were found in the calculated metal-ring distances for all. The most surprising result from this study was that the errors in calculated metal-ring distances obtained for cationic species are only half of those encountered for the neutral molecules. This fact points to correlation effects being the main cause for these errors, in contrast to calculation of spectroscopic data that is linked to relaxation effects. It is only when introducing correlation with the sophisticated modified coupled-pair functional (MCPF) approach that a satisfactory metal-ring distance of 1.68 Å is achieved.[106] The primary contribution to correlation effects was found to be dispersion interactions between the two cyclopentadienyl rings. Fairly decent agreement with experimental metal-ring distances has also been found using local-spin density methods.[107,108]

Two recent reports of density functional calculations on metallocenes indicate that this method has much promise for applications to metallocenes. A recent paper by Cook reported calculations at the MP2 level for ferrocene, ruthenocene, and osmocene.[109] Good agreement with experimental geometries was reached using effective core potentials for atoms of the second and higher rows. DFT calculations were also performed, and found to give numerical results comparable to MP2 using the same basis.

Weber reported calculations on ruthenocene[110] of the linear combination of Gaussian-type orbitals-model core potential-density functional (LCGTO-MCP-DF method — although the MCP option was not applied, — all the core and valence electrons were explicitly taken into account) employing the Vosko–Wilk–Nusair LSD exchange-correlation potential. The agreement between experimental and calculated ground and excited-state properties of ruthenocene was very good. Some of these results are summarized in Table 3. The quality of these predictions for ruthenocene are perfectly in line with similar results previously obtained using the same technique for organic molecules,[111,112] which showed that bond lengths are

**Table 3 Comparison Between Experimental and Calculated (LCGTO–DF) Ground and Excited State Properties of Ruthenocene**

| Parameter | Calculated | Experimental |
|---|---|---|
| ${}^1A_1 \rightarrow {}^3E_1$ absorption ($cm^{-1}$) | 26,900 | 27,000 |
| ${}^3E_1 \rightarrow {}^1A_1$ emission ($cm^{-1}$) | 19,000 | 17,000 |
| ${}^1A_1 \rightarrow {}^1E_{1/2}$ absorption ($cm^{-1}$) | 28,300–32,000 | 29,000–34,000 |
| Ru–Cp distance, ground state (Å) | 1.799 | 1.816 |
| Ru–Cp distance, excited state (Å) | 1.94 | 1.94–1.96 |
| Ru–Cp force constants [$f$(GS)/$f$(ES)] | 1.35 | 1.25 |

generally predicted within about 0.02 Å compared with experiment. The structure prediction of the Ru–Cp distance is significantly better than that obtained for the Fe–Cp distance in ferrocene using the same model and a one-electron basis set of similar quality. The authors suggested that the optimization of the LSD basis set for iron has not been performed using the same criteria as that of ruthenium, such that the problem is with the basis set. However, one must wonder if the DF calculation does a better job with ruthenocene than ferrocene because of the much more covalent bonding in the second and third row congeners. It will be interesting to see if better iron basis sets do lead to better results.

### 3.3 Summary

This section has explored the problems encountered when performing calculations on metallocenes, especially ferrocene. The large amount of relaxation and correlation effects encountered in these systems have caused the calculation of experimental observables to be especially troublesome. The computational technique used must be carefully chosen depending upon what type of information the calculation is to give. If spectroscopic results are to be investigated, calculations using $\Delta$SCF or transition state methods must be used to compensate for the large amount of relaxation energy that affects these results. If geometry optimization is to be performed, electron–electron correlation becomes important and methods that handle these correlation effects must be used.

Rather than being viewed as a failure, the computational difficulties for ferrocene have proven an exciting challenge for computational chemists, and have been the main driving forces for the development of computational tools capable of effectively describing inorganic systems.

## 4 COMPUTATIONAL STUDIES OF METAL–CARBONYL BONDING

This section will summarize some of the computational studies that have been reported investigating metal–carbonyl bonding. Literally thousands of papers have been written using various computational techniques to study some aspect of the structure or bonding in carbonyl complexes. This section will focus on the many studies concerning the synergistic nature of metal–carbonyl sigma and pi bonding in chromium hexacarbonyl.

### 4.1 Studies of Metal–Carbonyl Back-Bonding via X$\alpha$ Calculations

Early semi-empirical and approximate method calculations were very useful in the interpretation of the absorption and photoelectron spectra of metal carbonyls.[43,113] Some of the earliest calculations that addressed the relative importance of $\sigma$ and $\pi$ interactions to the overall metal carbonyl bond used the X$\alpha$ method. In 1972, Johnson and Wahlgren reported SW-X$\alpha$ calculations on $Ni(CO)_4$ and came to the conclusion that there was no evidence for any significant metal-to-ligand back-donation.[114] A later study of $Cr(CO)_6$ by Klemperer and Johnson came to a similar conclusion.[115] The analyses of the molecular orbitals in both of these papers were derived through the use of atomic sphere charges and orbital contour plots.

A more rigorous analysis of the results from X$\alpha$ calculations was carried out by Bursten and Fenske,[116] who used the projected X$\alpha$ (Px$\alpha$) method.[117] This method gives LCAO molecular orbitals, which allows the interpretative difficulties often associated with X$\alpha$ calculations to be circumvented. Their results for $Cr(CO)_6$ indicated that there is both significant CO $\sigma$ donation and $\pi$ acceptance in this molecule. The $e_g^*$ orbital was found to contain 33% Cr 3*d* character, indicative of strong CO 5$\sigma$ to Cr 3d donation. The highest occupied $t_{2g}$ orbital was found to contain 19% CO $2\pi^*$ character due to Cr to CO back-bonding. The total Mulliken orbital populations indicated that 1.42 e is donated to and 1.25 e is back-donated from the Cr 3*d* orbitals. Calculations on $Ni(CO)_4$ gave similar results.

The use of X$\alpha$ calculations to evaluate the bonding of metal carbonyls has also been exploited by Nash and Bursten to study actinide and transactinide metals.[118] These authors used nonrelativistic and fully relativistic DV-X$\alpha$ calculations to evaluate the electronic structure of a series of $M(CO)_6$ complexes, where M is Cr, W, U, or Sg (element 106). The periodic table would predict that Sg should behave chemically more like Cr, Mo, and W than like U. This study indicated that $U(CO)_6$ would suffer from an inherent instability, due to a high density of 5*f*-based states near the HOMO. However, $Sg(CO)_6$ would not have this electron problem. This study indicated that the transactinide Sg would be expected to behave chemically very similar to a group 6 transition metal, albeit with a much larger relativistic contribution. And indeed, recent experimental study of the reactivity of $^{265}Sg$ and $^{266}Sg$ atoms formed in nuclear fusion reactions found Sg to have many properties similar to Mo and W.[119]

### 4.2 *Ab Initio* Calculations

Many papers have appeared evaluating the $\sigma$ donation and $\pi$ back-bonding effects for metal carbonyls by *ab initio* methods.[120] While many of the details of these studies differ, the majority agree that $\pi$ back-bonding is more important than $\sigma$ donation. In addition, the metal $(n+1)s$ and $(n+1)p$ orbitals play only a small role in bonding, even in $Ni(CO)_4$ where there are no empty d orbitals.

Davidson has raised a question as to the validity of these *ab initio* results and the appropriateness of the overlap and frontier orbital models for describing the bonding of metal carbonyls.[121] He points out that an inherent error is introduced in *ab initio* calculations studying the interaction of two molecular fragments if limited basis sets are used. If calculations are performed on two fragments $A$ and $B$, where for example $A$ could be some metal containing fragment and $B$ is CO, their properties will be calculated with errors that could be eliminated by using better (larger) basis sets. When $A$ and $B$ are brought together to perform a calculation on an entire molecule using the same two fragment basis sets, the number of basis functions becomes that of $A+B$. Any superposition of the basis sets, in addition to describing the $A$–$B$ interaction, improves the calculated properties within each of the fragments. Because of this, the energy of interaction between $A$ and $B$ is *not* equal to the difference between the energies of the joint system and the sum of the energies of the separate fragments because of the changes in energy for $A$ and $B$ due to the superposition of the two basis sets. This difference in energy is referred to as the basis set superposition error (BSSE). The example given by Davidson specifically for metal carbonyls is that it is possible to introduce the CO orbitals into a calculation on the free metal atom without actually introducing the CO nuclei and electrons. The overlap with the metal orbitals would be the same in this case, but any energy lowering would be due to BSSE. If the metal orbitals were already optimal, no further improvement would be possible. If a complete basis set were used, the $\pi^*$ orbitals on CO would disappear.

Davidson has analyzed the bonding in $Cr(CO)_6$ using Hartree–Fock *ab initio* calculations in conjunction with the Morokuma partitioning scheme.[122] This study found that $Cr(CO)_6$ is unbound at the Hartree–Fock level of theory relative to six COs and Cr in its $d^5s$ ($^7S$) electronic ground state. The total metal–CO bond energy of $Cr(CO)_6$ was found to be approximately equal to the correlation energy. This makes interpretation of the bonding in terms of orbital interactions questionable. The authors emphasized the importance of the atomic promotion energy from the ground state to the valence excited state for a quantitatively correct understanding of the bond energy. They criticized the vocabulary of frontier orbital theory, which leads to the impression that the orbital overlap might be the driving force for the formation of a bond. The authors point out that the energy-lowering effect of "HOMO-LUMO" mixing is driven by the interactions between the nuclei and electrons, not by overlap of the occupied and unoccupied orbitals.

Davidson finds that the driving force for the bond is the electrostatic energy from the penetration of the $5\sigma$ electrons into the chromium valence shell. The increased shielding results in increased diffuseness of the $3d$ $t_{2g}$ orbitals and $\pi$ back-bonding into the carbonyls. The strong correlation energy between the $5\sigma$ electrons and the $3d$ electrons is an important factor leading to energy lowering terms that are able to compensate the promotion energy of the metal to the bonding state and the exchange repulsion of the $5\sigma$ electrons with the metal $3s$ and $3p$ electrons. Although the driving force is described differently than in the overlap and frontier orbital models of bonding, the synergism of the carbonyl $5\sigma$ and $2\pi$ roles is apparent, and orbital plots of the $t_{2g}$ and $e_g$ distributions that are obtained are qualitatively similar to those shown in the introduction to this chapter.

Agreement has not been reached as to the relative importance of BSSE to the results of *ab initio* calculations on metal–ligand interactions. One of the primary causes of BSSE is the use of truncated basis sets. Frenking and co-workers have addressed this issue by performing calculations using the charge-decomposition analysis (CDA) method and various large basis sets. In the CDA method the molecular orbitals of a complex are expressed in terms of the MOs of appropriately chosen fragments, similar to an LCAO approach. It was found that the choice of basis set did not greatly affect the relative amounts of $\sigma$ donation and $\pi$ back-bonding for metal carbonyls. For example, the donation and back-donation of $W(CO)_6$ are calculated to be 0.316 and 0.184 e, respectively, at the Møller-Plesset level using an effective core potential for the metal, while they are calculated to be 0.315 and 0.233 e, respectively, when Frenking's standard "basis set II"[123] is used.[124]

As pointed out by Frenking,[125] the strength of a good chemical model lies in the ability to explain and to predict *differences* in chemical behavior. The Dewar–Chatt–Duncanson model has been widely accepted in transition metal chemistry because a large class of complexes can be easily classified and interpreted in terms of synergistic $\sigma$ donation and $\pi$ back-donation between the metal and the ligands. To test the validity of this statement, the ability for *ab initio* calculations to agree with expected or experimentally determined trends in the amount of $\sigma$ donation and $\pi$ back-donation between a metal and different ligands has been evaluated using the CDA method for the complexes $M(CO)_5L$ (M = Cr, Mo, W) and $M(CO)_3L$ (M = Ni, Pd, Pt), where L is a variety of ligands including CO, SiO, CS, $N_2$, $NO^+$, etc. The CDA results show the *relative* amount of charge donation and back-donation between the ligand and the remaining fragment. Thus, it is possible, using this computational technique, to classify different ligands with regard to their donating and back-donating strength. The results for the CDA for the complexes $Cr(CO)_5L$ are shown in Table 4. The CDA results agree with the standard classification of the nature of these ligands.

**Table 4 Results of *Ab Initio* Charge-Decomposition Analysis for $Cr(CO)_5L$ Complexes**

| L in $Cr(CO)_5L$ | $\sigma$ Donation ($e^-$) | $\pi$ Back-Donation ($e^-$) |
|---|---|---|
| $NC^-$ | 0.482 | 0.038 |
| $CN^-$ | 0.604 | 0.089 |
| $H_2$ | 0.393 | 0.143 |
| $N_2$ | 0.166 | 0.173 |
| HCCH | 0.516 | 0.186 |
| SiO | 0.299 | 0.208 |
| $CH_2$ | 0.331 | 0.238 |
| $CF_2$ | 0.494 | 0.264 |
| **CO** | 0.460 | 0.314 |
| $CCH_2$ | 0.364 | 0.322 |
| $NO^+$ | 0.152 | 0.326 |
| CS | 0.310 | 0.345 |

**Table 5 Comparison of the Analysis of Spectroscopic Vibration Structure for $M(CO)_6$**

| | Ground State (IR) | | Absorption Spectra $t_{2g}(d) \rightarrow t_{1u}$ (CO $\pi^*$) | | | | UPS $^2T_{2g}$ Ionization | | | XPS $4f$ Ionization |
|---|---|---|---|---|---|---|---|---|---|---|
| Compound | $\nu_1$ (cm$^{-1}$) | $\nu_2$ (cm$^{-1}$) | $\nu_1$ (cm$^{-1}$) | $\nu_2$ (cm$^{-1}$) | $\Delta Q_1$ (Å) | $\Delta Q_2$ (Å) | $\nu_1$ (cm$^{-1}$) | $\nu_2$ (cm$^{-1}$) | $\Delta Q_2$ (Å) | $\nu_1$ (cm$^{-1}$) |
| $Cr(CO)_6$ | 2,118 | 390 | — | — | — | — | — | 310 | 0.14 | — |
| $Mo(CO)_6$ | 2,124 | 392 | 1,950 | 380 | 0.03 | 0.16 | — | — | — | — |
| $W(CO)_6$ | 2,124 | 420 | 1,840 | 410 | 0.04 | 0.15 | 2,090 | 387 | 0.10 | 2,220 |

### 4.3 "$\sigma + \pi$" Interactions

As this section has shown so far, most calculations on metal carbonyls have been interpreted in terms of the Dewar–Chatt–Duncanson model of bonding. This model involves the separation of $\sigma$ and $\pi$ bonding effects and, as for molecules such as metal hexacarbonyls, this separation is also a result of the symmetry of the molecules. However, this separation of $\sigma$ and $\pi$ bonding effects does start to break down in the analysis of specific metal–carbonyl bonds. One example of this that has been studied by relativistic Dirac-scattered-wave X$\alpha$ calculations, where the specific ligand–complex bond $(CO)_5Cr–CO$ was analyzed.[126]

The mixing of $\sigma$ and $\pi$ bonding effects is most apparent in the $t_{1u}$ orbital of $Cr(CO)_6$. For example, one component of this three fold degenerate orbital lies in the $xy$ plane of the molecule, and a contour diagram of this component is shown in Figure 8. In this plane, the $p_x$ orbital of the Cr atom has some $\sigma$ overlap with the carbonyls on the $x$ axis, and some $\pi$ overlap with the carbonyls on the $y$ axis. The overall effect upon the chemical bonding is that the $\sigma$ bonds of some ligands are mixed with the $\pi$ bonds of others. Any such mixing is dependent upon the importance of metal $p$ orbitals to the overall metal–carbonyl bonding.

As pointed out by Bersurker,[127] the $t_{1u}$ is a typical molecular orbital resulting from the general scheme for octahedral complexes. The $\sigma + \pi$ description emerges when one tries to consider the metal–ligand bond separately as a local property. Such a description, however, is useful in revealing the properties of coordinated ligands and ligand–ligand interactions.

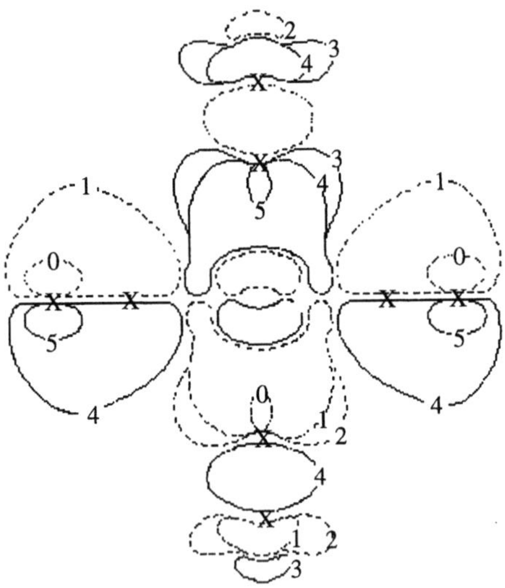

**Figure 8.** Contour plot of one component of the $t_{1u}$ orbital of $Cr(CO)_6$ as calculated by relativistic Dirac-scattered-wave X$\alpha$ calculations. (Reprinted with permission from Reference 133. Copyright 1985 American Institute of Physics.)

### 4.4 Summary

This section has given some indication of what can be learned from computational studies of metal carbonyls. While the Dewar–Chatt–Duncanson is a model for describing this bonding that may not be completely valid for computational methods, the results from computation studies of metal carbonyls and other complexes can be evaluated in terms of this model. The use of this model allows us to classify and interpret the relative amounts of bonding interactions between metals and various ligands.

## 5 VIBRATIONAL FINE STRUCTURE IN THE ELECTRONIC SPECTRA OF METAL HEXACARBONYLS

One important way that spectroscopy can give detailed bonding information on molecules is by the characterization of vibrational fine structure associated with transitions between electronic states of the molecule.[128] Excitation of an electron changes the electronic structure and bonding in the molecule and results in a shift of the equilibrium internuclear separations. If the geometry changes are sufficiently great, the most probable transitions from the initial electronic state are to excited vibrational levels of the final electronic state. When these vibrational levels are resolved in the electronic transition, the change in vibrational spacing from the initial state to the final state gives a measure of the change in vibrational frequencies and force constants associated with the excitation, and the intensity pattern of the transitions to the excited vibrational levels gives a measure of the change in equilibrium bond distances. In the molecular orbital model, these changes are due to the change of bonding (or antibonding) when an electron is removed from an orbital in the case of ionization, or transferred to a higher orbital in the case of excitation. Thus the vibrational pattern in an electronic spectrum shows the influence of an orbital electron on bond force constants and distances. In this section, we will review and compare the results from three reports of the observation of vibrational structure in spectroscopic studies of group 6 metal hexacarbonyls. Each of these studies allows us to gain unique bonding information about these compounds.

For these octahedral molecules the vibrational selection rules indicate that, in the absence of any vibronic or Jahn–Teller coupling, a vibration activated by an electronic excitation must have an $a_{1g}$ symmetry designation. There are two fundamental vibrations of $a_{1g}$ symmetry that can be derived from a normal mode analysis. The first of these, referred to as $\nu_1$, corresponds primarily to the symmetric C–O stretch. The frequency of this vibration in the ground state is 2,118 cm$^{-1}$ for $Cr(CO)_6$ and 2,124 cm$^{-1}$ for both $Mo(CO)_6$ and $W(CO)_6$.[129] The second vibration of $a_{1g}$ symmetry ($\nu_2$) corresponds primarily to the symmetric M–C stretch. The ground state frequency of $\nu_2$ is 390 cm$^{-1}$ for $Cr(CO)_6$, 392 cm$^{-1}$ for $Mo(CO)_6$, and 420 cm$^{-1}$ for $W(CO)_6$.

### 5.1 Electronic Absorption Spectra

The first observation of vibrational structure in electronic spectroscopy of metal hexacarbonyls was in the absorption spectra of $Mo(CO)_6$ and $W(CO)_6$.[130] The optical spectra of $Mo(CO)_6$ and $W(CO)_6$ measured in argon matrices at 10 K exhibit vibrational structure on the electronic absorption band at 286 nm (35,000 $cm^{-1}$) in the spectra of each molecule. This band is assigned to the allowed charge-transfer excitation $^1T_{1u} \leftarrow {}^1A_{1g}$, which in the molecular orbital designation is the transfer of an electron from the predominantly metal $d$ $t_{2g}$ orbital to the predominantly CO $\pi^*$ orbital of $t_{1u}$ symmetry. This type of "metal-to-ligand charge transfer" is also one type of transition that occurs upon photochemical labilization of a carbonyl ligand.

The 286-nm absorption of $Mo(CO)_6$ exhibits only one major progression in a 350-$cm^{-1}$ vibration, which is attributed to the $\nu_2$ vibration in the excited state. A progression in the higher-frequency $\nu_1$ is not evident. For $W(CO)_6$, a progression in $\nu_2$ is observable (ca. 400 $cm^{-1}$), and a second weak set of progressions appears to higher energy of the 286-nm band that is assigned to a progression in the other $a_{1g}$ mode, $\nu_1$.

A Franck–Condon analysis was carried out[131] to determine the distortions in the normal vibrational modes. From the band shape analysis, it is possible to calculate the amount of lengthening of the carbon–oxygen bond ($\Delta Q_1$) and metal-carbon bond ($\Delta Q_2$) that occur due to the electronic distortion. From the Franck–Condon analysis, it was concluded that for both $Mo(CO)_6$ and $W(CO)_6$ there is a large distortion in the metal–carbon bond lengths and significant weakening of these bonds. Some distortion of the C–O bond also appears, but it is not as large as for the M–CO bond.

The authors pointed out that, in terms of the crystal field theory description of charge transfer from the non-bonding metal $t_{2g}$ orbital to the carbon–oxygen anti-bonding $t_{1u}$ CO $\pi^*$ level, the small change in carbon–oxygen distance and the larger change in metal–carbon distance is surprising. The small distortion of the C–O bond is interesting in light of the analogous one-electron excitation of free CO.[132] For the $A^1\Pi \leftarrow X^1\Sigma^+$ transition of free CO, which corresponds to a $\sigma_n$ to $\pi^*$ transition, the CO stretching frequency is decreased by 655 $cm^{-1}$ and the bond is lengthened by 0.11 Å. Although crystal field theory does not account for the small change in carbon–oxygen distances and larger change in metal–carbon distances with this excitation of metal hexacarbonyls, this result is not surprising in terms of the delocalized picture of molecular orbital theory. In this picture the $t_{2g}$ metal orbitals are partially delocalized through backbonding to the carbonyls, so an electron in one of these orbitals already has some CO $\pi^*$ character. The $t_{1u}$ orbitals also have CO $\pi^*$ character, so the transition from the $t_{2g}$ to the $t_{1u}$ does not represent a full change of charge in the CO $\pi^*$ orbitals. The calculations in the paper support this view, showing that the electron charge on the carbonyls increases by a total of only 0.04 e on $t_{2g}$ to $t_{1u}$ excitation. The larger change in metal–carbon distances is also consistent with the molecular orbital description. The $t_{2g}$ orbitals are $\pi$ bonding with the carbonyls through interaction with the CO $\pi^*$ orbitals while the $t_{1u}$ orbitals are non-bonding with respect to the metal d orbitals, and slightly anti-bonding with the metal through the metal $p$ orbitals according to the authors. The metal–carbon bonding of the $t_{2g}$ set is reduced when an electron is removed from these orbitals. The vibrational fine structure gives an indication of the influence of one electron in the $t_{2g}$ orbitals on the metal–carbon bond distances, depending on the amount of anti-bonding in the $t_{1u}$ orbitals.

## 5.2 Valence Photoelectron Spectra of Transition Metal Hexacarbonyls

As seen in the previous section, the analysis of the vibrational structure observed on bands in absorption spectra can be complicated due to the fact that two different orbitals are involved in the transition. This problem does not arise when analyzing vibrational structure in photoelectron spectroscopy. In photoelectron spectroscopy, an electron is completely removed from a molecule, and the molecular distortions that occur do so simply due to removal of an electron from a particular orbital.

As discussed in Section 2, the photoelectron spectra of group 6 metal hexacarbonyls have been the subjects of many studies. A high resolution photoelectron study to observe vibrational fine structure on the valence ionizations was reported by Lichtenberger and Hubbard.[133] In this study, specific attention is given to interpretation of the fine structure observed on the ionization band corresponding to removal of an electron from the metal-based $t_{2g}$ orbitals. This predominantly metal-based orbital has pure $\pi$ symmetry with respect to the carbonyl ligands. The vibrational fine structure associated with this ionization thus can provide unique information on the $\pi$-bonding between the metal and carbonyl ligands.

The He I photoelectron spectra in the energy region associated with removal of an electron from the $t_{2g}$ orbital of $Cr(CO)_6$, $Mo(CO)_6$, and $W(CO)_6$ are shown in Figure 9. The spectrum of $Cr(CO)_6$ shows one band in this region, while that of

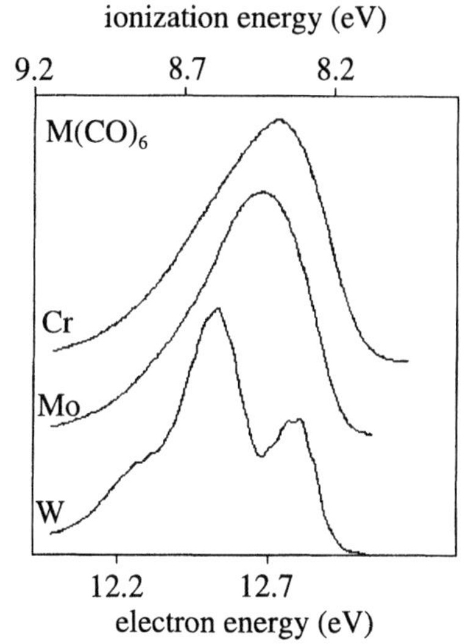


**Figure 9.** Photoelectron spectra of the metal-based ($t_{2g}$) ionizations of $Cr(CO)_6$, $Mo(CO)_6$, and $W(CO)_6$. (Reprinted with permission from Reference 133. Copyright 1982 American Chemical Society.)

$Mo(CO)_6$ contains a band that is slightly broader than that of $Cr(CO)_6$. In contrast, the spectrum of $W(CO)_6$ shows two well-separated bands of approximately 2 : 1 ratio. The band shape seen for $W(CO)_6$ is due to the large spin-orbit coupling of the W atom, which splits the sixfold degenerate $^2T_{2g}$ ion state ($t_{2g}^5$) into a fourfold degenerate $^2U'$ state and a twofold degenerate $^2E''$ state in the double group of O symmetry.

In addition to the structure present due to spin-orbit splitting, vibrational structure is also present. For the $^2U'$ ionization of $W(CO)_6$, two overlapping vibrational progressions are seen. A shoulder on the high ionization energy side of the $^2U'$ band is clearly observed at a separation of 2,090 $cm^{-1}$ and corresponds to the totally symmetric, primarily carbon–oxygen stretch. The other progression is 390 $cm^{-1}$ and corresponds to the totally symmetric metal–carbon stretching mode. In the ground state of this molecule the metal–carbon stretch is 426 $cm^{-1}$. The lowering of the metal–carbon stretching frequency when an electron in a metal orbital that is $\pi$ symmetry with respect to the carbonyls is removed indicates a reduction of the metal–carbonyl bond strength that is due to $\pi$ back-bonding from the metal to the ligand.

The intensity pattern observed also indicates that the metal–carbon bond distance in the positive ion is considerably different from that in the neutral molecule. The results of a Franck–Condon analysis of $W(CO)_6$ are illustrated in Figure 10. This is constructed with the correct scale between the vibrational potential wells of the

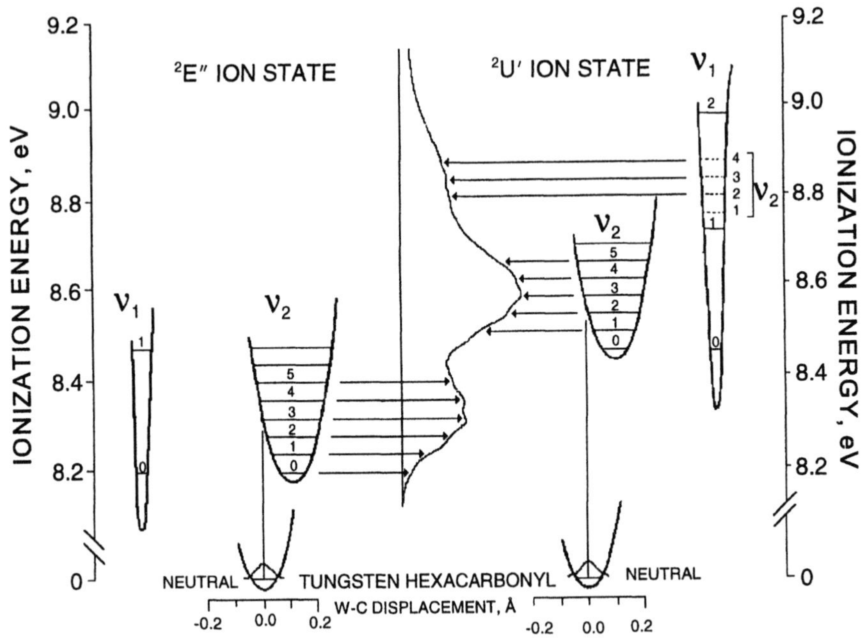


**Figure 10.** The $\nu_1$ and $\nu_2$ ($a_{1g}$) potential surfaces for the $W(CO)_6$ $^2E''$ and $^2U'$ ion states constructed from experimental data and ground-state normal mode analysis. The experimental ionization band is in the center of the figure. (Reprinted with permission from Reference 133. Copyright 1982 American Chemical Society.)

ground state and the positive ions and matches the vibrational quantum levels with the observed vibrational progressions in the ionization band. The $\nu_2$ ($a_{1g}$) normal mode potential well is plotted versus the internal coordinate representing the W–C bond distances. This Franck–Condon analysis indicates that the W–C bond distances in the positive ion are displaced 0.10 Å ($\Delta Q_2$) from the W–C bond distances in neutral $W(CO)_6$. This is additional indication of the metal–carbon bond-strengthening nature of an electron in the $t_{2g}$ orbital. Because the $t_{2g}$ orbital has only a $\pi$-symmetry interaction with the carbonyls, the direct deduction from these observations is that the metal orbitals are $\pi$ bonding with respect to the carbonyls. It is difficult to attribute the observed ionization structure to any significant changes in the $\sigma$ bonding. The 0.10 Å change in metal–carbon distances with removal of an electron from the $t_{2g}$ orbitals is comparable to the 0.15 Å change that was found for electron excitation from the $t_{2g}$ to the $t_{1u}$ orbitals from the previous section. The significance of the difference in these numbers is uncertain, but it suggests some metal–carbon antibonding character in the excited $t_{1u}$ orbitals, as suggested by the authors of that work.

For $Cr(CO)_6$, the $\nu_2$ frequency reduction is 60 $cm^{-1}$ from the ground-state value, again indicating a considerable decrease in Cr–C bond order upon ionization from the $t_{2g}$ orbital. This frequency reduction represents a 15% frequency reduction, as compared to a 10% frequency reduction for $W(CO)_6$. Also, the $\nu_2$ progression in the $Cr(CO)_6$ spectrum is roughly twice as long as the $W(CO)_6$ progression. This indicates a greater displacement of the metal–carbon distances for $Cr(CO)_6$. Franck–Condon analysis gives a Cr–C bond displacement of 0.14 Å. These observations indicate that the metal–carbon bond strength is more sensitive to removal of a $t_{2g}$ electron in the case of $Cr(CO)_6$ than in the case of $W(CO)_6$.

## 5.3 Core Level Ionizations of Metal Hexacarbonyls

Vibrational structure has also been observed on *core* level ionizations for metal carbonyls.[134] In particular, vibrational structure on the $4f$ ionization of W in $W(CO)_6$ has been observed. The region of the spectrum of $W(CO)_6$ that contains the ionizations corresponding to removal of an electron from the $4f$ orbitals is shown in Figure 11.

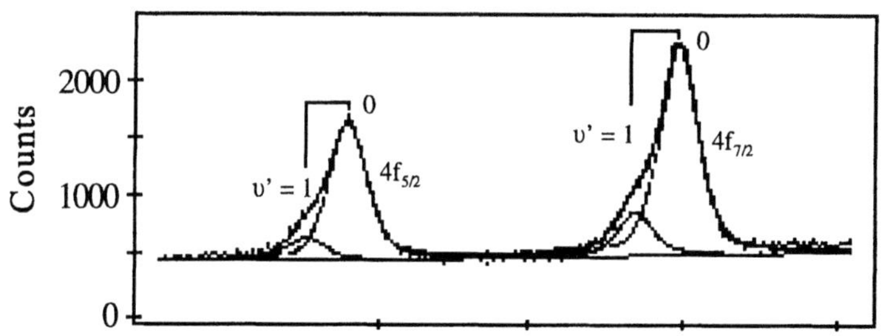


**Figure 11.** High resolution $4f$ photoelectron spectrum of $W(CO)_6$. (Reprinted with permission from Reference 134. Copyright 1995 American Chemical Society.)

The vibrational structure observed on a core ionization is not due to direct interaction of the ionized orbital with the ligand as for a valence ionization. Instead, the presence of vibrational structure on core ionizations can be understood by considering the core equivalent model of Jolly.[135] In this model, the properties of a molecular ion with a core hole are approximated by the isoelectronic molecule with a $Z^{+1}$ atom. The core equivalent species for core ionized $W(CO)_6$ would therefore be $[Re(CO)_6]^+$. The M–C bond length in $[Re(CO)_6]^+$ is 2.01 Å, which is smaller than the 2.07 Å M–C bond length in $W(CO)_6$. This difference in expected bond length between neutral and core ionized $W(CO)_6$, due to the deshielding effect, should lead to substantial M–C vibrational structure on the core ionizations. The widths of the ionizations are largely attributed to this vibrational structure. The vibrational splitting of 2,220 $cm^{-1}$ can be attributed to the C–O stretch. This is larger than that of the ground state of $W(CO)_6$ (2,126 $cm^{-1}$) and is closer to that of $[Re(CO)_6]^+$, which is about 2,200 $cm^{-1}$.[136] Little change in vibrational splitting is seen from the core equivalent ground state, indicating that little change in C–O bonding is occurring, although it is difficult to determine the precise vibrational spacing in the core ionization. The relative intensity of the vibrational shoulder on the core ionization is also much less than that on the valence $t_{2g}$ ionization, indicating a smaller change in CO bond distance with core ionization compared to valence ionization. The core ionization of the metal, which is effectively an increase in oxidation state localized at the metal, may have secondary effects, such as diminishing the ability of the valence $t_{2g}$ orbitals to backbond to the carbonyls, and this could contribute to the observation of a short vibrational progression.

### 5.4 Summary

The observation of vibrational structure in spectroscopic studies of metal carbonyls provides unique insight into the bonding for these types of molecules. Study of the electronic absorption spectra gives further information about the photo-excited states of metal–carbonyls, which is important for understanding their photoreactivity. The vibrational structure observed on the $t_{2g}$ based ionizations of metal hexacarbonyls gives direct experimental observation of the structural influence of metal to carbonyl back-bonding. The main power of this observation is the fact that photoelectron spectroscopy allows the separation, by symmetry, of metal–ligand $\sigma$ and $\pi$ interactions. The observation of vibrational structure on core ionizations in XPS is important for surface science studies of small absorbates on metals in which vibrational and ligand field contributions to the metal core spectra must be considered.

## 6 CLOSING

Understanding of the bonding and structure of metal carbonyls and metallocenes has continued to progress since the discovery of these classes of compounds. The development of this understanding is a wonderful illustration of the interplay of experiment and theory. Vibrational spectroscopy initially provided simple applications of the synergistic bonding aspects of the Dewar–Chatt–Duncanson model. Electronic excitation and ionization spectroscopies have provided direct information on the relationships of electron configurations and distributions to the bonding and stability of these molecules. The increasingly detailed information from the energies, transition probabilities and vibrational fine structures of these spectroscopies has provided increasingly specific challenges to methods of calculation and interpretation. The concepts of valence bond theory that have been widely successful for interpretation of the structure and behavior of organic molecules are wholly inadequate in accounting for the structure and spectroscopy of these molecules. Likewise, the basic principles of crystal field theory that have been applicable to understanding the properties and spectroscopy of classical coordination complexes are generally unsuccessful in accounting for these features of organometallic complexes. The experimental investigation of these compounds has generally paralleled the development of molecular orbital theory. The experimental results, computational methods, and conceptual models have not yet reached a completely common ground for the description of these molecules. Fundamental points of continued focus for metal carbonyls are the relative roles of carbonyl $\sigma$ and $\pi$ bonding in both the ground and excited or ion states, the driving forces and synergism of the bonding, and the relevance of the frontier orbital model. The orbital model becomes an even greater concern for the metallocenes, both in terms of the relative energies of excited and ion states, and in terms of the fundamental bonding and geometrical structures of the molecules. These are continuing areas of investigation. Increasing computational capabilities and the development of new computational approaches, such as the emerging density functional theories, promise interesting new developments in the future.

## ACKNOWLEDGMENTS

We wish to thank Dona Myers for assistance with the manuscript. Work in our laboratory is supported by the National Science Foundation, the Department of Energy, and the State of Arizona's Materials Characterization Program.

## REFERENCES

1. Schützenberger, M.P. *Anneles (Paris)* **1868**, *15*, 100.
2. Mond, L.; Langer, C.; Quincke, F. *J. Chem. Soc. (London)* **1890**, *57*, 749–753, reproduced in *J. Organomet. Chem.* **1990**, *383*, 1–6.
3. Abel, E. *J. Organomet. Chem.* **1990**, *383*, 11–20.
4. Mond, L. *J. Soc. Chem. Ind.* **1895**, *14*, 945–946.
5. Herrmann, W.A. *J. Organomet. Chem.* **1990**, *383*, 21–44 and references therein.
6. Brockway, L.O.; Cross, P.C. *J. Chem. Phys.* **1935**, *3*, 828–833.
7. Pauling, L. *The Nature of the Chemical Bond*, 3rd ed.; Cornell University Press: Ithaca, New York, 1960, 331–336 and references therein.

8. Richardson, J.W. in *Organometallic Chemistry*; Zeiss, H., Ed., Reinhold: New York, 1960, 12–20 and references therein.
9. Orgel, L.E. *An Introduction to Transition-Metal Chemistry*; Methuen: London, 1960, 135–143 and references therein.
10. Dewar, M.J.S. *Bull. Soc. Chim.* **1951**, *18*, C79.
11. Dewar, M.J.S.; Chatt, J.; Duncanson, L.A. *J. Chem. Soc.* **1953**, 2939.
12. Kealy, T.J.; Pauson, P.L. *Nature* **1951**, *168*, 1039–1040.
13. Miller, S.A.; Tebboth, J.A.; Tremaine, J.F. *J. Chem. Soc.* **1952**, 632–635.
14. Wilkinson, G.; Rosenblum, M.; Whiting, M.C.; Woodward, R.B. *J. Am. Chem. Soc.* **1952**, *74*, 2125.
15. For a historic recollection, see Wilkinson, G. *J. Organomet. Chem.* **1975**, *100*, 273–278.
16. a. Mingos, D.M.P.; Wales, D.J. *Introduction to Cluster Chemistry*; Prentice Hall: Englewood Cliffs, New Jersey, 1990.
    b. Adams, R.D. in *The Chemistry of Metal Cluster Complexes*; Shriver, D.F.; Kaesz, H.D.; Adams, R.D., Eds., VCH Publishers, Inc.: New York, 1990, 120–170.
    c. Lewis, J.; Raithby, P.R. *J. Organomet. Chem.* **1995**, *500*, 227–237.
    d. Masters, A.F.; Meyer, J.T. *Polyhedron* **1995**, *14*, 339–365.
    e. Beattie, J.K.; Masters, A.F.; Meyer, J.T. *Polyhedron* **1995**, *14*, 829–868.
17. Somorjai, G.A. *Introduction to Surface Chemistry and Catalysis*; Wiley-Interscience: New York, 1994.
18. Kaminsky, W. *Macromol. Chem. and Phys.* **1996**, *197*, 3907–3945.
19. Homielec, A.E.; Soares, J.B.P. *Prog. Polymer Sci.* **1996**, *21*, 651–706.
20. Rosenblum, M.; Nugent, H.M.; Jang, K.-S.; Labes, M.M.; Cahalane, W.; Klemarczyk, P.; Reiff, W.M. *Macromolecules* **1995**, *28*, 6330–6342.
21. Miller, J.S.; Epstein, A.J.; Reiff, W.M. *Chem. Rev.* **1988**, *88*, 201–220.
22. Lichtenberger, D.L.; Gruhn, N.E.; Renshaw, S.K. *J. Mol. Struct.* **1997**, *405*, 79–86.
23. Ehlers, A.W.; Dapprich, S.; Vyboishchikov, S.F.; Frenking, G. *Organometallics* **1996**, *15*, 105–117.
24. a. Albright, T.A.; Burdett, J.K.; Whangbo, W.-H. *Orbital Interactions in Chemistry*; Wiley-Interscience: New York, 1984.
    b. Cotton, F.A. *Chemical Applications of Group Theory*, 3rd Ed.; Wiley-Interscience: New York, 1990.
25. Lever, A.B.P.; Solomon, E.I. "Ligand field theory and the properties of transition metal complexes," in *Inorganic Electronic Structure and Spectroscopy, Vol. 1, Methodology* (this series), Solomon, E.I.; Lever, A.B.P., Eds.; Wiley-Interscience: New York, 1999.
26. Nakamoto, K. *Infrared and Raman Spectra of Inorganic and Coordination Compounds*, 4th ed.; Wiley-Interscience: New York, 1986.
27. a. Abel, E.W.; Stone, F. G.A. *Quart. Ref.* **1969**, *23*, 219.
    b. Haines, L.M.; Stiddard, M.H.B. *Adv. Inorg. Chem. Radiochem.* **1969**, *12*, 53.
    c. Braterman, P.S. *Metal Carbonyl Spectra*; Academic Press: New York, 1974.
    d. Bigorgne, M. *J. Organomet. Chem.* **1975**, *94*, 161.
    e. Kettle, S.F.A. *Topics Curr. Chem.* **1977**, *71*, 111.
    f. Horwitz, C.P.; Shriver, D.F. *Adv. Organomet. Chem.* **1984**, *23*, 219.
28. Jones, L.H.; McDowell, R.S.; Goldblatt, M. *Inorg. Chem.* **1969**, *8*, 2349.

29. Jones, L.H.; McDowell, R.S.; Goldblatt, M.; Swanson, B.I. *J. Chem. Phys.* **1972**, *57*, 2050.

30. Ottesen, D.K.; Gray, H.B.; Jones, L.H., Goldblatt, M. *Inorg. Chem.* **1973**, *12*, 1051.

31. a. Cotton, F.A.; Kraihanzel, C.S. *J. Am. Chem. Soc.* **1962**, *84*, 4432.
b. Cotton, F.A.; Kraihanzel, C.S. *Inorg. Chem.* **1963**, *2*, 533.
c. Cotton, F.A.; Kraihanzel, C.S. *Inorg. Chem.* **1964**, *3*, 702.

32. Timney, J.A. *Inorg. Chem.* **1979**, *18*, 2502.

33. Hall, M.B.; Fenske, R.F. *Inorg. Chem.* **1972**, *11*, 1619.

34. Fritz, H.P. *Adv. Organomet. Chem.* **1964**, *1*, 239.

35. Brunvoll, J.; Cyvin, S.J.; Schäfer, L. *J. Organomet. Chem.* **1971**, *27*, 107.

36. Wilkinson, G.; Pauson, P.L.; Cotton, F.A. *J. Am. Chem. Soc.* **1954**, *76*, 1970.

37. Bodenheimer, J.S.; Low, W. *Spectrochim. Acta* **1973**, *29A*, 1733.

38. Pavlik, I.; Klilorka, J. *Collect. Czech. Chem. Commun.* **1965**, *30*, 664.

39. Geoffrey, G.L.; Wrighton, M.S. *Organometallic Photochemistry*; Academic Press: New York, 1979.

40. Wrighton, M.; Gray, H.B.; Hammond, G.S. *Mol. Photochem.* **1973**, *5*, 165.

41. Wrighton, M.S.; Handeli, D.I.; Morse, D.L. *Inorg. Chem.* **1976**, *15*, 434–440.

42. Gray, H.G.; Beach, N.A. *J. Am. Chem. Soc.* **1963**, *85*, 2922–2927.

43. Beach, N.A.; Gray, H.B. *J. Am. Chem. Soc.* **1968**, *90*, 5713–5721.

44. Schreiner, A.F.; Brown, T.L. *J. Am. Chem. Soc.* **1968**, *90*, 3366–3374.

45. a. Bock, C.R.; Koerner von Gustorf, E.A. *Adv. Photochem.* **1977**, *10*, 221.
b. Geoffrey, G.L.; Wrighton, M.S. *Organometallic Photochemistry*; Academic Press: New York, 1979, 230–255.

46. Sohn, Y.S.; Hendrickson, D.N.; Gray, H.B. *J. Am. Chem. Soc.* **1971**, *93*, 3603–3612.

47. Rösch, N.; Johnson, K.H. *Chem. Phys. Lett.* **1974**, *24*, 179–184.

48. Green, J.C. *Struct. Bonding (Berlin)* **1981**, *43*, 37–112.

49. Green, J.C. *Acc. Chem. Res.* **1994**, *27*, 131–137.

50. Lloyd, D.R.; Schlag, E.W. *Inorg. Chem.* **1969**, *8*, 2544–2544.

51. Evans, S.; Green, J.C.; Green, M.L.H.; Orchard, A. F.; Turner, D.W. *Discuss. Faraday Soc.* **1969**, *47*, 112–120.

52. Folkesson, B.; Larsson, R. *J. Electron Spectrosc. Relat. Phenom.* **1990**, *50*, 267–276.

53. Cooper, G.; Green, J.C.; Payne, M.P.; Dobson, B.R.; Hillier, I.H. *Chem. Phys. Lett.* **1986**, *125*, 97–100.

54. Cooper, G.; Green, J.C.; Payne, M.P.; Dobson, B.R.; Hillier, I.H. *J. Am. Chem. Soc.* **1987**, *109*, 3836–3843.

55. Hubbard, J.L.; Lichtenberger, D.L. *J. Chem. Phys.* **1981**, *75*, 2560–2568.

56. Turner, D.W. in *Physical Methods in Advanced Inorganic Chemistry*; Hill, H.A.O. and Day, P., Eds.; Wiley-Interscience: London, 1968, 74–105.

57. Evans, S.; Green, M.L.H.; Jewitt, B.; Orchard, A.F.; Pygall, C.F. *J. Chem. Soc. Faraday, Trans. II* **1972**, *68*, 1847.

58. Cauletti, C.; Green, J.C.; Kelly, M.R. *J. Electron Spectrosc. Rel. Phenom.* **1980**, *19*, 327–353 and references therein.

59. Khvostenko, V.I.; Asfandiarov, N.L. *J. Electron Spectrosc. Rel. Phenom.* **1993**, *63*, 419–429 and references therein.

60. Lichtenberger, D.L.; Copenhaver, A.S. *J. Chem. Phys.* **1989**, *91*, 663–673.
61. Ryan, M.F.; Richardson, D.E.; Lichtenberger, D.L.; Gruhn, N.E. *Organometallics* **1994**, *13*, 1190–1199.
62. Lichtenberger, D.L.; Elkadi, Y.; Gruhn, N.E.; Hughes, R.P.; Curnow, O.J.; Zheng, X. *Organometallics*, **1997**, *16*, 5209–5217.
63. Cooper, G.; Green, J.C.; Payne, M.P. *Mol. Phys.* **1988**, *63*, 1031–1051.
64. Brennan, J.; Copper, G.; Green, J.C.; Payne, M.P.; Redfern, C.M. *J. Electron Spectrosc. Rel. Phenom.* **1993**, *66*, 101–115.
65. a. Böhm, M.C. *Inorg. Chim. Acta* **1982**, *62*, 171.
b. Green, J.C.; Payne, M.P.; Teuben, J.H. *Organometallics* **1983**, *2*, 203.
c. van Dam, H.; Oskam, A. *Trans. Met. Chem.* **1985**, *9*, 125.
d. Lichtenberger, D.L.; Darsey, G.P.; Kellogg, G.E.; Sanner, R.D.; Young, V.G. Jr.; Clark, J.R. *J. Am. Chem. Soc.* **1989**, *111*, 5019.
66. von Wald, G.A.; Taylor, J.W. *J. Electron Spectrosc. Rel. Phenom.* **1988**, *47*, 315–325.
67. Yates, B.W.; von Wald, G.A.; Taylor, J.W.; Grimm, F.A.; Tse, J.S. *Chem. Phys.* **1990**, *147*, 431–436.
68. Stucky, G.D.; Rye, R.R.; Jennison, D.R.; Kelber, J.A. *J. Am. Chem. Soc.* **1982**, *104*, 5951–5959.
69. Yencha, A.J. in *Electron Spectroscopy: Theory, Techniques and Applications*, vol. 5, Brundle, C.R.; Baker, A.D., Eds.; Academic Press: New York, 1984, 198–355.
70. Harada, Y.; Ohno, K.; Mutoh, H. *J. Chem. Phys.* **1983**, *79*, 3251–3255.
71. Masuda, S.; Harada, Y. *J. Chem. Phys.* **1992**, *96*, 2469–2474.
72. Higginson, B.R.; Lloyd, D.R.; Burroughs, P.; Gibson, D.M.; Orchard, A.F. *J. Chem. Soc. Faraday Trans. 2* **1973**, *69*, 1659.
73. Bozso, F.; Yates, J.T., Jr.; Metiu, J.A.H.; Martin, R.M. *J. Chem. Phys.* **1983**, *78*, 4256.
74. Sesselmann, W.; Woratschek, B.; Ertl, G.; Küppers, J.; Haberland, H. *Surf. Sci.* **1984**, *146*, 17.
75. Rocker, G.; Tochihara, H.; Martin, R.M.; Metiu, H. *Surf. Sci.* **1987**, *181*, 509.
76. Munakata, T.; Harada, Y.; Ohno, K.; Kuchitsu, K. *Chem. Phys. Lett.* **1981**, *84*, 6–8.
77. Masuda, S.; Aoyama, M.; Harada, Y. *J. Am. Chem. Soc.* **1990**, *112*, 6445–6446.
78. Prins, R. *Mol. Physics* **1970**, *19*, 603.
79. Koopmans, T. *Physica* **1933**, *1*, 104.
80. Jungwirth, P.; Stussi, D.; Weber, J. *Chem. Phys. Lett.* **1992**, *190*, 29.
81. Lichtenberger, D.L.; Copenhaver, A.S. in *Bonding Energetics in Organometallic Compounds*; Marks, T.J., Ed.; ACS Symposium Series 428; American Chemical Society: Washington, D.C., 1990, 84–99.
82. Shustorovich, E.M.; Dyatkina, M.E. *Zh. Strukt. Khim.* **1960**, *1*, 98.
83. Dahl, J.P.; Ballhausen, C.J. *Math.-Fys. Meddr.* **1961**, *33*, 5.
84. Coutière, M.-M.; Demuynck, J.; Veillard, A. *Theoret. Chim. Acta (Berlin)* **1972**, *27*, 281–287.
85. Calabro, D.C.; Lichtenberger, D.L. *Inorg. Chem.* **1980**, *19*, 1732.
86. Lichtenberger, D.L.; Calabro, D.C.; Kellogg, G.E. *Organometallics* **1984**, *3*, 1623–1630.
87. Bagus, P.S. *Phys. Rev. Sect. A* **1965**, *139*, 619.
88. Moser, C.M.; Nesbet, R.K.; Verhaegen, G. *Chem. Phys. Lett.* **1971**, *12*, 230.

89. Roher, M.-M.; Veillard, A. *J. Chem. Soc., Chem. Commun.* **1973**, 250.
90. Rohmer, M.-M.; Denuynck, J.; Veillard, A. *Theoret. Chim. Acta (Berlin)* **1974**, *36*, 93.
91. For example, see: Ohno, M.; von Neissen, W.; Schüle, J. *Chem. Phys.* **1991**, *158*, 1–7.
92. Bagus, P.S.; Walgren, U.I.; Almlöf, J. *J. Chem. Phys.* **1976**, *64*, 2324–2334.
93. a. Epstein, L.M. *J. Chem. Phys.* **1962**, *36*, 2731.
    b. Wertheim, G.K.; Herber, R.H. *J. Chem. Phys.* **1963**, *38*, 2106.
    c. Zahn, U.; Kienle, P.; Eicher, H. *Z. Phys.* **1962**, *166*, 220.
94. Slater, J.C.; Johnson, K.H. *Phys. Rev. B* **1972**, *5*, 844.
95. Baerends, E.J.; Ros, P. *Chem. Phys. Lett.* **1973**, *23*, 391–393.
96. Rösch, N.; Johnson, K.H. *Chem. Phys. Lett.* **1974**, *24*, 179–184.
97. Böhm, M.C. *Theoret. Chim. Acta (Berlin)* **1982**, *61*, 539–558 and references therein.
98. Böhm, M.C.; Gleiter, R. *Theoret. Chim. Acta (Berlin)* **1981**, *59*, 127.
99. a. Ecker, F.; Hohlneicher, G. *Theoret. Chim. Acta (Berlin)* **1972**, *25*, 289.
    b. Hohlneicher, G.; Ecker, F.; Cederbaum, L.S. in *Electron Spectroscopy*, Shirley, D.E., Ed.; North Holland Publishing Company: Amsterdam, 1972.
    c. Nerbrant, P.O. *Int. J. Quantum Chem.* **1975**, *9*, 901.
    d. Cederbaum, L.S.; Domcke, W. *Adv. Chem. Phys.* **1977**, *36*, 205.
    e. von Neissen, W.; Cederbaum, L.S.; Domcke, W.; Schirmer, J. in *Computational Methods in Chemistry*; Bargon, J., Ed.; Plenum Press: New York, 1980.
100. Lathan, W. A.; Curtiss, L.A.; Hehre, W.J.; Lisle, J.B.; Pople, J.A. *Prog. Phys. Org. Chem.* **1974**, *11*, 175.
101. Lüthi, H.P.; Ammeter, J.; Almlöf, J.; Korsell, K. *Chem. Phys. Lett.* **1982**, *69*, 540.
102. Lüthi, H.P.; Ammeter, J.H.; Almlöf, J.; Faergi, K., Jr. *J. Chem. Phys.* **1982**, *77*, 2002–2009.
103. Pietro, W.J.; Hehre, W.J. *J. Comput. Chem.* **1983**, *4*, 241.
104. Marynick, D.S.; Axe, F.U.; Kirkpartick, C.M.; Throckmorton, L. *Chem. Phys. Lett.* **1983**, *99*, 406.
105. Almlöf, J.; Faegri, K., Jr.; Schilling, B.E.R.; Lüthi, H.P. *Chem. Phys. Lett.* **1984**, *106*, 266–270 and references therein.
106. Park, C.; Almlöf, J. *J. Chem. Phys.* **1991**, *95*, 1829–1833.
107. Rösch, N.; Jörg, H. *J. Chem. Phys.* **1986**, *84*, 5967–5968.
108. Jungwirth, P.; Stussi, D.; Weber, J. *Chem. Phys. Lett.* **1992**, *190*, 29–35.
109. Cook, D.B. *Int. J. Quantum Chem.* **1995**, *53*, 309–319.
110. Daul, C.; Güdel, H.-U.; Weber, J. *J. Chem. Phys.* **1993**, *98*, 4023–4029.
111. Fournier, R.; Andzelm, J.; Salahub, D.R. *J. Chem. Phys.* **1989**, *90*, 6371.
112. Andzelm, J.; Wimmer, E. *J. Chem. Phys.* **1992**, *96*, 1280.
113. a. Hillier, I.H.; Saunders, V.R. *Mol. Phys.* **1971**, *22*, 1025.
     b. Caulton, K.G.; Fenske, R.F. *Inorg. Chem.* **1968**, *7*, 1273.
114. Johnson, K.H.; Wahlgren, U. *Int. J. Quantum Chem. Symp.* **1972**, *6*, 243.
115. Johnson, J.B.; Klemperer, W.G. *J. Am. Chem. Soc.* **1977**, *99*, 7132.
116. Bursten, B.E.; Freir, D. G.; Fenske, R.F. *Inorg. Chem.* **1980**, *19*, 1810–1811.
117. Bursten, B.E.; Fenske, R.F. *J. Chem. Phys.* **1977**, *67*, 3138.
118. Nash, C.S.; Bursten, B.E. *New J. Chem.* **1995**, *19*, 669–675.
119. Schädel, M. *Nature* **1997**, *388*, 55.

120. For a few examples, see reference 13 in the following reference.

121. Davidson, E.R.; Kunze, K.L.; Machado, F.B.C.; Chakravorty, S.J. *Acc. Chem. Res.* **1993**, *26*, 628–635 and references therein.

122. Kunze, K.L.; Davidson, E.R. *J. Phys. Chem.* **1992**, *96*, 2129.

123. Frenking, G.; Antes, I.; Böhme, M.; Dapprich, S.; Ehlers, A.W.; Jonas, V.; Neuhaus, A.; Otto, M.; Stegmann, R.; Veldkamp, A.; Vyboishchikov, S.F. in *Reviews in Computational Chemistry*; Liplowitz, K.B., Boyd, D.B., Eds.; VCH: New York, 1996, vol. 8.

124. Dapprich, S.; Frenking, G. *J. Phys. Chem.* **1995**, *99*, 9352.

125. Ehlers, A.W.; Dapprich, S.; Vyboishchokov, S.F.; Frenking, G. *Organometallics* **1996**, *15*, 105–117.

126. Arratia-Perez, R.; Yang, C.Y. *J. Chem. Phys.* **1985**, *83*, 4005–4014.

127. Bersurker, I.B. *Electronic Structure and Properties of Transition Metal Compounds: Introduction to the Theory*; Wiley-Interscience: New York, 1996, 224–227.

128. Bersurker, I.B. *Electronic Structure and Properties of Transition Metal Compounds: Introduction to the Theory*; Wiley-Interscience: New York, 1996, 285–292 and references therein.

129. Jones, L.H. *Spectrochim. Acta* **1963**, *19*, 329–338.

130. Trogler, W.C.; Desjardins, S.R.; Solomon, E.I. *Inorg. Chem.* **1979**, *8*, 2131–2136.

131. Wilson, R.B.; Solomon, E.I. *Inorg. Chem.* **1978**, *17*, 1729–1736.

132. Herzberg, G. *Molecular Spectra and Molecular Structure*; Van Nostrand-Rheinhold: New York, 1950, vol. 1.

133. Hubbard, J.L.; Lichtenberger, D.L. *J. Am. Chem. Soc.* **1982**, *104*, 2132–2138.

134. Hu, Y.-F.; Bancroft, G.M.; Liu, Z.; Tan, K.H. *Inorg. Chem.* **1995**, *34*, 3716–3723.

135. Jolly, W.; Hendrickson, D.M. *J. Am. Chem. Soc.* **1980**, *92*, 1863.

136. Braterman, P.S. *Metal Carbonyl Spectra*; Academic Press: London, 1975.

# 10 Spin Transition in Iron(II) Compounds

PHILIPP GÜTLICH AND HARTMUT SPIERING
Institut für Anorganische Chemie und Analytische Chemie
Johannes Gutenberg-Universität
Staudingerweg 9
D-55099 Mainz, Germany
E-mail: p.guetlich@uni-mainz.de

ANDREAS HAUSER
Département de Chimie Physique
Section de Chimie – Sciences II
Université de Genève
30, Quai E. Ansermet
CH-211 Genève 4, Switzerland



*Inorganic Electronic Structure and Spectroscopy, Volume II: Applications and Case Studies.*
Edited by E. I. Solomon and A. B. P. Lever.
ISBN 0-471-32682-8. 

## 1 INTRODUCTION

The electronic ground state properties of most transition metal compounds do not change appreciably with temperature. They are either diamagnetic or paramagnetic, depending essentially on the number of unpaired valence electrons. For octahedral complexes of $3d$ transition metal ions with electron configuration $3d^4$ up to $3d^7$ the electronic ground state may be either high-spin (HS) with maximum spin multiplicity or low-spin (LS) with minimum spin multiplicity depending on the ligand field (LF) strength as compared to the spin pairing energy. The HS state is adopted when weak-field ligands are coordinated, while the LS state is obtained when strong-field ligands are involved. If the LF strength in a particular complex is such that the zero-point energy difference between the HS and LS state becomes comparable to thermal energies, $k_B T$, then thermally induced spin transition or spin crossover becomes observable.

Thermal spin crossover was discovered more than sixty years ago by Cambi and coworkers for iron(III) trisdithiocarbamate complexes.[1–4] Later, many other transition metal complexes exhibiting thermal spin crossover have been found. Most of them are Fe(II), Fe(III),[5] and to a lesser extent Co(II).[6] Relatively few examples refer to Mn(II),[7–10] Mn(III),[11–13] Co(III),[14–16] Ni(II),[17–20] and Mo(II).[21] It is, however, iron(II) ($3d^6$) for which the phenomenon has been most widely studied, although the first example of an iron(II) spin crossover complex viz. $[Fe(phen)_2(NCS)_2]$ (phen = 1,10-phenantroline)[22–24] was communicated more than thirty years after Cambi's work. Not only is the number of iron(II) spin crossover systems relatively large, but the range of behavior exhibited by them is extremly diverse. It is principally through the study of iron(II) systems that our understanding of the nature of the spin crossover phenomenon has developed.[5,22–30]

When a thermally induced spin transition occurs, a strong temperature-dependence of the properties of the system will be observed, ranging from those associated with the pure LS state to those associated with the pure HS state. From the changes observed, it is possible to evaluate the fraction of molecules in each spin state, $\gamma_{HS}$ or $\gamma_{LS}$ ($\gamma_{HS} + \gamma_{LS} = 1$), at a particular temperature. The plot of $\gamma_{HS}$ vs. temperature is generally referred to as a *spin transition curve*. Such curves can be diagnostic

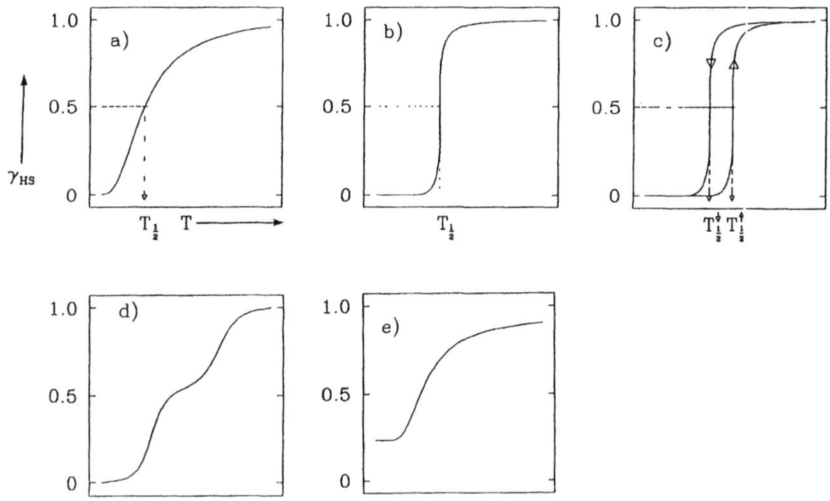


**Figure 1.** Classification of spin transition curves: (a) gradual, (b) abrupt, (c) discontinuous with hysteresis, (d) two-step transition, and (e) incomplete spin transition.

of the nature of the spin crossover. Spin transition curves for systems in solution closely resembles the one shown in Figure 1a. The transition occurs essentially at the molecular level and, in the absence of the constraints of lattice interactions, may be treated as an equilibrium involving thermal population of all vibronic levels of the two spin states. For systems in solution the spin transition is affected primarily by fine-tuning of the ligand field through ligand modification or replacement. For solid systems transition curves reveal the full range of behaviours shown schematically in Figures 1a–c. The transition may be gradual and continuous over an extended temperature range as in solution (Figure 1a), or it may be more abrupt and occur within a very narrow temperature range (Figure 1b). The transition may be associated with a thermal hysteresis loop (Figure 1c) or it may be a two-step process (Figure 1d). Finally, transitions may be incomplete at either or both extremes of the spin transition curve $\gamma_{HS}(T)$, as indicated in Figure 1e. The occurrence of hysteresis is often indicative of a structural phase transition accompanying the spin change. In this instance the transition is described as discontinuous, in contrast to those which do not show hysteresis and are classified as continuous. For the discontinuous transitions X-ray powder diffraction patterns characteristic of two phases are observed, the relative contributions changing with the progress of the transition. For continuous transitions the broad features of the pattern are retained over the entire transition region, a gradual displacement of the peaks being observed.[25]

The different manifestations of spin transitions in solid systems are subject to lattice forces, and thus the nature of the spin transition curve $\gamma_{HS}(T)$ can be strongly influenced by chemical modifications[24,29] which affect the lattice interactions. These include the nature of the anion associated with the complex cation, the extent and nature of solvation of the complex, variations in ligand substituents and actual ligand replacement. In addition, isotopic substitution within the complex salt, e.g., hydrogen by deuterium, being the most subtle modification possible, can influence the nature of the $\gamma_{HS}$ vs. $T$ plot. Finally, mechanical treatment, such as grinding the sample, can drastically change the features of the transition. Although the interactions considerably complicate the crossover behavior, they are in fact responsible for much of the current interest in the spin crossover phenomenon. This interest is focused on the potential for applications of systems, particularly undergoing discontinuous transitions as displayed in Figure 1c, in electronic and optical switching and memory devices.[31] The yearly publication rate in this highly innovative research area of transition metal chemistry has considerably increased in recent years and reaches 100 on average by now. By far, the majority of the many iron(II) systems, which have been shown to undergo a thermal spin transition, contain nitrogen-donor ligands bound to the metal ion in six-coordinate [Fe(II)$N_6$] species, but iron(II) complexes with [Fe(II)$N_4O_2$] and [Fe(II)$P_4X_2$] (X = halogen) cases have also been shown to exhibit thermal spin transition.[24,30]

It is not our purpose to present here a rigorous review article covering quantitatively the research work performed with spin crossover systems so far. We rather restrict ourselves to a few systems, such as the picolylamine, tetrazole and triazole complexes of iron(II), and focus on the problems connected with the questions of *why* and *how* thermal and optical spin transition in solid iron(II) complexes occur. We shall begin in the following chapter with ligand field theoretical considerations and thereby discuss the conditions for thermal spin crossover to occur in iron(II) complexes.

## 2 LIGAND–FIELD THEORETICAL CONSIDERATIONS

In an octahedral LF, the $^{2S+1}L$ terms of a free $d^6$ ion are split according to the energy level diagram[32,33] shown in Figure 2. The parameter along the abscissa is the LF strength $10Dq$, representing the splitting of the five metal $d$ orbitals into the $t_2$ and $e$ subsets under the influence of the ligands.

As mentioned in the introduction, iron(II) complexes with ligands exerting only a weak influence on the $d$ electrons of the central ion in relation to the electron–electron repulsion, as for example in $[Fe(H_2O)_6]^{2+}$, have a paramagnetic HS $^5T_2$ ground state with a $(t_2^4e^2)$ electron configuration stemming from the $^5D$ term of the free ion. With increasing LF strength, however, the LS $^1A_1$ state, with a $t_2^6$ electron configuration, emerging from the $^1I$ term of the free ion, is quickly lowered in energy, and from a critical LF strength $Dq$ onwards ($\approx$1700 cm$^{-1}$ in the energy level diagram of Figure 2), where the LF stabilisation exceeds the electron–electron repulsion, it, in turn, becomes the ground state. This is, for example, the case for the diamagnetic $[Fe(CN)_6]^{4-}$ complex.

For a proper appreciation of the importance of the crossover point $\Delta_{crit}$ and the phenomenon of a thermal spin transition in some iron(II) complexes, it is not sufficient to discuss the energy level diagram of Figure 2 on its own because for a given ligand-field strength it gives only the electronic energies of the excited states with

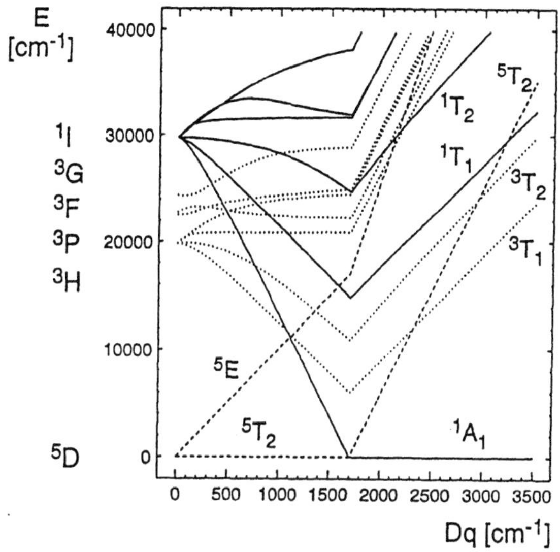


**Figure 2.** Energy level diagram for iron(II) with six $d$ electrons. The diagram is calculated using the free ion values for the Racah parameters, $B = 917\ \text{cm}^{-1}$ and $C = 4040\ \text{cm}^{-1}$.[32]

respect to the electronic energy of the corresponding ground state. It is, in fact, indispensable to look also at the potential energy surfaces[33] of the two states because the ligand-field strength for a given set of ligands is not only determined by the properties of those ligands. It is also a strong function of the metal–ligand distance $r(\text{M}-\text{L})$ according to

$$10Dq \sim r^{-n} \tag{1}$$

with $n = 5$ for charged ligands (point charge model) to 6 for neutral ligands (dipoles),[34] whereas the electron–electron repulsion within the $d$ shell as described by the Racah parameters[33] depends only weakly on $r(\text{M} - \text{L})$. In the most simple approach, the potentials are modelled along one single configurational coordinate.[35] For the ensuing discussion, the metal–ligand breathing mode, that is, the totally symmetric normal mode, is the most important configurational coordinate.

Usually $10Dq$ is known from optical spectroscopy at the equilibrium bond length of a particular state. Thus, in the configurational coordinate diagram, only the vertical energies of the other states with respect to that particular state are accessible experimentally. However, starting from a harmonic potential, for example, the $^1A_1$ state with an equilibrium bond length $r^0_{\text{LS}}$ and a corresponding value for $10Dq^{\text{LS}}$, the energies of the other states can easily be calculated as a function of $r(\text{M} - \text{L})$ using the proportionality

$$10Dq(r) = 10Dq^{\text{LS}}\left(\frac{r^0_{\text{LS}}}{r}\right)^n \tag{2}$$

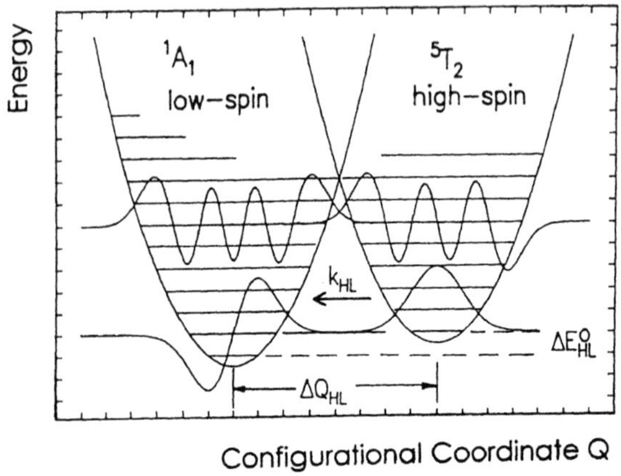


**Figure 3.** Configurational coordinate diagram along the totally symmetric breathing mode for the two lowest states in an octahedral $d^6$ system, the LS $^1A_1$, and the HS $^5T_2$ state. The condition for an entropy driven thermal spin transition is that $\Delta E^0_{\mathrm{HL}} \approx k_B T$.

A natural consequence of such a calculation is that for the other LF states the resulting potential wells are shifted both vertically and horizontally with respect to the reference state. In particular, the HS $^5T_2$ state is always shifted to substantially larger metal–ligand distances irrespective of the ligand, as shown in Figure 3. This is not unexpected in view of the electronic configuration of the HS state with two of the six electrons in the anti-bonding *e* orbitals as opposed to the LS state with all six electrons in the essentially non-bonding or due to $\pi$-back donation slightly bonding $t_2$ orbitals. An immediate consequence of this difference in equilibrium bond length $\Delta r^0_{\mathrm{HL}} = r^0_{\mathrm{HS}} - r^0_{\mathrm{LS}}$ between the LS and the HS state is that at $r^0_{\mathrm{HS}}$ the corresponding LF strength $10Dq^{\mathrm{HS}}$ is substantially smaller.

The vertical shift of the two potentials relative to each other depends very much on the specific properties of the ligands. For complexes with suitable ligands, the difference in zero-point energies between the two states, $\Delta E^0_{\mathrm{HL}} = E^0_{\mathrm{HS}} - E^0_{\mathrm{LS}}$, can fall within the range of thermal energies. In such a case, only the $^1A_1$ state is occupied at low temperatures, whereas at elevated temperatures a thermal population of the $^5T_2$ state occurs. Because the $^5T_2$ has a 15-fold electronic degeneracy and a higher vibrational density of states than the $^1A_1$ state, this entropy-driven spin transition may be almost quantitative. Conversely, for complexes with a HS ground state, the LS state will never be populated thermally to any appreciable extent even for very small energy gaps.

What then is the significance of the crossover point $\Delta_{\mathrm{crit}}$ in the energy level diagram? It has to be identified with the intercept of the two potentials in the configurational coordinate diagram of Figure 3. This point can never correspond to an equilibrium configuration of the ground state! Thus, for the small values of the zero-point energy difference, $\Delta E^0_{\mathrm{HL}}$ required for a spin crossover compound the critical LF strength is in between $10Dq^{\mathrm{HS}} < \Delta_{\mathrm{crit}} < 10Dq^{\mathrm{LS}}$.

Iron(II) spin crossover compounds mainly have sixfold nitrogen coordination. Typical values for $r^0_{LS}$ are in the range between 1.96 and 2.00 Å. Those for $r^0_{HS}$ lie between 2.16 and 2.20 Å and are typically longer by $\Delta r^0_{HL} \approx 0.16$–$0.22$ Å.[36–39] Thus for neutral ligands

$$\frac{10Dq^{LS}}{10Dq^{HS}} = \left(\frac{r^0_{HS}}{r^0_{LS}}\right)^6 \approx 1.75 \tag{3}$$

The range of LF strengths for which spin crossover compounds can be expected is rather narrow because of the condition that $\Delta E^0_{HL}$ has to be on the order of thermal energies. Therefore we expect

$$10Dq^{HS} < 11{,}000\ \text{cm}^{-1} \qquad \text{HS complexes}$$

$$\left.\begin{aligned} 10Dq^{HS} &\approx 11{,}000 - 13{,}000\ \text{cm}^{-1} \\ 10Dq^{LS} &\approx 19{,}000 - 21{,}000\ \text{cm}^{-1} \end{aligned}\right\} \quad \text{spin crossover complexes}$$

$$10Dq^{LS} > 21{,}000\ \text{cm}^{-1} \qquad \text{LS complexes}$$

Minor changes in the ligands and even in the second coordination sphere can shift the thermal spin transition to very different temperatures or even suppress it entirely.

## 3 PHYSICAL CHARACTERIZATION

Spin crossover in iron(II) compounds, ${}^1A_{1g} \rightleftharpoons {}^5T_{2g}$ is accompanied by large changes of structural and electronic properties. Thus, the spin transition is frequently indicated by a color change, from generally feebly colored in the HS state to more deeply colored in the LS state. Therefore, the optical spectra in the UV/visible region differ markedly for the HS and the LS states and are well suited to follow the spin state transition quantitatively. The magnetic behavior, of course, also changes drastically. A change from no unpaired electrons in the LS state to 4 unpaired electrons in the HS state, i.e., a change from diamagnetic or weakly paramagnetic to strongly paramagnetic behavior is easily verified by measuring the magnetic susceptibility. The increase in metal–ligand bond length and the concomitant weakening of the metal–ligand bands can be detected by vibrational spectroscopy, i.e., by a decrease in the metal–donor atom stretching frequencies, and directly from a crystal structure determination, e.g., by X-ray diffraction. Mössbauer spectroscopy is a very elegant tool for the investigation of spin transitions in solid iron compounds; the HS and LS states can unambigiously be assigned on the basis of two parameters: the isomer shift and the quadrupole splitting. Finally, thermoanalytical measurements can be employed to identify the spin transition and to explore the changes in enthalpy and entropy. In the following we shall illustrate the use of these techniques in spin crossover research.

### 3.1 Optical Spectroscopy

From the energy level diagram shown in Figure 2, one sees that for a $^5T_{2g}$ ground state only one spin-allowed $d$–$d$ transition, viz., $^5T_{2g} \rightarrow {^5E_g}$ in the NIR can be expected. For a $^1A_{1g}$ ground state two spin-allowed transitions, $^1A_{1g} \rightarrow {^1T_{1g}}$ and $^1A_{1g} \rightarrow {^1T_{2g}}$ are predicted, and these fall in the UV/visible region.

This is illustrated in Figure 4a and b where the single crystal absorption spectra of $[Fe(ptz)_6](BF_4)_2$ (ptz = 1-propyl-tetrazole) at 293 K (HS state) and 20 K (LS state), respectively, are depicted. The spectra show the relatively weak quintet-quintet band at 12,000 cm$^{-1}$ of the colorless crystal in the HS state and the more intense singlet-singlet excitations to the $^1T_{1g}$ and $^1T_{2g}$ states at 18,000 and 25,000 cm$^{-1}$, respectively, in the LS state. Also shown in Figure 4b, are the very weak spin-forbidden singlet-triplet transitions. These spectra can be used to verify the relation of Equa-

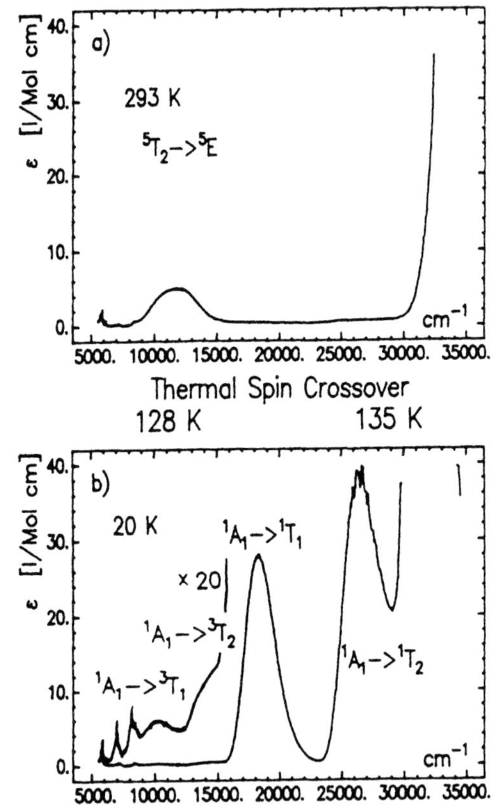


**Figure 4.** Single crystal UV/VIS absorption spectra along the crystallographic $c$-axis of the $R\bar{3}$ phase of $[Fe(ptz)_6](BF_4)_2$, showing the $d$–$d$ transitions (a) at 293 K, and (b) at 20 K (after rapidly cooling). The enlarged inset (×20) in (b) shows the weak singlet-triplet transitions. [All spectra show the molar absorption coefficient $\varepsilon$ [$\ell$/(mol cm) versus the wave number $\tilde{\nu}$ [cm$^{-1}$] (adapted from Reference 40).]

tion 3 giving the ratio of LF strength of the two states. The value of the LF strength in the HS state is given by:

$$10Dq^{\mathrm{HS}} = E(^5E) - E(^5T_2) = 11{,}760\ \mathrm{cm}^{-1}$$
$$10Dq^{\mathrm{LS}} \approx E(^1T_1) - E(^1A_1) + \frac{1}{4}\left[E(^1T_2) - E(^1T_1)\right] = 20{,}550\ \mathrm{cm}^{-1} \quad (4)$$

$10Dq^{\mathrm{LS}}/10Dq^{\mathrm{HS}} \approx 1.75$, which is almost identical to the value estimated in Section 2. The second equation is obtained setting $C = 4B$ and ignoring configuration interaction according to Reference 41. A more involved treatment using the full strong-field matrices for the electron–electron repulsion of Sugano et al.[32] gives essentially the same value.[42] From the area $A(T)$ of, e.g., the $^1A_{1g} \rightarrow {}^1T_{1g}$ band as a function of the temperature the spin transition curve can be obtained quantitatively according to:

$$\gamma_{\mathrm{HS}}(T) = 1 - \gamma_{\mathrm{LS}}(T) = 1 - \frac{A(T)}{A(T \rightarrow 0)} \quad (5)$$

As an example, Figure 5 shows the temperature dependence of the optical density OD of the $^1A_{1g} \rightarrow {}^1T_{1g}$ and the $^1A_{1g} \rightarrow {}^1T_{2g}$ absorption bands of $[Fe_{0.87}Zn_{0.13}(ptz)_6](BF_4)_2$.[43] The corresponding transition curve derived from the $^1A_{1g} \rightarrow {}^1T_{1g}$ band is plotted in Figure 6 together with the transition curves of other mixed crystals differing in their iron concentration. This study demonstrates the effect of metal dilution on the transition behavior: The transition curves become

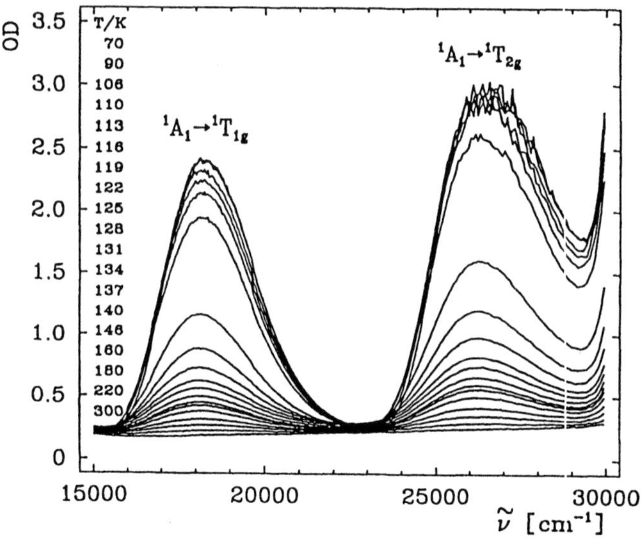


**Figure 5.** Temperature dependence of the optical density OD of the $^1A_1 \rightarrow {}^1T_1$ and the $^1A_1 \rightarrow {}^1T_2$ absorption bands of $[Fe_{0.87}Zn_{0.13}(ptz)_6](BF_4)_2$ (from Reference 43).

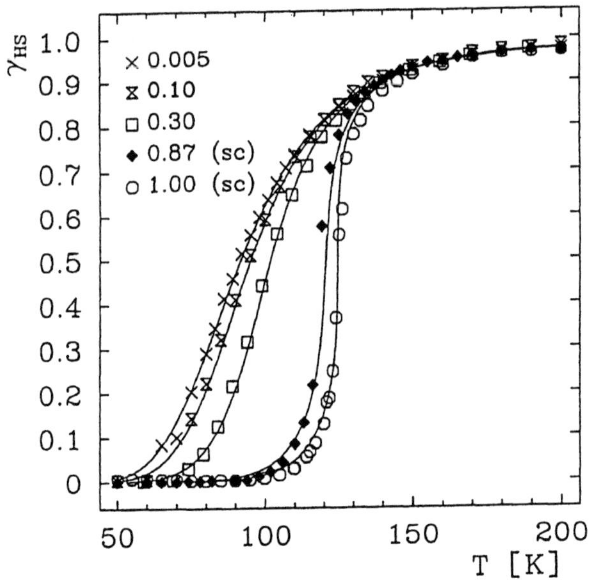


**Figure 6.** Transition curves $\gamma_{HS}(x, T)$ from optical absorption measurements on single crystals of $[Fe_xZn_{1-x}(ptz)_6](BF_4)_2$ in the structure $R\bar{3}$. In the case of $x = 0.87$ and 1.0 the structure was preserved by supercooling (sc) (from References 29,46).

more gradual and the transition temperature $T_{1/2}$ shifts to lower temperatures with decreasing iron concentration. This is clear evidence for a reduction of cooperative interactions and will be treated in more detail in Section 4.

When a neat single crystal of $[Fe(ptz)_6](BF_4)_2$ is cooled slowly, the spin transition is accompanied by a crystallographic phase change with the space group $R\bar{3}$ above and $P\bar{1}$ below the transition temperature,[44] and a hysteresis in the $\gamma_{HS}$ curve is observed with $T_{1/2}^{\downarrow} = 128$ and $T_{1/2}^{\uparrow} = 135$[45] (see Section 3.2). A rapidly cooled crystal, however, retains its high temperature structure $R\bar{3}$, but still exhibits the thermal spin transition as demonstrated in Figure 6 for the crystals with $x = 1.0$ and $x = 0.87$. Thus it is clear that in this case the thermal spin crossover phenomenon is not induced by the crystallographic phase change.

## 3.2 Magnetic Susceptibility

The magnetic behavior of any material may be characterized by measuring its susceptibility using a Faraday balance, Foner type, or SQUID magnetometer.[34,47] The susceptibility $\chi(T)$ of a spin crossover compound is determined by the temperature-dependent fractional contributions of the susceptibility of the two spin states according to:

$$\chi(T) = \gamma_{HS}(T) \cdot \chi_{HS} + \left(1 - \gamma_{HS}(T)\right) \cdot \chi_{LS} \tag{6}$$

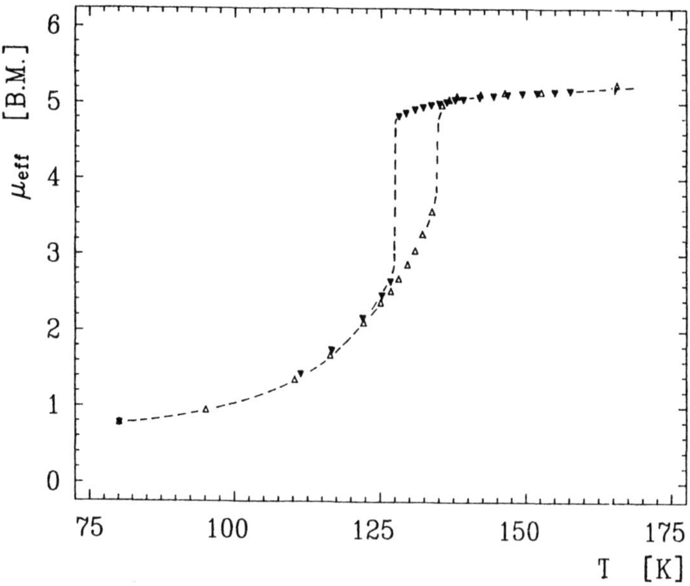


**Figure 7.** Effective magnetic moment $\mu_{eff}$ of the $^5T_{2g}$ state in $[Fe(ptz)_6](BF_4)_2$ as a function of temperature derived from magnetic susceptibility measurements on slowly cooling (▼) and heating (Δ) (from Reference 45).

After determination of the susceptibilities of the compound in the pure HS and LS states (at sufficiently high and low temperatures when the spin transition is complete) the spin state conversion curve $\gamma_{HS}$ is readily obtained. Alternatively, one can also derive the effective magnetic moment $\mu_{eff} = \sqrt{8\chi \cdot T}$ [34] and plot this as a function of temperature. Figure 7 shows the dramatic change of the effective magnetic moment of $[Fe(ptz)_6](BF_4)_2$ upon slow temperature variation. The above mentioned hysteresis of 7 K is clearly discernible.

The spin-only moment $\mu_{s.o.} = \sqrt{4S(S+1)}\mu_B$ of the HS state with $S = 2$ is 4.9 B.M. and that of the LS state with $S = 0$ is 0 B.M. However, due to orbital contributions the effective magnetic moments are slightly higher than the spin-only values. For a quantitative evaluation, it is thus neccessary to explicitely consider the effects of spin-orbit coupling. The influence of anisotropic thermal expansion on the spin-orbit coupling and the successive thermal depopulation of split $^5T_{2g}$ states in the case of relatively low transition temperatures ($T_{1/2} < 100$ K) may lead to a temperature dependence of the effective moment of the HS state.

### 3.3 Vibrational Spectroscopy

As mentioned above, a depletion of charge in the antibonding $e_g$ orbitals and simultaneous filling of the slightly bonding $t_{2g}$ orbitals causes a strengthening of the metal–donor atom bonds. This should be reflected in the vibrational spectrum in the region between ~250 and 500 cm$^{-1}$, where the metal–donor atom bond stretching frequencies of transition metal compounds usually occur. Vibrational spectroscopy was applied very early to spin crossover compounds and significant changes in stretching frequencies with the spin transition were observed.[48–50] The charge rearrangement accompanying the spin transition affects not only the stretching and deformation frequencies of the metal bonds, but can also induce changes in intra-ligand vibrational frequencies, e.g., the NCS stretching mode in $[Fe(phen)_2(NCS)_2]$, which can be used to monitor the spin transition.[49,50] All lattice vibrational modes may, in fact, be affected due to the lattice deformation accompanying the spin transition. Even the characteristic $\nu_4(F_2)$ band of the $BF_4^-$ anion, appearing at 524 $cm^{-1}$, is affected by the spin transition. Thus the anion, which is certainly not coordinated to the metal, "feels" the effect of the spin transition in the coordination center.

### 3.4 Mössbauer Spectroscopy

$^{57}$Fe Mössbauer spectroscopy is a particularly elegant and powerful tool in probing the different oxidation and spin states of iron compounds.[51] The isomer shift $\delta$ (usually given relative to $\alpha$-iron) and the quadrupole splitting $\Delta E_Q$, the two most important parameters derived from a Mössbauer spectrum, are different for iron(II)-HS and iron(II)-LS. This has been discussed in detail in a separate chapter of this series devoted to Mössbauer spectroscopy.[52] In the case of iron(II) the isomer shift $\delta$ is more positive for the HS state (ca. 1 mm/s) than for the LS state (0–0.3 mm/s) for the following reason: Due to the $\pi$-back donation of charge from the filled $t_{2g}$ orbitals of iron(II) in the LS state into the ligand $\pi^*$-orbitals, the $3d$-electron density decreases. As a consequence, the shielding effect of the $d$-electrons for $s$-electrons (mainly $3s$) towards the nuclear charge decreases, which in turn increases the $s$-electron density $|\Psi_A(0)|^2$ at the nucleus. As the relative change of the nuclear volume on going from the excited to the ground state of Fe,[57] $R_e^2 - R_g^2$ is negative, the higher electron density at the nucleus causes a more negative shift of $\delta$ in the LS state as compared to the HS state. $\pi$-back donation from the HS configuration is not as effective, hence the shielding effect of $3d$-electrons is greater and the $s$-electron density at the nucleus decreases, which in turn causes a more positive isomer shift.

Iron(II)-HS compounds usually have a large quadrupole splitting (near 3 mm/s) due to a large valence electron contribution to the electric field gradient (EFG); this arises from a noncubic distribution of the four $3d$ electrons over the $t_{2g}$ orbitals which are split (e.g., into $b_{2g}$ and $e_g$ orbitals in a ligand field of $D_{4h}$ symmetry) due to the Jahn–Teller distortion or noncubic coordination of the ligands. The lattice contribution arising from such an axial distortion is usually smaller than the valence electron contribution and often has an opposite sign. The quadrupole splitting of iron(II)-LS compounds is generally considerably smaller than that of iron(II)-HS compounds because the $t_{2g}$ orbitals, whether split or unsplit, are fully occupied by the six $3d$ electrons; the partial contributions from each orbital add up to a vanishing valence electron contribution, and what remains, e.g., in iron compounds with ligands of different kinds or with bi- or multidentate ligand molecules, is a lattice contribution to the EFG.

These typical Mössbauer parameters effectively give a fingerprint to each spin state and enable one to differentiate readily between the different spin states. When a thermally induced spin crossover occurs the intensity of the HS quadrupole doublet decreases and that of the LS doublet increases as the temperature is lowered. An example is presented in Figure 8, which shows a series of temperature-dependent $^{57}Fe$ Mössbauer spectra of $[Fe(phen)_2(NCS)_2]$ (phen = 1,10-phenanthroline), which, as stated earlier, is the first unequivocal spin crossover compound of iron(II) reported in the literature.[53–55] The room temperature spectrum (300 K) contains only the HS quadrupole doublet with a splitting close to 3 mm/s. As the temperature is lowered, the quadrupole doublet of the LS state appears with increasing intensity at the expense of the HS intensity. At 77 K, the spin transition to LS is complete. Most of the transition takes place in an abrupt fashion within a few Kelvin around 185 K. In fact, there is even a small hysteresis (0.2 K width) in the transition curve.[56]

The quantitative determination of the spin transition curve from a series of Mössbauer spectra over temperature is based on the evaluation of the areas $A_{HS}$ and $A_{LS}$ of the HS and LS resonance lines. These are proportional to the products $f_{HS} \cdot \gamma_{HS}$ and $f_{LS} \cdot \gamma_{LS}$, respectively, according to:

$$\frac{A_{HS}}{A_{HS} + A_{LS}} = \frac{\gamma_{HS} \cdot f_{HS}}{\gamma_{HS} \cdot f_{HS} + (1 - \gamma_{HS}) \cdot f_{LS}} \tag{7}$$

where $f_{HS}$ and $f_{LS}$ are the so-called Lamb-Mössbauer factors (LMF) for the two spin states. The LMF is a measure for the tightness of binding of the Mössbauer nuclide to its environment. They have been found to be practically equal for most spin crossover systems, particularly for those without structural changes.

## 3.5 Crystal Structure Determination

As pointed out above, a thermal spin transition is always accompanied by marked changes of the metal-donor atom bond lengths. This can be directly observed by X-ray diffraction on a single crystal or on a polycrystalline sample as a function of temperature. The metal–ligand bond lengths decrease by as much as ca. 10%, and the unit cell volumes decrease by ca. 3–4% when going from HS to LS.[57,58] The spin transition may also be accompanied by a change in the space group, i.e., a first-order crystallographic phase change. If this is the case, the spin state conversion curve shows hysteresis. However, the reverse is not necessarily true, i.e., the observation of hysteresis in the spin transition curve does not necessarily imply the occurrence of a crystallographic phase change. Examples are the spin crossover systems $[Fe(phen)_2(NCS)_2]$ and $[Fe(2\text{-pic})_3]Br_2{\cdot}EtOH$ (2-pic = 2-picolylamine), for which hysteresis in the function $\gamma_{HS}(T)$ has been measured,[56] but for which no change in the space-group was observed by variable temperature crystal structure determination.[36,59]

Lattice expansion with increasing fractions of HS molecules was observed for many examples by variable-temperature X-ray studies.[60] It finally emerged that the lattice volume is proportional to the fraction of HS molecules with a high degree

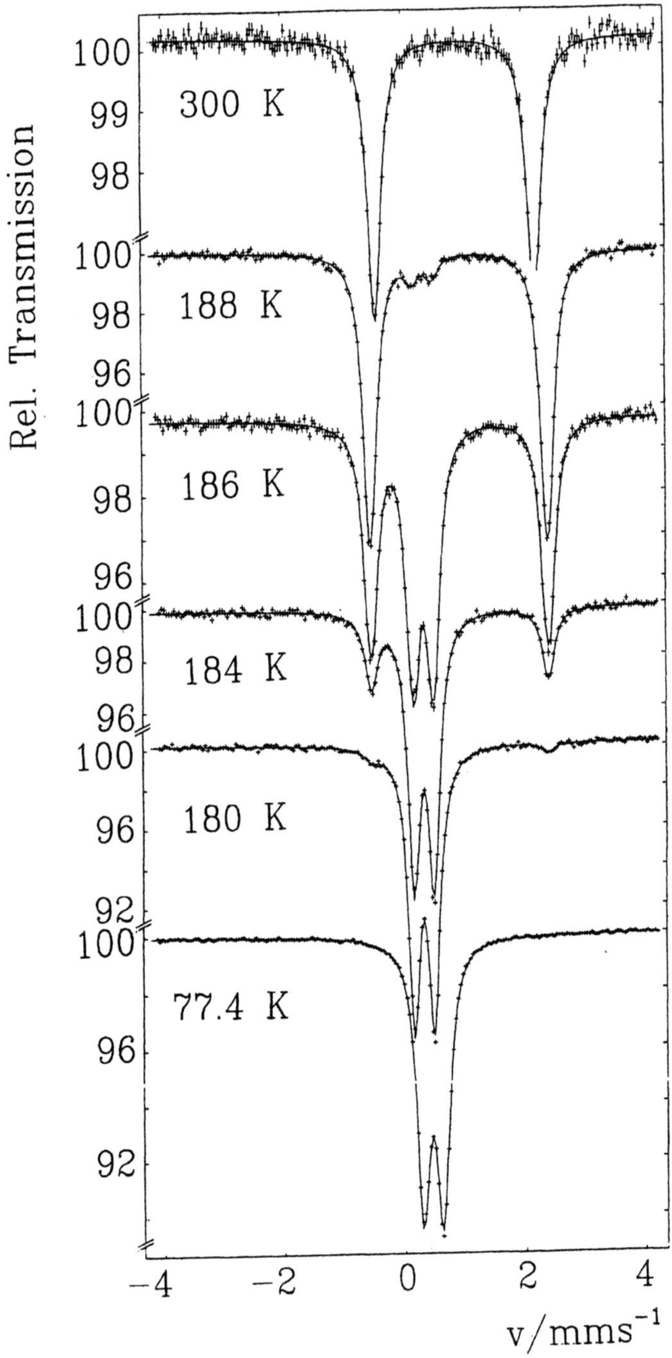


**Figure 8.** $^{57}$Fe Mössbauer spectra of $[Fe(phen)_2(NCS)_2]$ at different temperatures (from Reference 24) showing the abrupt spin transition at $T_{1/2} \simeq 185$ K.

of accuracy.[61] The total increase in volume $V(T) - V_{LS}(T_0)$ (relative to the unit cell volume in the LS state at a reference temperature $T_0$) is the sum of the linear volume increase due to thermal lattice expansion, $V_{LS}\alpha_v(T - T_0)$ ($\alpha_v$ is the linear expansion coefficient), and the volume increase due to the spin transition, $\Delta V_{HL} \cdot \gamma_{HS}(T)$ according to:

$$V(T) = V_{LS}(T_0)\big[1 + \alpha_v(T - T_0)\big] + \Delta V_{HL} \cdot \gamma_{HS}(T) \tag{8}$$

As the volume difference $\Delta V_{HL} = V_{HS} - V_{LS}$ between the HS and LS unit cell is temperature independent, we can neglect the small differences between the expansion coefficients of the pure HS and LS lattices. The $[Fe(2\text{-pic})_3]Br_2{\cdot}EtOH$ complex, being the first example of a spin crossover with a temperature hysteresis without a crystallographic change of structure,[36] and the anomalous spin transition in the analogous $[Fe(2\text{-pic})_3]Cl_2{\cdot}EtOH$ complex with a step in the transition curve are especially convincing because the $\gamma_{HS}(T)$ curves are exactly mirrored (that is, with the same step) in the temperature dependence of the unit cell volume $V$ and the lattice parameters $a$, $b$, and $c$ themselves, as shown in Figure 9. Irrespective of an overall volume decrease, certain distances in the lattice may well increase and thus cause a lattice deformation. This deformation can be adequately described by two tensors, an expansion tensor $\alpha$ and a temperature-independent tensor $\varepsilon$. The unit cell vectors $\mathbf{x}(T)$ at temperature $T$ are related to the vectors $\mathbf{x}(T_0)$ at temperature $T_0$ according to:

$$\mathbf{x}(T) = \big[1 + \alpha(T - T_0) + \varepsilon\gamma_{HS}(T)\big]\mathbf{x}(T_0) \tag{9}$$

Equation 8 is a special case of this equation with $\alpha_v = \text{trace}(\alpha)$ and $\Delta V_{HL}/V_{LS} = \text{trace}(\varepsilon)$. The curves in Figure 9 have been calculated with the experimental values for the HS fraction $\gamma_{HS}(T)$.

The bond-length change in the first coordination sphere of each spin-changing molecule is considered as the origin for this deformation. Such information has been of utmost importance for the development of a model to describe the cooperative interactions within the framework of elasticity theory (see Section 4).

It should be noted that EXAFS measurements have also been successfully applied to the determination of the metal-donor atom bond length changes in spin crossover complexes.[62–66] This method is very useful in those cases where X-ray diffraction becomes difficult or impossible because of highly dispersed particles.

### 3.6 Calorimetry

The spin transition can be related to a phase transition associated with a change of the Gibbs free enthalpy $\Delta G = \Delta H - T\Delta S$. The enthalpy change $\Delta H = H_{HS} - H_{LS}$ is typically 10 to 20 kJ mol$^{-1}$, and the entropy change $\Delta S = S_{HS} - S_{LS}$ is of the order of 50 to 80 J mol$^{-1}$ K$^{-1}$.[28] These quantities are, of course, temperature dependent. The thermally induced spin transition is an entropy driven process; the degree of freedom is much higher in the HS state than in the LS state. Approximately 25% of the total

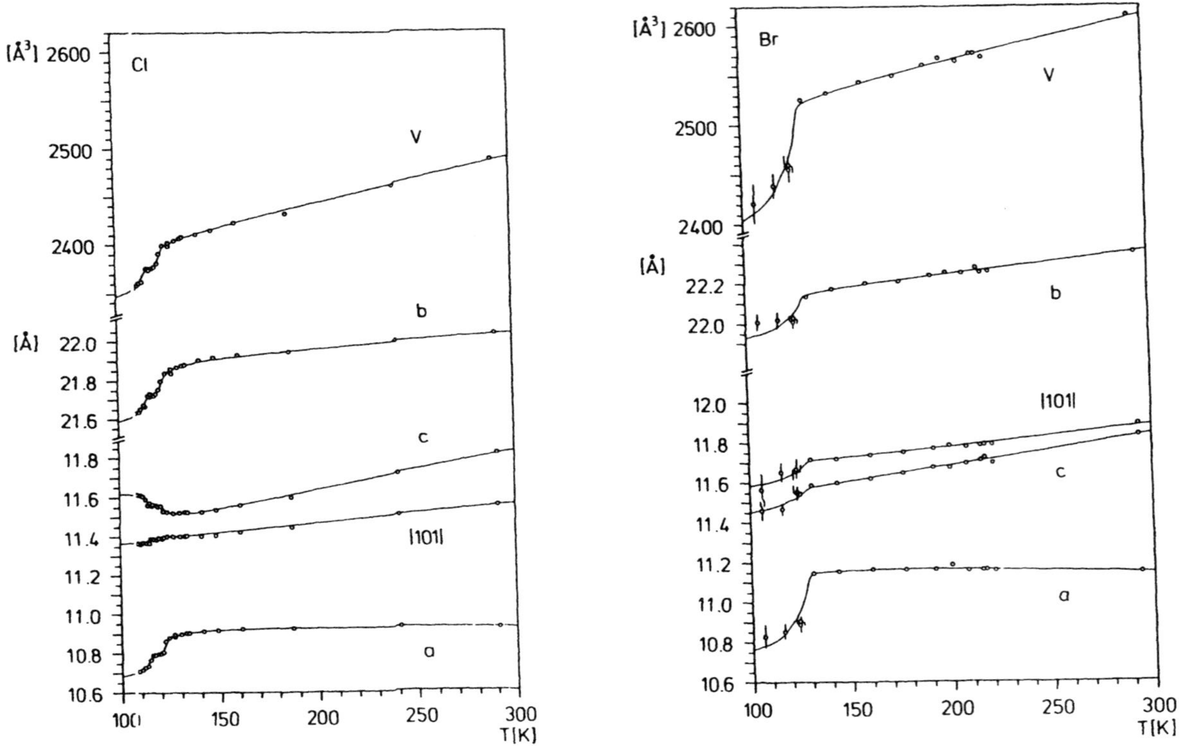


**Figure 9.** Temperature dependence of the unit cell volume $V[\text{Å}^3]$ and of the lengths of four vectors (**a**, **b**, **c**, $|\mathbf{a}+\mathbf{c}|$) in the monoclinic crystal system (space group $P2_1/c$ with $Z = 4$) for $[Fe(2\text{-pic})_3]Cl_2{\cdot}EtOH$ (left) and $[Fe(2\text{-pic})_3]Br_2{\cdot}EtOH$ (right). The solid curves were calculated from the observed values for the HS fraction $\gamma_{HS}(T)$. For the hysteresis in the $Br^-$ system, $\gamma_{HS}(T\downarrow)$ was plotted from values obtained on decreasing the temperature (from Reference 36).

entropy gain on going from LS to HS arises from the change in spin multiplicity, $\Delta S_{\mathrm{mag}} = R \cdot [\ln(2S+1)_{\mathrm{HS}} - \ln(2S+1)_{\mathrm{LS}}]$, and the major part originates from changes in the intra- and intermolecular vibrations.[67,68] An example of a specific heat measurement is given in Section 4.

# 4 MECHANISM OF SPIN CROSSOVER IN SOLIDS

Our present knowledge of the mechanism of spin crossover in solids results from investigations of the effect of metal dilution in mixed crystals. The basic idea behind these studies is the possibility of gradually reducing the cooperative interactions which are active in the neat crystal, but no longer in sufficiently diluted mixed crystals. In view of the observations that the shape and volume of the complex molecules and the lattice change significantly upon spin transition, it appeared to be a reasonable working hypothesis that the cooperative interactions are elastic in nature. In the following, we shall describe first the spin crossover phenomenon in highly diluted systems, and then the cooperative interactions in more concentrated systems.

## 4.1 Dilute Mixed Crystal Systems

Highly diluted spin crossover compounds, in which the concentration of Fe ions in the lattice positions of Zn, Co, or Mn ions is only 1–2%, show very simple transition behavior. The population of the energy levels in the isolated complexes is then described by the Boltzmann law, which determines the thermodynamic spin equilibrium. If $Z_{\mathrm{HS}}(T)$ and $Z_{\mathrm{LS}}(T)$ are the partition functions of the molecules in the HS and the LS states, respectively, the HS fraction as a function of temperature, $\gamma_{\mathrm{HS}}(T)$, is given by:

$$\gamma_{\mathrm{HS}}(T) = Z_{\mathrm{HS}}(T) / \left[ Z_{\mathrm{HS}}(T) + Z_{\mathrm{LS}}(T) \right] \tag{10}$$

Thus, the ratio of the partition functions of all HS and all LS energy levels is accessible experimentally by measuring the HS fraction $\gamma_{\mathrm{HS}}(T)$:

$$\frac{Z_{\mathrm{HS}}(T)}{Z_{\mathrm{LS}}(T)} = \frac{\gamma_{\mathrm{HS}}(T)}{1 - \gamma_{\mathrm{HS}}(T)} \tag{11}$$

Introducing the free energy per complex molecule $f_i = -k_B T \ln Z_i,\ i = \mathrm{HS, LS}$ the difference $\Delta f_{\mathrm{HL}}(T) = f_{\mathrm{HS}}(T) - f_{\mathrm{LS}}(T)$ is proportional to the logarithm of the ratio $Z_{\mathrm{HS}}/Z_{\mathrm{LS}}$ and in turn can be expressed by the HS fraction:

$$\Delta f_{\mathrm{HL}}(T) = -k_B T \ln \left[ \gamma_{\mathrm{HS}}(T) / (1 - \gamma_{\mathrm{HS}}(T)) \right] \tag{12}$$

The difference of the free energy of all molecules in the HS state and in the LS state of an ensemble of $N$ molecules, $N \cdot \Delta f_{\mathrm{HL}}(T)$, is therefore obtained from the transition curve. This consideration obviously assumes energy levels, electronic and vibrational levels, being independent of temperature, at least in the temperature range of the spin transition. For molecular crystals with much weaker intermolecular than intramolecular bonds this assumption is expected a good approximation. It is tested by incorporating the same complex into different host lattices which provide different amount of space in their metal lattice positions and/or have different elasticity constants. Because the volume of the complex as well as its shape change along with the spin transition, an elastic energy difference stored in the lattice will contribute to the free energy difference, such that $\Delta f_{\mathrm{HL}}(T)$ for two different host lattices should only differ by a constant energy contribution.

This simple shifting of the $\Delta f_{\mathrm{HL}}(T)$ curves has been measured for the spin crossover compound $[Fe(2\text{-pic})_3]Cl_2\cdot Sol$ (2-pic = 2-picolylamine, Sol = EtOH, MeOH). The compound was dissolved in the isomorphous solid Zn and Co compounds, and above *RT* in two liquid solvents. Figure 10 shows $\Delta f_{\mathrm{HL}}(T)$ for all

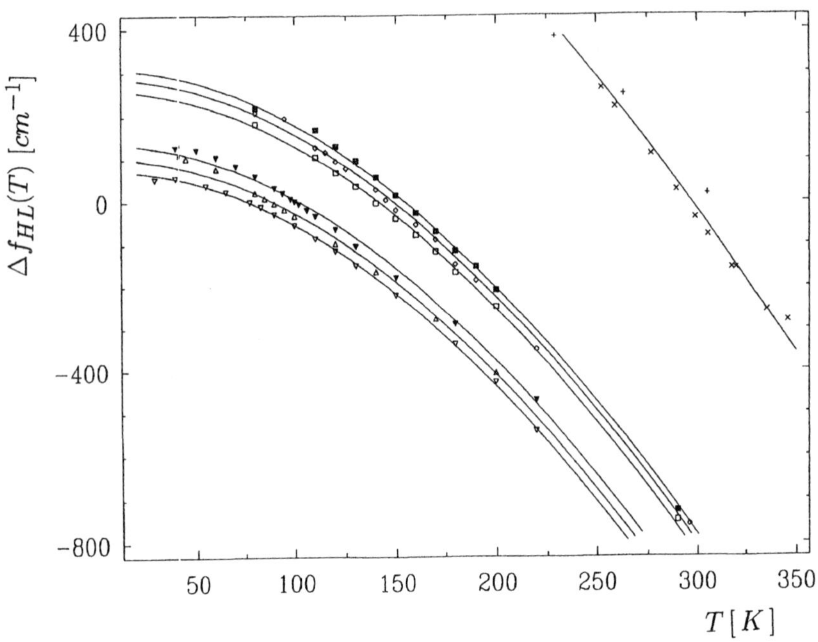


**Figure 10.** The difference between the free energies $\Delta f_{\mathrm{HL}}(T) = f_{\mathrm{HS}}(T) - f_{\mathrm{LS}}(T)$ of the highly dilute $[Fe(2\text{-pic})_3]^{2+}$ complexes in the HS and LS state in different lattices and in liquid solution. The values are derived from the spin equilibrium. ▽ $[Fe_{0.007}Zn_{0.993}(2\text{-pic})_3]Cl_2\cdot EtOH$; △ $[Fe_{0.02}Co_{0.98}(2\text{-pic})_3]Cl_2\cdot EtOH$ ($\Delta' = 20\ cm^{-1}$); ▼ $[Fe_{0.03}Zn_{0.97}(2\text{-pic-}ND_2)_3]Cl_2\cdot EtOD$ ($\Delta' = 86\ cm^{-1}$); □ $[Fe_{0.02}Zn_{0.98}(2\text{-pic})_3]Cl_2\cdot MeOH$ ($\Delta' = 255\ cm^{-1}$); ◇ $[Fe_{0.01}Co_{0.99}(2\text{-pic})_3]Cl_2\cdot MeOH$ ($\Delta' = 243\ cm^{-1}$); ■ $[Fe_{0.01}Zn_{0.99}(2\text{-pic-}ND_2)_3]Cl_2\cdot MeOD$ ($\Delta' = 282\ cm^{-1}$); × $[Fe(2\text{-pic})_3]^{2+}$ complex in acetonitrile/water, + $[Fe(2\text{-pic})_3]^{2+}$ complex in methanol/water ($\Delta' = 800\ cm^{-1}$); observed values for the solution are taken from Reference 69. The parameter of the solid lines fitted to the data differ by the energy shift $\Delta'$.

measured transitions. The shifts are related to the transition of $[Fe_{0.007}Zn_{0.993}(2\text{-pic})_3]Cl_2{\cdot}EtOH$ with the lowest transition temperature. Clearly, the curves do not intersect but are merely shifted by a specific quantity along the energy axis. One of the solid lines has been parameterized, the others were obtained by $\Delta f_{HL}(T) + \Delta'$.

The interpretation of the shift $\Delta'$ is evident. If we disregard the influence of the polar solvent on the ligand field strength, we have practically undisturbed complex molecules in the liquid solution. There is no elastic energy to overcome on going from the HS to the LS state (work against ambient pressure is minimal and can be neglected). Upon incorporation into a crystal lattice, the complete set of energy levels of the HS state is shifted downward by $\Delta'$, relative to the LS state. At ambient temperature the solution in this case[69] contains about the same number of HS and LS complex molecules. Because the HS complex molecules, along with the solvent molecules, fit best into the available structure, a HS crystal precipitates. Thus the lattice stabilizes the HS state by $\Delta'$ (downward shift by $\Delta'$). The zero temperature limit of the free energy difference $\Delta f_{HL}(T = 0) + \Delta'$, which the energy difference between the lowest HS level and LS level denoted by $\Delta E^0_{HL}$, plays also an important role in the dynamics of the spin transition (see Section 6).

Because a transition from the LS state to the HS state entails a volume increase, the LS state is stabilized by $\Delta = p\delta v_{HL}$ when an increasing external pressure $p$ is applied, where $\delta v_{HL}$ is the volume change of the host crystal per complex molecule. For example, the shift $\Delta'$ of $\Delta f_{HL}(T)$ was found to be ca. 80 $cm^{-1}$ for the system $[Fe_{0.03}Zn_{0.97}(2\text{-pic-}ND_2)_3]Cl_2{\cdot}EtOD$ at a pressure of $p = 1{,}350$ bar.[70] The LS state stabilization of $\Delta' = 80\ cm^{-1}$ corresponds to a volume change of $\delta v_{HL} = 80\ cm^{-1}/1{,}350\ bar = 11.9\ Å^3$. From variable-temperature X-ray measurements on the all-iron compound an increase of the unit cell volume going from the LS to the HS state of 49.3 $Å^3$ was found applying Equation 8 of Section 3.5.[71] Taking into account that there are 4 molecules per unit cell of this compound one obtains a change of crystal volume 12.3 $Å^3$ in good agreement with the value obtained from the pressure measurements.

## 4.2 Concentrated Systems

In concentrated systems, particularly, in mixed crystals with Fe concentrations of $0 \ll x \leqslant 1$ a deviation from the Boltzmann type population is observed. Such a deviation is attributed to an interaction between the spin crossover molecules leading to an increasing cooperativity of the transition with increasing $x$. The simplest approach for the free energy per complex molecule, $f_x$, is to consider a random mixture of HS, LS, and M complexes (M = Zn, Co, etc.) in a mixed crystal system:

$$\begin{aligned} f_x(T) = {} & x\big[\gamma_{HS}(T) f_{HS}(T) + (1 - \gamma_{HS}(T)) f_{LS}(T) - T s_{mix}(\gamma_{HS}(T))\big] \\ & + (1 - x) f_{Zn}(T) + x w\big(\gamma_{HS}(T), x, T\big) + f_G(T) \end{aligned} \tag{13}$$

Here $w(\gamma_{HS}(T), x, T)$ is the interaction energy per Fe complex molecule, and $f_G(T)$ is the free energy of the residual lattice, taken as independent of $\gamma_{HS}$. The mixing entropy between HS and LS molecules is $s_{mix} = -k_B[\gamma_{HS} \ln \gamma_{HS} + (1 - \gamma_{HS}) \ln(1 -$

$\gamma_{HS})]$, and contributions to the mixing entropy dependent on $x$ have been omitted. At infinite dilution the interaction energy per complex molecule must vanish, $w(\gamma_{HS}(T), x \to 0, T) \to 0$. Then the equilibrium condition $\partial f_{x\to 0}(T)/\partial\gamma_{HS} = 0$ yields the Boltzmann type population given in Equation 10.

From the temperature dependence of the spin equilibrium of a mixed crystal series we obtain an expression for $w$. Equation 14 follows from the equilibrium condition $\partial f_x(T)/\partial\gamma_{HS} = 0$:

$$\frac{\partial w(\gamma_{HS}, x, T)}{\partial\gamma_{HS}} = -k_B T \cdot \ln\frac{\gamma_{HS}(T)}{[1-\gamma_{HS}(T)]} - \Delta f_{HL}(T) \qquad (14)$$

$\Delta f_{HL}(T)$ is known from the transition curve for highly dilute lattices (see Figure 10). With the experimental data for $\gamma_{HS}(T)$[72,73] at concentration $x$ the derivative $\partial w(\gamma_{HS}, x, T)/\partial\gamma_{HS}$ of the interaction term is obtained.

In Figure 11, we have plotted the derivatives of $w$ for several concentrations $x$ as a function of $\gamma_{HS}$ in the metal diluted and deuterated system $[Fe_xZn_{1-x}(2\text{-pic-}ND_2)_3]Cl_2\cdot EtOD$. As a good approximation $\partial w/\partial\gamma_{HS}$ is a linear function of $\gamma_{HS}$ and turns out to be independent on temperature explicitly: $\partial w/\partial\gamma_{HS} = \Delta - 2\Gamma\gamma_{HS}$. Figure 12 shows that both constants, $\Delta$ and $\Gamma$, are proportional to the concentration: $\Delta(x)/x = 273\ \text{cm}^{-1}$ and $\Gamma(x)/x = 152\ \text{cm}^{-1}$.

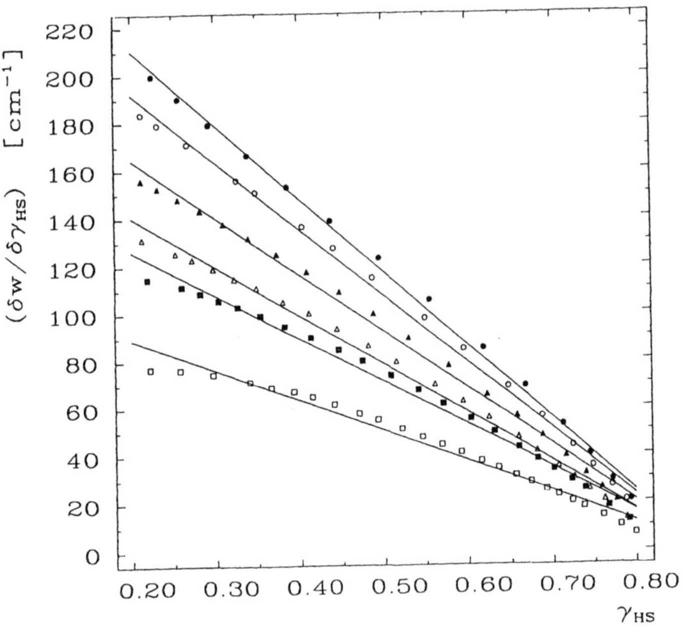


**Figure 11.** The partial derivative $\partial w/\partial\gamma_{HS}$ as a function of $\gamma_{HS}$ for different Fe concentrations $x$ in $[Fe_xZn_{1-x}(2\text{-pic-}ND_2)_3]Cl_2\cdot EtOD$ with linear regression lines (from References 72,73). (●): $x = 1.00$; (○): $x = 0.91$; ▲: $x = 0.78$; △: $x = 0.68$; ■: $x = 0.60$; and □: $x = 0.46$.

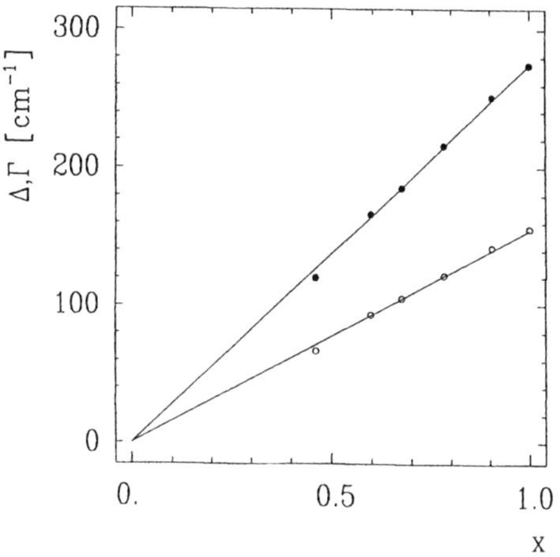


**Figure 12.** $\Gamma(x) = 152\ \text{cm}^{-1} \cdot x$ (●) and $\Delta(x) = 273\ \text{cm}^{-1} \cdot x$ (○) as linear functions of the Fe concentration $x$. The linear regression lines were made to pass through the origin, so that the condition $w(x = 0) = 0$ is met (from Reference 73).

This implies that $w$ fulfills the simplest possible phenomenological formulation, the Bragg–Williams approximation for a binary mixture with interactions between the components:

$$w(\gamma_{\text{HS}}, x) = \Delta(x)\gamma_{\text{HS}}(T) - \Gamma(x)\gamma_{\text{HS}}^2(T) \tag{15}$$

The formal integration constant, which can only be dependent on $T$ and $x$, is included in the lattice contribution $f_G(x, T)$ or taken to be part of this term a priori. The linear dependence of $\partial w/\partial\gamma_{\text{HS}}$ on $x$ as well as on $\gamma_{\text{HS}}$ was first observed for the $[Fe_xZn_{1-x}(2\text{-pic})_3]Cl_2 \cdot EtOH$ mixed crystal series[74] and later also verified for other mixed crystals.[75–77] The free energy of Equation 13 agrees well with observations from calorimetric measurements.[73]

## 4.3 Interpretation of the Interaction Term

We have already pointed out that the elastic interactions between the spin-changing complex molecules certainly play an important or even decisive role in the HS–LS transition in solid spin crossover compounds. This section will show that the parameters $\Delta(x)$ and $\Gamma(x)$ in the interaction term of Equation 15 for a mixed crystal series can be well understood on the basis of elasticity theory. We use a simplified picture of the crystal, in order to derive quantitative expressions. The HS, LS, and M complex molecules are considered as defects in their own crystal lattice into which they fit more or less well. The crystal lattice itself is taken as a homogeneous medium with certain elastic properties. As a further simplification, the elastic medium is assumed to be isotropic so that it can be characterized by just two elastic constants, the (elastic) bulk modulus $K$, and the Poisson ratio $\sigma$. $K$ is defined by the volume change $dV$ caused by a pressure change $dp$: $-KdV/V = dp$. $\sigma$ is connected to the shear properties of the elastic medium.

Here we look at the isotropic part of the defects only. A stress-free volume $v_0$ will be assigned to a metal ion lattice site in the homogeneous medium. The metal ion in this lattice position can be Fe(II)-HS, Fe(II)-LS, and the host metal ion M with their corresponding molecular volumes $v_{HS}$, $v_{LS}$, and $v_M$ in a mixed crystal of the $[Fe_xZn_{1-x}(2\text{-pic})_3]X_2\cdot Sol$ type. These molecular volumes can be smaller or larger than $v_0$, and for the following simple consideration are assumed to be incompressible, i.e., hard spheres: When a volume $v_\alpha$ ($\alpha$ = HS, LS, M) is placed into the medium of macroscopic volume $V$, a stress field is built up, because the volumes $v_{HS}$, $v_{LS}$, and $v_M$ are different from $v_0$. This stress field leads to an elastic energy stored in the lattice of the form[78]

$$E = \frac{1}{2}K(\gamma_0 - 1)(v_\alpha - v_0)^2/v_0 - \frac{1}{2}K\gamma_0(\gamma_0 - 1)(v_\alpha - v_0)^2/V \qquad (16)$$

The Eshelby constant $\gamma_0$ is related to the Poisson ratio $\sigma$ by $1 \leqslant \gamma_0 = 3(1 - \sigma)/(1 + \sigma) \leqslant 3$. The first term is the energy to be spent to increase/decrease the sphere $v_0$ to $v_\alpha$ in an infinite medium ($V \to \infty$). The second term in Equation 16 corrects for the surface effect of a finite volume $V$. It is very small since $V/v_0$ is of the order of Avogadro's number and vanishes for an infinite medium. The elastic energy for an isolated complex in the lattice is therefore given by the first term only. In a liquid solution, the inner degrees of freedom of the complex molecule alone are active; in the highly dilute mixed crystal, the difference in the elastic energies of the lattice between the molecules in the HS and the LS state is added. Within this simple picture the free energy difference between the spin state changing complex molecules in mixed crystals (Figure 10) is traced back to the differences in $K$, $\gamma_0$, and $v_0$. We shall not discuss this energy difference in the following but concentrate on the nature of the interaction itself.

In the finite crystal, the second term in Equation 16, although very small, gives rise to an interaction between the defects by the so-called *image pressure* or *chemical pressure* of defects in crystals. This interaction is understood when applying a simple result of elasticity theory derived by Eshelby.[78] The crystal volume $V$ changes upon incorporation of a molecule of volume $v_\alpha$ at the lattice site providing a volume $v_0$ by more than just $v_\alpha - v_0$, namely increased by the Eshelby constant $\gamma_0$.

$$\delta v_\alpha = \gamma_0(v_\alpha - v_0) \qquad (17)$$

The additional volume change $\Delta v = \delta v_\alpha - (v_\alpha - v_0) = (\gamma_0 - 1)(v_\alpha - v_0)$ is formally attributed to a pressure $p_I = -K\Delta v/V$. The second term in Equation 16 is interpreted as volume work against this pressure $p_I$.[78] Indeed it reads $-\frac{1}{2}p_I\delta v_\alpha$, where the factor 1/2 stems from the integration $\int p_I d(\delta v)$. The effect can be used for the creation of positive or negative pressures by replacing atoms in the crystal with smaller or larger alien atoms. This effect is somehow against intuition, namely incorporating a *smaller* atom, associated with free volume in the crystal, leads to an effective *positive* external (image) pressure, saying that any site in the crystal "feels" a lack of volume. Of course, the term "pressure" applies only for defects centered in a sphere, otherwise the stress field can no longer be considered homogeneous. Eshelby,[78] however, could show that $\delta v_\alpha$ remains valid irrespective of the form of the volume and the location of the defect. A homogeneous distribution of defects results in a homogeneous stress field, so again we can speak of a "pressure". That this pressure leads to an interaction between HS complexes is obvious. Starting from low temperatures where all molecules are in the LS state the first small fraction of molecules in the thermally populated HS state causes a negative pressure due to their volume increase. This negative pressure acting on all molecules in the crystal stabilizes their HS states and in turn increases the population of HS states, so that HS states favour further HS states as described by a ferromagnetic type of interaction.

Although the pressure is small for a single defect, for one mole of defects, it adds up to a finite entity. For $N$ defects in a crystal we obtain $E_{\text{total}}$ according to:

$$E_{\text{total}} = \frac{1}{2}K(\gamma_0 - 1)\sum_{i=1}^{N}(v_i - v_0)^2/v_0 - \frac{1}{2}K\gamma_0(\gamma_0 - 1)\left[\sum_{i=1}^{N}(v_i - v_0)\right]^2/V \quad (18)$$

The first term just sums over all sites whereas in the second term the total volume change of all sites enters. The denominator of the second term becomes proportional to $N^2$ so that it is no longer negligible. If we express the sum $\sum_{i=1}^{N}(v_i - v_0)$ by the fraction $x$ and $\gamma_{\text{HS}}$:

$$\frac{1}{N}\sum_{i=1}^{N}(v_i - v_0) = x\left[\gamma_{\text{HS}}(v_{\text{HS}} - v_0) + (1 - \gamma_{\text{HS}})(v_{\text{LS}} - v_0)\right] + (1 - x)(v_{\text{M}} - v_0) \quad (19)$$

and consider only the $\gamma_{\text{HS}}$-dependent terms in the elastic energy, we obtain an interpretation of $\Delta(x)$ and $\Gamma(x)$ within this phenomenological approach for the free energy as given in:[74]

$$\begin{aligned}\Delta(x) &= xK\gamma_0(\gamma_0 - 1)(v_{\text{HS}} - v_{\text{LS}})(v_{\text{M}} - v_{\text{LS}})/v_c \\ \Gamma(x) &= x\frac{1}{2}K\gamma_0(\gamma_0 - 1)(v_{\text{HS}} - v_{\text{LS}})^2/v_c\end{aligned} \quad (20)$$

$v_c = V/N$ denotes the volume per complex molecule. We see that $\Delta$ and $\Gamma$ depend linearly on $x$, as observed, and that all parameters can be experimentally determined. Note that $v_0$ is absent from the formulae. Thus, $\gamma_0(v_{\text{HS}} - v_{\text{LS}})/v_c$ is the relative volume increase on going from the LS to the HS state, that is, the trace $\varepsilon_s^{\text{HL}}$ of the $\varepsilon$ tensor $\varepsilon_s^{\text{HL}}$ derivable from a variable-temperature structure determination. We obtain $\varepsilon_s^{\text{ML}} = \gamma_0(v_{\text{M}} - v_{\text{LS}})/v_c$ from a comparison of the unit cell volumes of the host metal compound and the Fe compound in the LS state. $K$ and $\gamma_0$ are accessible as elasticity constants.

The spherical part of the defects causes only image pressure, which is ubiquitous, and therefore the interaction effectively extends to infinity. In the course of the development of this model, it has been found that the nonspherical part of the shape of the defects plays the more important role. It causes both long-range (image pressure type) and short-range interactions.[79,80] The latter ones are the origin of deviations from random distribution of spin states of the spin changing molecules over the lattice sites. As an example, it was found for the $[Fe(2\text{-pic})_3]Cl_2{\cdot}EtOH$ system that such short-range interactions favor a HS–LS neighborhood rather than the formation of HS–HS and LS–LS neighbors. The HS–LS correlations are responsible for the occurence of the stepwise spin transition in $[Fe(2\text{-pic})_3]Cl_2{\cdot}EtOH$,[81,82] despite the fact that all complexes are crystallographically equivalent at all temperatures.

## 5 PHOTOPHYSICAL PHENOMENA

### 5.1 Light-Induced Excited Spin State Trapping (LIESST)

In contrast to many other transition metal compounds, iron(II) coordination compounds are not famous for their photophysical properties. This is basically due to the fact that with the exception of a few HS systems as ZnS[83] or $KMgF_3$[84–86] doped with Fe(II), they do not luminesce and that therefore they were ignored by most spectroscopists. Of course, the fact that they do not luminesce immediately signifies that they do not possess any excited states of lifetimes larger than typical vibrational times, i.e., $>10^{-11}$ s. This can be understood on the basis of the large number of excited LF states shown in Figure 13, which provide a perfect ladder for efficient multi-phonon relaxation in the intermediate vibronic coupling regime.[35,87]

Iron(II) spin crossover compounds are special with respect to their photophysical properties because of the small zero-point energy difference $\Delta E^0_{HL}$ between the ${}^1A_1$ ground state and the first excited state, the ${}^5T_2$ state, and the extremely large bond length difference $\Delta r^0_{HL}$ between the two states. This leads to the astonishing phenomenon of a light-induced spin transition at temperatures well below the thermal spin transition: Irradiating, for example, a crystal of $[Fe(ptz)_6](BF_4)_2$ at 20 K into the spin allowed ${}^1A_1 \rightarrow {}^1T_1$ absorption band of the LS species at 514.5 nm (19,440 $cm^{-1}$) quickly results in a complete bleaching of the initially red crystal. The resulting spectrum is again typical for the HS state (see Figures 14a and b). The mechanism for this light-induced spin transition is straight forward. The initially excited ${}^1T_1$ state is very short lived (<1 ns[88]), and decays non-radiatively via several intersystem-crossing (ISC) steps either to the ${}^1A_1$ ground state or to the ${}^5T_2$ state. Because of the substantial energy barrier between the two states, the relaxation from the metastable ${}^5T_2$ state back to the ground state is very slow. In $[Fe(ptz)_6](BF_4)_2$ it is so slow, that even with quite low light intensities a quantitative population of the ${}^5T_2$ state can be achieved within a few minutes. In fact below 20 K, the system remains trapped in the HS state for more than ten days, and it is this trapping that gave rise to the acronym of "light-induced excited spin state trapping (LIESST)" for this phenomenon.[89,90] Only at temperatures above 50 K does a noticeable thermal relaxation set in.

A third important state in the mechanism of LIESST is the comparatively low-lying ${}^3T_1$ state.[91] It serves as intermediate state in the cascade of intersystem-crossing processes leading from the excited ${}^1T_1$ to the ${}^5T_2$ state. Because for spin

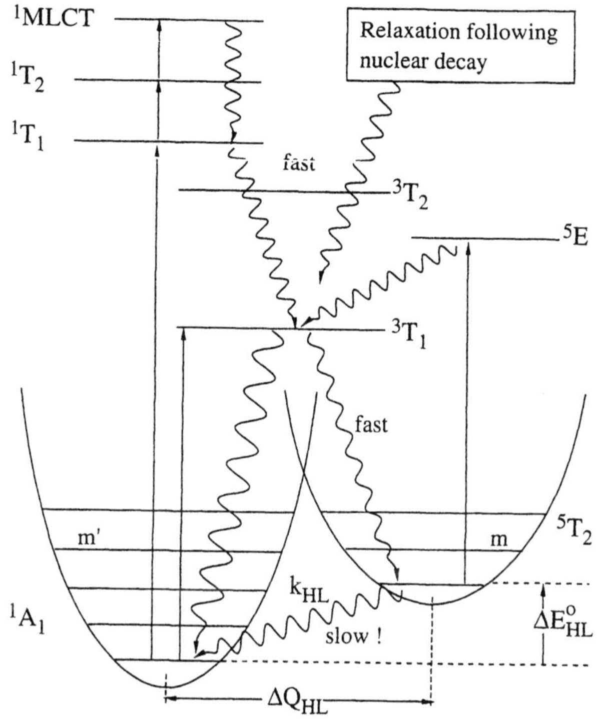


**Figure 13.** Configurational coordinate diagram for the HS and the LS state of a $d^6$ spin crossover compound. Higher excited ligand-field states are schematically included. Arrows show the mechanism of light and nuclear decay induced processes.

crossover complexes this triplet state lies below both the $^1T_1$ as well as the excited $^5E$ state, the reverse process, that is a light-induced HS $\rightarrow$ LS conversion via irradiation into the $^5T_2 \rightarrow {}^5E$ transition seems feasible. And indeed, if in a second step the colourless crystal is irradiated at 820 nm (12,200 cm$^{-1}$) with the temperature kept at 20 K, the red color returns. In the corresponding spectrum the typical LS absorption bands reach a saturation value of about 90% of their original intensity (see Figure 14c). This is due to the spectral overlap of the spin-allowed absorption band of the HS species with the spin-forbidden bands of the LS species, leading to a steady state type situation. This indicates that the $^3T_1$ indeed serves as intermediate state. And, in fact, narrow band excitation at 980 nm (10,200 cm$^{-1}$), that is right into the maximum of the spin-forbidden $^1A_1 \rightarrow {}^3T_1$ transition but well outside the $^5T_2 \rightarrow {}^5E$ band, results in a quantitative LS $\rightarrow$ HS conversion just as irradiation into the spin-allowed transitions of the LS species did.

LIESST and reverse LIESST can be followed quantitatively. In the region of spectral overlap the actual saturation value for the LS fraction depends critically upon

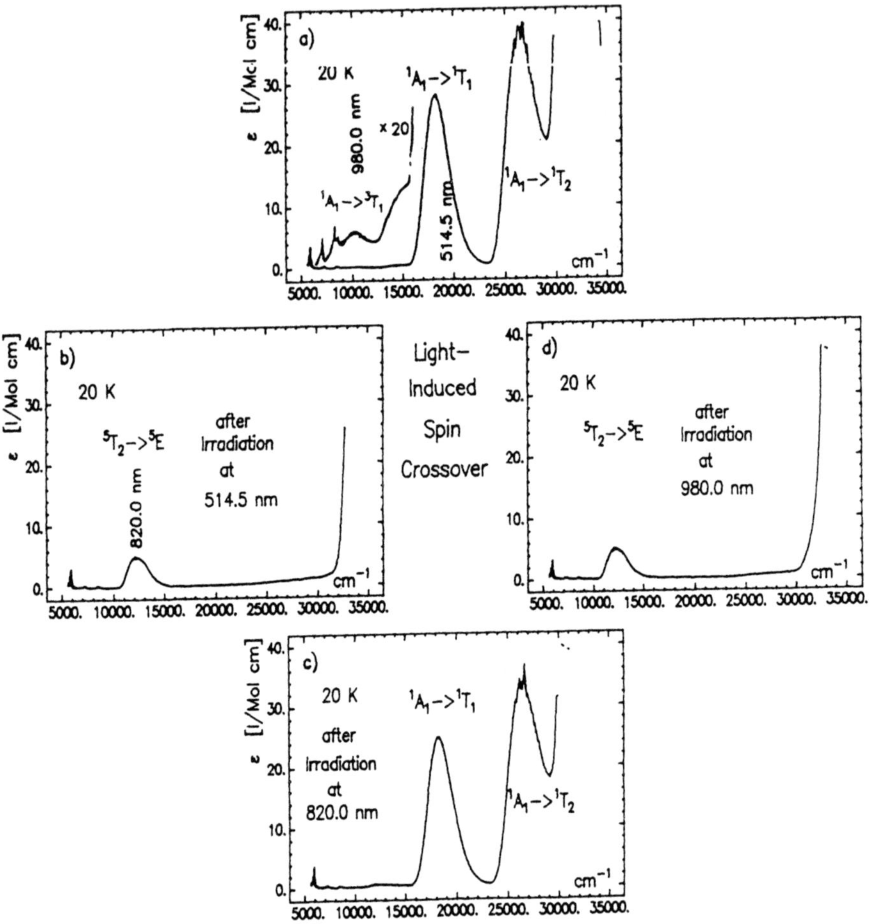


**Figure 14.** Single crystal absorption spectra of $[Fe(ptz)_6](BF_4)_2$ at 20 K: a) normal spectrum before irradiation, b) after irradiation into the spin-allowed transition of the LS state, ${}^1A_1 \rightarrow {}^1T_1$, at 514.5 nm, c) after subsequent irradiation into the maximum of the spin-allowed transition of the HS species, ${}^5T_2 \rightarrow {}^5E$, at 820 nm, and d) after irradition into the spin-forbidden transition, ${}^1A_1 \rightarrow {}^3T_1$, at 980 nm (adapted from Reference 91).

the exact irradiation wavelength and is determined by the relative absorption coefficients of the transitions involved as well as the quantum yields of various processes and the branching ratio at the ${}^3T_1$ state. Quantum yields for the first ISC step, that is, ${}^1T_1 \rightarrow {}^3T_1$ and ${}^5E \rightarrow {}^3T_1$ are on the order of unity in $[Fe(ptz)_6](BF_4)_2$. The branching ratio from ${}^3T_1$ to ${}^1A_1$ and ${}^5T_2$ is approximately $1:4$.[91]

LIESST is not restricted to neat $[Fe(ptz)_6](BF_4)_2$. Since its discovery,[89] the phenomenon has also been observed in the series of compounds of composition $[Fe(2\text{-pic})_3]X_2Sol$ ($X = Cl, Br$; Sol = MeOH, EtOH),[90] $[Fe(phen)_2(NCS)_2]$[92,93] and others,[94–96] as well as in diluted mixed crystals with various classical spin crossover complexes, among them $[Zn_{1-x}Fe_x(ptz)_6](BF_4)_2$,[97,98] $[M_{1-x}Fe_x(2\text{-pic})_3]Cl_2Sol$ (M = Mn, Zn)[99,100] and $[Zn_{1-x}Fe_x(mepy)_3tren](PF_6)$,[99] and for spin crossover complexes embedded in polymer matrices.[101,102] Although cooperative effects do influence the LIESST behaviour in a fascinating way (see Section 5.2), LIESST is therefore essentially a property of a single complex.

As mentioned in Section 3.1, in most of the above systems the *d–d* bands of the LS species are hidden by low-energy MLCT bands. These are much more intense for the LS state than for the HS state, and irradiation into the MLCT bands of the LS species results in an efficient light-induced LS → HS conversion, too.[99,102,103] Thus the LIESST scheme in Figure 13 may be expanded to include MLCT states. However, these MLCT transitions with their long tails into the near IR diminish the efficiency for reverse LIESST. For instance, in $[Zn_{1-x}Fe_x(2\text{-pic})_3]Cl_2{\cdot}EtOH$ irradiation at the maximum of the $^5T_2 \rightarrow {^5E}$ transition results in a steady state value for the LS fraction of 10% only.[104]

The lower limit of the relaxation rate constants for all ISC processes, with the exception of the $^5T_2 \rightarrow {^1A_1}$ relaxation, is $10^9\ s^{-1}$, because up to now, no transient populations of an intermediate state have been found using picosecond pulsed laser excitation.[88,105] The low-temperature lifetimes of the light-induced HS states in the above systems are between $10^1$ and $10^6$ s. The factors governing these lifetimes will be treated in detail in Section 6.

## 5.2 Light-Induced Bistability

For negative values of $\Delta E^0_{HL}$, that is with the HS state as the quantum mechanical ground state, no thermal spin transition is expected. In analogy to reverse LIESST it should, however, be possible to induce a HS → LS conversion by irradiation in the near IR. For this, the potential for the $^5E$ state has to be above the one for the $^3T_1$ state, otherwise the quantum efficiency for the first ISC step following the excitation drops to zero. This condition is only met for complexes whose LF strengths are not significantly below $10Dq^{HS}$ for spin crossover compounds.

Such a HS → LS conversion was first realised in the methyl substituted compound $[Fe(mtz)_6](BF_4)_2$ (mtz = 1-methyltetrazole).[106] In this system, the complexes are distributed over two non-equivalent lattice sites.[44] Complexes on the one site show both a thermal spin transition as well as the light-induced processes observed in the propyl derivative. Complexes on the other site remain in the HS state down to 10 K. Irradiation in the near IR results in a steady state population of the LS state of around 82% for complexes on both lattice sites.[107] The LS state on the second site is a metastable state only, and at temperatures above 50 K relaxation to the HS state sets in.

As shown schematicaly in Figure 15 an external pressure stabilizes the LS state. Thus for $\Delta E^0_{HL} < 0$, it is possible to induce the spin transition by applying a sufficiently high pressure.[108–110] In neat materials, $\Delta E^0_{HL}$ is a function of $\gamma_{LS}$. Thus, even if at $\gamma_{LS} = 0$, the HS state is the true ground state, an increasing internal pressure with an increasing LS fraction could stabilize the LS state sufficiently to result in it becoming the true ground state at $\gamma_{LS} = 1$. The only problem is how to get there. Such

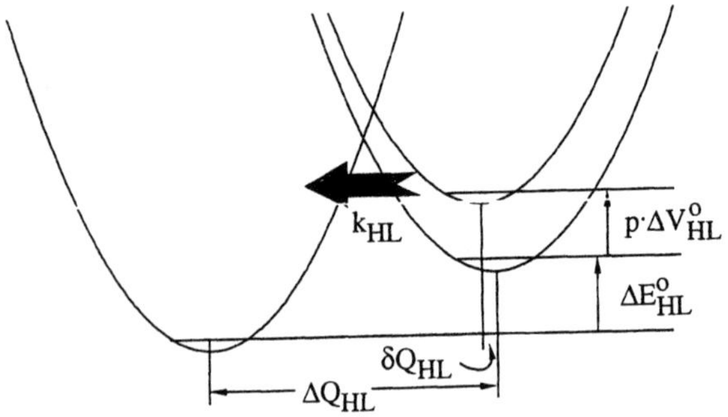


**Figure 15.** Schematic representation of the effect of external pressure on the potentials of the HS and the LS state. The main effect is an increase in energy difference by the work term $p\Delta v_{HL}$. The change in bond length difference is comparatively small for pressures below 1 kbar.

a system corresponds to a system with a hysteresis in the thermal spin transition, but as $T_c^{\downarrow} \rightarrow 0$ it has no reason to populate the $^1A_1$ state upon lowering of the temperature. However, at low temperatures, the LS state can be populated photophysically by irradiating into the spin-allowed *d–d* transition of the HS species. Initially, this LS state is just a metastable state, but as the light-induced HS $\rightarrow$ LS conversion proceeds, $\Delta E^0_{HL}$ may become positive by virtue of the increasing internal pressure.

The ethylderivative $[Fe(etz)_6](BF_4)_2$ (etz = 1-ethyltetrazole) of the tetrazole series provides a convincing example of such a light-induced bistability.[111] In this system, the iron(II) complexes sit on two nonequivalent lattice sites in a ratio of 2 : 1. Complexes on one site (denoted as site A) show a comparatively steep thermal spin transition with a transition temperature of 105 K, those on the other site (site B) stay in the HS state down to liquid helium temperatures (see Figure 16). Complexes on site A behave pretty much as $[Fe(ptz)_6](BF_4)_2$ with respect to LIESST. Those on site B can be converted almost quantitatively to the LS state at 20 K by irradiating at 820 nm, that is, into the $^5T_2 \rightarrow {}^5E$ transition of the HS species. A partial light-induced, steady-state type population of the LS state on the two sites results in an interesting relaxation behaviour at temperatures below 80 K. As expected, complexes on site A always fully relax to the LS state, irrespective of the initial LS fraction because for them $\Delta E^0_{HL}$ is always positive. For initial values of the total LS fraction $\gamma_{LS}$ of less than 0.3, complexes on site B relax to the HS state, and the system ends up with the same overall LS fraction of ≈2/3, as obtained by recording a straightforward thermal transition curve (see inset Figure 16). For initial values of $\gamma_{LS}$ larger than 0.3, however, site B complexes, too, relax to the LS state, so that the system ends up completely in the LS state. At temperatures below 80 K, the cooperative effects thus result in the postulated optical bistabilty.

As the temperature is raised to above 80 K, site B complexes go back to the HS state. This may be regarded as the thermal spin transition in heating mode of a system with a very large hysteresis, with the corresponding branch in cooling mode not being accessible in a simple temperature cycle.

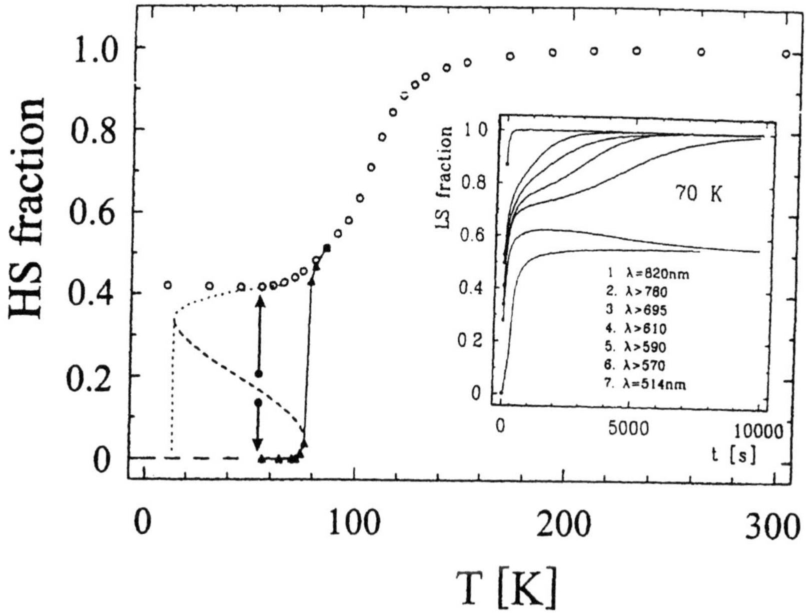


**Figure 16.** Thermal spin transition in $[Fe(etz)_6](BF_4)_2$ (○). Only complexes on one lattice site go to the LS state at low temperatures. Inset: Relaxation curves for various initial conditions, illustrating the light-induced bistability. For a total light-induced population of the LS state above a critical value, the system relaxes fully to the LS state at temperatures below 80 K (adapted from Reference 111).

## 6 RELAXATION DYNAMICS

Intersystem crossing (ISC), that is, the radiationless transition between electronic states of a system involving a change in spin quantum number, is one of the most simple yet universal radiationless processes in chemistry and physics. In very many photochemical and photophysical processes of technological relevance it constitutes the crucial first step, taking the system from the initially excited and usually short-lived state to the metastable state with a lifetime long enough for the processes of interest to actually have a chance of occurring. The three-level ruby laser,[112] for example, is based on the ISC step from the initially excited $^4T_2$ to the metastable $^2E$ state and the subsequent luminescence back to the $^4A_2$ ground state of Cr(III) doped into $Al_2O_3$. In a more recent application, the ISC process from an excited $^1$MLCT to the low-lying $^3$MLCT state of $[Ru(bpy)_3]^{2+}$ (bpy = 2,2′-bipyridine) lead to the development of a photovoltaic cell, making use of the resulting photoredox properties.[113]

Given the central role of ISC processes in photochemistry and photophysics, it is of paramount importance to arrive at a clear understanding of the parameters which govern their rates and their quantum efficiencies. Spin crossover compounds of first-row transition metal ions, especially of iron(II), provide the ideal systems to study the dynamics of ISC processes. The HS $\rightarrow$ LS relaxation is particularly suited for such a study, because there are no competing processes such as luminescence to contend with, allowing the process of interest to be followed over a wide temperature range and over many orders of magnitude of the relaxation rate constant.

## 6.1 Experimental Results

Initial studies on the HS $\rightarrow$ LS relaxation dynamics in spin crossover compounds were mostly performed at ambient temperatures in solution,[114] using such diverse methods as Raman temperature jump techniques,[115] ultrasonic relaxation,[116] and laser perturbation.[103,117] In solution only a limited temperature range is accessible and (except for laser perturbation) studies are restricted to systems with transition temperatures $T_{1/2}$ within this range, because these methods require both species of the unimolecular process to be present in a dynamic equilibrium. Despite the suggestion of Buhks et al.[118]—that the HS $\rightarrow$ LS relaxation should be treated as a nonadiabatic multiphonon process, i.e., a tunneling process during which the electronic energy of the initial state is spontanously transformed into vibrational energy in the final state—it continued to be treated classically, in all probability because Arrhenius plots of the observed rate constants usually gave straight lines over the limited temperature range. Table 1 gives the relaxation rate constant $k_{HL}$ at 295 K, and the corresponding experimental activation energies $E^a_{HL}$ and frequency factors $A_{HL}$, as well as the transition temperatures $T_{1/2}$ as compiled for a series of iron(II) spin crossover compounds in solution.[28,114] There is no obvious correlation between any of these quantities! In addition, Table 1 includes the results of laser perturbation experiments on three LS compounds.[88,119,120] The only thing to note at this stage is the comparatively small range of $10^6$–$10^8$ s$^{-1}$ spanned by $k_{HL}$ at room temperature for the spin crossover systems and the somewhat higher values for the LS systems.

Similar HS $\rightarrow$ LS relaxation rate constants and activation parameters as in solution were obtained for some iron(II) spin crossover systems at ambient temperatures in the solid state using lineshape analysis of Mössbauer spectra.[76,125] Unfortunately, this technique is restricted to the small range of rate constants from $10^6$–$10^8$ s$^{-1}$, and requires both species to be present in appreciable amounts, too. Thus, as in solution, only spin crossover compounds with $T_{1/2}$ around room temperature are suitable for such experiments (see Table 1).

With the discovery of the light-induced spin transitions at low temperatures a systematic study of the HS $\rightarrow$ LS relaxation over a much larger temperature interval, namely 10 to 300 K, and for compounds with $T_{1/2}$ as low as 70 K all the way to LS systems with $T_{1/2}$ well in excess of 300 K, became possible. It soon emerged that tunneling processes are very important, becoming dominant at temperatures below approximately 100 K.[99,130] Figure 17 shows the HS $\rightarrow$ LS relaxation rate constants for a series of dilute single crystals plotted as $\ln[k_{HL}(T)]$ versus $1/T$. The deviation from Arrhenius behavior is obvious. Table 2 gives the low temperature tunneling rate constants $k_{HL}(T \rightarrow 0)$ as well as the activation parameters for this series of complexes. In addition, it contains $T_{1/2}$ and, for comparison, $k_{HL}$ extrapolated to room

**Table 1 Transition Temperatures $T_{1/2}$, Activation Energies $E^a_{HL}$ and Frequency Factors $A_{HL}$, and $k_{HL}$ (295 K) for Some Spin Crossover Compounds in Solution and in the Solid State**

| Compound/ Solvent, Solid | $T_{1/2}$ [K] | $E^a_{HL}$ [cm$^{-1}$] | $A_{HL}$ [s$^{-1}$] | $k_{HL}$ (295 K) [s$^{-1}$] | Reference |
|---|---|---|---|---|---|
| $[Fe(pybimH)_3](BPh_4)_2$ in MeOH/acetone/$H_2O$ | 232 | 850 | $1.4 \cdot 10^8$ | $2 \cdot 10^7$ | 121,122 |
| $[Fe(phenmethoxa)_2)](BPh_4)_2$ in acetone | 250 | 930 | | $1.4 \cdot 10^6$ | 123 |
| $[Fe(paptH)_2]Cl_2$ in $H_2O$ | 264 | 1497 | $1 \cdot 10^{10}$ | $7 \cdot 10^6$ | 116,114 |
| $[Fe(pyimH)_3](BPh_4)_2$ in MeOH/acetone/$H_2O$ | 298 | 1135 | $2 \cdot 10^9$ | $1 \cdot 10^7$ | 121,122 |
| $[Fe(mepy)_2(py)tren](PF_6)_2$ in MeOH/acetone/$H_2O$ | 320 | | | $5 \cdot 10^6$ | 115 |
| $[Fe(biz)_2](ClO_4)_2$ in MeCN | ~300 | | | $1.9 \cdot 10^7$ | 103 |
| $[Fe(tpchxn)](ClO_4)_2$ in MeCN | 321 | 350 | $8 \cdot 10^8$ | $1.4 \cdot 10^7$ | 122 |
| $[Fe(tppn)](ClO_4)_2$ in DMF | 355 | 640 | $4 \cdot 10^8$ | $1.6 \cdot 10^7$ | 122 |
| $[Fe(HB(pz)_3)_2]$ in $CH_2Cl_2$/MeOH | 369 | 420 | $2 \cdot 10^8$ | $2 \cdot 10^7$ | 114 |
| $[Fe(ppa)_2](ClO_4)_2$ in acetone | 387 | 1320 | $4 \cdot 10^9$ | $6 \cdot 10^6$ | 103 |
| $[Fe(tpmba)](ClO_4)_2$ in DMF | 400 | 960 | $7 \cdot 10^9$ | $6.3 \cdot 10^7$ | 122 |
| $[Fe(mepy)(py)_2tren](PF_6)_2$ in MeOH/acetone/$H_2O$ | 450 | | | $5 \cdot 10^6$ | 115 |
| $[Fe(tptn)_3]^{2+}$ in $H_2O$ | LS | | | $6 \cdot 10^8$ | 88 |
| $[Fe(bipy)_3]^{2+}$ in $H_2O$ | LS | | | $1.2 \cdot 10^9$ | 119 |
| $[Fe(phen)_3]^{2+}$ in $H_2O$ | LS | | | $2 \cdot 10^9$ | 120 |
| $[Fe(2\text{-}pic)_3]Cl_2 \cdot MeOH$ | 150 | 1050 | $2 \cdot 10^9$ | $7 \cdot 10^6$ | 76 |
| $[Fe(2\text{-}pic)_3](PF_6)_2$ | 185 | 1058 | $1.3 \cdot 10^9$ | $6.6 \cdot 10^6$ | 124 |
| $[Fe(mtz)_6](PF_6)_2$ | 205 | | | $2.5 \cdot 10^6$ | 125 |
| $[Fe(mepy)_3tren](PF_6)_2$ | 220 | 420 | $3 \cdot 10^7$ | $3 \cdot 10^6$ | 126 |
| $[Fe(tpen)](ClO_4)_2$ | 420 | | | $10^8$ | 127 |

temperature. The latter ones again fall within the range of $10^6$–$10^8$ s$^{-1}$. In contrast to the room temperature data, $k_{HL}(T \to 0)$ spans 12 orders of magnitude, and, as shown in Figure 18, there is a clear correlation between $k_{HL}(T \to 0)$ and $T_{1/2}$.

## 6.2 Multiphonon Relaxation

Obviously, in order to arrive at a coherent understanding of the HS → LS relaxation dynamics, it is essential to consider the full wealth of experimental data simultaneously. In particular, since there are no fundamental differences between solution and solid state (in both, the spin transition is basically a unimolecular process albeit in-

**Table 2** **Transition Temperatures $T_{1/2}$, Activation Energies $E^{a}_{HL}$ and Frequency Factors $A_{HL}$ for the High Temperature Region, and Low Temperature Tunneling Rate Constants $k_{HL}(T \rightarrow 0)$ for Some Fe(II) Coordination Compounds in the Solid State and Embedded in Polymer Matrices. For Comparison, Values of $k_{HL}$ Extrapolated to Room Temperature are Included**

| Compound | $T_{1/2}$ [K] | $E^{a}_{HL}$ [$cm^{-1}$] | $A_{HL}$ [$s^{-1}$] | $k_{HL}(T \rightarrow 0)$ [$s^{-1}$] | $k_{HL}$ (295 K) [$s^{-1}$] | Reference |
|---|---|---|---|---|---|---|
| $[Zn_{1-x}Fe_x(ptz)_6)](BF_4)_2$ | 95 | 1100 | $5 \cdot 10^{7}$ | $5 \cdot 10^{-7}$ | $\sim 5 \cdot 10^{5}$ | 99 |
| $[Mn_{1-x}Fe_x(2\text{-pic})_3)]Cl_2 \cdot EtOH$ | 76 | | | $6 \cdot 10^{-6}$ | | 100 |
| $[Zn_{1-x}Fe_x(2\text{-pic})_3)]Cl_2 \cdot EtOH$ | 95 | | | $2.5 \cdot 10^{-5}$ | | 100 |
| $[Mn_{1-x}Fe_x(2\text{-pic})_3)]Cl_2 \cdot MeOH$ | 118 | 1120 | $2 \cdot 10^{8}$ | $2.5 \cdot 10^{-3}$ | $\sim 4 \cdot 10^{6}$ | 100 |
| $[Zn_{1-x}Fe_x(2\text{-pic})_3)]Cl_2 \cdot MeOH$ | 140 | | | $9 \cdot 10^{-3}$ | | 100 |
| $[Zn_{1-x}Fe_x(mepy)_2(py)tren](PF_6)_2$ | 210 | 837 | $5 \cdot 10^{8}$ | $1.4 \cdot 10^{-1}$ | $\sim 8 \cdot 10^{6}$ | 99 |
| $[Fe(mepy)_2)(py)tren]^{2+}$ in PMMA | 270 | | | $\sim 5 \cdot 10^{1}$ | | 99 |
| $[Fe(mepy)(py)tren]^{2+}$ in PMMA | 370 | | | $\sim 2 \cdot 10^{2}$ | | 99 |
| $[Zn_{1-x}Fe_x(py)_3tren](PF_6)_2$ | $> 400$ | 640 | $1 \cdot 10^{9}$ | $4 \cdot 10^{2}$ | $\sim 5 \cdot 10^{7}$ | 99 |
| $[Mn_{1-x}Fe_x(bipy)_3)](PF_6)_2$ | LS | | | $6 \cdot 10^{4}$ | | 128 |
| $[Zn_{1-x}Fe_x(bipy)_3](PF_6)_2$ | LS | 364 | $2 \cdot 10^{9}$ | $6 \cdot 10^{5}$ | $\sim 1 \cdot 10^{9}$ | 131 |
| $[Fe(phen)_3]^{2+}$ in Nafion | LS | | | $\sim 5 \cdot 10^{6}$ | | 131 |

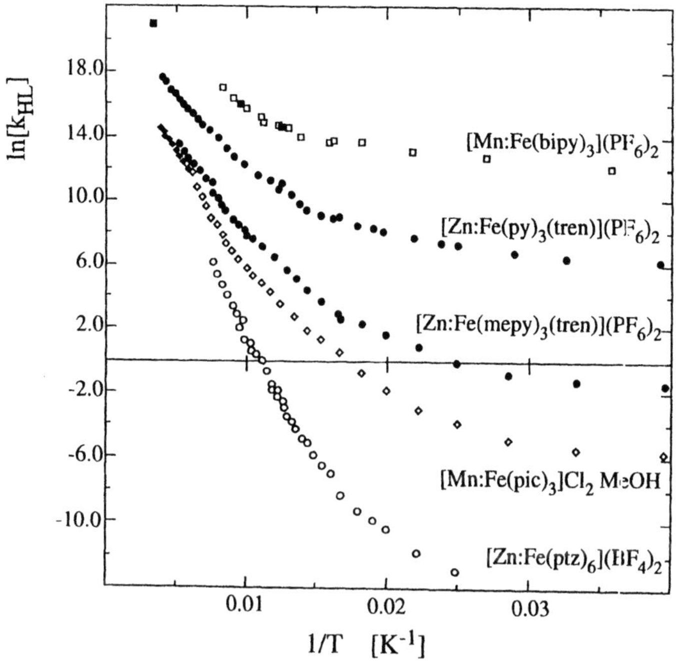


**Figure 17.** Experimental rate constants for the HS $\rightarrow$ LS relaxation plotted as $\ln[k_{HL}]$ versus $1/T$ for the diluted mixed crystals (○) $[Zn_{1-x}Fe_x(ptz)_6](BF_4)_2$, (◇) $[Mn_{1-x}Fe_x(2\text{-pic})_3]Cl_2\cdot MeOH$, (●) $[Zn_{1-x}Fe_x(mepy)_3tren](PF_6)_2$, (●) $[Zn_{1-x}Fe_x(py)_3tren](PF_6)_2$, (□) $[Mn_{1-x}Fe_x(bpy)_3](PF_6)_2$ using pulsed laser excitation. $x \approx 0.0005$ except for $[Zn_{1-x}Fe_x(ptz)_6](BF_4)_2$ for which $x \approx 0.1$. For comparison data for $[Fe(bpy)_3]^{2+}$ in solution (■) from Reference 119 and for $[Mn_{1-x}Fe_x(bpy)_3](PF_6)_2$ from the Mössbauer emission study of Reference 128 are included (■).

fluenced by the surrounding medium), the two should not be discussed separately! It is also obvious that a classical or semiclassical approach to the HS $\rightarrow$ LS relaxation is not adequate for its description. In the following, therefore, a brief outline of the quantum mechanical approach is given. Buhks et al.[118] describe the HS $\rightarrow$ LS relaxation in terms of "a radiationless nonadiabatic multiphonon process occurring between two distinct zeroth-order spin states characterized by different nuclear configurations". In order for such a process to occur, there has to be an electronic coupling between initial and final state. For iron(II) with $\Delta S = 2$, higher order spin-orbit coupling provides the required mixing of the two states. Furthermore, a reaction coordinate has to be defined, which describes the nuclear motion during the relaxation process. In the case of iron(II) spin crossover compounds it is mainly the metal–ligand bond length which is subject to large changes. Thus, the obvious choice for the reaction coordinate is the totally symmetric normal coordinate. This immediately leads to the single configurational coordinate model outlined in Volume I of this series,[35] where the potentials are taken to be displaced relative to each other by $\Delta Q_{HL}$ along a single normal coordinate $Q$ only. Using Fermi's Golden Rule and the Con-

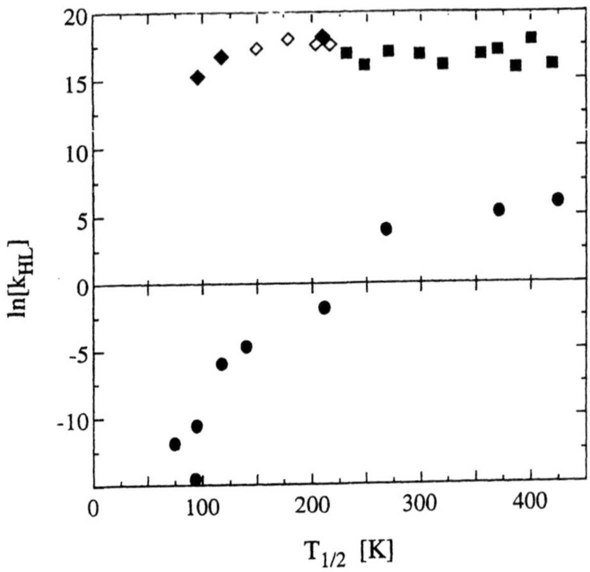


**Figure 18.** Experimental low-temperature tunneling rate constants $k_{HL}(T \rightarrow 0)$ as a function of transition temperature $T_{1/2}$ from Table 2 (●), and $k_{HL}$ at 295 K: solution data (■), solid state data from Mössbauer lineshape analysis (◇), and solid state data from pulsed laser excitation extrapolated to 295 K (◆) (from Reference 129).

don approximation the rate constant for the HS → LS relaxation can be expressed as:[132]

$$k_{HL} = \frac{2\pi}{\hbar^2 \omega} \beta^2 g_f F_n(T) \tag{21}$$

$\beta = \langle \Phi_{LS} | H_{so} | \Phi_{HS} \rangle$ is the electronic matrix element, which can be estimated to be $\approx$150 cm$^{-1}$ considering second order spin-orbit coupling via the low-lying $^3T_1$ state.[118] $\hbar\omega$ is the vibrational frequency of the accepting mode. Typical values for the totally symmetric metal–ligand stretch involving the bulky ligands of spin crossover compounds are around 250 cm$^{-1}$. $g_f$ is the electronic degeneracy of the final state, i.e., for the LS state as final state of the HS → LS relaxation $g_f = 1$. $F_n(T)$ is a thermally averaged Franck–Condon factor. In addition to being temperature dependent, $F_n(T)$ is a parametric function of $\Delta Q_{HL}$ and $\Delta E^0_{HL}$, i.e., the horizontal and the vertical displacements of the potential wells of the two states relative to each other. For harmonic potentials and, for simplicity, assuming equal force constants and vibrational frequencies in the two states, $F_n(T)$ takes the form:[132]

$$F_n(T) = \frac{\sum_m |\langle \chi_{m'} | \chi_m \rangle|^2 e^{-m\hbar\omega/k_B T}}{\sum_m e^{-m\hbar\omega/k_B T}} \tag{22}$$

The sum goes over all thermally populated vibrational levels of the initial state. In order to ensure energy conservation during the radiation process, the nuclear overlap integral $\langle \chi_{m'} | \chi_m \rangle$ have to be evaluated for $m' = m + n$, where the reduced energy gap $n = \Delta E^0_{\mathrm{HL}}/\hbar\omega$ gives the number of vibrational quanta required to bridge the difference in the zero-point energies of the two states. For $T \to 0$, $F_n$ simplifies to:

$$F_n(T \to 0) = \left|\langle \chi_n \mid \chi_0 \rangle\right|^2 = \frac{e^{-S} S^n}{n!} \tag{23}$$

where the Huang–Rhys factor $S = \frac{1}{2} f \Delta Q^2_{\mathrm{HL}}/\hbar\omega$ is a measure for the relative horizontal displacement of the potential wells. For iron(II) compounds with an $[\mathrm{FeN_6}]$ coordination, average values for $f$ and $\hbar\omega$ are $2 \times 10^5$ dyn/cm and 250 cm$^{-1}$, respectively. With $\Delta Q_{\mathrm{HL}} = \sqrt{6}\Delta r^0_{\mathrm{HL}} \approx 0.5$ Å, a value for $S$ of $\approx$45 has been estimated.[99] Whereas $S$ is not expected to vary to a large extent within the series of compounds, the zero-point energy difference and thus $n$ will vary all the way from $n < 1$ for spin crossover compounds having transition temperatures below 100 K to $n > 10$ for true LS compounds. Figure 19 shows $\ln[k_{\mathrm{HL}}]$ as a function of both $n$ and $T$, calculated according to Equations 21 and 22, with the above parameters.

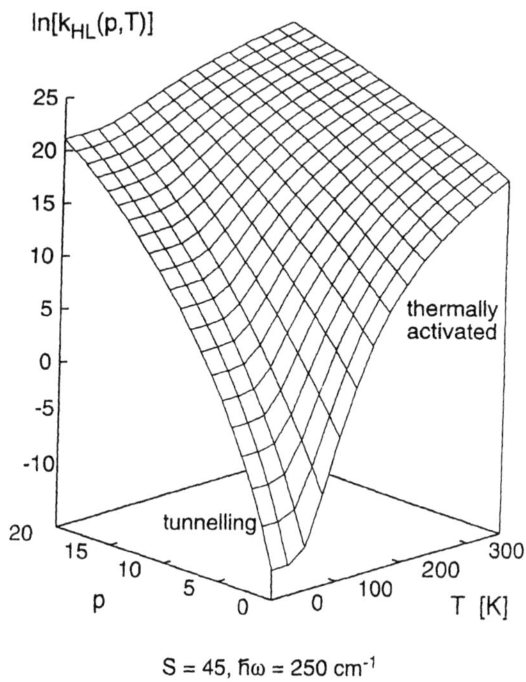


**Figure 19.** Parametric surface of $\ln[k_{\mathrm{HL}}(n, T)]$ calculated with a Huang–Rhys factor $S = 45$ and a vibrational frequency $\hbar\omega = 250$ cm$^{-1}$ according to Equations 21 and 22.

In the strong-coupling limit with $S \gg n$, the characteristic features of $k_{HL}(T)$ are temperature independent tunneling below $\approx 50$ K and a thermally activated process at elevated temperatures, in accordance with the experimental results. Furthermore, the low temperature tunneling rate $k_{HL}(T \rightarrow 0)$ is predicted to increase exponentially with increasing $n$ according to the inverse energy gap law:

$$k_{HL}(T \rightarrow 0) \propto \exp(a\,n) \tag{24}$$

with $a \approx \ln(S) \approx 3.8$ for $S \gg n$ and small values of $n$. At room temperature a much smaller dependence of $k_{HL}$ on $n$ is expected.

The vital parameters governing the temperature dependence and the low temperature tunneling rate constant are $S$ and $n$. As shown above, it is in principle straight forward to estimate $S$ from experimental data, but $n$ is somewhat more difficult to get at quantitatively. However, the transition temperature $T_{1/2}$ is certainly a measure for the zero-point energy difference, and as shown in Figure 18, the increase in $k_{HL}(T \rightarrow 0)$ with $T_{1/2}$ is indeed exponential, spanning 12 orders of magnitude between the spin crossover system $[Zn_{1-x}Fe_x(ptz)_6](BF_4)_2$ with $T_{1/2} < 100$ K and thus $n < 1$ to the LS system $[Zn_{1-x}Fe_x(bpy)_3](PF_6)_2$ with $T_{1/2} \gg RT$ and thus $n \geqslant 10$. And as predicted, the room temperature data is poorly correlated.

With $S \approx 45$ and $\hbar\omega \approx 250$ cm$^{-1}$ the classical activation energy would be approximately 2500 cm$^{-1}$. This is substantially larger than the experimental values. Thus even in the thermally activated region, the HS → LS relaxation has to be considered a nonadiabatic process, with tunnelling from thermally excited vibrational levels of the HS state.[118,129]

Of course, the single configurational coordinate model is rather crude. There are certainly additional normal modes which couple to the HS → LS relaxation. In particular, low-energy torsional modes have been suggested.[133–136] However, the 12 orders of magnitude increase in the low temperature tunneling rate constant are very difficult to explain on the basis of a low-frequency quasi-classical reaction coordinate. In addition, the pressure experiments presented in the next section show that the breathing mode is indeed the most important normal mode in the HS → LS relaxation.

### 6.3 External Pressure and Cooperative Effects

As for the thermal spin transition, an external pressure is expected to affect the HS → LS relaxation considerably. In the thermally activated region,

$$k_{HL}(T) \propto \exp\left(-\,p\Delta v^{\ddagger}_{HL}/k_B T\right) \tag{25}$$

where $\Delta v^{\ddagger}_{HL}$ is the activation volume. Experimental values for $\Delta v^{\ddagger}_{HL}$ of typically between $-10$ and $-20$ Å$^3$ both in solution[122,123] as well as in the solid state,[76,125] indicate that the volume of the transition state is approximately half way in between the volumes of the initial and the final state. At ambient temperatures the effect of an external pressure is to increase the HS → LS relaxation rate constant by typically a factor of 2. In the low temperature tunneling region the effect is much more dramatic. Figure 20 shows the pressure dependence of the HS → LS relaxation rate constant plotted as $\log[k_{HL}(p)/k_{HL}(p \to 0)]$ versus $p$ for the highly diluted $[Zn_{1-x}Fe_x(mepy)_3tren](PF_6)_2$ at 15 K.[137] There is an astonishing eight orders of magnitude increase in the low temperature tunneling rate constant between ambient pressure and 20 kbar, with an initial increase of one order of magnitude per kbar.

As mentioned in Section 4, the external pressure basically increases the zero-point energy difference $\Delta E^0_{HL}$. Thus, the pressure dependence of the reduced energy gap is given by:

$$n(p) = n(p \to 0) + \frac{\Delta v_{HL}}{\hbar\omega} p = n_0 + \beta_n p \tag{26}$$

With a value for $\Delta v_{HL}$ of 25 Å$^3$ as determined from the shift in the thermal equilibrium[125] and the standard value from the preceeding section for $\hbar\omega = 250\ \text{cm}^{-1}$, $\beta_n \approx 0.5$. In Figure 20, calculated curves of $\log[k_{HL}(p)/k_{HL}(p \to 0)]$ versus $p$ are shown, using Equation 25 for the Franck–Condon factor, Equation 26 for the reduced energy gap, the standard value for the Huang–Rhys factor $S$ of 45, and setting $n_0 = 0,\ 1$ and 3. The first and third value may be regarded as lower and upper limits for a spin crossover system, $n_0 = 1$ is typical for spin crossover compounds with

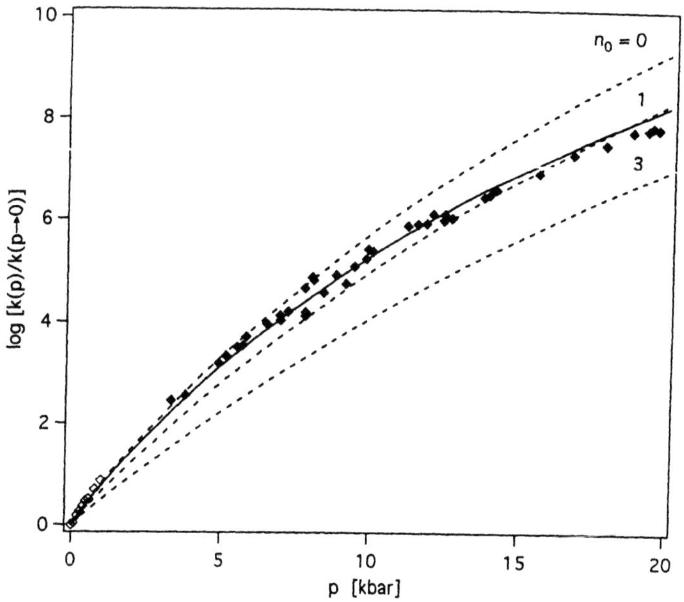


**Figure 20.** (◆) $\log[k_{HL}(p)/k_{HL}(p \to 0)]$ versus pressure $p$ for $[Zn_{1-x}Fe_x(mepy)_3tren](PF_6)_2$ at 15 K. (—) Calculated curves according to Equation 25 with a pressure dependent reduced energy gap $n$ as described in the text (from Reference 137).

transition temperatures of 100–200 K. The experimental curve falls well within the comparatively narrow band of the limiting curves predicted by this simple approach. The acceleration at low pressures is slightly larger than predicted; at high pressures it levels off more rapidly. These deviations must be due to the pressure dependence of $\Delta r_{HL}$, $\Delta v_{HL}$, and $\hbar\omega$, and thus of both $S$ and $n$, a full discussion of which is beyond the scope of this article.

Since in a simple minded approach, the cooperative effects may be pictured as an internal pressure which increases with increasing LS fraction, the HS $\rightarrow$ LS relaxation in neat spin crossover compounds is expected to deviate from a single exponential behavior. These deviations are substantial, as shown by the sigmoidal relaxation curves of neat $[Fe(ptz)_6](BF_4)_2$ in Figure 21. These are well described phenomenologically by a self-accelerating rate constant of the form:

$$k_{HL}(\gamma_{LS}) = k^0_{HL}\exp(\alpha\gamma_{LS}) \tag{27}$$

The initial rate constant as function of $T$ is within experimental error identical to the relaxation rate constant of the diluted mixed crystal system with the isomorphous zinc complex. This is not surprising because, as mentioned before, zinc(II) and HS iron(II) have very similar ionic radii. The self-acceleration factor $\alpha$ as function of $T$ is shown in the inset of Figure 21. At low temperatures $\alpha$ has a limiting value of approximately 5, above 50 K it decreases with $1/T$. Thus, at low temperatures the relaxation rate constant at the end of the relaxation is larger by a factor of 150 compared to the initial rate constant. In the diluted mixed crystal, the same increase in relaxation rate constant can be achieved with an external pressure of 2 kbar, which together with the value for $\Delta v_{HL}$ of $\approx 26$ Å$^3$ corresponds to an increase in zero-point energy difference of approximately 250 cm$^{-1}$.[138] Thus, relaxation measurements in concentrated systems and comparison with results obtained from mea-

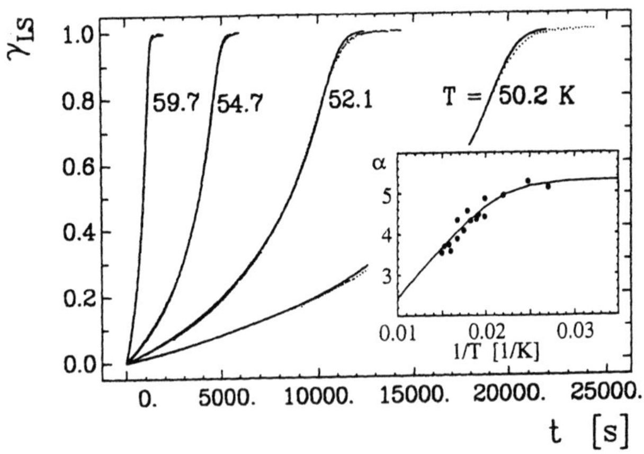


**Figure 21.** HS $\rightarrow$ LS relaxation curves of $[Fe(ptz)_6](BF_4)_2$ following a quantitative light induced population of the HS state. Full lines: best fit to the phenomenological dependence of $k_{HL}$ on the LS fraction according to Equation 27. Inset: self-acceleration factor $\alpha$ versus $T$.

surements under external pressure allow a direct and quantitative approach to the cooperative phenomena, too.

## 7 NUCLEAR DECAY-INDUCED SPIN CROSSOVER

The investigation of chemical and physical after-effects, associated with the nuclear decay in solid materials has been an active research field for decades in Nuclear and Radiochemistry. Chemical and physical after-effects refer primarily to the oxidation and spin states of the nucleogenic atom, bond rupture events in the nearby surroundings of the nucleogenic atom, and radiolysis effects in the coordinated ligand sphere. The nuclear decay-induced species are more or less stable, and the best way of studying the nature and the formation mechanism of such species is by virtue of spectroscopic techniques. A major obstacle hereby is caused by the low concentration of nucleogenic decay species of typically ca. $10^{-15}$ as an upper limit.

Mössbauer spectroscopy has proven to be an excellent tool for the investigation of after-effects in solid coordination compounds. Particularly the electron capture (EC) decay of $^{57}$Co into $^{57}$Fe, denoted by $^{57}$Co(EC)$^{57}$Fe, in cobalt or iron compounds has been widely explored.[139–141] The procedure is such that the $^{57}$Co-labelled compound is used as the Mössbauer source versus a single-line absorber material such as $K_4[Fe(CN)_6]$. The recorded spectrum delivers information on the chemical state of the nucleogenic $^{57}$Fe at ca. $10^{-7}$ s (which is approximately the lifetime of the 14.4 keV level of $^{57}$Fe; for details about Mössbauer Spectroscopy refer to Reference 52) after the nuclear decay. One has named this special kind of Mössbauer experiments Mössbauer Emission Spectroscopy or Mössbauer Source Experiments, in contrast to conventional Mössbauer Absorption Spectroscopy, where the sample under study is used as the Mössbauer absorber (or scatterer) and a suitable single-line matrix with the embedded radioactive precursor nuclide ($^{57}$Co in case of $^{57}$Fe spectroscopy) as the source.

Already in the early seventies one observed the formation of "anomalous" spin states of iron(II) in $FeN_6$ core complexes in Mössbauer "source experiments".[142] As sources we have used $^{57}$Co labelled coordination compounds versus $K_4[Fe(CN)_6]$ as absorber. We speak of "anomalous" spin states in the Mössbauer emission spectrum, if they do not exist as ground states at comparable temperatures in the corresponding iron coordination compounds studied by Mössbauer absorption spectroscopy.[143]

The system $[M(phen)_3](ClO_4)_2$ (M = $^{57}Fe_x/Co_{1-x}$, $^{57}Co_x/Co_{1-x}$ with $x$ = 0.001) was studied by Time Integral Mössbauer Emission Spectroscopy (TIMES) as an example of a strong field system. The corresponding iron complex (M = Fe) is a typical iron(II)-LS compound. Although the Co(II) analogue $[^{57}Co/Co(phen)_3]$ $(ClO_4)_2$ is HS with $^4T_{1g}$ ground state (in the approximation of $O_h$-symmetry), the $^{57}$Fe(II) complex molecules doped in the host lattice of $[Co(phen)_3](ClO_4)_2$ are LS with $^1A_{1g}$ ground state at all temperatures as in the pure iron compound. This can be seen from the Mössbauer absorption spectra on the left of Figure 22. If, however, the Co(II) compound is doped with $^{57}$Co and used as the Mössbauer source against the

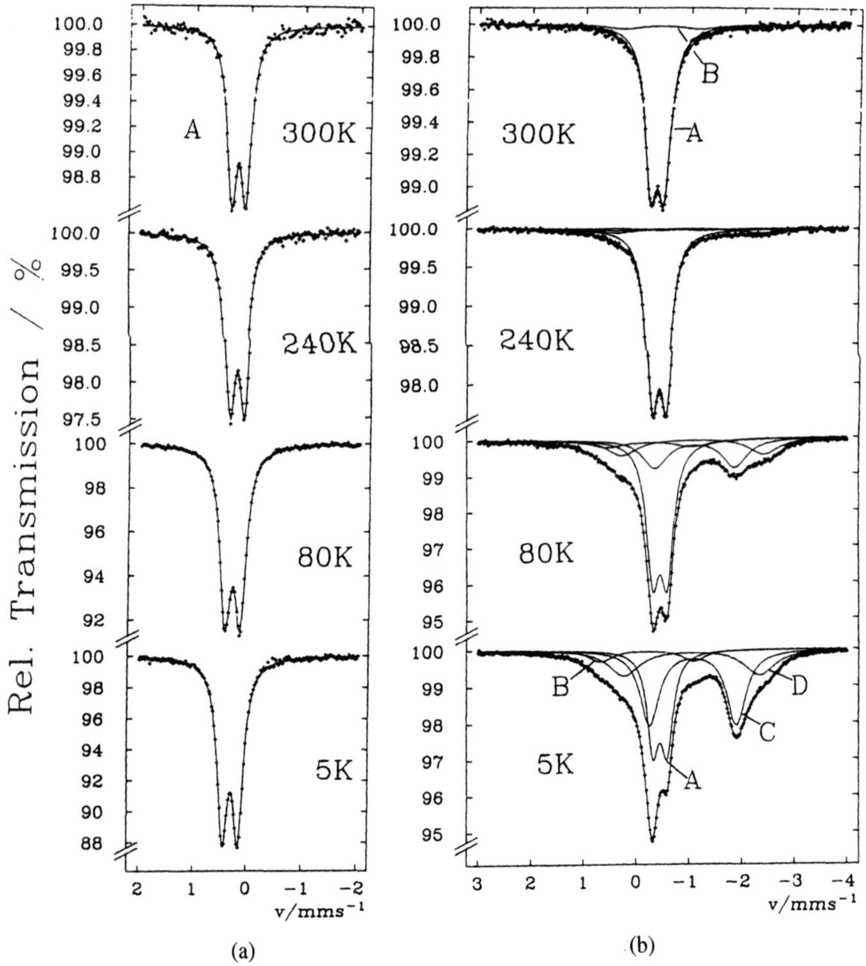


**Figure 22.** (a) $^{57}$Fe Mössbauer absorption spectra of $[^{57}Fe/Co(phen)_3](ClO_4)_2$ as a function of temperature versus $^{57}$Co/Rh (295 K) as source. (b) Time-integral $^{57}$Fe Mössbauer emission spectra of a $[^{57}Co/Co(phen)_3](ClO_4)_2$ source as a function of temperature versus $K_4[Fe(CN)_6]$ (295 K). Assignment: A: Fe(II)-LS, B: Fe(III)-LS, C: Fe(II)-$HS_1$, D: Fe(II)-$HS_2$. In (b), the velocity scale is inverted to allow a direct comparison with the absorption spectra in (a) (from Reference 141).

single-line absorber $K_4[Fe(CN)_6]$, emission spectra of the type shown on the right of Figure 22 are obtained.

At temperatures between 300 K and ca. 250 K, the emission spectra do not differ from the absorption spectra, which indicates that the electronic structure of the nucleogenic Fe atom is that of the Fe(II) ground state, viz. LS ($^1A_{1g}$) represented by signal A in Figure 22. At lower temperatures typical HS resonance lines C and D appear in addition to the LS ($^1A_{1g}$) signal, which have been assigned to the $^5T_{2g}$ state of an undistorted and a distorted environment of the nucleogenic iron, respectively. The resonance doublet B refers to iron(III)-LS and has practically unchanged intensity over the whole temperature range.

The compound $[Fe(phen)_2(NCS)_2]$ possesses a somewhat weaker ligand field and shows thermal spin transition as shown by the temperature dependent Mössbauer absorption spectra in Figure 8. The $^{57}$Co-labelled $[^{57}Co/Fe(phen)_2(NCS)_2]$ compound used as a Mössbauer source produces emission spectra which no longer exhibit Fe(II)-LS signals at any temperature. The HS quadrupole doublet of iron(II) is the dominant signal to be seen over the whole temperature range between 296 and 4.2 K. Similar behaviour was observed in the Mössbauer emission spectra of other $^{57}$Co-labelled cobalt complexes,[143–149] whose corresponding iron compounds all exhibit thermal spin crossover. The essential result from these Mössbauer source experiments is that, at a given temperature, the probability of trapping the HS state of the nucleogenic $^{57}$Fe(II) within the Mössbauer time window is considerably greater in the spin crossover compounds (intermediate field strength) than in strong field compounds. With our knowledge of the LIESST effect (generating metastable HS states by irradiation with an "external" light source), we are able to understand the mechanism of the nuclear decay-induced formation of metastable anomalous spin states, where the nuclear decay process $^{57}$Co(EC)$^{57}$Fe may be regarded as an "internal" molecular light source. By analogy to the LIESST effect, we have termed this phenomenon "Nuclear Decay-Induced Excited Spin State Trapping" (NIESST). The mechanism of NIESST is essentially the same as for LIESST (see Section 5.1), except for the initial excitation step.

The nuclear decay process $^{57}$Co(EC)$^{57}$Fe leads initially to highly ionized nucleogenic iron ions. However, at times before the Mössbauer technique enables one to observe metastable states, i.e., before ca. $10^{-8}$ s, there is fast electron recombination leading to iron(II) preferentially in the spin quintet states.[150] The rate of relaxation to the $^1A_1$ state is much faster in strong-field complexes than in spin crossover (intermediate field) complexes. The lifetime of the trapped $^5T_2$ state is in those strong-field systems studied so far comparable to the Mössbauer time window, such that relaxation can be observed.[128] The trapped $^5T_2$ state in spin crossover systems lives much longer than the Mössbauer time window. This corresponds to the observed correlation of the lifetime of the metastable LIESST state and the energy difference $\Delta E^\circ_{HL}$, which also depends on the ligand field strength (reduced energy gap law). In order to prove the validity of this mechanism we have constructed a coincidence spectrometer for time-differential Mössbauer emission spectroscopy (TDMES) with a resolution of ca. 3.5 ns in the time window of ca. 5 to 500 ns, in order to measure the lifetime of the metastable HS state of the nucleogenic iron as a function of temperature.[151] The TDME spectra prove that the intensity of the metastable HS state decreases the higher the temperature and the shorter the time interval elapsed after the nuclear decay. The lifetime of one of the HS states derived from the TDME measurements was found to be 390 ns (at 10 K), 205 ns (at 47 K), and 100 ns (at 100 K).[152] This compares well with results from optical spectroscopy on $[Fe(phen)_3]^{2+}$ embedded in a nafion foil, where a lifetime of 188 ns was determined for the $^5T_{2g}$ state at 50 K after

LIESST.[131] This agreement provides strong evidence for the validity of the proposed mechanism.

## 8 CONCLUDING REMARKS

Iron(II) spin crossover compounds constitute a model system for the demonstration of a number of elementary principles in physical inorganic chemistry and photophysics, and, in particular, they serve as a textbook example of applied ligand field theory. The phenomenon of spin crossover is basically due to the difference in the dependence of the ligand field strength and the electron–electron repulsion on the metal–ligand separation, which results in a double minimum type situation. The driving force is the increase in entropy as the HS state is populated at elevated temperatures.

Understanding the thermal spin transition thus involves concepts of classical and statistical thermodynamics. In a diluted system, such as for a spin crossover complex in solution or in a diluted mixed crystal, only the vibronic manifolds of the individual complex have to be considered. In concentrated compounds, the interactions of elastic origin between the spin changing complexes have to be taken into account. The key features of the resulting cooperative effects, namely the fact that in concentrated compounds transition curves are always steeper than in diluted systems, can already be explained on the basis of the comparatively simple level of the Bragg–Williams approximation. In this approximation, the interaction is treated as long-range interaction. The physical picture of an internal pressure, which increases with increasing fraction of complexes in the LS state, emerging from such a treatment can be justified to a certain extent. Indeed, as evidenced by X-ray crystallography, the unit cell dimensions decrease with increasing LS fraction. Thus the complexes are more densely packed and are thus subject to an increased chemical pressure. A direct comparison of the effects of an external pressure is also possible. Shifts of transition curves to higher temperatures as well as shifts in the maxima of ligand field absorption bands due to coopertive effects are compatible with those observed under external pressure in a number of cases.

The photophysical properties can be understood based on the principles of the electronic strucure of transition metal compounds and the resulting processes as outlined in various chapters in the two volumes of this series of monographs. Despite the fact that iron(II) spin crossover compounds do not luminesce, a number of extremely rapid intersystem crossing steps between higher ligand field states could unambiguously be identified, and the surprisingly long low-temperature life times of the light-induced metastable HS state could be rationalized quantitatively on the basis of the single configurational coordinate model in the strong vibronic coupling limit. The bond length difference between the HS and the LS state in iron(II) compounds is, in fact, the largest difference between any two ligand field states in coordination compounds, and together with the small driving force it is responsible for the long life time of the HS state. As a rule, the single configurational coordinate model only allows a rather qualitative estimate for the absolute values of intersystem crossing rate constants, basically due to the fact that a) the reaction coordinate is usually a linear combination of more than one normal coordinate and b) the parameters entering the theory are not known at the required level of accuracy. For spin crossover compounds the reaction coordinate is dominated by the totally symmetric breathing mode and is thus well defined, and the structural and energetic parameters can be accurately determined from independent experiments. The absence of competing relaxation processes such as luminescence allowed the determination of the HS–LS relaxation rate constants over a temperature range from 10 K to room temperature and for values from $10^{-6}$ $s^{-1}$ to $10^{8}$ $s^{-1}$, helping to establish the tunnelling nature of the process.

With this report, we hope to have established a coherent picture of the phenomenon of spin crossover in its most important aspects, the interdependence of thermodynamics and kinetics, cooperative effects, and photophysics. We hope to have shown that in order to arrive at such a picture, a combination of different experimental techniques as well as theoretical work are indispensable.

## ACKNOWLEDGMENTS

We thank all our students and collaborators, who, over the past two decades, have contributed towards our understanding of this class of fascinating compounds. Of course, we are indepted to our colleagues around the globe for enlightening discussions. We have paid them our respect with the list of references. Financial support was granted by the "Deutsche Forschungsgemeinschaft," the "Materialwissenschaftliches Forschungszentrum der Universität Mainz," the "Fonds der Chemischen Industrie," the "Bundesministerium für Bildung, Wissenschaft, Forschung und Technologie," and the "Schweizerischer Nationalfonds."

## REFERENCES

1. L. Cambi and A. Cagnasso, Atti Accad. Naz. Lincei **13**, 809 (1931).
2. L. Cambi and L. Szegö, Ber. dtsch. chem. Ges. **64**, 259 (1931).
3. L. Cambi, L. Szegö and A. Cagnasso, Atti Accad. Naz. Lincei **15**, 266 (1932).
4. L. Cambi, L. Szegö and A. Cagnasso, Atti Accad. Naz. Lincei **15**, 329 (1932).
5. R.L. Martin and A.H. White, Transition Met. Chem. (Chem. Soc. NY) **4**, 113 (1968).
6. J. Zarembowitch and O. Kahn, New J. Chem. **15**, 181–90 (1991).
7. J.H. Ammeter, R. Bucher and N. Oswald, J. Am. Chem. Soc. **96**, 7833 (1974).
8. J.H. Ammeter, J. Magn. Reson. **30**, 299 (1978).
9. J.H. Ammeter, L. Zoller, J. Bachmann, E. Baltzer, E. Gamp, R. Bucher and E. Deiss, Hel. Chim. Acta **64**, 1063 (1981).
10. N. Hebedanz, F.H. Köhler, G. Müller and J. Riede, J. Am. Chem. Soc. **108**, 3281 (1986).
11. M.E. Switzer, R. Wang, M.F. Rettig and A.H. Maki, J. Am. Chem. Soc. **96**, 7669 (1974).
12. P.G. Sim and J. Sinn, J. Am. Chem. Soc. **103**, 241 (1981).

13. H.J. Krokoszinski, C. Santandrea, E. Gmelin and K. Bärner, Phys. Status Solidi B **113**, 185 (1982).
14. W. Kläui, J. Chem. Soc., Chem. Commun., 700 (1979).
15. P. Gütlich, B.R. McGarvey and W. Kläui, Inorg. Chem. **19**, 3704 (1980).
16. W. Eberspach, N. El Murr and W. Kläui, Ang. Chem. Int. Ed. Engl. **21**, 915 (1982).
17. L. Sacconi, Pure Appl. Chem. **27**, 161 (1971).
18. R.H. Holm and M.J. O'Connor, Prog. Inorg. Chem. **14**, 241 (1973).
19. W. Kläui, K. Schmidt, A. Bockmann, P. Hofmann, H.R. Schmidt and P. Staufert, J. Organomet. Chem. **286**, 407 (1985).
20. H. Werner, B. Ulrich, U. Schubert, P. Hofmann and B. Zimmer-Gasser, J. Organomet. Chem. **297**, 27 (1985).
21. M.H. Chisholm, E.M. Kober, D.H. Ironmonger and P. Thornton, Polyhedron **4**, 1869 (1985).
22. E. König, Coord. Chem. Rev. **3**, 471 (1968).
23. H. A. Goodwin, Coord. Chem. Rev. **18**, 293 (1976).
24. P. Gütlich, Struct. Bond. (Berlin) **44**, 83 (1981).
25. E. König, G. Ritter and S.K. Kulshreshtha, Chem. Rev. **85**, 219 (1985).
26. H. Toftlund, Coord. Chem. Rev. **94**, 67 (1989).
27. P. Gütlich and A. Hauser, Coord. Chem. Rev. **97**, 1 (1990).
28. E. König, Structure and Bonding **76**, 51 (1991).
29. P. Gütlich, A. Hauser and H. Spiering, Angew. Chem. **106**, 2109 (1994). See also Angew. Chem. Int. Ed. Engl., **33**, 2024, (1994).
30. P Gütlich, J Jung and H.A. Goodwin, Spin Transition in Iron(II) Complexes, In *NATO-ASI-SERIES E: Applied Science*, Edited by E. Coronado, volume 321 (1996).
31. O. Kahn, J. Kröber and C. Jay, Advanced Materials **4**, 718–728 (1992).
32. S. Sugano, Y. Tanabe and H. Kamimura, *Multiplets of Transition-Metal Ions in Crystals*, volume 33 of *Pure and Applied Physics*, (Academic Press: New York, 1970).
33. E.I. Solomon and A.B.P. Lever, in *Inorganic Structure and Spectroscopy, Vol. I, Methodology* (this series), (John Wiley & Sons, Inc.: New York, 1999).
34. H.L. Schläfer and G. Gliemann, *Einführung in die Ligandenfeldtheorie*, (Akademische Verlagsgesellschaft: Wiesbaden, 1980).
35. H.U. Güdel and T. Brunold, in *Inorganic Structure and Spectroscopy, Vol. II, Applications and Case Studies* (this series), (John Wiley & Sons, Inc.: New York, 1999).
36. L. Wiehl, G. Kiel, C.P. Köhler, H. Spiering and P. Gütlich, Inorg. Chem. **25**, 1565 (1986).
37. M.A. Hoselton, L.J. Wilson and R.S. Drago, J. Am. Chem. Soc. **97**, 1722 (1975).
38. B.A. Katz and C. E. Strouse, J. Amer. Chem. Soc. **101**, 6214 (1979).
39. M. Mikami-Kido and Y. Saito, Acta Cryst. **B38**, 452 (1982).
40. A. Hauser, J. Chem. Phys. **94**, 2741 (1991).
41. A.B.P. Lever, "*Inorganic Electronic Spectroscopy*," *Studies in Physical and Theoretical Chemistry*, volume 33, (Elsevier: Amsterdam, 1984), p. 462.
42. A. Hauser, J. Chem. Phys. **94**, 2741 (1991).
43. J. Jung, *Two Aspects of Elasticity in the Spin Crossover Compound* $[Fe_xZn_{1-x}(ptz)_6](BF_4)_2$, Ph.D. thesis, D77, Johannes Gutenberg-Universität Mainz (1995).
44. L. Wiehl, Acta. Cryst. **B49**, 289 (1993).

45. E. W. Müller, J. Ensling, H. Spiering and P. Gütlich, Inorg. Chem. **22**, 2074 (1983).
46. J. Jung, G. Schmitt, A. Hauser, L. Wiehl, K. Knorr, H. Spiering and P. Gütlich, Z. Phys. B **100**, 523 (1996).
47. O. Kahn, *Molecular Magnetism*, (VCH Publisher, Inc.: New York, 1993).
48. J.H. Takemoto and B. Hutchinson, Inorg. Nucl. Chem. Lett. **8**, 769 (1972).
49. J.H. Takemoto and B. Hutchinson, Inorg. Chem. **12**, 705 (1973).
50. J.H. Takemoto, B. Streusand and B. Hutchinson, Spectrochim. Acta Part A **30**, 827 (1974).
51. P. Gütlich, R. Link and A.X. Trautwein, *Mössbauer Spectroscopy and Transition Metal Chemistry*, (Springer: Berlin, 1978).
52. P. Gütlich and J. Ensling, in *Inorganic Structure and Spectroscopy, Vol. I, Methodology* (this series), (John Wiley & Sons, Inc.: New York, 1998).
53. K. Madeja and E. König, J. Inorg. Nucl. Chem. **25**, 377 (1963).
54. W.A. Baker and H.M. Bobonich, Inorg. Chem. **3**, 1184 (1964).
55. E. König and K. Madeja, J. Chem. Soc., Chem. Commun. **3**, 61 (1966).
56. E. W. Müller, H. Spiering and P. Gütlich, Chem. Phys. Lett. **93**, 567 (1982).
57. M. Mikami, K. Konno and Y. Saito, Chem. Phys. Lett. **63**, 566 (1979).
58. M. Mikami, K. Konno and Y. Saito, Acta. Cryst. **B36**, 275 (1980).
59. B. Gallois, J.A. Real, C. Hauw and J. Zarembowitch, Inorg. Chem. **29**, 1152 (1990).
60. E. König, G. Ritter, J. Dengler and S.M. Nelson, Inorg. Chem. **26**, 3582 (1987).
61. E. Meissner, H. Köppen, H. Spiering and P. Gütlich, Chem. Phys. Lett. **95**, 163 (1983).
62. C. Cartier, P. Thuery, M. Verdaguer, J. Zarembowitch and A. Michalowicz, J. Phys., Colloq. **2**, C8/563 (1986).
63. P. Thuery, J. Zarembowitch, A. Michalowicz and O. Kahn, Inorg. Chem. **26**, 851–5 (1987).
64. H. Winkler, A. Sawaryn, A.X. Trautwein, A.S. Yousif, C. Hermes, H. Toftlund and R.H. Herber, Hyperfine Interact. **42**, 921 (1988).
65. E. König, G. Ritter, H. Grünsteudel, J. Dengler and J. Nelson, Inorg. Chem. **33**, 837 (1994).
66. N.V. Bausk, S.B. Erenburg, L.N. Mazalov and L.G. Lavrenova, J. Struct. Chem. **35**, 509 (1994).
67. M. Sorai and S. Seki, J. Phys. Soc. Jap. **33**, 575 (1972).
68. M. Sorai and S. Seki, J. Phys. Chem. Solids **35**, 555 (1974).
69. H.L. Chum, J.A. Vanin and M.I.D. Holanda, Inorg. Chem. **21**, 1146 (1982).
70. C.P. Köhler, R. Jakobi, E. Meissner, L. Wiehl, H. Spiering and P. Gütlich, J. Phys. Chem. Solids **51**, 239 (1990).
71. E. Meissner, H. Köppen, C.P. Köhler, H. Spiering and P. Gütlich, Hyperfine Interact. **36**, 1 (1987).
72. Jakobi R., *Thermoanalytische Untersuchungen an den Mischkristallen* $[Fe_xZn_{1-x}$(2-pic -$ND_2)_3]Cl_2 \cdot$ EtOD, Master's thesis, Johannes Gutenberg-Universität, Mainz, 1985.
73. R. Jakobi, H. Spiering, L. Wiehl, E. Gmelin and P. Gütlich, Inorg. Chem. **27**, 1823 (1988).
74. H. Spiering, E. Meissner, H. Köppen, E.W. Müller and P. Gütlich, Chem. Phys. **68**, 65 (1982).

75. I. Sanner, E. Meissner, H. Köppen, H. Spiering and P. Gütlich, Chem. Phys. **86**, 227 (1984).
76. P. Adler, H. Spiering and P. Gütlich, Inorg. Chem. **26**, 3840 (1987).
77. R. Jakobi, H. Spiering and P. Gütlich, J. Phys. Chem. Solids **53**, 267 (1992).
78. J.D. Eshelby, *Solid State Phys.*, volume 3, p. 79, (Academic: New York, 1956).
79. N. Willenbacher and H. Spiering, J. Phys. C. **21**, 1423 (1988).
80. H. Spiering and N. Willenbacher, J. Phys.: Condens. Matter **1**, 10089 (1989).
81. H. Köppen, E.W. Müller, C.P. Köhler, H. Spiering, E. Meissner and P. Gütlich, Chem. Phys. Lett. **91**, 348 (1982).
82. T. Kohlhaas, H. Spiering and P. Gütlich, Z. Physik B. **102**, 455 (1997).
83. D. Walsh and J. Lange, Phys. Rev. B **23**, 8 (1981).
84. G.A. Ham, F.S. Slack, Phys. Rev. B **4**, 777 (1971).
85. M. Skowronski and Z. Liro, J. of Luminescence **24**, 253 (1981).
86. E.E. Vogel, O. Mualin, H.J. Schulz and M. Thiede, Z. Phys. Chem. (Munich) **201**, 373 (1997).
87. J.F. Endicott, *Inorganic Structure and Spectroscopy, Vol. I, Methodology* (this series), (John Wiley & Sons, Inc.: New York, 1998).
88. J.K. McCusker, K.N. Walda, R.C. Dunn, J.D. Simon, D. Magde and D.N. Hendrickson, Am. Chem. Soc. **114**, 6919 (1992).
89. S. Decurtins, P. Gütlich, C.P. Köhler, H. Spiering and A. Hauser, Chem. Phys. Lett. **105**, 1 (1984).
90. S. Decurtins, P. Gütlich, C.P. Köhler and H. Spiering, J. Chem. Soc., Chem. Commun., 430 (1985).
91. A. Hauser, Coord. Chem. Rev. **111**, 275 (1991).
92. S. Decurtins, P. Gütlich, K.M. Hasselbach, A. Hauser and H. Spiering, Inorg. Chem. **24**, 2174 (1985).
93. R.H. Herber, Inorg. Chem. **26**, 173 (1987).
94. P. Poganiuch and P. Gütlich, Inorg. Chem. **26**, 455 (1987).
95. A.W. Addison, S. Burman, C.G. Wahlgren, O.A. Rajan, T.M. Rowe and E. Sinn, J. Chem. Soc. Dalton Trans., 2621 (1987).
96. D.C. Figg and R.H. Herber, Hyperfine Interact. **62**, 99 (1990).
97. A. Hauser, P. Gütlich and H. Spiering, Inorg. Chem. **25**, 4245 (1986).
98. A. Hauser, Chem. Phys. Lett. **124**, 543 (1986).
99. A. Hauser, A. Vef and P. Adler, J. Chem. Phys. **95**, 8710 (1991).
100. A. Vef, U. Manthe, P. Gütlich and A. Hauser, J. Chem. Phys. **101**, 9362 (1994).
101. K.-U. Baldenius, A.K. Campen, H.-D. Hönk and A.J. Rest, J. Mol. Struct. **157**, 295 (1987).
102. A. Hauser, J. Adler and P. Gütlich, Chem. Phys. Lett. **152**, 468–72 (1988).
103. J.J. McGarvey and I. Lawthers, J. Chem. Soc., Chem. Commun. 906 (1982).
104. H. Romstedt, H. Spiering and A. Hauser, J. Chem. Phys. Solids **59**, 265 (1998).
105. J.K. McCusker, K.N. Walda, R.C. Dunn, J.D. Simon, D. Magde and D.N. Hendrickson, Inorg. Chem. **32**, 394 (1993).
106. P. Poganiuch, S. Decurtins and P. Gütlich, J. Am. Chem. Soc. **112**, 3270 (1990).

107. R. Hinek, P. Gütlich and A. Hauser, Inorg. Chem. **33**, 567 (1994).
108. H.G. Drickamer and C.W. Frank, *Electronic Transitions and the High Pressure Chemistry and Physics of Solids*, (Chapman and Hall: London, 1973).
109. J.K. McCusker, M. Zvagulis, H.G. Drickamer and D.N. Hendrickson, Inorg. Chem. **28**, 1380 (1989).
110. C. Roux, J. Zarembowitch, J.-P. Itie, A. Polian and M. Verdaguer, Inorg. Chem. **35**, 574 (1996).
111. R. Hinek, H. Spiering, P. Gütlich and A. Hauser, Chemistry—A European Journal **2**, 1435 (1996).
112. T.H. Mayman, Nature **187**, 493 (1960).
113. M. Grätzel, Comments Inorg. Chem. **12**, 93 (1991).
114. J.K. Beattie, Adv. Inorg. Chem. **32**, 1 (1988).
115. E.V. Dose, M.A. Hoselton, N. Sutin, M.F. Tweedle and L.J. Wilson, J. Am. Chem. Soc. **100**, 1141 (1978).
116. J.K. Beattie, R.A. Binstead and R.J. West, J. Am. Chem. Soc. **1008**, 3044 (1978).
117. I. Lawthers and J. J. McGarvey, J. Am. Chem. Soc. **106**, 4280 (1984).
118. E. Buhks, G. Navon, M. Bixon and J. Jortner, J. Am. Chem. Soc. **102**, 2918 (1980).
119. A.D. Kirk, P.E. Hoggard, G.B. Porter and M.W. Windsor, Chem. Phys. Lett. **37**, 199 (1976).
120. A.J. Street, D.M. Goodall and R.C. Greenhow, Chem. Phys. Lett. **56**, 326 (1978).
121. J.J. McGarvey, I. Lawthers, K. Heremans and H. Toftlund, J. Chem. Soc. Chem. Commun. 1575 (1984).
122. J.J. McGarvey, I. Lawthers, K. Heremans and H. Toftlund, Inorg. Chem. **29**, 252 (1990).
123. J. DiBenedetto, V. Arkle, H.A. Goodwin and P.C. Ford, Inorg. Chem. **24**, 455 (1985).
124. P. Adler, A. Hauser, A. Vef, H. Spiering and P. Gütlich, Hyperfine Interact. **47–48**, 343 (1989).
125. P. Adler, P. Poganiuch and H. Spiering, Hyperfine Interact. **52**, 47 (1989).
126. P. Adler, H. Spiering and P. Gütlich, J. Phys. Chem. Solids **50**, 587 (1989).
127. H.R. Chang, J.K. McCusker, H. Toftlund, S.R. Wilson, A.X. Trautwein, H. Winkler and D.N. Hendrickson, J. Am. Chem. Soc. **112**, 6814 (1990).
128. A. Deisenroth, S. Hauser, H. Spiering and P. Gütlich, Hyperfine Interact. **93**, 1573 (1994).
129. A. Hauser, Comments Inorg. Chem. **17**, 17 (1995).
130. C.L. Xie and D.N. Hendrickson, J. Am. Chem. Soc. **109**, 6981 (1987).
131. A. Hauser, Chem. Phys. Lett. **173**, 507 (1990).
132. G.F. Imbush and C.J. Donnelly, volume 249 of *NATO ASI Series B* (Plenum Press: New York, 1991).
133. K.F. Purcell, J. Am. Chem. Soc. **101**, 5147 (1979).
134. L.G. Vanquickenborne and K. Pierloot, Inorg. Chem. **20**, 3673 (1981).
135. J.J. McGarvey, H. Toftlund, A.H.R. Al-Obaidi, K.P. Taylor and S.E.J. Bell, Inorg. Chem. **32**, 2469 (1993).
136. J.K. McCusker, A.L. Rheingold and D.N. Hendrickson, Inorg. Chem. **35**, 2100 (1996).
137. W. Wang, I.Y. Chan, S. Schenker and A. Hauser, J. Chem. Phys. **106**, 3817 (1997).
138. J. Jeftic and A. Hauser, Chem. Phys. Lett. **248**, 458 (1996).

139. J.M. Friedt and J. Danon, Atomic Energy Review **18**, 893 (1980).

140. P. Bonville, C. Garcin, A. Gerard, P. Imbert and G. Jehanno, Phys. Rev. B **32**, 4239, 4310 (1981).

141. H. Sano and P. Gütlich, Hot Atom Chemistry in Relation to Mössbauer Emission Spectroscopy, in *Hot Atom Chemistry*, Edited by T. Matsuura, pp. 265–302 (Kodanski: Tokyo, 1984).

142. P. Gütlich, J. Jung and H.A. Goodwin, Spin Transitions in Iron(II) Complexes, volume 321 of *NATO-ASI-Series E: Applied Sciences*, pp. 327–378 (Plenum Press: New York, 1996).

143. J. Ensling, P. Gütlich, K.M. Hasselbach and B.W. Fitzsimmons, Chem. Phys. Lett. **42**, 232 (1976).

144. J. Ensling, B.W. Fitzsimmons, P. Gütlich and K.M. Hasselbach, Angew. Chemie **9**, 637 (1970).

145. J. Fleisch and P. Gütlich, Chem. Phys. Lett. **42**, 237 (1976).

146. J. Fleisch and P. Gütlich, Chem. Phys. Lett. **45**, 29 (1977).

147. J. Fleisch, P. Gütlich and H. Köppen, Radiochem. Radioanalyt. Lett. **42**, 279 (1980).

148. P. Gütlich, H. Köppen and H.G. Steinhäuser, Chem. Phys. Lett. **74**, 3 (1980).

149. C. Hennen, Master's thesis, D77, Johannes Gutenberg-Universität Mainz (1986).

150. H. Spiering, M. Alflen, P. Gütlich, A. Hauser, C. Hennen, U. Manthe and F. Tuczek, Hyperfine Interact. **52**, 113 (1990).

151. R. Albrecht, M. Alflen, P. Gütlich, Zs. Kajcsos, R. Schulze, H. Spiering and F. Tuczek, Nucl. Instr. Meth. **A257**, 209 (1987).

152. R. Grimm, P. Gütlich, E. Kankeleit and R. Link, J. Chem. Phys. **67**, 5491 (1977).

# 11 Neutron and Optical Spectra of Magnetically Ordered Crystals

PETER DAY
The Royal Institution
21 Albemarle Street
London W1X 4BS, UK
E-mail: pday@ri.ac.uk



## 1 INTRODUCTION

From its earliest days, inorganic chemistry developed with an emphasis on the behavior of individual molecular entities called complexes, which might be charged or neutral, but whose structural and electronic properties remain the same whether they

*Inorganic Electronic Structure and Spectroscopy, Volume II: Applications and Case Studies.*
Edited by E. I. Solomon and A. B. P. Lever.
ISBN 0-471-32682-8. 

were isolated from one another in solution or packed together in a crystal lattice. For example, to a very good approximation, the ligand field transitions of octahedral $Ni(H_2O)_6^{2+}$ do not shift when a dilute aqueous solution of $Ni^{2+}$ is concentrated up to the point where crystals of, say $NiSO_4 \cdot 7H_2O$ are formed. Furthermore, from neutron diffraction measurements on aqueous solutions of $Ni^{2+}$, isotopically substituted in turn on nickel, oxygen and hydrogen, we know that the octahedral geometry of $Ni(H_2O)_6^{2+}$, and in particular the Ni–O bond length, is all but identical in the solution and crystalline phases. The physical properties of such crystals, for example magnetic and optical properties, are very close to the sum of the properties of the constituents, and the same is true even when the crystal contains two kinds of complex ion, e.g., $[Cr(NH_3)_6][FeCl_6]$.

In particular, the approach to the electronic structure of transition metal compounds known as ligand field theory was so successful in rationalizing the number and energy distribution of the lower lying excited states that it focused attention completely on the properties of single ions surrounded by their immediate neighbors. For example, because the visible spectrum of $Ni^{2+}$ in aqueous solution is so similar to that of NiO or of $Ni^{2+}$ dissolved in MgO, theoretical treatments of the spectrum of NiO rightly take as their starting point a single $Ni^{2+}$ surrounded by six oxide ions and view the optical properties of the solid lattice as a degenerate superposition of the $10^{23}$ single ion excited states.

In the early 1960s, instances began to be noticed where this generalization was far from borne out. Indeed, some spectacular cases such as Prussian Blue, formed from colorless or pale yellow $Fe(H_2O)_6^{3+}$ and yellow $Fe(CN)_6^{4-}$, had been known for two and a half centuries, and had even been commented on 80 years ago. The physical origins of what has been called the "non-additivity of ionic colors"[1] are multifarious, ranging from small shifts and intensity changes in ligand field transitions to shifts of many thousands of reciprocal centimeters in intense allowed transitions, and the appearance of new absorption bands in the solid which have no direct analogue in the spectra of the constituent ions isolated from one another in dilute solution.[2,3] In this chapter, we deal with some of the consequences of magnetic exchange interactions on the optical spectra of inorganic compounds.

In transition metal complexes in which covalency is not very marked, i.e., with metal ions in the +2 or +3 oxidation states and O- or N-donor ligands, magnetic exchange is usually weak and the exchange energy is a small fraction of the energy separation between the ground and lowest excited ligand field levels. Magnetic exchange therefore has only a small effect on the energies of the ligand field transitions, so high resolution and low temperatures are needed to observe them. However, the effect on the intensity, especially of spin-forbidden transitions, is much more spectacular. To measure and interpret the spin excitations that occur within the electronic ground states of magnetically ordered crystals, and which are also important in determining the intensities of the optical transitions, we need another form of spectroscopy, not otherwise dealt with in this book, namely neutron spectroscopy. This chapter therefore includes an introduction to inelastic neutron scattering.

## 2 PAIRS OF EXCHANGE-COUPLED METAL IONS IN CRYSTALS

A starting point in understanding the effect of magnetic exchange coupling on optical spectra is to look at pairs of coupled ions in a diamagnetic host crystal. One impetus to these studies has been the interest in laser activity in transition metal doped crystals, and the first clear observation of cooperative magnetic effects on ligand field excited states came from emission studies on ruby.[4] For example, the $^2E \rightarrow {}^4A_2$ emission of octahedrally coordinated $Cr^{3+}$ is split into two components by the trigonal distortion of the sites in $Al_2O_3$, and close examination showed that these were also accompanied by other weak lines. These so-called "sidebands" were extremely unusual in having intensities relative to the main bands which were roughly proportional to the $Cr^{3+}$ ion concentration. Such bands must be attributed to new states originating in randomly distributed near neighbor pairs of $Cr^{3+}$. At very low $Cr^{3+}$ concentrations (below 0.5 atomic %) the intensity dependence was accurately linear, but became steeper at higher concentrations, possibly because the pairs were stealing excitation from single ion transitions. The extent to which the sideband emission of the pairs is separated from the single ion emission evidently depends on the size of the pairwise exchange interaction, and often has the order of magnitude 50 $cm^{-1}$. Corresponding effects in the absorption spectrum of highly doped ruby were also observed. For example the intensity of the band system near 3400 Å varied as $[Cr^{3+}]^2$ but the one near 4770 Å as $[Cr^{3+}]$.[5]

A particularly striking example of the change in selection rules from single ions to pairs is the case of $Co^{2+}$ as a substitutional impurity in the hexagonal layer crystal $CdBr_2$.[6] A sharp, well-resolved band system near 560 nm arises from the spin-forbidden ligand field transition $^4T_{1g}(F) \rightarrow {}^2T_{1g}(H)$. The ground term is split in first order by spin-orbit coupling, leaving an $E'_g$ component lying at the lowest level. The upper state is likewise split into $E'_g$ and $U'_g$ components and the small trigonal distortion from the octahedral field further splits the latter into $U'_g(\pm 3/2)$ and $U'_g(\pm 1/2)$. Transitions from $E'_g(^4T_{1g})$ to the two $U'_g$ components are observed as sharp, purely electronic absorption lines (Figure 1). In a uniaxial crystal, one can distinguish between absorption due to the electric or magnetic dipole mechanisms by measuring three spectra, with the electric (***E***) and magnetic (***M***) vectors of the incident light parallel or perpendicular to the unique $c$ axis as follows: $E \perp c$, $M \perp c$ (axial); $E \perp c$, $M \parallel c(\sigma)$; and $E \parallel c$, $M \perp c(\pi)$. If the axial and $\sigma$ spectra are the same, the electric dipole operator provides the dominant transition mechanism, whereas if the axial spectrum is the same as the $\pi$ spectrum, it is the magnetic dipole mechanism that dominates. Following this recipe, Figure 1 shows immediately that, in a crystal of $CdBr_2$ containing about 1% $Co^{2+}$, the mechanism of the $E'_g(^4T_{1g}) \rightarrow U'_g(\pm 1/2 \pm 3/2)$ transitions is purely magnetic dipole. In striking contrast, however, if one measures the spectra of a $CdBr_2$ crystal in which about 10% of the $Cd^{2+}$ is substituted by $Co^{2+}$, one finds that it is the $\sigma$ spectrum that resembles the axial one, so that the mechanism has changed to electric dipole. Not only that, but the spectrum is much more complicated, with at least five bands instead of two. That the multiplicity of bands arises from splitting of the ground state manifold as well

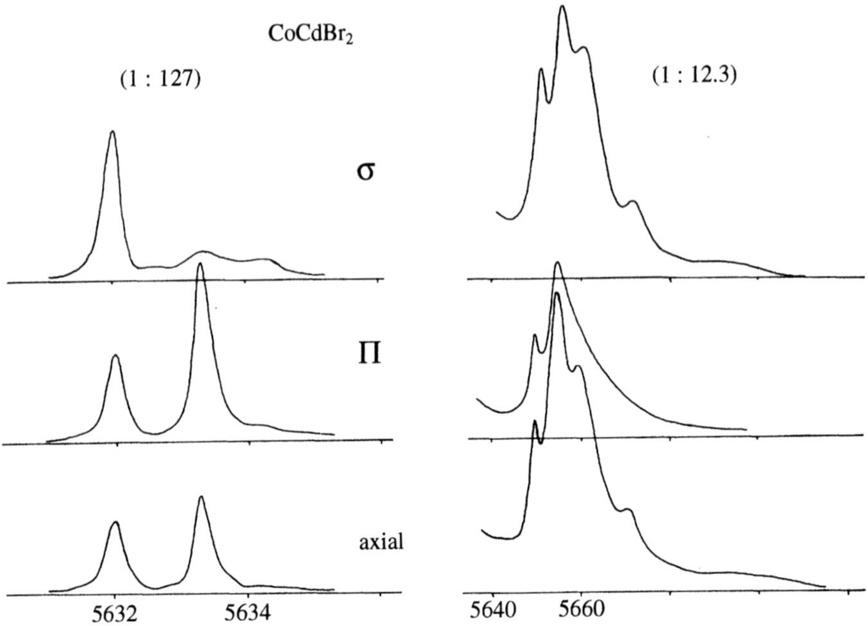


**Figure 1.** Linearly polarized absorption spectra of crystals of $CdBr_2$ containing $2^+$ different doping levels of $Co^{2+}$ in 563–566 nm region: (a) 1% $Co^{2+}$; and (b) 10% $Co^{2+}$. Reproduced by permission from Reference 6.

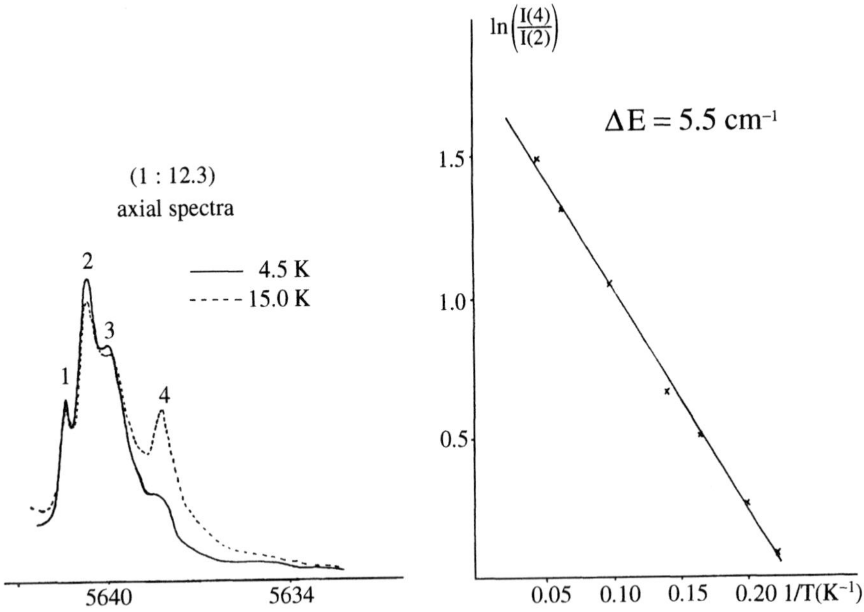


**Figure 2.** Temperature dependence of the absorption spectra of a crystal of $CdBr_2$ doped with 10% $Co^{2+}$ in the 563–566 nm region: (a) the spectrum at 4.5 and 15 K; and (b) temperature variation of intensity of band 4. Reproduced by permission from Reference 6.

as the excited state can be verified by measuring the temperature dependence of the spectrum. Figure 2(a) shows that the band labeled "4" increases rapidly in intensity with increasing temperature, so it originates from a thermally populated upper level within the electronic ground state. Its energy separation from the lowest level can be found by following the intensity quantitatively, as illustrated in Figure 2(b). Since the lowest level of the single-ion spectrum is a Kramers doublet, the further splitting must be due to exchange.

Related work on cooperative magnetic effects on the spectra of pairs of $Mn^{2+}$ and $Ni^{2+}$ in $KZnF_3$ was reported in a series of elegant papers by Ferguson, Guggenheim, and Tanabe.[7] The intensities of the pair spectra were an order of magnitude higher than those of isolated ions, so that even in crystals containing as little as 1 atomic % Mn, absorption by pairs dominated the spectrum. Thus the bands at 400 and 330 nm in $KMnF_3$ are sixteen times as intense as the corresponding bands in the spectrum of $Mn(H_2O)_6^{2+}$.

As in the case of the $^6A_{1g} \rightarrow {}^4T_{1g}$ transition of $MnF_2$, it was shown that the transitions $^6A_{1g} \rightarrow {}^4E_g$, $(^4G)$ of $Mn^{2+}$ doped in $KZnF_3$ were electric-dipole in character and had temperature independent oscillator strengths.[7] The particularly sharp $^4A_{1g}(^4G)$ component at 400 nm under high resolution consisted of four bands, whose relative intensities were temperature dependent in the manner shown in Figure 3.

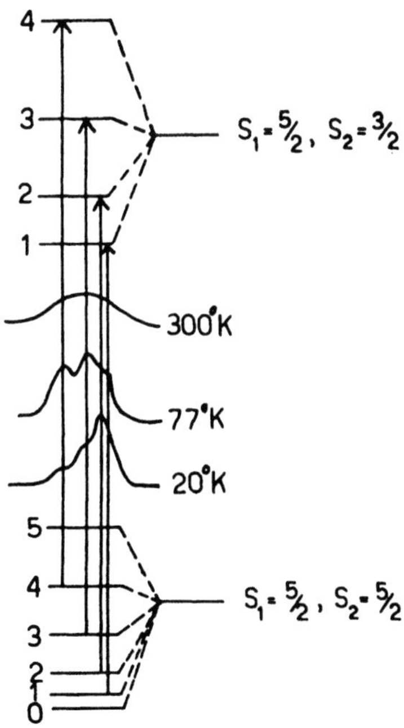


**Figure 3.** The exchange split energy levels of pairs of $Mn^{2+}$ in $KZnF_3$ corresponding to the $^6A_{1g} \rightarrow {}^4A_{1g}(^4G)$ transition near 400 nm and the temperature dependence of the resulting absorption spectrum. Reproduced with permission from Reference 7.

A pairwise exchange Hamiltonian $J S_a S_b$ leads to the energy levels

$$(J/2)[S(S+1) - S_a(S_a+1) - S_b(S_b+1)] \tag{1}$$

so that the energy separations between the four components should be 2 K, 5 K, 9 K, in terms of an apparent coupling constant. The actual separations were found to be 4.7, 6.5 and 10.7 $cm^{-1}$. All except the lowest energy line are hot bands, and disappear at 4 K.

Similar effects to those observed when pairs of magnetic ions are coupled in a diamagnetic host are of course to be expected in pure lattices of magnetic ions in the antiferromagnetic state, i.e., below the Neel temperature. A few years after the ruby sidebands had been detected, it was noticed that the spin-forbidden electric-dipole forbidden $^6A_{1g} \rightarrow {}^4T_{1g}$ absorption band in the 4 K spectrum of $MnF_2$ had a much greater intensity than would have been anticipated. The electronic origin line, which would be only magnetic-dipole allowed, would be expected to be extremely weak, though of course accompanied by vibronic sidebands of the usual type. These, on the basis of experience with other systems will be about 300 $cm^{-1}$ distant from the electronic origin, i.e., one quantum of an asymmetric stretch within

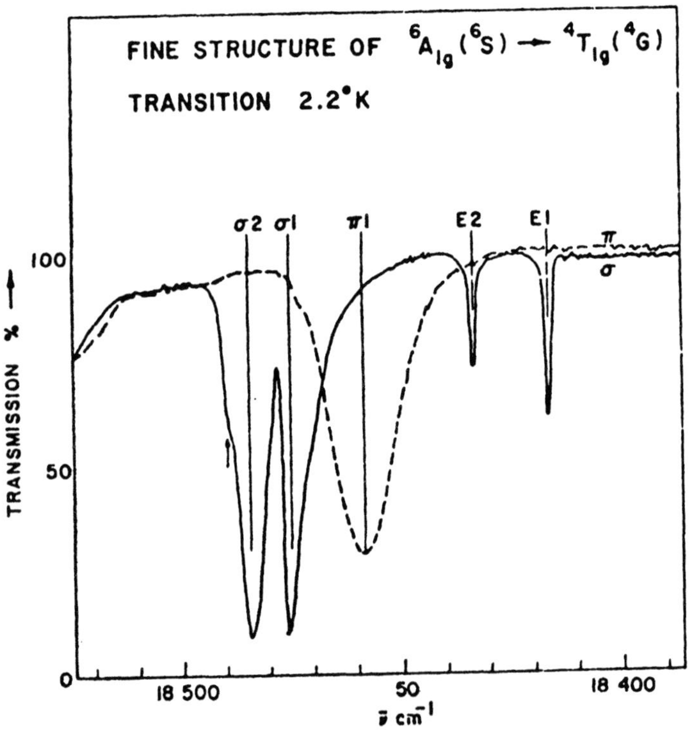


**Figure 4.** The linearly polarized $^6A_{1g} \rightarrow {}^4A_{1g}(^4G)$ absorption spectrum of a crystal of $MnF_2$ in the 540–543 nm region. Reproduced by permission from Reference 9.

the local $MnF_6^{4-}$ octahedron. In fact, at 4 K the ${}^6A_{1g}({}^6S) \rightarrow {}^4T_{1g}({}^4G)$ transition of $MnF_2$ consists of two sharp weak magnetic-dipole lines (Figure 4), identified as such by their polarisations, but the bulk of the intensity is in three broad electric-dipole lines about 50 $cm^{-1}$ to higher energy.[9] Apart from their small separation from the pure magnetic-dipole lines, the electric-dipole lines do not have the temperature dependence of intensity which would be associated with vibrationally assisted transitions. They are therefore apparently electric-dipole allowed, and their intensity derives from magnetic coupling between the ions in the lattice. Pictorially one could say that while one ion makes the electronic transition, accompanied by a reduction in the spin from $5/2$ to $3/2$, a neighboring ion undergoes a spin-flip which keeps the total spin-projection on a given axis constant. However, we cannot speak about exciting one particular $Mn^{2+}$ electronically, or flipping the spin of one identifiable neighbor. In fact, there are two sublattices of parallel spins, aligned antiparallel to one another since it is within the sublattices as a whole that the electronic and spin excitations occur, we must examine the kinds of excitations that arise in an infinite lattice.

## 3 SPIN EXCITATIONS IN MAGNETICALLY ORDERED CRYSTALS

In order to see how these different kinds of excited state arise in a magnetically ordered crystal we shall look more closely at a simplified model situation, that of a chain of spins coupled ferromagnetically. In fact, as we shall see later, there are some quite good examples of this situation in real life. A classical picture of the fully ordered ground state in such a system is shown in Figure 5(a), and an excited state, in which we have flipped the spin on one ion, in Figure 5(b). The most general form

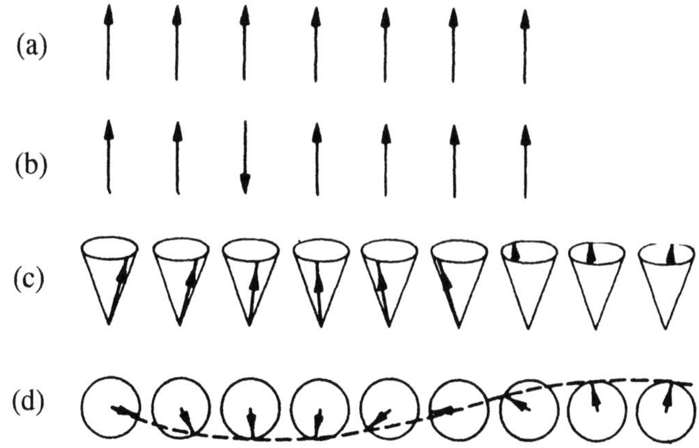


**Figure 5.** (a) The classical ground state of a ferromagnetic chain; (b) a classical spin excitation; and (c) a spin wave.

of the magnetic exchange Hamiltonian coupling two neighboring spins 1 and 2 is

$$\mathcal{H} = -J\left[\gamma(S_{1x}S_{2x} + S_{1y}S_{2y}) + S_{1z}S_{2z}\right] \tag{2}$$

which enables us to define how many degrees of freedom the spin vector has, i.e., in how many directions it is allowed to point. This is also called the spin dimensionality $d$. Thus when $\gamma = 1$ in Equation 2, $d = 3$ and $x, y, z$ are equivalent. That case defines the Heisenberg exchange Hamiltonian. On the other hand if $\gamma = \infty$, $d = 2$ and the spins are constrained to the $xy$-plane: This is called the XY model. Finally, if $\gamma = 0$ the spins can only be in one direction, $z(d = 1)$, defining the Ising model.

If we adopt the Heisenberg model, the total exchange energy of the fully aligned chain of $N$ spins is

$$U = -J\sum_{l=1}^{N} S_l \cdot S_{l+1} \tag{3}$$

If we treat the spins as classical, then $S_l \cdot S_{l+1} = S^2$ so the exchange energy of the ground state becomes

$$U_0 = -JNS^2 \tag{4}$$

while the energy of the first excited state is

$$U_1 = U_0 + 4JS^2 \tag{5}$$

Evidently, the classical picture is over naive because it is necessary to take into account the proper quantum mechanical view of the situation, which starts from the proposition that if one spin in the chain has been reversed, we have no means of knowing which one, and therefore that a correct wavefunction for the first excited state must be one that allows the reversed spin to be on any atom in the chain, with equal probability. For insulating materials, such as the example of NiO that we started with, the ground and lower excited states all arise from the same Heitler–London configuration, e.g., $Ni^{2+}O^{2-}Ni^{2+}\ldots$. Thus, for the linear chain ferromagnet, where the spin wave functions of each cation can be $\alpha$ or $\beta$ (spin up or down), the ground state

$$\varphi_0 = \alpha_1\alpha_2\ldots\alpha_N \tag{6}$$

and the first excited state is

$$\varphi_l = \alpha_1\alpha_2\ldots\alpha_{l-1}\beta_l\alpha_{l+1}\ldots\alpha_N \tag{7}$$

Because all the ions $l$ are identical, and to preserve translational symmetry, the correct wave function is

$$\psi_{\mathrm{k}} = \sum_{l} c_l^{\mathrm{k}} \varphi_l \tag{8}$$

i.e., a superposition of all the $\varphi_l$ in which the spin on atom $l$ has been reversed.

With periodic boundary conditions, i.e., $N$ atoms forming a ring, the coefficients $c_l^k$ are

$$c_l^k = \exp\left[-\mathrm{i}\left(\frac{2\pi}{N}\right)kl\right] \tag{9}$$

so that

$$\psi_k = \left(\frac{1}{N^{1/2}}\right)\sum_{l=1}^{N} \varphi_l \exp\left[-\mathrm{i}\left(\frac{2\pi}{N}\right)kl\right] \tag{10}$$

where $k$ can have the values 0, 1, ... $N$.

In pictorial terms, the significance of Equation 10 is that the spin flip, which was confined to one atom in the chain ($l$ in Equation 7) is now distributed over the whole chain. Instead of a single spin being reversed by 180° with respect to all the others, the overall decrease of one in the number of spins pointing in the 'up' direction is achieved by a small deviation from the 'up' direction at every site, as shown in Figures 5(c) and 5(d). It is easy to see, then, why such an excited state is called a "*spin wave*."

Now, to calculate the spectrum, we need to know the energies of the states $E(k)$, which are given by the diagonal element of the wavefunction in Equation 10 with the Hamiltonian of the system. To calculate the energies $E(k)$ associated with the $\psi_k$, the spin dependent part of the total Hamiltonian can be considered to be made up to two terms.[11] The first arises from anisotropy, which constrains the spins to a plane or axis, as in the XY and Ising cases already mentioned. This may be the result of the low symmetry of the ligand field around the ion in question (in which case it is called "single ion anisotropy") or it may arise from the magnetic exchange interaction. Either way, it mimics the effect of a large static applied field that lifts the $(2S + 1)$ fold degeneracy of the ground state with spin $S$. The second term takes account of the magnetic exchange interaction, assumed to be Heisenberg-like. In all,

$$\mathcal{H} = -g\mu_B H_A \sum_{l} S_l^z - \sum_{l>m} J_{lm} \mathbf{S}_l \cdot \mathbf{S}_m \tag{11}$$

The energy difference between the ground and the first excited state arising from the anisotropy term is just $g\mu_B H_A$. To calculate the energy difference arising from the second term, we note that the Heisenberg Hamiltonian can be written in two parts using the raising and lowering operators, i.e.,

$$\begin{aligned}\mathcal{H}''_{\text{ex}} &= -\sum J_{lm} S_l^z S_m^z \\ \mathcal{H}^{\perp}_{\text{ex}} &= -J_{lm}\left(S_l^+ S_l^- + S_l^- S_m^+\right)\end{aligned} \tag{12}$$

where

$$\begin{aligned}S^z|sm\rangle &= m\hbar|sm\rangle \\ S^{\pm}|sm\rangle &= \left[s(s+1) - m(m \pm 1)\right]^{1/2}\hbar|sm \pm 1\rangle\end{aligned} \tag{13}$$

The parallel exchange term yields an energy $2J$ that is independent of $k$, while that for the perpendicular term depends on $k$. For example $S_l^+ S_{l+1}^-$ acting on the function where the $\beta$-spin is on site $(l+1)$, moves the $\beta$ spin to site $l$, i.e., to the left:

$$\ldots\alpha_{l-1}\alpha_l\beta_{l+1}\alpha_{l+2}\ldots \rightarrow \ldots\alpha_{l-1}\beta_l\alpha_{l+1}\alpha_{l+2}\ldots \tag{14}$$

Adding the left and right shift terms, the contribution to the energy of the lowest excited state from the parallel term is

$$J\left[\exp\left(\frac{2\pi ik}{N}\right) + \exp\left(\frac{2\pi ik}{N}\right)\right] = -J\cos\left(\frac{2\pi k}{N}\right) \tag{15}$$

In consequence, the total excitation energy from the fully aligned ground state to the proper quantum mechanically defined excited state corresponding to the flipping of one spin is the sum of all these contributions, namely

$$E(k) = g\mu_B H_A + J\left[1 - \cos\left(\frac{2\pi k}{N}\right)\right] \tag{16}$$

The energy of the spin wave depends on its wavelength, or the inverse of the wavelength, which is defined by the parameter $k$. In fact, $k$ is related to the reciprocal of the wavelength because, if $k = 0$ the wavelength of the spin wave is infinite, which corresponds to all the $\phi_l$ in Equation 10 having the same value. In other words, there is no spin deviation at all. Then the energy of excitation comes from the anisotropy and the parallel exchange terms:

$$E(0) = g\mu_B H_A + J \tag{17}$$

With increasing k the wavelength of the spin deviation, as exemplified by Figure 5(d), decreases. The shortest wavelength that one could envisage is when the direction of adjacent spins are all antiparallel, in which case it is $2a$, where $a$ is the distance between the ions in the chain. In that case the maximum energy of a spin wave is $g\mu_B H_A + 4J$. To define energies between these two limits it is convenient to define the quantity $k = 2\pi k/Na$, so that the last term in Equation 16 becomes $\cos(ka)$.

If the lattice is three-dimensional, up to three independent length parameters $a$, $b$, $c$ are needed to define the repeat distance between ions, in which case $k$ is a vector quantity, called the wave vector $\boldsymbol{k}$.

The result for a one-dimensional chain can be generalized to three dimensions as:

$$E(k) = g\mu_B H_A + SzJ(1 - \gamma_k) \tag{18}$$

where

$$z\gamma_k = \sum_n \exp(-i\boldsymbol{k}\boldsymbol{r}_n) \tag{19}$$

$z$ being the number of nearest neighbors $n$ and $\boldsymbol{r}$ their vector displacement from the origin. When the system has a centre of symmetry (the most common situation), Equation 19 simplifies to

$$z\gamma_k = \sum_n \cos(\boldsymbol{k}\boldsymbol{r}_n) \tag{20}$$

Without going into details, the corresponding expression for a one-dimensional antiferromagnet is

$$E(k)^2 = g\mu_B H_A + J\left[1 - \cos^2\left(\frac{2\lambda k}{N}\right)\right] \tag{21}$$

Since the electronic spectra that we are considering in this chapter are mostly measured at low temperature the most important spin wave states for our purposes are those with the lowest energy, that is, the ones with smallest $k$. Under those circumstances Equations 16 and 21 simplify, so that for a ferromagnet $E(k) \propto k^2$ while for an antiferromagnet $E(k) \propto k$, a difference which is crucially important when we come to examine the way in which spin waves couple to electronic excitations to create the optical spectra that we actually observe.

## 4 MEASURING SPIN EXCITATIONS

In the previous section, we saw how the energy needed to create a spin deviation in a magnetically ordered crystal is defined in terms of the anisotropy, the exchange constant and the wavelength (wavevector) of the deviation. Now we want to consider how to measure it. The first thing to notice is that when we want to measure the energies of spin waves as a function of their wavevector we cannot use any form of optical spectroscopy. This is because photons, having zero mass, carry no momentum and hence only induce transitions in which the wavevector is conserved, i.e., $\Delta\boldsymbol{k} \sim 0$. Put another way, the wavelength of visible (or even infrared) light is so much longer than the unit cell repeat distance that neighboring unit cells are excited essentially in phase with one another. However, help is at hand in the form of another particle, the neutron.[12] Neutrons have mass, and hence, momentum. The equivalent wavelength of a neutron with a thermal energy corresponding to room temperature is 1.5 Å, that is, of the same order as interatomic spacings. Furthermore, neutrons have spin, so they interact with unpaired electron spins in solids. To measure the energy $E(\boldsymbol{k})$ of a spin wave as a function of $\boldsymbol{k}$, we therefore let neutrons with a wavevector $\boldsymbol{k}_\mathrm{n}$ impinge on a crystal, where they are scattered by interaction with a spinwave of wavevector $\boldsymbol{k}$. Conservation of momentum requires that the outgoing neutrons have a new wavevector $\boldsymbol{k}'_\mathrm{n}$ such that

$$\boldsymbol{k}_\mathrm{n} = \boldsymbol{k}'_\mathrm{n} + \boldsymbol{k} \tag{22}$$

At the same time the law of conservation of energy applies, so

$$\frac{\hbar^2 \boldsymbol{k}_\mathrm{n}^2}{2M_\mathrm{n}} = \frac{\hbar^2 \boldsymbol{k}_\mathrm{n}'^2}{2M_\mathrm{n}} + E(\boldsymbol{k}) \tag{23}$$

where $M_\mathrm{n}$ is the mass of the neutron.

The apparatus used to make this kind of measurement was developed many years ago by Brockhurst,[13] quite soon after neutron beams started to be used as probes in condensed matter research. It is called a triple axis spectrometer, and while the technical details have evolved greatly over the last 40 years, the principle of the technique remains unchanged. A schematic view of the arrangement is shown in Figure 6(a) and a picture of an actual spectrometer in Figure 6(b). The incident wavevector of the neutron beam $\boldsymbol{k}_\mathrm{n}$ is chosen by selecting a monochromatic wavelength out of the "white" beam emerging from a nuclear reactor. This is accomplished by exploiting a particularly intense reflection from a single crystal, e.g., Si(111) or Cu(110). The beam is then scattered from the sample crystal and the energy which it has gained or lost in the scattering process is determined using reflection from an analyzing crystal. There is a strong analogy between the process just described and that occurring in Raman spectroscopy, where a photon is scattered at a higher or lower frequency than that of the incident beam, with the difference corresponding to a vibrational (or sometimes electronic) excitation mode. The beauty of inelastic neutron scattering, however, compared with inelastic photon scattering is the access it gives to $E(\boldsymbol{k})$ at different values of $\boldsymbol{k}$.

### 4.1 Ferromagnetic Chains

As a simple practical example we take the one-dimensional ferromagnet $CsFeCl_3$, which has the hexagonal perovskite structure consisting of chains of $FeCl_6$ octahedra sharing opposite faces. The chains are separated by the $Cs^+$ so near-neighbor exchange along the chains dominates the inter-chain interaction (Figure 7). The ground state of a high spin $Fe^{2+}$ ion ($3d^6$) in an octahedral ligand field is $^5T_{2g}$. However, since this state has both orbital and spin angular momentum its degeneracy is further lifted in first order by spin-orbit coupling. In addition, there is a minor trigonal

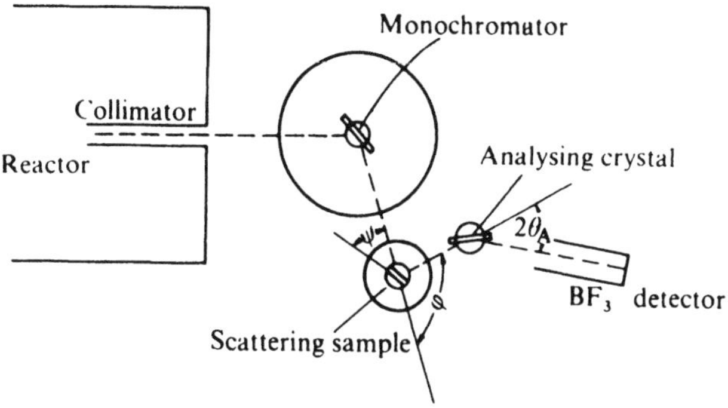


(a)

(b)

**Figure 6.** The neutron triple axis spectrometer: (a) schematic view; and (b) the IN8 spectrometer at the Institut Lane-Langevin, Grenoble.

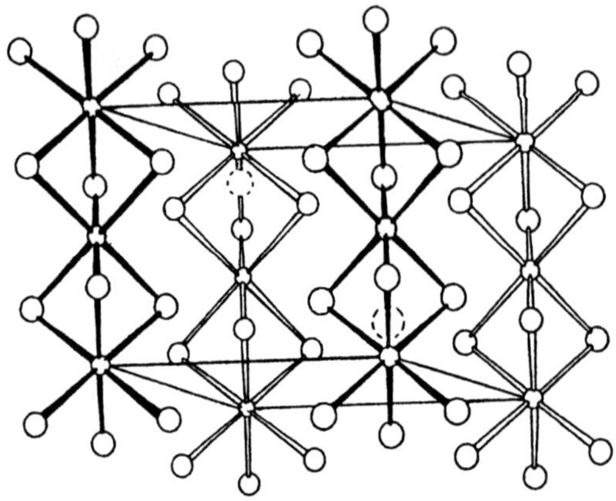

**Figure 7.** The crystal structure of $CsFeCl_3$.

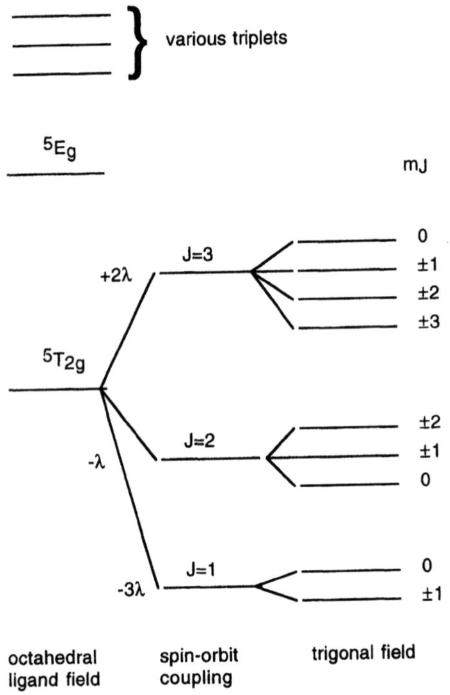


**Figure 8.** Energy levels of high spin $Fe^{2+}$ in an octrahedral ligand field with a small trional distortion, in the presence of first order spin-orbit coupling.

distortion. The relevant Hamiltonian is therefore

$$\mathcal{H} = \sum_i \left[ \lambda \boldsymbol{L}_i \cdot \boldsymbol{S}_i + \delta \left( \boldsymbol{L}_{zi}^2 - \frac{2}{3} \right) \right] - \sum_{i>j} J_{ij} \boldsymbol{S}_i \cdot \boldsymbol{S}_j \tag{24}$$

The resulting energy level scheme is shown in Figure 8.

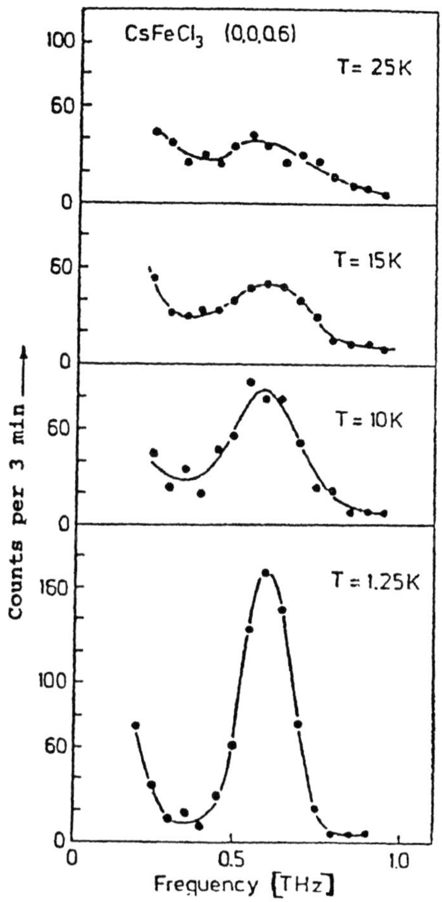


**Figure 9.** Frequency dependence of inelastic neutron scattering in $CsFeCl_3$ at different temperatures. Reproduced by permission from Reference 14.

The variation of the spin wave energy with wavevector $\boldsymbol{k}_c$ along the chain is given by Equation 16 as

$$E(\boldsymbol{k}_c) = E(0) + J(1 - \cos \pi \boldsymbol{k}_c) \tag{25}$$

The relative ordering of the lowest lying $m_J = 0$ and $m_J = \pm 1$ levels depends on the sign of the trigonal field component of the ligand field, i.e., whether the $FeCl_6$ octahedron are trigonally elongated or compressed. In fact it is the latter, so the $m_J = 0$ level is lowest.

Inelastic neutron scattering measurements have been carried out on $CsFeCl_3$.[14] As the temperature is lowered a peak begins to appear, and becomes narrower with decreasing temperature. The frequency dependence of the scattering is shown in Figure 9.

The width of the observed peak depends in part on the resolution of the spectrometer, of course, but it is also governed by the degree of magnetic order in the chain, that is, the mean distance between irregularities in the relative orientations of neighboring moments. At absolute zero, this distance (called the correlation length) will be infinite, i.e., all the moments in the chain are perfectly aligned with respect to one another. However, as the temperature is raised the moments become increasingly disordered. Quantitatively, the peakwidth is inversely proportional to the correlation length.

Turning to the wavevector dependence, Figure 10 shows the energy of the spin wave measured along the c-direction, i.e., parallel to the chains of $Fe^{2+}$. An excellent fit is obtained to Equation 25 with $E(0) = 5.4(3)\ \mathrm{cm}^{-1}$ and $J = 4.79(7)\ \mathrm{cm}^{-1}$. Finally, the ordering of the sublevels in the groundstate is verified by applying a magnetic field. The single excitation found in zero field is split, as illustrated in Figure 11, showing conclusively that the upper component is the degenerate one.

## 4.2 Ferromagnetic Layers

A two-dimensional lattice in which an octahedral arrangement of halide ions is also subject to a trigonal distortion occurs in the $MX_2$ compounds, which have either the $CdCl_2$ or $CdI_2$ structures. All the dihalides of the $3d$ elements have one of these structures or the other, though in the cases of $CrX_2$ and $CuX_2$ the structure is distorted by the Jahn–Teller effect. The magnetic properties of the $MX_2$ compounds are also quite unusual, making them excellent prototypes to exemplify the connections between magnetic order and optical spectra. They were in fact the first examples to be discovered of "metamagnets," that is, antiferromagnetic compounds which can nevertheless be ferromagnetically saturated in quite modest applied fields. The reason for this apparently strange behavior lies in the layer structure: The exchange interaction between nearest neighbors within the layers is ferromagnetic while the interaction between layers is antiferromagnetic, but much weaker. It is worth examining the magnetic excitations in some key examples of this type of compound in detail because it turns out that we can explain the behavior of their optical spectra quite quantitatively from the nature of the spin wave spectra.

As an example of an easy-axis system we take $FeCl_2$, and of an easy-plane system, $CoBr_2$.[15] The cubic single-ion ground states of the metal ions in these two compounds are respectively $^5T_{2g}$ and $^4T_{1g}$, i.e., they have both spin and orbital degeneracy. In both cases, the three-fold orbital degeneracy mimics that of an atomic $P$ state so we can define a pseudo-angular momentum quantum number $L = 1$. Combining this with the spin of $S = 2$ in the case of $FeCl_2$ and $S = 3/2$ in the case of $CoBr_2$, spin-orbit coupling leads to states with $J = 1, 2, 3$ for the former and $1/2$, $3/2$ for the latter, with the $J = 1$ and $J = 1/2$ components lying lowest in energy. Further, the $J = 1$ ground state of $Fe^{2+}$ is split by the trigonal ligand field distortion in the $CdCl_2$ structure, though of course the Kramers degeneracy of the $J = 1/2$ ground state of $Co^{2+}$ cannot be so lifted, as we saw in Section 2. In the effective spin-Hamiltonian formalism, we may therefore write the exchange Hamiltonians of

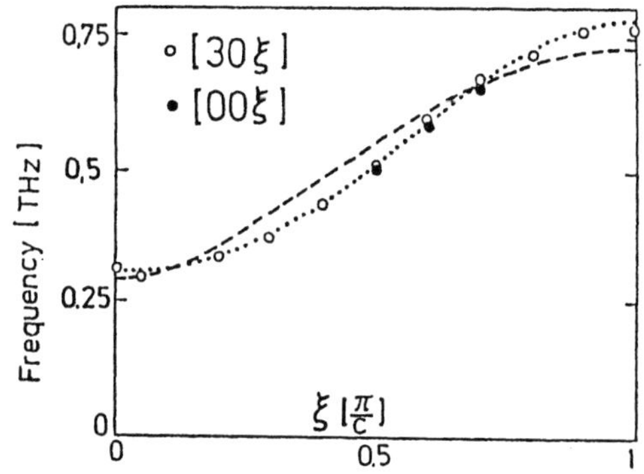


**Figure 10.** Variation of the spin wave frequency in $CsFeCl_3$ with wavevector along the chain axis, $K_c$ at 1.25 K. The dotted line is a fit to Equation 25. Reproduced by permission from Reference 14.

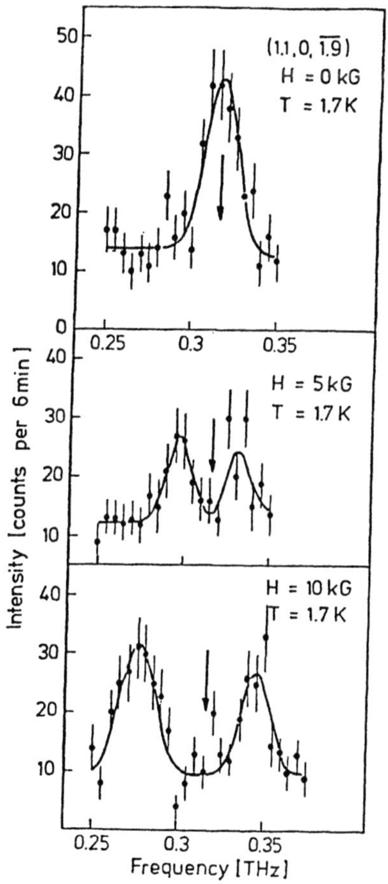


**Figure 11.** Zeeman splitting of a spin wave in $CsFeCl_3$ with $K_c = 0.1$, as a function of magnetic field. The arrows indicate the frequency in zero field. Reproduced by permission from Reference 14.

the two ions as:

$$\begin{aligned} Fe^{2+}: \mathcal{H}_{ex} &= \sum_{NN}(-J_1 S_i \cdot S_j - D_1 S_{iz} S_{jz}) + \sum_{NNN}(J_2 S_i \cdot S_j + D_2 S_{iz} S_{jz}) \\ &\quad - \sum_i \Delta\left[(S_{iz})^2 - \frac{1}{3}S(S+1)\right] \end{aligned} \tag{26}$$

$$Co^{2+}: \mathcal{H}_{ex} = \sum_{NN}(-J_1' S_i' \cdot S_j' + D_1' S_{ix}'' S_{jx}'') + \sum_{NNN}(J_2' S_i' \cdot S_j' - D_2' S_{ix}'' S_{jx}'') \tag{27}$$

where, in the former, the effective spin of the ground state is $S = 1$ and in the latter $S'' = 1/2$. In Equations 26 and 27, the subscripts 1 and 2 indicate intra- and inter-plane exchange, respectively, $J(J')$ and $D(D')$ are isotropic and anisotropic exchange parameters, and $\Delta$ is the trigonal field anisotropy parameter. The $z$ axis is the axis of preferred spin orientation, and only nearest-neighbor intra-(NN) and inter-plane (NNN) interactions are included. The magnetic properties of $NiCl_2$ can also be treated by a Hamiltonian of the form of Equation 27, but with $D_1 = D_2 = 0$. The latter approximation arises because the $^3A_{2g}$ ground state of $Ni^{2+}$ is split only in second-order by the trigonal field and spin-orbit interaction. In fact, the experimentally determined anisotropy is very small ($D_1 = D_2 = 0.40$ K for $NiCl_2$) and on this basis we assume that an isotropic Hamiltonian can be used as a first approximation of the $Ni^{2+}$ case.

The Hamiltonian equation 26, applicable to $FeCl_2$, is that of a simple two-sublattice anti-ferromagnet with the anisotropy field directed along the preferred spin axis. From Equations 18 and 20, the spin wave energy is

$$E_{sw} \sim S(A + z_1 J_1 + z_2 J_2 - J_1 z_1 \gamma_{k1}) \tag{28}$$

This is just what would be expected for a two-dimensional ferromagnet with strong axial anisotropy, for which the dominant interactions are ferromagnetic and in-plane. The spin-waves on each sublattice are doubly degenerate, and at the same time the spin-wave modes on the two sublattices of the antiferromagnet are degenerate in the absence of an external field.

In Hamiltonian equation 27 the anisotropy field directed along the $x$ axis makes the solution more difficult. In this case the degeneracy of the spin-waves on each sublattice is lifted by the anisotropy field, and Lines[17] gives the energies as

$$E_{sw}^{(1)} = S''\left[(\mu + \lambda)(\mu - \lambda + \rho_1 + \rho_2)\right]^{1/2} \tag{29a}$$

$$E_{sw}^{(2)} = S''\left[(\mu - \lambda)(\mu + \lambda + \rho_1 - \rho_2)\right]^{1/2} \tag{29b}$$

where

$$\mu = z_1 J_1' + z_2 J_2' - \sum_{NN} J_1' \exp\left[i\boldsymbol{k}.\delta(1)\right] = z_2 J_2' + z_1 J_1'(1 - \gamma_{k1}) \tag{30a}$$

$$\lambda = \sum_{\text{NNN}} J_2' \exp\left[i\boldsymbol{k}.\delta(2)\right] = z_2 J_2' \gamma_{k2} \tag{30b}$$

$$\rho_1 = \sum_{\text{NN}} D_1' \exp\left[i\boldsymbol{k}.\delta(1)\right] = z_1 D_1' \gamma_{k1} \tag{30c}$$

$$\rho_2 = \sum_{\text{NNN}} D_2' \exp\left[i\boldsymbol{k}.\delta(2)\right] = z_2 D_2' \gamma_{k2} \tag{30d}$$

Neglecting the anisotropy terms $\rho_1$ and $\rho_2$, as in the $Ni^{2+}$ case, makes the modes degenerate again. We are then back to the case of the isotropic antiferromagnet, i.e., to the expression used for $FeCl_2$, though now with very small anisotropy energy. In the same way as in Equation 20, the summations of Equations 30, can also be simplified when the crystal is centrosymmetric.

As we noted already, only the lower-energy spinwave states, i.e., those near the zone centre, are populated at low temperature. Then, for $FeCl_2$, the spin-wave energy expressed by Equation 28 simplifies to

$$E_{\text{sw}} \sim A'\left(1 + \alpha' k^2\right) \tag{31}$$

There is an energy gap at $k = 0$ where magnitude depends on the anisotropy parameter $A'$. For $CoBr_2$, the two spin-wave energy relations of Equation 29 take rather different forms. When $J_1' > D_1' > J_2' > D_2'$, as in the $CoX_2$ compounds,

$$E_{\text{sw}}^{(1)} \sim \left(B_1 + \gamma k^2\right)^{1/2} \sim B_1'(1 + \gamma' k^2) \tag{32}$$

In Equation 32, $B_1'$ is determined by $D_1'$ and $J_2'$. We see that Equation 32 has the same form as Equation 31 for $FeCl_2$. As $k$ tends to zero, the energy of $E_{\text{sw}}^{(1)}$ tends to a finite limit governed by the anisotropy field, and the dependence of the energy on $k$ is quadratic, a result typical of ferromagnetic systems.

At small $k$ becomes the second spin-wave mode, $E_{\text{sw}}^{(2)}$.

$$E_{\text{sw}}^{(2)} \sim \left(B_2 k^2 + \ldots\right)^{1/2} \sim B_2' k \tag{33}$$

As $k$ tends to zero, the energy of this second mode also tends to zero, and further, its energy is linear in $k$, a characteristic usually associated with antiferromagnetic ordering.

The spinwave spectrum of $CoCl_2$within the layers, determined by inelastic neutron scattering[18] results on $CoCl_2$ clearly show the two modes with the behavior predicted by Equations 32 and 33 (Figure 12). If the inter-plane coupling is small compared with that within the planes, the energies of $E_{\text{sw}}^{(1)}$ and $E_{\text{sw}}^{(2)}$ become degenerate and assume the form typical of a two-dimensional easy-plane ferromagnet with no energy gap and a linear dependence of energy on $k$.

Turning to the nickel dihalides, in view of their orbital singlet ground states, the appropriate Hamiltonian is that of an isotropic antiferromagnet, i.e., either Equation 26 with $D_1 = D_2 = \Delta = 0$ or Equation 27 with $D_1' = D_2' = 0$. For very

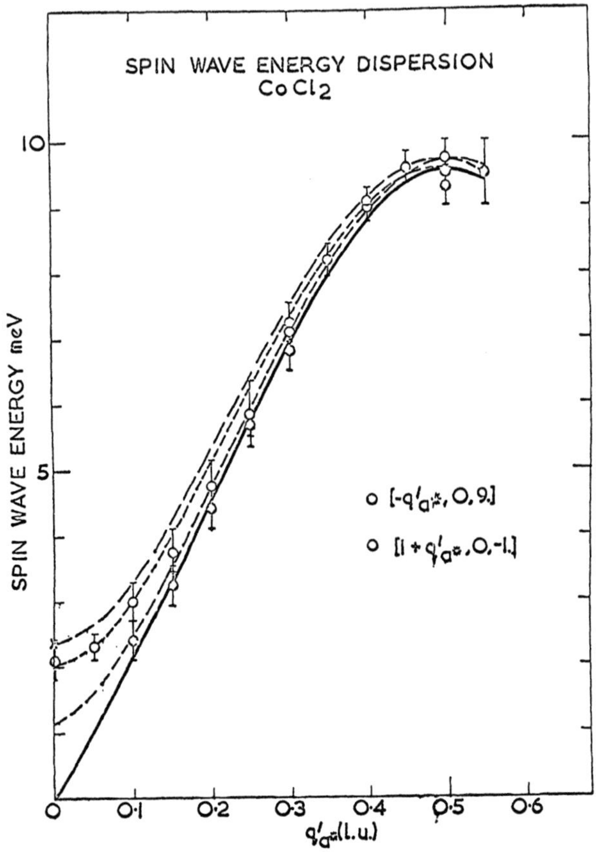


**Figure 12.** Spin wave energies in $CoCl_2$ at 4.5 K. The lines are fits to Equations like 32 and 33. Reproduced by permission from Reference 18.

small $k$, $E_{sw}$ is proportional to $k$. On the other hand, for larger values of $k$, as $J_2''$ becomes smaller and smaller relative to $J_1''$, $E_{sw}$ again becomes proportional to $k^2$. Once again, this prediction is beautifully borne out by the inelastic neutron scattering results on $NiCl_2$.[19]

Finally, let us summarize the three types of behavior for the spin waves in two-dimensional ferromagnets:

(a) Easy-axis with large anisotropy, the best example of which is $FeCl_2$: There is a single spin wave whose energy varies as $k^2$ for small $k$ and at $k = 0$ has a finite energy, equal to 17.2 cm$^{-1}$ in $FeCl_2$ at 5 K.

(b) Easy-plane with large anisotropy, examples being $CoCl_2$ and $CoBr_2$. There are two spin waves: one varying as $k^2$ for small $k$, and tending to a finite energy at $k = 0$, and the other varying as $k$ and tending to zero energy at $k = 0$. It is the latter which is most important at low temperature.

(c) Easy-plane with small anisotropy, found in $NiCl_2$ and $NiBr_2$, as well as layer compounds with $K_2NiF_4$ structure such as $Rb_2CrCl_4$. Here the spin wave energy varies as $k^2$ for small $k$ but with a small or negligible energy at $k = 0$.

## 5 ELECTRONIC EXCITED STATES IN MAGNETICALLY ORDERED CRYSTALS

As we illustrated above for the case of pairs of magnetic cations doped into a diamagnetic host lattice, the magnetic exchange interaction between near neighbors provides a mechanism for overcoming the spin selection rule in ligand field transitions, so that formally spin-forbidden transitions which might be expected to be extremely weak often appear as electric-dipole allowed, with moderate intensity. A similar situation is found in undiluted transition metal compounds, where there is a lattice of magnetic cations infinite in one, two or three dimensions. The difference from the case of discrete pairs or finite clusters, however, is that the magnetic exchange interaction splits the electronic ground state, not into discrete energy levels, but into a continuous spectrum of spin waves whose energies $E(\boldsymbol{k})$ are a function of wavevector $\boldsymbol{k}$. In a similar way, the electronic excited states are not localized on individual ions either. Because the ions are indistinguishable, any quantum mechanically correct description of the excited state must allow the excitation to reside on every ion with equal probability. The wavefunctions describing such an excited state take the same functional form as the ones we used to describe the spin excitations in an infinite lattice, i.e., Equations 6 and 10, though the functions $\alpha$ and $\beta$ have a different significance. In the case of electronic excitation they denote the ground and excited state electronic wavefunctions. Building on the notion of particle-wave duality, wavefunctions like those of Equations 7 and 10 are often thought of in terms of pseudo-particles: spin waves are called *magnons* and the analogous electronic excitation wave is called an *exciton*. In place of the picture we used to describe the discrete pair of magnetic ions, where an electronic excitation on one ion couples to a spin flip on another, we therefore envisage that absorbing a photon simultaneously creates an exciton and a magnon. Consequently the absorption bands in the optical spectra of magnetically ordered compounds are called exciton-magnon combination bands.

Figure 13 illustrates the difference between the mechanisms of spin-conservation in exciton-magnon transitions available to antiferromagnets and ferromagnets. They are called "hot" and "cold," a nomenclature due to Shinagawa and Tanabe.[21] This denotes the fact that the "hot" mechanism requires the photon to annihilate a thermally populated $\Delta M_S = -1$ magnon to compensate the decrease in the overall spin projection of the crystal brought about by the creation of a $\Delta M_S = -1$ exciton. On the other hand, the "cold" mechanism, which, at the absolute zero of temperature, would only be available to an antiferromagnet, invokes the creation of a $\Delta M_S = +1$ magnon on the sublattice opposite to that on which the $\Delta M_S = -1$ exciton is created. At temperatures far removed from absolute zero both mechanisms can con-

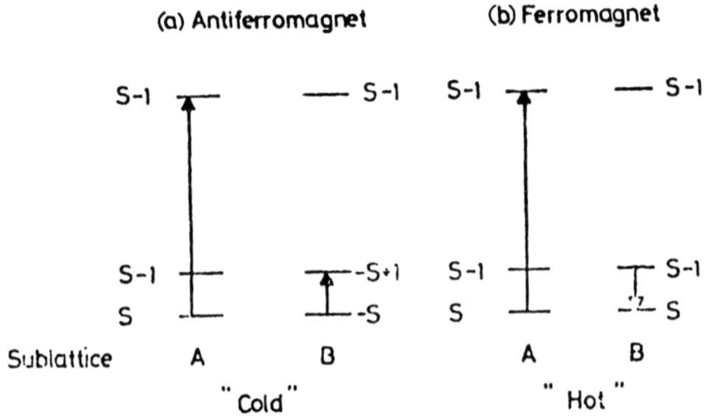


**Figure 13.** Exciton-magnon combinations in antiferromagnets (left hand side) and ferromagnets (right hand side). A and B denote the two sublattices.

tribute to the electric-dipole intensity in antiferromagnets but at low temperature one anticipates that the optical spectrum of an antiferromagnet will be dominated by "cold" bands, and of a ferromagnet by "hot" bands. Spectacular differences between the light absorption properties of the two kinds of magnet should thus become apparent.

The incident photon creates an exciton with a wavevector of $+\boldsymbol{k}$ and a magnon with a wavevector of $-\boldsymbol{k}$. To obtain the total intensity resulting from either mechanism one must sum contributions for all values of $\boldsymbol{k}$, and thus important terms are the $\boldsymbol{k}$-dependent electric-dipole transition moment $M_k$, which depends on the symmetry of both the exciton and the magnon and on the crystal symmetry, and the thermal population of magnons $\langle n_k \rangle$ having the same wavevector $|k|$ as the corresponding exciton. For the "cold" mechanism we have

$$I(\omega) = \sum_k I_k(\omega) = \sum_k |M_k|^2 (1 - \langle n_k \rangle) \cdot \delta(h\omega - E_k - \lambda_k) \tag{35}$$

where $E_k$ is the energy of the magnon and $\lambda_k$ that of the exciton. Correspondingly, for the "hot" mechanism,

$$I(\omega) = \sum_k I_k(\omega) = \sum_k |M_k|^2 \langle n_k \rangle \cdot \delta(h\omega - E_k - \lambda_k) \tag{36}$$

Let us now consider the various terms in Equations 35 and 36. At low temperatures only low energy magnons are thermally populated, so only $M_k$ having small $\boldsymbol{k}$ need to be considered. For a centrosymmetric system the leading terms in $M_k$ are proportional to $\sin(\boldsymbol{k}.\delta)$, $\delta$ being a vector connecting the ion in question with the neighbors included in the summation Equation 19. For small $\boldsymbol{k}$ this term is therefore proportional to $\boldsymbol{k}$. At thermal equilibrium the average value of $n_k$ is given by the Planck distribution:

$$\langle n_{\boldsymbol{k}} \rangle = \left[ \exp\left( \frac{E_{\mathrm{sw}}}{k_B} \right) - 1 \right]^{-1} \tag{37}$$

and the total number of magnons excited at temperature $T$ is

$$\sum_{\boldsymbol{k}} \langle n_{\boldsymbol{k}} \rangle = \int D(\omega) \langle n(\omega) \rangle \, \mathrm{d}\omega \tag{38}$$

where $D(\omega)$ is the number of magnon modes per unit frequency.

To translate Equations 35–38 into predictions of temperature dependence we need to know how the energy of the magnon $E_{\mathrm{sw}}$ varies with $\boldsymbol{k}$. This in turn depends on whether we have an easy-axis or easy-plane system, and on whether the single-ion anisotropy, which determines the gap in the spin-wave spectrum at $\boldsymbol{k} = 0$, is large or small. The appropriate expressions for $E(\boldsymbol{k})$ were summarized in the previous section. In evaluating Equation 38 (note that the integral may be taken from 0 to $\infty$, since in the low-temperature limit $\langle n(\omega) \rangle \rightarrow 0$ as $\omega \rightarrow \infty$). Further,

$$D(\omega) = \frac{V}{(2\pi)^3} \int \frac{\mathrm{d}S_\omega}{|\mathrm{d}\omega/\mathrm{d}k|} \tag{39}$$

where $V$ is the volume of the sample and $\mathrm{d}S_\omega$ is an element of area on the surface of constant frequency in $k$-space.[22] If we assume, for the two-dimensional lattices we are considering, that only magnons propagating within the layers are important, we can write $\int \mathrm{d}S_\omega \propto 2\pi k$ and also, if $\omega \propto |k|^m$ where $m$ is some power, then $(\mathrm{d}\omega/\mathrm{d}k) \propto mk^{m-1}$. For the two-dimensional easy-axis ferromagnet, we noted above that $m = 2$ at small $k$, so $D(\omega)$ becomes a constant, independent of $k$. In contrast, in the two-dimensional easy-plane system, the low-energy mode has $m = 1$, giving $D(\omega) \propto k$. In the following section we compare the predictions based on these simple considerations with the form of temperature variation of $I(\omega)$ actually observed in two dimensional transition metal dihalides.

## 6 EXAMPLES OF EXCITON–MAGNON COMBINATION BANDS

Armed with the information about the energy dispersion and density of the magnons and excitons, we can now look at some examples of the optical absorption spectra of magnetically ordered compounds.

### 6.1 Metamagnetic Transition Metal Dihalides

**6.1.1** ***Easy Axis*** Figure 14 shows the axial absorption spectrum of $FeCl_2$ in the 427 nm region.[15] Band $A$ is clearly a hot band, the relative intensities of the other two varying only very slightly over this range of temperature. Band $B$, which is

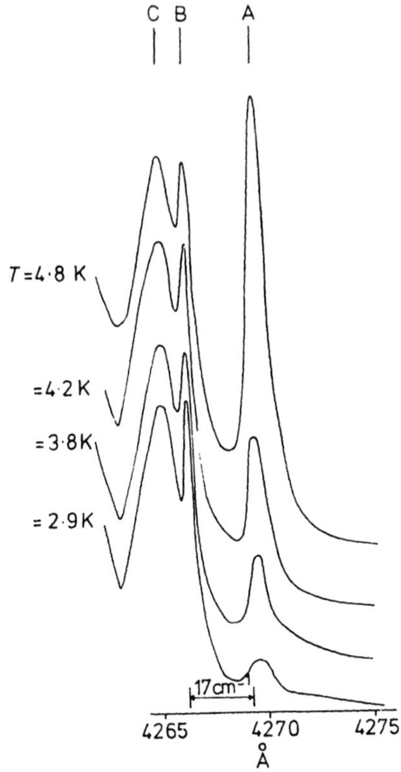


**Figure 14.** Absorption spectrum of $FeCl_2$ in the 427 nm region measured between 2.9 and 4.8 K. Reproduced with permission from Reference 15.

much narrower than either $A$ or $C$, occurs some 17 $cm^{-1}$ higher than band $A$. From inelastic neutron scattering, Birgeneau et al.[20] found that the gap in the spin-wave dispersion of $FeCl_2$ at the zone center was 17.2 $cm^{-1}$ at 5 K, so we assign band $B$ as the exciton line corresponding to the hot exciton–magnon band $A$. Band $C$ is a cold exciton–magnon combination band whose intensity comes from zero-point spin fluctuations in the two-dimensional spin system brought about by the weak antiferromagnetic coupling between the layers.

From equations 31, 36, 38 and 39, recalling at the same time that, for the two-dimensional easy-axis ferromagnet, $D(\omega)$ is independent of $k$ at small $k$, and neglecting the small inter-plane interactions, we have

$$\sum_k I_k(\omega) = \int_0^\infty k^2 \exp\left(\frac{-E_{\mathrm{sw}}}{k_B T}\right) \mathrm{d}\omega = \int_0^\infty \omega' \exp\left[\frac{-(A' + \hbar\omega)}{k_B T}\right] \mathrm{d}\omega' \qquad (38)$$

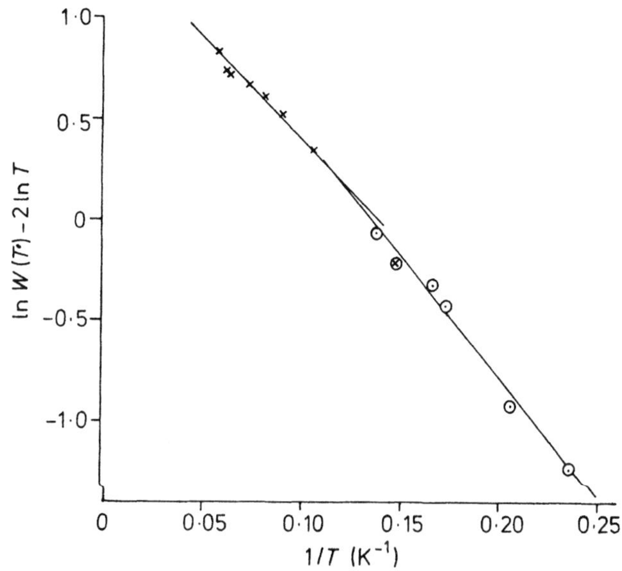


**Figure 15.** The intensity ($W$) of band $A$ in Figure 14 as a fraction of temperature. The crosses are from Reference 15 and the circles from Reference 23. They are normalized together at the point labeled X. Reproduced with permission from Reference 15.

or, putting $x = \omega'/T$,

$$\sum_k I_k(\omega) = C \exp\left(\frac{-A'}{k_B T}\right) T^2 \int_0^\infty x \exp(-ax)\, \mathrm{d}x = W(T) \tag{39}$$

The integral in $x$ is a definite integral and $A'$ is the energy gap in the spin-wave dispersion. It follows from Equation 39 that a plot of $\ln W(T) - 2\ln T$ against $1/T$ should be a straight line whose slope depends on $A'$. Figure 15 shows the temperature dependence of the intensity of band $A$ plotted according to this function. The two independent sets of data[15,23] included define values of $A'$ of 16 and 19(1) $\text{cm}^{-1}$, in good agreement with the zone center gap of 17.2 $\text{cm}^{-1}$ measured by inelastic neutron scattering.

**6.1.2** ***Easy Plane; Large Anisotrophy*** A good example of a compound which behaves as an easy-plane ferromagnet with large single ion anisotropy is $CoBr_2$. At 494 nm, there is a narrow absorption band associated with a quartet-to-doublet transition, whose intensity, like that of the 427 nm band in $FeCl_2$, increases with increasing temperature.[15] Assuming that the exciton concerned is coupling with the low energy spin-wave $E_{\text{sw}}^{(2)}$ mentioned above (Equation 33) the $k$-dependences of the electric-dipole transition moment, the magnon density of states, etc., yield an expression for the integrated intensity:

$$\sum_k I_k(\omega) = \int_0^\infty k^2\left(\frac{1}{k}\right)\left[\exp\left(\frac{E_{\mathrm{SW}}^{(2)}}{k_B T}\right) - 1\right]^{-1} \mathrm{d}\omega$$
$$= \int_0^\infty \omega^2 \exp\left[\exp\left(\frac{\hbar\omega}{k_B T}\right) - 1\right]^{-1} \mathrm{d}\omega \quad (40)$$

Letting $x = \omega/T$:

$$\sum_k I_k(\omega) = CT^3 \int_0^\infty x^2[\exp(ax) - 1]^{-1}\,\mathrm{d}x = CT^3 \quad (41)$$

Experimentally, over the temperature interval 4–18 K, the straight line giving the best fit to a log–log plot of band area *versus* temperature has a slope of 2.8(2).

**6.1.3** ***Easy Plane; Small Anisotrophy*** The third class of two-dimensional ferromagnets, easy-plane systems with small anisotropy, may be exemplified first by $NiBr_2$. The ground state is $^3A_{2g}$ and although no inelastic neutron scattering data are available for $NiBr_2$, in $NiCl_2$, which has the same structure, an anisotropy constant $D$ of only 0.40 K has been deduced.[16] In the visible spectrum of $NiBr_2$ there is a weak, highly structured absorption band near 610 nm,[24] the electronic origin region of which is shown in Figure 16. Once again we have a band increasing strongly in

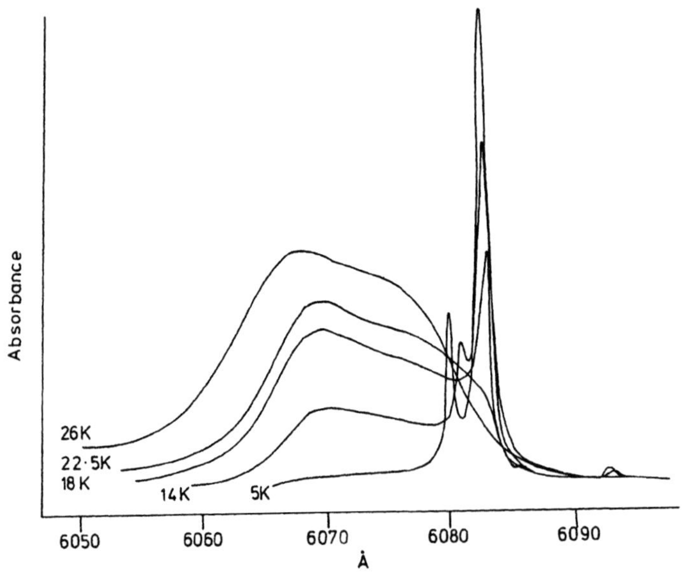


**Figure 16.** The absorption spectrum of $NiBr_2$ from 605–609 nm measured from 5–26 K. Reproduced with permission from Reference 24.

intensity with increasing temperature, and a log–log plot of the band area *versus* temperature from 4–20 K is a good straight line, this time with a slope of 2.0. As we remarked earlier, $E_{sw} \propto k^2$ in this case so

$$\sum_k I_k(\omega) \propto \int_0^\infty k^2\left[\exp\left(\frac{E_{sw}}{k_B T}\right) - 1\right]^{-1} d\omega$$

$$\propto T^2 \int_0^\infty x[\exp(ax) - 1]^{-1}\,dx \propto T^2 \tag{42}$$

where $x$ is $\omega/T$.

An unusual feature of Figure 16 is the pair of very sharp lines which appear on the long wavelength side of the hot exciton–magnon line. Their intensity decreases with increasing temperature till they finally vanish at about 22 K. The Neel temperature of $NiBr_2$ is 52 K, but there is evidence from antiferromagnetic resonance, and neutron diffraction, that a further magnetic phase transition takes place at 22 K, which leads to a small canting of the spins out of the plane, of the layer. That introduces a small component of antiferromagnetic coupling and thus of "cold" exciton–magnon intensity. However, its effect on the major "hot" band is only slight. When a magnetic field is applied to the crystal parrallel to the layers, the sharp cold lines switch off abruptly. The magnitude of the applied field needed to do this varies with temperature, so a plot of critical field *versus* temperature is in essence a magnetic phase diagram. More details about the very unusual (indeed unique) spin structure of $NiBr_2$ are given in Reference 25.

## 7 CONCLUSIONS

In this chapter, we have described how the spins are excited in infinite lattices showing long range magnetic order. Like the electronic energy levels in metals and semiconductors, the spin states in magnetically ordered compounds are not localized on single identifiable ions, but have to be considered as delocalized throughout the crystals. The corresponding spin waves, or magnons, can be excited by inelastically scattering neutrons, or optically when combined with electronic excitations (or excitons) such as arise from ligand field transitions. In the latter case, when the energies of the electronic transitions put them in the visible region of the spectrum, the consequences can be quite spectacular. Since combining magnons with excitons provides a mechanism to overcome the spin selection rule, nominally spin-forbidden ligand field transitions can become allowed, and so increase in intensity by several orders of magnitude. The mechanism depends on the magnon states being thermally populated, so the optical absorption bands change intensity in a most striking way as the temperature is varied. In the interests of brevity and clarity, only a small group of examples has been considered in detail, but the principles described apply to all inorganic solids where long range magnetic order is found. As such they form an important element in the spectroscopy of inorganic materials.